Historical Geography and Geosciences

This book series serves as a broad platform for contributions in the field of Historical Geography and related Geoscience areas. The series welcomes proposals on the history and dynamics of place and space and their influence on past, present and future geographies including historical GIS, cartography and mapping, climatology, climate history, meteorology and atmospheric sciences, environmental geography, hydrology, geology, oceanography, water management, instrumentation, geographical traditions, historical geography of urban areas, settlements and landscapes, historical regional studies, history of geography and historic geographers and geoscientists among other topically related areas and other interdisciplinary approaches. Contributions on past (extreme) weather events or natural disasters including regional and global reanalysis studies also fit into the series.

Publishing a broad portfolio of peer-reviewed scientific books Historical Geography and Geosciences contains research monographs, edited volumes, advanced and undergraduate level textbooks, as well as conference proceedings. This series appeals to scientists, practitioners and students in the fields of geography and history as well as related disciplines, with exceptional titles that are attractive to a popular science audience.

If you are interested in contributing to this book series, please contact the Publisher.

Matija Zorn · Olga Pelcer-Vujačić ·
Peter Mikša
Editors

Discourses on Mountains of Montenegro and Slovenia

Springer

Editors
Matija Zorn
ZRC SAZU Anton Melik Geographical Institute
Ljubljana, Slovenia

Olga Pelcer-Vujačić
Historical Institute
University of Montenegro
Podgorica, Montenegro

Peter Mikša
Department of History
Faculty of Arts
University of Ljubljana
Ljubljana, Slovenia

ISSN 2520-1379 ISSN 2520-1387 (electronic)
Historical Geography and Geosciences
ISBN 978-3-031-99016-8 ISBN 978-3-031-99017-5 (eBook)
https://doi.org/10.1007/978-3-031-99017-5

This work was supported by ZRC SAZU Anton Melik Geographical Institute and the Slovenian Research and Innovation Agency's program no. P6-0101 (Geography of Slovenia).

This Springer imprint is published by the registered company Springer Nature Switzerland AG
The registered company address is: Gewerbestrasse 11, 6330 Cham, Switzerland

Preface

Montenegro and Slovenia are among the youngest independent countries in the world and also among the smallest. Because they are barely visible as dots on the globe, the average person at the global level knows very little or nothing about them. The editors of the volume *Discourses on Mountains of Montenegro and Slovenia* hope that this volume will contribute at least a little to raising Montenegro's and Slovenia's profiles, especially with regard to their mountains, thereby drawing attention to their natural, cultural, and symbolic treasures.

The subject matter of this book is divided into 29 chapters, which were written by 45 authors (22 from Slovenia, 19 from Montenegro, two from Serbia, and two from Germany) representing 18 different institutions (nine from Slovenia, six from Montenegro, two from Serbia, and one from Germany), and the volume therefore has a strong interdisciplinary character.

The chapters, not counting the introduction and the conclusion (Part VII), are grouped into six parts based on their content. Part I is devoted to a geographical overview of the mountains in both counties and consists of four chapters. Part II deals with human activity in the mountains from prehistory up to the Middle Ages and consists of five chapters. Part III is devoted to the decline of the cultural landscape in the mountains as a consequence of socioeconomic changes and consists of three chapters. Part IV, the longest part, with seven chapters, deals with the importance of mountains as an identity feature in both countries. Part V, with four chapters, highlights mountain cultural heritage, and Part VI, also with four chapters, highlights some aspects of the natural environment in the mountains of Montenegro and Slovenia.

The editors of the volume represent three institutions that have a long tradition in mountain research. The **ZRC SAZU Anton Melik Geographical Institute** (Ljubljana, Slovenia) was founded in 1946 by the Slovenian Academy of Sciences and Arts. In 1976 it was named after Slovenia's greatest geographer, academy member Anton Melik (1890–1966). Since 1981, the institute has been part of the Research Centre of the Slovenian Academy of Sciences and Arts (ZRC SAZU). Since the very beginning, the institute's main task has been to conduct basic and applied geographical research on Slovenia and its landscapes and to prepare basic geographical texts about Slovenia as a country and as part of the world. The institute has six organizational units: the Geographical Museum, Geographical Library, Department of Geosciences, Department of Human Geography, Department of Geoinformatics, and Physical Geography Laboratory. It is the seat of the Slovenian

Government's Commission for the Standardization of Geographical Names. The institute takes part in many Slovenian and international projects, holds research conferences, trains junior researchers, and participates in researcher exchange programs. It publishes the SCIE and SCOPUS-indexed journal *Acta geographica Slovenica.*

The **Historical Institute of the University of Montenegro** (Podgorica, Montenegro) was founded in 1948; it is one of the oldest research institutions in Montenegro and one of the founders of the University of Montenegro (1974). The institute deals with research on and treatment of topics related to Montenegrin and general history and cultural heritage. It publishes studies, monographs, paper collections, document collections, and bibliographies, it edits and publishes the journal *Istorijski zapisi* (*Les Écrits Historiques*), and it organizes research meetings and conferences. It is involved in educating and training research staff and, from 2012 to 2016, it organized and carried out master's programs in history, the history of culture, and historical anthropology. The institute was involved in several successful international and national projects concerning cultural heritage and digital humanities, such as Montenegro on the Political and Cultural Map of Europe: CLIO MAP (GIS mapping of cultural heritage and a mobile tourism application).

The **Department of History at the University of Ljubljana's Faculty of Arts** (Ljubljana, Slovenia) was founded in 1919. The University of Ljubljana's Faculty of Arts is a research-intensive faculty in the humanities and social sciences, and it is the largest teaching and research institution in the social sciences and humanities in Slovenia. The Department of History is one of the founding departments of the Faculty of Arts. With more than four hundred students enrolled, it is also one of the largest departments. The department consists of six units: ancient history, medieval history and auxiliary sciences of history, early modern history, nineteenth-century history, contemporary history, and history education. The department is one of the most insightful and internationally renowned research groups in historiography and the humanities in general in Slovenia. Department members have contributed to the development of the Slovenian historiographic language and developed own methodological approaches and regularly contribute their findings to teaching practice and to the public. Department members participate in national and international projects, organize international conferences, host lecturers, and visit institutions abroad.

Ljubljana, Slovenia — Matija Zorn
Podgorica, Montenegro — Olga Pelcer-Vujačić
Ljubljana, Slovenia — Peter Mikša
July 2024

Acknowledgements

The content of this volume is partly based on the results of research programs co-funded by the Slovenian Research and Innovation Agency. The authors of the chapters acknowledge financial support from Slovenian Research and Innovation Agency program no. P6-0101 (Geography of Slovenia) for chaps. 1, 3–5, 12, 17, 22, 26, and 29, program no. P6-0235 (Slovene History) and project no. J6-60101 (The Struggle for the Mountains: Slovene National Appropriations of the Alps, 19th–21st Centuries) for chaps. 1, 15, and 19, and program no. P5-0408 (Heritage on the Margins: New Perspectives on Heritage and Identity Within and Beyond National) for chaps. 12 and 22.

In addition, the authors of chaps. 6, 11, 20, and 24 acknowledge financial support from the Ministry of Science of Montenegro during the bilateral project The Role of Mountains in Determining the Identity of Slovenia and Montenegro (2018–2021).

We would like to thank all the reviewers for their time and effort and their insightful comments on the chapters because these led to the improvement of the manuscripts. In this regard, we thank Jelena Beloica, Predrag Djurović, Pablo Dominguez, Andrea Carteny, Andraž Čarni, Maruša Goluža, Maja Gori, Jana Horvat, Borut Juvanec, Domen Kušar, Peter Mikša, Janez Mlinar, Gojko Nikolić, Darko Ogrin, Olga Pelcer-Vujačić, Hrvoje Petrić, Olivera Popović, Nika Razpotnik Visković, Slavica Stamatović Vučković, Slobodan Vukićević, and Matija Zorn.

Matija Zorn
Olga Pelcer-Vujačić
Peter Mikša

Contents

Editors and Contributors

About the Editors

Matija Zorn was born in 1975 in Kranj, Slovenia. In 2001, he received a bachelor's degree in geography and history at the University of Ljubljana's Faculty of Arts, and he received his doctorate in geography from the same institution in 2007. In 2001 he was employed by the Anton Melik Geographical Institute at the Research Centre of the Slovenian Academy of Sciences and Arts (ZRC SAZU) (Ljubljana, Slovenia). He served as the institute's assistant director from 2007 to 2018, and since 2018 he has been its director. From 2008 to 2022, he headed the institute's Department of Physical Geography.

His research interests focus on physical geography, especially geomorphology and the geography of natural disasters, and he also works on land degradation, environmental history, historical geography, and geographic information systems. Since 2011 he has been a lecturer at the Faculty of Humanities at the University of Primorska (Koper, Slovenia). He heads projects, advises junior researchers, and is the coeditor of several book series. Since 2011 he has been the Editor-in-Chief of the SCOPUS-indexed journal *Geografski vestnik* (Geographical Bulletin) and the Area Editor for physical geography for the SCIE and SCOPUS-indexed journal *Acta geographica Slovenica*. Currently he is the Chair of the Commission on Land Degradation and Desertification (COMLAND) of the International Geographical Union.

Olga Pelcer-Vujačić was born in 1977 in Belgrade, Serbia. In 2001 she received a bachelor's degree in history at the University of Belgrade's Faculty of Philosophy, and she received her doctorate in ancient history from the same institution in 2015. From 2003 until 2013, she was employed by the Faculty of Philosophy in Belgrade. Since 2013 she has worked at the University of Montenegro's Historical Institute. She served as the institute's deputy director for international relations from 2020 to 2023.

Her research interests focus on ancient history, epigraphy, ancient cultural heritage, and digital humanities. She has led and actively participated in several national and international projects concerning the Balkans in antiquity, epigraphy and cultural heritage, and GIS. During her undergraduate and graduate education, she studied in Oxford and London (in 2001 and 2003), was a scholarship holder of the CEEPUS program in Austria (University of Graz, Department of Southeast European History) in May 2005, and was a DAAD scholarship holder for a two-month research visit at the University of Heidelberg, Germany (June–July 2017). Among other things, she conducted research at institutions in Cologne (Institute of Classical Studies at the University of Cologne), Rome (German Archaeological Institute, CNR-ISMA, French School of Rome, American Academy in Rome), and Athens (French School of Athens, American School of Classical Studies). She is a member of ICOM Montenegro and the Society for Ancient Studies. Since 2023 she has been the Editor-in-Chief of the journal *Istorijski zapisi* (*Les Écrits Historiques*).

Peter Mikša was born in 1977 in Celje, Slovenia. In 2007 he received a bachelor's degree in history at the University of Ljubljana's Faculty of Arts, where he also received his doctorate in history in 2013. In 2009 he was employed at the faculty's Department of History, where he has worked ever since. In 2023 he was named an Associate Professor of Slovenian contemporary history.

His research interests focus on contemporary Slovenian history, especially on border studies, the Rapallo border, Second World War borders in Slovenia (1941–1945), the history of sports, local history, the history of tourism, and the Alpine region. He has authored several books, especially on Second World War borders and mountaineering, one of his most influential works being *Zgodovina slovenskega alpinizma* (The History of Slovenian Alpinism). He also leads projects, such as the basic research project. The Struggle for the Mountains: Slovene National Appropriations of the Alps, 19th–21st Centuries, the target research project Supplementing the Register of Hidden War Graves and the Register of War Graves with Data on Hidden War Graves in Caves, and the bilateral project The Role of Mountains in Determining the Identity of Slovenia and Montenegro. He is also the Editor of several historical and mountaineering publications, the most important being the journal *Retrospektive*, for which he has been serving as the Editor-in-Chief since 2018. He is the head of the Center for Public History at the University of Ljubljana's Faculty of Arts.

Contributors

Kornelija Ajlec Department of History, Faculty of Arts, University of Ljubljana, Ljubljana, Slovenia

Vladimir Bakrač Faculty of Philosophy, University of Montenegro, Nikšić, Montenegro

Slavko Burzanović Historical Institute, University of Montenegro, Podgorica, Montenegro

Goran Ćeranić Faculty of Philosophy, University of Montenegro, Nikšić, Montenegro

Pavle Cikovac Department of Geography, University of Munich, Munich, Germany

Mirela Djurović Faculty of Philosophy, University of Belgrade, Belgrade, Serbia

Predrag Djurović Faculty of Geography, University of Belgrade, Belgrade, Serbia

Alenka Fikfak Faculty of Architecture, University of Ljubljana, Ljubljana, Slovenia

Primož Gašperič ZRC SAZU Anton Melik Geographical Institute, Ljubljana, Slovenia

Matjaž Geršič ZRC SAZU Anton Melik Geographical Institute, Ljubljana, Slovenia

Miljan Gogić Historical Institute, University of Montenegro, Podgorica, Montenegro

Janez P. Grom Faculty of Architecture, University of Ljubljana, Ljubljana, Slovenia

Mauro Hrvatin ZRC SAZU Anton Melik Geographical Institute, Ljubljana, Slovenia

Olga Jokanović Biotechnical Faculty, University of Montenegro, Podgorica, Montenegro

Danilo Kalezić Historical Institute, University of Montenegro, Podgorica, Montenegro

Aleksandra Kapetanović EXPEDITIO Center for Sustainable Spatial Development, Kotor, Montenegro

Janez Kermavnar Department of Forest Ecology, Slovenian Forestry Institute, Ljubljana, Slovenia

Tatjana Koprivica Historical Institute, University of Montenegro, Podgorica, Montenegro

Carola Küfmann Department of Geography, University of Munich, Munich, Germany

Ivan Laković Historical Institute, University of Montenegro, Podgorica, Montenegro

Kristijan Lavtižar Faculty of Architecture, University of Ljubljana, Ljubljana, Slovenia

Jelena Lazarević Biotechnical Faculty, University of Montenegro, Podgorica, Montenegro

Špela Ledinek Lozej ZRC SAZU Institute of Slovenian Ethnology, Ljubljana, Slovenia

Peter Mikša Department of History, Faculty of Arts, University of Ljubljana, Ljubljana, Slovenia

Božena Miljić University Donja Gorica, Podgorica, Montenegro

Gojko Nikolić Department of Geography, Faculty of Philosophy, University of Montenegro, Nikšić, Montenegro

Marija Ogrin ArheoAlpe zavod za kulturo, izobraževanje in turizem Bohinj, Bohinjska Bistrica, Slovenia

Matej Ogrin Department of Geography, Faculty of Arts, University of Ljubljana, Ljubljana, Slovenia

Gašper Oitzl Department of History and Applied Arts, National Museum of Slovenia, Ljubljana, Slovenia

Olga Pelcer-Vujačić Historical Institute, University of Montenegro, Podgorica, Montenegro

Drago Perko ZRC SAZU Anton Melik Geographical Institute, Ljubljana, Slovenia

Miloš B. Petričević Museums of Kotor, Kotor, Montenegro

Primož Pipan ZRC SAZU Anton Melik Geographical Institute, Ljubljana, Slovenia

Katarina Polajnar Horvat ZRC SAZU Anton Melik Geographical Institute, Ljubljana, Slovenia

Olga Radonjić Faculty of Philosophy, University of Montenegro, Nikšić, Montenegro

Miomirka Rakonjac Faculty of Philosophy, University of Montenegro, Nikšić, Montenegro

Obrad Samardžić Faculty of Philosophy, University of Montenegro, Nikšić, Montenegro

Rade Šarović Faculty of Philosophy, University of Montenegro, Nikšić, Montenegro

Mateja Šmid Hribar ZRC SAZU Anton Melik Geographical Institute, Ljubljana, Slovenia

Aleš Smrekar ZRC SAZU Anton Melik Geographical Institute, Ljubljana, Slovenia

Domen Svetlin Envirodual d.o.o., Slovenske Konjice, Slovenia

Maja Topole ZRC SAZU Anton Melik Geographical Institute, Ljubljana, Slovenia

Mimi Urbanc ZRC SAZU Anton Melik Geographical Institute, Ljubljana, Slovenia

Predrag Živković Faculty of Philosophy, University of Montenegro, Nikšić, Montenegro

Matija Zorn ZRC SAZU Anton Melik Geographical Institute, Ljubljana, Slovenia

1 Discourses on Mountains of Montenegro and Slovenia: An Introduction

Matija Zorn, Olga Pelcer-Vujačić, and Peter Mikša

Abstract

Montenegro and Slovenia have been associated with mountains for centuries. Over the course of history, the two countries have largely followed different paths, but some geographical, cultural, and historical features of their mountains have nonetheless been similar. This duality (i.e., their similarities on the one hand and differences on the other) was one of the main inspirations for this volume, which is the first multidisciplinary work on geographical, cultural, historical, and some other aspects of mountains in Montenegro and Slovenia. It contains twenty-nine chapters, written by forty-five authors. The title of the volume, with "discourses" in the plural, emphasizes the diversity of the topics discussed, which all shed light on the mountains in the two countries. Chronologically, the volume deals with topics ranging from prehistory to the present, spatially covering most of Montenegro's and Slovenia's mountainous areas.

Keywords

Mountains · Nature · Culture · Identity · Montenegro · Slovenia

1.1 Introduction

"Mountains were the arena of relatively independent social, economic, and ideological development, where specific forms of settlement, modes of subsistence, and social structures developed" (Pelisiak et al. 2018:4). This is something that all mountainous areas have in common, including those in Montenegro and Slovenia, which are discussed in this volume. "It is true to say that mountain people throughout the world—beyond their cultural, religious or political differences—easily feel at one. From the close ties with their environment, which is often their prime resource, to the community traditions that are deeply rooted thanks to the age-old need of mutual help, via their knowledge of nature" (Debarbieux and Rudaz 2008:509). Despite these similarities, mountainous areas have developed their own special local features that contribute to their diversity (Mountain … 2016).

"Mountain peoples are sometimes said to share a specific type of relation with the natural environment" (Debarbieux and Rudaz 2008:508), which

M. Zorn (✉)
ZRC SAZU Anton Melik Geographical Institute, Ljubljana, Slovenia
e-mail: matija.zorn@zrc-sazu.si

O. Pelcer-Vujačić
Historical Institute, University of Montenegro, Podgorica, Montenegro

P. Mikša
Department of History, Faculty of Arts, University of Ljubljana, Ljubljana, Slovenia
e-mail: peter.miksa@ff.uni-lj.si

M. Zorn et al. (eds.), *Discourses on Mountains of Montenegro and Slovenia*,
Historical Geography and Geosciences, https://doi.org/10.1007/978-3-031-99017-5_1

helped shape the local communities and people's identity (see Chaps. 14, 15, 16, and 17 of this volume). People living in mountainous areas have historically developed practices that are adapted to the challenges and opportunities presented by their environment (e.g., transhumance and terraced slopes). These adaptations, associated with isolation or peripherality, can give rise to differences in belief systems, building materials, architecture, clothing, cuisine, customs, music, or other cultural practices. Today, this rich cultural heritage (see Chaps. 21, 22, 23 and 24) affects the tourist appeal of mountainous areas and thus also their globalization, with all its advantages and disadvantages (Price 2015).

Over the course of history, Montenegro and Slovenia largely followed different paths, but some cultural, historical (see Chaps. 6, 7, 8, 9 and 10), and geographical features (see Chaps. 2, 3, 4, 5, and 29) of their mountains have nonetheless been similar. This duality (i.e., similarities on the one hand and differences on the other) was one of the main inspirations for this volume. Montenegro and Slovenia have been associated with mountains for centuries, mainly thanks to scientists, travelers (tourists), or artists (see Chap. 18; Mikša and Zorn 2016), and this volume continues this tradition.

Mountainous areas cover between 26.4% and 30.5% of the global land surface (excluding Antarctica; Sayre et al. 2018). In Montenegro, areas above an elevation of 500 m cover over four-fifths of the country, and areas above an elevation of 1,000 m cover nearly three-fifths (see Chaps. 2 and 29). In Slovenia, areas above 500 m in elevation cover over half of the country, and areas above 1,000 m account for over a tenth; just under three-fifths of the country are defined as mountainous areas (see Chaps. 3, 4, 5 and 29).

Mountains were (and still are) often perceived as peripheral areas (Pelisiak et al. 2018), which are nonetheless home to roughly a tenth (i.e., 14%) of the world population (Ehrlich et al. 2021); estimates vary between less than 5% (or approximately 0.34 billion) and over 31% (or approximately 2.29 billion) of the word population (Thornton et al. 2022), and the corresponding share in Europe is just under a fifth (Europe's ... 2010). Montenegro and Slovenia are no exception in this regard (e.g., around a quarter of Slovenia's population lives in the mountains; see Chap. 5).

Many lowland regions across the globe depend on resources originating in mountainous regions. Projections show that, by the mid-twenty-first century, nearly a quarter of the world's lowland population, or approximately 1.5 billion people, will depend on water coming from mountainous areas alone (Viviroli et al. 2020). Therefore, it is not surprising that mountains have been defined as "the world's water tower" (Price 2015:32), which, according to some estimates, supplies half of the world's population with fresh water for drinking, domestic use, irrigation, industry, and hydropower. Forty percent of the world's irrigation is supported by flows originating from large mountain systems (Mountains ... 2014). There are also various mineral resources (e.g., nearly half of the world's supplies of silver originate in the mountains of western North America, nearly a third of the world's supplies of copper come from the Chilean part of the Andes, and half of the world's supplies of tungsten originate in the mountains of southern China; Price 2015) and, first and foremost, forest. Forests in mountainous areas account for over a fourth of all global forests (Price 2015). In Europe, forests cover 41% of mountainous areas (Europe's ... 2010), and forest cover in Montenegro and Slovenia is significantly above average (see Chap. 25). In mountainous areas, wood is an important building material (see Chap. 21) and the primary source of energy. In addition, forests have an important protective function against natural hazards (e.g., avalanches, rockfalls, or storms), and they also hold significant stores of carbon (Price 2015).

Mountainous areas are inseparably connected with leisure and tourism (e.g., mountain hiking and skiing), as are many protected areas (see Chap. 26). Globally, these cover just under one-sixth of mountains (Price 2015). In Europe, Natura 2000 sites cover 43% of mountainous areas (Europe's ... 2010). Today, mountains account for 15% to 20% of the global tourism market (Ng 2022). In this regard, Montenegro and Slovenia differ significantly, with overnights at Montenegrin mountain resorts accounting for only 2% of all overnight stays, whereas in Slovenia overnights at mountain resorts account for nearly a third of all overnights (see Chap. 29).

In terms of the "benefits" provided by mountains, today ecosystem services come to the fore. "Provisioning services come from agricultural and forestry systems; natural ecosystems; and rivers, which provide water and hydroelectricity. Regulating services relate particularly to climate, air quality, water flow, and the minimisation of natural hazards. Cultural services are associated with tourism, recreation, aesthetics, protected areas and locations of religious importance" (Europe's … 2010).

In the past, mountainous areas were associated with certain social characteristics (Gams 2001), the most prominent among which include a rich (fragmented) ethnic and linguistic composition associated with the difficulty in traversing mountains, a delay in innovations, based on which mountainous areas were defined as backward, seasonal movement of people and their livestock (to higher elevations in summer and to valleys in winter; see Chap. 8), which used to be vital for people to make a living and now largely remains an special ethnological feature, and recent extensive rural depopulation resulting in land abandonment and the overgrowth and deterioration of the cultural landscape (see Chaps. 11, 12 and 13), and the loss of habitat diversity. In general, mountainous areas are considered biodiversity hotspots (Price 2015; see Chaps. 27 and 28).

Mountains have always served as areas of separation but also connection (Pelisiak et al. 2018), and their protective role has been gratefully acknowledged by those that have lived there or by those that sought refuge there.

However, changes in mountain landscapes are not only associated with social processes; they are also affected by global climate change (Hock et al. 2019). The most evident changes in the global mountains include the melting of glaciers, including in Montenegro (Gachev 2020) and Slovenia (Zorn et al. 2020), and the movement of plants to higher elevations. Other changes include the disappearance of permafrost (and the consequent increase in rockfall hazards and threats to infrastructure), a general decline in low-elevation snow cover (and the consequent impact on winter tourism), changes in the volume and seasonality of river discharge (see Chap. 5) and the impact on the occurrence of floods and droughts, and an increase in fire hazards and potential pest outbreaks (e.g., bark beetle outbreaks; Price 2015; Thornton et al. 2022).

1.2 Montenegro, Slovenia, and Mountains

Montenegro's mountainous identity is highlighted in its very name (literally, 'black mountain'), first attested in the eleventh century (Rastoder 2003). Moreover, Mounts Lovćen, Durmitor, and Prokletije (see Chap. 2) are among the most familiar features associated with Montenegro in general, and they are also the locations of national parks bearing the same names. In addition, Mount Lovćen (see Chaps. 20 and 24) was depicted on the coat of arms of the Socialist Republic of Montenegro during the Yugoslav era.

Mount Lovćen and its foothills are a historical area where the medieval Crnojević noble family ruled the Principality of Zeta (see Chap. 9), which encompassed parts of present-day Montenegro (Andrijašević and Rastoder 2006). Although it was not the only one, this area is the most familiar stronghold of the liberation struggle of Montenegrin and highland tribes against Ottoman rule.

The strategic importance of Mount Lovćen attracted the interest of military and diplomatic circles in the second half of the nineteenth century, and especially at the beginning of the twentieth century (see Chap. 20). When it was captured by Austria-Hungary in January 1916, this marked the end of the Kingdom of Montenegro (Andrijašević and Rastoder 2006), both factually and symbolically.

Mount Lovćen, with the Njegoš mausoleum located on one of its peaks (*Jezerski vrh*, 1,659 m; see Chap. 24), stands out as an important part of national cultural heritage. This memorial, built in 1845, changed its architectural and artistic characteristics twice, following the political changes that Montenegro experienced in the twentieth century (Simic 1997).

Although Mounts Lovćen and Durmitor (see Chaps. 13 and 16) receive the most attention, the mountainous areas (see Chap. 2) of Prokletije Mountains, Komovi Mountains, Kuči Mountains, Bjelasica Mountains (see Chap. 21), Mount Sinjajevina (see Chap. 8), the Morača Mountains, and the Maganik mountain range in the continental part of the country, as well as Mounts Rumija (see Chap. 14) and Orjen in the coastal part, have equally important and distinguishing attributes of recognizable natural and cultural-historical domains. Their cultural heritage (see Chap. 21) is multilayered and rich, and it is closely connected with the transhumant nature of Montenegrin society.

Mountains have left a special mark on people in **Slovenia**, which they continue to identify with (see Chap. 15; Šaver 2005). Mount Triglav (2,864 m), as the highest mountain in the Julian Alps and in Slovenia, is the most prominent in this regard (see Chap. 19). It is depicted on the national coat of arms and on the national flag. The mountain was already portrayed on the coat of arms of the Socialist Republic of Slovenia during the Yugoslav era. It is the most powerful national symbol for the Slovenians (Mikša 2021).

With the advent of the Renaissance and the development of the natural sciences, the Slovenian Alpine landscape, like other areas in the Alps, became a place for research, especially its geology, flora, and fauna (Mikša and Zorn 2016).

During the strengthening of Slovenian national consciousness in the nineteenth century (Slovenian territory was then part of the Austrian Empire), the symbolism of the mountain landscape and Alpine culture became one of the central points in the formation of the Slovenians as a nation. In the late nineteenth century, hiking clubs were established on Slovenian territory, first a German–Austrian one in 1874 and later a Slovenian one in 1893 (Mikša and Ajlec 2015). At that time, the "battle" between Slovenians and the German–Austrians for primacy in Slovenia's mountains, which took place in the form of establishing hiking trails and building mountain lodges (i.e., "marking" and "controlling" the mountains), also contributed to Slovenian national consciousness (see Chap. 19; Mikša and Zorn 2018).

After the First World War, Slovenians became part of a new Slavic state (i.e., the Kingdom of Serbs, Croats, and Slovenes, later named Yugoslavia) with an identity already formed, in which the mountains and especially Mount Triglav played an important role. This was also reflected in culture (e.g., in the first Slovenian feature film, *V kraljestvu Zlatoroga* 'In the Kingdom of the Goldhorn', filmed in the Julian Alps with an ascent of Mount Triglav; released in 1931) and in the popularity of alpinism, alpine skiing, or ski jumping. Slovenians retained their mountain identity even after the Second World War in communist Yugoslavia. Mountain activities strengthened Slovenian identity and the profile of the nation, also in the form of climbing the highest mountains in the world (Mikša and Zorn 2024). It is important to mention the large-scale phenomenon of mountain hiking, mountain literature, touring, alpine skiing, and alpinism, and the continuity of Mount Triglav as a prominent symbol of the Slovenians (see Chaps. 15 and 19).

1.3 Content and Scope of This Volume

This volume is the first multi-authored inter- and multidisciplinary work on geographical, cultural, historical, and other aspects of mountains in Montenegro and Slovenia. It contains twenty-nine chapters, written by forty-five authors: geographers, historians, archaeologists, architects, sociologists, ethnologists, and biotechnology and forestry specialists, representing eighteen institutions from Montenegro, Slovenia, Serbia, and Germany. The chapters were prepared for publication in the years 2022 to 2024.

The title of this volume, with "discourses" in the plural, emphasizes the diversity of the topics discussed, which all shed light on the mountains in the two countries. Chronologically, the volume deals with topics ranging from prehistory to the present, spatially covering most of Montenegro's and Slovenia's mountainous areas (Fig. 1.1).

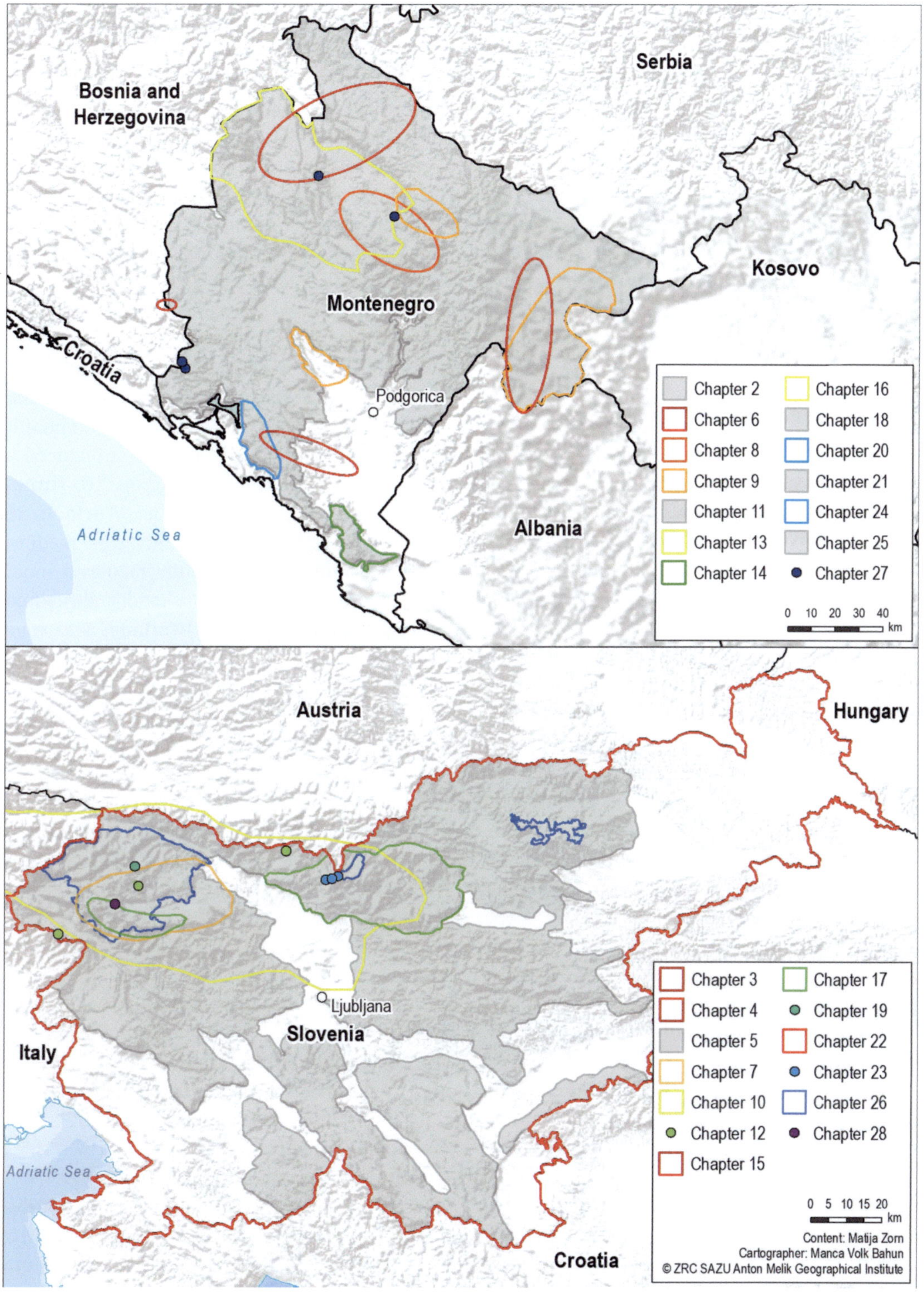

Fig. 1.1 Spatial coverage of chapters in the volume

The chapters address various research topics, forming separate parts of the book. Following the introductory chapter by the volume editors, **Part I** provides a general geographical overview of mountains in Montenegro and Slovenia. **Part II** focuses on prehistory and the Middle Ages in the mountains of the two countries, and **Part III** on changes in the cultural landscape. **Part IV** discusses the relations between identity and mountains, **Part V** is dedicated to cultural heritage, and **Part VI** focuses on the natural environment. The volume ends with a conclusion as **Part VII**.

Part I: Geographical Overview

This part includes four chapters, one discussing Montenegro's mountains and three focusing on Slovenia's mountains. In Chap. 2, **Mountains of Montenegro**, Predrag and Mirela Djurović present the division and spatial distribution of mountains in Montenegro, as well as their physical-geographical features. This is followed by three chapters discussing the mountains of Slovenia. In Chap. 3, **The Definition of Mountainous Areas in Slovenia**, Drago Perko defines mountainous areas in Slovenia. The definition of mountainous areas in Slovenia from this chapter is used in several other chapters in the volume (e.g., in Chaps. 4, 5, 17, and 26). In Chap. 4, **Geomorphometry of Slovenia's Mountainous Surface**, Drago Perko uses some basic geomorphometric indicators, such as surface height, slope, and aspect, to compare Slovenia's mountainous and non-mountainous areas, as well as to introduce a geomorphometric landform typology of Slovenia. In Chap. 5, **Mountains of Slovenia**, Mauro Hrvatin and Matija Zorn present the basic physical-geographical and human-geographical features of Slovenia's mountains.

Part II: Prehistory and the Middle Ages

This part includes five chapters, three discussing Montenegro's mountains and two dealing with Slovenia's mountains. In Chap. 6, **Continuity of Mountain Settlements in Montenegro**, Olga Pelcer-Vujačić provides an overview of mountain settlements from prehistory to Roman times. In Chap. 7, **Human Presence in the Eastern Julian Alps from the Copper Age to the Early Middle Ages**, Marija Ogrin presents evidence of the human presence in the Slovenian Alps from prehistory to the early Middle Ages. In Chap. 8, **Prehistory of High-Altitude Landscape in Montenegro: Mount Sinjajevina Case Study**, Miloš B. Petričević presents recent archaeological investigations at high elevations that shed new light on the human capacity to survive in extreme and isolated environments, which also provide insight into the prehistoric origins of transhumant pastoralism in Montenegro. In Chap. 9, **Mountains as Medieval Man's Habitat in Montenegrin Territory**, Miljan Gogić discusses the formation of isolated medieval communities and their way of life, which was influenced by natural and social factors. In Chap. 10, **Human Activity in the Slovene Alpine World in the Middle Ages**, Gašper Oitzl presents medieval mountain colonization and the medieval mountain economy (e.g., animal husbandry, agriculture, hunting, mining and metalworking, and transport).

Part III: The Downfall of Cultural Landscape

This part includes three chapters, two discussing Montenegro's mountains and one addressing Slovenia's mountains. In Chap. 11, **Abandoning Traditional Identity—Depopulation of the Mountain Areas of Montenegro Throughout the Second Half of the 20th Century**, Ivan Laković discusses the socioeconomic circumstances that led to the large-scale depopulation of mountain areas in Montenegro and, along with this, the loss of traditional (mountain) identity. In Chap. 12, **Mountain Pastures in the Slovenian Alps: Their Role in Shaping and Sustaining the Cultural Landscape**, Mimi Urbanc, Špela Ledinek Lozej, and Mateja Šmid Hribar focus on mountain pastures and their role in shaping and maintaining the cultural landscape. In Chap. 13, **Population Migrations from Poddurmitor Region (Consequences and Alternative Development Paths)**, Rade Šarović and Goran Ćeranić discuss issues related to out-migration in Durmitor National Park, and they also present an alternative development model to change this trend.

Part IV: Identity and the Mountains

This part includes seven chapters, four discussing Montenegro's mountains and three focusing on Slovenia's mountains. In Chap. 14, **The Archetypal Discourse of the Mountain Rumija as a Source of Montenegrin Identity and the Significance of its Existence through the Character and Work of Prince Vladimir**, Miomirka Rakonjac, Predrag Živković, and Vladimir Bakrač discuss the geoidentity of Mount Rumija and its symbolic importance as the origin of Montenegrin statehood and spirituality. In Chap.15, **Slovenian Mountain Attachment as a Hallmark of a Nation**, Peter Mikša and Matej Ogrin present the relationship that Slovenians have with the mountains, and how mountains have become an integral part of national identity, with Mount Triglav being the most important national symbol. In Chap.16, **Spatio-Temporal Landscape of Durmitor and Drobnjak: Its Role in Creation of Highland Community's Local Identity and Its Toponyms**, Obrad Samardžić, Rade Šarović, and Olga Radonjić discuss the spatio-temporal landscape of Mount Durmitor and the area of Drobnjak, and its role in shaping the local identity; they also discuss the various origins of local toponyms. In Chap. 17, **Geographical Names and Place Identity in the Slovenian Alps**, Matjaž Geršič, Primož Gašperič, and Matija Zorn focus on the origin of geographical names in the Slovenian Alps, their standardization, and their role connected to national identity. In Chap. 18, **Depiction of the Montenegrin Mountains as Seen in British, German, and French Travelogues: Geosophy in Montenegro in the 19th and Beginning of the 20th Century**, Božena Miljić and Danilo Kalezić address the experiences of foreign travel writers encountering Montenegro's mountains and the highlanders that lived there—that is, how nineteenth- and early twentieth-century foreign travelers understood the cultural identity of Montenegro's mountainous communities. In Chap. 19, **Mount Triglav and Its Role among Slovenes**, Peter Mikša and Kornelija Ajlec explain why Mount Triglav, the highest Slovenian peak, which is also depicted on the national coat of arms and the national flag, is the most prominent Slovenian national symbol. In Chap. 20, **"The Issue of Mt. Lovćen" in World War I and Italy**, Slavko Burzanović discusses the military and political significance of Montenegro's "holy" Mount Lovćen during the First World War, along with Austro-Hungarian and Italian territorial ambitions in Montenegro.

Part V: Cultural Heritage

This part includes four chapters, two covering Montenegro's mountains and two discussing Slovenia's mountains. In Chap. 21, **Examples of Immovable Cultural Heritage of Montenegrin Mountains**, Aleksandra Kapetanović focuses on the traditional architecture of temporary seasonal mountain settlements known as *katuni*, and especially on various types of huts and their development. In Chap. 22, **Alpine Heritage in Slovenia: Between Authorisation and Imagination**, Špela Ledinek Lozej, Primož Pipan, and Maja Topole discuss the presence of Alpine heritage in official registers and compare the presence of different heritage units (natural and cultural) in the Slovenian Alps with other Slovenian macroregions. In Chap. 23, **Recognition of Spatial Properties for the Location Assessment and Design of Bivouacs in the Slovenian Mountains**, Kristijan Lavtižar, Alenka Fikfak, and Janez P. Grom identify the spatial properties that define bivouac placement according to their basic rescue functions and the natural characteristics and values of the site. In Chap. 24, **The Chapel of Njegoš and the Mausoleum on Mt. Lovćen**, Tatjana Koprivica discusses the history of the construction of the chapel and mausoleum on Mount Lovćen dedicated to Petar II Petrović Njegoš, who ruled Montenegro between 1830 and 1851; Mount Lovćen is a symbol of Montenegrin statehood and national identity.

Part VI: The Natural Environment

This part includes four chapters, two covering Montenegro's mountains and two discussing Slovenia's mountains. In Chap. 25, **Mountain Forests in Montenegro: Usage and Public Perception**, Jelena Lazarević, Olga Jokanović, and Ivan Laković discuss the importance of

Montenegro's forests as a natural resource, and they also point out the threats connected with exploitation; they also discuss the public perception of forests. In Chap. 26, **Mountain Environment Protection and Protected Mountain Areas in Slovenia**, Aleš Smrekar, Katarina Polajnar Horvat, and Matija Zorn discuss the history of natural environment conservation in Slovenia, with an emphasis on protected mountainous areas. In Chap. 27, **Temperature Conditions and Vegetation in Dinaric Frost Hollows**, Matej Ogrin, Gojko Nikolić, Domen Svetlin, Pavle Cikovac, and Carola Küfmann discuss the temperature and vegetation conditions in high mountain topographic depressions (i.e., frost hollows) in Montenegro's mountains, which act as cold air pools. In Chap. 28, **Frost Hollows as Interesting Areas of Geo- and Biodiversity in the Alps**, Domen Svetlin, Matej Ogrin, and Janez Kermavnar similarly discuss the climatological and vegetation characteristics of frost hollows in the Slovenian Alps, as well as their role in terms of geodiversity and biodiversity.

Part VII: Conclusion

In the concluding Chap. 29, **Between Slovenia and Montenegro**, Drago Perko and Matija Zorn sum up the geographical and historical connections between Montenegro and Slovenia.

Acknowledgements Support for producing this volume included funding from the Slovenian Research and Innovation Agency's program no. P6-0101 (Geography of Slovenia) and program no. P6-0235 (Slovene History).

References

Andrijašević ŽM, Rastoder Š (2006) Istorija Crne Gore: od najstarijih vremena do 2003 [History of Montenegro: from the earliest times to 2003]. Centar za iseljenike Crne Gore, Podgorica

Debarbieux B, Rudaz G (2008) Linking mountain identities throughout the world: the experience of Swiss communities. Cult Geogr 15(4):497–517. https://doi.org/10.1177/1474474008094619

Ehrlich D, Melchiorri M, Capitani C (2021) Population trends and urbanisation in mountain ranges of the world. Land 10. https://doi.org/10.3390/land10030255

Europe's ecological backbone: Recognising the true value of our mountains (2010) EEA, Copenhagen

Gachev EM (2020) Small glaciers in the Dinaric Mountains after eight years of observation: on the verge of extinction? Acta Geogr Slov 60(2):191–211. https://doi.org/10.3986/AGS.8092

Gams I (2001) O Afganistancih in drugih gorjancih [On the Afghanistans and other highlanders]. Geografski Vestnik 73(2):33–41

Hock R, Rasul G, Adler C, Cáceres B, Gruber S, Hirabayashi Y, Jackson M, Kääb A, Kang S, Kutuzov S, Milner A, Molau U, Morin S, Orlove B, Steltzer H (2019) High mountain areas. In: IPCC special report on the ocean and cryosphere in a changing climate. Cambridge University Press, Cambridge, pp 131–202. https://doi.org/10.1017/9781009157964.004

Mikša P (2021) Od gore do simbola: Triglav in njegova vloga pri Slovencih [From a mountain to a symbol: Mount Triglav and its role among Slovenes]. Geografija v šoli 29(2):6–14. https://doi.org/10.59132/geo/2021/2/6-14

Mikša P, Ajlec K (2015) Slovensko planinstvo = Slovene mountaineering. Planinska zveza Slovenije, Ljubljana

Mikša P, Zorn M (2016) The beginnings of the research of Slovenian Alps. Geografski Vestnik 88(2):103–131. https://doi.org/10.3986/GV88206

Mikša P, Zorn M (2018) The "battle" for the mountains: Germans versus Slovenes in the South-Eastern alps in the late 19th and early 20th century. Adv Environ Res 65:199–227

Mikša P, Zorn M (2024) The 'Slovenian' Everest 1979: a small nation and the highest mountain in the world. In: Gilchrist P, Hansen P, Westaway J (eds) Other Everests: one mountain, many worlds. University of Manchester, Manchester, pp 178–191. https://doi.org/10.7765/9781526179173.00016

Mountains as the water towers of the world: a call for action on the sustainable development goals (SDGs) (2014) https://www.grida.no/publications/242. Accessed 7 Jul 2024

Mountain cultures: celebrating diversity and strengthening identity (2016) FAO, Rome. https://openknowledge.fao.org/server/api/core/bitstreams/3183ea2e-521d-4cde-bb59-5336904c70c0/content. Accessed 7 Jul 2024

Ng SL (2022) Bibliometric analysis of literature on mountain tourism in Scopus. J Outdoor Recreat Tour 40. https://doi.org/10.1016/j.jort.2022.100587

Pelisiak A, Nowak M, Astaloş C (2018) Introduction. In: Pelisiak A, Nowak M, Astaloş C (eds) People in the mountains: current approaches to the archaeology of mountainous landscapes. Archaeopress, Oxford, pp 1–5

Price M (2015) Mountains: a very short introduction. Oxford University Press, Oxford. https://doi.org/10.1093/actrade/9780199695881.001.0001

Rastoder Š (2003) A short review of the history of Montenegro. In: Bieber F (ed) Montenegro in transition: problems of identity and statehood. Nomos Verlagsgesellschaft, Baden-Baden, pp 107–137

Sayre R, Frye C, Karagulle D, Krauer J, Breyer S, Aniello P, Wright DJ, Payne D, Adler C, Warner H, VanSistine DP, Cress J (2018) A new high-resolution map of world mountains and an online tool for visualizing and comparing characterizations of global mountain distributions. Mt Res Dev 38(3):240–249. https://doi.org/10.1659/MRD-JOURNAL-D-17-00107.1

Simic A (1997) Montenegro: beyond the myth. In: Danopoulos CP, Messas KG (eds) Crises in the Balkans. Routledge, New York

Šaver B (2005) Nazaj v planinski raj: alpska kultura slovenstva in mitologija Triglava [Back to the mountain paradise: The alpine culture of Slovenia and the mythology of Triglav]. Fakulteta za družbene vede, Ljubljana

Thornton JM, Snethlage MA, Sayre R, Urbach DR, Viviroli D, Ehrlich D, Muccione V, Wester P, Insarov G, Adler C (2022) Human populations in the world's mountains: spatio-temporal patterns and potential controls. Plos One 17(7). https://doi.org/10.1371/journal.pone.0271466

Viviroli D, Kummu M, Meybeck M, Kallio M, Wada Y (2020) Increasing dependence of lowland populations on mountain water resources. Nat Sustain 3:917–928. https://doi.org/10.1038/s41893-020-0559-9

Zorn M, Komac B, Carrey A, Hrvatin M, Ciglič R, Lyons B (2020) The disappearing cryosphere in the southeastern Alps: introduction to special issue. Acta Geogr Slov 60(2):109–124. https://doi.org/10.3986/AGS.9396

Part I
Geographical Overview

Mountains of Montenegro

2

Predrag Djurović and Mirela Djurović

Abstract

Montenegro is a mountainous country where more than 50% of the land is more than 1,000 m above sea level. There are two mountain systems: the Dinaric Alps and the Prokletije Alps. In the Dinaric Alps the following mountain zones stand out: the Western Mountain zone (coastline), the Mid-Mountain zone and the Northeastern zone. The Prokletije Alps are divided into two zones by the valley of the Lim River and display different orographic, geological and other characteristics in comparison to mountains in the Dinaric system. In the Western and Mid-Mountain zone of the Dinaric Alps, carbonate rocks of Mesozoic age that exceed 2 km in thickness in some places prevail. The Northeastern zone consists of igneous and metamorphic rocks. The Prokletije Alps are very heterogeneous in composition, playing home to a high geo- and biodiversity, as is the case in the overall mountainous area in Montenegro. The mountains have a diverse climate with extreme values of distinct climate elements depending on the distance from the sea and elevation. Maximal annual precipitation on slopes in the Western Mountain zone exceeds 9,000 mm, whilst the average annual precipitation is above 5,000 mm. The mean air temperature in the highest parts of Montenegro is about 0 °C. The domination of carbonate rocks in mountains prejudices their hydrological properties. Hence, the mountains are dry and commonly waterless. However, in many places in the mountainous areas there are allogenic rivers that have incised canyons of over 1,000 m in depth. The basins of a number of recent mountain lakes were formed by glacial erosion, while at Mt. Durmitor the recent glacier–Debeli Namet–still survives. In the mountain relief, karst relief forms dominate. These include a number of poljes, uvalas, dry valleys and frequent occurrences of holokarst. Glaciokarst reliefs are also noteworthy and typical in mountain areas. There are a number of caves whose depth exceeds 1,000 m and reach a maximal length of 17 km. At the highest levels there are preserved tool marks of Pleistocene glaciation displayed by a number of cirques, basins and moraines. These forms owe their origin primarily to cirque, valley and plateau glaciers that covered more than 50% of the mountain area during the most intensive glaciation. Diverse recent periglacial and colluvial reliefs have also evolved on the mountains. A striking geodiversity has supported the development of a rich biodiversity in the

P. Djurović
Faculty of Geography, University of Belgrade, Belgrade, Serbia
e-mail: geodjura@gmail.com

M. Djurović (✉)
Faculty of Philosophy, University of Belgrade, Belgrade, Serbia
e-mail: mirela.djurovic@gmail.com

M. Zorn et al. (eds.), *Discourses on Mountains of Montenegro and Slovenia*,
Historical Geography and Geosciences, https://doi.org/10.1007/978-3-031-99017-5_2

mountainous areas, which host a great variety of endemic and relic flora and fauna taxa. In order to protect and conserve the rich biodiversity, there are four national parks and a number of protected areas in the mountainous areas of Montenegro.

Keywords

Physical geography · Mountains · Glaciation · Karst · Dinaric Alps · Prokletije Alps · Montenegro

2.1 Introduction

Mountains and mountainous areas have been the subject of geomorphological, climate, biological, ecological and many other studies for a long period of time (e.g., Milojević 1937; Kapos et al. 2000; Comes and Kadereit 2003; Körner 2003; Aeschlimann et al. 2004; Owens and Slaymaker 2004; Barry 2008). However, morphological diversity and different genesis, along with different local perceptions of mountainous landscapes, means there is no concise universal definition of mountains (Gerrard 1990; Byers et al. 2013). According to the number of criteria, two main approaches for defining mountains can be distinguished: one-criterion and multiple-criterion delineation. One of the most frequent criteria used to delineate mountains in the simplest way is elevation (one-criterion delineation) (Ives et al. 1997). As the boundary values range from one country to another, the comparing areas designated as mountainous across different countries is often difficult. In order to more precisely delineate mountains, steepness has been added to the other criteria. Although the both criteria are based on quantitative relief patterns, distinguishing mountainous areas from the surrounding region is hard to approach practically, particularly in the case of plateaus and high and very steep slopes at lower altitudes. Biogenic indices, such as the effects on human beings caused by the decreased concentration of atmospheric oxygen (above 2,500 m a.s.l.), mapping of vegetation (forest) cover or the determination of the upper limit of forest growth (Troll 1973; Tranquillini 1979) displays high variations, and consequently does not provide satisfactory results in the assessment of a distinctive methodology for mountains, i.e., the delineation of their frontiers.

The implementation of new methods and techniques, and the introduction of more parameters (multiple-criterion delineations) in defining mountains has led to them being more precisely distinguished from surrounding areas (Kapos et al. 2000). Previous measures of the percentage mountainous areas in the world–such as 20% (Louis 1975) and 36% (Fairbridge 1968)–were later corrected to 35.8 million km^2, or 24% (Kapos et al. 2000), thanks to the introduction of more criteria. The delineation of mountainous regions in Europe began in the second half of the twentieth century (Price et al. 2004; Castelein et al. 2006). Aside from absolute relief elevation, which represents the main criterion for mountain delineation (with bordering values that differ from country to country), other specific parameters, such as climate conditions that are similar to mountainous climates, were also included (EC 2004). The application of variable criteria in mountain delineation gave rise to imprecise insights into proportions (in percentages) of mountain areas across different countries.

The aforementioned general problem in defining mountains along with the complexity and difficulties in establishing the borders of mountains (Rasemann et al. 2004) could not be avoided during attempts to outline mountainous areas in Montenegro. The overall elevation zoning of the relief of Montenegro (Table 2.1) was performed

Table 2.1 Elevation zoning on Montenegro (Mladenović 1984)

Elevation zone	Spatial area in km^2	%
0–200 lowland relief	1,425	10.3
200–500 hilly relief	791	5.5
500–1000 low mountainous relief	3,763	27.2
1000–1500 mid-mountainous relief	5,378	38.5
1500–2000 mountainous relief	2,455	17.4
> 2000 high-mountainous relief	159	1.1

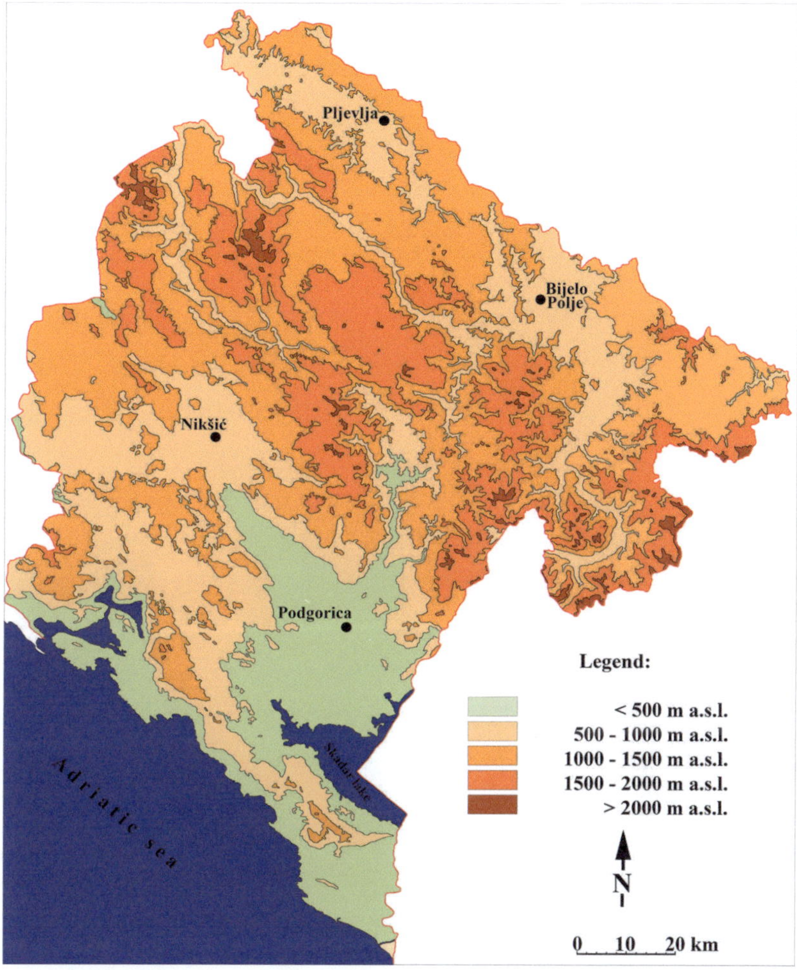

Fig. 2.1 The hypsometric map of Montenegro. (Author of the map Predrag Djurović)

through the frame of evaluating an elevation map of the Socialist Federal Republic of Yugoslavia made for military-geographic purposes (Mladenović 1984). According to just one criterion (absolute height), six elevation zones were identified. The first two consider lowland and hilly landforms (from 0 to 500 m a.s.l.), while the other four refer to low, moderate-height, above-moderate and high mountains (Fig. 2.1). Such a schematic division does not involve specific patterns of mountains in the given area (coastal mountains, mountains on a mountainous plateau, etc.). This division should be employed with a certain amount of care, as it refers to general relief within the distinct elevation zone, not only to mountains as morphological forms of interest. Relief forms above 500 m a.s.l. occupy 11,755 km^2 or 84.2% of the territory of Montenegro, which is 13,821 km^2 in total. The highest point in Montenegro is at its border with Albania (Mt. Bjelič, summit Maja Kolata, 2,528 m a.s.l.), while the peak named Bobotov kuk on Mt. Durmitor reaches 2,523 m a.s.l. (Osnovna … 1980).

2.2 Division and Spatial Distribution of Mountains

Two vast mountain ranges are recognised in Montenegro: the Dinaric Alps in the northwest, and the Prokletije Alps in the southeast (Fig. 2.2).

Fig. 2.2 Spatial distribution of mountain systems, mountain zones, mountains and mountain lakes. (Author of the map Predrag Djurović)

These ranges are distinguished by their geological, relief, climatic, biological and some additional differences. The boundary between these two ranges is not always clearly displayed.

The mountains of the Dinaric system (Fig. 2.2) occupy the largest part of the territory of Montenegro, extending from northwest to southeast, which follows the main direction of this mountain range. Within it are three discernible

Fig. 2.3 Toponyms of mountains. (Author of the map Predrag Djurović)

zones: coastal, mid- and external. The Western Mountain zone spreads from the Adriatic Sea to the valley of the Bojana River on the border with Bosnia and Herzegovina. It stretches like a narrow belt approximately 100 km long and from 10 to 20 km wide, displaying the characteristic NW–SE direction of the Dinaric Alps. Mountains with steep slopes that abruptly and almost vertically rise above the coastline are found abundantly in this zone. Notable mountains are

Fig. 2.4 Mountains of Montenegro: **a** Mt. Lovćen, **b** Mt. Bjelasica, **c** Mt. Vojnik, **d** Mt. Mali and Veliki Žurim, **e** Mt. Bogićevica, **f** Mt. Karanfili, **g** Mt. Komovi and **h** Mt. Hajla. (Photos by Predrag Djurović)

Orjen (1,894 m a.s.l.), Lovćen (1,749 m a.s.l.; Fig. 2.4a), Rumija (1,594 m a.s.l.) and Sozina (780 m a.s.l.). The Mid-Mountain zone (Fig. 2.2) comprises the largest part of the mountainous area in Montenegro. It extends from the border with Bosnia and Herzegovina to the valley of the Morača River and the Podgorička Basin. This zone represents the highest part of the wholly Dinaric system. Its main morphological feature is that mountains rise from 800 to 1,000 m above the old Mio-Pliocene plains, which were referred to as the area of "plains and hills" (Cvijić 1926). Notable mountains in this zone are: Somina (1,598 m a.s.l.), Njegoš (1,721 m a.s.l.), Prekornica (1,927 m a.s.l.), Golija (1,942 m a.s.l.), Pivska planina (1,776 m a.s.l.), Lebršnik (1,985 m a.s.l.), Vojnik (1,998 m a.s.l.; Fig. 2.4c), Treskavac (2,094 m a.s.l.), Maganik (2,139 m a.s.l.), Stožac (2,141 m a.s.l.), Žurim (2,030 m a.s.l.; Fig. 2.4d), Lola (2,129 m a.s.l.), Sinjajevina (2,277 m a.s.l.), Volujak (2,336 m a.s.l.), Bioč (2,397 m a.s.l.), Bjelasica (2,139 m a.s.l.; Fig. 2.4b) and Durmitor (2,523 m a.s.l.). The northeastern (outer) mountain zone extends from Tara Canyon in the east to the border with Serbia, and from the border with Bosnia and Herzegovina in the northeast to Bjelopoljska Basin and to the canyon of the Ibar River. It includes the following mountains: Gradina (1,303 m a.s.l.), Lisa (1,503 m a.s.l.), Kovač (1,530 m a.s.l.), Lisac (1,769 m a.s.l.), Obzir (1,869 m a.s.l.) and Ljubišnja (2,238 m a.s.l.) (Fig. 2.3).

Studies employed by Cvijić at the end of the 19th and beginning of the twentieth century (Cvijić 1899, 1913), along with later research through the frame of various natural branches, confirmed that Prokletije needs to be distinguished as an independent mountain range (Milojević 1937; Bošnjak 1938; Bešić et al. 1985; Lakušić et al. 1985; Petrović 1985; Djurović and Djurović 2016). Mountains in the Prokletije Alps (Fig. 2.3) commonly extend perpendicular to the direction of the Dinaric Alps and display significant geological and geomorphological differences, including different vegetation cover, etc. The mountain system is located in the southeastern part of Montenegro, between the valley of the Morača River and Kolašinska, Mojkovača and Bjelopoljska Basin, and the canyon of the Ibar River. In the south it spreads until the border with Albania. The Prokletije Alps are divided by the deep valleys of the rivers Grnčar, Ljuča and Lim in the Southwestern zone: Žijovo (2,131 m a.s.l.), Crna planina, Komovi (2,487 m a.s.l.; Fig. 2.4g), Zeletin (2,126 m a.s.l.), Greben (2,196 m a.s.l), Visitor (2,211 m a.s.l.,.) and the Southeastern zone: Trojan (2,190 m a.s.l.), Karanfili (2,119 m a.s.l.; Fig. 2.4f), Bjelič 2,528 m a.s.l.), Bor (2,106 m a.s.l.), Bogićevica (2,374 m a.s.l.; Fig. 2.4e), Mokra planina (1,968 m a.s.l.), Hajla (2,403 m a.s.l.; Fig. 2.4h), Žljeb (2,356 m a.s.l.), and Suva planina (2,098 m a.s.l.) (Fig. 2.3).

2.3 Geological Background

Mountains in Montenegro are predominantly composed of a variety of Mesozoic carbonate rocks (limestone and dolomite), and to a lesser extent by carbonate and clastic rocks of Cenozoic and Paleozoic age (sandstones, flysch, schists, phyllite, conglomerates, etc.). Igneous rocks (diabase, dacite, andesite and tuffs) are more or less abundant in a small number of mountains (Radojičić 1996) (Fig. 2.5).

Mountains in the Dinaric Alps are commonly composed of carbonate rocks of Mesozoic age (Bešić 1975, 1980). A small geological difference can be noted between the distinguished mountain zones. Marginal parts of the Western Mountain zone are made of flysch and limestones of Cretaceous-Eocene age, while elsewhere in this zone, Mesozoic limestones from several hundred meters to over 1,500 m thick prevail. Mesozoic carbonate rocks also dominate in the construction of mountains within the Mid-Mountain zone, where their thickness occasionally exceeds 2,000 m. Carbonate rocks are dissected by long, narrow belts of Paleogene flysch in zones of thrust faults. Igneous rocks occur sporadically in this zone in the foothills of several mountains, e.g., Sinjajevina, Volujak, Pivska planina and Durmitor (Mirković 1983). The Northeastern mountain zone displays a heterogeneous composition due to the presence of rocks of various ages and composition. Carbonate rocks are also very

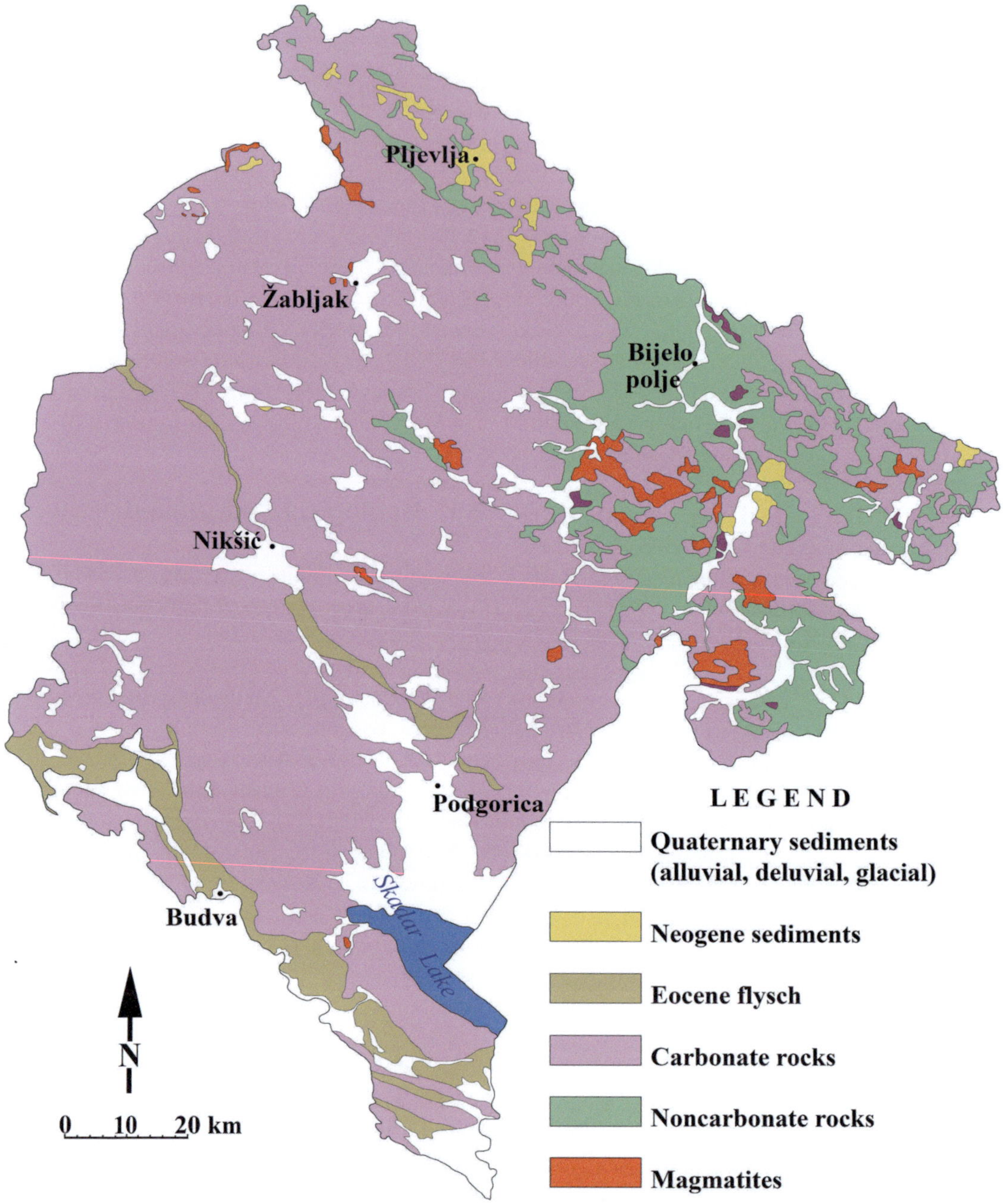

Fig. 2.5 Generalized geologic map (modified after Mirković et al. 1985). (Author of the map Predrag Djurović)

common in this zone, but cover much smaller areas and rarely exceed 1,000 m in thickness (near the canyon of the Tara River). The "mosaic texture" revealed by the frequent alternation of carbonate and non-carbonate rocks is a striking geological feature of this zone. Non-carbonate rocks include Triassic sandstones, andesite and dacite; Paleozoic schists and sandstones, as well as Neogene sediments, which outcrop in intramontane spaces. In southeastern parts of this zone, towards the boundary with the Prokletije Alps, large complexes of Permian sandstones, schists and marlstones are present in the construction of mountains.

The Prokletije Alps display a very heterogeneous geological composition. The highest parts of the mountains are composed of Mesozoic carbonate rocks, which range in thickness from a few tens to several hundreds of meters (e.g., Bjelasica, Komovi, Visitor, Hajla), but are notably lesser in abundance than in the Dinaric Alps. Marginal parts of these mountains are composed of Jurassic schists, diabase and marlstone, and of Triassic dacite, andesite and tuff. Thick limestone complexes (1,000–1,500 m) are specific only for Mt. Bjelič, which differs significantly in composition from the other mountains in this zone. Some mountains in this zone lack carbonate rocks, consisting either of Jurassic schists, diabase and marlstone and Triassic dacite, andesite and tuff (Mt. Greben), or of Paleozoic phyllites and Permian–Triassic conglomerates (e.g., Bogićevica).

2.4 Climate Characteristics

The climate characteristics of mountains in Montenegro are governed by three main factors: elevation, distance from the sea and the dissection of the mountainous area. Continuous and systematic measurements of the main climate parameters (precipitation and air temperature) are not adequate for considering the whole mountainous area due to the small number of meteorological stations and their locations at lower altitudes. For that reason, the climate on the mountains was analysed using different modeling. Instrumental measurements at high altitudes were applied within the Dinaric Alps, precisely within the Western Mountain zone on Mt. Orjen–at the meteorological station in Crkvice (937 m a.s.l.), and within the Mid- Mountain zone on Mt. Durmitor–at the meteorological station in Žabljak (1,450 m a.s.l.). Instrumental measurements in the Prokletije Alps were performed at lower altitudes.

According to results obtained by instrumental measurements, the annual average air temperature on the eastern slopes of Mt. Durmitor was 4.7 °C during the period from 1961–1990 (Djurović 2011), or 4.6 °C (Burić et al. 2011, 2012). The annual average air temperature fluctuates over time, thus between 1960 and 1980, a temperature drop and minimal value of 3.7 °C was recorded (Djurović 2011). Afterwards, the increase in annual average temperature led to a maximum of 7.1 °C in 2014 (data of Republic Hydro-meteorological Service of Montenegro). The average annual air temperature in the highest parts of Mt. Durmitor was established in terms of vertical temperature gradients (1958–1993), being 1.6 °C at 2,050 m a.s.l., i.e., 0.1 °C at 2,550 m a.s.l. (Djurović 2011).

The lack of stations at higher elevations doesn't allow for the precise analysis of precipitation in mountains. The annual average precipitation on the eastern slopes of Mt. Durmitor (Žabljak, 1,450 m a.s.l.) is 1,453 mm (1958–1993) (Djurović 2011), i.e., 1,473 mm (1961–1990) (Burić et al. 2012). The estimated annual precipitation in the highest part of Durmitor, based on precipitation gradients at 11 precipitation stations, ranges from 2,200 to 2,600 mm (Djurović 2011). In marginal areas of Mt. Durmitor during months with positive air temperature (eight months), 67.1% of the annual amount of precipitation falls, while the remaining 32.9% falls during months with negative average air temperatures (four months). According to estimates applied to the areas above 2,150 m a.s.l. this mountain receives about 53% of its precipitation (1,400 mm) during a period of negative air temperatures (six months) and 47% (1,200 mm) during periods of positive mean monthly temperatures (six months) (Djurović 2011).

Instrumental measurements at Mt. Orjen showed an average annual precipitation of 4,604 mm between 1961 and 1990. A maximal value of 9,079 mm was recorded in 2010 (Burić et al. 2016). Such an amount is extreme for Europe itself, not only for mountains in Montenegro. Additional extreme value and record rains fell for 30 h on this mountain, totalling 719 mm (from November 30 to December 1, 2010).

2.5 Hydrological Characteristics

The hydrological characteristics of mountains are strongly determined by their geological composition, climate and relief. The aforementioned domination of carbonate rocks in the Dinaric Alps, regardless of a large precipitation rate, resulted in sparse or no water courses.

In the Western Mountain zone, there is no larger surface runoff. The broadest area of the mountains is heavily karstified and without any surface flow. Intermittent flows occur in karst poljes (Figs. 2.3 and 2.10), such as Grahovsko, Dragaljsko, Njeguško (Fig. 2.10b) and Cetinjsko (Fig. 2.10a) poljes, which are situated in marginal mountain locations (Djurović and Djurović 2015). Periodical flows that occur on mountainsides oriented towards the Adriatic Sea display torrential pattern due to steep slope gradients. Groundwater flows are also characteristic for the Mid-Mountain zone due to prevailing carbonate bedrocks. Mountain water courses belong either to the Adriatic (Morača with a tributary to Zeta) or to the Black Sea watershed (Tara and Piva). Groundwater appears on the surface in lower and marginal mountain levels in the form of powerful karst springs: Pivsko oko, Vukovo vrelo, Vidrovan and Gornjepoljski vir in Nikšićko polje, Ljutica (Fig. 2.6d) in the canyon of the Tara River, Glava Zete, Pivsko oko—20 m^3/s (Piva Canyon) (Prohaska et al. 2004). Intermittent springs (Vidov potok (Fig. 2.6a), Šavnička glava and Zaslapnica) and estavelles (Gornjopoljski vir (Fig. 2.6b)) are hydrological rarities in this mountainous zone (Fig. 2.6). These springs contribute to the formation of large rivers, such as the Morača (Adriatic Sea watershed), whose mean annual discharge was 95 m^3/s in 1983 and 255 m^3/s in 1979, and was 156 m^3/s from 1961 to 2001 (Knežević 2009). The average annual discharge of the Piva River is 75 m^3/s, while the discharge of the Tara River is 79 m^3/s (Prohaska et al. 2004). Rivers and their tributaries incised some of the deepest canyons in Europe (the Tara Canyon –1,341 m, Mrtvica –1,247 m, Piva –1,034 m, Morača –1,168 m, Kruševački potok –1,008 m) (Djurović and Petrović 2007). Surface water courses are more significant in the Northeastern Mountain zone due to the larger extent of noncarbonate rocks. This zone is drained by the Tara River, and partly by the Lim and Ćehotina rivers (average annual discharge 18 m^3/s) (Prohaska et al. 2004). More suitable for surface flows are the Prokletije Alps. The Lim River (average annual discharge 112 m^3/s) and the Ibar River span to the Black Sea watershed, whereas the Cijevna River drains into the Adriatic Sea. In the foothill of mountains composed of carbonate rocks, there are powerful karst springs, such as Alipašini izvori (Fig. 2.4e), Glava Ibra, Savino oko (Fig. 2.6f), etc. In places, karst springs emerge from caves, e.g., Glava Bistrice (Djurović and Lješević 1994), Juriška vrela etc. (Fig. 2.6).

Most lakes in Montenegro are located in mountainous zones (Fig. 2.2) (Stanković 1975), commonly at their highest levels and being of small surface area and depth (Fig. 2.7). The lake Zeleni Vir sits at 2,028 m a.s.l., while the Jablan Lake sits at 1,791 m a.s.l. (Mt. Durmitor). The lake Hridsko (Mt. Bogićevica) sits at 1,970 m a.s.l., and Visitorsko Lake (Visitor) is at 1,735 m a.s.l. Mt. Stožac features the following lakes: Kapetanovi at 1,678 m a.s.l. and Manito Lake at 1,773 m a.s.l. A small number of lakes, like Plavsko in the foothills of Prokletije Alps (1.9 km^2 at 906 m a.s.l.) and Crno in the foothills of Mt. Durmitor (0.515 km^2 at 1,416 m a.s.l.) is more extensive. The depth of the lakes is negligible, commonly only reaching a few meters. The exceptions are Kapetanovo (–37 m) and Crno (–49 m).

The lake basins are of very diverse origin, having been created by a strong interplay of different processes (glacial, karst, colluvial, cryonivation, etc.) that took place on the mountains of Montenegro from the Pleistocene to nowadays. Basins were predominantly shaped by glacial erosion hence they are located at the highest levels of Pleistocene cirques (Zeleni vir, Veliko and Malo Škrčko Lake (Fig. 2.7b), Jablan Lake at Mt. Durmitor; Veliko Stabansko and Trnovačko Lakes at Volujak Mt. and Lebršnik; Lake Manito at Mt. Stožac). Lakes also occur in lower, marginal mountain regions and at the bottom of Pleistocene valleys (Crno, Modro (Fig. 2.7e), Valovito, Zminje and Pošćenska Lakes (Fig. 2.7b) at Mt. Durmitor; Malo

Fig. 2.6 Karst spring of Montenegro: **a** Vidov potok—estavelle and intermittent springs, **b** Gornjepoljski vir—estavelle, **c** Perućica, **d** Ljutica, **e** Alipašini izvori springs and **f** Savino oko. (Photos by Predrag Djurović)

Fig. 2.7 Mountain lakes: **a** Pošćenska Lakes, **b** Veliko and Malo Škrčko Lakes, **c** Ursulovačko Lake, **d** Ropojansko Lake, **e** Modro Lake and Hridsko Lake. (Photos by Predrag Djurović)

Stabanjsko Lake at Mt. Bioč, Kapetanovo Lake at Mt. Stožac and Ropojansko Lake (Fig. 2.7d) in the Prokletije Alps). Lake basins that developed within intramoraine depressions, such as Vražje and Riblje at Mt. Durmitor, Ursulovačko Lake (Fig. 2.7c) at Mt. Bjelasica, Plavsko Lake in the Prokletije Alps and Biogradsko Lake at Mt. Bjelasica, are placed in terminal basins. Cryonivation processes also contributed to the development of lake basins. Examples are lakes like Hridsko (Fig. 2.7f) on Mt. Bogićevica, which occupies a previous rock glacier (Gachev et al. 2019), and Svrablje at Mt. Durmitor, which occupies a prior protalus rampart. The karst process is particularly significant regarding water balance in lakes. Disappearing streams in a number of lakes

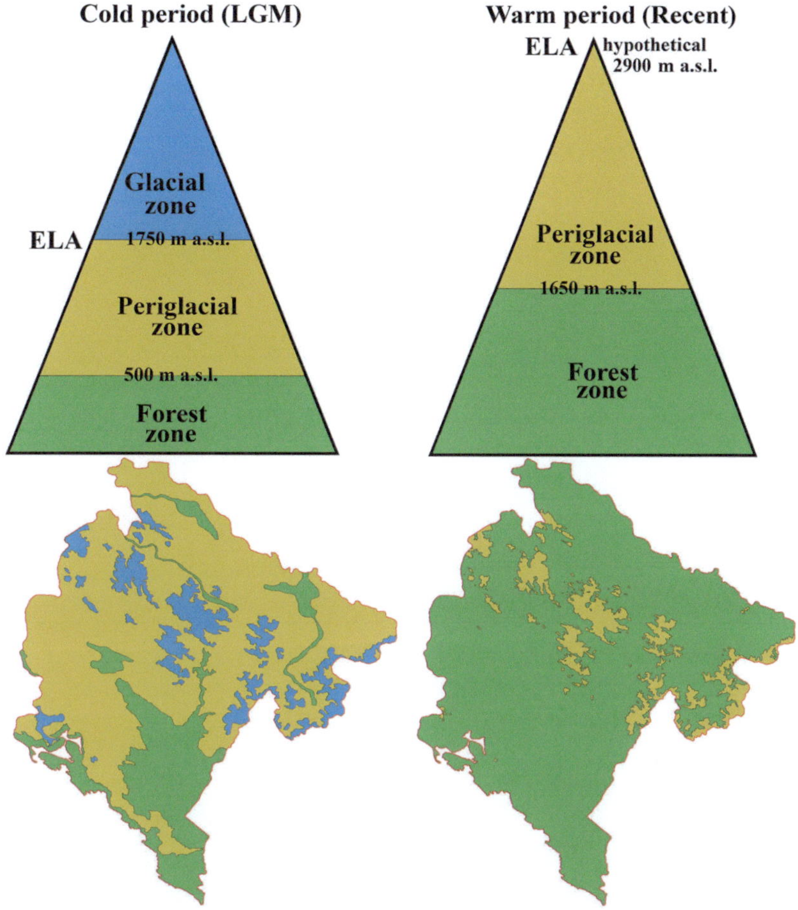

Fig. 2.8 Elevation zones influenced by nival, nivation and glacial processes during Quaternary (LGM—last glacial maximum, ELA—equilibrium line altitude; modified after Milivojević and Djurović 2010)

strongly influence the maximal water level. The most important disappearing stream is in the lake Crno jezero, from which water flows away in opposite directions, i.e., to the adjacent rivers Komarnica and Tara.

The recent glacier named Debeli Namet certainly represents the most amazing occurrence in the mountainous area of Montenegro (Fig. 2.9a). It is located within the Mid-Mountain zone of the Dinaric Alps, at the Mt. Durmitor. This glacier is a subject of intensive explorations since the end of the last century to nowadays (Nicod, 1968; Djurović, 1996a, 2009, 2012a; Veselinović et al. 1997; Kern et al. 2006; Hughes 2007, 2008; Grünewald and Scheithauer 2010; Gachev 2017). The glacier is positioned at the highest level of Pleistocene cirque Velika Kalica, at altitudes ranging from 2,050 to 2,160–2,170 m a.s.l. Its length varies from 320 to 350 m. It is fed from an area of about 0.144 $km^2$and it occupies an area of 0.056 km^2 at the end of the accumulation period, while at the end of the ablation period it spatially ranges from 0.025 to 0.032 km^2, with minimal ice thickness of 8 m. The Debeli Namet glacier has not experienced significant changes in its aerial extent over the past 50 years despite the negative impact of increasing temperature, because it is fed additionally by avalanches coming from high surrounding ridges, not only by direct precipitation (Djurović 2012a).

2.6 Characteristics of Mountain Relief

The relief of mountains in Montenegro was formed under the permanent influence of low temperatures of varying strength (cryogenic process), snow

Fig. 2.9 Pleistocene glacial morphology, periglacial and colluvial relief: **a** glacier Debli Namet at Mt. Durmitor, **b** valley Grebaja at Mt. Karanfili, **c** nunatak Stožina at Mt. Durmitor, **d** moraina at Mt. Žurim, **e** sliding blocks at Mt. Bogićevica, **f** protalus rampart at Mt Žurim, **g** sea of stones at Mt. Komovi and **h** talus slope at Mt. Komovi (Photos by Predrag and Mirela Djurović)

Fig. 2.10 Poljes of Montenegro: **a** Cetinjsko, **b** Njeguško, **c** Ivanje and **d** Bitinsko. (Photos by Predrag and Mirela Djurović)

(nival process), snowpatches (nivation process) and ice (glacial process) over the entire Quaternary (Milivojević and Djurović 2010). Hence, the boundaries between the surfaces shaped by the aforementioned processes changed and shifted upward or downward depending on the intensity of the warming or cooling climate. Consequently, the area subjected to distinct processes could reduce or enlarge (Fig. 2.8). These processes took part on mountains that are commonly composed of carbonate rocks, supporting chemical processes to be involved in the geomorphological evolution of mountains. Besides the climate and lithology, the intensive neotectonic movements, particularly the ascension of terrain, played a significant role in the creation of the mountain relief. The recent vertical growth of mountains ranges from 2 to 4 mm/year (Ivanović 1991). The current mountain relief in Montenegro resulted from the individual or combined action of the aforementioned processes, allowing several genetic relief types to be distinguished: Pleistocene glacial, karst, periglacial and colluvial relief.

2.6.1 Pleistocene Glacial Relief

Taking into consideration studies devoted to the glaciation of the mountains in Montenegro carried out by a number of authors for over a century (e.g., Cvijić 1903, 1913, 1917; Milojević 1937; 1951; Riđanović 1966; Djurović 1996a, 2009, 2012a, 2012b, 2020; Milivojević 2007; Milivojević et al. 2008; Stepišnik et al. 2009; Hughes et al. 2010; 2011; Stepišnik and Žebre 2011; Petrović 2014), comprising the existing differences in conceptual approach and methodology in between, it could be concluded that the mean equilibrium-line altitude (ELA) during the most extensive glaciation in the Pleistocene was above 1,750 m a.s.l. on the mountains in Montenegro, covering approximately 8% of the territory of Montenegro.

Morphological tool-marks of Pleistocene glaciation are preserved on all high mountains in Montenegro, either as erosional (cirque, glacial valley, glacial shoulders, nunataks) or as accumulative landforms (lateral and terminal moraine) (Fig. 2.9). These forms developed during three stages (Djurović 2009), which correlated with the MIS 12 (470–420 ka), the MIS 6 (190–130 ka) and the MIS 5d–2 (110–11.7 ka) (Hughes et al. 2011). According to morphological marks, four glacier types have been reconstructed: valley, plateau (piedmont), karst and regenerated glacier. Preglacial planes supported the generation of the extensive and relatively thin plateau glaciers comprising specific landforms: hummocky moraines (Djurović 2020). The formation of glaciers on the carbonate basement allowed the generation of a specific type (Cvijić 1899, 1913). The most extensive glaciation centers were on Mt. Durmitor (54% of its area covered by glaciers during the MIS 12 and 34% of the area in the MIS 5) (Djurović 2011; Hughes et al. 2011), Mt. Žurim, Mt. Stožac, Mt. Lola (Hughes et al. 2011; Djurović 2020), Mt. Orjen (Hughes et al. 2010), the Prokletije Alps (Milivojević et al. 2008), and Mt. Žijovo (Petrović 2014). Gathering actions of glacial and karst processes led to the development of distinct glaciokarst relief on all mountains consisting of carbonate rocks (Stepišnik and Žebre 2011; Djurović 2012a, 2012b, 2020; Petrović 2014; Žebre and Stepišnik 2014).

2.6.2 Periglacial and Colluvial Relief

The warmer climate during the Holocene led to the dominant influence of recent periglacial processes in the area previously subjected to Pleistocene glaciation (Fig. 2.9). Low temperatures and long retention of snow caused more intensive weathering of rocks and the specific transport and accumulation of such derived material, leading to the formation of a number of periglacial and colluvial relief forms above the forest line (Djurović 2011). More extensive areas exposed to *solifluction* occur on mountains of noncarbonate composition, as well as on impure carbonate rocks wherever the formation of thicker pedological cover is possible. Solifluction movements were noted in almost all dolines carved in carbonate flysch (Mt. Durmitor, Mt. Žurim, Mt. Stožac etc.). *Grass terraces* reflect the disequilibrium between soil and vegetation cover attained by the removal and shearing of elongated soil lanes, causing the appearance of stepped soil terraces. *Sliding blocks* are widespread on mountains composed of noncarbonate rocks, including those made up of marly limestones. A *sea of stones* appears at the bottom of nearly all Pleistocene cirques covered by rocky blocks of significant size. Large amounts of snow falling near ridges allowed the formation of a *nivation niche*, which resembles a saddle on ridge slopes. One of the largest forms of periglacial-colluvial process is *protalus rampart*. It is formed by the gravitational sorting of rock material in response to long retention of snow in the talus base. Two generations of it were recognised—one recent and another ascribed to the Little Ice Age (Djurović 2011, 2012b, 2020). Of similar dimensions are *rock glaciers*. Their occurrence is exclusively related mountains composed of non-carbonate rocks. One such case is at Mt. Bogićevica, where the basin of Hridsko Lake formed within the accumulation of rock glaciers (Gachev et al. 2019). Intensive frost decay at the base of rocky ridges led to a large accumulation of *talus*, as in the case of Komovi, where it reaches a length of 1.8 km and up to 400 m in width.

2.6.3 Karst Relief

Karst relief is the most abundant relief in Montenegro, as well as on the mountains. Pure limestones of significant thickness, high elevation and dissected terrain enabled the supreme development of subsurface and surface karst relief. Possessing characteristics of both, the karst relief in Montenegro consists of holokarst, i.e., represents the north-easternmost part of the Dinaric karst. Karst poljes, as the most prominent landforms in surface karst relief, occur in marginal mountain regions (Fig. 2.3). The mountainous area of Montenegro contains 14 of the country's 16 recognised karst poljes, which are distributed at elevations from 600 to 1,450 m a.s.l. (Djurović and Djurović 2015). Karst poljes are small of size, from 1 to 3 km^2, maximally about 5 km^2 (Fig. 2. 10), excluding Nikšićko polje (66.5 km^2) (Radojičić 1952–1953, 1996) and Brezna (24 km^2) (Lješević 1975). Intermittent (periodic) streams in a number of them are common. The exception is Nikšičko polje, which is permanently drained by the Zeta River, the longest intermittent river in Montenegro. The evolution of the majority of poljes was influenced by glacial processes and the formation of moraines, either in poljes themselves or by the entrainment of re-transported moraine material (Djurović and Djurović 2015, 2020; Adamson et al. 2016). Dolines are particularly numerous within polygonal karst, which evolved on distinct mountains e.g., Pivska planina (Lješević 2005), Sinjajevina, Lovćen. In the highest mountainous parts, the snow retention within dolines over the year and the occurrences of kettle-shaped (*kotlič*) dolines are common (Djurović 2011). In these mountains the density of dolines is 80–120 per km^2. The bottoms of many dolines are filled with different types of sediments. Dolines whose bottoms are filled with aeolian sediments are particularly noticeable in mountainous regions above 2400 m a.s.l. (Durmitor) (Djurović et al. 2022).

In re-shaped Pleistocene cirques in the highest mountains and in basinal regions uvalas formed. A few types of uvalas were distinguished depending on the transformation rate of Pleistocene cirques into recent uvalas on Mt. Durmitor and Mt. Žijovo (Djurović et al. 2010; Djurović 2011). Glacial uvalas are placed at the highest levels and are strongly affected by cryo-nivation and colluvial processes. In many cases, the morphology of the glacial phase is preserved and significant elements of karst relief are lacking. Glacial-karst uvalas appear either in lower altitudes, or at the same altitudes as the former do, but in notably warmer climate conditions. Such uvalas display the main morphological and hydrological features of karst, such as dolines and karren, whereas intermittent springs derived from snowmelt disappear. Glacial-fluvial uvalas are situated at lower elevations than the two former types. They formed after previous basins with dolines, blind valleys, springs and intermittent rivers. Parts of them were exposed to permanent or temporary flooding (lakes) (Fig. 2.11).

In the formation of the recent relief, the cryo-nivation process is of minor importance (Djurović et al. 2010). The influence of the glacial process on recent karst morphology could be inferred from the development of karst perched block with frequent karst pedestals (Djurović and Djurović 2011). The annual lowering of the topographic surface under chemical erosion of about 0.014 mm per year, i.e., 13.57 mm over 1,000 years, has been determined by studying pedestals on Durmitor (1,850 m a.s.l.) (Djurović and Djurović 2011). The number of microkarst landforms (karren and kamenitza) of various shapes and dimensions is considerable. The long retention of snow has a strong influence on the evolution and specific morphology that distinguish them from similar forms at low altitudes (Djurović 1996b).

The most significant caves in Montenegro are located in the mountainous area (Fig. 2.12). High altitudes and the prevailing of vertical water circulation in karst regions led to the domination of vertical caves (Djurović 2017). Close to the pit entrance snow corks are common up to a few tens of meters thick (Gavrilović 1963; Djurović 2011). A few mountains are outstanding in their number and depths of pits: Orjen, Lovćen, Durmitor, Maganik and Bjelič. In total, 22 caves were explored at depths from 250 to 500 m, 11 caves range from 500 to 1,000 m in depth, and one cave exceeds 1,000 m in depth (Djurović and Djurović 2021). The deepest pit is the Iron Deep on Maganik, with the depth of 1,162 m (Adamec et al. 2013; Dvořák and Baldík 2013; Dvořák 2014). A smaller number of caves have a considerable length, e.g., Pećina nad Vražjim Firovima Cave, which is 17.5 km long (Sirotek 2005, 2006, 2015).

Fig. 2.11 Karst relief: **a** karren, Bolj at Mt. Durmitor, **b** glaciokarst at Mt. Lovćen: pleistocene cirk (1), moraine (2), glacial valley (3), Njeguško polje (4), **c** dolines on Krnovo plateau (Dinaric Alps/Mid-Mountain zone), **d** Lokvice glacial uvalas at Mt. Durmitor, **e** Ledena pećina Cave at Mt. Durmitor and **f** Lipska Cave (Dinaric Alps/Mid-Mountain zone). (Photos by Predrag Djurović)

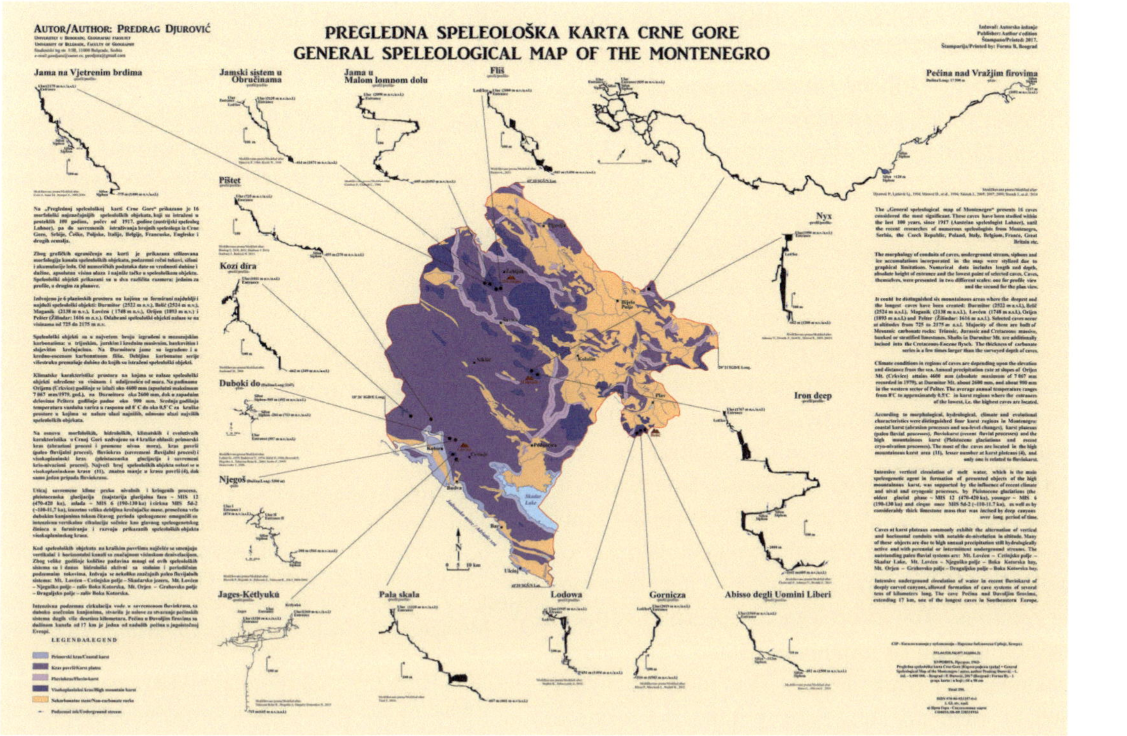

Fig. 2.12 Distribution of karst and the most significant caves (Djurović 2017)

2.7 Geodiversity, Biodiversity and Mountain Protection

Geodiversity refers to the overall abiotic diversity of an area where representative objects should be distinguished as geoheritage sites. According to the universal classification proposed by ProGeo (Wimbledon 1996, 1999), through a unique methodology of recognition, valorisation and systematisation, representative objects were distinguished within a great number of geodiversity categories (Djurović and Mijović 2006), and a preliminary inventory of geoheritage sites in Montenegro was created (Djurović and Djurović 2010). Geodiversity objects are classified into nine main categories: objects of historical-geological and stratigraphical heritage, structural objects and objects of petrological heritage, geomorphological objects, those of neotectonic activity, speleological geoheritage sites, objects of hydrological-hydrogeological geoheritage, and pedological and archeological geoheritage sites. Within the entire area of Montenegro, there are 217 recognised geoheritage sites, the majority of them (170) are located within the mountainous area (Djurović and Djurović 2010). Geoheritage sites could not always be clearly spatially ascribed to mountains, as their extent may be solely within mountainous areas, or may spread to lowland regions (and vice-versa). Such cases can be found among structural, historical-geological, stratigraphical, and petrological and neotectonic sites. Distinct groups of geomorphological sites are either completely (paleoglacial and periglacial relief) or mostly (karst relief and speleological objects) linked to mountains. The lowest parts of canyons, which deeply incise mountains (1,000 m), sit at lower elevations. Much of the hydrological-hydrogeological geoheritage sites (very powerful springs) are connected to the mountainous region in terms of their functional properties, although they are located at lower altitudes. On the contrary, the majority of lakes are distributed within mountains.

It has been estimated after almost two centuries of investigations (Rohlena 1942; Pulević 2006) that in Montenegro, there are more than 3,600 plant species and subspecies (Stešević and Caković 2013). Such high diversity results from complex and dynamic relief, variable climate, heterogeneous soil cover, and turbulent geological and biological history etc. (Stevanović and Stevanović 1995). From a flora perspective, the mountainous area in Montenegro includes regions above 700 m a.s.l. (Fuštić and Đuretić 2000) and is divided into four sub-regions: central, eastern, northern and western (Stešević and Caković 2013). Of the total number of vascular plants in Montenegro, 10.4% of the taxa consist of Balkan endemites numbering 147 species from 46 families. The chorological spectrum reflects the domination of mountainous and alpine floral endemism in Montenegro. The most abundant Balkan endemic taxa are within the eastern (224) and northern subregion (205), and the least abundance is recorded in the mid-region (60) (Vuksanović et al. 2016). Glacial and interglacial periods during the Pleistocene allowed for the migration of high-mountain flora from the Dinaric Alps and the Balkan Mountains, and the Alps and the Carpathians (Kramp et al. 2009). A high percentage of Balkan endemic taxa in the lowest mountain parts and in deep canyons within the eastern and northern mountainous subregion display an increasing number of Balkan endemic taxa, thanks to the influence of the Mediterranean climate and thermophytic ecological elements (Turrill 1929). Mt. Durmitor in the northern subregion, along with the canyons of the Piva and Tara rivers, represents one of the most important refuge centers of endemic alpine flora in the whole Balkan Peninsula (Stevanović and Lakušić 1996). The mountain ridge of Prokletije (eastern subregion) is noteworthy for having a large number of endemic taxa (Stevanović and Stevanović 1995). The endemic flora of Prokletije links the ancient orophytic flora of the Alps and the Carpathians with the Oromediterranean Tertiary flora, hence displaying characteristics of both (Bulić et al. 2010).

References

Adamec V, Baldik V, Dvořák Z (2013) "Česká" jeskyně hluboká přes jeden kilometr (Maganik Range, Černá Hora) ["Czech" cave over one kilometre deep (Maganik Range, Montenegro)]. Czech Speleol Soc Speleofórum 23:45–50

Adamson KR, Woodward JC, Hughes PD (2016) Middle Pleistocene glacial outwash in poljes of the Dinaric karst. In: Feinberg JM, Gao Y, Alexander EC (eds) Caves and karst across time. Geological Society of America, Boulder, pp 247–262. https://doi.org/10.1130/2015.2516(20)

Aeschlimann D, Lauber K, Moser DM, Theurillat JP (2004) Flora Alpina. Verlag Haupt, Bern

Barry RG (2008) Mountain weather and climate. Cambridge University Press, Cambridge

Bešić Z (1975) Geologija Crne Gore—Stratigrafija i facijalni sastav Crne Gore [Geology of Montenegro—Stratigraphy and facial structure of Montenegro] Društvo za nauku i umjetnost Crne Gore, Titograd

Bešić Z (1980) Geologija Crne Gore [Geology of Montenegro]. Društvo za nauku i umjetnost Crne Gore, Titograd

Bešić Z, Perović Z, Mihajlović R (1985) Geološki sastav terena opštine Plav, Prokletije—zaštita i unapredjenje plavskog dijela Prokletija [The geological composition of the terrain of the municipality of Plav, Prokletije—protection and improvement of the Plav part of Prokletije]. Opština Plav, Plav

Bošnjak K (1938) Nekoliko novosti iz flore gornjeg Ibra i Prokletija. [Some news from the flora of the upper Ibar and Prokletije]. Glasnik Hrvatskog Prirodoslovnog Društva 49–50:22–28

Bulić Z, Stevanović V, Lakušić D, Hadžiablahović S (2010) Biljni svijet NP Prokletije. [Flora of National park Prokletije]. Časopis Nacionalnih Parkova Crne Gore 6:18–23

Burić D, Ducić V, Luković J (2011) Kolebanje klime u Crnoj Gori u drugoj polovini XX i početkom XXI vijeka [Climate fluctuation in Montenegro in the second half of 20th century and the beginning of 21st century]. CANU, Podgorica

Burić M, Micev B, Mitrović L (2012) Atlas klime Crne Gore [Climate atlas of Montenegro]. Leksikografski centar, CANU, Podgorica

Burić D, Ducić V, Doderović M (2016) Poplave u Crnoj Gori krajem 2010. godine sa osvrtom na kolebanje proticaja Morače [Floods in Montenegro at the end of 2010 with reference to the fluctuation of the Morača flow] Glasnik odjeljena prirodnih nauka CANU 21:47–64

Byers AC, Price LW, Price MF (2013) Introduction to Mountains. In: Price MF, Byers AC, Friend DA, Kohler T, Price LW (eds) Mountain geography: physical and human dimensions. University of California Press, Berkeley, pp 1–10

Castelein A, Dinh TTV, Mekouar MA, Villeneuve A (2006) Mountains and the law—emerging trends. FAO Legislative Study 75. FAO, Rome

Comes HP, Kadereit JW (2003) Spatial and temporal patterns in the evolution of the flora of the European alpine system. Taxon 52(3):451–462. https://doi.org/10.2307/3647445
Cvijić J (1899) Glacijalne i morfološke studije Bosne, Hercegovine i Crne Gore [Glacial and morphological studies of Bosnia, Herzegovina and Montenegro]. Glas Srpke kraljevske akademije 57. Srpska kraljevska akademija, Beograd
Cvijić J (1903) Novi rezultati o glacijalnoj eposi Balkanskog poluostrva [New results about the glacial epoch of the Balkan Peninsula]. Glas Srpke kraljevske akademije 65. Srpska kraljevska akademija, Beograd
Cvijić J (1913) Ledeno doba u Prokletijama i okolnim planinama [Ice age in Prokletije and the surrounding mountains]. Glas Srpke kraljevske akademije 91. Srpska kraljevska akademija, Beograd
Cvijić J (1917) L'epoque glaciaire dans la péninsule balkanique. Annales De Géographie 26(141):189–218
Cvijić J (1926) Geomorfologija II [Geomorphology II]. Državna štamparija Kraljevine Srba, Hrvata i Slovenaca, Beograd
Djurović M, Djurović P (2011) Kraški stolovi—fenomen visokoplaninskog kraškog reljefa Durmitora. [Karst tables—the phenomenon of highe-mountain karst landscape of Mt. Durmitor]. In: Zaštita prirode u XXI vijeku: zbornik referata. Zavod za zaštitu prirode Crne Gore, Žabljak, pp 349–354
Djurović M, Djurović P (2015) The impact of glacial processes in the evolution of the poljes of Montenegro. Acta Geobalcanica 1(1):25–31. https://doi.org/10.18509/AGB.2015.03
Djurović M (2020) Genesis of the pleistocene hummocky moraines in the Moračke Planine mountains (Montenegro). Glasnik Srpskog Geografskog Društvo 100(2):25–43. https://doi.org/10.2298/GSGD2002025D
Djurović M, Djurović P (2021) Review of the most significant caves in Montenegro. Acta Carsologica 50(1):11–27. https://doi.org/10.3986/ac.vi.7182
Djurović M, Gajić V, Djurović P (2022) Filling of dolines with aeolian sediments in the high-mountain karst of the Dinarides (Mt. Durmitor, Montenegro). J Mt Sci 19(7):1886–1900. https://doi.org/10.1007/s11629-021-7274-5
Djurović P (1996a) Visokoplaninski kras Durmitora – geomorfološka studija [Alpine karst of Mt. Durmitor – geomorphological study]. PhD Thesis, University of Belgrade, Beograd
Djurović P (1996b) Specifični oblici mikrokrasa na severnim padinama Bolja (Durmitor) [Specific forms of micro-karst on the northern slopes of Bolja (Mt. Durmitor)]. In: Lješević M (ed) Priroda Nacionalnog parka Durmitor. Univerzitet u Beogradu, Crnogorska akademija nauka i umetnosti, Nacionalni park "Durmitor", Beograd, pp 108–114
Djurović P (2009) Reconstruction of the Pleistocene glaciers of mount Durmitor in Motenegro. Acta Geogr Slov 49(2):263–289. https://doi.org/10.3986/AGS49202
Djurović P (2011) Reljef Durmitora [The relief of Mt. Durmitor]. In: Zaštita prirode u XXI vijeku: zbornik referata. Zavod za zaštitu prirode Crne Gore, Žabljak, pp 159–168
Djurović P (2012a) The Debeli Namet glacier from the second half of the 20th century to the present. Acta Geogr Slov 52(2):277–301. https://doi.org/10.3986/AGS52201
Djurović P (2012b) Visokoplaninski kras Durmitora. [Alpine karst of Mt. Durmitor]. Geografski fakultet, Beograd
Djurović P (2017) General speleological map of Montenegro. Belgrade
Djurović P, Djurović M (2010) Objekti geonasledja—reprezenti geodiverziteta Crne Gore [Geoheritage sites—reprezentatives of geodiversity of Montenegro]. In: Geoekologija—XXI vijek, teorijski i aplikativni zadaci: zbornik referata. Filozofski fakultet, Žabljak, pp 508–517
Djurović P, Djurović M (2016) Physical geographic characteristics and sustainable development of the mountain area in Montenegro. In: Zhelezov G (ed) Sustainable development in mountain regions: Southeastern Europe. Springer, Cham. https://doi.org/10.1007/978-3-319-20110-8_8
Djurović P, Lješević M (1994) Speleološki objekti Djalovića klisure i njihov značaj za razumevanje cirkulacije vode u krasu Peštera [Speleological sites of the Djalovića gorge and their significance for understanding the circulation of water in the karst of Pešter]. Zbornik radova Odbora za kras i speleologiju. SANU, Beograd, pp 109–131
Djurović P, Mijović D (2006) Geonasledje Srbije – reprezent njenog ukupnog geodiverziteta [Geoheritage of Serbia—Representative of its total geodiversity]. Zbornik Radova—Geografski Fakultet Univerziteta u Beogradu 54:4–18
Djurović P, Petrović A (2007) Large canyons of Dinaric and Prokletije mountains region of Montenegro. Geogr Panonica 11:14–18. https://doi.org/10.5937/GeoPan0711014D
Djurović P, Petrović A, Simić S (2010) Uticaj pleistocene glacijacije na morfološku raznovrsnost uvala Durmitora i Žijova [The overall impact of Pleistocene glaciation on morphological diversity of uvalas at Durmitor and Žijovo]. Glasnik Srpskog Geografskog Društva 90(1):17–34
Dvořák Z, Baldík V (2013) Czech discoveries in the Maganik Mts., Montenegro. In: Filippi M, Bosák P (eds) Proceedings of the 16th international congress of speleology. Czech Speleological Society, Brno, pp 74–77
Dvořák Z (2014) Iron deep pokračuje—Maganik, Černá Hora [Iron deep continued—Maganik, Montenegro]. Speleofórum 33:72–77
EC [European Commission] (2004) Mountain areas in Europe: analysis of mountain areas in EU

member states, acceding and other European countries. Directorate-General for Regional Policy, Brussels

Fairbridge RW (1968) Mountain and hilly terrain, mountain systems; mountain types. In: Fairbridge RW (ed) Encyclopedia of geomorphology. Reinhold, New York, pp 745–761

Fuštić B, Đuretić G (2000) Zemljišta Crne Gore [Soils of Montenegro]. Univerzitet Crne Gore, Biotehnički Institut, Podgorica

Gachev E (2017) The unknown southernmost glaciers of Europe. In: Godone D (ed) Glacier evolution in a changing world. InTech, Rijeka, pp 77–102. https://doi.org/10.5772/intechopen.68899

Gachev E, Djurović P, Stoyanov K, Bezinska G, Ivanov M, Petrov S (2019) Otnosno obrazuvaneto na Hridskoto ezero v planinata Bogičevica, Černa gora [About the formation of Hridsko lake in Bogićevica Mountains, Montenegro]. Izvestija Na Blgarskoto Geografsko Družestvo 40:3–9

Gavrilović D (1963) Snežnice na Lovćenu [The snowmelt in Lovćen Mt.]. Zbornik Radova Geografskog Inst PMF 10:7–68

Gerrard AJ (1990) Mountain environments: an examination of the physical geography of mountains. Belhaven, London

Grünewald K, Scheithauer J (2010) Europe's southernmost glaciers: response and adaptation to climate change. J Glaciol 56(195):129–142. https://doi.org/10.3189/002214310791190947

Hughes PD (2007) Recent behaviour of the Debeli namet glacier, Durmitor, Montenegro. Earth Surf Process Land 32(10):1593–1602. https://doi.org/10.1002/esp.1537

Hughes DP (2008) Response of a Montenegro glacier to extreme summer heatwaves in 2003 and 2007. Geogr Ann Ser B 90(4):259–267. https://doi.org/10.1111/j.1468-0459.2008.00344

Hughes PD, Woodward JC, van Calsteren PC, Thomas LE, Adamson KR (2010) Pleistocene ice caps on the coastal mountains of the Adriatic sea. Quat Sci Rev 29:27–28. https://doi.org/10.1016/j.quascirev.2010.06.032

Hughes PD, Woodward J, van Calsteren P, Thomas L (2011) The glacial history of the Dinaric Alps, Montenegro. Quat Sci Rev 30:3393–3412. https://doi.org/10.1016/j.quascirev.2011.08.016

Ivanović S (1991) Zemljotresi fenomeni prirode [The earthquakes phenomena of nature] CANU, Građevinski fakultet, Titograd

Ives JD, Messerli B, Spiess E (1997) Mountains of the world—a global priority. In: Messerli B, Ives JD (eds) Mountains of the world: a global priority. Parthenon, New York, London

Kapos V, Rhind EM, Price MF, Ravilious C (2000) Developing a map of the world's mountain forests. In: Price MF, Butt N (eds) Forests in sustainable mountain development: a state of knowledge report for 2000. CAB International, Wallingford, pp 4–9

Kern Z, Surányi G, Molnár M, Nagy B, Balogh D (2006) Investigation of natural perennial ice deposits of Durmitor Mts, Montenegro. In: Zelinka J (ed) Proceedings of the 2nd international workshop on ice caves. Liptovsky Mikuláš, pp 70–73

Knežević M (2009) Studija—vodni režim rijeke Morače i Skadarskog jezera [The study—water regime of the Morača River and Skadar Lake]. WWF, Podgorica

Körner C (2003) Alpine plant life: functional plant ecology of high mountain ecosystems. Springer, Berlin, Heidelberg. https://doi.org/10.1007/978-3-642-18970-8

Kramp K, Huck S, Niketić M, Tomović G, Schmitt T (2009) Multiple glacial refugia and complex postglacial range shifts of the obligatory woodland plant *Polygonatum verticillatum* (Convallariaceae). Plant Biol 11(3):392–404. https://doi.org/10.1111/j.1438-8677.2008.00130.x

Lakušić R, Vučković M, Markišić H (1985) Specifičnosti Prokletija [The specifics of the Prokletije Alps]. Prokletije—zaštita i unapredjenje Plavskog dijela Prokletije. Republički zavod za zaštitu prirode, Titograd, pp 91–102

Lješević M (1975) Brezansko polje [Brezansko polje]. Glasnik Srpskog Geografskog Društva 55(1):103–113

Lješević AM (2005) Karst Pive [Karst of Piva]. CANU, Podgorica

Louis H (1975) Neugefasstes Höhendiagramm der Erde. Bayerischer Akademie der Wissenschaften, München

Mladenović T (1984) Visinska struktura reljefa zemljišta u SFR Jugoslaviji [Elevation structure of land relief in SFR Yugoslavia]. Zbornik Radova Vojnogeografskog Inst 4:67–75

Milojević BŽ (1937) Visoke planine u našoj kraljevini [High mountains in our kingdom]. Državna štamparija Kraljevine Jugoslavije, Beograd

Milojević BŽ (1951) Durmitor—regionalno-geografska istraživanja [Mt. Durmitor—regional-geographical research]. Zbornik Radova SANU 9:1–74

Milivojević M, Djurović P (2010) Glacijalna i periglacijalna morfologija—indikator promene klime na planinama Crne Gore [Glacial and periglacial morphology—an indicator of climate change on the mountains of Montenegro]. In: Geoekologija XXI vijeka: teorijski i aplikativni zadaci: zbornik referata. Filozofski fakultet, Nikšić, pp 536–543

Milivojević M (2007) Glacijalni reljef na Volujku sa Bioćem i Maglićem [Glacial relief of Mts. Maglić, Bioč and Volujak]. Geografski institut „Jovan Cvijić", SANU, Beograd

Milivojević M, Menković Lj, Ćalić J (2008) Pleistocene glacial relief on the central part of the Mt. Prokletije (Albanian Alps). Quat Int 190:112–122. https://doi.org/10.1016/j.quaint.2008.04.006

Mirković M (1983) Geološki sastav i tektonika planina Durmitora, Pivske planine i Volujka [Geology and tectonics of Durmitor, Piva and Volujak mountans]. Zavod za geološka istraživanja, Titograd

Mirković M, Živaljević M, Đokić V, Perović Z, Kalezić M, Pajović M (1985) Geološka karta Crne Gore 1:200.000 [Geological map of Montenegro 1:200,000]. RSIZ za geološka istraživanja, Titograd

Nicod J (1968) Premières recherches de morphologie karstique dans le massif du Durmitor. Meditérranée 3:187–216
Osnovna topografska karta Crna Gora 1:25.000 [Basic topographic map Montenegro 1:25,000], sheet Dragobilja, sheet Durmitor north (1980). Vojnogeografski institute, Republička geodetska uprava SR Crne Gore, Titograd
Owens NP, Slaymaker O (2004) Mountain geomorphology. Oxford University Press, Oxford
Petrović J (1985) Prokletije—geomorfološke odlike [Prokletije Alps—geomorphological features] In: Prokletije—zaštita i unapredjenje Plavskog dijela Prokletija. Republički zavod za zaštitu prirode, Titograd, pp 39–50
Petrović A (2014) A reconstruction of the Pleistocene maximum in the Žijovo range (Prokletije mountains, Montenegro). Acta Geogr Slov 54(2):255–269. https://doi.org/10.3986/AGS54202
Price MF, Lysenko I, Gloersen E (2004) Delineating Europe's Mountains. Revue De Géographie Alpine 92(2):75–86
Prohaska S, Simić Z, Orlić A, Ristić V (2004) Osnovne hidrografsko-hidrološke karakteristike sliva Drine i hidrometeorološki podaci [Major hydrographic and hydrological feature of the Drina River basin and hydrometeorological information]. Vodoprivreda 36:3–21
Pulević V (2006) Botaničari i Crna Gora [Botanists and Montenegro]. Prirodnjački muzej Crne Gore, Podgorica
Radojičić B (1952–1953) Nikšićko polje, geomorfološka promatranja [Nikšićko polje, geomorphological observations]. Geografski glasnik 14–15:71–86
Radojičić B (1996) Geografija Crne Gore prirodna osnova [The geography of Montenegro natural basis]. Univerzitet Crne Gore, Filozofski fakultet, Nikšić
Rasemann S, Schmidt J, Schrott L, Dukau R (2004) Geomorphometry in mountain terrain. In: Bishop M, Shroder JF (eds) Geographic information science and mountain geomorphology. Springer, Heidelberg, pp 101–103
Riđanović J (1966) Orjen—La montagne Dinarique. Geografski institut, Zagreb
Rohlena J (1942) Conspectus florae montenegrinae. Česká botanická společnos, Praha
Sirotek J (2005) Expedice Medúza 2004 [Medúza expedition 2004]. Speleofórum 24:71–73
Sirotek J (2006) Expedice Medúza 2005 [Medúza expedition 2005]. Speleofórum 25:29–31
Sirotek J (2015) Medúza 2014 [Medúza 2014]. Speleofórum 34:81–83
Stanković S (1975) Planinska jezera Crne Gore [Mountain lakes of Montenegro]. Društvo za nauku i umjetnost Crne Gore, Titograd
Stepišnik U, Ferk M, Kodelja B, Medenjak G, Mihevc A, Natek K, Žebre M (2009) Glaciokarst of western Orjen, Montenegro. Cave Karst Sci 36(1):21–28
Stepišnik U, Žebre M (2011) Glaciokras Lovčena [Glaciokarst of Lovćen]. E-GeograFF 2. Znanstvena založba Filozofske fakultete, Ljubljana. https://doi.org/10.4312/9789612374402
Stešević D, Caković D (2013) Katalog vaskularne flore Crne Gore 1 [Catalogue of vascular flora of Montenegro 1]. CANU, Podgorica
Stevanović V, Stevanović B (1995) Osnovni klimatski, geološki i pedološki činioci biodiverziteta kopnenih ekosistema Jugoslavije [Basic climatic, geological and pedological factors of biodiversity of the terrestrial ecosystems of Yugoslavia]. In: Stevanović V, Vasić V (eds) Biodiverzitet Jugoslavije sa pregledom vrsta od međunarodnog značaja. Biološki fakultet, Ecolibri, Beograd, pp 75–95
Stevanović V, Lakušić D (1996) Florističke i florogenetske karakteristike visokoplaninske endemične flore Durmitora [Floristic and florogenetic characteristics of the high-mountain endemic flora of Mt Durmitor]. In: Lješević M (ed) Priroda Nacionalnog parka "Durmitor". Geografski fakultet, Beograd, pp 206–219
Tranquillini W (1979) Physiological ecology of the Alpine timberline: tree existence at high altitudes with special reference to the European Alps. Springer, Berlin. https://doi.org/10.1007/978-3-642-67107-4
Troll C (1973) The upper timberlines in different climatic zones. Arct Alp Res 5(3):A3–A18
Turrill WB (1929) The plant life of the Balkan Peninsula: a phytogeographical study. Clarendon Press, Oxford
Veselinović D, Kovačević D, Marković D, Vukmirović Z, Rajšić S, Stanković S, Stanković A, Djordjević D (1997) Sadržaj metala i radioaktivnost leda sa ostataka lednika Debeli namet na Durmitoru [Metal content and radioactivity of ice from the remains of the Debeli namet glacier on Durmitor]. Rizik tehnoloških sistema i životna sredina: zbornik radova. Univerzitet u Nišu, Niš, pp 112–116
Vuksanović S, Tomović G, Niketić M, Stevanović V (2016) Balkan endemic vascular plants of Montenegro—critical inventory with chorological and life-forms analyses. Willdenowia 46(3):387–397. https://doi.org/10.3372/wi.46.46307
Wimbledon WAP (1996) National site selection, a stop on the road to a Europen geosites list. Geol Balc 26(1):15–27
Wimbledon WAP (1999) Geosites—an international union of geological sciences initiative to conserve our geological heritage. Pol Geol Inst Spec Pap 2:5–8
Žebre M, Stepišnik U (2014) Reconstruction of late Pleistocene glaciers on Mount Lovćen, Montenegro. Quat Int 353:225–235. https://doi.org/10.1016/j.quaint.2014.05.006

The Definition of Mountainous Areas in Slovenia

3

Drago Perko

Abstract

Slovenia is very diverse in terms of landscapes. Most differences between them refer to landforms. Flatlands account for barely a tenth of Slovenia, and the remainder comprises various forms of dynamic terrain, from rounded low hills on the edges of the Pannonian and Mediterranean areas to high and steep mountains in the Alps. Taking into account all landscape typologies of Slovenia, more than half of the country is mountainous, primarily comprised of mountains, hills, and high plateaus. Nearly all these landforms belong to three landscape types: Alpine mountains (15% of Slovenia), Alpine hills (23%), and Dinaric plateaus (19%), which together cover a total of 11,531 km^2. Slovenia's highest Alpine mountain is Mount Triglav (2,864 m) in the Julian Alps, its highest Alpine hill is Mount Porezen (1,630 m), which is part of the Cerkno Hills, and its highest Dinaric plateau is the Snežnik Plateau with Big Mount Snežnik (*Veliki Snežnik*, 1,796 m). Beyond these three landscape types, there are only two other elevations that could partially be considered mountainous: Mount Boč (978 m) in the Pannonian part of Slovenia and Mount Slavnik (1,028 m) in its Mediterranean part.

Keywords

Mountain · Hill · Plateau · Alps · Dinaric Alps · Landform · Landscape

3.1 Mountains and Related Terms in Slovene

Slovenia is a country with a distinctly dynamic terrain, which is why Slovenian uses numerous expressions for convex and concave landforms. The most common terms referring to convex forms are *grič* 'low hill', *hrib* 'high hill', and *gora* 'mountain'. Their plural forms are *griči*, *hribi*, and *gore*, and their collective nouns are *gričevje*, *hribovje*, and *gorovje*. Collective nouns predominate over plural forms in Slovenian geographical names; for example, *Posavsko hribovje* 'Sava Hills' rather than *Posavski hribi*. Slovenian also uses a common term for basic convex landforms: *vzpetina* 'elevation'.

Differences between the terms *grič*, *hrib*, and *gora* are not very precisely defined. Usually, a *grič* is an elevation with a 200 m difference between its foot and top, a *hrib* is an elevation with a difference of up to 600 m between its foot and top, or an elevation that rises up to the tree line, and a *gora* is an elevation with a difference of over 600 m between its foot and top or an elevation that rises

D. Perko (✉)
ZRC SAZU Anton Melik Geographical Institute, Ljubljana, Slovenia
e-mail: drago.perko@zrc-sazu.si

M. Zorn et al. (eds.), *Discourses on Mountains of Montenegro and Slovenia*, Historical Geography and Geosciences, https://doi.org/10.1007/978-3-031-99017-5_3

beyond the tree line (Badjura 1953; Demek 1976; Gams 1984, 1985, 1986, 1987; Natek 1993; Perko 1992, 2001).

In western Slovenia, a *grič* is also called a *brdo* and in eastern Slovenia it is called a *gorica* (literally 'little mountain'), which usually denotes a low hill with vineyards.

Other common Slovenian terms for convex and concave landforms are *dolina* 'valley', *podolje* 'elongated lowland', *kotlina* 'basin', *ravnina* 'flat, plain', *ravnik* 'corrosion plain', *planota* 'plateau', *višavje* 'highlands', and *nižavje* 'lowlands'.

Slovenian also distinguishes between the following:

- *razgibano površje* 'rough terrain', which primarily includes mountains, high hills, low hills, and plateaus, versus *nerazgibano površje* 'level terrain', which includes plains;
- *gorato površje* 'mountainous terrain', which primarily includes mountains, high hills, and high plateaus, versus *negorato površje* 'non-mountainous terrain', which primarily comprises plains, low hills, and low plateaus; and
- *gorsko površje* 'mountain terrain', which only includes mountains, versus *negorsko površje* 'non-mountain terrain', which includes all other landforms.

If Slovenia, which is located at the intersection between the Alps, the Pannonian Basin, the Dinaric Alps, and the Mediterranean (Perko et al. 2021), is graphically depicted as a square (Fig. 3.1), the northwestern part of its central section is covered in Alpine mountains, high plateaus, and high hills with intermittent basins and valleys, and the southeastern part of its central section is occupied by Dinaric high plateaus, high hills, and elongated lowlands in between. The Alps and Dinaric Alps gradually turn into Pannonian low hills and plains toward the northeastern corner of the square and into Mediterranean low hills and low plateaus toward its northwestern corner.

Mountainous areas in Slovenia usually primarily include mountains, high hills, and

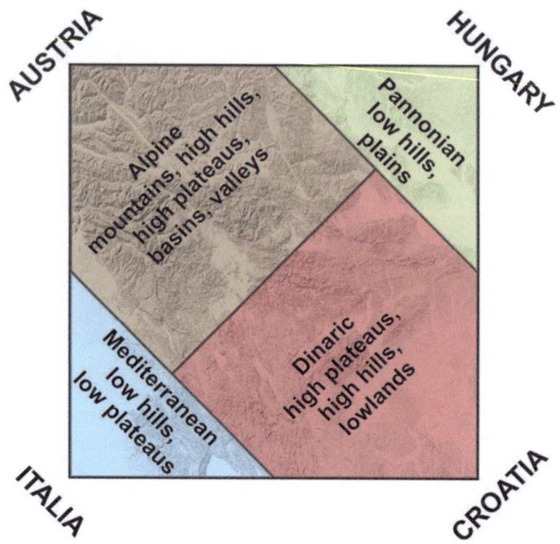

Fig. 3.1 Simplified schematic representation of Slovenia's landform and landscape types

high plateaus with small intermittent concave landforms (e.g., the Bohinj Basin in the Julian Alps). They do not include large basins (e.g., the Ljubljana Basin), plains (e.g., the Drava Plain), lowlands (e.g., the Inner Carniola Lowland), corrosion plains (e.g., White Carniola), and low plateaus (e.g., the Karst Plateau) with small intermittent convex landforms, such as Mount Boč in the middle of the Pannonian lowland.

3.2 Landscape Typologies

Landscape typologies can be used to delineate Slovenia's mountainous terrain because they all heavily rely on the dynamic nature of terrain, which is the main reason for landscape differences in Slovenia. Geographers have produced five landscape typologies for all of Slovenia to date (Perko et al. 1996; 2015; 2021; Melik 1946; Ilešič 1958; Špes et al. 2002). Due to the different methodologies used for defining these types, it is primarily the first four that are suitable for delineating mountainous terrain.

The first landscape typology of Slovenia was produced by Anton Melik (1946), who divided territory in terms of geomorphology, rocks, and climate into eight basic landscape types (Perko et al. 2021):

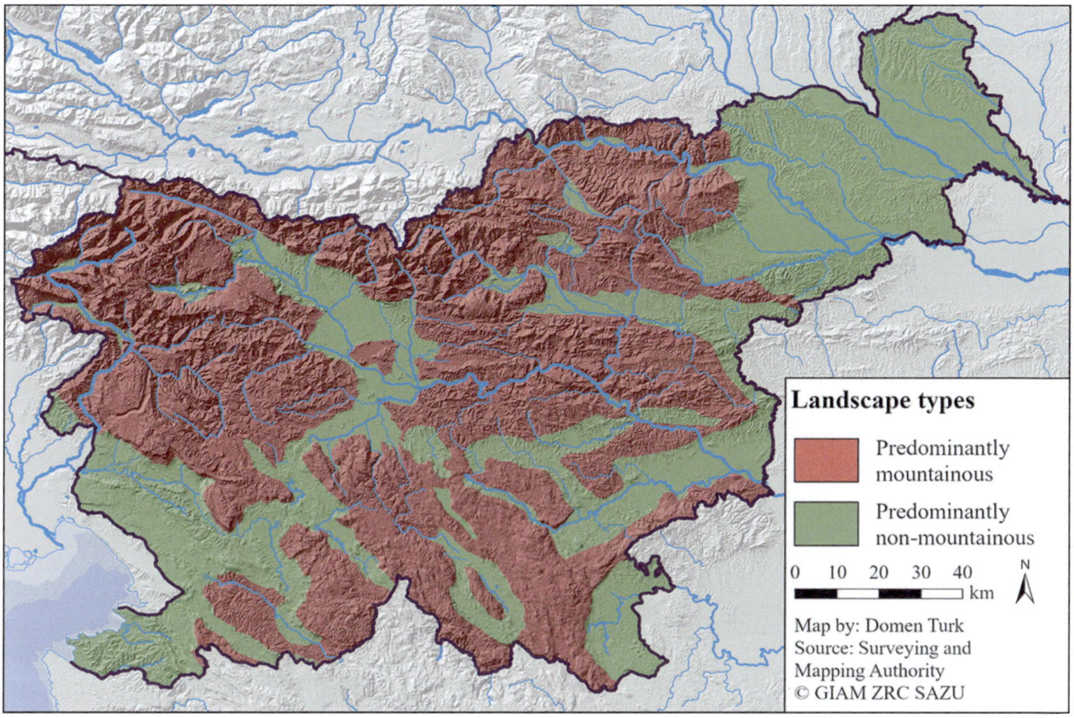

Fig. 3.2 Predominantly mountainous types in Melik's landscape typology of Slovenia

1. high Alpine mountains;
2. low Alpine mountains;
3. high hills;
4. low hills;
5. plains and wide valleys;
6. Dinaric high plateaus;
7. karst hills with low plateaus and lowlands; and
8. karst poljes.

The following types may primarily be deemed mountainous: Type 1 (11.2% of Slovenia), Type 2 (7.9%), Type 3 (23.8%), and Type 6 (14.9%), which account for a total of 57.8% of Slovenia (Fig. 3.2).

The second landscape typology was produced by Svetozar Ilešič (1958). The basic landscape types in his division were (Perko et al. 2021):

1. western Alpine landscapes;
2. eastern Alpine landscapes;
3. western sub-Alpine landscapes;
4. eastern sub-Alpine landscapes;
5. northeastern sub-Alpine landscapes;
6. sub-Pannonian landscapes;
7. sub-Pannonian–sub-Dinaric landscapes;
8. karst landscapes of inner Slovenia;
9. sub-Mediterranean–sub-Alpine landscapes;
10. sub-Mediterranean landscapes; and
11. sub-Mediterranean–Dinaric landscapes.

The following types may primarily be considered mountainous: Type 1 (8.2% of Slovenia), Type 2 (2.9%), Type 3 (11.6%), Type 4 (14.5%), Type 5 (6.5%), and Type 8 (20.2%), which account for a total of 63.9% of Slovenia (Fig. 3.3).

The third landscape typology was produced half a century after Melik in 1996 by Drago Perko (Perko 1998; Perko and Ciglič 2020a), who divided the territory in terms of mostly physical landscape components into nine landscape types using a geographic information system (Perko et al. 2021):

1. Alpine mountains;
2. Alpine hills;
3. Alpine plains;

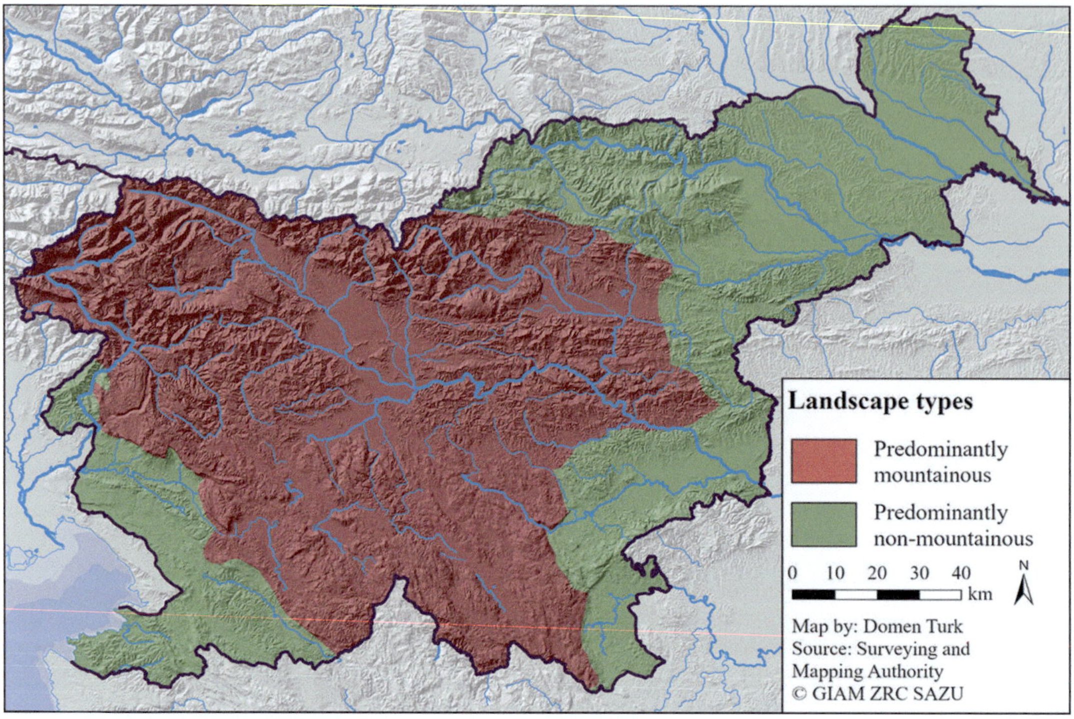

Fig. 3.3 Predominantly mountainous types in Ilešič's landscape typology of Slovenia

4. Pannonian hills;
5. Pannonian plains;
6. Dinaric plateaus;
7. Dinaric lowlands;
8. Mediterranean hills; and
9. Mediterranean plateaus.

The following types may primarily be considered mountainous: Type 1 (15.1% of Slovenia), Type 2 (23.0%), and Type 6 (18.8%), which account for a total of 56.9% of Slovenia (Fig. 3.4).

Metka Špes et al. (2002) divided Slovenia into 223 landscape ecological units and combined them into thirteen landscape types (Perko et al. 2021):

1. mountains;
2. wide valleys in mountains, high hills, and karst;
3. high karst plateaus and high hills of carbonate rocks;
4. high hills of non-carbonate rocks;
5. inter-mountain basins;
6. low hills in inner Slovenia;
7. plains and wide valleys in low hills in inner Slovenia;
8. karst poljes and lowlands;
9. low karst of Inner Carniola and Lower Carniola;
10. low karst of White Carniola;
11. Karst and Podgorje Karst plateaus;
12. low hills in the Littoral; and
13. wide valleys and plains in the Littoral.

The following types may primarily be considered mountainous: Type 1 (8.3% of Slovenia), Type 2 (3.6%), Type 3 (17.7%), Type 4 (22.4%), and Type 9 (5.4%), which account for a total of 57.4% of Slovenia (Fig. 3.5).

In 2013, Drago Perko et al. produced several typologies with a different number of types, based on the spatial overlap of 195 landform units, 938 lithological units, and sixty-five vegetation units in the geographical information system.

The most generalized typology includes twenty-four landscape types (Perko et al. 2015):

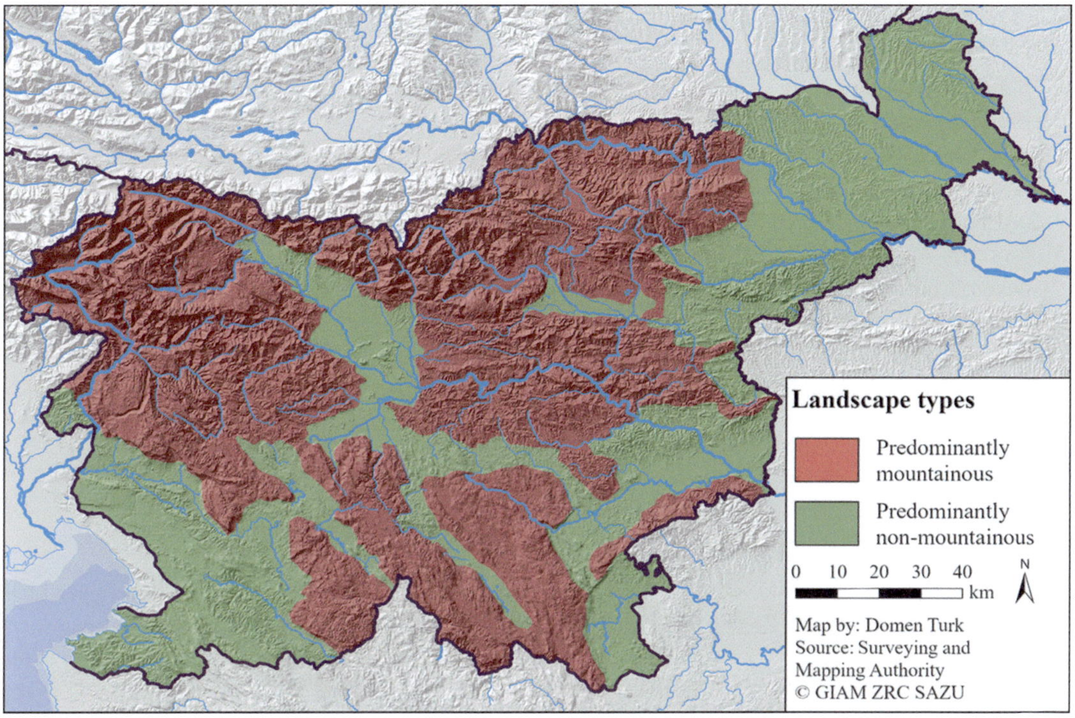

Fig. 3.4 Predominantly mountainous types in Perko's landscape typology of Slovenia

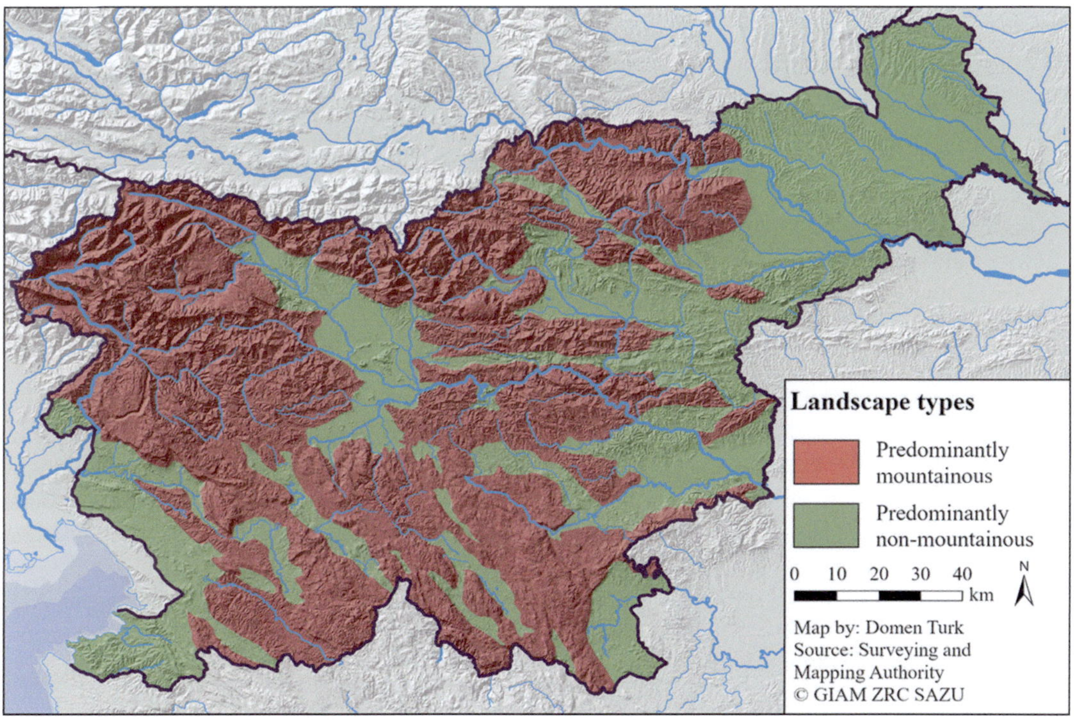

Fig. 3.5 Predominantly mountainous types in Špes et al.'s landscape typology of Slovenia

1. plains + non-carbonate sediments + European hornbeam, oak, red pine;
2. plains + non-carbonate sediments + beech, chestnut, various oaks;
3. plains + carbonate sediments + European hornbeam, oak, red pine;
4. other plains;
5. low hills + non-carbonate sediments + European hornbeam, oak, red pine;
6. low hills + non-carbonate sediments + beech, chestnut, various oaks;
7. low hills + fine clastic rocks + beech, chestnut, various oaks;
8. low hills + flysch + downy oak, sessile oak;
9. low hills + flysch + beech, chestnut, various oaks;
10. low hills + carbonate rocks + downy oak, sessile oak;
11. low hills + carbonate rocks + European hornbeam, oak, red pine;
12. low hills + carbonate rocks + beech;
13. low hills + carbonate rocks + beech, fir;
14. low hills + carbonate rocks + beech, chestnut, various oaks;
15. other low hills;
16. high hills + coarse clastic rocks + beech;
17. high hills + carbonate rocks + beech;
18. high hills + carbonate rocks + beech, fir;
19. high hills + carbonate rocks + beech, European hophornbeam;
20. high hills + metamorphic and igneous rocks with tuffs + beech, chestnut, various oaks;
21. other high hills;
22. mountains + carbonate rocks + beech;
23. mountains + carbonate rocks + fir, spruce, highland vegetation; and
24. other mountains.

The following types may primarily be considered mountainous: Type 16 (1.8% of Slovenia), Type 17 (8.7%), Type 18 (7.1%), Type 19 (3.5%), Type 20 (2.1%), Type 21 (14.9%), Type 22 (2.5%), Type 23 (1.8%), and Type 24 (5.3%), which account for a total of 47.7% of Slovenia (Fig. 3.6).

The largest share of mountainous terrain is included in Ilešič's typology (i.e., 63.9%) and the smallest in that of Perko et al. (i.e., 47.7%); roughly all five landscape typologies define approximately half of Slovenia as predominantly mountainous (Table 3.1).

The correlation between the landscape typologies or the spatial overlap of individual typologies calculated using contingency tables shows that all typologies correlate most highly with Perko's typology from 1996 (Perko et al. (2021). Therefore, a decision was made to define and treat the following three landscape types from his typology (Fig. 3.7) as mountainous areas in Slovenia: Alpine mountains, Alpine hills, and Dinaric plateaus, which account for a total of 56.9% of Slovenia.

3.3 Mountainous Regions

Together with the landscape typology, Perko et al. also produced a geographical regionalization of Slovenia with four macroregions and forty-eight mesoregions (Kladnik 1996; Perko 1998; Perko and Ciglič 2020b). This division relies on four major European geographical units intersecting in Slovenia: the Alps, the Pannonian Basin, the Dinaric Alps, and the Mediterranean.

The Alps are the highest and largest mountain chain in Europe. They cover about 200,000 km^2 and run from France in the west to Austria in the east. The Alps are more than 1,200 km long and up to 250 km wide. The southeastern part of the Alps extends into Slovenia. Slovenia's Alps macroregion lies in northern and northwestern Slovenia, covering 42.1% of the country.

The Dinaric Alps are the southeastern continuation of the Alps. They are 700 km long and up to 200 km wide, and they cover less than half the area of the Alps. The northwestern part of the Dinaric Alps extends into Slovenia. Slovenia's Dinaric Alps macroregion lies in the southern and southeastern part of the country, encompassing 28.2% of Slovenia.

The Pannonian Basin lies east of the Alps and north of the Dinaric Alps. It runs about 600 km from north to south and 700 km from east to west, and it is almost twice as large as the Alps. Only the southwestern margin of the Pannonian Basin extends into Slovenia. Slovenia's Pannonian Basin macroregion lies in

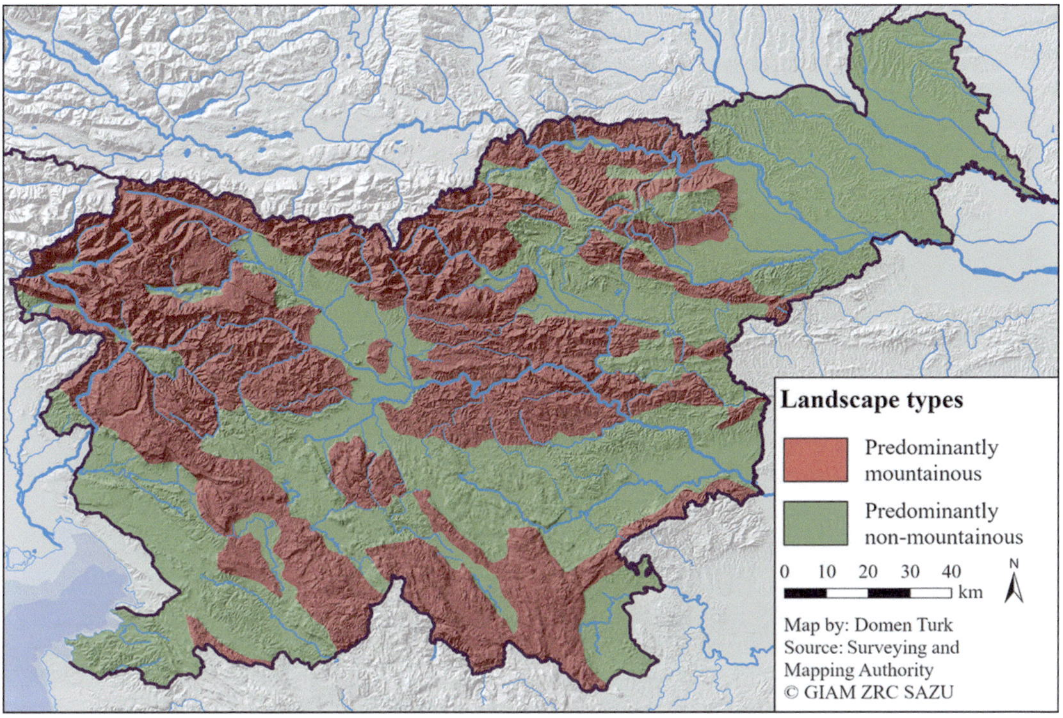

Fig. 3.6 Predominantly mountainous types in Perko et al.'s landscape typology of Slovenia

Table 3.1 Comparison of landscape macrotypologies of Slovenia

Author(s)	Year	No. of basic types	Average basic type size (km^2)	No. of mountainous types	Proportion of no. of mountainous types (%)	Area of mountainous types (km^2)	Proportion of area of mountainous types (%)
Melik	1946	8	2,534	4	50.0	11,709	57.8
Ilešič	1958	11	1,843	6	54.5	12,958	63.9
Perko	1996	9	2,253	3	33.3	11,532	56.9
Špes et al	2002	13	1,559	5	38.5	11,634	57.4
Perko et al	2013	24	845	9	37.5	9,671	47.7

the eastern and northeast part of the country, covering 21.2% of Slovenia.

The Mediterranean region surrounds the Mediterranean Sea. Its arm known as the Adriatic Sea stretches almost 700 km between the Italian Apennines to the southwest and the Dinaric Alps to the northeast. With an area of 132,000 km^2, the Adriatic Sea is somewhat larger than the Dinaric Alps. The extreme northern margin of the Mediterranean extends into Slovenia. Slovenia's Mediterranean macroregion lies along the Gulf of Trieste and its hinterland in southwestern Slovenia, encompassing 8.6% of the country.

The typology and regionalization are interconnected in the sense that the borders of macroregions and mesoregions overlap with those of the landscape types (Perko et al. 2021). The area covered by the Alps macroregion (42.1% of Slovenia) is the same as the total area covered by the Alpine mountains, Alpine hills, and Alpine plains landscape types. The area covered by the Pannonian Basin macroregion (21.2%) corresponds to the total area covered by the Pannonian hills and Pannonian plains landscape types, the area covered by the Dinaric Alps (28.2%) corresponds to the total area covered by Dinaric

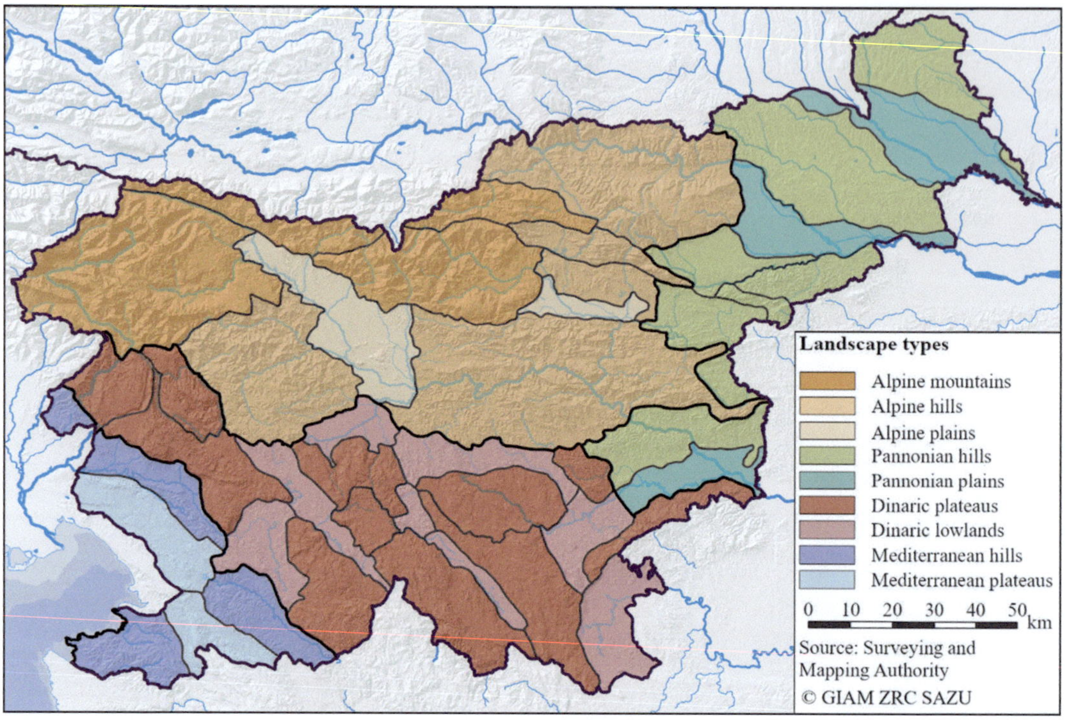

Fig. 3.7 Perko's landscape typology. (Perko 1998; Perko et al. 2021)

plateaus and Dinaric lowlands, and the area covered by the Mediterranean region (8.5%) corresponds to that of Mediterranean hills and Mediterranean plateaus combined (Perko 1998; Perko et al. 2021).

Mountainous regions (Fig. 3.8, Table 3.2) are the ones that are part of Alpine mountains, Alpine hills, or Dinaric plateaus (Fig. 3.7).

3.4 Regions Within Alpine Mountains

Alpine mountains cover 15.1% of Slovenia. They rise in the country's northwest, in the corner between Italy and Austria, and are divided into four regions (Perko 1998; Perko et al. 2015):

- *Julijske Alpe* 'Julian Alps' (Figs. 3.9, 3.10, 3.11 and 3.12);
- *Zahodne Karavanke* 'Western Karawanks' (Fig. 3.13);
- *Vzhodne Karavanke* 'Eastern Karawanks' (Fig. 3.14); and
- *Kamniško-Savinjske Alpe* 'Kamnik–Savinja Alps' (Figs. 3.15 and 3.16).

The Karawanks extend along the Slovenian–Austrian border in an east–west direction (highest peak: Mount Stol, 2,236 m). They are divided into two regions: the Western Karawanks and the Eastern Karawanks. South of them, the Kamnik–Savinja Alps (highest peak: Mount Grintovec, 2,558 m) descend into the Ljubljana Basin. The Skuta Glacier, the southeasternmost glacier in the Alps, lies below nearby Mount Skuta (2,533 m). Continuing from Italy south of the Sava River are the Julian Alps, the most magnificent Slovenian mountain range. This is where Mount Triglav (2,864 m), Slovenia's highest peak, rises. The remnants of the Triglav Glacier lie on its northeastern slopes, and Triglav National Park extends around the mountain. Located nearby are two glacial lakes: Lake Bohinj (328 hectares),

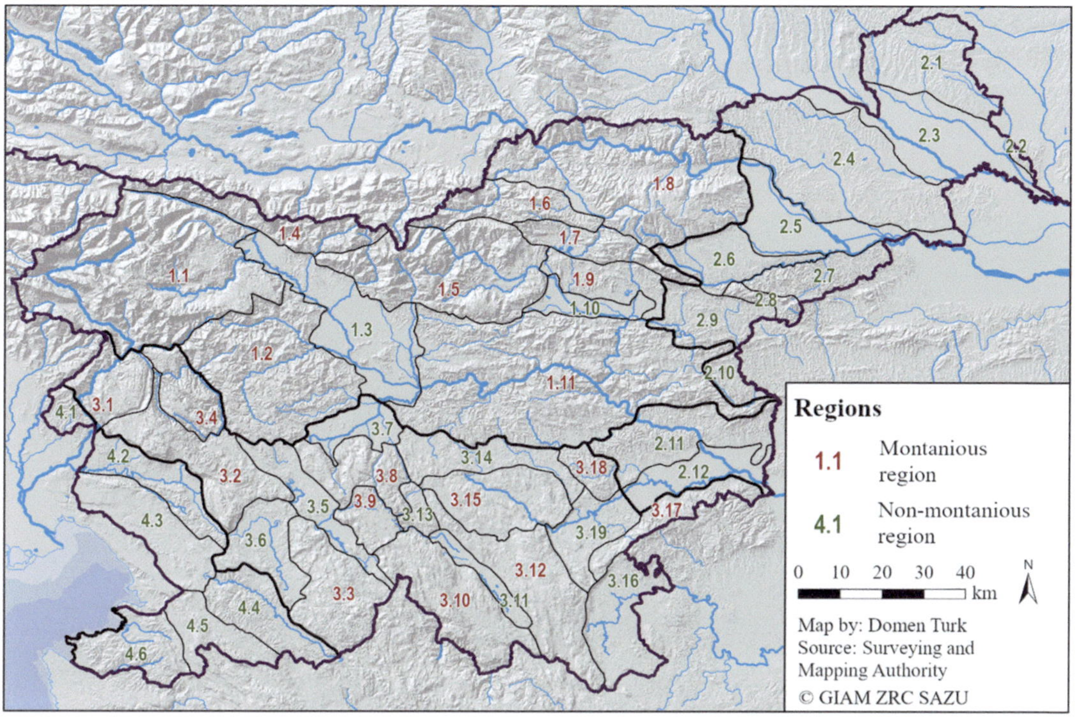

Fig. 3.8 Predominantly mountainous regions in Perko et al.'s geographical regionalization of Slovenia. (Perko 1998)

Slovenia's largest natural lake (Fig. 3.11), and cosmopolitan tourism-oriented Lake Bled (145 hectares; Fig. 3.13). The Julian Alps are the divide between the Adriatic Sea to the south, which the emerald-color Soča River empties into (Fig. 3.12), and the Black Sea to the east, which Slovenia's largest river, the Sava, flows toward.

3.5 Regions Within the Alpine Hills

Alpine hills cover 23.0% of Slovenia. They surround the Alpine mountains in a wide arc to the south and east, and they are divided into five regions (Perko 1998; Perko et al. 2015):

- *Cerkljansko, Škofjeloško, Polhograjsko in Rovtarsko hribovje* 'Cerkno, Škofja Loka, Polhov Gradec, and Rovte Hills' (Fig. 3.17);
- *Pohorje, Strojna in Kozjak* 'Pohorje, Strojna, and Kozjak Hills' (Fig. 3.18);
- *Velenjsko in Konjiško hribovje* 'Velenje and Konjice Hills';
- *Ložniško in Hudinjsko gričevje* 'Ložnica and Hudinja Hills' (Fig. 3.19); and
- *Posavsko hribovje* 'Sava Hills' (Fig. 3.20).

Extending furthest to the west are the Cerkno, Škofja Loka, Polhov Gradec, and Rovte Hills, and furthest to the north are the Pohorje, Strojna, and Kozjak Hills, which are the southern outcroppings of the Central Eastern Alps, where the Drava River has cut a deep valley. From there, the Velenje and Konjice Hills extend to the south, where many years of lignite mining resulted in subsidence, forming artificial lakes. Further to the south, the hills transition into the lower Ložnica and Hudinja Hills, which form the northern edges of the Celje Basin. To the south of this basin, the Sava Hills extend from Ljubljana in the west to the Croatian border in the east, in a stretch nearly 100 km long, making it the largest region in Slovenia (1,909 km^2).

Table 3.2 English and Slovenian names of predominantly mountainous regions and predominantly non-mountainous regions (Fig. 3.8)

No. in Fig. 3.8	English names	Slovenian names
	macroregion	***makroregija***
	mountainous regions	*gorata regija*
	non-mountainous regions	*negorata regija*
	Alps	***Alpe***
1.4	Western Karawanks	*Zahodne Karavanke*
1.6	Eastern Karawanks	*Vzhodne Karavanke*
1.5	Kamnik–Savinja Alps	*Kamniško-Savinjske Alpe*
1.1	Julian Alps	*Julijske Alpe*
1.2	Cerkno, Škofja Loka, Polhov Gradec, and Rovte Hills	*Cerkljansko, Škofjeloško, Polhograjsko in Rovtarsko hribovje*
1.6	Sava Hills	*Posavsko hribovje*
1.7	Velenje and Konjice Hills	*Velenjsko in Konjiško hribovje*
1.8	Pohorje, Strojna, and Kozjak Hills	*Pohorje, Strojna in Kozjak*
1.9	Ložnica and Hudinja Hills	*Ložniško in Hudinjsko gričevje*
1.3	Sava Plain	*Savska ravan*
1.10	Savinja Plain	*Savinjska ravan*
	Pannonian Basin	***Panonska kotlina***
2.1	Goričko	*Goričko*
2.2	Lendava Hills	*Lendavske gorice*
2.4	Slovenian Hills	*Slovenske gorice*
2.6	Dravinja Hills	*Dravinjske gorice*
2.7	Haloze	*Haloze*
2.8	Mount Boč and the Macelj Ridge	*Boč in Macelj*
2.9	Voglajna and Upper Sotla Hills	*Voglajnsko in Zgornjesotelsko gričevje*
2.10	Central Sotla Hills	*Srednjesotelsko gričevje*
2.11	Krško, Senovo, and Bizeljsko Hills	*Krško, Senovsko in Bizeljsko gričevje*
2.3	Mura Plain	*Murska ravan*
2.5	Drava Plain	*Dravska ravan*
2.12	Krka Plain	*Krška ravan*
	Dinaric Alps	***Dinarsko gorovje***
3.1	Kambreško and Banjšice Plateaus	*Kambreško in Banjšice*
3.2	Trnovo Forest Plateau, Mount Nanos, and Hrušica Plateau	*Trnovski gozd, Nanos in Hrušica*
3.4	Idrija Hills	*Idrijsko hribovje*
3.3	Javornik Hills and Snežnik Plateau	*Javorniki in Snežnik*
3.6	Pivka Lowland and Mount Vremščica	*Pivško podolje z Vremščico*
3.5	Inner Carniola Lowland	*Notranjsko podolje*
3.8	Krim Hills and Menišija Plateau	*Krimsko hribovje in Menišija*
3.9	Bloke Plateau	*Bloke*
3.10	Big Mountains, Mount Stojna, and Mount Gotenica	*Velika gora, Stojna in Goteniška gora*
3.11	Ribnica–Kočevje Lowland	*Ribniško-Kočevsko podolje*
3.12	Little Mountains, Kočevje Rog Plateau, and Mount Poljane	*Mala gora, Kočevski rog in Poljanska gora*
3.13	Velike Lašče Region	*Velikolaščanska pokrajina*
3.7	Ljubljana Marsh	*Ljubljansko barje*
3.19	Novo Mesto Region	*Novomeška pokrajina*
3.14	Lower Carniola Lowland	*Dolenjsko podolje*
3.18	Radulja Hills	*Raduljsko hribovje*
3.15	Dry Carniola and Dobrepolje	*Suha krajina z Dobrepoljem*
3.16	White Carniola	*Bela krajina*
3.17	Gorjanci Hills	*Gorjanci*
	Mediterranean	***Sredozemlje***
4.1	Gorica Hills	*Goriška brda*
4.2	Vipava Hills	*Vipavska brda*
4.3	Karst Plateau	*Kras*
4.4	Brkini Hills and Reka Valley	*Brkini in dolina Reke*
4.5	Podgorje Karst Plateau, Čičarija Plateau, and Podgrad Lowland	*Podgorski kras, Čičarija in Podgrajsko podolje*
4.6	Koper Hills	*Koprska brda*

Fig. 3.9 Julian Alps are the northwesternmost region in Slovenia. Triglav, the highest mountain in Slovenia (2,864 m), stands out the most. The view from the Karawanks also shows Lake Bled on the left, the Sava River in front, and the town of Jesenice on the right. (Photo by Aleš Krivec, Shutterstock.com)

Fig. 3.10 Planica is a picturesque glacial valley on the north side of the Julian Alps. It is the cradle of ski flying, one of Slovenia's most popular winter spectator sports. The picture shows the valley in winter with the facilities of the new Nordic Center. (Photo by Planica Nordic Centre, GIAM ZRC SAZU archive)

3.6 Regions Within the Dinaric Plateaus

Dinaric plateaus cover 18.8% of Slovenia. They lie in the southern part of the country; they are almost entirely made of limestone and dolomite, and covered in forest.

They are divided into eleven regions (Perko 1998; Perko et al. 2015):

- *Kambreško in Banjšice* 'Kambreško and Banjšice Plateaus';
- *Trnovski gozd, Nanos in Hrušica* 'Trnovo Forest Plateau, Mount Nanos, and Hrušica Plateau' (Fig. 3.21);
- *Idrijsko hribovje* 'Idrija Hills';
- *Javorniki in Snežnik* 'Javornik Hills and Snežnik Plateau' (Fig. 3.22);
- *Krimsko hribovje in Menišija* 'Krim Hills and Menišija Plateau' (Fig. 3.23);
- *Bloke* 'Bloke Plateau';
- *Velika gora, Stojna in Goteniška gora* 'Big Mountains, Mount Stojna, and Mount Gotenica' (Fig. 3.24);

Fig. 3.11 Lake Bohinj is a picturesque glacial lake located among the alpine plateaus in the southwestern part of the Julian Alps. With an area of 328 hectares, it is Slovenia's largest natural lake. Its tributary, the Savica River, is one of the two headwaters of the Sava River. (Photo by Bojan Erhartič, GIAM ZRC SAZU archive)

Fig. 3.12 Source of the Soča River is west of Mount Triglav in the Julian Alps. It is known for its emerald color. It is the most important river in Slovenia's Adriatic catchment. Its upper valley is one of the best-preserved natural areas in the Alps. (Photo by Bojan Erhartič, GIAM ZRC SAZU archive)

- *Mala gora, Kočevski rog in Poljanska gora* 'Little Mountains, Kočevje Rog Plateau, and Mount Poljane';
- *Raduljsko hribovje* 'Radulja Hills';
- *Suha krajina in Dobrepoljem* 'Dry Carniola and Dobrepolje'; and
- *Gorjanci* 'Gorjanci Hills'.

In the northwesternmost part of the Dinaric region, the Alpine mountains come into contact with the region of the Kambreško and Banjšice high plateaus, which toward the southeast, beyond the deeply cut Soča Valley, continues into the plateau-like Idrija Hills region and the Trnovo Forest, Nanos, and Hrušica high Dinaric plateau

Fig. 3.13 Western Karawanks (highest peak: Stol, 2,236 m) are a beautiful backdrop to glacial Lake Bled (145 ha) in a moraine landscape on the eastern edge of the Julian Alps. It is known for its castle above a rock wall and islet with a church. (Photo by Bojan Erhartič, GIAM ZRC SAZU archive)

Fig. 3.14 Of all the mountains in Slovenia, the Eastern Karawanks extend the farthest east, where scenic Mount Saint Ursula (*Uršlja gora*, 1,699 m) especially dominates. At its top stands Saint Ursula's Church, a television transmitter, and a mountain lodge (view from Pohorje). (Photo by Bojan Erhartič, GIAM ZRC SAZU archive)

region. This series of western Dinaric plateaus ends in the region of the Javornik Hills and Snežnik Plateau, which extends to the border with Croatia.

The central part of the Dinaric plateaus is composed of a series of plateaus of varying elevations extending nearly from Ljubljana in the north to the Kolpa River in the south. They are separated by elongated lowlands. All these series of Dinaric plateaus and lowlands in between run in the Dinaric direction from the northwest to the southeast; the only exception is the easternmost region of the Dinaric plateaus—that is, the Gorjanci Hills—which predominantly run in a southwest–northeast direction. They form the border with Croatia (Perko and Ciglič 2020b).

Fig. 3.15 Kamnik–Savinja Alps (highest peak Grintovec, 2,558 m) rise between the Karawanks to the north and the edge of the Ljubljana Plain to the south. A picture taken in a rapeseed field near the town of Kranj shows a mountain range in their western part. (Photo by Bojan Erhartič, GIAM ZRC SAZU archive)

Fig. 3.16 In the eastern part of the Kamnik–Savinja Alps lies the picturesque glacial Logar Valley. Since 1987, it has been protected as a landscape park, stretching for 24 km^2. The Savinja River, the largest tributary of the Sava in Slovenia, has its source here. (Photo by Bojan Erhartič, GIAM ZRC SAZU archive)

Fig. 3.17 Cerkno Hills lie southeast of the Julian Alps. Their highest peak, Porezen (1,630 m), rises on the watershed between the Black Sea and the Adriatic catchments. The Sava tributaries flow to the east and the Soča tributaries to the other sides. (Photo by Bojan Erhartič, GIAM ZRC SAZU archive)

Fig. 3.18 Plateau-like Pohorje Hills are about 50 km long and 20 km wide south of the Drava River and west of Maribor (highest peak: Black Peak, *Črni vrh*, 1,543 m). Among its natural attractions are also the small Lovrenc Lakes, lying among conifers. (Photo by Bojan Erhartič, GIAM ZRC SAZU archive)

3.7 Missing Mountains

The selection of the three landscape types (i.e., Alpine mountains, Alpine hills, and Dinaric plateaus) from Perko's typology (Perko et al. 2015, 2021, 2026), which are expected to cover all predominantly mountainous areas in Slovenia, may seem approximate at first glance, but in fact only two elevations that at least partly correspond to the definition of a mountain in Slovenia (i.e., the difference in elevation between its foot and top is over 600 m, or its peak is above the tree line) remain outside these three landscape types.

Fig. 3.19 North of the town of Celje, into which the Savinja River flows from the Alps, the mountains quickly turn into hills and then sink below the plain. The area on the southeastern Alpine edge is characterized by an mix of Alpine and Pannonian features. (Photo by Bojan Erhartič, GIAM ZRC SAZU archive)

Fig. 3.20 Sava Hills along the Sava River are the most extensive Slovenian region between Ljubljana in the west and the meandering Sotla, a border river, in the east (highest peak: Mount Kum, 1,220 m). The snowy peaks of the Kamnik–Savinja Alps rise in the back. (Photo by Matevž Lenarčič, GIAM ZRC SAZU archive)

The first is Mount Boč (978 m) in Pannonian Slovenia, which actually does not rise beyond the tree line, but it markedly stands out from the surroundings and the difference in elevation between its foot and top is over 600 m (Fig. 3.25). The town of Poljčane to the north of Mount Boč lies 716 m lower than its peak, and Rogaška Slatina to the south of the mountain lies 754 m lower than its peak.

Fig. 3.21 Nanos Plateau (highest peak: Dry Peak, *Suhi vrh*, 1,313 m) is a typical Dinaric high plateau. It dominates west of Postojna. At the southern foot of the Nanos Plateau, the freeway branches to Gorizia to the west and Trieste to the south (both in Italy). (Photo by Matevž Lenarčič, GIAM ZRC SAZU archive)

Fig. 3.22 Cerknica Polje (front) with intermittent Lake Cerknica is surrounded by forested Dinaric plateaus. The Javornik Hills rise southwest of the polje (right rear), and Snežnik (highest peak: Big Mount Snežnik, *Veliki Snežnik*, 1,796 m) stands out in the southwest (left rear). (Photo by Bojan Erhartič, GIAM ZRC SAZU archive)

The second is Mount Slavnik (1,028 m) in Slovenia's Mediterranean region, which dominates its surroundings, even though the difference in elevation between its foot and top is smaller than 600 m (Fig. 3.26). The village of Materija north of Mount Slavnik lies only 521 m lower than its peak, and the village of Podgorje south of the mountain lies 520 m lower than its peak. The top is not forested, but the natural tree line was originally higher because farmers cleared the area for pastures.

Fig. 3.23 South of the Ljubljana Marsh, a series of Dinaric plateaus begins with the Krim Hills (highest peak: Krim, 1,107 m), stretching across central southern Slovenia, crossing the Slovenian–Croatian border and sinking steeply into the Adriatic Sea in the Gulf of Rijeka. (Photo by Bojan Erhartič, GIAM ZRC SAZU archive)

Fig. 3.24 Kolpa is a border river between Slovenia and Croatia. It carved deep into carbonate Dinaric plateaus populated by brown bears (*Ursus arctos*). The plateaus continue along the Adriatic Sea all the way to Albania. The view is from Slovenia to Croatia. (Photo by Jure Tičar, GIAM ZRC SAZU archive)

Fig. 3.25 Mount Boč (978 m) is a geological extension of the Karawanks. It rises east from Slovenia's continuous mountainous surface. A lookout and television tower are located at the top. Since 1992, the 22 km^2 surrounding area has been protected as a nature park. (Photo by Matej Gabrovec, GIAM ZRC SAZU archive)

Fig. 3.26 From the sea, Mount Slavnik (1,028 m) looks like a mighty mountain, but from the mainland only an ordinary hill. It offers a beautiful view of the Adriatic Sea and the Alps. Due to endemic animal species, it has been protected as a natural monument since 1992. (Photo by Špela Čonč, GIAM ZRC SAZU archive)

Slovenians have also named many other elevations across the country *gora* 'mountain', even though they do not meet the Slovenian criteria. One example is *Šmarna gora* (Mount Saint Mary, 669 m) north of Ljubljana, which rises 369 m above the Sava River and is among the most popular hiking destinations in Slovenia.

Thus, it can be concluded that Slovenia consists of three predominantly mountainous landscape types (Alpine mountains, Alpine hills, and Dinaric plateaus), which include all areas that can be defined as mountainous, and six predominantly non-mountainous landscape types (Alpine plains, Pannonian hills, Pannonian plains, Dinaric lowlands, Mediterranean hills, and Mediterranean plateaus), which include most of the plains, elongated lowlands, low plateaus, and low hills.

Acknowledgements This study was conducted as part of the research program The Geography of Slovenia (P6-0101), financed by the Slovenian Research and Innovation Agency.

References

Badjura R (1953) Ljudska geografija [People's Geography]. Državna založba Slovenije, Ljubljana

Demek J (1976) Handbuch der geomorphologischen Detailkartirung. Verlag Ferdinand Hirt, Wien

Gams I (1984) Metodologija geografske razčlenitve ozemlja [Methodology of the geographical territorial division]. Geografski Vestnik 56:75–82

Gams I (1985) Problemi sodobnega raziskovanja gorskih sistemov [Problems of the contemporaneous research of mountain systems]. Geografski Vestnik 57:65–72

Gams I (1986) Za kvantitativno razmejitev med pojmi gričevje, hribovje in gorovje [For the quantitative defination of the notions low hills, hills and mountains]. Geografski Vestnik 58:77–82

Gams I (1987) Omejitev alpskega sveta v Sloveniji [Delimitation of the Alpine world in Slovenia]. Geografski Vestnik 59:13–20

Ilešič S (1958) Problemi geografske rajonizacije ob primeru Slovenije [On the problems of delimitation and classification of geographical regions according to the example of Slovenia]. Geografski Vestnik 29–30:83–140

Kladnik D (1996) Naravnogeografske členitve Slovenije [Natural-geographical divisions of Slovenia]. Geografski Vestnik 68:123–159

Melik A (1946) Prirodnogospodarska sestava Slovenije [Naturally-economic structure of Slovenia]. Geografski Vestnik 18:3–22

Natek K (1993) Tipi površja v Sloveniji 1 [The types of terrain in Slovenia 1]. Geografski Obzornik 40(4):26–31

Perko D (1998) The regionalization of Slovenia. Geografski Zbornik 38:11–57

Perko D, Hrvatin M, Ciglič R (2015) A methodology for natural landscape typification of Slovenia. Acta Geogr Slov 55(2):235–270. https://doi.org/10.3986/AGS.1938

Perko D, Hrvatin M, Ciglič R (2017) Determination of landscape hotspots in Slovenia. Acta Geogr Slov 57(1):7–29. https://doi.org/10.3986/AGS.4618

Perko D, Ciglič R, Hrvatin M (2021) Landscape macrotypologies and microtypologies of Slovenia. Acta Geogr Slov 61(3):7–89. https://doi.org/10.3986/AGS.10384

Perko D, Ciglič R (2020a) Slovenia's landscapes. In: Perko D, Ciglič R, Zorn M (eds) The geography of Slovenia: small but diverse. Springer, Cham, pp 211–225. https://doi.org/10.1007/978-3-030-14066-3_14

Perko D, Ciglič R (2020b) Slovenia's regions. In: Perko D, Ciglič R, Zorn M (eds) The geography of Slovenia: small but diverse. Springer, Cham, pp 227–243. https://doi.org/10.1007/978-3-030-14066-3_15

Perko D, Ciglič R, Hrvatin M (2026) Slovenia's landscape and landform diversity. In: Zorn M, Ciglič R, Perko D (eds) Landscapes and landforms of Slovenia. Springer, Cham

Perko D (1992) Zveze med reliefom in prebivalstvom 1880–1981 v Sloveniji [Relation between relief and population 1880–1981 in Slovenia]. PhD Thesis, University of Ljubljana, Ljubljana

Perko D (2001) Analiza površja Slovenije s stometrskim digitalnim modelom reliefa [Analysis of the surface of Slovenia using the 100-meter digital elevation model]. Geografija Slovenije, vol 3. Založba ZRC, Ljubljana. https://doi.org/10.3986/9789612544973

Špes M, Cigale D, Lampič B, Natek K, Plut D, Smrekar A (2002) Študija ranljivosti okolja: metodologija in aplikacija [Environmental vulnerability study: methodology and application]. Geographica Slovenica 35:3–150

4 Geomorphometry of Slovenia's Mountainous Surface

Drago Perko

Abstract

Geomorphometry is the science of measuring land surfaces. The basic geomorphometric indicators include surface height, slope, and aspect. Their values were calculated for all of Slovenia and separately for its mountainous and non-mountainous surface, using a 5-m digital elevation model. Slovenia's average surface height is 556.4 m, its average surface slope is 16.5°, and its average surface aspect measured from the south (0°) to the north (180°) is 84.8°. The average height of its non-mountainous surface is 320.0 m and that of its mountainous surface is 735.5 m, the average slope of its non-mountainous and mountainous surface is 9.5° and 21.8°, respectively, and the average aspect of its non-mountainous and mountainous surface is 81.2° and 87.5°, respectively. Slovenia's lowest point is at 0 m on the coast of the Gulf of Trieste and the lowest point of its mountainous surface is at 54 m at the bottom of the Soča Valley north of Nova Gorica. The highest point of Slovenia and its mountainous surface is at 2,864 m at the top of Mount Triglav. Geomorphometric indicators help design landform typologies. Slovenian geographers have produced five so far: the first in 1935 and the last in 2019. The one created in 1992 is the only typology with a distinct geomorphometric character and the first computer-designed one. It divides Slovenia into 195 geomorphometric units and seven geomorphometric types. Among all Slovenian regions, Haloze has the roughest surface and the Mura Plain has the least rough surface, and among the mountainous regions, the roughest surface is characteristic of the Idrija Hills and the least rough surface can be found in the Dry Carniola and Dobrepolje region.

Keywords

Digital elevation model · Height · Slope · Aspect · Geomorphometric unit · Geomorphometric type · Landform · Landscape

4.1 Introduction

Geomorphometry, or the science of measuring the Earth's landforms (from Greek *gê* 'Earth', *morphḗ* 'form', and *metréō* 'I measure' or *métron* 'measure') is a branch of geomorphology that primarily measures landforms but also explains their connections with other landscape elements. It relies largely on geographic information systems and digital elevation models. The basic geomorphometric indicators are surface slope and aspect. The former shows changes in surface

D. Perko (✉)
ZRC SAZU Anton Melik Geographical Institute, Ljubljana, Slovenia
e-mail: drago.perko@zrc-sazu.si

M. Zorn et al. (eds.), *Discourses on Mountains of Montenegro and Slovenia*, Historical Geography and Geosciences, https://doi.org/10.1007/978-3-031-99017-5_4

height relative to the horizontal plane, and the latter relative to the vertical plane.

A 5 m digital elevation model was used to conduct a geomorphometric analysis of Slovenia's surface. The model was produced in 2023 based on a 0.5-m digital elevation model that the Ljubljana-based company GDi d.o.o. designed in 2021 using data from the point cloud obtained through the laser scanning of Slovenia commissioned by the Slovenian Ministry of Natural Resources and Spatial Planning. The indicators were calculated separately for Slovenia and its mountainous and non-mountainous surface. The following three landscape types (defined by Perko in Chap. 3 of this book), which together account for 57% of Slovenia, were characterized as mountainous surfaces: Alpine mountains, Alpine hills, and Dinaric plateaus. Five landscape types were categorized as non-mountainous surfaces: Alpine plains, Pannonian hills, Pannonian plains, Dinaric lowlands, Mediterranean hills, and Mediterranean plateaus (Perko et al. 2021).

4.2 Geomorphometric Indicators

Because the faces of landforms are two-dimensional and the space in which they are located is three-dimensional, each face of a landform can be analyzed in terms of its geomorphometric characteristics relative to the horizontal and vertical plane, especially distance, inclination (describing spatial deviation in distance), and curvature (describing spatial changes in inclination; Perko 2001, 2007; Hrvatin and Perko 2002).

Based on this, six geomorphometric characteristics of landform faces and points can be determined (Perko 2007):

- **Vertical distance** or distance from the horizontal plane. In the geographical sense, this is the same as surface height, which, as a third dimension in addition to longitude and latitude, gives the precise location of all features on the Earth's surface. Its basic unit is the meter.
- **Horizontal distance** or distance from the vertical plane; in geographical terms, this is the distance between two faces (or points) on the Earth's surface. It is measured in meters.
- **Vertical inclination** refers to the change in distance relative to the horizontal plane. It is usually measured in degrees. In the mathematical sense, it corresponds to the first derivative of vertical variation in height, and, in the geographical sense, it is the same as surface slope.
- **Horizontal inclination** refers to the change in distance relative to the vertical plane. It is usually measured in degrees. In the mathematical sense, it corresponds to the first derivative of horizontal variation in height, and, in the geographical sense, it is the same as surface orientation or aspect.
- **Vertical curvature** (also profile curvature or slope curvature) refers to the change in inclination relative to the horizontal plane. It is measured in radians per meter. In the mathematical sense, it corresponds to the first derivative of a vertical change in inclination or the second derivative of a vertical change in distance, and, in the geographical sense, it corresponds to surface concavity or convexity relative to the horizontal plane and indicates the direction of the maximum slope.
- **Horizontal curvature** (also plan curvature, aspect curvature, or contour curvature) refers to the change in surface inclination relative to the vertical plane. It is measured in radians per meter. In the mathematical sense, it corresponds to the first derivative of a horizontal change in inclination or the second derivative of a horizontal change in distance, and, in the geographical sense, it corresponds to surface concavity or convexity relative to the vertical plane and is thus perpendicular to the direction of maximum slope (i.e., parallel to contour lines and surface aspects).

However, it is not only the geomorphometric characteristics of the landforms' faces that is important, but also the geomorphometric characteristics of the faces of the neighboring landforms. Hence, in addition to the six geomorphometric

Table 4.1 Overview of the basic geomorphometric and geographical indicators of landform characteristics and their spatial variability

Geomorphometric characteristic	Vertical distance	Horizontal distance	Vertical inclination	Horizontal inclination	Vertical curvature	Horizontal curvature
Geographical characteristic	Surface height	–	Surface slope	Surface aspect	–	–
Geomorphometric variability	Spatial variability of vertical distance	Spatial variability of horizontal distance	Spatial variability of vertical inclination	Spatial variability of horizontal inclination	Spatial variability of vertical curvature	Spatial variability of horizontal curvature
Geographical variability	Surface height variability	–	Surface slope variability	Surface aspect variability	–	–
Surface roughness (variability) coefficient	Height coefficient	–	Slope coefficient	Aspect coefficient	–	–

indicators relating to surface distance, inclination, and curvature, a further six geomorphometric indicators of their spatial variability (Table 4.1) can be used to determine surface variability or roughness.

Thus, for example, the vertical inclination or surface slope indicator also includes a surface slope variability indicator, which shows the frequency or speed at which surface slopes change.

The coefficient of variation is among the most suitable statistical indicators for calculating spatial variability. It is the ratio of the standard deviation to the mean of the digital elevation model cells analyzed multiplied by 100, and it shows how much the standard deviation differs from the mean in percentages (Perko 2001). The greater the ratio, the more variable and rougher the surface.

Geomorphometric indicators are important not only for conducting landform analyses, but also for producing a wide range of landform classifications, including landform typologies.

4.3 Landform Typologies

So far, Slovenian geographers have produced five landform typologies covering all of Slovenia.

The first one was produced by Anton Melik. He primarily took macro landforms and surface height into account to define the basic types, along with rock types to define the subtypes (Fig. 4.1). He distinguished between sixteen types and twelve subtypes (Melik 1935):

1. High mountains;
2. Low mountains with high mountain ridges (with three subtypes);
3. Low mountains (with three subtypes);
4. High hills (with three subtypes);
5. Low hills (with three subtypes);
6. Pannonian carbonate–shale hills above 550 m;
7. Pannonian carbonate–shale hills below 550 m;
8. Carbonate plateaus above 400 m;
9. Carbonate plateaus below 400 m;
10. Mediterranean flysch high hills;
11. Mediterranean flysch low hills;
12. Pannonian Paleogene hills;
13. Pannonian Neogene hills;
14. Hills in basins and Dinaric lowlands;
15. Pannonian Pleistocene terraces; and
16. Plains.

Types 1 through 8 in particular can be characterized as mountainous areas; they cover a total of 55.2% of Slovenia.

Nearly sixty years later, Drago Perko produced the first computer-designed landform typology. Using the relief coefficient, which is based on the spatial variability of surface height, slope, and aspect (see the following section), he initially defined eight types (Perko 1992, 2001), which he then reduced to seven (Perko 2007; Perko et al. 2015):

1. Gentle plains (areas with a predominant relief coefficient between 0.0 and 0.9);

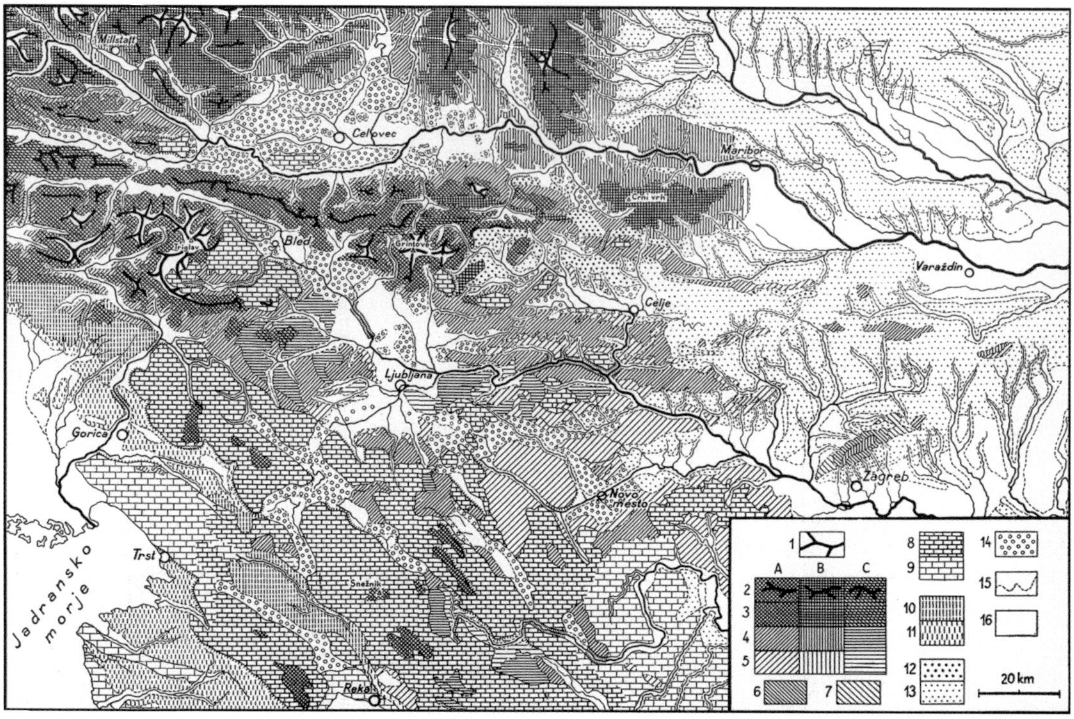

Fig. 4.1 Melik's landform typology was the first such typology of Slovenia. (Melik 1935, GIAM ZRC SAZU archive)

2. Rough plains (areas with a predominant relief coefficient between 1.0 and 1.9);
3. Gentle low hills (areas with a predominant relief coefficient between 2.0 and 5.9);
4. Rough low hills (areas with a predominant relief coefficient between 6.0 and 9.9);
5. Gentle high hills (areas with a predominant relief coefficient between 10.0 and 15.9);
6. Rough high hills (areas with a predominant relief coefficient between 16.0 and 19.9); and
7. Gentle and rough mountains (areas with a predominant relief coefficient of 20.0 or more).

Types 5, 6, and 7, in particular, can be characterized as mountainous areas. They cover 33.0%, 5.1%, and 9.6% of Slovenia, respectively, or altogether 47.7% of the country.

A year later, Karel Natek divided Slovenia into eight types, using surface height difference and rock type as the main criteria (Natek 1993):

1. Plains (level areas with a surface height difference up to 30 m, mainly on fluvial and fluvioglacial sediments);
2. Low hills (ridge-and-valley-dissected areas with a surface height difference between 30 and 200 m, mostly on Miocene and Pliocene rock);
3. High hills (ridge-and-valley-dissected areas with a surface height difference over 200 m, extending to the alpine tree line);
4. Mountains (areas above the alpine tree line, mostly alpine karst);
5. Low karst (predominantly level areas or low hills and depressions between them at lower elevations than the surrounding landscape);
6. High karst (high plateaus and hills with a surface height difference over 200 m, ridges running in a Dinaric direction, and karst depressions between them);
7. Low fluviokarst (karst areas with predominantly fluvio-denudational landforms at lower elevations than the surrounding landscape); and
8. High fluviokarst (mostly hills on dolomite rock, with predominantly fluvio-denudational landforms).

Types 3, 4, 6, and 8, in particular, can be characterized as mountainous areas. They cover 35.1%,

4.3%, 14.9%, and 8.8% of Slovenia, respectively, or altogether 63.1% of the country.

In 1998, Matej Gabrovec and Mauro Hrvatin produced a genetic and a geomorphometric landform typology of Slovenia. Both include six types. The geomorphometric typology is primarily based on surface height difference and macro landforms, and it comprises the following types (Gabrovec and Hrvatin 1998):

1. Plains;
2. Low hills (a surface height difference of up to 300 m);
3. High hills (a surface height difference between 300 and 1,000 m);
4. Mountains (the peaks and ridges extend over the tree line up to approximately 1,700 m);
5. Low plateaus (up to an elevation of 700 m); and
6. High plateaus (above 700 m in elevation; the peaks may even extend over 1,000 m in elevation).

Types 3, 4, and 6, in particular, can be characterized as mountainous areas. They cover 25.8%, 8.7%, and 22.4% of Slovenia, respectively, or altogether 56.9% of the country.

In 2019, Hrvatin et al. improved this typology, dividing Slovenia into seven types, primarily based on surface height, slope, and dominant landforms (Perko et al. 2024):

1. Plains;
2. Low hills;
3. High hills;
4. Mountains;
5. Corrosion plains;
6. Low plateaus; and
7. High plateaus.

Types 3, 4, and 7, in particular, can be characterized as mountainous areas. They cover 25.0%, 8.3%, and 22.3% of Slovenia, respectively, or altogether 55.6% of the country (Fig. 4.2).

Natek's typology contains the largest share of mountainous surface (63.1%) and Perko's typology contains the smallest share (47.7%), but roughly all five landform typologies define approximately half of Slovenia as predominantly mountainous (Table 4.2).

4.4 Perko's Geomorphometric Landform Typology

Among the five typologies, only the one produced by Perko is distinctly geomorphometric. It relies on three geomorphometric indicators that show spatial variations in surface height and slope, calculated from a 100-m digital elevation model: the surface height coefficient, the surface slope coefficient, and the relief (combined) coefficient or surface roughness coefficient, or simply the height, slope, and relief coefficients (Perko 1992, 2001, 2002).

Perko first used the *Idrisi* geographic information system (now *Terraset*) to calculate the following parameters for each cell based on its height and the heights of its eight neighboring cells:

- The height coefficient, which is the ratio between the standard deviation of surface height of nine neighboring cells and the average surface height of all cells multiplied by 100;
- The slope coefficient, which is the ratio between the standard deviation of surface slopes of nine neighboring cells and the average surface slopes of all cells multiplied by 100; and
- The relief or surface roughness coefficient, which is the geometric mean of the height and slope coefficients.

The lowest relief coefficient established was 0, the highest was 111.5, and the average was 9.3. The share of cells with a relief coefficient below 1 was 11.0%, that with a relief coefficient below 10 already rose to 59.1%, and the share of cells with a relief coefficient below 20 even reached 93.7%.

After reviewing the relief coefficients of the most typical Slovenian plains, low hills, high hills, and mountains, Perko logically combined their values into the following eight classes:

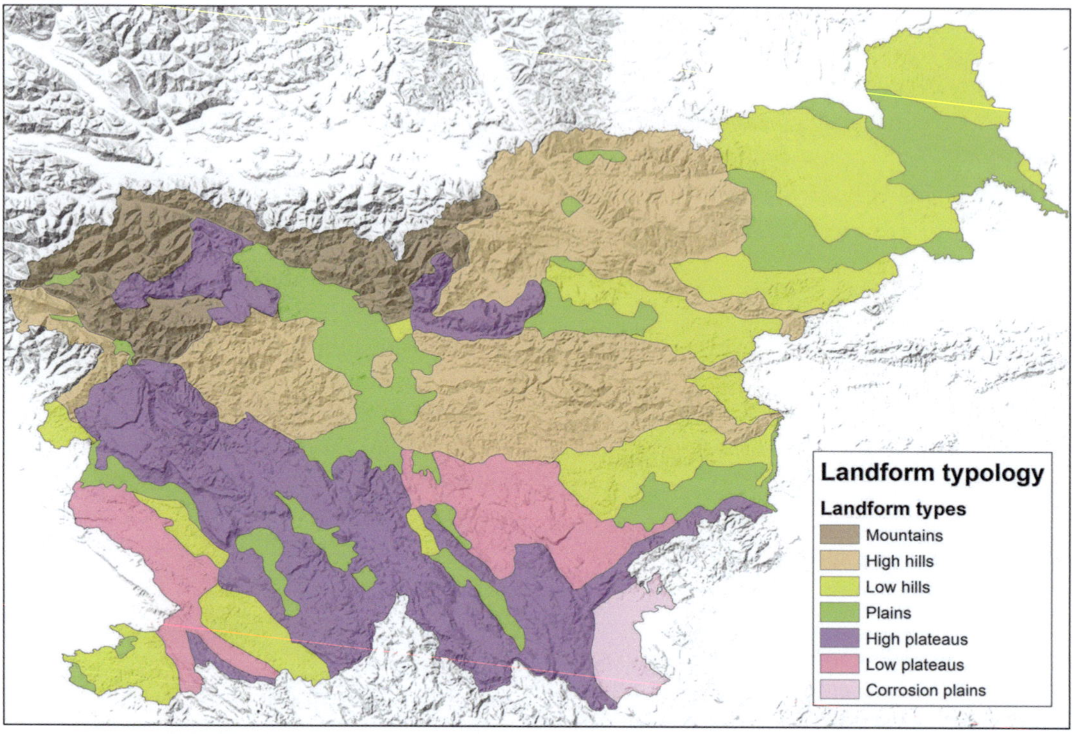

Fig. 4.2 Hrvatin et al.'s landform typology is the last such typology of Slovenia. (Perko et al. 2026)

Table 4.2 Comparison of landform macrotypologies of Slovenia

Author(s)	Year	No. of basic types	Average basic type size (km^2)	No. of mountainous types	Proportion of no. of mountainous types (%)	Area of mountainous types (km^2)	Proportion of area of mountainous types (%)
Melik	1935	16	1,267	8	50.0	11,189	55.2
Perko	1992	7	2,896	3	42.9	9,671	47.7
Natek	1993	8	2,534	4	50.0	12,791	63.1
Gabrovec and Hrvatin	1998	6	3,379	3	50.0	11,535	56.9
Hrvatin et al	2019	7	2,896	3	42.9	11,282	55.6

- Surfaces with relief coefficients between 0.0 and 0.9 (11.0% of Slovenia);
- Surfaces with relief coefficients between 1.0 and 1.9 (2.8% of Slovenia);
- Surfaces with relief coefficients between 2.0 and 5.9 (20.4% of Slovenia);
- Surfaces with relief coefficients between 6.0 and 9.9 (24.9% of Slovenia);
- Surfaces with relief coefficients between 10.0 and 15.9 (26.4% of Slovenia);
- Surfaces with relief coefficients between 16.0 and 19.9 (8.2% of Slovenia);
- Surfaces with relief coefficients between 20.0 and 39.9 (6.1% of Slovenia); and
- Surfaces with relief coefficients of 40.0 or more (0.2% of Slovenia).

He then filtered the layer with the relief coefficient classes three times using the modus inside of a moving 3 × 3 cell square window, obtaining

195 larger and spatially separate homogenous cores with similar values. He printed these cores on a 1:250,000 scale map and manually plotted their boundaries, mostly along geomorphological boundaries and larger watercourses. In the end, he combined these 195 manually delineated geomorphometric units into eight geomorphometric landform types (Perko 1992, 2001).

The eight landform types were the following:

1. Gentle plains (areas with predominance of the first class or relief coefficients between 0.0 and 0.9, comprising twenty-one units and covering 9.5% of Slovenia);
2. Rough plains (areas with relief coefficients predominantly between 1.0 and 1.9, comprising twenty-nine units and covering 5.8% of Slovenia);
3. Gentle low hills (areas with relief coefficients predominantly between 2.0 and 5.9, comprising forty-two units and covering 24.0% of Slovenia);
4. Rough low hills (areas with relief coefficients predominantly between 6.0 and 9.9, comprising twenty-two units and covering 12.9% of Slovenia);
5. Gentle high hills (areas with relief coefficients predominantly between 10.0 and 15.9, comprising forty-one units and covering 31.8% of Slovenia);
6. Rough high hills (areas with relief coefficients predominantly between 16.0 and 19.9, comprising eleven units and covering 5.1% of Slovenia);
7. Gentle and rough mountains (areas with relief coefficients predominantly of 20.0 or more, comprising twenty-four units and covering 9.5% of Slovenia); and
8. Large valleys (large areas between mountains and high hills with relief coefficients predominantly below 10.0, which were able to be drawn on a 1:250,000 map, comprising five units and covering 1.4% of Slovenia).

When a 25-m digital elevation model became available for all of Slovenia, Perko used a moving 11 × 11 cell square window to recalculate the height, slope, aspect, and relief coefficients (Perko 2007), reducing the number of types to seven (Fig. 4.3) by logically assigning the large valleys type to other types (Perko et al. 2015). The analysis of differences between the 100 m and 25 m digital elevation models showed that the average values of individual geomorphometric indicators for Slovenia and its landscape types and regions remained almost the same, but their minimum and maximum values increased (Hrvatin and Perko 2005).

Initially, Perko (1992, 2001) used the height and slope coefficients (i.e., two vertical surface roughness indicators) to calculate the relief coefficient, whereas the second time (Perko 2007) he used the slope and aspect coefficients (i.e., one vertical and one horizontal surface roughness indicator) because a combination of the two better shows their mutual or combined surface roughness relative to the horizontal and vertical plane.

Of the 195 geomorphometric units defined (Perko 2007), the Haloze rough low hills (i.e., the western part of the Haloze region) have the highest relief coefficient (46.9) and the Ravensko and Dolinsko Plain (i.e., the eastern part of the Mura Plain region in northeastern Slovenia) has the lowest relief coefficient (10.0), thus constituting the roughest and the least rough geomorphometric units in Slovenia.

Of the seven geomorphometric types defined (Perko 2007), rough high hills have the highest relief coefficient (37.8), and rough plains have the lowest (23.7).

In terms of the mountainous geomorphometric types alone, the most typical representative of the type *mountains* is the Western Karawanks region (Fig. 4.4), where this type covers 92.8% of the region, and the most typical representative of the type *rough high hills* is the Velenje and Konjice Hills region (Fig. 4.5), where this type covers 43.6% of the region. The most typical example of the geomorphometric type *gentle high hills* is the Big Mountains, Mount Stojna, and Mount Gotenica region, where this type covers 95.0% of the region, followed closely by the Mount Boč and Macelj Ridge region (Fig. 4.6), where this type accounts for 94.1% of the region, even though it already lies outside the mountainous area (Perko 2007).

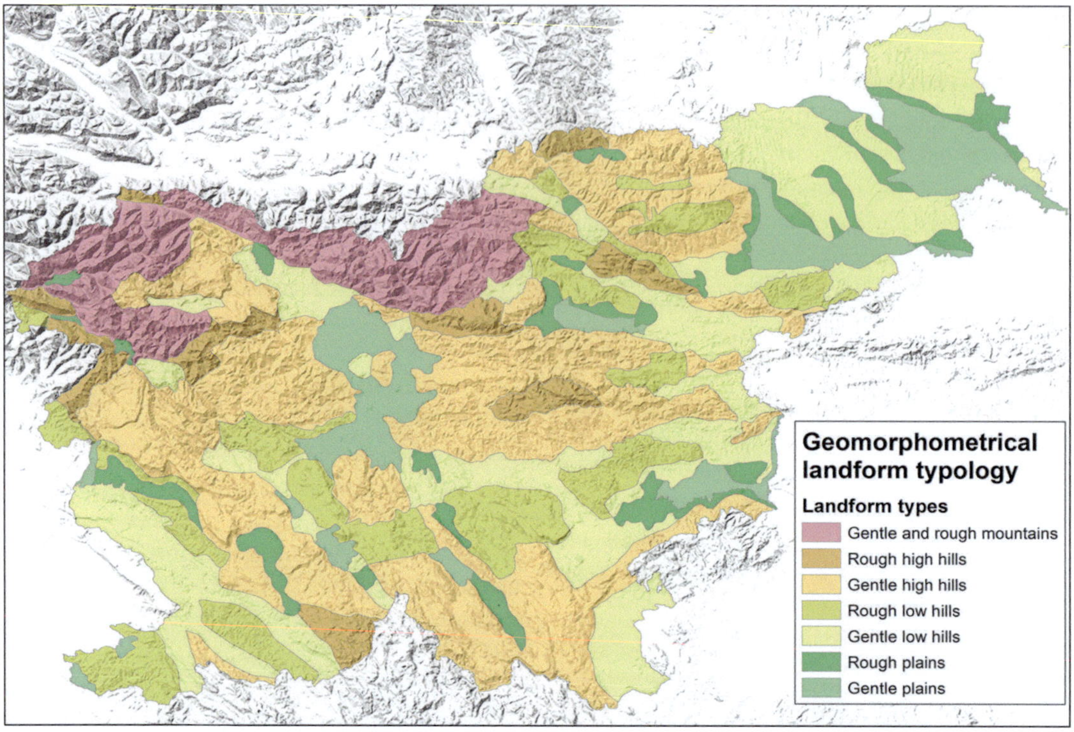

Fig. 4.3 Perko's geomorphometric landform typology. (Perko et al. 2015)

Fig. 4.4 Western Karawanks region is the most typical representative of the geomorphometric type *mountains*, which covers over 90% of the region, accounting for the largest share in any of the regions defined. The rest of the region belongs to the type *rough high hills*. The border between Slovenia and Austria runs along the ridges, with Mount Stol (2,236 m) as the highest peak. (Photo by kato08, Adobe Stock)

Fig. 4.5 Velenje and Konjice Hills region is the most typical representative of the geomorphometric type *rough high hills*. Second in terms of the share of the area covered in this region is the type *rough low hills*. The ridges of the Paški Kozjak Mountains and the Konjice Mountains (*Konjiška Gora*) with Mounts Basališče (1,272 m) and Stolpnik (1,012 m) as their highest peaks rise to the east of the Velenje Basin and its subsidence lakes resulting from coal mining operations in the past. (Photo by Marjan Garbajs, GIAM ZRC SAZU archive)

Fig. 4.6 Mount Boč and Macelj Ridge region along the eastern edges of Slovenia's mountainous region is the second most typical example of the geomorphometric type *gentle high hills*. The types *rough low hills* and *gentle low hills* are also common, but they only account for 6% of the region. Its highest peak is Mount Boč (978 m), which offers good views of the Slovenian mountains to the west and the non-mountainous areas to the east. (Photo by Matej Lipar, GIAM ZRC SAZU archive)

A comparison of selected indicators for the mountainous and non-mountainous types from the geomorphometric landform typology (Perko 2007), the latest landform typology produced (Perko et al. 2026), and the landscape typology of Slovenia (Perko 1998; Perko et al. 2021) is presented in Tables 4.3, 4.4, and 4.5.

4.5 Geomorphometric Indicators of Slovenia's Landscape Types

Based on Perko's geomorphometric typology, the following seven geomorphometric indicators were selected to present the geomorphometric

Table 4.3 Some basic characteristics of Slovenian mountainous and non-mountainous geomorphometric types (Perko 1992, 2001; Perko et al. 2015; data sources: 5-m digital elevation model; Gabrovec 1996; Perko et al. 2015, 2017, 2020; Ministry of Agriculture, Forestry, and Food, and Ministry of Natural Resources and Spatial Planning of the Republic of Slovenia)

Variables	Mountainous geomorphometric types (M)	Non-mountainous geomorphometric types (N)	All geomorphometric types	Ratio (M:N)
Area (km^2)	9,671.1	10,602.3	20,273.4	0.9
Area (%)	47.7	52.3	100.0	0.9
Mean elevation (m)	773.8	358.0	556.4	2.2
Mean inclination (°)	23.0	10.6	16.5	2.2
Insolation (MJ/m^2)	3,879.5	4,134.6	4,012.8	0.9
Average landscape diversity	0.1686	0.1567	0.1624	1.1
Fields (%)	1.1	16.4	9.1	0.1
Vineyards (%)	0.2	1.8	1.0	0.1
Orchards (%)	0.9	2.1	1.5	0.4
Meadows and pastures (%)	15.3	23.7	19.7	0.6
Forests (%)	76.5	47.0	61.1	1.6
Built-up areas (%)	2.6	7.9	5.4	0.3
Barren areas (%)	3.4	1.0	2.2	3.4
Most frequent landscape type	Alpine hills 35.7%; Alpine mountains 28.8%; Dinaric plateaus 27.9%	Pannonian hills 26.1; Dinaric lowlands 14.8%; Pannonian plains 12.2%; Alpine hills 11.4%; Dinaric plateaus 10.5%	Alpine hills 23.0%; Dinaric plateaus 18.8%; Alpine mountains 15.1%; Pannonian hills 14.8	
Most frequent landform type	High hills 44.2%; high plateaus 38.8%; mountains 16.8%	Low hills 36.7%; plains 29,8%; low plateaus 15.0%	High hills 25.1%; high plateaus 22.3%; low hills 19.3%; plains 15.6%	
Most frequent rock type	Limestone 33.6%; dolomite 18.3%	Clay and silt 22.7%; limestone 20.4%	Limestone 26.7%; clay and silt 13.1%; dolomite 12.4%	
Most frequent vegetation type	Beech 42.0%; beech and fir 17.6%; beech, chestnut, and various oaks 14.0%; beech and hophornbeam 12.3%	Beech, chestnut, and various oaks 34.9%; beech 18.1%; hornbeam 10.9%	Beech 29.5%; beech, chestnut, and various oaks 24.9%; beech and fir 10.5%	

Table 4.4 Some basic characteristics of Slovenian mountainous and non-mountainous landform types (Hrvatin et al. 2019; data sources: 5 m digital elevation model; Gabrovec 1996; Perko et al. 2015, 2017; Perko and Ciglič 2020; Ministry of Agriculture, Forestry, and Food, and Ministry of Natural Resources and Spatial Planning of the Republic of Slovenia)

Variables	Mountainous landform types (M)	Non-mountainous landform types (N)	All landform types	Ratio (M:N)
Area (km^2)	11,280.9	8,992.5	20,273.4	1.3
Area (%)	55.6	44.4	100.0	1.3
Mean elevation (m)	753.2	309.5	556.4	2.4
Mean inclination (°)	22.1	9.5	16.5	2.3
Insolation (MJ/m^2)	3,897.4	4,157.8	4,012.8	0.9
Average landscape diversity	0.1702	0.1525	0.1624	1.1
Fields (%)	1.2	19.1	9.1	0.1
Vineyards (%)	0.2	2.1	1.0	0.1
Orchards (%)	1.0	2.2	1.5	0.5
Meadows and pastures (%)	17.3	22.7	19.7	0.8
Forests (%)	74.8	43.8	61.1	1.7
Built-up areas (%)	2.5	9.0	5.4	0.3
Barren areas (%)	3.0	1.1	2.2	2.7
Most frequent landscape type	Alpine hills 37.0%; Dinaric plateaus 27.6%; Alpine mountains 26.6%	Pannonian hills 30.5; Dinaric lowlands 15.4%; Pannonian plains 14.4%; Mediterranean hills 10.5%	Alpine hills 23.0%; Dinaric plateaus 18.8%; Alpine mountains 15.1%; Pannonian hills 14.8	
Most frequent geomorphometric type	High gentle hills 59.1%; mountains 17.3%	Low gentle hills 47.7%; level plains 21.3%; low rugged hills 17.5%; rugged plains 13.3%	High gentle hills 33.0%; low gentle hills 23.9%; low rugged hills 13.0%	
Most frequent rock type	Limestone 32.0%; dolomite 18.5%	Clay and silt 25.7%; limestone 20.0%; sandstone and marl (flysch) 10.6%;	Limestone 26.7%; clay and silt 13.1%; dolomite 12.4%	
Most frequent vegetation type	Beech 41.0%; beech and fir 18.1%; beech, chestnut, and various oaks 14.8%; beech and hophornbeam 11.7%	Beech, chestnut, and various oaks 36.8%; beech 15.1%; hornbeam 12.7%; hornbeam and pedunculate oak 11.0%	Beech 29.5%; beech, chestnut, and various oaks 37.7%; beech and fir 10.5%	

characteristics of Slovenia's mountainous and non-mountainous landscape types and regions (Tables 4.6 and 4.7): surface height, surface slope, surface aspect, height coefficient to indicate surface height diversity, slope coefficient to indicate surface slope diversity, aspect coefficient to indicate surface aspect diversity, and relief coefficient to indicate the overall surface diversity.

All landscape types and many regions extend to Slovenia's border, which in places distinctly cuts off only certain surface aspects. Therefore, surface aspect is an appropriate indicator only for comparing landscape units that lie entirely inside Slovenia, whereas it can be misleading in other cases and is therefore omitted in general comparisons. A good example is the Karawanks on the Slovenian–Austrian border, which extend from the west toward the east, and so the northern surface aspects predominate on the Austrian side and the southern aspects predominate on the Slovenian side.

Table 4.5 Some basic characteristics of Slovenian mountainous and non-mountainous landscape types (Perko 1998; Perko et al. 2021; data sources: 5-m digital elevation model; Gabrovec 1996; Perko et al. 2015, 2017; Perko and Ciglič 2020; Ministry of Agriculture, Forestry, and Food, and Ministry of Natural Resources and Spatial Planning of the Republic of Slovenia)

Variables	Mountainous landscape types (M)	Non-mountainous landscape types (N)	All landscape types	Ratio (M:N)
Area (km^2)	11,531.4	8,742.0	20,273.4	1.3
Area (%)	56.9	43.1	100.0	1.3
Mean elevation (m)	735.5	320.0	556.4	2.3
Mean inclination (°)	21.8	9.5	16.5	2.3
Insolation (MJ/m^2)	3,885.9	4,180.3	4,012.8	0.9
Average landscape diversity	0.1701	0.1522	0.1624	1.1
Fields (%)	1.6	19.0	9.1	0.1
Vineyards (%)	0.2	2.2	1.0	0.1
Orchards (%)	1.1	2.1	1.5	0.5
Meadows and pastures (%)	16.9	23.4	19.7	0.7
Forests (%)	74.3	43.6	61.1	1.7
Built-up areas (%)	3.0	8.6	5.4	0.3
Barren areas (%)	3.0	1.0	2.2	3.0
Most frequent geomorphometric type	High gentle hills 51.8%; low rugged hills 12.8%	Low gentle hills 46.4%; level plains 20.7%; low rugged hills 13.4%; rugged plains 11.1%	High gentle hills 33.0%; low gentle hills 23.9%; low rugged hills 13.0%	
Most frequent landform type	High hills 41.2%; high plateaus 33.5%; mountains 14.4%	Low hills 40.1%; plains 31.9%; low plateaus 12.7%	High hills 25.1%; high plateaus 22.3%; low hills 19.3%; plains 15.6%	
Most frequent rock type	Limestone 32.2%; dolomite 17.4%	Clay and silt 25.0%; limestone 19.4%; sandstone and marl (flysch) 11.3%; marl 10.4%	Limestone 26.7%; clay and silt 13.1%; dolomite 12.4%	
Most frequent vegetation type	Beech 42.9%; beech and fir 16.5%; beech, chestnut, and various oaks 16.0%; beech and hophornbeam 11.3%	Beech, chestnut, and various oaks 36.8%; hornbeam 13.1%; beech 11.9%; hornbeam and pedunculate oak 11.2%	Beech 29.5%; beech, chestnut, and various oaks 24.9%; beech and fir 10.5%	

Among the remaining six geomorphometric indicators selected, Alpine mountains show the highest average values for all indicators except the aspect and relief coefficients, and the Pannonian plains show the lowest average values for all indicators except the aspect coefficient.

Among mountainous landscape types, Alpine hills have the highest aspect coefficient (40.4) because they are more dissected by river valleys than the more uniform Alpine mountains (33.5). Among non-mountainous landscape types, Pannonian plains have the highest aspect coefficient (56.0) because the finely terraced fluvial deposits change the surface aspect quickly, even though that may not be so obvious due to the small height differences. Because the finely dissected karst surface also contributes to a rapid alternation of aspect, Dinaric plateaus, Dinaric lowlands, and Mediterranean plateaus also have a higher average aspect coefficient than Alpine mountains.

Table 4.6 Some basic morphometric indicators of Slovenian mountainous (in bold) and non-mountainous (in italic) landscape types (Perko 1998; Perko et al. 2021; data sources: 5-m digital elevation model and Ministry of Natural Resources and Spatial Planning of the Republic of Slovenia; Perko 2007)

Landscape types	Area in km^2	Surface height average	Surface slope average	Surface aspect average	Surface height coefficient	Surface slope coefficient	Surface aspect coefficient	Surface relief coefficient
Alpine mountains	3,062	1053.79	28.24	84.12	6.74	45.61	33.51	37.49
Alpine hills	4,660	582.02	21.10	87.30	4.23	39.94	40.42	38.88
Alpine plains	819	373.14	5.69	66.88	0.98	16.90	44.98	23.09
Pannonian hills	2,995	288.47	12.01	83.65	2.10	28.16	44.03	33.33
Pannonian plains	1,297	195.79	1.65	70.19	0.21	4.62	55.98	13.01
Dinaric plateaus	3,809	667.34	17.49	90.37	3.44	34.70	38.05	34.90
Dinaric lowlands	1,897	402.67	9.59	85.61	1.59	23.28	46.39	30.25
Mediterranean hills	1,061	305.22	14.40	84.41	2.71	31.45	42.36	34.61
Mediterranean plateaus	673	425.73	10.58	91.18	1.90	24.83	42.59	30.57
Slovenia	20,273	556.43	16.52	84.76	3.35	32.38	41.38	33.50

Alpine hills have the highest average relief coefficient (37.5, compared to 33.5 for all of Slovenia, 37.2 for mountainous areas, and 28.6 for non-mountainous areas) among all landscape types, which means they constitute the roughest parts of Slovenia in terms of geomorphometry; the opposite applies to Pannonian plains (13.0; Table 4.6).

4.6 Geomorphometric Indicators of Slovenia's Regions

Among the twenty mountainous regions defined, the Western Karawanks (Fig. 4.7; for the locations of individual regions, see Fig. 3.8. and Table 3. 2 in Chap. 3 of this book) have the highest average surface height (1,186.0 m, compared to 556.4 m for all of Slovenia) and the highest average surface slope (29.4°, compared to 16.5° for all of Slovenia), and the Julian Alps (Fig. 4.8) have the highest surface height coefficient (7.2, compared to 3.4 for all of Slovenia) and the highest surface slope coefficient (48.1, compared to 32.4 for all of Slovenia), which means that the surface height and slope change the most rapidly in this region. The Ložnica and Hudinja Hills have the highest surface aspect coefficient (46.0), whereas among the non-mountainous regions the Mura Plain has an even higher one (57.7). The Idrija Hills (Fig. 4.9) have the highest relief coefficient (40.7, compared to 33.5 for all of Slovenia), which means that, in terms of geomorphometry, they constitute the roughest mountainous region in Slovenia. The highest relief coefficient overall (45.8) has, however, been established for the non-mountainous region of Haloze (Fig. 4.10), which makes it morphometrically the roughest region in Slovenia (Table 4.7); among other things, this is also reflected in higher landslide hazard (Zorn and Komac 2008).

The Mura Plain (Fig. 4.11) has the lowest relief coefficient in Slovenia (11.1), which makes it the least rough region of all in terms of geomorphometry. The lowest relief coefficient among

Table 4.7 Some basic morphometric indicators of Slovenian mountainous (in bold) and non-mountainous (in italic) regions (Perko 1998; Perko et al. 2021; data sources: 5-m digital elevation model and Ministry of Natural Resources and Spatial Planning of the Republic of Slovenia; Perko 2007)

Regions	Area in km^2	Surface height average	Surface slope average	Surface aspect average	Surface height coefficient	Surface slope coefficient	Surface aspect coefficient	Surface relief coefficient
Julian Alps	1,542	1107.24	28.91	87.11	7.21	48.05	32.67	37.76
Kamnik–Savinja Alps	889	957.26	27.06	84.48	6.26	43.18	34.11	36.91
Western Karawanks	331	1186.01	29.35	65.13	6.74	43.20	33.05	36.77
Eastern Karawanks	300	918.82	27.02	88.70	5.78	42.89	36.59	38.65
Cerkno, Škofja Loka, Polhov Gradec, and Rovte Hills	979	646.50	23.60	88.01	4.74	42.24	40.71	40.35
Sava Hills	1,909	483.02	20.72	87.96	4.11	40.55	41.18	39.52
Ložnica and Hudinja Hills	244	363.35	15.26	81.32	2.61	35.72	45.97	39.16
Velenje and Konjice Hills	241	625.38	22.22	82.46	4.55	42.51	41.06	40.41
Pohorje, Strojna, and Kozjak	1,287	713.28	20.67	87.81	4.26	37.62	37.90	36.49
Sava Plain	676	393.32	6.12	65.11	1.07	18.29	43.72	23.97
Savinja Plain	143	277.98	3.66	75.22	0.55	10.33	50.94	18.95
Goričko	492	274.87	7.29	78.47	1.36	18.15	39.19	25.41
Lendava Hills	17	225.53	11.18	83.65	2.09	27.82	36.75	30.94
Slovenian Hills	1,034	268.16	9.58	83.83	1.66	24.26	45.07	31.05
Dravinja Hills	281	317.62	9.32	82.39	1.59	22.56	43.80	29.26
Haloze	241	316.59	22.48	94.99	3.38	44.55	48.82	45.76
Mount Boč and the Macelj Ridge	73	461.97	25.07	84.09	4.80	45.06	40.48	41.66
Voglajna and Upper Sotla Hills	293	307.05	13.94	84.58	2.40	33.01	46.47	38.06
Central Sotla Hills	101	304.62	14.19	94.70	2.73	33.32	42.55	35.78
Krško, Senovo, and Bizeljsko Hills	461	275.37	14.88	80.51	2.73	35.52	44.11	38.10
Mura Plain	602	183.92	1.28	65.48	0.14	3.00	57.67	11.06
Drava Plain	425	234.81	1.59	69.60	0.20	4.82	56.17	13.54
Krka Plain	270	160.81	2.58	81.61	0.41	7.93	51.92	16.52
Kambreško and Banjšice Plateaus	272	549.47	22.30	91.13	4.70	39.96	37.06	37.27
Idrija Hills	239	647.59	28.14	90.82	5.96	46.19	37.80	40.70
Trnovo Forest Plateau, Mount Nanos, and Hrušica Plateau	508	830.70	20.66	87.99	4.02	39.78	37.63	37.32
Javornik Hills and Snežnik Plateau	458	967.61	16.62	90.47	3.37	35.02	38.23	35.56

(continued)

Table 4.7 (continued)

Regions	Area in km^2	Surface height average	Surface slope average	Surface aspect average	Surface height coefficient	Surface slope coefficient	Surface aspect coefficient	Surface relief coefficient
Krim Hills and Menišija Plateau	300	633.20	17.38	96.53	3.48	36.35	37.48	35.41
Bloke Plateau	144	743.37	14.35	88.60	2.68	31.70	42.58	35.09
Big Mountains, Mount Stojna, and Mount Gotenica	553	735.80	17.31	86.28	3.41	34.77	37.07	34.42
Little Mountains, Kočevje Rog Plateau, and Mount Poljane	584	611.62	14.52	86.76	2.68	28.40	35.82	30.68
Dry Carniola and Dobrepolje	424	399.06	11.43	90.16	2.09	25.76	39.45	30.34
Radulja Hills	116	373.06	16.67	88.71	3.11	36.83	45.57	39.93
Gorjanci Hills	212	469.21	17.07	108.88	3.32	35.69	39.71	36.51
Pivka Lowland and Mount Vremščica	297	607.09	9.90	86.52	1.81	24.65	43.71	31.36
Inner Carniola Lowland	270	571.38	10.28	81.36	1.56	24.40	46.33	30.84
Ribnica–Kočevje Lowland	113	489.24	6.83	83.44	0.84	14.90	49.72	24.65
Velike Lašče Region	71	591.08	15.76	90.84	2.82	35.43	45.53	39.09
Ljubljana Marsh	180	296.78	3.25	78.25	0.55	10.76	56.30	18.55
Lower Carniola Lowland	318	363.43	10.05	86.69	1.88	26.32	43.35	32.16
Novo Mesto Region	259	280.58	11.12	94.99	1.98	26.42	43.25	32.22
White Carniola	388	231.66	10.09	83.82	1.44	22.89	47.67	31.54
Gorizia Hills	83	232.20	19.31	81.32	3.79	37.84	42.42	38.97
Vipava Hills	310	215.07	12.54	80.41	2.34	26.84	41.47	31.15
Brkini Hills and Reka Valley	342	525.36	15.87	87.45	2.75	34.37	45.03	38.13
Koper Hills	327	179.12	13.38	85.83	2.76	31.16	40.40	33.11
Karst Plateau	429	333.64	9.66	90.54	1.72	23.10	43.31	29.65
Podgorje Karst Plateau, Čičarija Plateau, and Podgrad Lowland	245	587.25	12.20	92.30	2.22	27.88	41.33	32.18
Slovenia	20,273	556.43	16.52	84.76	3.35	32.38	41.38	33.50

the mountainous regions (30.3) is typical of Dry Carniola and Dobrepolje (Fig. 4.12, Table 4.7).

Fig. 4.7 Western Karawanks have the highest average surface height and slope among all Slovenian regions. The Košuta Ridge, the longest Slovenian mountain ridge, is also part of this region. Left: the highest peak of the Košuta Ridge, Mount Košutnikov Turn (2,133 m), right: Mount Veliki Vrh (2,088 m). (Photo by Bor Rojnik / Wirestock Creators, Adobe Stock)

Fig. 4.8 Due to significant surface height differences, the Julian Alps have the highest average height and slope coefficients among all Slovenian regions. The photograph shows a view of the Soča Valley and Kobarid toward Big Mount Kanin (*Visoki Kanin*, 2,587 m) on the Slovenian–Italian border, at the southern edges of the Julian Alps. (Photo by Mny-Jhee, Adobe Stock)

4.7 Geomorphometric Indicators of Slovenia's Mountainous and Non-Mountainous Areas

The main geomorphometric differences between Slovenia's mountainous and non-mountainous areas (Table 4.8) show that, despite its smallness, Slovenia is a diverse country not only in terms of landscapes (Ciglič and Perko 2013) but also geomorphometry. Moreover, its landscape diversity is largely the result of geomorphometric diversity (Perko et al. 2015, 2017, 2024).

The average **surface height** of all of Slovenia is 556.4 m, of its non-mountainous areas is 320.0 m, which is 42% less, and that of its mountainous areas is 735.5 m, which is 32% more than the average surface height of all of Slovenia and 130% more than that of its non-mountainous areas. The Western Karawanks have the highest average surface height (1,186.0 m) among the mountainous regions, and the Ložnica and Hudinja Hills have the lowest (363.4 m), and so the ratio between the two is slightly more than 3:1.

Slovenia's average surface height is nearly 300 m lower than the average land surface height

Fig. 4.9 Idrija Hills have the highest average relief coefficient among Slovenian mountainous regions. Its significant surface roughness is primarily the result of numerous creeks, the Idrija Fault, and rapid rock alternation. The photograph shows Spodnja Idrija at the confluence of the Idrijca and Kanomljica rivers. (Photo by Taljat, Adobe Stock)

Fig. 4.10 Haloze has the highest average relief coefficient among all Slovenian regions and thus forms the roughest part of Slovenia. Its wooded western section is the roughest geomorphometric unit. The region lies southwest of the Drava River along the Slovenian–Croatian border. (Photo by Marjan Garbajs, GIAM ZRC SAZU archive)

of the Earth, which is 841 m (Perko 2001). With an average surface height of 830.7 m, the Trnovo Forest Plateau, Mount Nanos, and Hrušica Plateau region most closely approaches this average in Slovenia.

Among the 100-m elevation zones, the largest share of Slovenia's total surface (18.2%) lies in the zone between 200 and 300 m, the largest share of mountainous areas (13.5%) lies in the 500–600-m zone, and the largest share of non-mountainous areas (36.3%) lies in the 200–300-m zone. A total of 6.5% of all of Slovenia lies above an elevation of 1,200 m, which is roughly the upper settlement limit in Slovenia (Perko 2001), along with 11.4% of mountainous areas. There are no non-mountainous areas above that line. Among the regions, the Western Karawanks account for the largest share in that zone (45.7%; Figs. 4.13 and 4.14).

There are also great differences among the landscape types. Alpine mountains extend across all 100-m elevation zones above an elevation of 100 m, and the Pannonian plains lie almost entirely within the 200–300-m elevation zone. Alpine mountains account for over 80% of Slovenia's areas above an elevation of 1,200 m and Mediterranean hills

Fig. 4.11 Mura Plain has the lowest average relief coefficient, which makes it the least rough part of Slovenia and its breadbasket. The eastern part of the region between the meandering Mura River and its tributary Ledava from the Austrian to the Croatian border is the least rough geomorphometric unit in Slovenia. (Photo by Jože Pojbič, GIAM ZRC SAZU archive)

Fig. 4.12 Dry Carniola and Dobrepolje is the Slovenian mountainous region with the lowest average relief coefficient. Dry Carniola is a plateau-like area cut through by the Krka Valley, and Dobrepolje is an oblong dry karst polje to its west. The photograph shows Žužemberk, the largest town in the region. (Photo by Stepo, Adobe Stock)

cover over 90% of all of Slovenia's areas below an elevation of 100 m.

The average **surface slope** of Slovenia is 16.5°, of its non-mountainous areas is 9.5°, which is 42% less, and of its mountainous areas is 21.8°, or 32% more than Slovenia's total surface average and 129% more than the average of its non-mountainous areas. The Western Karawanks have the highest average surface slope (29.4°) among Slovenian regions, and the Bloke Plateau has the lowest (14.4°). The ratio between the two is 2:1, which is lower than the ratio

Table 4.8 Some basic morphometric indicators of Slovenian mountainous (in bold) and non-mountainous (in italic) areas (Perko 1998; Perko et al. 2021; data sources: 5-m digital elevation model and Ministry of Natural Resources and Spatial Planning of the Republic of Slovenia; Perko 2007)

Areas	Area in km^2	Surface height average	Surface slope average	Surface aspect average	Surface height coefficient	Surface slope coefficient	Surface aspect coefficient	Surface relief coefficient
Mountainous areas	11,531	735.47	21.80	87.47	4.64	39.71	37.80	37.20
Non-mountainous areas	8742	320.04	9.53	81.18	1.66	22.70	46.09	28.63
Slovenia	20,273	556.43	16.52	84.76	3.35	32.38	41.38	33.50

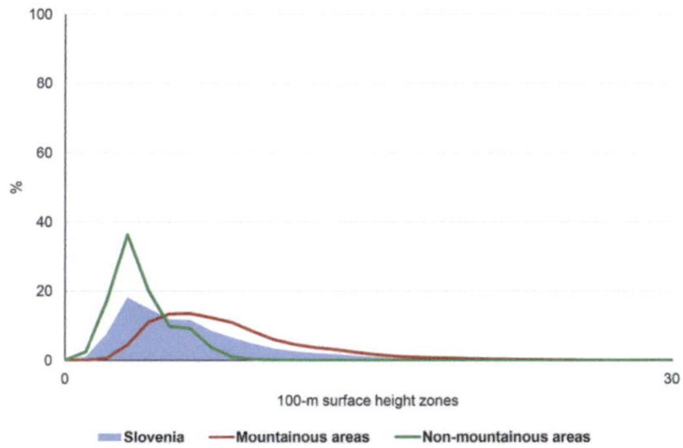

Fig. 4.13 Surface height frequency distribution across thirty 100-m elevation zones for all of Slovenia and its mountainous (red) and non-mountainous areas (green)

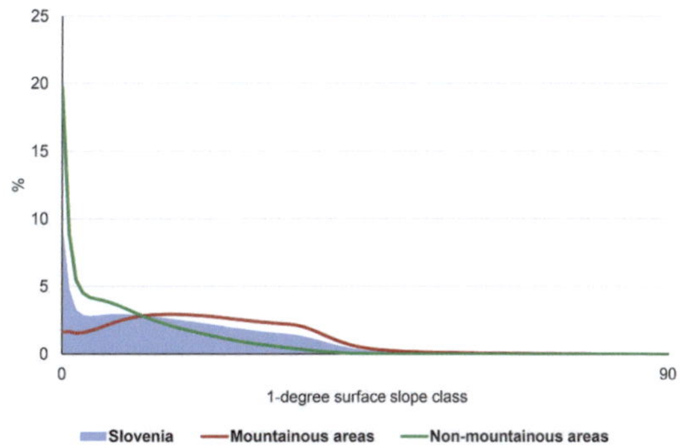

Fig. 4.15 Surface slope frequency distribution across ninety one-degree categories for all of Slovenia and its mountainous (red) and non-mountainous areas (green)

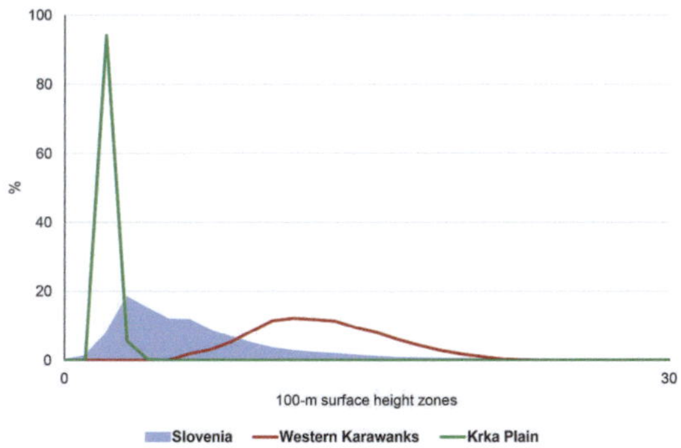

Fig. 4.14 Surface height frequency distribution across thirty 100-m elevation zones for all of Slovenia, the highest region on average (red), and the lowest region on average (green)

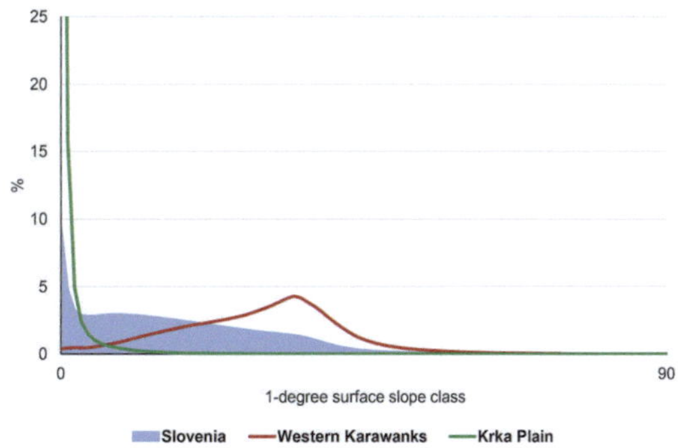

Fig. 4.16 Surface slope frequency distribution across ninety one-degree categories for all of Slovenia, the steepest region on average (red), and the flattest region on average (green)

between the highest and lowest average surface heights.

Among the one-degree slope categories, the largest share of Slovenia's total surface lies in the 0°–1° category (9.4%), the largest share of its mountainous areas lies in the 15°–16° category (2.9%), and the largest share of its non-mountainous areas lies in the 0°–1° category

(19.7%). A total of 56.8% of all of Slovenia lies in areas with slopes over 12°, where modern agriculture is no longer possible; the share of mountainous surface in that category is 75.5% and that of non-mountainous surface is 32.3%. Among the regions, the Western Karawanks account for the largest share (92.0%; Figs. 4.15 and 4.16).

There are also great differences between individual landscape types in terms of surface slope. The surface slope of over 80% of the Pannonian plains is below 2°, which accounts for over a third of all of Slovenia. The corresponding share of Alpine mountains is below 3%, which means that the ratio between the two landscape types is as high as 32:1.

The average **surface aspect** of all of Slovenia is 84.8°, of its non-mountainous areas is 81.2° or 4% less, and of its mountainous areas is 87.5° or 3% more than the average surface aspect of all of Slovenia and 8% more than that of its non-mountainous areas. The differences are thus small. Taking account of only those regions that lie entirely within Slovenia and are not cut off by the border in certain parts (e.g., the northern aspects of the Western Karawanks or the southern aspects of the Gorjanci Hills), the Krim Hills and Menišija Plateau region has the highest average surface aspect (96.5°), or the most northern aspects, among all the mountainous regions. The predominance of northern aspects over southern ones is mainly due to the region's extensive northern slopes, which descend from the mountainous areas in the north toward the flat Ljubljana Marsh, which then gently transitions into the Bloke Plateau to the south. The Ložnica and Hudinja Hills region, whose numerous ridges descend toward the Savinja Plain in the south, has the lowest average surface aspect (81.3°). The ratio between the two is only slightly higher than 1.

The average **surface height coefficient** is 3.4 for all of Slovenia, 1.7 for its non-mountainous areas (50% lower), and 4.6 for its mountainous areas (39% higher than the average coefficient for all of Slovenia and 180% higher than that for its non-mountainous areas). The Julian Alps, where the surface heights change the most rapidly, have the highest average surface height coefficient (7.2) among the mountainous regions, and the Dry Carniola and Dobrepolje region has the lowest (2.1), making the ratio between the two higher than 3:1.

The average **surface slope coefficient** is 32.4 for all of Slovenia, 22.7 for its non-mountainous areas (30% lower), and 39.7 for its mountainous areas (23% higher than the average coefficient for all of Slovenia and 75% higher than that for its non-mountainous areas). The Julian Alps have the highest average surface slope coefficient (48.1) among the mountainous regions, and the Dry Carniola and Dobrepolje region has the lowest (25.8), making the ratio between the two slightly less than 2:1.

The average **surface aspect coefficient** is 41.4 for all of Slovenia, 46.1 for its non-mountainous areas (11% higher), and 37.8 for its mountainous areas (9% lower than the average coefficient for all of Slovenia and 18% lower than that for its non-mountainous areas). The Ložnica and Hudinja Hills region has the highest average surface aspect coefficient (46.0) among the mountainous regions, and the Little Mountains, Kočevje Rog Plateau, and Mount Poljane region has the lowest (35.8), making the ratio between the two barely higher than 1.

A comparison between the surface slope and aspect coefficients shows that the surface aspect coefficient for Slovenia and all its landscape types except the Alpine mountains is higher that the slope coefficient. This means that in most of Slovenia the surface aspects change more rapidly than the surface slopes, or that Slovenia's surface aspect variability is greater than its surface slope variability. This is largely contributed to by the dense river system on non-karst terrain and the corroded character of karst terrain. In non-mountainous areas, the aspect coefficient is twice as much as the slope coefficient, whereas in mountainous areas the difference is almost negligible, with the aspect coefficient only 5% lower than the slope coefficient.

The **surface relief coefficient** combines the surface slope and aspect coefficients. It is the most synthetic of all the geomorphometric indicators used, combining the surface slope and aspect variability, and showing the total surface roughness. The average surface relief coefficient is 33.5 for all of Slovenia, 28.6 for its non-mountainous areas (or 15% lower), and 37.2 for its mountainous areas (or 11% higher than the average coefficient for all of Slovenia and 30% higher than that for its non-mountainous areas).

The Idrija Hills have the highest average surface relief coefficient (40.7) among the mountainous regions, and the Dry Carniola and Dobrepolje region has the lowest (30.3), making the ratio between the two slightly higher than 1. Haloze has the highest average surface relief coefficient (45.8) among the non-mountainous regions, and the Mura Plain has the lowest (11.1), and so the ratio between the two is just slightly over 4:1.

The ratios between the mountainous regions with the highest and lowest indicators thus range from 3:1 for surface height to nearly 1:1 for surface slope and aspect. Taking account of mountainous and non-mountainous regions together, the ratios between the highest and lowest indicators are significantly higher, such as, for example, higher than 7:1 for surface height, where the highest average surface height among the mountainous regions is 1,186.0 m recorded for the Western Karawanks, and the lowest among the non-mountainous regions is 160.8 m recorded for the Krka Plain.

Comparing all geomorphometric units, regions, landscape types, and geomorphometric types, the highest average surface relief coefficients were recorded for the western Haloze geomorphometric unit (45.9), the Haloze region (45.8), the Alpine hills landscape type (38.9), and the rough high hills geomorphometric type (37.8). These are thus areas at various scales with the roughest surface in Slovenia.

Acknowledgements This study was conducted as part of the research program The Geography of Slovenia (P6-0101), financed by the Slovenian Research and Innovation Agency.

References

Ciglič R, Perko D (2013) Europe's landscape hotspots. Acta Geogr Slov 53(1):117–139. https://doi.org/10.3986/AGS53106

Gabrovec M (1996) Solar radiation and the diverse relief of Slovenia. Geografski Zbornik 36:47–68

Gabrovec M, Hrvatin M (1998) Površje [Surface]. In: Fridl J, Kladnik D, Orožen Adamič M, Perko D (eds) Geografski atlas Slovenije. DZS, Ljubljana, pp 80–83

Hrvatin M, Perko D (2002) Ugotavljanje ukrivljenosti površja z digitalnim modelom višin in njena uporabnost v geomorfologiji [Determination of surface curvature by digital elevation model and its application in geomorphology]. In: Podobnikar T, Perko D, Krevs M, Stančič Z, Hladnik D (eds) Geografski informacijski sistemi 2001–2002. Založba ZRC, Ljubljana, pp 65–72. https://doi.org/10.3986/9616358650

Hrvatin M, Perko D (2005) Differences between 100-meter and 25-meter digital elevation models according to types of relief in Slovenia. Acta Geogr Slov 45(1):7–31. https://doi.org/10.3986/AGS45101

Melik A (1935) Slovenija: geografski opis [Slovenia: geographical description]. Slovenska matica, Ljubljana

Natek K (1993) Tipi površja v Sloveniji 1 [The types of terrain in Slovenia 1]. Geografski Obzornik 40(4):26–31

Perko D (2002) Določanje vodoravne in navpične razgibanosti površja z digitalnim modelom višin [Determination of horizontal and vertical surface roughness by digital elevation model]. Geografski Vestnik 74(2):85–98

Perko D, Hrvatin M, Ciglič R (2015) A methodology for natural landscape typification of Slovenia. Acta Geogr Slov 55(2):235–270. https://doi.org/10.3986/AGS.1938

Perko D, Hrvatin M, Ciglič R (2017) Determination of landscape hotspots of Slovenia. Acta Geogr Slov 57(1):7–29. https://doi.org/10.3986/AGS.4618

Perko D, Ciglič R, Hrvatin M (2021) Landscape macrotypologies and microtypologies of Slovenia. Acta Geogr Slov 61(3):7–89. https://doi.org/10.3986/AGS.10384

Perko D, Ciglič R (2020) Slovenia's landscapes. In: Perko D, Ciglič R, Zorn M (eds) The geography of Slovenia: small but diverse. Springer, Cham, pp 211–225. https://doi.org/10.1007/978-3-030-14066-3_14

Perko D, Ciglič R, Hrvatin M (2026) Slovenia's landscape and landform diversity. In: Zorn M, Ciglič R, Perko D (eds) Landscapes and landforms of Slovenia. Springer, Cham

Perko D (1992) Zveze med reliefom in gibanjem prebivalstva 1880–1981 v Sloveniji [Relation between relief and population change 1880–1981 in Slovenia]. PhD Thesis, University of Ljubljana
Perko D (1998) The regionalization of Slovenia. Geografski Zbornik 38:11–57
Perko D (2001) Analiza površja Slovenije s stometrskim digitalnim modelom reliefa [Analysis of the surface of Slovenia using the 100-meter digital elevation model]. Geografija Slovenije, vol 3. Založba ZRC, Ljubljana. https://doi.org/10.3986/9789612544973
Perko D (2007) Morfometrija površja Slovenije [Morphometry of Slovenia's surface]. Georitem, vol 3. Založba ZRC, Ljubljana. https://doi.org/10.3986/9789612545345
Zorn M, Komac B (2008) Zemeljski plazovi v Sloveniji [Landslides in Slovenia]. Georitem, vol 8. Založba ZRC, Ljubljana. https://doi.org/10.3986/9789612545505

Mountains of Slovenia

5

Mauro Hrvatin and Matija Zorn

Abstract

Slovenia's mountainous areas comprise Alpine mountains, Alpine hills, and Dinaric plateaus, which together cover over half of the country. Its mountainous landscapes lie at the intersection of the geotectonic units of the Dinarides, Southern Alps, and Eastern Alps. Limestone and dolomite predominate, although there are also extensive areas in the Alpine hills composed of poorly permeable clastic, igneous, and metamorphic rock. In terms of geomorphological development, mountainous areas can be divided into glacial, fluvio-denudational, and karst landscape types. Glacial landscapes are limited to areas that were exposed to glacial erosion and accumulation during the cold Pleistocene periods, and fluvio-denudational landscapes are typical of areas made of poorly permeable rocks and characterized by surface runoff. Karst landscapes developed on carbonate rocks. They are characterized by intense chemical dissolution of the bedrock and subterranean karst hydrology. The high mountains feature a combination of glacial and karst landscapes, or glaciokarst. Due to the Alpine–Dinaric barrier, Slovenia's climate is wetter than average. Over the past decades, the average air temperature has been rising rapidly, which causes frequent droughts in the summer and reduces the duration and depth of snow cover in the winter. Climate change has already resulted in significantly lower discharges of mountain rivers and altered discharge regimes. Water conditions largely depend on the rock composition. Nearly two-thirds of mountainous areas is karst and almost devoid of surface water flows, while elsewhere the river system is highly branched. The predominant soils on carbonate rocks are Rendzina and Chromic Cambisol, whereas Dystric Cambisol predominates on non-carbonate rocks. Forest is the natural vegetation and the predominant land-use type in all mountainous areas. The dominant tree is beech, which creates numerous forest communities. Conifers dominate at higher elevations, especially spruce and larch, which grow up to the tree line between 1,550 and 1,900 m. Late-medieval colonization of higher elevations was a key stage in the permanent colonization of mountainous areas, extending even above 1,000 m. In the Alpine mountains, it was the valleys that were usually settled, and in the Alpine hills people also settled the ridges and slopes. The upper limit of settlement in Slovenia's mountainous areas varies; the highest settlements can be found in the Eastern Karawanks, even at elevations above

M. Hrvatin · M. Zorn (✉)
ZRC SAZU Anton Melik Geographical Institute, Ljubljana, Slovenia
e-mail: matija.zorn@zrc-sazu.si

M. Hrvatin
e-mail: mauro.hrvatin@zrc-sazu.si

M. Zorn et al. (eds.), *Discourses on Mountains of Montenegro and Slovenia*, Historical Geography and Geosciences, https://doi.org/10.1007/978-3-031-99017-5_5

1,300 m. The traditional economic activity is forestry.

Keywords

Rocks · Surface · Climate · Hydrology · Soil · Vegetation · Settlement · Alpine mountains · Alpine hills · Dinaric plateaus · Slovenia

5.1 Introduction

Slovenia's mountainous areas (as defined by Perko in Chap. 3 of this volume; Fig. 5.1) cover 11,531.1 km^2 or 56.9% of the country. They can be divided into:

- Alpine mountains (3,061.8 km^2, 26.5% of mountainous areas, 15.1% of the entire country);
- Alpine hills (4,660.0 km^2, 40.4% of mountainous areas, 23.0% of the entire country); and
- Dinaric plateaus (3,809.3 km^2, 33.0% of mountainous areas, 18.8% of the entire country).

Alpine mountains (for basic morphometric indicators, see Table 4.6 in Chap. 4 of this volume; Fig. 5.1) include the Julian Alps, the Kamnik–Savinja Alps (Fig. 5.2), and the Karawanks. The first two mountain ranges are very imposing, but they are dissected by deep, even glacially transformed valleys that separate individual mountain groups. They are characterized by very steep slopes, which in many places turn into picturesque rockfaces. Prominent karst and wooded plateaus lie on their edges, such as the Pokljuka, Mežakla, and Jelovica plateaus on the edges of the Julian Alps and the Velika Planina, Dleskovec, Golte, Menina, and Dobrovlje plateaus on the edges of the Kamnik–Savinja Alps; rocky karst plateaus (referred to as *podi* in Slovenian) can also be found closer to the center of these mountains, above the tree line. The Karawanks are a distinctly elongated and slightly lower mountain range stretch along the Austrian border. The glacial valley of Belska kočna (Vellacher Kotschna) divides the range into the Western and Eastern Karawanks (Kladnik 1998a).

Because limestone predominates in the rock composition of the Alpine mountains, surface streams are rare and are limited to the deeply cut valleys. The permeable surface gave rise to high-mountain karst, in which karst landforms are closely interwoven with glacial and periglacial ones (Kladnik 1998a).

The extensive **Alpine hills** (for basic morphometric indicators, see Table 4.6 in Chap. 4 of this volume; Fig. 5.1) are composed of several landscapes: the Cerkno (Fig. 5.3), Škofja Loka, Polhov Gradec, and Rovte Hills in western Slovenia, the extensive Sava Hills in central and eastern Slovenia, and the Ložnica and Hudinja Hills and the Velenje and Konjice Hills to the northeast. The Pohorje (Fig. 5.3), Strojna, and the Kozjak Hills extend farthest to the northeast; these are the southern spurs of the Central Alps separated by the deeply cut Drava, Mislinja, and Mežica valleys. The surface is heavily dissected into valleys and ridges with an elevation difference of over 200 m per square kilometer. The slopes are mostly long, steep, and wooded, and in many places they are broken up by shelves and side ridges with sufficient space for isolated farms and small hamlets. The rock composition is very diverse. Carbonate rocks predominates in the western hills, with sandstone and shale exposed only in the lower parts. In the Sava Hills, narrow belts of higher limestone hills and slightly lower shale and sandstone hills follow one another from north to south, whereas hills made of volcaniclastic, igneous, and metamorphic rock dominate the Savinja and Drava basins (Kladnik 1998a).

Dinaric karst **plateaus** (for basic morphometric indicators, see Table 4.6 in Chap. 4 of this volume; Fig. 5.1) extended in what is nearly a contiguous belt from the central Soča Valley in the west to the Kolpa River in southeastern Slovenia. The Kambreško and Banjšice Plateaus lie in the northwesternmost part of the Dinaric region, separated by the deeply cut Soča Valley. The high Dinaric Trnovo Forest (*Trnovski gozd*; Fig. 5.4), Nanos, and Hrušica Plateaus extend to the southeast from there. These transition into the more rugged Idrija Hills in the north and the high

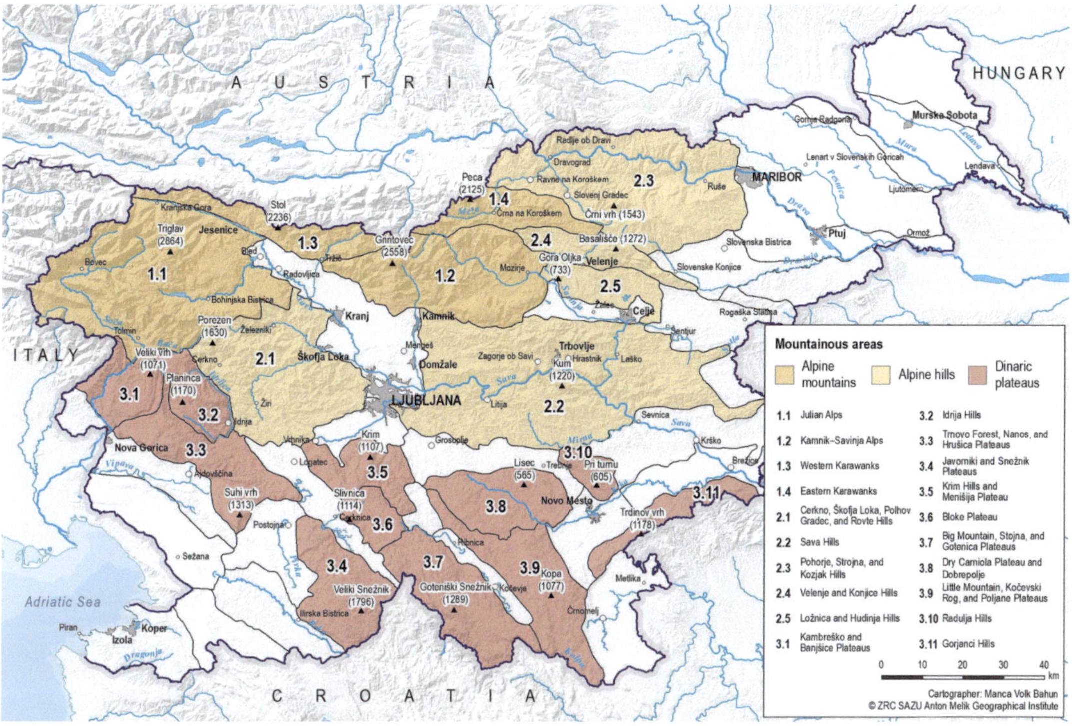

Fig. 5.1 Slovenia's mountainous areas with the Alpine mountains, Alpine hills, and Dinaric plateaus marked separately. (Adapted from Perko and Ciglič 2020a, b)

Dinaric Javorniki (Fig. 5.4) and Snežnik Plateaus to the southeast. East of the Inner Carniola Lowland (*Notranjsko podolje*), a series of slightly lower plateaus extends from the Menišija Plateau and the Krim Hills in the north via the Bloke Plateau to the Big Mountain (*Velika gora*), Stojna, and Gotenica (*Goteniška gora*) Plateaus in the south. The plateau of Dry Carniola (*Suha krajina*) rises above the upper Krka Valley, followed by Little Mountain (*Mala gora*), Kočevski Rog, and Poljane (*Poljanska gora*) Plateaus to the south. The Radulja Hills extend east of the Lower Carniola Lowland (*Dolenjsko podolje*), and the Gorjanci Hills rise above the lower reaches of the Krka River (on its right bank). The Gorjanci Hills extend in a predominantly southwest–northeast direction, which makes them an exception among the Dinaric landscapes. Karst plateaus are heavily broken up by countless karst depressions, conical hills, and rugged rocky terrain. Thick fir–beech and beech forests grow in very unfavorable conditions on these plateaus because there is very little fertile soil on the surface and rainwater drains quickly through the thick limestone layers into the karst underground, only resurfacing in the powerful karst springs below the plateaus (Kladnik 1998b).

5.2 Rocks

Slovenia's mountainous areas lie at the intersection of the Dinarides, Southern Alps, and Eastern Alps. Even though today these geotectonic units are close together and even overlap in places, they have different geological histories. They all belong to the Adriatic microplate, which separated from the African plate during the Mesozoic. While moving north, it collided with the Eurasian plate during the Neogene, causing the Alps to rise. The two plates have continued to approach each other since this collision, which is still the main factor affecting Slovenia's tectonic structure (Novak 2016; Hrvatin et al. 2020; Fig. 5.5).

Fig. 5.2 The two highest mountain ranges in Slovenia. Top: Mount Triglav (2,864 m), the highest peak in Slovenia and the Julian Alps, with the Vodnik Lodge (1,817 m) in the foreground. Bottom: the central part of the Kamnik–Savinja Alps with Mount Grintovec (2,558 m) as its highest peak. (Photo by Matija Zorn)

The borders between the geotectonic units run along the Periadriatic (Seam), Labot, Ljutomer, and Sava faults, the thrust front of the Southern Alps, and the external front of the thrust zone of the External Dinarides (Celarc and Placer 2016).

The **Dinarides** are divided into the External and Internal Dinarides, and a transitional zone in between. Slovenia only includes the External Dinarides with the transitional zone. The External Dinarides have a thrust and nappe structure, which formed at the end of the Eocene and the beginning of the Oligocene. It is composed of extensive nappes made of Mesozoic carbonate rocks thrust horizontally on top of one another. The thrusting took place from the northeast toward the southwest, and the nappe units slid across a base made of Paleozoic rocks. In the front part of the mountain range, they are thrust over Eocene flysch layers. Typical examples of thrust tectonics in the External Dinarides include the Trnovo (Fig. 5.4) and Hrušica nappes, and the Snežnik thrust block (Placer 1999, 2008; Celarc and Placer 2016).

In terms of paleogeography, the **Southern Alps** are part of the Dinarides, from which they separated during the Miocene. They are delimited by the Periadriatic (Seam), Labot, and Ljutomer faults to the north, and the Southern Alpine thrust front and the Sava Fault to the south. They are

Fig. 5.3 Alpine hills. Top: the Cerkno Hills with Mount Porezen (1,630 m) as their highest peak. Bottom: the wooded Pohorje Hills, whose highest parts rise above 1,500 m. (Photo by Matija Zorn)

made of the Upper Triassic rocks of the Julian Carbonate Platform and the Mesozoic rocks of the Slovenian Basin. The Southern Alps comprise the Julian and Kamnik–Savinja Alps, the Western Karawanks, the Paški Kozjak and Boč hills, the Alpine hills along the Bača and Selška Sora rivers, and part of the Škofja Loka Hills. The thrust sheets of the Southern Alps formed predominantly between the Eocene and Middle Oligocene. Due to the north–south thrust direction, the ridges of the mountains and hills run largely from west to east (Placer 2008).

The **Eastern Alps** encompass the Precambrian and Early Paleozoic metamorphosed rocks and the Permian and Mesozoic sedimentary rocks north of the Periadriatic (Seam) Fault. In Slovenia, the Eastern Alps include the Pohorje, Strojna, and Kozjak Hills and the Eastern Karawanks with Mounts Peca and Uršlja gora (Placer 2009). The Eastern Alps are made of extensive nappes that formed in the Alpine orogeny between the Cretaceous and the Neogene. In addition to its nappe structure another important feature is plutonism, which is divided into Periadriatic tonalite intrusions of Oligocene age and Pohorje granodiorite pluton with dacitic sills and dikes of Miocene age (Fodor et al. 2008; Placer 2008).

Fig. 5.4 Dinaric plateaus. Top: the thrust structure of the Trnovo Forest Plateau rises to 1,000 m above the Vipava Valley (at approximately 100 m). Bottom: the rounded tops of the Javorniki Plateau, rising above 1,200 m. (Photo by Matija Zorn)

In many places, the geological structure is clearly reflected in the landforms. This especially applies to large geological structures, such as thrusts, folds, and faults. In some places, the thrust fronts take the form of sudden transitions in terrain. The steep limestone slopes of the Trnovo Forest Plateau above the flysch Vipava Valley form the front of the Trnovo thrust (Fig. 5.4). Folded terrain most often takes the form of parallel elongated sequences of ridges and valleys, such as in the Sava Hills. In many places, river valleys are cut into the tectonically damaged rocks along the faults. The Soča and Idrijca rivers cut part of their valleys along the Idrija fault.

The rocks (Figs. 5.6 and 5.7, Table 5.1) shaping the surface of Slovenia's mountainous areas formed from at least the Early Paleozoic to the Quaternary. The oldest metamorphic rocks are probably even of Precambrian age (Novak 2016). Sedimentary rocks predominate, covering 78.1% of the surface. Especially common among these are limestone and dolomite (50.3%) as carbonate rocks, whereas quartz sandstone and conglomerate, shale, and flysch and volcanoclastic rocks predominate among the noncarbonate rocks (27.8%). Loose sediments cover the river valleys and certain slopes, covering 12.0% of the territory. Metamorphic rocks account

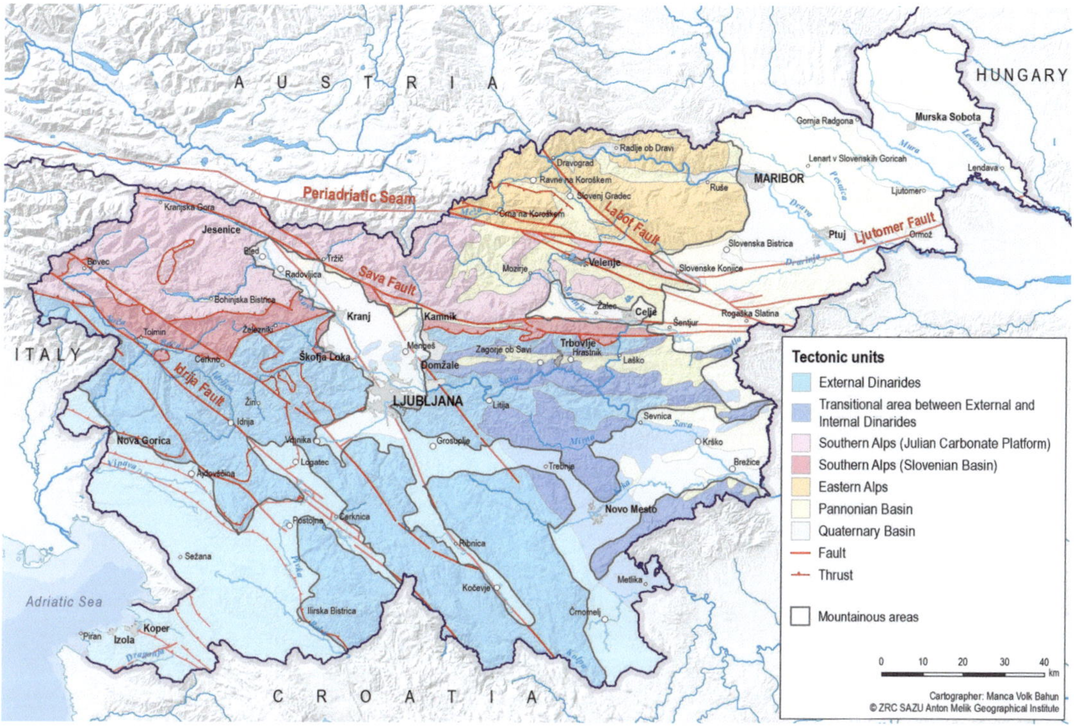

Fig. 5.5 Tectonic structure of Slovenia's mountainous areas. (Adapted from Poljak 2007; Celarc and Placer 2016)

for 7.1% of the territory, and they are predominantly clustered in the northeast of mountainous areas. Igneous rocks account for the smallest share, covering only 2.8% of the territory. Intrusive igneous rocks can only be found on the surface along the Periadriatic Seam and in the Pohorje Hills, and volcanic rocks are dispersed across small patches in the Alpine hills and mountains.

The distribution of individual rock types is fairly uneven. Alpine mountains are predominantly made of limestone and dolomite, whereas carbonate gravel, rubble, and till have accumulated in large quantities in the valleys. Shale, quartz sandstone, quartz conglomerate, and dolomite dominate the Alpine hills of Central Slovenia. Standing out in this respect are the Alpine hills in northeastern Slovenia, which are made of metamorphic and igneous rocks. Nearly all Dinaric plateaus are made of limestone and dolomite. Due to their diversity, rocks play an important role in Slovenian landscapes. They are what largely determines today's landforms, network of rivers and creeks, soil types, distribution of plant species, and human activities in the landscape (Natek and Natek 1998).

5.3 Landscape Types

In terms of their geomorphological development (genesis), one can distinguish between glacial, fluvio-denudational, and karst landscapes in Slovenia's mountainous areas (Figs. 5.8, 5.9 and Table 5.2). Glacial landscapes are limited to areas that were exposed to glacial erosion and accumulation during the cold Pleistocene periods, and fluvio-denudational landscapes are typical of areas made of poorly permeable rocks and characterized by surface runoff. Karst landscapes developed on carbonate rocks. They are characterized by intense chemical dissolution of the bedrock and subterranean karst hydrology (Hrvatin 2004).

Some areas feature a combination of genetic landscape types. For example, glaciokarst is typi-

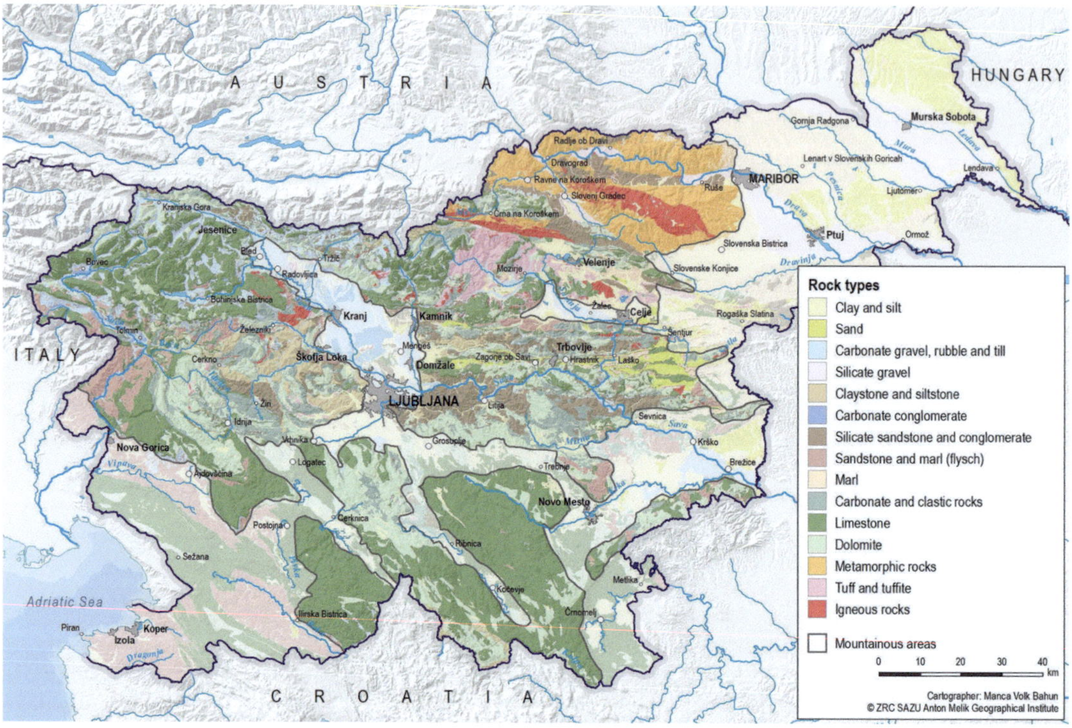

Fig. 5.6 Rock types in Slovenia's mountainous areas. (Adapted from Litostratigrafska ... 2011; Perko et al. 2015)

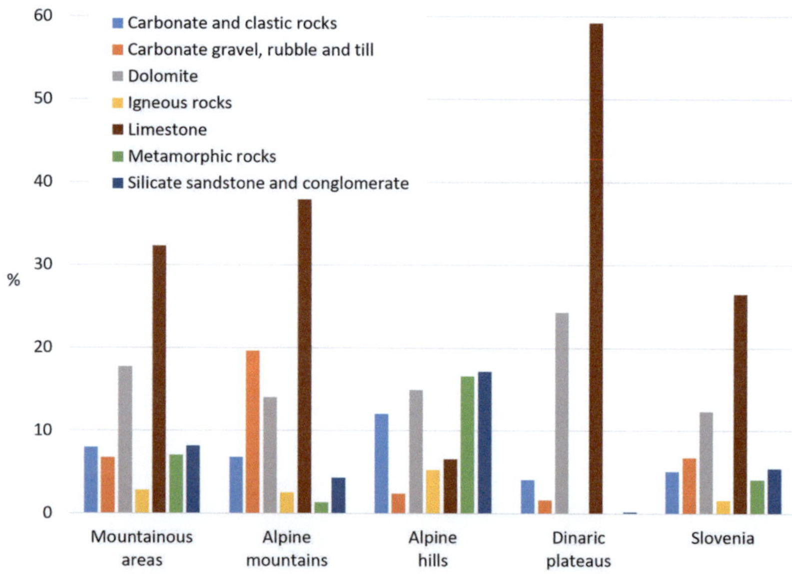

Fig. 5.7 Predominant rock types in Slovenia's mountainous areas. (Based on Table 5.1)

cally found in mountains that were reshaped by glaciers during the Pleistocene and where the chemical dissolution of carbonate rocks now predominates. In addition to landforms that more or less clearly belong to a specific genetic landscape type, there is a variety of those that developed through the interaction of various geomorphological processes and are therefore referred to as polygenetic. The most typical among these are peaks, ridges, and slopes of various

Table 5.1 Rock types in Slovenia's mountainous areas. (Based on Fig. 5.6)

Rock type	Mountainous area			*Alpine mountains*				*Alpine hills*				*Dinaric plateaus*				Slovenia	
	Area (km^2)	% of mountainous areas	% of Slovenia	Area (km^2)	% of Alpine mountains	% of mountainous areas	% of Slovenia	Area (km^2)	% of Alpine hills	% of mountainous areas	% of Slovenia	Area (km^2)	% of Dinaric plateaus	% of mountainous areas	% of Slovenia	Area (km^2)	%
Carbonate and clastic rocks	926.07	8.03	4.57	209.71	6.85	1.82	1.03	559.66	**12.01**	4.85	2.76	156.69	4.11	1.36	0.77	1,040.52	5.13
Carbonate conglomerate	24.83	0.22	0.12	19.98	0.65	0.17	0.10	4.38	0.09	0.04	0.02	0.47	0.01	0.00	0.00	152.49	0.75
Carbonate gravel, rubble and till	777.93	6.75	3.84	602.47	**19.68**	5.23	2.97	112.80	2.42	0.98	0.56	62.66	1.65	0.54	0.31	1,368.59	6.75
Clay and silt	473.83	4.11	2.34	20.47	0.67	0.18	0.10	353.92	7.60	3.07	1.75	99.44	2.61	0.86	0.49	2,628.61	**12.97**
Shale, claystone and siltstone	337.32	2.93	1.66	44.90	1.47	0.39	0.22	252.15	5.41	2.19	1.24	40.27	1.06	0.35	0.20	348.77	1.72
Dolomite	2,051.72	**17.80**	**10.12**	430.80	**14.07**	3.74	2.13	695.11	**14.92**	6.03	3.43	925.81	**24.31**	8.03	4.57	2,504.06	**12.36**
Igneous rocks	326.10	2.83	1.61	77.15	2.52	0.67	0.38	247.64	5.31	2.15	1.22	1.31	0.03	0.01	0.01	332.24	1.64
Limestone	3,723.66	**32.30**	**18.37**	1,160.50	**37.91**	**10.07**	5.73	307.96	6.61	2.67	1.52	2,255.20	**59.22**	**19.56**	**11.13**	5,378.79	**26.54**
Marl	113.31	0.98	0.56	10.28	0.34	0.09	0.05	101.55	2.18	0.88	0.50	1.48	0.04	0.01	0.01	980.92	4.84
Metamorphic rocks	812.30	7.05	4.01	40.29	1.32	0.35	0.20	772.00	**16.57**	6.70	3.81	0.00	0.00	0.00	0.00	831.16	4.10
Sand	143.75	1.25	0.71	0.90	0.03	0.01	0.00	142.63	3.06	1.24	0.70	0.22	0.01	0.00	0.00	913.81	4.51
Sandstone and marl (flysch)	472.76	4.10	2.33	118.69	3.88	1.03	0.59	103.73	2.23	0.90	0.51	250.34	6.57	2.17	1.24	1,443.65	7.12
Silicate gravel	130.08	1.13	0.64	8.10	0.26	0.07	0.04	121.97	2.62	1.06	0.60	0.00	0.00	0.00	0.00	930.93	4.59
Quartz sandstone and conglomerate	941.95	8.17	4.65	133.70	4.37	1.16	0.66	800.07	**17.17**	6.94	3.95	8.19	0.21	0.07	0.04	1,106.23	5.46
Tuff and tuffite	273.59	2.37	1.35	183.26	5.99	1.59	0.90	84.22	1.81	0.73	0.42	6.11	0.16	0.05	0.03	304.81	1.50
Total	11,529.19	100.00	56.89	3,061.20	100.00	26.55	15.11	4,659.80	100.00	40.42	22.99	3,808.20	100.00	33.03	18.79	20,265.58*	100.00

* The data in the data layer used do not entirely correspond to Slovenia's actual size: 20,273 km^2
Those covering more than 10% are in bold

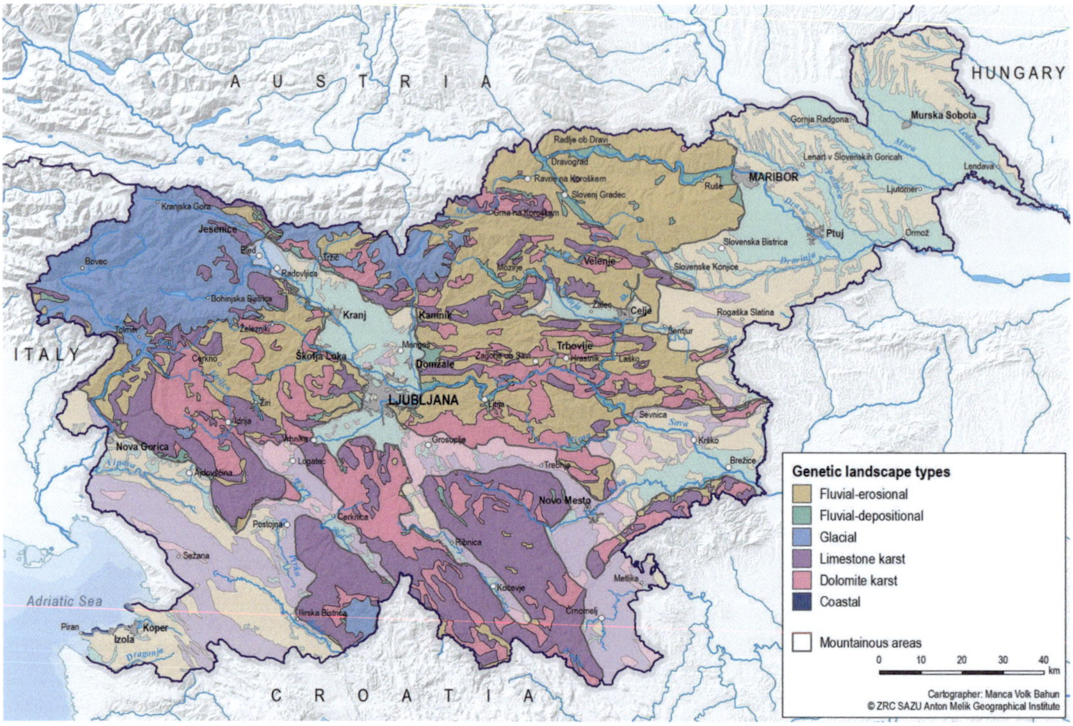

Fig. 5.8 Genetic landscape types in Slovenia's mountainous areas. (Adapted from Gabrovec and Hrvatin 2018)

Fig. 5.9 Predominant genetic landscape types in Slovenia's mountainous areas. (Based on Table 5.2)

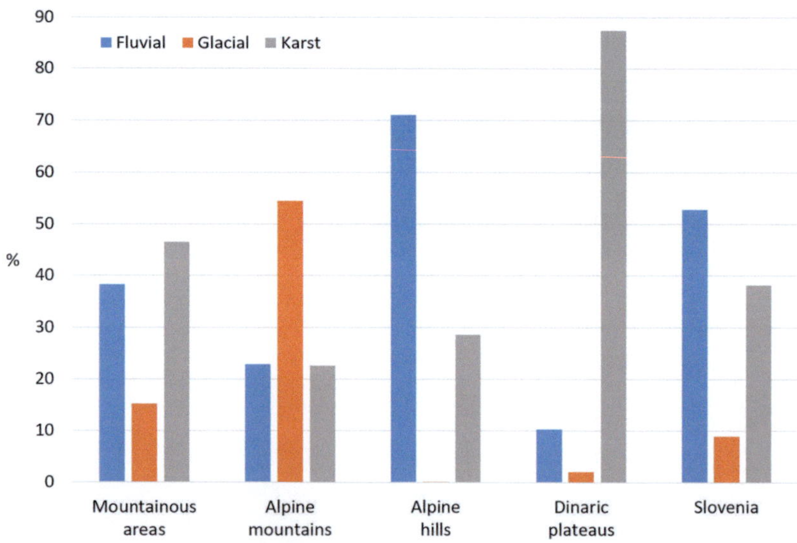

Table 5.2 Genetic landscape types in Slovenia's mountainous areas. (Based on Fig. 5.8)

Type, *subdivision*	**Description**	**Mountainous areas**			*Alpine mountains*				*Alpine hills*				*Dinaric plateaus*				**Slovenia**	
		Area (km^2)	% of mountainous areas	% of Slovenia	Area (km^2)	% of Alpine mountains	% of mountainous areas	% of Slovenia	Area (km^2)	% of Alpine hills	% of mountainous areas	% of Slovenia	Area (km^2)	% of Dinaric plateaus	% of mountainous areas	% of Slovenia	Area (km^2)	%
Fluvial	Dominates on poorly permeable rocks with surface drainage	4,409.40	**38.24**	**21.75**	701.21	**22.90**	6.08	3.46	3,313.16	**71.10**	**28.73**	**16.34**	395.03	**10.37**	3.43	1.95	10,712.21	**52.85**
Erosion	Characteristic of mountainous and hilly landscapes with valleys and ridges	3,947.31	**34.23**	**19.47**	617.58	**20.17**	5.36	3.05	3,015.70	**64.72**	**26.15**	**14.88**	314.03	8.24	2.72	1.55	7,110.76	**35.08**
Deposition	Characteristic of flat landscapes and plains	462.09	4.01	2.28	83.63	2.73	0.73	0.41	297.46	6.38	2.58	1.47	81.00	2.13	0.70	0.40	3,601.45	**17.77**
Glacial		1,761.04	**15.27**	8.69	1,668.61	**54.50**	**14.47**	8.23	9.94	0.21	0.09	0.05	82.49	2.16	0.72	0.41	1,825.10	9.00
Karst	Characterized by chemical weathering of carbonate rocks (limestone and dolomite), circular surface depressions, and subsurface drainage	5,361.25	**46.49**	**26.45**	691.89	**22.60**	6.00	3.41	1,336.82	**28.69**	**11.59**	6.59	3,332.54	**87.47**	**28.90**	**16.44**	7,731.44	**38.14**
Limestone		3,102.09	**26.90**	**15.30**	477.41	**15.59**	4.14	2.36	429.26	9.21	3.72	2.12	2,195.42	**57.62**	**19.04**	**10.83**	4,874.24	**24.05**
Dolomite		2,259.16	**19.59**	**11.15**	214.48	7.01	1.86	1.06	907.56	**19.48**	7.87	4.48	1,137.12	**29.85**	9.86	5.61	2,857.20	**14.10**
Coastal	Limited to a narrow Adriatic coastal belt																1.75	0.01
Total		11,531.69	100.00	56.89	3,061.70	100.00	26.55	15.10	4,659.93	100.00	40.41	22.99	3,810.06	100.00	33.04	18.80	20,270.50*	100.00

* The data in the data layer used do not entirely correspond to Slovenia's actual size: 20,273 km^2
Those covering more than 10% are in bold

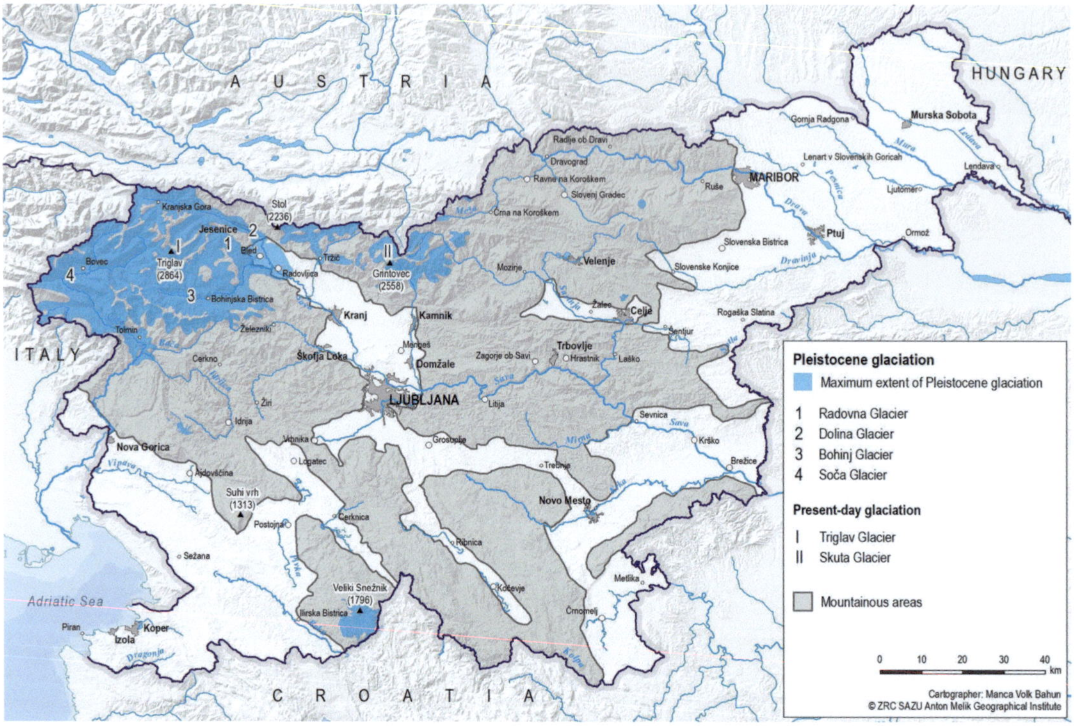

Fig. 5.10 Pleistocene glaciation of Slovenia's mountainous areas. (Adapted from Šifrer 1998; Ferk et al. 2017)

shapes (Gabrovec and Hrvatin 1998; Zorn et al. 2020a, b).

5.3.1 Glacial Landscapes

Glacial landscapes cover over 15.3% of Slovenia's mountainous areas (Fig. 5.10). They formed through glacial erosion and accumulation during the cold periods of the Pleistocene. They are typical of the Alpine mountains, such as the Julian Alps and the Kamnik–Savinja Alps, and partly also the Karawanks, the Pohorje Hills among the Alpine hills, and certain high Dinaric plateaus, such as the Snežnik and the Trnovo Forest plateaus (Ferk et al. 2017). They make up over half of the Alpine mountains, only 0.2% of the Alpine hills, and just over 2% of the Dinaric plateaus.

During the Last Glacial Maximum, the Julian Alps were covered by an ice cap above an approximate elevation of 1,300 m, from which glaciers crept down through the valleys, extending far below the snow line. The greatest volume of ice masses accumulated in the Bohinj Valley, into which ice flowed down from the ridges and high-mountain plateaus on its north, west, and south sides. At Bohinjska Bistrica, the ice masses were as thick as 800 m. From the Bohinj Valley, the glacier descended through the valley of what is now the Sava Bohinjka River all the way down to the Bled–Radovljica Basin (Melik 1930; Šifrer 1969).

Two other glaciers descended toward that same basin. The first crept down the Radovna Valley and the other down the Sava Dolinka (Dolina) Valley. The Radovna Glacier was fed by ice masses coming down from the Krma and Kot high-mountain valleys, as well as from the northern part of the Pokljuka Plateau. The Dolina Glacier accumulated ice masses from the Planica, Pišnica, and Vrata high-mountain valleys, and to a smaller degree also from the Western Karawanks (Gams 1992; Ferk et al. 2017).

Great volumes of ice moved along the Soča Valley. According to an older interpretation, the Soča Glacier reached down to Tolmin or even

Most na Soči (being almost 70 km long) during its maximum extent (Šifrer and Kunaver 1978). However, the glacier's characteristics and the model of its possible dimensions show that the size is most likely exaggerated (being maybe half the size), at least when referring to the periods of most recent major Alpine glaciations, and that during that time the main ice flow in the valley must not have extended farther than the Bovec Basin (Bavec et al. 2004).

Only minor glaciers developed in the Kamnik–Savinja Alps. Ice masses descended from beneath the central peaks and ridges down into the deeply cut valleys, extending further down along them, but these glaciers were not long. They only filled the upper parts of the Kokra (Meze 1974), Kamniška Bistrica (Šifrer 1961), and Savinja valleys (Meze 1966), which still have a high-mountain character, and so they did not extend beyond the mountain range.

As a slightly lower and narrow mountain range, the Karawanks were covered by less ice. In their eastern part, small glaciers developed on Mounts Peca and Olševa (Gams 1970), and in the central part there were several glaciers on the slopes of Mount Košuta to the south and in the area of the Zelenica between Mounts Begunjščica, Vrtača, and Stol, from where ice masses descended down the Završnica Valley (Melik 1932; Šifrer 1969).

During the ice ages, the Pohorje Hills were too low to develop an ice cap, but nonetheless several small cirque glaciers developed in the shady depressions on the northern slopes of the main ridge (Natek 2007; Obu 2011).

In the Dinaric Alps, ice mainly accumulated on the extensive plateaus south of Mount Snežnik. These extended over 1,400 m and thus above the permanent snow line. From them, the ice flowed toward Padežnica hollow and from there down the valley onto Gomance piedmont polje, with the ice-covered area extending contiguously all the way to Mount Risnjak in Croatia. Moraine embankments of various ages suggest the area was exposed to multiple glaciations (Šifrer 1959).

The Pleistocene glaciation of the Trnovo Forest Plateau affected at least 8 km^2. The ice cap was approximately 4,8 km^2 in size, and its maximum thickness was approximately 180 m in the karst depression known as *Smrekova draga*. Outlet glaciers fell across a steep transition in the terrain in the form of seracs. After nearly 4 km, the glacier descending down the plateau into the Trebuša Valley stopped at an elevation of around 350 m (Kodelja et al. 2013; Žebre et al. 2013).

The glaciers heavily reshaped the surface. The debris they carried with them eroded many slopes and widened the highest karst plains. Cirques formed under the rocky ridges high up in the mountains. Large cirques include the *Za Akom* and *Pod srcem* cirques above Gozd-Martuljek in the Julian Alps, and the Okrešelj Cirque above the headwall of the Logar Valley in the Kamnik–Savinja Alps. Glaciers filled the former river valleys, transforming them into glacial troughs with a wide, flat bottom and steep slopes. The most typical glaciated valleys include the Trenta, Lepena, Vrsnik, Možnica, Bavšica, Zadnjica, Lower and Upper Bohinj, Voje, Tamar, Vrata, Kot, Krma, Radovna, and Triglav Lakes valleys in the Julian Alps, the Makek (*Makekova kočna*), Ravne (*Ravenska kočna*), Logar (*Logarska dolina*), Matk (*Matkov kot*), Roban (*Robanov kot*), and Kamniška Bistrica valleys in the Kamnik–Savinja Alps, and the Završnica, Draga, and Podljubelj valleys in the Karawanks (Gabrovec and Hrvatin 1998; Kladnik 1998a).

The interglacial, warmer periods of the Quaternary were characterized by significant deepening of the valleys, which began when the glaciers started to recede. The downcutting of rivers was so pronounced that any new deposition during each of the following ice ages no longer reached the height of the previous one. Thus, six main Quaternary terraces formed down the valleys, with each lower one younger than the previous one and hence less weathered and karstified on the surface. Each consecutive terrace is covered by a thinner layer of weathered material, the gravel in it is increasingly less consolidated and, in the youngest terrace from the most recent ice age, it is even completely fresh (Šifrer 1998).

During the ice ages, the nearby non-glaciated periglacial areas were also exposed to profound surface transformation. Due to extremely low temperatures, forest almost completely disappeared. This was followed by strong mechanical weathering of rocks and solifluction (i.e., slow downslope movement of water-saturated debris). Significant amounts of periglacial debris can be found in the Lower Soča Valley and below the karst plateaus, from the Banjšice and Trnovo Forest plateaus past the Nanos, Hrušica, Javorniki, and Snežnik plateaus, and onward toward the southeast (Šifrer 1998).

Recent geomorphological research has provided new insights into periglacial landforms. In the Pohorje Hills, research has revealed numerous cryoplanation terraces and nivation hollows, as well as the presence of ploughing rocks, blockstreams, and blockfields (Natek 2007; Obu 2011). In the Alpine mountains, detailed laser scanning of the surface has revealed many relict rock glaciers (Fig. 5.11) and nival moraines (Colucci et al. 2016; Triglav Čekada et al. 2016).

Currently, there are only two glaciers left in Slovenia: the Triglav Glacier (Fig. 5.12) in the Julian Alps and the Skuta Glacier in the Kamnik–Savinja Alps. Both are subject to long-term monitoring and hence important for studying past and present environmental changes. The Triglav Glacier lies on the relatively sunny northeastern side of Mount Triglav, at an elevation between 2,400 and 2,500 m. In the second half of the nineteenth century, it still covered nearly forty-six hectares. By 1946, when regular monitoring started, it shrank to fourteen hectares, and it now covers less than half a hectare (Gabrovec et al. 2013, 2014; Triglav Čekada and Zorn 2020). The Skuta Glacier is a cirque glacier, less than a hectare in size, which lies in a year-round shady area at an elevation between 2,000 and 2,100 m (Pavšek 2007; Triglav Čekada et al. 2020). For a long time, the view predominated that both glaciers formed during the Little Ice Age and hence cannot be considered remnants of the Pleistocene glaciation (Ferk et al. 2017). However, the latest uranium–thorium dating of subglacial carbonate deposits in the Triglav Glacier area (Lipar et al. 2021) has surprisingly shown that the sediments were deposited during the Last Glacial Maximum and the Younger Dryas. Exposed subglacial carbonate deposits tend to weather quickly and they erode completely within a few decades (Lipar et al. 2025). Their presence, therefore, indicates that the Triglav Glacier has most likely remained continuously present throughout the Holocene.

Fig. 5.11 Relict rock glaciers can be found in the Alpine mountains, such as the one in the *Beli potok* basin in the Julian Alps with a mean elevation of 1,040 m. Top: lidar image. (*Source* Surveying and Mapping Authority of the Republic of Slovenia) Bottom: steplike surface between two "arches" of the rock glacier. (Photo by Matija Zorn)

5.3.2 Fluvio-Denudational Landscapes

The fluvio-denudational landscape type constitutes nearly four-tenths of Slovenia's mountainous areas. It is the most common in the Alpine hills, covering nearly two-thirds of the total area. In addition, it constitutes just under a fourth of the Alpine mountains and only one-tenth of the Dinaric plateaus. These areas are largely made of poorly permeable rocks, and they are primarily characterized by various slope processes (Komac and Zorn 2007), from slow creeping and leaching

Fig. 5.12 The Triglav Glacier at the end of the nineteenth century (top) and in 2012 (bottom). (GIAM ZRC SAZU archive)

of weathered material to large landslides and rockfalls, and water erosion.

Due to exogenic processes the yearly sediment productions in the Alpine mountains is 1,100 to 1,300 m^3/km^2, and in the Alpine hills 400 to 700 m^3/km^2 (Natek and Natek 1998). Especially in the Alpine hills slumps and landslides are frequent on slopes covered by a thick layer of weathered material and soil; these are triggered in great numbers (up to several dozen per km^2) during heavy rain (Gabrovec 1990). Extensive landslides are common in the Škofja Loka, Idrija, Cerkno, and Sava hills (Komac and Zorn 2007). Rockfalls occasionally develop on the mountain cliffs and steep rocky slopes of mountain valleys (Zorn 2002). In recent decades, they have been especially common in the mountains above the Soča Valley. A significant number of rockfalls developed during the 1976, 1998, and 2004 earthquakes. During the earthquake on April 12, 1998, which had a magnitude of 5.8 on the Richter scale, over a hundred rockfalls of various sizes were triggered, and several million cubic meters of material were moved. During the earthquake on July 12, 2004, which had a magnitude of 4.9 on the Richter scale, thirty-eight rockfalls were recorded in roughly the same area but, because of the smaller earthquake magnitude, they were smaller and primarily only superficial (Vidrih et al. 2001; Komac and Zorn 2007).

An important factor in the transformation of mountainous surfaces is various watercourses. The most important feature of mountain rivers and creeks is their distinctly torrential character, which shows most clearly in the over tenfold difference between their maximum and average discharges. Between 1991 and 2020, the maximum discharge on the Paka River measured at Šoštanj (137 m^3/s) was 59.9 times higher than average (2,29 m^3/s), the maximum discharge on the Kokra River measured at Kokra was 43.9 times higher than average, and on the Koritnica River at Kal–Koritnica it was 43.3 times higher. A difference of at least 30 times was also recorded on the Kamniška Bistrica River at Kamnik, on the Savinja River at Laško, on the Sora River at Suha, and on the Meža River at Otiški Vrh.

Flash floods are of short duration, but they are extremely violent. They are caused by relatively short but intense rainfall during summer thunderstorms or heavy rain in the fall. They occur along hundreds of small torrents in the mountains and hills, and along certain large rivers, such as the Meža, Mislinja, Savinja, Tržiška Bistrica, Kamniška Bistrica, and Sora. The water rises very fast, transports large amounts of gravel, which it deposits on the fans or floodplains, and then recedes within a few hours (Brilly et al. 1999). The increasing damage from floods in recent years has primarily been the result of ill-considered spatial planning in flood-prone areas (Natek and Natek 1998; Komac et al. 2008).

Alpine mountains and hills are very sensitive to erosion, and so forest is of exceptional importance in these areas. It retains and absorbs water and hinders its runoff but, first and foremost, it prevents the formation of large volumes of unconsolidated erosion debris that surface runoff could transport into the torrents (Horvat 2002).

5.3.3 Karst Landscapes

Karst landscapes are the predominant landform in Slovenia's mountainous areas, covering nearly half of their total area. Taking into account that most previously glaciated areas are also karstified, mountain karst constitutes over 60% of Slovenia's mountainous areas. Most karst landscapes can be found on Dinaric plateaus, where they cover nearly nine-tenths of the total area. Nearly a quarter of the Alpine mountain terrain has a karst character but, if glaciokarst is also taken into account (Fig. 5.13), karst landscapes make up over three-quarters of the Alpine mountains. They also constitute nearly three-tenths of the Alpine hills.

The main characteristics of karst landscapes include intense chemical dissolution of carbonate rocks and underground karst hydrology. Based on the hardness or mineralization of karst waters, it was established that the karst surface in Slovenia lowers by 2 to 10 mm every century; however, on the Trnovo Forest Plateau and Snežnik Plateau and the edges of the Julian Alps, where annual precipitation is up to 3,000 mm, it most likely lowers by more than 10 mm per century locally (Gams 2004). The discovery of unroofed caves has confirmed intense lowering of the surface due to corrosion (Mihevc 2001).

Based on the type of carbonate rock exposed to karstification, one can distinguish between limestone and dolomite karst. **Limestone karst landscapes** typically consist of karst depressions and conical hills (Figs. 5.4 and 5.14). Dolines are the most common karst depressions; in flat areas their density can be exceptional, even over 500 per km^2. There are no dolines at the bottoms of karst poljes and steep slopes; they are also less common in slightly inclined areas (Mihevc and Mihevc 2021). Large karst depressions include uvalas and karst poljes. The best-known ones are located in the lowlands outside the mountainous areas; however, some can also be found in mountainous areas, such as the Dobrepolje–Struge, Bloke and Globodol karst poljes (on Dinaric plateaus), as well as Velo and Malo polje (in Alpine mountains) below Mount Triglav (Fig. 5.13).

Karst hydrology resulted in the formation of many caves and shafts. As of July 2022, the register of the Slovenian Speleological Association included 14,695 karst cavities, 9,956 of which are located in mountainous areas (Kataster ... 2022). With a total length of 43 km, the Migovec Cave System is the longest cave in Slovenia, and the Rombon Cave System is already over 20 km long (both in the Julian Alps). The deepest shafts (i.e., deeper than 1,000 m) to date have been explored on Mounts Rombon and Kanin (the Julian Alps), and on the Dleskovec Plateau (the

Fig. 5.13 Because of the predominantly carbonate bedrock, most Alpine mountains are karstified, and they were also reshaped by glaciers. Top: glaciokarst in the Triglav Lakes Valley in the Julian Alps. (Photo by Bojan Erhartič, GIAM ZRC SAZU archive) Bottom: the Velo polje (right) and Malo polje (left) in the Julian Alps are high-mountain karst poljes. (Photo by Matija Zorn)

Kamnik–Savinja Alps). The Čehi 2 Shaft (1,505 m deep) on the Rombon Plateau is currently the fourteenth-deepest shaft in the world. Collapse dolines formed in places as a result of collapsed cave roofs; they are especially common in the catchment areas of the karst sources of the Ljubljanica and Krka rivers. The Pekel collapse doline between the Little Mountain (*Mala gora*) and the Kočevski Rog Plateau is over 100 m deep, and the Dol collapse doline near Mala Ilova Gora has a volume of 12.7 million m^3, which ranks it among the largest in Slovenia (Stepišnik 2010).

In karst studies, conical karst hills (Fig. 5.4) are mostly typical of tropical areas, and less attention is devoted to those in temperate climate zones. These are landforms that rise, either in isolation or from a common base, several tens of meters high. The hills predominantly have a round base, and they vary in height and slope inclination. It was initially believed that the conical hills typical of Dinaric karst landscapes were a relic of the tropical climate during the Pliocene (Habič 1968, 1980), but more recent research has shown that there are two types of isolated conical hills, reflecting two different formation

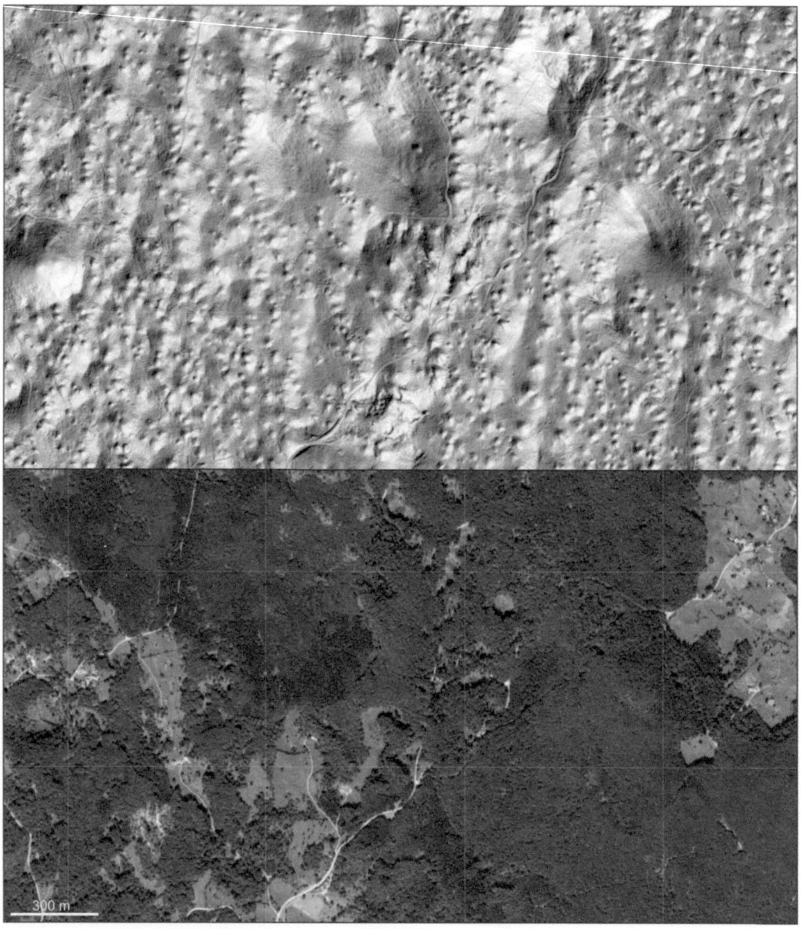

Fig. 5.14 Dinaric plateaus (Dinaric karst landscapes) are characterized by dolines and conical hills (top), which are mostly covered in forest (bottom); the area of the Trnovo cadastral municipality on the Trnovo Forest Plateau at an elevation between 800 and 900 m. (*Source* Surveying and Mapping Authority of the Republic of Slovenia)

mechanisms. The first type is corrosion hills, which, due to their varied rock permeability, lower more slowly than their surroundings. The second type is isolated hills, which can be tectonically defined as vertical uplifts. The types have different morphometric characteristics, with corrosion hills being lower, gentler, and covering smaller areas than the tectonic hills (Stefanovski and Repe 2019).

In **dolomite karst landscapes**, surface karst features are less common, and so their karst character is less pronounced than in the limestone type. Large karst caves are also relatively rare there. Shallow dry valleys, or dells, are typical (Komac 2006). In addition to the chemical dissolution of rock, erosion and denudation are also common on dolomite, which is why in many places dolomite karst landscapes resemble fluvio-denudational landscapes. This landscape type is often referred to as fluviokarst (Komac 2004). Because of the poor mechanical resistance or brittleness of the dolomite, badlands are common in this landscapes.

From a regional perspective, Slovenia's mountain karst landscapes are divided into three main units, which are spatially well separated, but they also differ in a number of natural geographical and other characteristics. Alpine karst landscapes are typical of the Alpine mountains, several small isolated karst areas can be found in the Alpine hills, and Dinaric karst landscapes are common on high Dinaric plateaus. This typology was applied in the outline of Slovenia's speleological characteristics (Gams 1965) and in discussing the hydrographic characteristics of Slovenian karst areas (Habič 1969).

Alpine karst landscapes primarily developed on the Triassic limestone and dolomite of the Julian Alps, Kamnik–Savinja Alps, and Karawanks. They typically have a deep vadose zone perforated with shafts, some of which over 1,000 m deep (Mihevc et al. 2016). Some shafts are connected with subhorizontal passages, forming long cave systems. At the foot of the mountains that rise up to 2,864 m (e.g., Mount Triglav), there are many karst springs. Because the mountains were covered in ice several times during the Pleistocene, the configuration of the terrain is the result of the interaction between glacial and karst processes. High-mountain plateaus and exposed rock areas above the tree line are characterized by limestone pavements—that is, glacially smoothed and heavily karstified rock blocks on top of thick layers of limestone (Kunaver 1983). Another common feature is snow kettles: round rock hollows with vertical walls, in which snow lingers most of the year. In the past, their formation was associated with the corrosive effects of snow melt, but now the predominant belief is that snow kettles are the remnants of former subglacial swallow holes (Stepišnik 2020).

Isolated karst landscapes formed on limestone and dolomite patches surrounded by poorly permeable non-carbonate rock. These patches are small, only a few tens of square kilometers in size, and dispersed across the Alpine hills (Habe 1972). Karst features tend to be less pronounced in these areas, and they are often described as fluviokarst due to the combined interaction of fluvial and karst processes there (Mihevc et al. 2016). Thicker soils that formed through the weathering of allogenic alluvial sediments are common on the surface and in the dolines. Therefore, small suffosion dolines are frequent in this karst landscape type (Mihevc and Mihevc 2021).

Dinaric karst landscapes cover a contiguous karst area in western and southern Slovenia. They are mostly made of Jurassic and Cretaceous limestone and dolomite (Habič 1975). Diverse surface karst features, such as minor karst landforms, various depressions, and conical hills (Figs. 5.4 and 5.14), appear across the entire area. Dolines are the predominant feature (Fig. 5.14). In the past, many dolines were adapted for agricultural use and are thus referred to as "cultivated dolines." High Dinaric karst landscapes consist of plateaus at elevations between 900 and 1,400 m, with individual peaks rising even significantly higher than that. The highest peak, Mount Snežnik, rises to 1,796 m. The high plateaus receive from 1,500 to 3,000 mm of precipitation a year. Because their surface is heavily karstified, water immediately sinks underground, and so there is no surface runoff. The vadose zone is several hundred meters deep and criss-crossed by many shafts (Mihevc et al. 2016). The percolating water resurfaces in the powerful karst springs below the plateaus. The most prominent among them are the Vipava, Ljubljanica, and Krka springs.

5.4 Climate

The climate in Slovenia's mountainous areas (Figs. 5.15 and 5.16, Table 5.3) is the result of various climate factors. Slovenia's position in the middle latitudes relatively close to the Atlantic and the influence of western air circulation are two important factors. Its position on the edge of the Adriatic or in a transitional zone between the Mediterranean Basin and the Eurasian continent also plays a great role. Due to its highly dynamic terrain, elevation is also very important (Ogrin 1996; Ogrin et al. 2023).

Because of its location in the middle latitudes, Slovenia has a temperate and humid climate. Great weather variability is typical, resulting from frequent changes in air masses of various characteristics. The weather is the most stable during the expansion of the Azores or Siberian High. The Azores High causes hot and muggy weather in summer and the Siberian High causes dry and cold weather in winter, with atmospheric inversion and fog in the basins and valleys. When the cold fronts pass through, they bring showers and storms, making the weather deteriorate briefly. Long deteriorations occur with the passing of the Mediterranean cyclones, which bring heavy precipitation and primarily snow in the mountains (Ogrin 2004).

The following three climate types can be distinguished in Slovenia's mountainous areas:

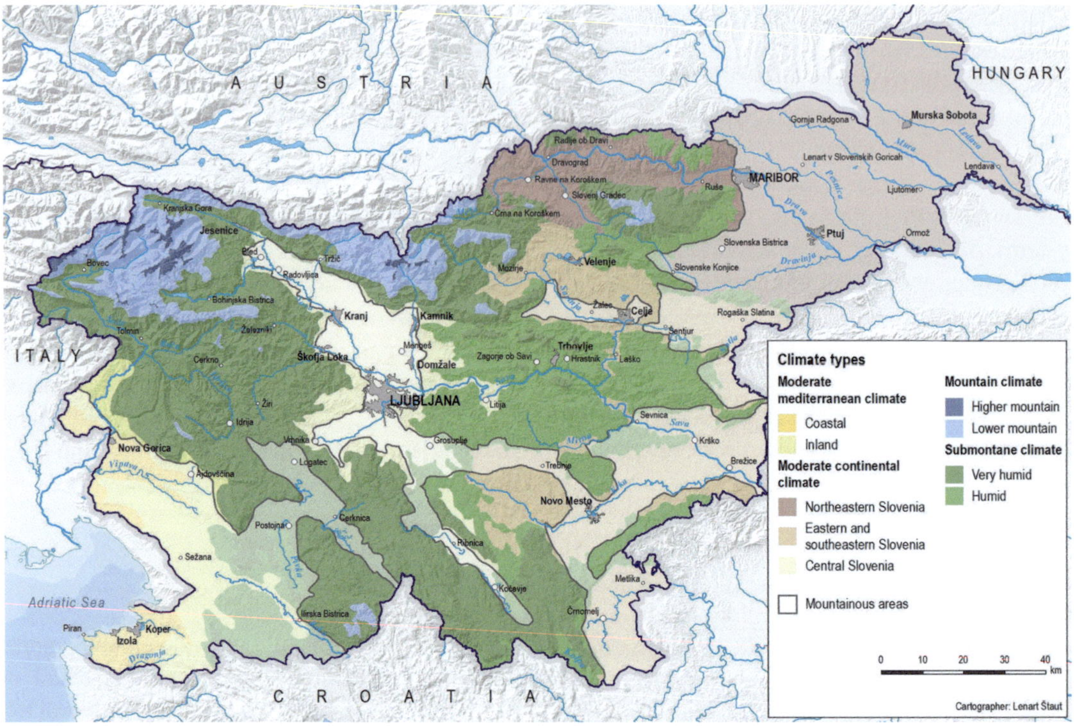

Fig. 5.15 Climate types in Slovenia's mountainous areas. (Adapted from Ogrin et al. 2023)

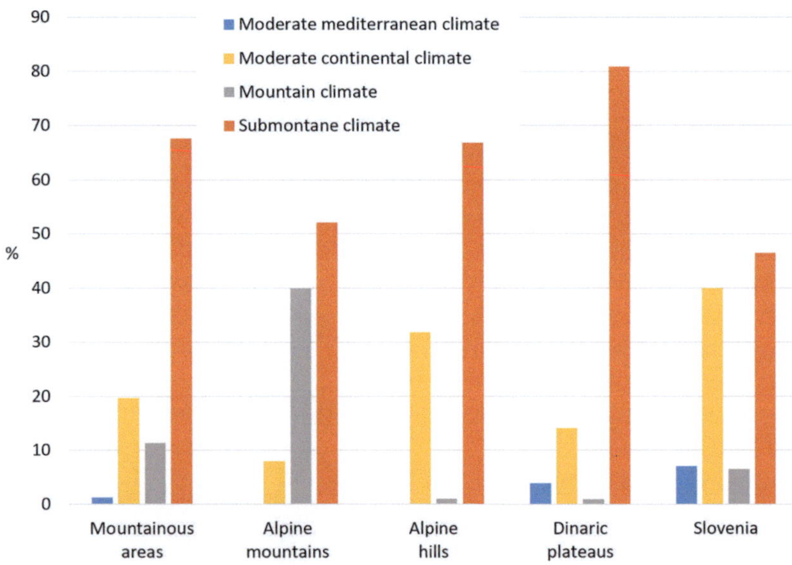

Fig. 5.16 Predominant climate types in Slovenia's mountainous areas. (Based on Table 5.3)

mountain, submontane, and moderate continental climate. Just over a tenth of the area covered by mountainous areas has a mountain climate, two-thirds has a submountain climate, and a fifth has a moderate continental climate.

With elevation, the air temperature usually decreases, precipitation increases, the duration and thickness of snow cover increases, the winds increase, and the vegetation period shortens. Therefore, altitudinal climatic–vegetation zones

Table 5.3 Climate types in Slovenia's mountainous areas. (Based on Fig. 5.15)

Climate types		**Mountainous areas**			*Alpine mountains*				*Alpine hills*				*Dinaric plateaus*				**Slovenia**	
		Area (km^2)	% of mountainous areas	% of Slovenia	Area (km^2)	% of Alpine mountains	% of mountainous areas	% of Slovenia	Area (km^2)	% of Alpine hills	% of mountainous areas	% of Slovenia	Area (km^2)	% of Dinaric plateaus	% of mountainous areas	% of Slovenia	Area (km^2)	%
Moderate mediterranean climate	Coastal	1.09	0.01	0.01	0.00	0.00	0.00	0.00	0.00	0.00	0.00	0.00	1.09	0.03	0.01	0.01	487.09	2.40
	Inland	147.91	1.28	0.73	0.00	0.00	0.00	0.00	0.00	0.00	0.00	0.00	147.91	3.88	1.28	0.73	953.01	4.70
Moderate continental climate	Northeastern Slovenia	766.25	6.65	3.78	77.29	2.52	0.67	0.38	688.97	14.79	5.98	3.40	0.00	0.00	0.00	0.00	3,823.89	**18.87**
	Eastern and southeastern Slovenia	995.99	8.64	4.91	120.38	3.93	1.04	0.59	453.25	9.73	3.93	2.24	422.35	**11.09**	3.66	2.08	2,785.88	**13.75**
	Central Slovenia	504.71	4.38	2.49	46.30	1.51	0.40	0.23	341.67	7.33	2.96	1.69	116.73	3.06	1.01	0.58	1,500.67	7.40
Mountain climate	High mountains	101.72	0.88	0.50	101.72	3.32	0.88	0.50	0.00	0.00	0.00	0.00	0.00	0.00	0.00	0.00	101.72	0.50
	Low mountains	1,212.29	**10.51**	5.98	1,122.20	**36.66**	9.73	5.54	50.45	1.08	0.44	0.25	39.63	1.04	0.34	0.20	1,212.29	5.98
Submontane climate	Very humid	5,362.12	**46.50**	**26.46**	1,546.89	**50.53**	**13.42**	7.63	1,244.67	**26.71**	**10.79**	6.14	2,570.56	**67.48**	**22.29**	**12.68**	6,013.98	**29.67**
	Humid	2,438.56	**21.15**	**12.03**	46.58	1.52	0.40	0.23	1,880.80	**40.36**	**16.31**	9.28	511.18	**13.42**	4.43	2.52	3,388.41	**16.72**
Total		11,530.64	100.00	56.89	3,061.36	100.00	26.55	15.11	4,659.81	100.00	40.41	22.99	3,809.47	100.00	33.04	18.80	20,266.95*	100.00

*The data in the data layer used do not entirely correspond to Slovenia's actual size: 20,273 km^2

Those covering more than 10% are in bold

are one of the main characteristics of mountainous areas. The **mountain climate**, typical of the Alps and the highest parts of the Pohorje Hills and the Snežnik Plateau, is the coldest and most humid climate type in Slovenia, with a long-lasting and thick snow cover, which in average winters is deeper than 150 cm. The average temperature of the coldest month is below −3 °C and annual precipitation mostly exceeds 1,600 mm. Areas with a mountain climate in western Slovenia are wetter, with annual precipitation over 2,500 mm and a precipitation maximum in the late fall. Areas in the east receive less precipitation, and summer is the wettest part of the year. Winter has the least precipitation. Due to global warming, the mountain climate zone has been shrinking rapidly over the past decades (Ogrin et al. 2023).

The **submontane climate** is typical of the foothills of the Alps and most of the Alpine hills and Dinaric karst plateaus. It is a transitional climate between the mountain and moderate continental climate on the eastern side of the Alpine–Dinaric barrier and the mountain and moderate Mediterranean climate on the southwestern side of the barrier. The average January temperatures mostly range between 0 and −3 °C, and the average July temperatures are between 16 and 20 °C. Due to the Alpine–Dinaric barrier, the climate is wetter than average, with the least precipitation in winter. Due to lower elevations and higher temperatures, snow cover is less reliable than in mountain climate. Precipitation and temperature conditions form the basis for dividing the submontane climate into very humid, found in the central, highest, coldest, and wettest part of the barrier, and humid, found in the lower and slightly warmer peripheral areas on the continental and coastal sides of the barrier (Ogrin et al. 2023).

The **moderate continental climate** is found in the low-lying areas of the Alpine hills in northeastern Slovenia and in Dry Carniola (*Suha krajina*). The highest average annual temperature amplitude of over 20 °C is typical, along with high maximum summer temperatures. Annual precipitation is below 1,400 mm. Even though the proportion of precipitation is higher in the warm half of the year, summers approach arid conditions due to the high temperatures. Frost is common in winters, along with colder periods when daily temperatures remain below zero. Spring frost is also relatively common, and summer hot spells are often interrupted by storms (including with hail and strong winds), which cause major damage to agriculture (Ogrin et al. 2023).

Over the past decades, the average air temperature in Slovenia has been rising faster than the global and Alpine average. The number of hot days is increasing, droughts are becoming increasingly more frequent, winters are becoming increasingly milder, and extreme weather events, such as heatwaves, are more pronounced and they last longer (Dolinar and Vertačnik 2010). Over the past sixty years, average annual temperatures in Slovenia's mountainous areas have increased by over 2 °C, irrespective of elevation (Table 5.4). In contrast, annual precipitation has been decreasing by as much as several hundred millimeters, or 10% to 20% (Table 5.5). A declining trend has been particularly evident in the number of days with snow cover. In the high mountains, at an elevation of about 2,500 m, the changes have not (yet) been so significant (the station on Mount Kredarica), whereas at elevations close to 1,000 m the number of days with snow cover has already declined considerably, by even more than forty days or approximately 30% a year. At even lower elevations, the decline has been even greater, exceeding fifty days or 50% a year (Table 5.6).

Due to the increasing air temperatures and diminishing snow cover, the discharges of Slovenia's mountain rivers have also decreased significantly (Sect. 5.5), and their discharge regimes have also changed considerably (Kobold and Ulaga 2010; Hrvatin and Zorn 2017, 2020, 2024). Because of the diminished snow cover, water supplies will have decreased by at least a third by the end of the twenty-first century (Dolinar and Vertačnik 2010).

5.5 Hydrology

Water conditions in Slovenia's mountainous areas largely depend on the rock composition of individual landscapes. The river system comprises a multitude of rivers, creeks, and torrents with a total

Table 5.4 Changes in average annual temperature in mountainous areas, 1961–2020 (Hrvatin and Zorn 2017, 2020, 2024; Arhiv ... 2024b)

Region (ID on Fig. 5.1)	Temperature station	Elevation (m)	1961 trend value (°C)	2020 trend value (°C)	1961–2020 trend difference (°C)	1961–2020 trend difference (%)*
Julian Alps (1.1)	**Kredarica**	2,513	−2.26	−0.04	**2.22**	**0.82**
Kamnik–Savinja Alps (1.2)	**Krvavec**	1,742	2.59	4.36	**1.77**	**0.64**
Western Karawanks (1.3)	**Rateče**	864	5.19	7.65	**2.46**	**0.88**
Polhov Gradec Hills (2.1)	**Topol pri Medvodah**	662	7.87	10.48	**2.61**	**0.93**
Sava Hills (2.2)	**Lisca**	947	6.06	8.62	**2.56**	**0.92**
Pohorje Hills (2.3)	**Šmartno pri Slovenj Gradcu**	444	7.06	9.67	**2.61**	**0.93**
Banjšice Plateau (3.1)	**Čepovan**	604	7.88	10.16	**2.28**	**0.81**
Idrija Hills (3.2)	**Vojsko**	1,070	5.19	7.70	**2.51**	**0.90**
Nanos Plateau (3.3)	**Nanos**	896	6.43	9.00	**2.57**	**0.92**
Bloke Plateau (3.6)	**Nova Vas**	720	6.23	8.94	**2.71**	**0.97**

*The trend difference as a percentage is calculated based on the absolute (Kelvin) temperature scale

Table 5.5 Changes in annual precipitation in mountainous areas, 1961–2020 (Hrvatin and Zorn 2020, 2022, 2024; Arhiv ... 2024b)

Region (ID on Fig. 5.1)	Precipitation station	Elevation (m)	1961 trend value (mm)	2020 trend value (mm)	1961–2020 trend difference (mm)	1961–2020 trend difference (%)
Julian Alps (1.1)	**Kredarica**	2,513	1,940.31	2,204.00	**263.69**	**13.59**
Kamnik–Savinja Alps (1.2)	**Solčava**	639	1,627.07	1,506.89	**−120.18**	**−7.39**
Western Karawanks (1.3)	**Javorniški Rovt**	939	1,949.68	1,939.68	**−10.00**	**−0.51**
Eastern Karawanks (1.4)	**Podpeca**	942	1,436.68	1,494.32	**57.64**	**4.01**
Sava Hills (2.2)	**Kal pri Krmelju**	509	1,228.33	1,133.16	**−95.17**	**−7.75**
Pohorje Hills (2.3)	**Mislinja**	622	1,248.93	1,189.11	**−59.82**	**−4.79**
Strojna Hills (2.3)	**Strojna**	940	1,119.74	1,042.30	**−77.44**	**−6.92**
Idrija Hills (3.2)	**Vojsko**	1,070	2,540.49	2,228.16	**−312.33**	**−12.29**
Trnovo Forest Plateau (3.3)	**Lokve**	946	2,546.21	2,275.60	**−270.61**	**−10.63**
Hrušica Plateau (3.3)	**Hrušica**	872	2,222.23	1,776.49	**−445.75**	**−20.06**
Menišija Plateau (3.5)	**Pokojišče**	716	1,656.06	1,387.09	**−268.97**	**−16.24**
Bloke Plateau (3.6)	**Nova Vas**	720	1,482.03	1,535.02	**52.99**	**3.58**
Big Mountain Plateau (3.7)	**Grčarice**	782	1,731.75	1,796.95	**65.20**	**3.76**
Gotenica Plateau (3.7)	**Trava**	520	1,582.08	1,552.43	**−29.65**	**−1.87**
Poljane Plateau (3.9)	**Sinji Vrh**	365	1,361.52	1,407.54	**46.02**	**3.38**

Table 5.6 Changes in the number of days with snow cover in mountainous areas, 1961–2020 (Hrvatin and Zorn 2020, 2022, 2024; Arhiv ... 2024b)

Region (ID on Fig. 5.1)	Precipitation station	Elevation (m)	1961 trend value (days)	2020 trend value (days)	1961–2020 trend difference (days)	1961–2020 trend difference (%)
Julian Alps (1.1)	**Kredarica**	2,513	267.7	259.5	**−8.14**	**−3.04**
Kamnik–Savinja Alps (1.2)	**Solčava**	639	106.2	43.0	**−63.21**	**−59.52**
Western Karawanks (1.3)	**Javorniški Rovt**	939	135.3	83.8	**−51.44**	**−38.02**
Eastern Karawanks (1.4)	**Podpeca**	942	125.0	77.8	**−47.20**	**−37.76**
Sava Hills (2.2)	**Kal pri Krmelju**	509	59.3	31.2	**−28.11**	**−47.40**
Pohorje Hills (2.3)	**Mislinja**	622	111.0	66.3	**−44.70**	**−40.25**
Strojna Hills (2.3)	**Strojna**	940	126.4	59.8	**−66.56**	**−52.67**
Idrija Hills (3.2)	**Vojsko**	1,070	153.0	94.0	**−59.00**	**−38.56**
Trnovo Forest Plateau (3.3)	**Lokve**	946	111.8	48.9	**−62.93**	**−56.29**
Hrušica Plateau (3.3)	**Hrušica**	872	95.6	60.2	**−35.40**	**−37.03**
Menišija Plateau (3.5)	**Pokojišče**	716	71.5	30.2	**−41.30**	**−57.76**
Bloke Plateau (3.6)	**Nova Vas**	720	107.2	62.8	**−44.41**	**−41.41**
Big Mountain Plateau (3.7)	**Grčarice**	782	91.1	52.7	**−38.35**	**−42.12**
Gotenica Plateau (3.7)	**Trava**	520	111.5	72.8	**−38.72**	**−34.72**
Poljane Plateau (3.9)	**Sinji Vrh**	365	76.5	39.0	**−37.55**	**−49.05**

length of nearly 15,300 km. The average density is 1.32 km of watercourses per km^2, which is only slightly lower than the Slovenian average. Surface streams are not evenly distributed, because nearly two-thirds of the mountainous areas are karst landscapes and nearly without surface streams. The density of streams is especially low (i.e., below 0.05 km per km^2) on some karst plateaus, such as the Javorniki (Fig. 5.4), Snežnik, Kočevski Rog, Poljane, and Dry Carniola plateaus. Outside karst landscapes, the river system is highly branched and, in some areas of the Alpine hills (e.g., the Cerkno, Polhov Gradec, Rovte, Pohorje, Strojna, and Kozjak hills), the density of streams exceeds 2.25 km per km^2.

The divide between the Black Sea and the Adriatic runs across Slovenia. Rivers drain to the Black Sea several hundred kilometers from four-fifths of Slovenia's mountainous areas, and to the nearby Adriatic from only one-fifth of this areas. The Black Sea watershed comprises the Drava, Sava, and Kolpa basins, and the Adriatic watershed includes the Soča and Reka basins.

The mountainous **river system** is characterized by short watercourses, with only seventeen rivers longer than 20 km. The Soča River is the longest, measuring 91 km in the mountainous part of Slovenia, followed by the Savinja (70 km) and Drava (67 km); the only two other rivers longer than 50 km in this part of Slovenia are the Sava (61 km) and Idrijca (58 km). In terms of water volume, the Drava is in first place, with an average annual discharge of 262 m^3/s at Dravograd, where it enters Slovenia from Austria. Lagging only slightly behind is the Sava at Radeče (230 m^3/s), and the discharge of the Soča at Solkan (94 m^3/s) is already less than half that. An average annual discharge of over 50 m^3/s is also typical of the Kolpa River, the Savinja River has a discharge of over 40 m^3/s, and the Ljubljanica and Idrijca over 20 m^3/s.

The runoff coefficients, which show the percentage of water flowing into rivers, vary considerably across the mountainous areas. The differences are primarily due to precipitation volume and regime, and rock and topographic characteristics. The highest values (i.e., over 70%) can be observed in the Upper Soča and Bohinj valleys (the Julian Alps). Runoff coefficients between 60 and 70% are typical of the rest of the Alpine mountains and the western parts of the Alpine hills and Dinaric plateaus. Values between 45 and 60% predominate in the eastern areas of the Alpine hills and Dinaric plateaus (Frantar 2008).

Even a few decades ago, most mountain rivers still had a very pronounced snow–rain discharge regime. Due to snow retention, the minimum discharges were in winter, and the primary discharge maximum was in spring when the snow melted. A secondary discharge minimum was in summer, due to stronger evapotranspiration, followed by a secondary discharge maximum in the fall resulting from heavy rain (Hrvatin 1998). The past decades have seen notable decreases in the discharge and gradual changes in the discharge regimes of all mountain rivers. The reasons for this include higher average temperatures, which accelerate evaporation, a smaller volume of annual precipitation, a shorter duration and smaller amount of snow cover, and an increase in the share of wooded areas. In nearly all rivers, the fall discharge maximum has already exceeded the spring maximum, and the summer discharge minimum is already very close to the winter minimum. Because of the shorter duration and smaller amount of snow cover, and stronger evapotranspiration, there has been a decrease in the late spring and early summer discharges, along with an increase in discharges between October and December, which indicates that winter is "running late" (Hrvatin and Zorn 2017, 2020, 2024).

For example, changes in the discharge regimes from 1961 to 1990 and from 1991 to 2020 (Arhiv ... 2024a) can be clearly seen on the Meža River (Fig. 5.17). Its November rain maximum has already significantly exceeded its spring high-flow conditions, which are primarily the result of snow melting in the mountains and hills. At the same time, due to summer droughts, its August minimum has already exceeded its winter minimum, which is the result of snow retention.

Natural **lakes** are rare and small in Slovenia's mountainous areas. In terms of origin, they can be divided into glacial and karst lakes. The largest and best-known glacial lakes are Lake Bohinj and Lake Bled (Fig. 5.18). Lake Bohinj covers an area

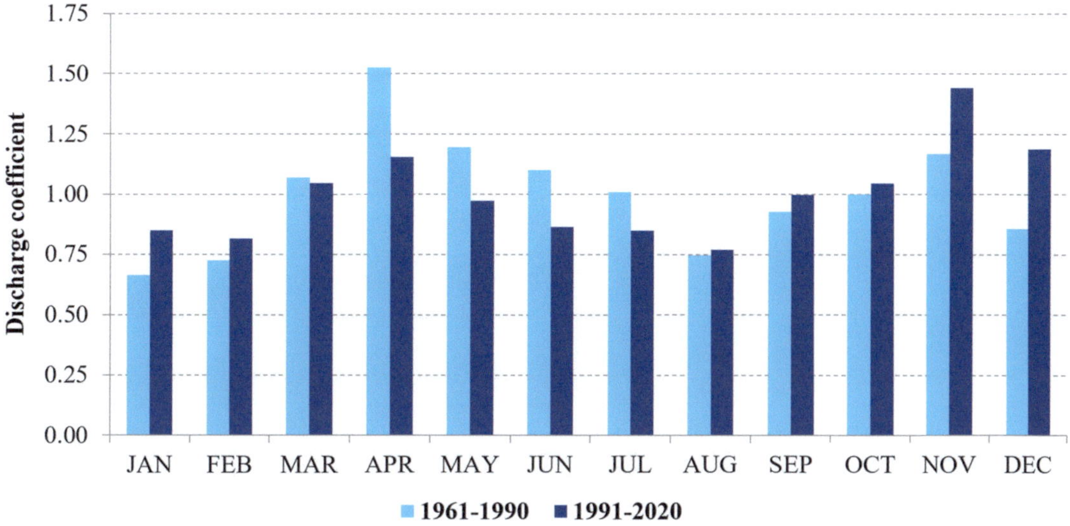

Fig. 5.17 Changes in the discharge regime of the Meža River at the Otiški Vrh gauging station, 1961–1990 and 1991–2020

of 3.18 km^2 and has a maximum depth of 44.5 m. It is a drainage lake with a karst catchment area. Its main tributary is the Savica River, which collects karst groundwater from the Triglav Lakes Valley area (the Julian Alps). The lake's watershed is part of Triglav National Park and almost completely unpopulated, and so the lake is in very good ecological condition (Remec-Rekar and Bat 2004; Remec-Rekar 2016).

Lake Bled is smaller (area: 1.40 km^2, maximum depth: 30.6 m). It is without a major freshwater inflow, and its catchment is considerably urbanized, burdened by traffic, and exploited for agriculture, fishing, and tourism. Eutrophication advanced heavily in the second half of the twentieth century, but the lake's ecological condition was significantly improved though various measures (Remec-Rekar and Bat 2004; Remec-Rekar 2016).

Both lakes are popular tourist destinations. The mean monthly temperature of Lake Bled in summer exceeds 22 °C, and the temperature of Lake Bohinj is 2 to 3 °C lower (Bat 1997). There are several other small glacial lakes in the high mountains of the Julian Alps, the best known among them being the Triglav Lakes, Križ Lakes (*Kriška jezera*), and Krn Lakes (*Krnska jezera*) (Brancelj 2002).

Marshes are also rare and small, just like lakes, and they are categorized as wetlands with permanent or intermittent water retention. They are overgrown with hygrophilous plants, which, due to the shortage of air, do not decompose into humus, but turn into peat in the process of coalification. In their original form, peat bogs now only remain on the Pokljuka Plateau (the Šijec, Veliko Blejsko barje, and Goreljek bogs) and Jelovica Plateau (the Ledina and Za blatom bogs) in the Julian Alps, and in the Pohorje Hills (the Lovrenc and Ribnica bogs; Bat 1997; Beltram 2004; Vreš et al. 2014).

Snowfields and glaciers are important but very changeable elements of high-mountain landscapes. A comparison between their condition as reported in various historical sources and the current situation shows that the number and size of once frequent and large mountain snowfields are decreasing significantly (Triglav Čekada and Ajdova 2023). The Triglav Glacier (Fig. 5.12) and Skuta Glacier mentioned above have also been shrinking rapidly over the past decades (Gabrovec et al. 2014).

Groundwater, which in Slovenia's mountainous areas primarily includes karst water, is significantly greater in volume than surface water. Karst water feeds the sources of the Soča, Sava, Idrijca, Vipava, Ljubljanica, Krka, and

Fig. 5.18 The largest and best-known glacial lakes in Slovenia are Lake Bled (top; Western Karawanks with their highest peak, Mount Stol (2,236 m), background) and Lake Bohinj (bottom). (Photo by Matija Zorn)

Kolpa rivers and many of their tributaries. Some karst springs, such as Malenščica, Podroteja, and Mrzlek springs, are important for the water supply of entire regions, supplying three-quarters of all available water reserves during droughts (Kranjc 1998).

5.6 Soil and Vegetation

In Slovenia's mountainous areas, the impact of pedogenesis factors, including rock composition, climate, topography, water, plant and animal species, and human factors, changes greatly over short distances, which is why the **soil types** are relatively diverse (Figs. 5.19 and 5.20, Tables 5.7 and 5.8). The most common types are Rendzina and Chromic Cambisol, which formed on limestone and dolomite, and together they cover over half of mountainous areas, or even over four-fifths of Dinaric plateaus. On steep slopes, they are mostly overgrown with forest, and on flat, non-rocky surfaces they are covered in meadows. Because the soil is shallow, fields are rare (Vidic et al. 2015; Vrščaj 2017). Dystric Cambisol predominates on non-carbonate rocks; it covers just under a third of mountainous areas, but over half of Alpine hills.

Forests are the **natural vegetation** in all of Slovenia's mountainous areas except in the high mountains above the alpine forest line and in very wet habitats. Over the centuries, people burned and cut down forests to obtain space for pastures and fields, and by using forest resources they also radically changed the forest composition. Only a few fully natural forests remain today, such as the highly protected forest reserves in the Kočevje region and remote mountains, but most

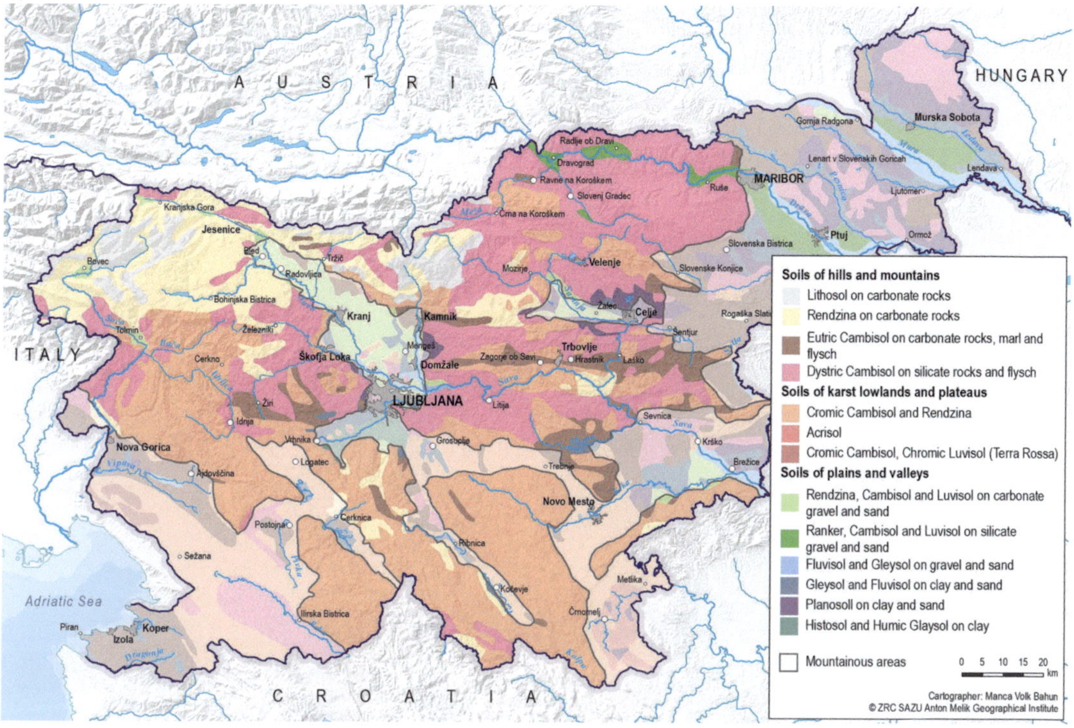

Fig. 5.19 Pedogeographical soil types in Slovenia's mountainous areas. (Adapted from Lovrenčak 1998)

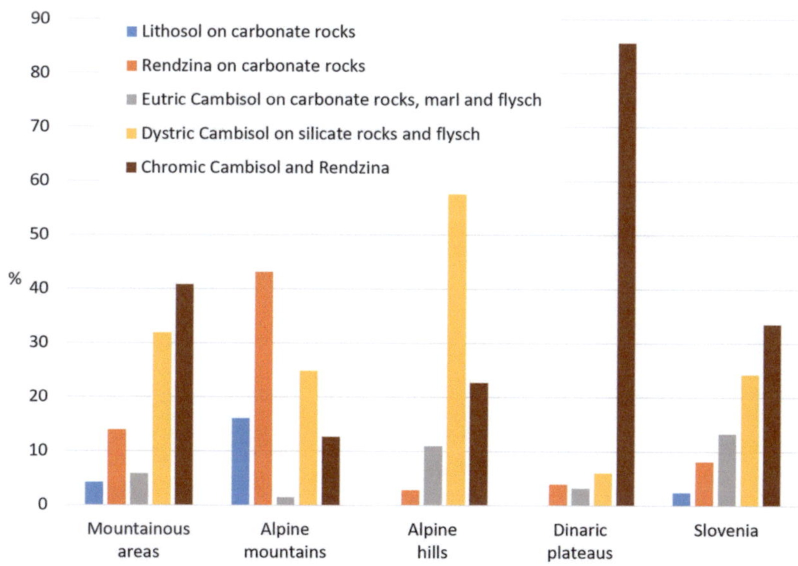

Fig. 5.20 Predominant pedogeographical soil types in Slovenia's mountainous areas. (Based on Table 5.7)

Table 5.7 Pedogeographical soil types in Slovenia's mountainous areas. (Based on Fig. 5.19)

Soil type		Mountainous areas			*Alpine mountains*				*Alpine hills*				*Dinaric plateaus*				Slovenia	
		Area (km^2)	% of mountainous areas	% of Slovenia	Area (km^2)	% of Alpine mountains	% of mountainous areas	% of Slovenia	Area (km^2)	% of Alpine hills	% of mountainous areas	% of Slovenia	Area (km^2)	% of Dinaric plateaus	% of mountainous areas	% of Slovenia	Area (km^2)	%
Soils of plains and valleys	Rendzina, Cambisol, and Luvisol on carbonate gravel and sand	89.26	0.77	0.44	51.08	1.67	0.44	0.25	26.35	0.57	0.23	0.13	11.83	0.31	0.10	0.06	640.57	3.16
	Ranker, Cambisol, and Luvisol on silicate gravel and sand	107.28	0.93	0.53	0.01	0.00	0.00	0.00	107.26	2.30	0.93	0.53	0.00	0.00	0.00	0.00	443.39	2.19
	Fluvisol and Gleysol on gravel and sand	12.13	0.11	0.06	0.00	0.00	0.00	0.00	11.25	0.24	0.10	0.06	0.89	0.02	0.01	0.00	437.54	2.16
	Gleysol and Fluvisol on clay and sand	49.44	0.43	0.24	0.01	0.00	0.00	0.00	27.41	0.59	0.24	0.14	22.02	0.58	0.19	0.11	805.50	3.97
	Planosol on clay and sand	102.17	0.89	0.50	4.06	0.13	0.04	0.02	98.12	2.11	0.85	0.48	0.00	0.00	0.00	0.00	1,030.26	5.08
	Histosol and Humic Gleysol on clay	6.00	0.05	0.03	0.00	0.00	0.00	0.00	0.00	0.00	0.00	0.00	6.00	0.16	0.05	0.03	105.09	0.52

(continued)

Table 5.7 (continued)

Soil type		Mountainous areas			*Alpine mountains*				*Alpine hills*				*Dinaric plateaus*				Slovenia	
		Area (km^2)	% of mountainous areas	% of Slovenia	Area (km^2)	% of Alpine mountains	% of mountainous areas	% of Slovenia	Area (km^2)	% of Alpine hills	% of mountainous areas	% of Slovenia	Area (km^2)	% of Dinaric plateaus	% of mountainous areas	% of Slovenia	Area (km^2)	%
Soils of hills and mountains	Lithosol on carbonate rocks	491.81	4.26	2.43	491.81	**16.06**	4.26	2.43	0.00	0.00	0.00	0.00	0.00	0.00	0.00	0.00	492.05	2.43
	Rendzina on carbonate rocks	1,605.10	**13.92**	7.92	1,321.86	**43.18**	**11.46**	6.52	133.36	2.86	1.16	0.66	149.88	3.93	1.30	0.74	1,656.57	8.17
	Eutric Cambisol on carbonate rocks, marl, and flysch	680.49	5.90	3.36	43.89	1.43	0.38	0.22	512.99	**11.01**	4.45	2.53	123.61	3.24	1.07	0.61	2,705.96	**13.35**
	Dystric Cambisol on silicate rocks and flysch	3,679.88	**31.91**	**18.15**	762.13	**24.89**	6.61	3.76	2,686.00	**57.64**	**23.29**	**13.25**	231.75	6.08	2.01	1.14	4,914.75	**24.25**
Soils of karst plains and plateaus	Chromic Cambisol and Rendzina	4,708.22	**40.83**	**23.23**	386.60	**12.63**	3.35	1.91	1,057.59	**22.69**	9.17	5.22	3,264.04	**85.67**	**28.30**	**16.10**	6,814.36	**33.62**
	Acrisol	0.00	0.00	0.00	0.00	0.00	0.00	0.00	0.00	0.00	0.00	0.00	0.00	0.00	0.00	0.00	155.45	0.77
	Chromic Cambisol, Chromic Luvisol (Terra Rossa)	0.00	0.00	0.00	0.00	0.00	0.00	0.00	0.00	0.00	0.00	0.00	0.00	0.00	0.00	0.00	69.09	0.34
Total		11,531.79	100.00	56.89	3,061.45	100.00	26.55	15.10	4,660.32	100.00	40.41	22.99	3,810.02	100.00	33.04	18.80	20,270.59*	100.00

*The data in the data layer used do not entirely correspond to Slovenia's actual size: 20,273 km^2
Those covering more than 10% are in bold

Table 5.8 Basic characteristics of the main pedogeographical soil types in Slovenia's mountainous areas. (Adapted from Vidic et al. 2015; Zorn et al. 2020a, b)

Groups of pedogeographical soil types	ID on Fig. 5.19	Pedogeographical soil type	Basic characteristics
Soils of hills and mountains	Lithosol on carbonate rock	Lithosol/Lithic Leptosols (Calcaric)	**Horizons**: (A)–R **Texture**: — **pH**: 7–8 **Parent material**: limestone and dolomite **Land use**: not suitable for agriculture and forestry **Comments**: incipient soil development
	Rendzina on carbonate rock	Rendzina/Rendzic Leptosols	**Horizons**: Ah–R **Texture**: silty loam **pH**: 5.5–7.0 **Parent material**: limestone and dolomite **Land use**: forest, alpine pasture **Comments**: grassland above the forest line
	Eutric Cambisol on carbonate rock, marl, and flysch	Eutric Cambisol	**Horizons**: A–Bv–C **Texture**: silty clay, clay **pH**: 5.5–6.5 **Parent material**: marly limestone **Land use**: arable land, orchards, vineyards, meadows, forest **Comments**: limitation in soil workability can be expected due to high clay content **Horizons**: A–Bv–C, A–BCa–C **Texture**: silty clay loam, silty clay **pH**: 5.5–8 **Parent material**: Eocene flysch **Land use**: vineyards, orchards, arable land, meadows, forest **Comments**: land use adapted to Mediterranean and sub-Mediterranean conditions
	Dystric Cambisol on silicate rocks and flysch	Dystric Cambisol	**Horizons**: A–Bv–C **Texture**: sandy loam **pH**: 4–5 **Parent material**: igneous rocks **Land use**: forest, grassland, arable land **Comments**: arable land limited due to slope conditions, liming required **Horizons**: A–Bv–C **Texture**: silty clay loam, silty clay **pH**: 4.5–5.5 **Parent material**: non-calcareous and low calcareous flysch and decalcificated marlstone **Land use**: grassland, arable land, forest **Comments**: limitation in soil workability can be expected due to texture, liming required
Soils of karst plains and plateaus	Chromic Cambisol and Rendzina	Chromic Cambisol and Rendzina/ Chromic Cambisol and Rendzic Leptosols	**Horizons**: Ah–C, A–Brz–C **Texture**: silty loam, silty clay loam, silty clay **pH**: 5.5–7 **Parent material**: limestone and dolomite **Land use**: forest, meadows **Comments**: possible surface rockiness, possible soil water deficiency

(continued)

Table 5.8 (continued)

Groups of pedogeographical soil types	ID on Fig. 5.19	Pedogeographical soil type	Basic characteristics
	Acrisol	Acrisol	**Horizons**: A–E–Bt–C **Texture**: silt, silty clay loam **pH**: 3.5–4.2 **Parent material**: limestone **Land use**: forest, grassland, arable land **Comments**: abundant liming is required as well as selection of acid-tolerant plants
	Chromic Cambisol, Chromic Luvisol (Terra Rossa)	Chromic Cambisol (Terra Rossa)	**Horizons**: A–Brz–C, A–E–Bt–C **Texture**: silty loam, clay loam **pH**: 4.5–6 **Parent material**: limestone **Land use**: vineyards, grassland, forest **Comments**: limitations in vineyard land use due to the variable solum thickness **Horizons**: A–Brz–C **Texture**: clay loam, silty clay loam **pH**: 5–6 **Parent material**: limestone with chert **Land use**: vineyards, grassland, forest **Comments**: limitations in vineyard land use due to the variable solum thickness and the amount of skeleton
		Luvisol	**Horizons**: A–E–Bt–C **Texture**: silty loam, silty clay loam **pH**: 4–5.5 **Parent material**: limestone and dolomite **Land use**: grassland, forest, arable land **Comments**: abundant liming is required, tillage limitations due to variable solum thickness

other forests are also well preserved (Natek and Natek 2008).

In 2020, forests covered three-quarters of Slovenia's mountainous areas (Grafični ... 2020). They account for 83.3% of Dinaric plateaus and 74.5% of Alpine mountains, and the smallest share (68.1%) can be found in the somewhat more populated Alpine hills.

The predominant tree species in Slovenian forests is beech. At elevations between 500 and 600 m, it forms a wide variety of forest communities. Submontane beech forests combined with European hornbeam, sessile oak, and linden predominate on carbonate rocks up to elevations of 700 m and, above them, montane beech forests mixed with sycamore, elm, and maple extend up to elevations of 900 m. Mountain beech forests, which may already include spruce and fir in addition to deciduous trees, extend to 1,400 to 1,500 m in elevation, and the highest elevations above 1,500 m are covered in subalpine beech forests. On non-carbonate rock, thermophilic beech forests mixed with European hop-hornbeam, manna ash, and whitebeam predominate on sunny slopes, whereas elsewhere acidophilic beech forests combined with sessile oak and sweet chestnut and, in shady locations, spruce are the most common (Marinček and Čarni 2002; Figs. 5.21 and 5.22, Table 5.9).

Communities of Dinaric fir and beech forests, which contain large volumes of timber-growing stock and are among Slovenia's most important commercial forests, stand out among the mixed forests (Kordiš 1993).

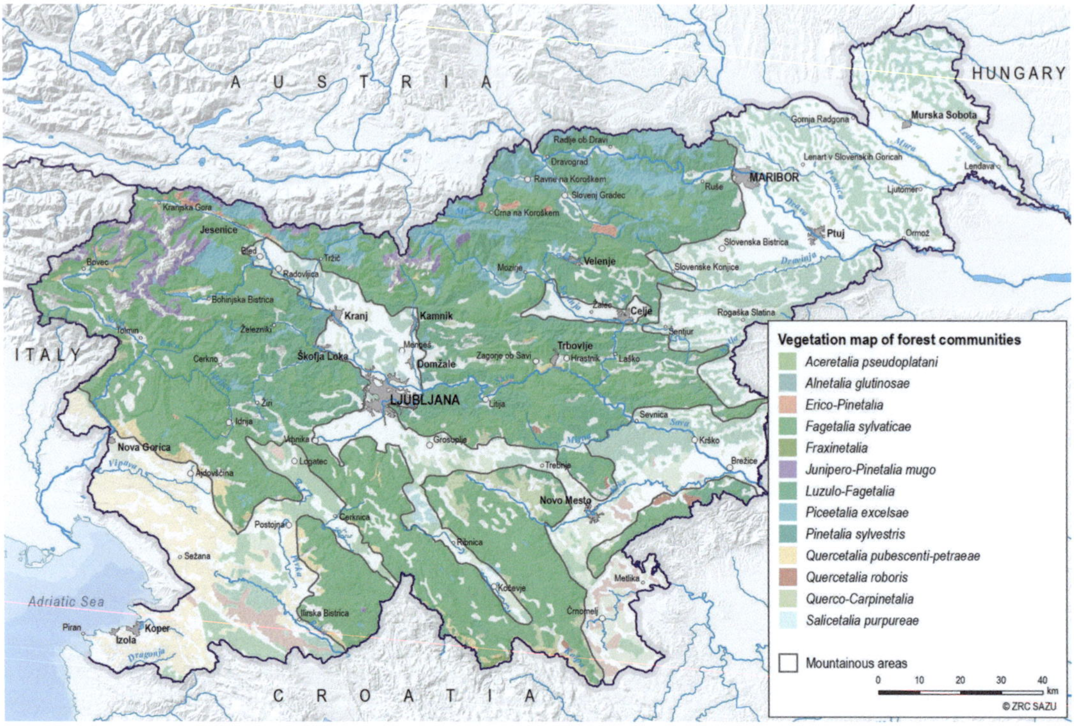

Fig. 5.21 Forest communities in Slovenia's mountainous areas. (Adapted from Čarni et al. 2002; Šilc et al. 2020)

Most Slovenian spruce forests are secondary forests growing at the site of former acidophilic beech forests. After beech forests were cleared, pastures were first arranged at their site in the Alpine hills and at lower elevations in the Alpine mountains. These were later largely reforested with spruce. The only natural spruce forests are the acidophilic spruce forests at higher Alpine elevations (between 1,400 and 1,600 m) and in the cold air pools of high Dinaric plateaus. In Alpine Slovenia, spruce forests extend up to the alpine forest line and, as elevation increases, they also include more larch (Kaligarič 2004).

Low temperatures, short vegetation periods, strong and persistent winds, and long-lasting snow cover prevent tree growth in the highest parts of the Alpine mountains and Dinaric plateaus. Because of these factors and anthropogenic changes, the alpine forest line in Slovenia varies between 1,550 and 1,900 m. A belt of mountain pine grows above this line, followed by highly diverse communities of high-mountain vegetation at even higher elevations, which becomes increasingly sparse and discontinuous with elevation (Lovrenčak 2003).

5.7 Settlement and Land Use

Archeological research shows that humans have been present in the Slovenian mountains for a very long time. Tens of thousands of years ago, during the Paleolithic, hunters and gatherers found shelter in caves in the mountains, such as Potočka zijalka (1,675 m) on Mount Olševa, Medvedova jama (1,525 m) on Mount Mokrica (both in the Kamnik–Savinja Alps), or Divje babe (448 m) in the Idrija Hills. The finds of weapons from the Bronze Age are the first accumulated evidence of people visiting the Slovenian high mountains (Cevc 2006). Moreover, the finds indicate that routes over mountain passes were used since ancient times; for example, the passes in the Karawanks or those in the Julian Alps connecting the Soča and Bohinj

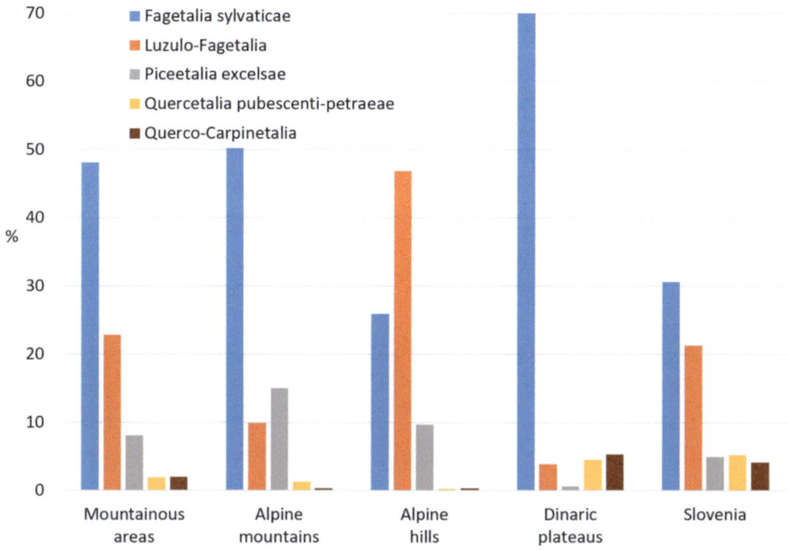

Fig. 5.22 Predominant forest communities in Slovenia's mountainous areas. (Based on Table 5.9)

Table 5.9 Predominant forest communities in Slovenia's mountainous areas (in %). (Based on Fig. 5.21)

Forest community	Mountainous areas	*Alpine mountains*	*Alpine hills*	*Dinaric plateaus*	Slovenia
Fagetalia sylvaticae (beech forests on nutrient-rich soil)	**48.10**	**50.73**	**25.98**	**73.05**	**30.63**
Luzulo-Fagetalia (acidophilic beech forests)	**22.83**	9.89	**46.87**	3.82	**21.29**
Piceetalia excelsae (mountain spruce and fir forests)	8.10	**15.03**	9.70	0.59	4.91
Quercetalia pubescenti-petraeae (thermophilic deciduous forests)	1.92	1.25	0.24	4.51	5.25
Querco–Carpinetalia (oak and hornbeam forests)	1.97	0.33	0.35	5.27	4.15

Those covering more than 10% are in bold

valleys (Horvat 2013). These routes became especially important with the expansion of trade in the Middle Ages; for example, the routes over the Ljubelj (Loibl) Pass (1,370 m), Jezerski Vrh (Seeberg) Saddle (1,218 m), and Koren (Wurzen) Pass (1,073 m) in the Karawanks (today all on the border between Slovenia and Austria), or over the Predel (Predil) Pass (1,156 m) in the Julian Alps (today on the border between Slovenia and Italy; Kosi 1998; Mikša and Zorn 2016). The development of road network also had an important impact on settlement pattern (Senegačnik 2012).

Archeological finds also show that during Roman times the Slovenian high mountains were already used for grazing. The number of herdsmen's huts increased during the decline of the Roman Empire, when people withdrew from the exposed areas in the valleys to the mountains. The presence of Alpine dairy farming proper was already reported from the High Middle Ages onward (Horvat 2013).

The first-stage colonization of higher elevations between the thirteenth and fifteenth centuries, resulting from the agrarian overpopulation of flatland areas, and the second-stage colonization that took place in the sixteenth century were key to the **settlement** of Slovenia's mountainous areas. The former affected hilly and wooded areas, where the number of settlements was even larger than today in some places, and the latter affected partly the extensive wooded areas and partly the surroundings of old settlements, completing the scale of today's settlement. In the late Middle Ages, colonization was intense up to elevations of 1,000 m and, in the eastern part of the mountainous areas, even up to 1,300 m. In the Alpine mountains, people usually settled the valleys, whereas in the Alpine hills they also settled the ridges and slopes. Forestry and animal husbandry were the main economic activities. Most alpine pastures were also created during that period; there were over six hundred of them, according to various sources (Grafenauer 1991; Kosi 1997; Kladnik 1998a; Senegačnik 2012). The numerous iron ore deposits in the eastern Julian Alps and certain parts of the Karawanks led to the expansion of ironworking as early as the sixteenth century. The industry died away by the end of the nineteenth century. Today, the mountainous areas play an important role in tourism (Fig. 5.23) and recreation. Animal husbandry is still a significant economic activity in these areas, along with forestry-related activities, especially on the Dinaric plateaus (Kladnik 1998b; Senegačnik 2012).

The upper limit of settlement in Slovenia's mountainous areas varies (Fig. 5.24). The elevation of permanently settled areas is the highest in the Alpine mountains, extending above 1,300 m in the Eastern Karawanks between Mounts Raduha, Olševa, and Peca. The highest farm in Slovenia, the Bukovnik (Bukovc) Farm, stands below Mount Raduha (2,062 m), at an elevation of 1,327 m. The Kamnik–Savinja Alps are settled above an elevation of 1,000 m, whereas the Western Karawanks and the eastern Julian Alps are settled up to an elevation of 1,000 m (Fig. 5.25). Settlements above the Selška Sora Valley on the southern edges of the Julian Alps, on the border with the Škofja Loka Hills, can be found above 1,000 m (Fig. 5.25). Zgornje Danje is among the highest settlements in Slovenia (1,120 m). In the Alpine hills, the area of the Pohorje, Strojna, and Kozjak is settled up to a maximum elevation of approximately 1,200 m on Mount Košenjak (1,522 m) in the northwest, and below 900 m in the eastern Pohorje Hills. In the Velenje and Konjice Hills, settlements reach a maximum elevation of approximately 1,000 m, and in the Sava Hills they extend up to approximately 800 m. In the Škofja Loka and Cerkno Hills, farms can be found at elevations between 900 and 1,000 m, or even higher in the Davča area (1,100 m). On the Dinaric plateaus, such as the Banjšice and Trnovo Forest plateaus, the highest settlements are located at 900 m and 1,000 m, and in the Idrija Hills they can even be found above 1,000 m (e.g., Vojsko, which stands at 1,077 m). The highest settlements on the Dinaric plateaus of southern Slovenia are at elevations between 800 and 900 m (Gams 1960).

In terms of population, the Alpine hills stand out among Slovenia's mountainous areas, with a population density of seventy-six people per km^2 (Table 5.10), which is below Slovenia's average (approximately hundred people per km^2), but significantly higher than the population density on the sparsely populated Dinaric plateaus (seventeen people per km^2) and in the Alpine mountains (thirty people per km^2). Dinaric plateaus also lag far behind in terms of settlement size, with settlements having an average of eighty residents, in contrast to settlements in the Alpine hills, which have an average population of over two hundred, and those in the Alpine mountains with an average population of just under three hundred. The higher population density of Alpine hills may relate to their central locations and the consequent development of main roads and industrialization (Senegačnik 2012). Over the past century, there has been a significant decline in population especially on the Dinaric plateaus, whereas in the Alpine hills the population has increased (Table 5.10).

Fig. 5.23 Slovenia's mountainous areas have an important tourism function, but some places are already overburdened by tourism, such as the Rogla ski resort (approximately 1,500 m) in the Pohorje Hills in summer. (Photo by Matija Zorn)

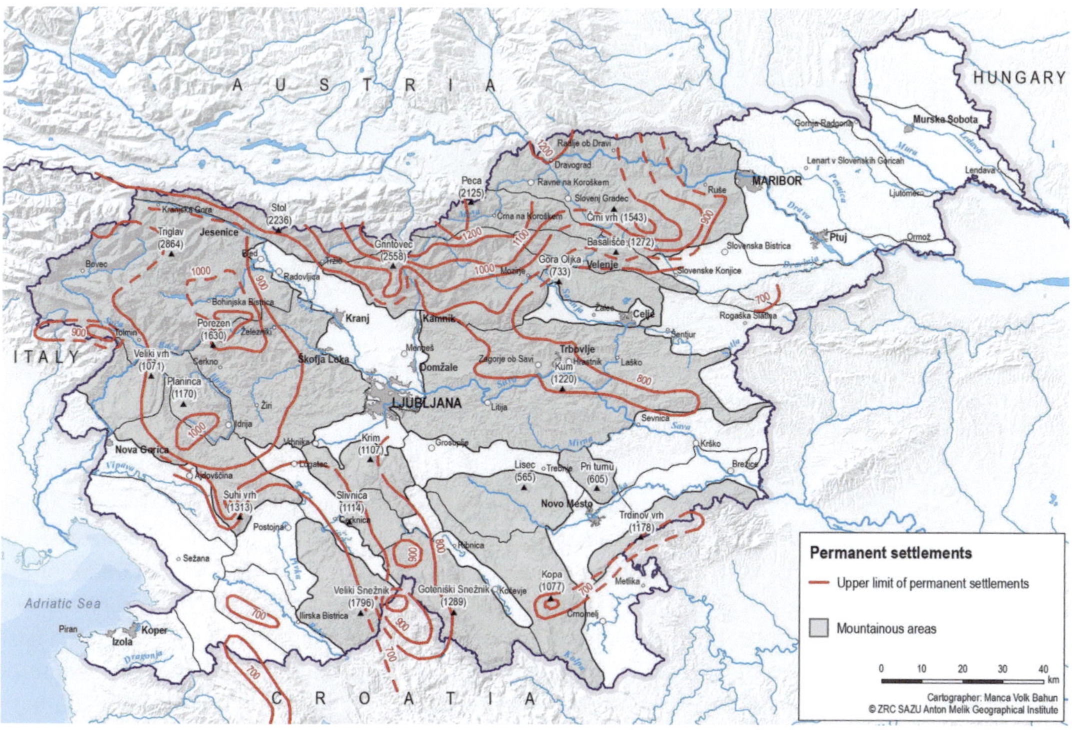

Fig. 5.24 Upper limit of settlement. (Adapted from Gams 1960)

Fig. 5.25 The maximum elevation of Gorjuše (top) on the Pokljuka Plateau in the Julian Alps is 1,080 m, and that of Ravne (bottom) in the Selška Sora Valley on the southern edges of the Juian Alps, on the border with the Škofja Loka Hills, is 1,060 m. (Photo by Matija Zorn)

In terms of **land use**, forests predominate heavily in Slovenia's mountainous areas. They cover over four-fifths of Dinaric plateaus and, on some of them, even over 90% of land. They cover three-quarters of Alpine mountains, and they account for the smallest share in the Alpine hills (they cover just over two-thirds of land), although that is still above Slovenia's average (just over three-fifths of the country). The presence of fields is greatly below Slovenia's average (i.e., just under a tenth of land), accounting for only 0.5% in Alpine mountains, 1.3% on Dinaric plateaus, and 3.1% in Alpine hills. The share of built-up areas in Slovenia's mountainous areas is also significantly below average, especially in Alpine mountains and on Dinaric plateaus. These two regions also have a below-average share of orchards, in contrast to Alpine hills, which have an above-average share. There are almost no vineyards in Slovenia's mountainous areas (Table 5.11 and Fig. 5.26).

Table 5.10 Basic information on the settlement of mountainous areas (Orožen Adamič et al. 1995; Perko and Orožen Adamič 1998; Perko and Ciglič 2020b; Perko et al. 2021; SiStat 2024)

ID on Fig. 5.1	Landscape type/region	Population in 1931	% of population in 1931	Population in 1961	% of population in 1961	Population in 1991	% of population in 1991	Population in 2011	% of population in 2011	Population density in 2011 (people per km^2)	Number of settlements in 2011	Settlement density in 2011 (number per 100 km^2)	Average size of settlement in 2011 (population)	Largest settlement (elevation, m)	Largest settlement in 2023 (population)	Highest peak (elevation, m)
	Alpine mountains	82,379	5.91	89,621	5.63	92,224	4.69	89,581	4.38	29.26	303	9.90	296	Jesenice (584)	13,702	Triglav (2,864)
1.1	Julian Alps	34,480	2.47	30,642	1.93	30,664	1.56	29,020	1.42	18.82	136	8.82	213	Tolmin (200)	3,228	Triglav (2,864)
1.2	Kamnik–Savinja Alps	20,352	1.46	20,098	1.26	20,916	1.06	21,657	1.06	24.37	114	12.83	190	Mozirje (340)	2,328	Grintovec (2,558)
1.3	Western Karawanks	16,994	1.22	26,586	1.67	29,466	1.50	28,447	1.39	85.90	25	7.55	1,138	Jesenice (584)	13,702	Stol (2,236)
1.4	Eastern Karawanks	10,553	0.76	12,295	0.77	11,178	0.57	10,457	0.51	34.90	28	9.35	373	Mežica (495)	3,127	Peca (2,125)
	Alpine hills	260,837	18.66	292,985	18.41	339,024	17.24	354,244	17.29	76.02	1,528	32.79	232	Velenje (390)	25,235	Porezen (1,630)
2.1	Cerkno, Škofja Loka, Polhov Gradec, and Rovte Hills	39,255	2.81	34,954	2.20	39,343	2.00	45,251	2.21	46.25	287	29.33	158	Žiri (492)	3,743	Porezen (1,630)
2.2	Sava Hills	121,713	8.71	132,142	8.30	136,848	6.96	142,184	6.94	74.47	857	44.89	166	Trbovlje (287)	13,678	Kum (1,220)
2.3	Pohorje, Strojna, and Kozjak Hills	66,146	4.73	83,773	5.26	98,652	5.02	99,510	4.85	77.34	218	16.94	456	Ravne na Koroškem (394)	7,160	Črni vrh (1,543)
2.4	Velenje and Konjice Hills	15,992	1.14	23,027	1.45	42,329	2.15	41,522	2.03	172.05	51	21.13	814	Velenje (390)	25,235	Basališče (1,272)
2.5	Ložnica and Hudinja Hills	17,731	1.27	19,089	1.20	21,852	1.11	25,777	1.26	105.49	115	47.06	224	Vojnik (270)	2,460	Gora Oljka (733)
	Dinaric plateaus	102,613	7.34	77,143	4.84	63,992	3.24	65,565	3.21	17.21	809	21.24	81	Idrija (325)	5,848	Veliki Snežnik (1,796)
3.1	Kambreško and Banjšice Plateaus	15,271	1.09	11,644	0.73	9,328	0.47	8,108	0.40	29.84	50	18.40	162	Kanal (103)	1,070	Veliki vrh (1,071)

(continued)

Table 5.10 (continued)

ID on Fig. 5.1	Landscape type/region	Population in 1931	% of population in 1931	Population in 1961	% of population in 1961	Population in 1991	% of population in 1991	Population in 2011	% of population in 2011	Population density in 2011 (people per km^2)	Number of settlements in 2011	Settlement density in 2011 (number per 100 km^2)	Average size of settlement in 2011 (population)	Largest settlement (elevation, m)	Largest settlement in 2023 (population)	Highest peak (elevation, m)
3.2	Idrija Hills	14,715	1.05	12,712	0.80	11,628	0.59	10,881	0.53	45.54	31	12.97	351	Idrija (325)	5,848	Planinca (1,170)
3.3	Trnovo Forest, Nanos, and Hrušica Plateaus	7,940	0.57	5,919	0.37	4,534	0.23	4,662	0.23	9.17	30	5.90	155	Črni Vrh (683)	699	Suhi vrh (1,313)
3.4	Javorniki and Snežnik Plateaus	0	0	0	0	0	0	21	0	0.05	1	0.22	21	Snežnik (1,000)	17	Veliki Snežnik (1,796)
3.5	Krim Hills and Menišija Plateau	6,185	0.44	4,745	0.30	4,691	0.24	7,042	0.34	23.46	82	27.31	86	Kamnik pod Krimom (351)	1,001	Krim (1,107)
3.6	Bloke Plateau	4,533	0.32	3,293	0.21	2,424	0.12	2,380	0.12	16.53	80	55.57	30	Nova vas (720)	273	Slivnica (1,114)
3.7	Big Mountain, Stojna, and Gotenica Plateaus	12,266	0.88	7,884	0.50	5,564	0.28	5,413	0.26	9.79	133	24.04	41	Sodražica (550)	871	Goteniški Snežnik (1,289)
3.8	Dry Carniola Plateau and Dobrepolje	18,466	1.32	15,336	0.96	13,362	0.68	14,349	0.70	33.87	166	39.18	86	Žužemberk (220)	1,056	Lisec (565)
3.9	Little Mountain, Kočevski Rog, and Poljane Plateaus	8,501	0.61	3,054	0.19	1,730	0.09	1,571	0.08	2.69	75	12.84	21	Velike Poljane (637)	151	Kopa (1,077)
3.10	Radulja Hills	6,796	0.49	5,637	0.35	4,979	0.25	5,485	0.27	47.36	74	63.89	74	Mokronog (250)	737	Pri turnu (605)
3.11	Gorjanci Hills	7,940	0.57	6,919	0.43	5,752	0.29	5,653	0.28	26.69	87	41.08	65	Orehovec (219)	229	Trdinov vrh (1,178)
	Slovenia	1,397,650	100	1,591,523	100	1,965,986	100	2,050,189	100	101.13	6,030	29.74	340	Ljubljana (298)	287,076	Triglav (2,864)

Table 5.11 Land use in mountainous areas (%) (Perko and Ciglič 2020b)

ID on Fig. 5.1	Landscape type/ region	Fields	Vineyards	Orchards	Meadows and pastures	Forests	Built-up areas
	Alpine mountains	0.53	0.00	0.49	**12.68**	**74.38**	1.64
1.1	Julian Alps	0.29	0.00	0.30	**11.38**	**70.49**	1.46
1.2	Kamnik–Savinja Alps	0.84	0.00	0.77	**14.83**	**76.58**	1.65
1.3	Western Karawanks	0.18	0.00	0.21	**11.60**	**80.71**	2.15
1.4	Eastern Karawanks	1.23	0.00	0.95	**14.16**	**80.93**	1.98
	Alpine hills	3.11	0.17	2.18	**21.74**	**68.12**	4.03
2.1	Cerkno, Škofja Loka, Polhov Gradec, and Rovte hills	1.53	0.00	1.13	**21.85**	**72.28**	2.92
2.2	Sava Hills	3.42	0.19	2.33	**23.11**	**66.24**	4.15
2.3	Pohorje, Strojna, and Kozjak Hills	2.96	0.20	2.12	**18.51**	**71.37**	3.93
2.4	Velenje and Konjice Hills	2.80	0.19	2.95	**21.22**	**65.90**	5.69
2.5	Ložnica and Hudinja Hills	8.04	0.48	4.76	**28.16**	**51.28**	6.46
	Dinaric plateaus	1.26	0.24	0.61	**12.87**	**83.21**	1.47
3.1	Kambreško and Banjšice Plateaus	0.80	0.01	1.15	**15.98**	**78.86**	2.38
3.2	Idrija Hills	0.46	0.00	0.76	**15.88**	**80.30**	1.94
3.3	Trnovo Forest, Nanos, and Hrušica Plateaus	0.51	0.02	0.25	**11.30**	**86.25**	1.03
3.4	Javorniki and Snežnik Plateaus	0.12	0.00	0.02	9.08	**90.16**	0.17
3.5	Krim Hills and Menišija Plateau	0.95	0.00	0.60	**12.43**	**83.26**	2.57
3.6	Bloke Plateau	1.37	0.00	0.68	**27.26**	**68.79**	1.73
3.7	Big Mountain, Stojna, and Gotenica Plateaus	0.35	0.00	0.37	9.70	**88.31**	1.01
3.8	Dry Carniola Plateau and Dobrepolje	4.03	0.22	0.88	**20.05**	**72.23**	2.37
3.9	Little Mountain, Kočevski Rog, and Poljane Plateaus	0.29	0.03	0.32	6.50	**92.23**	0.54
3.10	Radulja Hills	7.19	2.27	2.74	**22.72**	**61.03**	3.98
3.11	Gorjanci Hills	3.56	2.56	1.53	**14.45**	**75.32**	2.49
	Slovenia	9.71	0.94	1.76	**18.47**	**61.90**	5.01

Those covering more than 10% are in bold

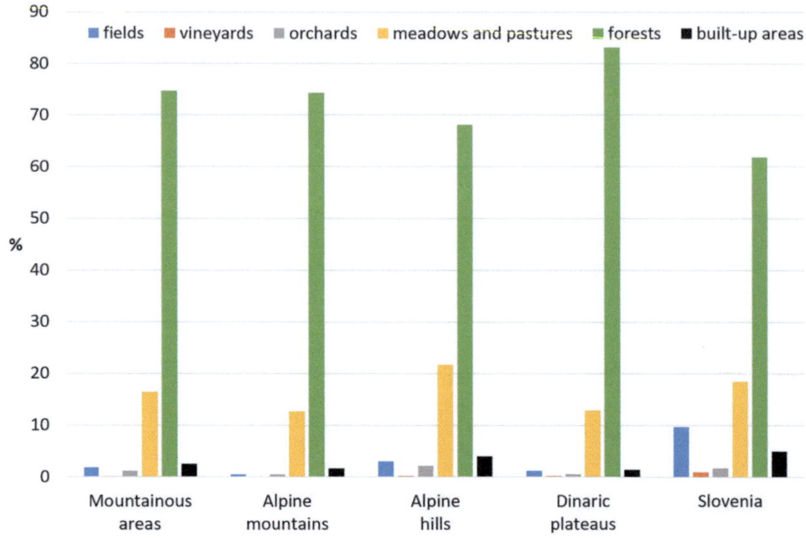

Fig. 5.26 Land use in Slovenia's mountainous areas. (Based on Table 5.11)

Acknowledgements This study was conducted as part of the research program The Geography of Slovenia (P6-0101) and research project Subglacial Carbonate Deposits - A New Source for Studying the Presence of Glaciers in a Glaciokarstic Environment (J6-3141), both financed by the Slovenian Research and Innovation Agency.

References

Arhiv hidroloških podatkov [Archive of hydrological data] (2024a). Agencija Republike Slovenije za okolje, Ljubljana. http://www.arso.gov.si/vode/podatki/arhiv/hidroloski_arhiv.html. Accessed 17 May 2024

Arhiv meteoroloških podatkov [Meteorological data archive] (2024b). Agencija Republike Slovenije za okolje, Ljubljana. http://meteo.arso.gov.si/. Accessed 17 May 2024

Bat M (1997) Slovenija (vodovje, površinske vode) [Slovenia (waters, surface waters)]. In: Voglar D (ed) Enciklopedija Slovenije, vol 11. Mladinska knjiga, Ljubljana, pp 325–327

Bavec M, Tulaczyk SM, Mahan SA, Stock GM (2004) Late quaternary glaciation of the Upper Soča River region (Southeastern Julian Alps, NW Slovenia). Sed Geol 165(3–4):265–283. https://doi.org/10.1016/j.sedgeo.2003.11.011

Beltram G (2004) Wetlands. In: Uhan J, Bat M (eds) Water treasures of Slovenia. Slovenian Environment Agency, Ljubljana, pp 46–53

Brancelj A (ed) (2002) High-mountain lakes in the Eastern part of the Julian Alps. Založba ZRC, Ljubljana. https://doi.org/10.3986/9616358588

Brilly M, Mikoš M, Šraj M (1999) Vodne ujme: varstvo pred poplavami, erozijo in plazovi [Water disasters: protection against floods, erosion and landslides]. Fakulteta za gradbeništvo in geodezijo, Ljubljana.

Celarc B, Placer L (2016) Geotectonic setting of Slovenia. In: Novak M, Rman N (eds) Geological atlas of Slovenia. Geological Survey of Slovenia, Ljubljana, pp 8–9

Cevc T (ed) (2006) Človek v Alpah: desetletje (1996–2006) raziskav o navzočnosti človeka v slovenskih Alpah [Man in the Alps: A decade (1996–2006) of research on human presence in the Slovenian Alps]. Založba ZRC, Ljubljana

Colucci RR, Boccali C, Žebre M, Guglielmin M (2016) Rock glaciers, protalus ramparts and pronival ramparts in the South-Eastern Alps. Geomorphology 269:112–121. https://doi.org/10.1016/j.geomorph.2016.06.039

Čarni A, Marinček L, Seliškar A, Zupančič M (2002) The vegetation map of forest communities of Slovenia, Scale 1:400,000. Institute of biology ZRC SAZU, Ljubljana

Dolinar M, Vertačnik G (2010) Spremenljivost temperaturnih in padavinskih razmer v Sloveniji [Temperature and precipitation variability in Slovenia]. In: Cegnar T (ed) Okolje se spreminja: podnebna spremenljivost Slovenije in njen vpliv na vodno okolje. Agencija Republike Slovenije za okolje, Ljubljana, pp 37–40

Ferk M, Gabrovec M, Komac B, Zorn M, Stepišnik U (2017) Pleistocene glaciation in Mediterranean Slovenia. In: Hughes PD, Woodward JC (eds) Quaternary glaciation in the Mediterranean mountains. Spec Publ Geol Soc London 433. Geological Society, London, pp 179–191. https://doi.org/10.1144/SP433.2

Fodor LI, Gerdes A, Dunkl I, Koroknai B, Pécskay Z, Trajanova M, Horváth P, Vrabec M, Jelen B, Balogh K, Frisch W (2008) Miocene emplacement and rapid cooling of the Pohorje pluton at the Alpine-Pannonian-Dinaridic junction, Slovenia. Swiss

J Geosci 101(Supplement 1):255–271. https://doi.org/10.1007/s00015-008-1286-9
Frantar P (2008) The water balance for the 1971–2000 period. In: Frantar P (ed) Water balance of Slovenia 1971–2000. Slovenian Environment Agency, Ljubljana, pp 71–79
Gabrovec M (1990) Uporaba digitalnega modela reliefa pri proučevanju usadov na primeru doline Lahomnice [Using a digital elevation model to study slumps in the Lahomnica valley]. In: Natek K (ed) Geomorfologija in geoekologija. ZRC SAZU, Ljubljana, pp 181–185
Gabrovec M, Hrvatin M (1998) Površje [Surface]. In: Fridl J, Kladnik D, Orožen Adamič M, Perko D (eds) Geografski atlas Slovenije. DZS, Ljubljana, pp 80–83
Gabrovec M, Hrvatin M, Komac B, Ortar J, Pavšek M, Topole M, Triglav Čekada M, Zorn M (2014) Triglavski ledenik [The Triglav Glacier]. Geografija Slovenije, vol 30. Založba ZRC, Ljubljana. https://doi.org/10.3986/9789610503644
Gabrovec M, Kladnik D (1997) Some new aspects of land use in Slovenia. Geografski Zbornik 37:8–64
Gabrovec M, Ortar J, Pavšek M, Zorn M, Triglav Čekada M (2013) The Triglav glacier between the years 1999 and 2012. Acta Geogr Slov 53(2):257–293. https://doi.org/10.3986/AGS53202
Gams I (1960) O višinski meji naseljenosti, ozimine, gozda in snega v slovenskih gorah [On the extreme upper limit of rural settlements, winter wheat, forest and snowline in Slovenia]. Geografski Vestnik 32:59–69
Gams I (1965) Speleological characteristics of the Slovene karst. Naše Jame 7:41–50
Gams I (1970) Geomorfološke in klimatske razmere v jugovzhodni Koroški [Geomorphological and climatic conditions in south-eastern Carinthia]. In: Jugovzhodna Koroška: zbornik 8. zborovanja slovenskih geografov. Geografsko društvo Slovenije, Ljubljana, pp 27–59
Gams I (1992) Prispevek k mladokvartarni geomorfologiji v Zgornjesavski dolini [A contribution to the young-quaternary geomorphology in the Upper Sava Valley]. Geografski Zbornik 32:5–49
Gams I (2004) Kras v Sloveniji v prostoru in času [Karst in Slovenia in space and time]. Založba ZRC, Ljubljana
Grafenauer B (1991) Kolonizacija [Colonization]. In: Javornik M (ed) Enciklopedija Slovenije, vol 5. Mladinska knjiga, Ljubljana, pp 203–205
Grafični podatki RABA za celo Slovenijo [Graphical data LAND USE for Slovenia] (2020). Ministrstvo za kmetijstvo, gozdarstvo in prehrano, Ljubljana. http://rkg.gov.si/. Accessed 17 Nov 2021
Habe F (1972) Nekatere speleološke značilnosti osamljenega krasa Slovenije [Some speleological features of the isolated karst of Slovenia]. Naše Jame 13:45–53
Habič P (1968) Kraški svet med Idrijco in Vipavo [The karstic region between the Idrijca and Vipava rivers]. Dela SAZU, vol 21. SAZU, Ljubljana
Habič P (1969) Hidrografska rajonizacija krasa v Sloveniji [Hydrographic regionalisation of karst in Slovenia]. Krš Jugoslavije 6:79–91
Habič P (1975) Razlike med alpskim in dinarskim krasom [Differences between Alpine and Dinaric karst]. Naše Jame 17:77–84
Habič P (1980) Nekatere značilnosti kopastega krasa v Sloveniji [Some characteristics of cone karst in Slovenia]. Acta Carsologica 9:5–25
Horvat A (2002) Varstvo pred erozijo [Protection against erosion]. In: Ušeničnik B (ed) Naravne nesreče in varstvo pred njimi. Uprava Republike Slovenije za zaščito in reševanje, Ljubljana, pp 533–541
Horvat J (2013) Visokogorje od kamene dobe do novega veka: Kako smo ljudje osvajali gorski svet [High mountains from the stone age to the modern age: how we humans conquered the mountainous world]. Planinski Vestnik 113(3):4–6
Hrvatin M (1998) Discharge regimes in Slovenia. Geografski Zbornik 38:59–87
Hrvatin M (2004) Relief. In: Orožen Adamič M (ed) Slovenia—a geographical overview. Založba ZRC, Ljubljana, pp 27–30. https://doi.org/10.3986/961650049X
Hrvatin M, Tičar J, Zorn M (2020) Rock and tectonic structure of Slovenia. In: Perko D, Ciglič R, Zorn M (eds) The geography of Slovenia: small but diverse. Springer, Cham, pp 23–34. https://doi.org/10.1007/978-3-030-14066-3_2
Hrvatin M, Zorn M (2017) Trendi pretokov rek v slovenskih Alpah med letoma 1961 in 2010 [River discharge trends in Slovenian Alps between 1961 and 2010]. Geografski Vestnik 89(2):9–35. https://doi.org/10.3986/GV89201
Hrvatin M, Zorn M (2020) Climate and hydrological changes in Slovenia's mountain regions between 1961 and 2018. Ekonomska i Ekohistorija 16:201–218
Hrvatin M, Zorn M (2022) Climate change impacts on hydrology in the Mediterranean part of Slovenia. In: Leal Filho W, Manolas E (eds) Climate change in the Mediterranean and Middle Eastern region. Springer, Cham, pp 85–118. https://doi.org/10.1007/978-3-030-78566-6_4
Hrvatin M, Zorn M (2024) Climate and discharge trends, and flood hazard in Slovenia's Dinaric Karst region since the mid-twentieth century. In: Fuerst-Bjeliš B, Mrgić J, Petrić H, Zorn M, Zwitter Ž (eds) Environmental histories of the Dinaric Karst. Environmental History, vol 17. Springer, Cham, pp 339–373. https://doi.org/10.1007/978-3-031-56089-7_12
Kaligarič M (2004) Rastlinske združbe [Plant communities]. In: Zych B, Mihelač Š (eds) Narava Slovenije. Mladinska knjiga, Ljubljana, pp 183–197
Kataster jam Jamarske zveze Slovenije [Cave registry of Speleological Association of Slovenia]. Ljubljana. https://kataster.jamarska-zveza.si/. Accessed 4 August 2022
Kladnik D (1998a) Alpski svet [Alpine region]. In: Perko D, Orožen Adamič M (eds) Slovenija: pokrajine in ljudje. Mladinska knjiga, Ljubljana, pp 33–52
Kladnik D (1998b) Dinarski svet [Dinaric region]. In: Perko D, Orožen Adamič M (eds) Slovenija: pokrajine in ljudje. Mladinska knjiga, Ljubljana, pp 295–309

Kobold M, Ulaga F (2010) Hidrološko stanje voda in podnebna spremenljivost [Hydrological conditions and climate variability]. In: Cegnar T (ed) Okolje se spreminja: podnebna spremenljivost Slovenije in njen vpliv na vodno okolje. Agencija Republike Slovenije za okolje, Ljubljana, pp 43–56

Kodelja B, Žebre M, Stepišnik U (2013) Poledenitev Trnovskega gozda [Glaciation of Trnovski gozd]. E-Geograff, vol 6. Znanstvena založba Filozofske fakultete, Ljubljana. https://doi.org/10.4312/9789612375614

Komac B (2004) Dolomitni kras ali fluviokras? [Dolomite karst or fluviokarst]. Geografski Vestnik 76(1):53–60

Komac B (2006) Dolec kot značilna oblika dolomitnega površja [Dell as a characteristic dolomite landform]. Geografija Slovenije, vol 13. Založba ZRC, Ljubljana. https://doi.org/10.3986/9789612545208

Komac B, Natek D, Zorn M (2008) Geografski vidiki poplav v Sloveniji. Geografija Slovenije, vol 20. Založba ZRC, Ljubljana. https://doi.org/10.3986/9789612545451

Komac B, Zorn M (2007) Pobočni procesi in človek [Slope processes and man]. Geografija Slovenije, vol 15. Založba ZRC, Ljubljana. https://doi.org/10.3986/9789612545307

Kordiš F (1993) Dinarski jelovo bukovi gozdovi v Sloveniji [Dinaric fir-beech forests in Slovenia]. Biotehniška fakulteta, Ljubljana

Kosi M (1997) Slowenien—ein Alpenland: geographische und historische Einführung. Histoire des Alpes 2:11–28

Kosi M (1998) Potujoči srednji vek [The travelling Middle Ages]. Založba ZRC, Ljubljana

Kranjc A (1998) Kraške vode [Karst water systems]. In: Fridl J, Kladnik D, Orožen Adamič M, Perko D (eds) Geografski atlas Slovenije. DZS, Ljubljana, pp 92–93

Kunaver J (1983) Geomorfološki razvoj Kaninskega pogorja s posebnim ozirom na glaciokraške pojave [Geomorphology of the Kanin mountains with special regard to the glaciokarst]. Geografski Zbornik 22:197–346

Lipar M, Zorn M, Ferk M, Cof K, Čretnik J (2025) The susceptibility of subglacial carbonates to frost weathering. Earth Surf Process Landf 50(5). https://doi.org/10.1002/esp.70055

Lipar M, Martín-Pérez A, Tičar J, Pavšek M, Gabrovec M, Hrvatin M, Komac B, Zorn M, Zupan Hajna N, Zhao JX, Drysdale RN, Ferk M (2021) Subglacial carbonate deposits as a potential proxy for a glacier's former presence. The Cryosphere 15(1):17–30. https://doi.org/10.5194/tc-15-17-2021

Litostratigrafska karta Slovenije [Lithostratigraphic map of Slovenia] (2011). Geološki zavod Slovenije, Ljubljana

Lovrenčak F (1998) Prsti [Soils]. In: Fridl J, Kladnik D, Orožen Adamič M, Perko D (eds) Geografski atlas Slovenije. DZS, Ljubljana, pp 114–115

Lovrenčak F (2003) Osnove biogeografije [Fundamentals of biogeography]. Filozofska fakulteta, Ljubljana

Marinček L, Čarni A (2002) Komentar k vegetacijski karti gozdnih združb Slovenije v merilu 1: 400.000 [Commentary to the vegetation map of forest communities of Slovenia in a scale 1:400,000]. Založba ZRC, Ljubljana

Melik A (1930) Bohinjski ledenik [The Bohinj Glacier]. Geografski Vestnik 5–6:1–39

Melik A (1932) O diluvialni poledenitvi v Karavankah [On Pleistocene glaciation in the Karawanks]. Geografski Vestnik 8:89–101

Meze D (1966) Gornja Savinjska dolina, nova dognanja o geomorfološkem razvoju pokrajine [Upper Savinja valley, new insights into the geomorphological evolution of the landscape]. Dela SAZU, vol 20. SAZU, Ljubljana

Meze D (1974) Porečje Kokre v pleistocenu [The river basin of Kokra during the pleistocene period]. Geografski Zbornik 14:5–101

Mihevc A (2001) Speleogeneza Divaškega krasa [Speleogenesis of the Divača Karst]. Založba ZRC, Ljubljana. https://doi.org/10.3986/9789610503224

Mihevc A, Gabrovšek F, Knez M, Kozel P, Mulec J, Otoničar B, Petrič M, Pipan T, Prelovšek M, Slabe T, Šebela S, Zupan Hajna N (2016) Karst in Slovenia. Bol Geol Min 127(1):79–97

Mihevc A, Mihevc R (2021) Morphological characteristics and distribution of dolines in Slovenia, a study of a lidar-based doline map of Slovenia. Acta Carsologica 50(1):11–36. https://doi.org/10.3986/ac.v50i1.9462

Mikša P, Zorn M (2016) The beginnings of the research of Slovenian Alps. Geografski Vestnik 88(2):103–131. https://doi.org/10.3986/GV88206

Natek K (2007) Periglacial landforms in the Pohorje mountains. Dela 27:247–263. https://doi.org/10.4312/dela.27.247-263

Natek K, Natek M (1998) Slovenija, geografska, zgodovinska, pravna, politična, ekonomska in kulturna podoba Slovenije [Slovenia, geographical, historical, legal, political, economic and cultural identity of Slovenia]. Mladinska knjiga, Ljubljana

Natek K, Natek M (2008) Slovenija, portret države [Slovenia, portrait of a country]. Natek in ostali, Ljubljana

Novak M (2016) An outline of the geology of Slovenia. In: Rman N, Novak M (eds) 70 geological wonders of Slovenia. Geological Survey of Slovenia, Ljubljana, pp 8–11

Obu J (2011) Periglacialne in ledeniške oblike v zahodnem delu Pohorja [Periglacial and glacial landforms in western part of Pohorje mountains]. Dela 35:55–71. https://doi.org/10.4312/dela.35.55-71

Ogrin D (1996) Podnebni tipi v Sloveniji [The climate types of Slovenia]. Geografski Vestnik 68:39–56

Ogrin D (2004) Vreme in podnebje [Weather and climate]. In: Zych B, Mihelač Š (eds) Narava Slovenije. Mladinska knjiga, Ljubljana, pp 73–101

Ogrin D, Ogrin M, Repe B, Svetlin D, Štaut L (2023) Climate classification of Slovenia based on data from the period 1991–2020. Dela 59:47–89. https://doi.org/10.4312/dela.59.5-89

Orožen Adamič M, Perko D, Kladnik D (eds) (1995) Krajevni leksikon Slovenije [Lexicon of place names of Slovenia]. DZS, Ljubljana

Pavšek M (2007) Ledenik pod Skuto kot pokazatelj podnebnih sprememb v slovenskem delu Alp [The Skuta glacier as an indicator of climate changes in Slovenian part of the Alps]. Dela 28:207–219. https://doi.org/10.4312/dela.28.207-219

Perko D, Ciglič R (2020a) Slovenia's landscapes. In: Perko D, Ciglič R, Zorn M (eds) The geography of Slovenia: small but diverse. Springer, Cham, pp 211–225. https://doi.org/10.1007/978-3-030-14066-3_14

Perko D, Ciglič R (2020b) Slovenia's regions. In: Perko D, Ciglič R, Zorn M (eds) The geography of Slovenia: small but diverse. Springer, Cham, pp 227–243. https://doi.org/10.1007/978-3-030-14066-3_15

Perko D, Ciglič R, Hrvatin M (2021) Landscape macrotypologies and microtypologies of Slovenia. Acta Geogr Slov 61(3):7–89. https://doi.org/10.3986/AGS.10384

Perko D, Hrvatin M, Ciglič R (2015) A methodology for natural landscape typification of Slovenia. Acta Geogr Slov 55(2):235–270. https://doi.org/10.3986/AGS.1938

Perko D, Orožen Adamič M (eds) (1998) Slovenija: pokrajine in ljudje [Slovenia: landscapes and people]. Mladinska knjiga, Ljubljana

Placer L (1999) Tektonska zgradba [Tectonic structure]. In: Voglar D (ed) Enciklopedija Slovenije, vol 13. Mladinska knjiga, Ljubljana, pp 214–215

Placer L (2008) Principles of the tectonic subdivision of Slovenia. Geologija 51(2):205–217. https://doi.org/10.5474/geologija.2008.021

Placer L (2009) Tectonic subdivision of Slovenia. In: Pleničar M, Ogorelec B, Novak M (eds) The geology of Slovenia. Geological Survey of Slovenia, Ljubljana, pp 43–60

Poljak M (2007) Structural-tectonic map of Slovenia 1:250,000, Explanatory booklet. Geološki zavod Slovenije, Ljubljana

Remec-Rekar Š (2016) Ocena stanja jezer v Sloveniji v letu 2015 [Assessment of the state of lakes in Slovenia in 2015]. Agencija Republike Slovenije za okolje, Ljubljana

Remec-Rekar Š, Bat M (2004) Lakes. In: Uhan J, Bat M (eds) Water treasures of Slovenia. Slovenian Environment Agency, Ljubljana, pp 39–45

Senegačnik J (2012) Slovenija in njene pokrajine [Slovenia and its regions]. Modrijan, Ljubljana

SiStat (2024) Prebivalstvo po spolu in po starosti, občine in naselja, Slovenija, letno [Population by sex and by age, municipalities and settlements, Slovenia, annually]. Statistični urad Republike Slovenije, Ljubljana. https://pxweb.stat.si/SiStatData/pxweb/sl/Data/-/05C5003S.px. Accessed 15 Apr 2024

Stefanovski S, Repe B (2019) Proučevanje vzpetin kopastega krasa s pomočjo digitalnega modela višin [Study of cone karst pinnacles with the use of digital elevation model]. Dela 52:141–160. https://doi.org/10.4312/dela.52.141-160

Stepišnik U (2010) Udornice v Sloveniji [Collapse dolines in Slovenia]. E-GeograFF, vol. 1. Znanstvena založba Filozofske fakultete, Ljubljana. https://doi.org/10.4312/9789612379841

Stepišnik U (2020) Fizična geografija krasa [Physical geography of karst]. Znanstvena založba Filozofske fakultete, Ljubljana

Šifrer M (1959) Obseg pleistocenske poledenitve na Notranjskem Snežniku [The extent of the Pleistocene glaciation on Snežnik]. Geografski Zbornik 5:27–83

Šifrer M (1961) Porečje Kamniške Bistrice v pleistocenu [The river basin of Kamniška Bistrica during the pleistocene period]. Dela SAZU, vol 10. SAZU, Ljubljana

Šifrer M (1969) Kvartarni razvoj Dobrav na Gorenjskem [The quaternary development of Dobrave in Upper Carniola (Gorenjska)]. Geografski Zbornik 11:99–221

Šifrer M (1998) Površje v kvartarju [Relief in the Quaternary]. In: Fridl J, Kladnik D, Orožen Adamič M, Perko D (eds) Geografski atlas Slovenije. DZS, Ljubljana, pp 78–79

Šifrer M, Kunaver J (1978) Poglavitne značilnosti geomorfološkega razvoja Zgornjega Posočja [The relevant characteristics of the geomorphological development of Upper Soča Valley]. In: Zgornje Posočje: zbornik 10. zborovanja slovenskih geografov. Geografsko društvo Slovenije, Ljubljana, pp 29–34

Šilc U, Vreš B, Čelik T, Gregorič M (2020) Biodiversity of Slovenia. In: Perko D, Ciglič R, Zorn M (eds) The geography of Slovenia: small but diverse. Springer, Cham, pp 109–124. https://doi.org/10.1007/978-3-030-14066-3_7

Triglav Čekada M, Ajdova L (2023) Površina snežišča v Skednju pod Prisojnikom iz fotografij [Snowfield in Skedenj under Prisojnik from images]. Raziskave s področja geodezije in geofizike 2022:95–103

Triglav Čekada M, Barbo P, Pavšek M, Zorn M (2020) Changes in the Skuta Glacier (southeastern Alps) assessed using non-metric images. Acta Geogr Slov 60(2):175–190. https://doi.org/10.3986/AGS.7674

Triglav Čekada M, Barborič B, Zorn M, Ferk M (2016) Lasersko skeniranje Slovenije in akumulacijske reliefne oblike v slovenskem visokogorju [Laser scanning of Slovenia and accumulation landforms in the Slovenian high mountains]. Raziskave s področja geodezije in geofizike 2015:51–65

Triglav Čekada M, Zorn M (2020) Thickness and geodetic mass balance changes for the Triglav Glacier (southeastern Alps) from 1952 to 2016. Acta Geogr Slov 60(2):155–173. https://doi.org/10.3986/AGS.7673

Vidic NJ, Prus T, Grčman H, Zupan M, Lisec A, Kralj T, Vrščaj B, Rupreht J, Šporar M, Suhadolc M, Mihelič R, Lobnik F (2015) Soils of Slovenia with soil map 1:250,000. Publications Office of the European Union, Luxembourg. https://doi.org/10.2788/88750

Vidrih R, Ribičič M, Suhadolc P (2001) Seismogeological effects on rocks during the 12 April 1998 upper Soča Territory earthquake (NW Slovenia). Tectonophysics 330(3–4):153–175. https://doi.org/10.1016/S0040-1951(00)00219-5

Vreš B, Gilčvert Berdnik D, Seliškar A (2014) Rastlinstvo življenjskih okolij v Sloveniji [Vegetation of natural environments in Slovenia]. Pipinova knjiga, Ljubljana
Vrščaj B (2017) An overview of soils of Slovenia. In: Vrščaj B, Repe B, Simončič P (eds) The soils of Slovenia. Springer, Dordrecht, pp 77–133. https://doi.org/10.1007/978-94-017-8585-3_5
Zorn M (2002) Rockfalls in Slovene Alps. Geografski Zbornik 42:123–160
Zorn M, Breg Valjavec M, Komac B, Volk Bahun M, Hrvatin M (2020) Soils of Slovenia. In: Perko D, Ciglič R, Zorn M (eds) The geography of Slovenia: small but diverse. Springer, Cham, pp 91–107. https://doi.org/10.1007/978-3-030-14066-3_6
Zorn M, Ferk M, Lipar M, Komac B, Tičar J, Hrvatin M (2020) Landforms of Slovenia. In: Perko D, Ciglič R, Zorn M (eds) The geography of Slovenia: small but diverse. Springer, Cham, pp 35–57. https://doi.org/10.1007/978-3-030-14066-3_3
Žebre M, Stepišnik U, Kodelja B (2013) Sledovi pleistocenske poledenitve na Trnovskem gozdu [Traces of Pleistocene glaciation on Trnovski gozd]. Dela 39:157–170. https://doi.org/10.43121/dela.39.9.157-170

Part II

Prehistory and the Middle Ages

Continuity of Mountain Settlements in Montenegro

6

Olga Pelcer-Vujačić

Abstract

The knowledge of prehistoric and ancient sites in the high-altitude Montenegrin landscape is fairly limited, and there is no comprehensive overview in the modern literature. Nevertheless, there are many remains of prehistoric activity in the mountains and caves, for example, around Cetinje (mountain Lovćen, cave Koronjina), Crvena Stijena near Petrovići, Nikšić (perhaps the most significant prehistoric archaeological site in Montenegro), and prehistoric petroglyphs at Prokletije. Another common form of archaeological site in Montenegrin mountains and widespread in the entire Balkan area are *tumuli* (characteristic tombstones built by forming an earthen or stone mound above one or more tombs) occurring in the Neolithic, and as the dominant burial form of the Bronze and Iron Age, they lasted until Roman domination of the Balkans. An important ancient Roman settlement is Municipium S, Komini near Pljevlja, at approximately 780 m above sea level. It was a prosperous inland town in the Roman province of Dalmatia, with distinguished magistrates, notable trade, and diverse religious beliefs. Other Illyrian and Roman settlements, hillforts, military camps, and villas are dispersed in the mountainous area in the north of Montenegro. Many sites are still to be thoroughly researched. This would be an opportunity to observe the creation of settlements in high-altitude zones and the initial dispersal of this cultural development.

Keywords

Prehistory · Mountains · Caves · Ancient settlements · Montenegro

6.1 Introduction

Nomen est omen, as the Roman author Plautus said (Plautus 2011:526) and one might think that writing about the continuity of settlements in the Montenegrin mountains is going to be an easy task. The first mentions of *Crna Gora*, or Latin equivalent *Montenegro* were during the Middle Ages. The first document mentioning the name *Crna Gora* is the charter by King Milutin to the monastery of St. Nicholas in Vranjina in 1296 (Stjepčević and Kovijanić 1953:231–233). The documents from the Kotor archive from 1397 and Venetian archive from 1435 are naming Montenegro, or in the case of the latter, specifically *katuns* of Montenegro (Božić 1979:150–155; Stjepčević and Kovijanić 1953:231–233). It started as a general term for a mountainous region rich in dark forests, similar to other such places in

O. Pelcer-Vujačić (✉)
Historical Institute, University of Montenegro, Podgorica, Montenegro
e-mail: olgapelcer@gmail.com

M. Zorn et al. (eds.), *Discourses on Mountains of Montenegro and Slovenia*,
Historical Geography and Geosciences, https://doi.org/10.1007/978-3-031-99017-5_6

the Balkans, but soon enough became a distinctive toponym (Blagojević 1999:57–77).

As the science of archaeology is undergoing constant change, our general understanding of human populations in the late Pleistocene and the first half of the Holocene has increased considerably in recent decades, owing to both new theoretical approaches and an increase in the number of researchers (Gassiot Ballbè et al. 2017:76; Vander Linden and Silva 2021). Technical and methodological innovations have actively contributed to such a development.

The archaeology of mountain areas in southern Europe is undergoing a kind of revolution that is very recent and yet cannot be adequately explained by general changes in archaeological science. In the previous period, research on prehistoric sites in the highlands of Montenegro has not attracted the special attention of Montenegrin archaeologists, with only periodic research and mentions (Ćulafić 1952:166–170; Parović-Pešikan 1980; Šćepanović 1996:111–123; Lazić and Cerović 2002:191–198). The prejudice that mountain areas are a space of marginal and occasional human activity influenced the fact that the notion of high mountain archaeology never gained importance within the framework of Montenegrin national archaeology.

Generally, the most important environmental factor in the early human settlement was water—rivers, lakes, and inland seas are good sources of fresh water. The second environmental factor was topography. Mountains and deserts were less friendly to human settlement. Steep mountains were hard to cross. Their jagged peaks and rocky land made farming difficult. This is the reason why many prehistoric settlements in Montenegro were used for a very short period of time as a habitat for a small community of people. Even in regions with similar mountain dispositions, such as the one comprised in present Slovenia, there are no documented larger Palaeolithic settlements at higher altitudes due to severe climate during the last glacial (Ogrin 2018:39–47).

The signs of anthropocentric disturbance of the alpine environment structures suggest a more stable and ubiquitous human presence, probably largely associated with the development of mobile herding practices. Pastoral transhumance occurs in a certain socio-political milieu, in which specialization and separation between the agricultural and pastoral sectors have occurred (Chang and Tourtellotte 1993:261). Therefore, the socio-political dynamics by which members of peripheral populations have held pivotal positions as middlemen and traders in environmentally diverse regions may also serve as a useful analogy for examining what roles pastoralist specialists might have had in the development of early complex societies (Chang and Tourtellotte 1993:261). There is still no definitive, based on solid evidence, answer to the question in which period of prehistory certain areas of Montenegro were inhabited, as well as the answer to the question of which communities were settled and finally where these communities came from. The analysis of certain elements of material culture tries to reach a more or less acceptable solution, which is mainly based on the observed similarities of certain forms of material culture contained in the layers of prehistoric sites in Montenegro, with forms of cultural artifacts from prehistoric sites in Europe, Asia, and Africa. It has been stated several times that the special geomorphology of Montenegro and favourable climatic conditions, which prevailed during the last interglacial and glacial periods, enabled the first human communities to settle in Montenegro, possibly at the end of the Early Palaeolithic, but certainly at the beginning of the Middle Palaeolithic. They continued their development throughout the Neolithic, including of course the Mesolithic, in which the process of gradual transformation of a man as an exclusive hunter and gatherer into a man who is a farmer and cattle breeder begins (Marković 2019:79–141).

However, this chapter is a modest overview of previous archaeological research of prehistoric sites, it doesn't offer more on the interpretation of continuities and discontinuities that are very relevant in the entire Balkan region. Some results still allow us to glimpse at the cultural and population permeations that took place during prehistory in Montenegro, starting from the Early Stone Age—Palaeolithic, then through the Mesolithic, Neolithic, Eneolithic, Bronze and Iron Ages, all the way to Roman domination.

6.2 Prehistoric Settlements

6.2.1 Cetinje Valley

The area of Cetinje (Fig. 6.1), namely Cetinjsko Polje and its surroundings, even during prehistory had a role in transferring cultural impulses from the seaside to the hinterland (Marković 1994a:20). On many peaks, there are early fortified settlements known as *gradine* as well as burial mounds dated mostly to the Iron Age, but with modest archaeological findings leading to the period of the Illyrians. Erdeljanović (1978:235–236) describes some visible remains of *gradine* (fortified settlements) and some *gomile* (burial mounds). Among them, there is one *gradina* above the Cetinje monastery. Remains of these burial mounds could be found in Orlov krš, Njeguši, and Krstac (some of the findings are displayed in the Historical Museum in Cetinje). The archaeological material dates to the early Iron age (Leković 1980:81–88). One of these burial mounds was just in front of Vlaška crkva (Vlaška church) in Cetinje, discovered and demolished during the levelling of the terrain. Several burials with inhumed remains and iron tools were discovered (Erdeljanović 1978:242).

The cave Koronjina is the best-known site close to Cetinje and Lovćen. Koronjina is a small *abri* (shelter) about 3.5 km north-east of Cetinje on the road to Bokovo, below the Dobroštak hill. The opening of the *abri* faces south-west, towards Cetinjsko Polje, 6 m wide and 4.5 m high. The maximum depth of the cave is 3.2 m; therefore, the total internal area does not exceed 15 m^2. The site testifies to the existence of the first human communities in the area of the Cetinje field. In addition to this prehistoric site on the territory of the municipality of Cetinje, there are prehistoric archaeological finds in the Vezačka cave near Rvaši. During short excavations done in the summer 1977 (Marković 1994a:19), 88 fragments of pottery and three animal horns were removed from the cultural layer, and only one of them showed traces of processing. Archaeological finds from Koronjina can be framed within the impresso complex (older Neolithic, about 12,000 years ago) which flourished in the entire Adriatic coast and its hinterland. It can be assumed that the Koronjina cave was used as a shelter for small communities, for a rather short period of time, which is probably the result of an unfavourable location (Anonymous 1913; Marković 1985:12–30; Marković 1994b:240–241; Marković 2006:220). Archaeological finds discovered during this excavation are stored in the Historical Museum in Cetinje.

Another new significant project concentrated on a site close to both Cetinje and Skadar lake, Vezačka cave, is an ERC project (lead by Marc Vander Linden) EUROFARM, active in Montenegro from 2012 to 2017, researching the transmission of farming and associated technological innovations (landscape use, pottery, lithics) in the western Balkans during the 6th and 5th millennium BC through a combination of new data collection (archaeological fieldwork, access to museum collections and literature survey), analysis and agent-based modelling (Gazivoda et al. 2019:121–124; Vander Linden and Silva 2021; Gaastra et al. 2022:44–60).

Due to rock configuration, Lovćen was not adequate in terms of rich, arable land or vast pastures, so there are no traces of continued settlement or archaeological sites. It is impossible to say which Illyrian tribe lived around Lovćen in the Hellenistic period, as there is no archaeological evidence, but one can assume that several smaller Illyrian tribes under the influence of Autariati were located there (Marković 1994b:245).

6.2.2 Crvena Stijena

The rock shelter of Crvena Stijena in Petrovići (Fig. 6.2), near Nikšić, is one of the most important Palaeolithic sites in south-eastern Europe with the richest stratigraphy and the largest number of cultural layers. Its 20-m-deep sequence of archaeological deposits spans the Middle Palaeolithic through the Bronze Age. The Middle Palaeolithic deposits themselves, which cover an astonishing 12 m in depth, contain one of the longest records of Neanderthal occupation in the region and one of the deepest rock shelter sequences in Europe. Since its discovery in

Fig. 6.1 Map of Montenegro with settlement locations

1954, the site has been the subject of two major research projects; the data they have produced have helped make it a critical type-site for the Palaeolithic in the Balkans. The site has been excavated from 1954–1963 and again since 2004 (Basler 1965/1966:107–111; Basler 1975; Whallon 2017).

Crvena Stijena is a large opening in a cliff composed of limestone and dolomite which is stained red by iron oxides, giving it the name 'Red Rock'. It is 26 m wide, 15 m high, and approximately 25 m deep, from front to back (Morley 2017:82–131). It is 700 m above sea level, facing south/south-west towards the Trebišnjica

Fig. 6.2 Crvena Stijena. (Photo by Colin McFadden)

river in Bosnia, and is sheltered from the strongest and coldest winds, those coming from the north (Ćulafić 2017:11–12). The oldest stone artifacts are about 180,000 years old, and the findings on the existence of prehistoric man are over 100,000 years old.

Palaeolithic man, whose existence depended on certain hunting species of animals and edible fruits, was forced to follow the migrations of flora and especially fauna, so today it is difficult to determine from which direction the first inhabitants of present-day Montenegro arrived (for recent discoveries Monnier et al. 2020:71–108; general global overview Boone 2002:6–25). The sequence of lithic industries has been studied by numerous lithic specialists (Dogandžić and Đuričić 2017: 68–192; Mihailović and Whallon 2017:36–49; Morley 2017:82–131; Mihailović 2020:47–62) and has provided valuable information about the nature of the Middle and Upper Palaeolithic in the Balkans, and its rich faunal record has revealed key information about the hunting behaviour of the people who lived here in relation to changing environments through time.

The record of the Palaeolithic occupation of Crvena Stijena provides an outstanding opportunity to closely examine changing Neanderthal adaptations. The excellent preservation of faunal remains and hearths signifies that the site has enormous potential for revealing more information about Neanderthal subsistence strategies, including hunting and butchery activities; the use of plants for food, fuel, and other purposes; how they made, managed, and used fire; how they organized their activities at the rock-shelter; and how they used the local landscape to obtain resources (Monnier et al. 2020:71–108).

6.2.3 Prokletije

One of the greatest features of the mountain range Prokletije in terms of cultural and historical heritage are prehistoric drawings—one carved on a stone slab below the top of Popadija (2,057 m above sea level), another at the site Vezirova brada. These drawings are mostly briefly mentioned, with scant data. This engraving, according to some accounts, dates from the Middle Bronze Age, around 1600 BC (Mijović 1992:11–23).

Near the village of Gusinje, on the eastern slope of mount Vezirova Brada (North Prokletije), drawings carved into the rock were found on the steep slope. They consist of a fairly large number of crosses and a human figure. The crosses vary in height from 5 to 15 cm. Their ends are rounded and they are arranged without any apparent order. The human figure, 17.5 cm high, is executed schematically. The head is in profile; the arms widen perpendicular to the body. The latter is disproportionately elongated and slightly curved. We distinguish the phallus between the short legs, spread diagonally. It seems that the right hand, bent in the elbow, supports an object hanging from the hip (sword or dagger). In the outstretched left hand is a bow with an arrow ready to be launched. The crosses and the archer were likely engraved simultaneously. In Grbaja, located a few kilometres south of Gusinje, other carved figures were found in the rock, but belonged, according to some information obtained, to modern times (Vasić 1972:131–135).

Another engraving on a rock, broken off the top of the Maja Popadija Mountain in the Prokletije Mountain range, on the border with Albania, has been discovered by local residents at the altitude of 1,700 m. The block of stone, on which the drawings have been found, lies in a ravine. Its dimensions are approximately 2 × 1.70 m. It is intact and lies in an almost horizontal position. The stone has the colour of *gres-lie-de-vin* (sandstone with iron-oxides). The engraving technique is pricking, most probably with the sharp tip of some flint. This pricking has probably been performed along an originally drawn line. Not many pricks have been made along the line, although it is hard to say to what degree damages affected the quality of the engravings found. The originality of these engravings lies in the geometrics of the figures, presented in the form of a circle (heads and female sexual organs) and a straight and interrupted line (all parts of the bodies and one head in the form of a triangle). These engravings represent a whole family with a mother who carries a child under each arm (Fig. 6.3). The figure of the mother is the biggest and it may thus be concluded that the main motive of this group is evoking the mother's role, i.e., the cult of the Big Mother. While all male figures are marked by a phallus that presents as an extension of the spine, the female sexual organ is presented as a big circle. The arms and legs of all figures but one, are spread up to the elbows or knees, at which point they are suddenly curved and lowered downward. They end with feet or hands stretched out aside. Only one figure has legs in the form of an arch and a wavy arm. Some figures are unfinished, and one of them—although it seems that the creator(s) started engraving it as the other ones—resembles a distorted handle as if it were a kind of tool or an unfinished animal.

Fig. 6.3 Petroglyphs in Prokletije. (Photo by Miloš Petričević)

One figure lies on the ground with spread arms in the form of a cross. It seems that this figure represents a deceased or killed family member around which all members have gathered, jumping and walling in grief. The female figures reminded Mijović (1992:11–23) of mourners, a well-known from the Montenegrin funeral folklore. Preliminarily, he dated these engravings in the medium Bronze Age, somewhere between 1800 and 1400 BC. In addition to the cave drawings in white color on the rocks of Lipci near Risan (Bay of Kotor), Žlijeb near Višegrad (Bosnia), Vezirova Brada near Gusinje (Montenegro), as well as to Tren in Albania and the river bed of the Komarnica river (Montenegro), the group of engravings at Maja Popadija represents a complete regional group that is extremely interesting and significant to science (Mijović 1992:11–23).

Soon a completely new analysis and findings will be published and probably shed some more light on these interesting findings (Petričević 2022).

6.2.4 Odmut

Odmut Cave is a well-known prehistoric site and is located at the mouth of the Vrbica River in Piva, at the foot of Kulina Hill at an altitude of 558 m. The opening of the cave, 20 m wide and 14 m high, faces south-east. The cultural layer is over 4 m thick and is divided into strata: Odmut I (Mesolithic), Odmut II (Early Neolithic), Odmut III (Late Neolithic), Odmut IV-VI (Eneolithic), and Odmut VII (Bronze Age). The research was conducted from 1972 to 1974, and geological surveys were performed in 1975 (Marković 2006:71). It is important to note that the results of earlier research show that the area of Piva, in relation to other areas of Montenegro, was inhabited much later, at the very beginning of the Holocene, at the beginning of the Mesolithic (Marković 2019:89–90).

The archaeological material found in the next three layers (Odmut II-V) confirms a longer stage of cultural development in the Piva area, the duration of which, using the ^{14}C method, was determined between 4,900 and 2,500 BC. In this period, the Odmut cave is no longer a permanent residence, but an occasional seasonal habitat of excellent shepherds and hunters. Ceramics with both coarse and fine textures were found in stratum II, and a percentage of coarse, monochrome ceramics is connected to the younger phase of Odmut IIb. Coarse ceramics are insufficiently cleaned, poorly baked and very often the outer surfaces are coated with a red coating and sometimes the inner ones as well. Fine, monochrome ceramics are better refined, better fired, with slightly thinner walls, most often smoothed surfaces with sometimes a reddish coating, and sometimes their outer surfaces are intentionally roughened. The shape of finer ceramics is the same as of the coarse ones, only of slightly smaller dimensions (Marković 1985:31–45). A total of 5 bone tools were found in both layers, and they were mostly awls made of the bones of larger mammals (Marković 1985:31–45). Among the found animal bones from stratum II, the presence of both domestic and wild animals was stated, but there was a statistically higher number of wild animals. The most commonly hunted animal is deer, followed by ibex, chamois, wild boar, bear, and marten. The most common domestic animals are goats and sheep, followed by cattle and pigs. Based on these data, it is concluded that animal husbandry played a secondary role in the diet of the inhabitants of the Odmut cave (Marković 1985:31–45). Based on the analysis of ceramic shapes and ways of decorating, it can be concluded that the older layer of Odmut IIa belongs to the Starčevo culture and that the influence from the Adriatic came to these parts somewhat later, the layer of Odmut IIb. Considering the influences of both the continental (Starčevo culture) and the Adriatic Neolithic (impresso culture), we can assume that it is a local culture (Marković 2006:71–75).

The Odmut III-V culture is related to the Middle Neolithic culture of the Danube (Vinča culture) and the southern Adriatic (Danila and Hvar culture). The last period in the early history of Piva is related to the Odmut VI-VII layers, dated by the ^{14}C method, between 2,500 and 1,800 BC, whose culture marks a complete break with local traditions. It is of foreign origin, that is, it is related to the nomads who in the first centuries of the third millennium of the old era moved from the southern Russian steppes and reached the core of the Balkan peninsula around 2500 BC. Their pottery differs from that of the previous layers, being made from uncleaned earth and regularly poorly fired. Fauna remains show that the inhabitants of Odmut at this time owned large herds of small livestock and that they hunted wild boars, bears, beavers, and rabbits (Marković 1974:7–13; Srejović, 1974:3–7; Marković 1985:31–45).

The Odmut cave was abandoned around 1,800 BC. New knowledge and skills were transferred to the lively ethno-cultural movements in the area of Piva. New residents knew the value of new raw materials—copper, gold, and silver. With weapons made of metal, shepherds became warriors, they left the caves and shelters under the rocks and built the first settlements in other areas of Piva.

6.2.5 General Overview of Other Prehistoric Settlements in Montenegro

Changes in the cold, harsh climate, with warm periods during the Pleistocene, must have affected the life of Palaeolithic men who were forced to adapt to these changes and resort to gradual migrations in search of those areas that provided them with more favourable conditions for existence. Noticeable landscape transformations occurred within individual lifetimes and on a generational basis during the Late Upper Palaeolithic (late Pleistocene) and Mesolithic (early Holocene). The search for more encouraging conditions was most often influenced by the impact of climate change on the animals, which were also forced to slow migrations in search of more favourable conditions for their survival and development (Pilaar Birch and Vander Linden 2018:177–191).

For now, the fact is that the settlement in certain areas of Montenegro was carried out at the very beginning of the Middle Palaeolithic, although the possibility that it happened earlier is not ruled out, as evidenced by only the palaeontological site Trlica near Pljevlja, which belongs to the Lower Pleistocene. Archaeological finds place this site in a somewhat younger phase of the older Neolithic, i.e., Starčevo culture, which shows that the oldest Neolithic settlement in the continental part of Montenegro was not formed at the very beginning of Neolithic development, but somewhat later, thus chronologically lagging behind the Neolithic Montenegrin coast (Marković 1985:76–77).

Another remarkable northern site is Beran krš located in Beranselo, approximately 2.5 km north-west of Berane. Archaeological research on this site was conducted in 1961 and 1975 (Marković 1974:7–12; Marković 1985). A house of a Neolithic type has been discovered (with the hearthstone), as well as parts of tools and furnishings, all dated in the middle of the 4th millennium BC (3600s BC). An altar of baked clay was also discovered, considered evidence of the Neolithic idolatry.

Relying on the results of archaeological research during 1970s (Marković 1974:7–12; Marković 1985), it is possible to indicate to a certain level the basic characteristics of the next epoch, the Bronze Age of Montenegro, and to a certain extent to consider cultural and population permeations during this prehistoric epoch. It is important to emphasize that for a more comprehensive view of the development of the Bronze Age, with special reference to cultural and population permeations, we must take into account the geographical and geomorphological characteristics of Montenegro. Certain specifics of these characteristics occur during the Bronze Age, which necessitates for the separation of larger territorial and cultural units with similar or identical manifestations, which culturally bring them closer to neighbouring areas—the central Adriatic, Hercegovina, the central Balkans, but also the distant, eastern Mediterranean. The data available today indicate that the Bronze Age of Montenegro was developing with all the features that characterize it (Marković 2006). Thus, in addition to the characteristic forms of material culture, forts appear as the basic type of settlement, although caves were still used for housing, as evidenced by the findings from the youngest stratum of Crvena stijena, the youngest 28 strata of the Odmut cave, and the youngest layer of the Vranjaj cave, while for burial, as the basic form of grave constructions, tombstones are used (Marković 2019:110).

At the end of the Late Bronze Age, and at the beginning of the Iron Age in the area from south-eastern Bosnia and south-western Serbia to central Albania, including Montenegro, a larger community was formed, marked by similar or identical manifestations of material and spiritual culture, known collectively as the Illyrians. As early as the beginning of the sixth century AD, this community of tribes was exposed to direct influences from developed centres, which led to major changes in social and economic relations and to the gradual separation of certain ethnic groups, namely Illyrian tribes, which can now be culturally and geographically separated. Cultural ties are being established with near and far areas—the continental regions of the Balkan Peninsula, the middle Danube region, Greece, and the Apennine Peninsula (Marković 2019:136).

6.3 Hellenistic and Roman Period

During the Late Iron Age, the Illyrians reached their highest rise in socio-economic relations, creating the most important socio-political organization—the Illyrian state. The emergence of new elements in material and spiritual culture during this period influenced the change of almost all previously established relations. We have no specific Illyrian sites in high altitude regions—their main settlements were on the coast. There are sporadic findings in a low mountainous area in the north, such as materials found in a grave in Lisijevo Polje that clearly show that an independent tribe, which at the end of the sixth century BC developed a culture with unique characteristics, lived in the Polimlje Region. Another interesting site is Tumbarice located in the village of Donja Rzanica, approximately 5 km east of Berane. Archaeological researches were conducted in 2008, and the remains of objects and fortresses with two sets of walls, which cover an area of 1.5 hectares, were discovered. Based on the discovered archaeological material, it can be concluded that humans continually lived on this locality from the Copper and Bronze Age, through the Roman period, until the Middle Ages. There are several more ancient Roman sites, some with remains of Roman housing near Bijelo Polje, or Lužac near Berane, where archaeological researches were conducted during the 1960s. Researches showed that a Roman fortress (castrum), with a series of ancillary facilities, existed in Lužac near Berane (Garašanin and Garašanin 1967:236–237).

6.3.1 Municipium S

Municipium S is, in addition to Risinium and Doclea, one of the most important urban centres of the Roman age in the south-eastern part and remote part of the Roman province of Dalmatia. The name of this ancient place is unknown, it is recorded on the epigraphic monuments in an abbreviated form as a *Municipium S.* (Šašel and Šašel 1978:1702), or *S [.] LO* (CIL III 8308). Today it seems that there is no doubt that it should be resolved as *municipium Splonistarum* (Loma 2002:168–170; Mirković 2012:29–30; Mirković 2013:59–61).

The Roman *Municipium S* was located in the northern mountain part of Montenegro, in an unurbanized area, away from the main roads. In Roman times as well as today, in this area settlements arose in the valleys of short, small rivers, such as Ćehotina or in the valley of the great river Lim. This municipium in today's village Komini is situated in the valley of a small river Vezišnica, a tributary of Ćehotina, surrounded by high mountains. It was not connected to any main road, but it was connected through the pass Jabuka with modern Kolovrat near Prijepolje in the Lim valley. From here the roads led north to Posavina and west to the Adriatic Sea. In the vicinity of modern Pljevlja, there were two larger Roman settlements—Komini, the place of municipium, and a smaller one, up the hill St. Ilija, on the southern end of the valley where an earlier local/native settlement was.

The Roman town evolved on the northern edge of a big valley. In the middle of the valley there are two hills, Veliki and Mali Pliješ rich with Bronze Age findings and on the south is St. Ilija hill (Evans 1885:32–37), inhabited since prehistoric times. High mountain ranges close the valley, but the town was not isolated. It was connected to the main roads toward the Neretva River and the Adriatic Sea, as well as Posavina, by vicinal roads.

Due to excavation and geo-magnetic research archaeologists (Mirković 2013:55; Cozzolino et al. 2020:187) estimate that the Roman settlement covered an area of about 40 ha. There were two large necropolises, dated first to fourth century. Many inscriptions and architectural parts were used in later centuries, building and rebuilding new communities (Mirković 2013:55).

The Roman city existed from the late first to the fourth century AD as evidenced by epigraphic monuments. In two areas of necropolis, about 700 burials with stone grave stelae, many of which bear inscriptions, were excavated. The tombs were mentioned by Arthur Evans in the mid-1800s (Evans 1885), and studied thoroughly between 1964 and 1975 (Cermanović-Kuzmanović 1966:4, 77–84; Cermanović-Kuzmanović 1981:43–52; Cermanović-Kuzmanović 1998; Mirković 2012). The discovery of a large quantity of pottery probably produced locally has led to the hypothesis of the existence of a ceramic workshop in *Municipium S* as testified by the discovery of mortars and bricks (Vujović and Cvijetić 2010:105–112). The most valuable finding is a glass vase, a rare example of diatretum, locked in a decorative cage produced by an important Cologne workshop and dated to the fourth century AD.

There is no information, however, on the public, administrative and residential centre of the Roman city. The lack of information on size and the urban conception, the location of its ramparts, gates, and routes of access roads causes a great concern to this day since only a small urban area was covered by archaeological research so far. Two large residential complexes in a rather peripheral area of the city came to light, one with a hypocaust. In the same area, an apsidal building has emerged which houses three altars with inscriptions. The excavation of this building is still ongoing by the Centre for Conservation and Archaeology of Montenegro (Cozzolino et al. 2020:187).

As previously stated, the municipium was probably founded in the late 1st to early second century and it seems that most of the inhabitants immigrated from elsewhere. Nevertheless, there is a strong native Illyrian component displayed in names documented in the inscriptions and shows a strong early presence of the indigenous elite in the ranks of municipal officials. The change in the dynamics of local elites changed the ways culture and identities were constructed on a local level. There was no strict ethnic division among the population in the *Municipium S*. Illyrian ancestry was not an obstacle for a marriage with a Roman citizen. Illyrian names are also attested frequently in families possessing Roman citizenship (Mirković 2012:44). Among many, particularly informative could be an inscription from Otilovići, where the father is named *Pletor*, mother *Maximina*, the daughter has a proper Roman name *Statia Fuscinia*, and the son has the Roman name *Victorinus* (Martinović 20001:266):

D(is) M(anibus) S(acrum)|Pleto|r(i) Maxi|mina vi|va sibi f(ecit) e|t Victor|ino et Stat|ie (!) Fuscinie f|lie (!) car(issimae) pi|ent(issimae) b(ene) m(erenti) marito.

We have another example from this region of a family that shows strong ethnic identity even after the *Constitutio Antoniniana*, with its members bearing Illyrian *cognomina* (third century) (Chastagnol et al. 1985:742):

D(is) M(anibus) s(acrum)/Aur(eliae) Sce/nutae/q(uae) v(ixit) a(nnos) XXXV/Aur(elius) Plares coni(ugi)/k(arissimae) et Lavius et/ Delmanna f(iliae) p(ientissimae)/et sibi vivi p(osuerunt)

In this inscription, Illyrian *cognomina* are *Scenuta, Plares*, and *Lavius* (Rendić-Miočević 1948:9; Alfoldy 1969:228, 289, 267; Mirković 2013:104). The name Delmanna is not previously attested, but Mirković (2012:48) classifies it as a personal name derived from an ethnic (Chastagnol et al. 1985:742).

Another member of the civic elite in *Municipium S* is quite interesting (Martinović 2011:211):

Sexto/Aur(elio) Lupi/ano Lupi/filio princip(i)/ decuriones/collegae et pop/ulares et pere/grini incolae/civi optimo ob/merita pos(uerunt)/epulo dedi/cata/l(oco) d(ato) d(ecreto) d(ecurionum).

Sextus Aurelius Lupianus, Lupi filius, is a full Roman citizen, *civis Romanus optimo iure.* He was probably of peregrine origin; his family was granted citizenship possibly during the reign of Marcus Aurelius or Caracalla. His Romanized peregrine status was perhaps emphasized by his *cognomen* deriving from *lupus*, an important symbol in Roman legend. He was designated *princeps* in the inscription (as well as the sons of Epicadus in the region of Docleates). He perhaps had a special task to fulfil during his office, as *princeps* in a city (Mirković 2012:42). His colleagues, members of *populares*, as well as *peregrini incolae* (thus showing the concord in the city), set up a statue in gratitude for all his excellent deeds and organized a banquet in his honour (Loma 2002:168–170).

Going through all the published inscriptions from *Municipium S* (Fig. 6.4), the percentage of Illyrian/native names is especially high, around 30%, (30 out of 100 names). Miroslava Mirković (2013) made a very thorough analysis of all the native names in *Municipium S*, as well as their onomastic connections with other Dalmatian tribes (Pelcer-Vujačić 2015:69–78). In this region, we can observe several layers of ethnic identities.

Fig. 6.4 Inscriptions of the *Municipium S.* (Photo by Olga Pelcer-Vujačić)

Also, Illyrian naming practices were preserved here much longer than in Doclea and coastal cities, even after the *Constitutio Antoniniana*. Some of the most notable examples of the persistence of Illyrian cognomina are Aplo, Vendo, and Turus (Rendić-Miočević 1948:34; Katičić 1962:100; Alföldy 1969:153, 315, 351, 323).

One can assume, as Džino (2010:180) suggested, that local elites gradually perceived their indigenous ancestry and Roman identity as standing side by side and were equally proud of both, combining them into a "*specific, regional kind of 'Romanness'*".

In the northern mountainous area of Montenegro, there is also one remarkable site, close to *Municipium S*. Valuable archaeological testimonies about the inhabitants of Lever Tara from the prehistoric period are provided by tumuli (grave mounds), from the Žugića gumno site, along the very bank of the Tara River. Near Gradina there is an altar, with a carved votive inscription, dedicated to the oriental deity Mithras (whose cult was respected throughout the Roman Empire in the first and second centuries; generally: Clauss 2012; Gordon 2015:451–455; in Dalmatia: Zotović 1973).

The rock with the ancient altar (Fig. 6.5) is popularly called *Preslica*, so the site itself is known by that name. It consists of a semi-circular empty niche and a votive inscription in Latin, carved in six lines next to it (CIL III 13,849):

Invicto/Aug(usto) Aur(elius)/[A]rgyria/[n]us proc(urator)/[Aug(usti)?] Orfito/[c]o(n)s(ule) l(aetus) l(ibens) m(erito) po(suit)

The given consul is probably Virius Orfitus, so the inscription is dated to 270 AD, but the relief was probably set up earlier (Loma 1998:193–202). In this inscription, although the sun is not explicitly mentioned, it is clear that it is the god Mithras, who was regularly called SOL INVICTUS (Invincible Sun) in the inscriptions. This sanctuary at the Preslica site in Lever Tara is unique in eastern Dalmatia, although there is one inscription dedicated to the same deity in Municipium S (CIL III 12715). Monuments of Mithraism have mostly been found on the main provincial roads, rare specimens have been found outside the route of these roads. Mithreum is located next to the road that in ancient times connected Pljevlja (Municipium S) and Nikšić (Anderba). The god Mithras is a Persian deity of light, revered throughout the Roman Empire. The Mithraic religion is based on the dualistic teaching of Mazdaism about the constant struggle between the principles of good (Ahura Mazda) and evil (Ariman). It was

Fig. 6.5 Mithreum in Lever Tara. (Photo by Olga Pelcer-Vujačić)

not until the time of the Roman Empire, in the time of Vespasian in 79 AD, that the cult of Mithras became popular and spread throughout the ancient world. Adherents of Mithra's religion were soldiers, freedmen, as well as slaves, while women were excluded from Mithra's cult. His cult reached its peak in the middle of the third century, and already in the second half of the second century, it was accepted by the elite of Roman society (Zotović 1973:12). Mithraism took on the role of the leading cult at the time of the predominance of monotheism when the Roman religion moved in that direction. It is fascinating to see a monument dedicated to such an important syncretic and international deity amid the Montenegrin mountains.

6.4 Conclusion

Human presence in the area of the Montenegrin mountains can be traced back to the Palaeolithic, Mesolithic, and during the Bronze Age people inhabited the highlands more often. Occasional outposts in the region of Berane and Pljevlja seem to confirm this. Any attempt to penetrate into a deeper analysis of cultural and population permeations and settlement continuation that took place during prehistory and early historic periods on the territory of Montenegro should directly emphasize that we are faced with relatively modest possibilities since a more serious approach to solving these problems is mainly based on the results of archaeological research, which are still modest in scope in Montenegro. This brief overview shows that during different phases the structure, types and hierarchy of ancient mountain settlements underwent considerable changes. These changes also concerned the types, numbers, and character of central places, which are difficult to differentiate according to modern categories like hamlet, village or city. Nevertheless, it seems sensible that making a tumulus needs effort and implies stronger social hierarchy (see different approaches in Borgna and Müller Celka 2012). Numerous prehistoric settlements in the Montenegrin mountains were used for a very short time as a habitat for a small community of people, indicating more extensive transhumance in the region. Later mountain settlements, during the early historic periods, are founded later than the ones on the coast, culturally more diverse and closer to the valleys and rivers. In the highlands, it seems that the vicinity of Pljevlja has the longest continuity of settlements from Stone Age to the Roman municipium. Much needed further research on these mountainous sites should be focused on possible demographic information that can be gathered from settlement size. Institutions (where there are some) can be studied in the form of public or civic structure types (market, palace, or temple forms), portable material culture (coins, religious paraphernalia), and any textual records. It could also be said that some ancient concepts of sacred geography in the Montenegrin mountains and the religious legitimization of political and economic power are still reflected in modern folklore. Nevertheless, these notions are rather general, and this connection remains just a suggestion. A rigorous assessment of all these different dimensions will provide the basis for undertaking a comprehensive analysis of settlement persistence between settlements in Montenegro and across regions.

References

Alföldy G (1969) Die Personennamen in der römischen Provinz Dalmatia. Winter, Heidelberg

Anonymous (1913) Važan pronalazak u Crnoj Gori [Important find in Montenegro]. Cetinjski glasnik 70:4, 16 Aug 1913

Basler Đ (1965/1966) Crvena stijena [Crvena Stijena]. Starine Crne Gore 3(3–4):107–111

Basler Đ (ed) (1975) Crvena stijena: zbornik radova [Crvena Stijena: collection of papers]. Zajednica kulturnih ustanova, Nikšić

Blagojević M (1999) Crne Gore i Crna Gora [Black forests and Montenegro]. In: Srednovjekovna istorija Crne Gore kao polje istraživanja III. Istorijski institut Crne Gore, Podgorica, pp 57–77

Borgna E, Müller Celka S (eds) (2012) Ancestral landscape: Burial mounds in the Copper and Bronze Ages

(Central and Eastern Europe—Balkans—Adriatic—Aegean, 4th-2nd millennium B.C.). Maison de l'Orient et de la Méditerranée Jean Pouilloux, Lyon

Boone JL (2002) Subsistence strategies and early human population history: an evolutionary ecological perspective. World Archaeol 34(1):6–25. https://doi.org/10.1080/00438240220134232

Božić I (1979) Katuni Crne Gore [Montenegrin Katuns]. In: Nemirno pomorje XV veka [Restless seas of 15th century]. Srpska književna zadruga, Beograd, pp 150–155

Cermanović-Kuzmanović A (1966) Rezultati arheoloških istraživanja na području Municipijuma S… kod Prijepolja—selo Komini [Results of the archaeological research in the area of Municipium S... near Prijepolje—village Komini]. Materijali 4:77–83

Cermanović-Kuzmanović A (1981) Arheološka istraživanja antičkih nekropola u selu Komini kod Pljevalja [Archaeological research of an ancient necropolis in the village of Komini near Pljevlja]. Starinar 31:43–52

Cermanović-Kuzmanović A (1998) Komini—Municipium S... Nekropole [Komini—Municipium S... Necropolae]. Centar za arheološka istraživanja, Filozofski fakultet, Beograd

Chastagnol A, Gagé J, Leglay M (1985) L'année épigraphique. Presses universitaires de France, Paris

Chang C, Tourtellotte PA (1993) Ethnoarchaeological survey of pastoral transhumance sites in the Grevena region, Greece. J Field Archaeol 20(3):249–264. https://doi.org/10.2307/530052

Clauss M (2012) Mithras. Kult und Mysterium. Philipp von Zabern, Mainz

Cozzolino M, Baković M, Borovinić N, Galli G, Gentile V, Jabučanin M, Mauriello P, Merola P, Živanović M (2020) The contribution of geophysics to the knowledge of the hidden archaeological heritage of Montenegro. Geosciences 10(5). https://doi.org/10.3390/geosciences10050187

Ćulafić G (2017) Geographical context: The Local topographic position of Crvena Stijena. In: Whallon R (ed) Crvena Stijena in cultural and ecological context: Multidisciplinary archaeological research in Montenegro. Montenegrin Academy of Sciences and Arts, National Museum of Montenegro, Podgorica, pp 11–21

Ćulafić V (1952) Arheološko bogatstvo doline Lima [Archaeological treasure of the Lim valley]. Istorijski Zapisi 5(1–3):166–170

Dogandžić T, Đuričić Lj (2017) Lithic production strategies in the middle paleolithic of the southern Balkans. Quat Int 450:68–102. https://doi.org/10.1016/j.quaint.2017.03.011

Džino D (2010) Illyricum in Roman politics 229BC–AD 68. Cambridge University Press, Cambridge

Erdeljanović J (1978) Stara Crna Gora [Traditional Montenegro]. Slovo Ljubve, Beograd

Evans A (1885) Antiquarian researches in Illyricum. Archaeologia 49:1–167

Gaastra JS, de Vareilles A, Vander Linden M (2022) Bones and seeds: an integrated approach to understanding the spread of farming across the western Balkans. Environ Archaeol 27(1):44–60. https://doi.org/10.1080/14614103.2019.1578016

Garašanin D, Garašanin M (1967) Istorija Crne Gore I [History of Montenegro]. Redakcija za istoriju Crne Gore, Titograd

Gassiot Ballbè E, Mazzucco N, Clemente Conte I, Rodríguez Antón D, Obea Gómez L, Quesada Carrasco M, Díaz Bonilla S (2017) The beginning of high mountain occupations in the Pyrenees: human settlements and mobility from 18,000 cal BC to 2000 cal BC. In: Catalan J, Ninot JM, Mercè Aniz M (eds) High mountain conservation in a changing world. Springer, Cham, pp 75–105. https://doi.org/10.1007/978-3-319-55982-7_4

Gazivoda D, Vander Linden M, Mihailović D (2019) Sondažna arheološka istraživanja u Vezačkoj pećini u Rvašima, opština Cetinje [Sondage archeological excavations in the Vezačka cave in Rvaši, Cetinje Municipality]. Glasnik Narodnog Muzeja Crne Gore 14:121–124

Gordon R (2015) From Mithra to Roman Mithras. In: Stausberg M, Vevaina YSD, Tessmann A (eds) The Wiley Blackwell companion to Zoroastrianism. Wiley, Chichester, pp 451–455. https://doi.org/10.1002/9781118785539.ch28

Katičić R (1962) Die illyrischen Personennamen in ihrem südöstlichen Verbreitungsgebiet. Živa Antika 12:95–120

Lazić M, Cerović M (2002) Prošlost u kanjonu Tare [The past in the Tara Canyon]. Istorijski Zapisi 75(3–4):191–198

Leković V (1980) Prilog proučavanju kulture gvozdenog doba u Crnoj Gori [Contribution to the study of the iron age culture in Montenegro]. Godišnjak Centra Za Balkanološka Ispitivanja 18:81–88

Loma S (1998) Zum Mithrasdenkmal im Taratal (CIL III 13849). Starinar 49:193–202

Loma S (2002) Princeps i peregrini incolae u municipiumu S(plonistarum?) [Princeps and Peregrini Incolae in Municipium S (plonistarum?)]. Živa Antika 52:168–170

Marković Č (1974) The stratigraphy and chronology of Odmut cave. Archaelogia Iugoslavica 15:7–12

Marković Č (1985) Neolit Crne Gore [Neolithics in Montenegro]. Filozofski fakultet, Beograd

Marković Č (1994a) Tragovi materijalne kulture do osnivanja Cetinja [Traces of material culture till the foundation of Cetinje]. In: Cetinje 1482–1982. CANU, Cetinje, pp 19–25

Marković Č (1994b) Tragovi materijalne kulture ranijih epoha na prostorima Nacionalnog parka "Lovćen" [Traces of material culture of earlier epochs in the area of the National Park "Lovćen]. In: Nacionalni park "Lovćen"—prirodna i kulturna dobra. CANU, Podgorica

Marković Č (2006) Arheologija Crne Gore [Archaeology of Montenegro]. CID, Podgorica

Marković Č (2019) Kulturna i populaciona prožimanja u praistoriji Crne Gore [Cultural and population permeations in the prehistory of Montenegro]. In: Martinović N (ed) Paradigma diskursa kulture u Crnoj Gori. CANU, Podgorica, pp 79–141

Martinović JJ (2011) Antički natpisi u Crnoj Gori = Corpus inscriptionum latinarum et graecarum Montenegri [Ancient Inscriptions from Montenegro]. Matica crnogorska, Podgorica

Mihailović D (2020) Push-and-pull factors of the Middle to Upper Paleolithic transition in the Balkans. Quatern Int 551:47–62. https://doi.org/10.1016/j.quaint.2019.10.010

Mihailović D, Whallon R (2017) Crvena Stijena revisited: the Late Mousterian assemblages. Quat Int 450:36–49. https://doi.org/10.1016/j.quaint.2016.12.026

Mijović P (1992) Bronzanodopske gravire u Maji Popadiji na Prokletijama [Bronze age engravings in Maja Popadija on Prokletije]. Glasnik Odeljenja Umjetnosti CANU 11:11–23

Mirković M (2012) Municipium S: A Roman town in the central Balkans, Komini near Pljevlja, Montenegro. BAR International Series 2357. Archaeopress, Oxford

Mirković M (2013) Municipium S: rimski grad u Kominima kod Pljevalja [Municipium S: A Roman town in Komini near Pljevlja]. Filozofski fakultet, Beograd

Monnier G, Tostevin G, Pajović G, Borovinić N, Baković M (2020) Nova istraživanja paleolitskog nalazišta Crvena Stijena, istorijski kontekst [New research at the Paleolithic site of Crvena Stijena within its historical context]. Istorijski Zapisi 93(1–2):71–108

Morley MW (2017) The geoarchaeology of Crvena Stijena: site formation processes, palaeoenvironments and hominin activity. In: Whallon R (ed) Crvena Stijena in cultural and ecological context: multidisciplinary archaeological research in Montenegro. Montenegrin Academy of Sciences and Arts, National Museum of Montenegro, Podgorica, pp 82–131

Ogrin M (2018) Najstarejše postojanke v Vzhodnih Julijskih Alpah [The oldest settlements in the Eastern Julian Alps]. In: Zorn M, Mikša P, Lačen Benedičič I, Ogrin M, Kunstelj A (eds) Triglav 240. Založba ZRC, Ljubljana, pp 39–47. https://doi.org/10.3986/9789610500841

Parović-Pešikan M (1980) Planinsko zaleđe Rizinijuma [Risinium mountain hinterland]. Arheološki institut, Beograd, Nikšić

Pelcer-Vujačić O (2015) Prilog proučavanju ilirskog identiteta u rimskim naseljima na tlu Crne Gore [Some notes on Illyrian identity in Roman settlements in Montenegro]. Istorijski Zapisi 88(3–4):69–78

Petričević M (2022) Visokoplaninski prostor Crne Gore u svjetlu novih arheoloških otkrića. Prilog proučavanju petroglifskih prikaza u Prokletijama [The high mountain area of Montenegro in the light of new archaeological discoveries. A contribution to the study of petroglyphic representations in Prokletije]. In: Raspopović R, Kujović D, Koprivica T, Laković I, Pelcer-Vujačić O (eds) Crna Gora na političkoj i kulturnoj mapi Evrope - CLIO MAP. Istorijski institut UCG, Podgorica, pp 298–307

Pilaar Birch SE, Vander Linden M (2018) A long hard road... Reviewing the evidence for environmental change and population history in the eastern Adriatic and western Balkans during the Late Pleistocene and Early Holocene. Quatern Int 465:177–191. https://doi.org/10.1016/j.quaint.2016.12.035

Plautus (2011) The Merchant. The Braggart soldier. The Ghost. The Persian. Harvard University Press, Cambridge MA

Rendić-Miočević D (1948) Ilirska onomastika na latinskim natpisima Dalmacije [Illyrian onomastics on Latin inscriptions of Dalmatia]. Slobodna Dalmacija, Split

Šašel A, Šašel J (1978) Inscriptiones latinae quae in Iugoslavia inter annos MCMLX et MCMLXX repertae et editae sunt [Latin inscriptions recorded and published in Yugoslavia between 1960 and 1970]. Narodni muzej, Ljubljana

Šćepanović S (1996) Tragovi starih naseobina na Bjelasici [Traces of older settlements at Bjelasica]. Istorijski Zapisi 69(2):111–125

Srejović D (1974) The Odmut cave—a new facet of the Mesolithic culture of the Balcan Peninsula. Archaeologia Iugoslavica 15:3–7

Stijepčević I, Kovijanić R (1953) Prvi pomeni Crne Gore u kotorskim spomenicima [The first mentions of Montenegro in the documents from Kotor]. Istorijski Zapisi 6(3):231–233

Vander Linden M, Silva F (2021) Dispersals as demographic processes: testing and describing the spread of the Neolithic in the Balkans. Philos Trans R Soc B 376(1816). https://doi.org/10.1098/rstb.2020.0231

Vasić R (1972) Crteži na steni u Gusinju [Rock drawings in Gusinje]. Starinar 23:131–135

Vujović MB, Cvijetić J (2010) Mortaria from Komini—Municipium S (Montenegro). Glasnik Srpskog Arheološkog Društva 26:105–112

Whallon R (ed) (2017) Crvena Stijena in cultural and ecological context: Multidisciplinary archaeological research in Montenegro. Montenegrin Academy of Sciences and Arts, National Museum of Montenegro, Podgorica

Zotović Lj (1973) Mitraizam na tlu Jugoslavije [Mithraism on the territory of Yugoslavia]. Arheološki institut, Beograd

Human Presence in the Eastern Julian Alps from the Copper Age to the Early Middle Ages

7

Marija Ogrin

Abstract

People have inhabited the high mountains of the Eastern Julian Alps on a more permanent basis ever since the transition from the Copper to the Bronze Age (late 4th/early 3rd millennia BC), with more notable habitation remains recorded in later periods, particularly the Late Bronze Age, Roman period, Late Antiquity and Early Middle Ages. Archaeological investigations have revealed sites located in grassland areas suitable for pasturing and areas with iron ore deposits that attracted mining activities.

Keywords

Alpine archaeology · Prehistory · Antiquity · Middle ages · Bohinj · Eastern Julian Alps

7.1 Introduction

Since 1996 archaeological investigations have brought to light a number of new archaeological sites in the high mountains of Slovenia, which challenged the view held until very recently that it was an area only sparsely inhabited until the Middle Ages (Horvat 2020).

M. Ogrin (✉)
ArheoAlpe zavod za kulturo, izobraževanje in turizem Bohinj, Bohinjska Bistrica, Slovenia
e-mail: arheo.alpe@gmail.com

Until recently, the scarce archaeological evidence from the Julian Alps included a stray find of a bronze dagger and pottery sherds unearthed during trial trenching at the Lipanca alp (Šribar 1955:328; Šinkovec 1995:99, Pl. 28:194), two flakes found in a fireplace dated between 6500 and 3500 BC at the V Lazu alp, in the cave of Jama za skalami, as well as prehistoric pottery from Zgornji Povden (Turk and Šercelj 1988:13). In the wider area of Slovenia, the best-known research was the excavations that Srečko Brodar conducted between 1928 and 1935 in the cave of Potočka zijalka, on Mt. Olševa in the eastern Karawanks (Brodar and Brodar 1983; Horvat 2020:13), but also his 1954 discovery of a Palaeolithic cave site in Mokriška jama cave in the Kamnik-Savinja Alps (Brodar 1959).

The late 20th century marks the beginning of a systematic archaeological exploration of the high mountains in Slovenia, which was certainly influenced by the discovery of the Ötztal Ice Man – human mummy trapped in ice in the Ötztal Alps on the border between Austria and Italy (Kothieringer et al. 2015:177). On the initiative of Tone Cevc, a group formed (Bizjak 2020:86–88) that gradually grew into a broad team of experts in alpine archaeology. Their systematic research brings ever new evidence on human presence at high altitudes. Until 2020, some 80 archaeological sites have been found in the Slovenian high mountains (Horvat 2020:12), 37 of which in the Julian Alps. The results of the archaeological investigations thus far have proven occasional human

M. Zorn et al. (eds.), *Discourses on Mountains of Montenegro and Slovenia*,
Historical Geography and Geosciences, https://doi.org/10.1007/978-3-031-99017-5_7

presence in the Upper Palaeolithic and Mesolithic (roughly between the 38th and the 7th millennium BC), and a more permanent habitation from the late 4th millennium BC to the present day, with possible interruptions.

7.2 Presentation of the Study Area

The chapter briefly presents the results of alpine archaeological investigations in the Eastern Julian Alps, where excavations primarily focused on the alps of the Bohinj area. The Eastern Julian Alps extend from the Upper Soča Valley (*Zgornje Posočje*) to the Sava Plain near Bled (Perko and Orožen Adamič 1998:34–54), in which Bohinj is a basin enclosed with the Jelovica Plateau, the Bohinj–Tolmin ridge (*Spodnje Bohinjske gore*), the Triglav mountain range (*Triglavsko pogorje*), as well as the Komna and Pokljuka Plateaus (Fig. 7.1).

Bohinj was already inhabited in the Bronze Age, some three thousand years ago, though permanent habitation only began in the Iron Age. The earliest settlements have been unearthed at Ajdovski gradec, Gradišče above Lepence and Dunaj near Jereka, for which the goods from the graves found in Bitnje, Lepence and Žlan (Gabrovec 1966:243–249; Gabrovec 1974:287–318) indicate most intensive habitation in the 6th and 5th centuries BC. The settlements on Ajdovski gradec and Dunaj near Jereka were also populated in the Late Iron Age, the Roman period and Late Antiquity. The shore of Lake Bohinj where the church of John the Baptist now stands was already inhabited in the Iron Age, while a sanctuary stood there between the 1st and the 4th century, abandoned in the 5th century. Dating to the Early Middle Ages are the cemeteries at Podonjce and Žale near Srednja vas that begin in the 8th century (Šmid 1908:17–26), possibly associated with the gradual settlement of Slavic populations.

Fig. 7.1 Archaeological sites in Bohinj and the surrounding high mountains. (Author of the map Mateja Rihtaršič, authors archive)

In addition to livestock farming, the economic activities significant in Bohinj from the Iron Age to the late 19th century were mining and iron-working (Mohorič 1969; Gabrovec 1974). The villages in the lowland had and still have their economic hinterlands in the highland.

7.3 Methods

The first step in alpine research is the identification of potential archaeological sites suitable for human habitation that takes into account location, accessibility, vegetation cover, local morphology and proximity to a water source.

The following step is a detailed topographic survey of the identified potential archaeological sites that includes test-pitting and collecting charcoal samples. This allows for a precise location of the site and detection of possible man-made earthwork and architectural remains. The typo-chronological attribution of the recovered small finds and radiocarbon analyses of the charcoal samples, in some cases also other analyses, then provide the dating of a site. In the last twenty years, the group of alpine researchers has visited the mountains on innumerable occasions and discovered prehistoric, Roman-period, Late Antique and early medieval archaeological sites in the Julian Alps (Horvat 2020:14–22), some of them multi-period. Finally, the identification phase is followed by archaeological excavation at select locations.

Between 2004 and 2019 (Ogrin 2006; Horvat 2020), test-pitting in the high mountains of the Eastern Julian Alps was conducted at Pečana, Klek, Dolga planja on Prvi Vogel, Poljanica and Kal on Zadnji Vogel, Poljanica on Lepa Komna and Bošinka or Vodene rupe above Dedno polje (Fig. 7.1). Most sites are located several hours of walking away from the nearest mountain huts or roads, which makes excavations all the more challenging. In spite of their remoteness, some alpine sites are endangered due to the expansion of ski slopes or intensive alpine farming (Prvi Vogel, Klek, Pečana) and investigations are all the more necessary in order to effectively protect the archaeological sites in the high mountains.

7.4 Chronological Outline

7.4.1 Stone Age—Mesolithic (8th Millennium BC)

The earliest traces of human presence date some eight thousand years back, to the time of the Middle Stone Age or Mesolithic, when subsistence was based on hunting and gathering, and people sought shelter in caves and rock shelters. One such rock shelter is Jama za skalami near the V Lazu Alp (Fig. 7.1), where two stone flakes and a fireplace have been unearthed (Turk and Šercelj 1988:13). Several Mesolithic sites are known in the Tolmin area to the southwest of Bohinj: Čadrg–Laze, Kašina, Pretovč and Zaprikraj (Mlinar and Turk 2016:30–35).

7.4.2 Copper and Bronze Ages (30th–9th Century BC)

The earliest permanent settlements in Slovenia date to the Late Stone Age, also known as Neolithic, and the Copper Age, but they are located in the lowland (Teržan and Črešnar 2014; Turk et al. 2020), while there are no known contemporary traces of human presence in the high mountains. Towards the end of the Copper Age, or Eneolithic, and the beginning of the Early Bronze Age, i.e. at the transition from the 4th to the 3rd and in the first half of the 3rd millennium BC, the south-eastern Alpine area was relatively sparsely inhabited, even in the lowland. Denser settlement in central Slovenia is only perceptible from the Middle and Late Bronze Age onwards (15th–9th century BC).

In the high mountains of Bohinj, as many as three sites date to the first half of the 3rd millennium BC, namely Poljanica on Lepa Komna, as well as Kal and Poljanica on Zadnji Vogel (Fig. 7.1).

The first one, in the south-eastern part of the Poljanica Alp (Fig. 7.2), was discovered and investigated in 2005 and 2006 (Ogrin 2006:99). Different earthen layers yielded the remains of a hearth, bones, numerous pottery sherds and over a hundred stone flakes, cores and tools. These

Fig. 7.2 Poljanica on Lepa Komna. The archaeological site is marked with an arrow. (Photo by Marija Ogrin)

include blades of quartz, as well as two arrowheads that indicate the use of the bow. The animal bone remains belong to the family of horned—hoofed animals, though the poor condition of the bone material prevents a more precise identification as to the species of sheep, goat or chamois. Pine, spruce and larch wood was used as firewood. The pottery remains include sherds decorated with cordons bearing impressions and stabbing (Fig. 7.7:3), as well as a handle (Fig. 7.7:1). The form and decoration of the pottery, coupled with the radiocarbon dates of the charcoal samples suggest that the site was inhabited in two periods, first towards the end of the Copper and beginning of the Bronze Age (29th–26th century BC) and again in the Late Bronze Age (10th–8th century BC).

The grassy basin of Kal on the Zadnji Vogel Alp was also inhabited very early. A pit was unearthed on a ridge in the centre of this basin that people had dug 1.8 m deep into a natural rock crevice (Ogrin 2006:100–101). It contained great amounts of charcoal and several pottery sherds decorated with fingernail impressions and notches. The charcoal from the bottom of the pit dates to the late 4th/early 3rd millennium BC, that from higher-lying layers to the 16th–13th century BC, showing that Kal was inhabited towards the end of the Copper/beginning of the Bronze Age and again in the Middle Bronze Age.

Several pits dug into a gravelly bedrock have been investigated at Poljanica on Zadnji Vogel, less than an hour's walk from Kal. The largest one, measuring 1.2 m in upper diameter and 2 m in depth, contained an interchanging succession of layers of charcoal and loam that yielded sherds of pots with an everted (Fig. 7.7:2) or cylindrical rim (Fig. 7.3). Similar vessels are known from the Parte–Iščica site in the Ljubljana Marsh (*Ljubljansko barje*), where they are dated to the first half of the 3rd millennium BC (Velušček et al. 2000:85, 99, Pl. 1: 1–12, Pl. 2: 1–12). The radiocarbon analyses of the charcoal samples from this pit indicate the same dating.

7.4.3 Iron Age (8th–1st Century BC)

In the late 9th and early 8th centuries BC, the knowledge of iron extraction spread into the south-eastern Alpine area and with it the use of the metal (Teržan and Črešnar 2014; Turk et al. 2020). Iron enabled the production of tools and weapons, which made it one of the most significant raw materials in the following centuries.

Fig. 7.3 Poljanica on Zadnji Vogel. Rim sherd of a pot. (Photo by Marija Ogrin, drawing by Tamara Korošec)

The rich deposits of iron ore attracted Iron Age people to settle in Bohinj. Despite the numerous settlements and cemeteries from this period known in the lowland of Bohinj, as well as the Bled area (to the east) and the neighbouring Soča Valley (*Posočje*; to the west), only two sites are known in the high mountains: Dolga planja on Prvi Vogel (Ogrin 2020:62) and Bošinka above Dedno polje (Horvat 2020:16).

Dolga planja is a grassy valley on the Prvi Vogel Alp, between Šija and Kratki plaz, where the archaeological site is located in immediate proximity to a ski slope. The investigated archaeological remains include a line of stones, interpreted as the stone foundations of a building, sherds of prehistoric pottery, an iron knife and a bronze brooch (Ogrin 2020:62–63). The brooch survives in two fragments, one with a decorated disc guard and the other foot with catchplate (Fig. 7.7:6). It belongs to Type X Certosa brooches characteristic of the Eastern Alpine costume in the 5th and 4th centuries BC (Teržan 1975:311). The knife and pottery are probably contemporary with the brooch. The habitation layer also revealed numerous pieces of iron ore. The recovered evidence suggests that Dolga planja was inhabited in the late part of the Early Iron Age.

A brooch of the same type was also found at Bošinka above Dedno polje. The Early Iron Age habitation at the site is corroborated by the layer with prehistoric pottery and charcoal, the latter radiocarbon dated to the 8th–6th century BC (Horvat 2020:16).

7.4.4 Roman Period (1st–4th Century)

Bohinj with part of the Upper Carniola (*Gorenjska*) region became integral parts of the Roman Empire in the middle of the 1st century AD.

Human presence in the high mountains increased substantially in this period (Horvat 2013). Sites revealed numerous bronze Roman brooches that fastened people's clothing.

The Klek Alp, located on the northern edge of the Pokljuka Plateau, revealed vast areas of chronologically unattributed mining pits. The earliest habitation remains here were found in the north part of the alp, at Sedlo (Ogrin 2006:103), where two bronze brooches from the 1st or 2nd century (Riha 1979:72–73) came to light. A bronze brooch was unearthed at Dolga planja on Prvi Vogel (Ogrin 2020:63), with two knobs on the bow (Fig. 7.7:7; the Almgren 236c type), such as were typically worn by both men and women in the south-eastern Alpine area in the 1st and 2nd centuries (Garbsch 1965:26–43). Similar brooches were found at Poljanica on Lepa Komna, Kal (Fig. 7.7:8), Poljanica on Zadnji

Fig. 7.4 Poljanica on Zadnji Vogel. Bronze Roman brooch, 1st century AD. (Photo by Marija Ogrin)

Vogel (Fig. 7.4), Pečana and Bošinka above Dedno polje. In addition to the brooch, a small bronze bell with the inscription of VRS was also found at Bošinka (Ogrin 2006:100–104; Horvat 2020:18–22).

7.4.5 Late Antiquity and the Early Middle Ages (5th–9th Century)

The decline of the Roman Empire in the 5th century was followed by turbulent times. In the 7th century, Slavs gradually began to settle the area (Turk et al. 2020) and the contact between the indigenous population and the newcomers is also observable in the high mountains.

The sites with Roman finds largely continue to be inhabited into the Late Roman period and the Early Middle Ages.

The remains of the stone foundations belonging to a hut from the 7th–9th century have been investigated at Dolga planja, on the Prvi Vogel Alp (Fig. 7.5). The hut interior measured 2.8 m in width and up to 4 m in length, its foundations were 0.6–0.8 m thick. They presumably supported the hut's wooden superstructure. The pottery sherds recovered at the site largely comprise jars, some decorated with brushing and some with an incised wavy line, both typical of the south-eastern Alpine area in Late Antiquity and the Early Middle Ages (Modrijan and Milavec 2011:197–200).

Stone foundations of early medieval buildings also came to light at Kal on Zadnji Vogel (Ogrin 2006:102), while early medieval pottery with everted rims and wavy line decoration (Fig. 7.7:4) was unearthed at the Pečana Alp on the Jelovica Plateau.

Two sites came to light at Klek. One is located in the southern part of the alp and yielded pottery and bronze brooches that indicate habitation in Late Antiquity (Ogrin 2006:104). The other lies at Na sedlu and was inhabited in the Roman period and again in the Early Middle Ages. Found in the same layers as diagnostic pottery sherds were also pieces of pisolitic iron ore. The pottery, mostly jars, is decorated with cordons bearing notches or stabs, sometimes appearing in combination with an incised wavy line (Fig. 7.6). Similar sherds were also unearthed on the Krstenica Alp (Lavrenčič 2020:70) and in the settlement at Pristava below Bled Castle. Pottery with such decoration is characteristic of the 9th and 10th centuries (Pleterski 2008:Pl. 25).

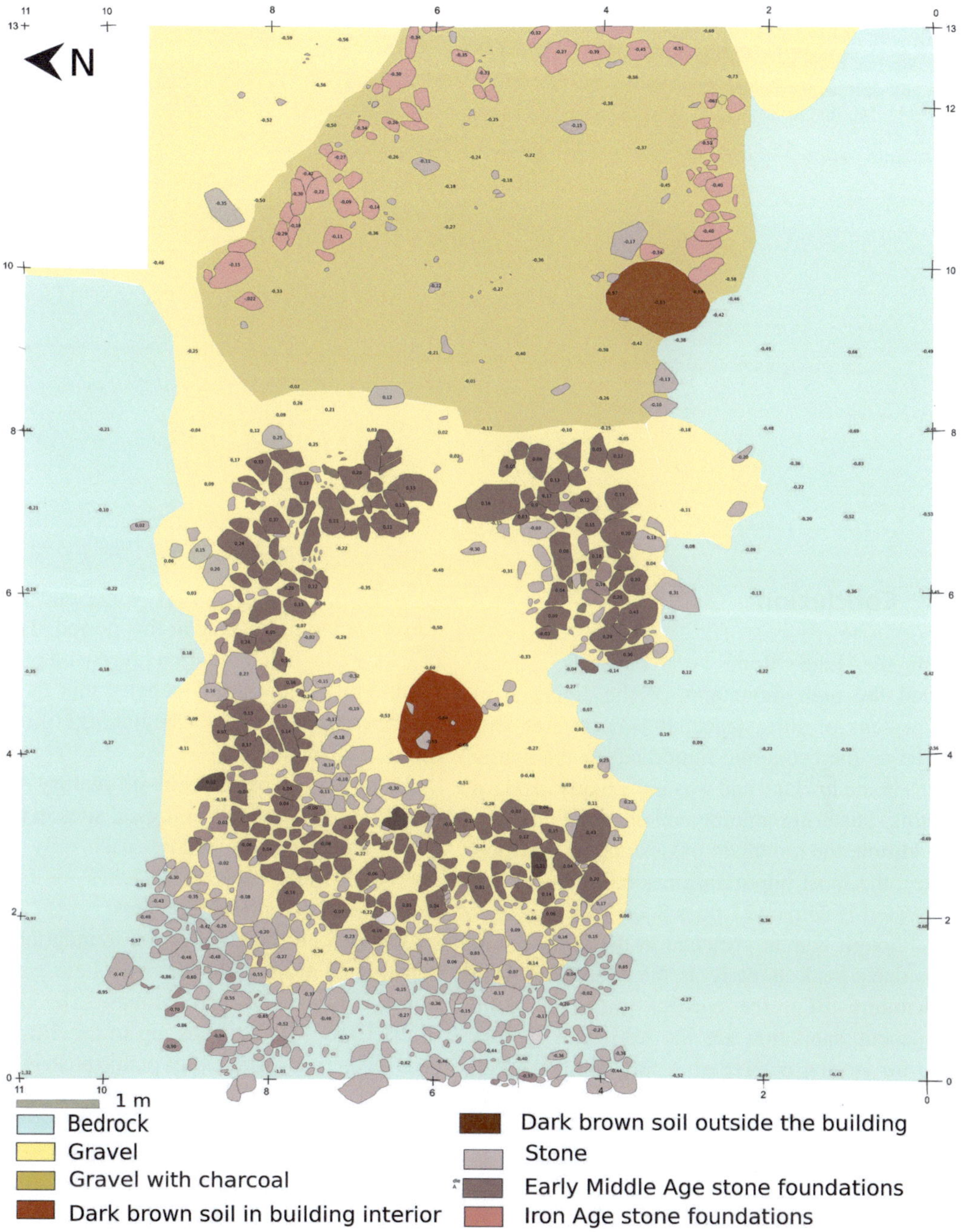

Fig. 7.5 Dolga planja on Prvi Vogel. Foundations of buildings from the Early Iron Age and the Early Middle Ages. (Digital plan by Marija Ogrin)

Fig. 7.6 Klek. Pottery sherds with incised wavy lines, straight horizontal lines and decorated cordon. (Photo by Marija Ogrin)

7.5 Conclusion

From the Late Stone Age onwards, people visited the high mountains of the Julian Alps for reasons of subsistence. It was an area that served as the economic hinterland to the settlements in the lowland and was most likely used for hunting, pasturing, mining and charcoal burning. In addition, the mountain passes hosted transport routes, the most important ones being those that connected Bohinj and Upper Soča Valley.

Although human presence in the high mountains dates back as early as the Mesolithic (8th millennium BC), the earliest traces of more permanent habitation are the sherds of pottery used for storing or preparing food from the late 4th/first half of the 3rd millennium BC. These were discovered at Poljanica on Lepa Komna Plateau, Kal and Poljanica on Zadnji Vogel, all located in grassland areas. The results of the research done on contemporary alpine sites in neighbouring regions suggest these traces can be attributed to the activities of the earliest shepherds (Carrer 2015:8; Korthieringer et al. 2015:190, 191, 194). Because of their advantageous location, the same areas were again inhabited in the Late Bronze Age.

People in the Early Iron Age were attracted to Bohinj for its rich deposits of iron ore. Two alpine sites, Dolga planja on Prvi Vogel and Bošinka above Dedno polje, date to this period. The finds of iron ore in their habitation layers and presumed mining pits in proximity suggest that these sites can be associated with the beginnings of opencast mining of iron ore.

In the Roman period (1st–4th century AD), the number of known archaeological sites in the high mountains increases, though each only yielded few small finds, such as brooches for fastening clothes, and no building remains.

Wooden buildings on stone foundations were constructed at high elevations in Late Antiquity. Some of these buildings continued to be inhabited in the Early Middle Ages, up to the 9th century (Kal on Zadnji Vogel, Dolga planja on Prvi Vogel). A continuity from Late Antiquity to the Early Middle Ages has also been observed at Pečana and Klek, where the layers with pottery also contained pieces of iron ore that suggest mining activities.

From the early medieval period onwards, most of the alps were inhabited, all the way to the foot of Mt Triglav. The intensity of human presence varied in time, influenced by climate, as well as economic, cultural and political changes in the lowland, whereas the type of activities witnessed little change through centuries and included hunting, pasturing, later also dairy farming and mining.

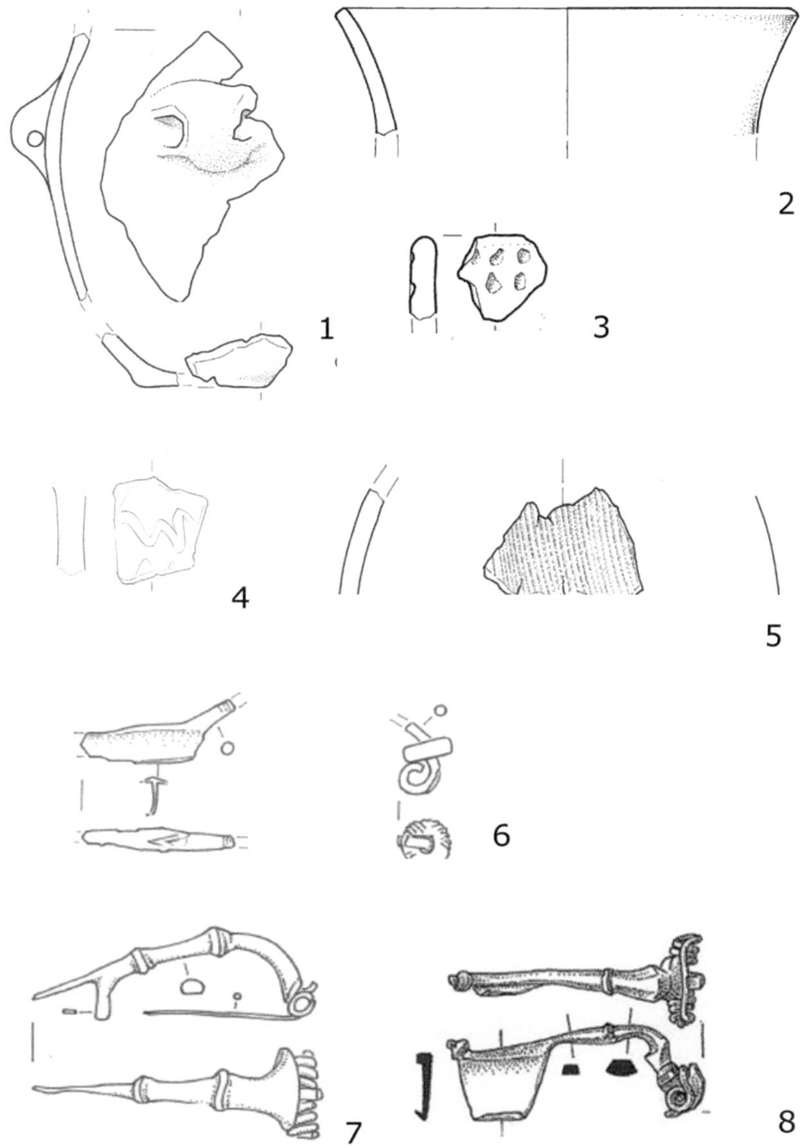

Fig. 7.7 Poljanica on Lepa Komna: 1, 3; Poljanica on Zadnji Vogel: 2, 5; Pečana: 4; Dolga planja on Prvi Vogel: 6, 7; Kal on Zadnji Vogel: 8; 1–5 pottery; 6–8 bronze

Acknowledgements I wish to thank Jana Horvat, from the ZRC SAZU Institute of Archaeology, for reviewing the contribution and Andreja Maver for translating it into English.

References

Bizjak J (2020) Po sledeh prvih rudarjev v visokogorju slovenskih Alp: V spomin Miranu Bremšaku (1961–2017) [Traces of early miners in the high mountain regions of the Slovenian Alps: In memory of Miran Bremšak (1961–2017)]. In: Ogrin M, Horvat J, Bizjak J, Langus K (eds) Življenje v Alpah: življenje v Alpah nekoč in danes – od prve obljudenosti visokogorskega sveta pred več tisoč leti do trajnostnega razvoja turizma danes. ArheoAlpe, Bohinjska Bistrica, pp 85–92

Brodar M (1959) Mokriška jama, nova visokoalpska aurignaška postaja v Jugoslaviji [Mokriška jama, a new Aurignacian site in the high mountains of Yugoslavia]. Razprave SAZU, Classis IV: Historia Naturalis et Medicina 5:416–419

Brodar S, Brodar M (1983) Potočka zijalka [Potočka zijalka Cave]. Dela SAZU 24. SAZU, Ljubljana

Carrer F (2015) Herding strategies, dairy economy and seasonal sites in the Southern Alps: Ethnoarchae-

ological inferences and archaeological implications. J Mediterr Archaeol 28(1):3–22. https://doi.org/10.1558/jmea.v28i1.27499
Gabrovec S (1966) Latensko obdobje na Gorenjskem [The La Tène period in the Gorenjska region]. Arheološki Vestnik 17:243–270
Gabrovec S (1974) Halštatske nekropole v Bohinju [The Hallstatt necropoleis in Bohinj]. Arheološki Vestnik 25:287–305
Garbsch J (1965) Die norisch-pannonische Frauentracht im 1. und 2. Jahrhundert. Müncher Beiträge zur Vor- und Frühgeschichte 11. Beck, München
Horvat J (2013) Roman sites in the high altitude areas of Slovenia. In: Magnani S (ed) Le aree montane come frontere: spazi d'interazione e connettività: atti del convegno internazionale. Aracne, Roma, pp 141–153
Horvat J (2020) Arheološke sledi v slovenskem visokogorju [Archaeological traces in Slovene mountain areas]. In: Ogrin M, Horvat J, Bizjak J, Langus K (eds) Življenje v Alpah: življenje v Alpah nekoč in danes – od prve obljudenosti visokogorskega sveta pred več tisoč leti do trajnostnega razvoja turizma danes. ArheoAlpe, Bohinjska Bistrica, pp 13–25
Korthieringer K, Walser C, Dietre B, Reitmaier T, Haas JN, Lambers K (2015) High impact: early pastoralism and environmental change during the Neolithic and Bronze Age in the Silvretta Alps (Switzerland/Austria) as evidenced by archaeological, palaeoecological and pedological proxies. Zeitschrift für Geomorphologie 59(Suppl 2):177–198. https://doi.org/10.1127/zfg_suppl/2015/S-59210
Lavrenčič L (2020) Sledi prvih koristnikov planine Krstenica [Traces of the first human activities at Krstenica Alp]. In: Ogrin M, Horvat J, Bizjak J, Langus K (eds) Življenje v Alpah nekoč in danes – od prve obljudenosti visokogorskega sveta pred več tisoč leti do trajnostnega razvoja turizma danes. ArheoAlpe, Bohinjska Bistrica, pp 66–72
Mlinar M, Turk M (2016) Prapoti skoz praproti: arheološka topografija dolin Tolminke in Zadlaščice [Ancient ways through winding woodland: archaeological topography of the valleys of the Tolminka and Zadlaščica waterways]. Katalog razstave. Tolminski muzej, Tolmin
Modrijan Z, Milavec T (2011) Poznoantična utrjena naselbina Tonovcev grad pri Kobaridu: Najdbe = Late Antique fortified settlement Tonovcov grad pri Kobaridu: finds. Opera Instituti Archaeologici Sloveniae 24. Založba ZRC, Ljubljana. https://doi.org/10.3986/9789612545871
Mohorič I (1969) Dva tisoč let železarstva na Gorenjskem, 1. del: Doba samoniklega gospodarstva: od pradob do ugasnitve plavžev in opustitve rudarjenja na Karavankah [Two thousand years of ironmaking in Gorenjska Region, Part 1: The age of independent economy: from prehistoric times to the shutdown of the blast furnaces and the abandonment of mining in Karavanks]. Mladinska knjiga, Ljubljana
Ogrin M (2006) Arheološke raziskave v Julijskih Alpah, Bohinj in Blejski kot [Archaeological investigations in the Julian Alps and the Bohinj and Bled areas]. In: Cevc T (ed) Človek v Alpah: desetletje (1996–2006) raziskav o navzočnosti človeka v slovenskih Alpah. Založba ZRC, Ljubljana, pp 96–110
Ogrin M (2020) Prve postojanke na Spodnjih Bohinjskih gorah – Dolga planja na Voglu [Early high-altitude sites in the Lower Bohinj Mountains –Dolga planja on Vogel]. In: Ogrin M, Horvat J, Bizjak J, Langus K (eds) Življenje v Alpah: življenje v Alpah nekoč in danes – od prve obljudenosti visokogorskega sveta pred več tisoč leti do trajnostnega razvoja turizma danes. ArheoAlpe, Bohinjska Bistrica, pp 59–65
Perko D, Orožen Adamič (eds) (1998) Slovenija, pokrajine in ljudje [Slovenia: landscapes and people]. Mladinska knjiga, Ljubljana
Pleterski A (2008) Zgodnjesrednjeveška naselbina na blejski Pristavi: najdbe = Frühmittelalterliche Siedlung Pristava in Bled: Funde. Opera Instituti Archaeologici Sloveniae 14. Založba ZRC, Ljubljana
Riha E (1979) Die römischen Fibeln aus Augst und Kaiseraugst. Forschungen in Augst 3. Augst
Šinkovec I (1995) Katalog posameznih kovinskih najdb bakrene in bronaste dobe = Catalogue of individual metal finds from the eneolithic and bronze ages. In: Teržan B (ed) Depojske in posamezne kovinske najdbe bakrene in bronaste dobe na Slovenskem = Hoards and individual metal finds from the eneolithic and bronze ages in Slovenia. Katalogi in monografije 29. Narodni muzej, Ljubljana, pp 29–127
Šmid V (1908) Altslovenische Gräber Krains: Das Gräberfeld in Wocheiner Mitterdorf. Carniola 1:17–26
Šribar V (1955) Najdiščno poročilo o sondah na planini Lipanca [Report on the trial trenching conducted on the Lipanca Alp]. Geografski Zbornik 3:321–345
Teržan B (1975) Certoška fibula [Die Certosafibel]. Arheološki Vestnik 25:317–536
Teržan B, Črešnar M (2014) Absolutno datiranje bronaste in železne dobe na Slovenskem = Absolute dating of Bronze and Iron Ages in Slovenia. Katalogi in monografije 40. Narodni muzej, Ljubljana
Turk I, Šercelj A (1988) Najstarejši dokazani obisk Ratitovca [Earliest evidence of visits to Mt. Ratitovec]. Loški Razgledi 35:11–18
Turk M, Velušček A, Pavlin P, Tecco Hvala S, Grahek L, Horvat J, Šašel Kos M, Modrijan Z, Pleterski A (2020) The history of Slovenia: archaeological evidence from prehistory to the Slavs. In: Perko D, Ciglič R, Zorn M (eds) The geography of Slovenia: small but diverse. Springer, Cham, pp 127–141. https://doi.org/10.1007/978-3-030-14066-3_8
Velušček A, Čufar K, Levanič T (2000) Parte-Iščica, arheološke in dendrokronološke raziskave [Parte–Iščica, archaeological and dendrochronological investigations]. Arheološki Vestnik 51:83–107

8 Prehistory of High-Altitude Landscape in Montenegro: Mount Sinjajevina Case Study

Miloš B. Petričević

Abstract

Montenegrin high-altitude landscapes have often been regarded as isolated and generally marginal environments. Our knowledge and understanding of prehistoric activity in zones of High Karst Unit is still limited. Recent archaeological research has identified the highest altitude archaeological sites in Montenegro close to 2,000 m a.s.l. A group of previously unknown burial mounds and hillforts has been discovered using remote sensing techniques. Chronological dating of archaeological sites at the mountain Sinjajevina cannot be provided without archaeological excavation. When burial mounds at Sinjajevina were erected remains an open question. New archaeological investigations in high-altitude zones shed new light on human capacity to survive in extreme and isolated environments, but on the other hand tell us about prehistoric origins of transhumant pastoralism in Montenegro.

Keywords

Prehistory · High-altitude archaeology · Remote sensing · Burial mound · Hillfort · Montenegro

M. B. Petričević (✉)
Museums of Kotor, Kotor, Montenegro
e-mail: milos.petricevic@muzejikotor.me

8.1 Introduction

The high-altitude region, also known as Montenegrin Hills (*Crnogorska brda*), extends over the largest part of northeastern Montenegro (Bešić et al. 1967:12). This is the area of high mountains such as Durmitor, Sinjajevina, Bjelasica, Visitor, Vojnik, Lola and Kapa Moračka. Sinjajevina is a spacious mountain area stretching south-eastwards from Drobnjačka Jezera (Milojević 1935:76). The mountain boundaries are morphologically finely defined due to the surrounding deep river valleys. This the largest Montenegrin mountain in terms of base area is located between the canyons of Tara River in the north-east and Morača River in the south-west. It is bordered with valleys of Bukovica and Tušinja on the north-west and west side, respectively. Stretching north-westward to the Durmitor plain, Sinjajevina creates a vast expanse due to which it is often referred to as the Balkan's biggest pasture (Fig. 8.1).

Sinjajevina's ultra-prominent peaks are: Torna (2,217 m), Vranova Glava (2,215 m), Jablanov Vrh (2,203 m) and Gradišta (2,174 m). This plain is a part of the largest limestone terrain from the Middle to Late Jurassic age (Fuštić and Đuretić 2000:55). It was created by secondary formation through erosion-accumulative processes of fluvial, karst and glacial erosion. The predominant geomorphological features of this karst landscape are dolines which were partly reshaped by glacial erosion (Lipovac 1987:28).

M. Zorn et al. (eds.), *Discourses on Mountains of Montenegro and Slovenia*,
Historical Geography and Geosciences, https://doi.org/10.1007/978-3-031-99017-5_8

Fig. 8.1 Red rectangle shows the extent of research area (*Source* ESRI). (Map by Miloš B. Petričević)

The investigation of prehistoric sites in the high-altitude areas of Montenegro has not really been the national archaeologists' focus of interest. Moreover, it could be said that their presence was considered as highly unlikely. The prejudice that mountain regions are an area of marginal and occasional human activity has contributed to the fact that high-altitude archaeology was never regarded as an important constituent of the national archaeology. Both low-level of knowledge and remoteness of Montenegrin high-altitude areas still make it an unknown to Montenegrin archaeology, especially in terms of prehistory. Precisely these areas, where occasional and scarce settlements *katuni* (Laković 2017) of the remaining Montenegrin herders are located, represent a challenge to discovering new archaeological sites.

In his works published in the middle of the 1970s, eminent Montenegrin geographer Milisav Lutovac was the first to make claims of possible prehistoric activities in Montenegrin high-altitude areas. Besides describing the location, natural and social characteristics of Bjelasica, the author geographically identifies and describes the remains of a settlement which, according to him, leads to "a distant antiquity". The data relating to the existence of *Illyrian* settlement and necropolises above Bardov Dol (Lutovac 1975:17) are of crucial importance. According to the description, it can be concluded that there were low houses with an extremely prolonged base and dry-stone foundations, and adjacent single graves. The data presented by Milisav Lutovac (1975) have not been confirmed through the aerial photographic interpretation. The only trace of a prehistoric activity has been identified at Bardova Kosa which is a stone burial mound (*tumulus*) with a diameter of 15 m at over 1,900 m a.s.l. This is currently the height maximum in terms of vertical distribution of prehistoric *tumuli* in Montenegro.

Discoveries of necropolises in the high mountain areas are not uncommon. For instance, the presence of several necropolises, whose age is yet to be determined, have been recorded in Prokletije (near Lake Ropojansko), in *katun* Popadija and at the foot of Maja e Can Peak. According to the local tradition, these necropolises date from the pre-Slavic epoch, i.e. Illyrian period. Lutovac (1976) sees the presence of necropolises in high-altitude areas as a relict of a ceaseless battle for resources, i.e. battle for ownership of pastures. Their abundance is the evidence of a high-level exploitation of the remote high-altitude zones as an important constituent of maintaining economic stability of herder communities.

Nikola Lipovac (1987) made the first direct confirmation of the presence of prehistoric remains by identifying them as "piles of rocks" and "ramparts" on Sinjajevina. He devoted particular attention to the site located at the foot of Stražnica containing remains which are, presumably, the foundations of former settlements (Lipovac 1987:177). The unearthing of a Bronze Age axe head in Gornja Bijela near Boan (Lipovac 1987:101) represents an important testament of life in isolated and hardly accessible mountains in this region. The most important archaeological discovery in the high-altitude karst areas are petroglyphs in Prokletije. The first time a wider scientific audience came to know about the presence of petroglyphs in Prokletije was through the works of Rastko Vasić (1972) and Pavle Mijović (1992). The ongoing research related to the high-altitude zones of Prokletije has led to the discovery of a previously unknown petroglyph site close to 2,000 m a.s.l. and a sundial, possibly prehistoric (Petričević and Vodolazhskaya 2021).

The above stated bibliographic data, as well as accidental archaeological records, have indicated possible organized exploitation of high-altitude areas by a transhumant prehistoric community. For thousands of years, high-altitude areas have been a place of living, a stage of many fights and often a place of death (Stojanović 1976), which is testified by recent archaeological finds in Sinjajevina.

The attention of the national archaeology has been, by default, aimed at discovering new sites in close or distant proximity to already known or explored sites. Unfortunately, such spatial centrism is a dominant component of the process of discovering new sites. It appears that the main criterion in the field survey strategy is closely related to the nature of landscape morphology and belief that significant former anthropogenic activity should not be expected in high-altitude areas (Walsh et al. 2006; Della Casa and Walsh 2007). Such tendency, as a result of evident subjectivity of a researcher, leads to a reductive and harmful generalization of a presumed geographical distribution of archaeological sites.

The focus of modern Montenegrin archaeological practice during the process of field survey is still placed on verbal data which might confirm the presence of archaeological sites. Researchers, as a rule, collect indicative information from members of local community which they subsequently try to validate on field. However, it remains unclear in what way they are supposed to explore the areas which are nowadays unpopulated or seasonally populated. In 1948 Montenegro had the biggest agrarian population in the former Yugoslavia (75.9%). Planned industrialization of this once socialist republic had led to its rapid deagrarization and deruralization (Gluščević 2004:355). As industry became the dominant branch of economic activity, it suppressed agriculture and livestock farming, thus causing migrations of rural population to urban areas. This led to a drastical decline in transhumant pastorialism and the abandonment of *katuni* in high-altitude areas. Today Sinjajevina represents an area with a tendency of complete abandonment.

Conducting a field survey in hardly accessible area spreading over a large and immense surface is complicated in terms of identifying the extent of former human activity. The specific geomorphic features of Sinjajevina, whose surface is covered with numerous dolines, make it hard to carry out field surveys and identify possible archaeological indicators. As the height above mean sea level increases, the probability of there being an archaeological site decreases as its mostly ephemeral nature reduces the visibility of archaeological records. Accordingly, this vast karst terrain was used to test the possibility of remote sensing in high-altitude areas aimed at collecting basic data on the amount of archaeological sites in this area.

In accordance with contemporary methodological and technological innovations, landscape archaeology will adopt new methods of data collection creating in such manner their sensible interrelatedness in attempt to successfully and rationally explore larger landscape units (Novaković 2008:35). Archaeological site detection does not recognize a universal technique and, in most cases, the specific qualities and availability of information about a particular area affects its choice. In this particular case, aerial photo interpretation technique was used and this proved to be extremely effective in detecting archaeological sites in hardly accessible areas. The gathering of data on the specifics of archaeological areas has been conducted through interpreting historical maps and toponomastics, historic aerial photographs, digital aerial photographs, data acquired using unmanned aerial vehicle (UAV) and targeted field survey. In order to study the remains of human activity in the area, aerial photogrammetry was used to record several representative sites to create technical documentation and preserve data on the state of discovered sites. Photogrammetry enables the reconstruction of the position, orientation and size of an artifact based on photographs (Kraus 2011:1). The primary purpose of photogrammetric modelling is to reconstruct a three-dimensional digital replica of an artefact or context from a series of photographs which represent a source of spatial information (Luhmann et al. 2011:2).

8.2 Toponymy

In the area of Sinjajevina the local tradition mostly exists as a relict of collective memory formed after the turmoil in the second half of the nineteenth century and is, consequently, highly intertwined with the landscape in close vicinity to *katuni.* The lore surrounding piles stones and fortified hilltops near Okrugljak Katun represents a reflection of a younger local tradition. The same mentions that Turks were buried under a "pile of rocks", while the so-called *šančevi* (*šanac*, singular form) belong to a respected Baig family called Mušović, whose members were captains from Nikšić and Kolašin (Konjević 2017:591). Well until 1878 Sinjajevina was under the rule of Turkish feudal masters from Mušović and Kalić families (Lutovac 1976:35). There is a memory of a newer epoch in the backdrop of this type of orality, which is also partly mirrored in the prehistoric landscape. Based on conversations with local people in Okrugljak Katun, data about the presence of, as they say, the "oldest buildings" and cemeteries at the foot of Mušovića Šanac was gathered, but it was not possible to determine the age of the sites based on these testimonies.

The toponymy of the area is mostly represented by the recent layer of Slavic toponymy layered with loanwords. Therefore, oronym *Šanac* should be directly linked to the German *schanze*, i.e. high profile fortification (Skok 1974:381). The equivalent to the noun *schanze* in Montenegrin language is a ditch, i.e. trench which was an integral part of defensive architecture. Interestingly, in Montenegro the toponym *Šanac* is predominately used as an oronym which leads us to assume that a new equivalent word for the term for a fortified hilltop. In certain cases, the toponym *Šanac* can be equalized with a walled, fenced, i.e. fortified hilltop, or simply a hillfort. Geographical names, besides them being primary the target of linguistic research, can be the object of archaeological research (Tomičić 2009:230). The potential and validation of cartographic sources as the means to discover archaeological remains have not yet been completely recognized within the framework of archaeological practices in Montenegro. With global informatization and development of geoinformation systems the cartography, as a detection tool, is growing in significance. Computerized cartography enables a more efficient and faster localization of toponomastic indicators which can help us to discover new archaeological sites. By analysing the registry of names of places located in the area of Sinjajevina, it was possible to identify several toponyms which could denote the presence of archaeological remains (e.g. *Gomile*—piles, *Gradišta*—hillforts, *Gradište*—hillfort, *Jerinina kula*—Jerina's tower, *Stražnica*—fortification, *Šanac*—ditch).

Remote sensing of the high-altitude areas of Montenegrin Hills begun in 2017 and encom-

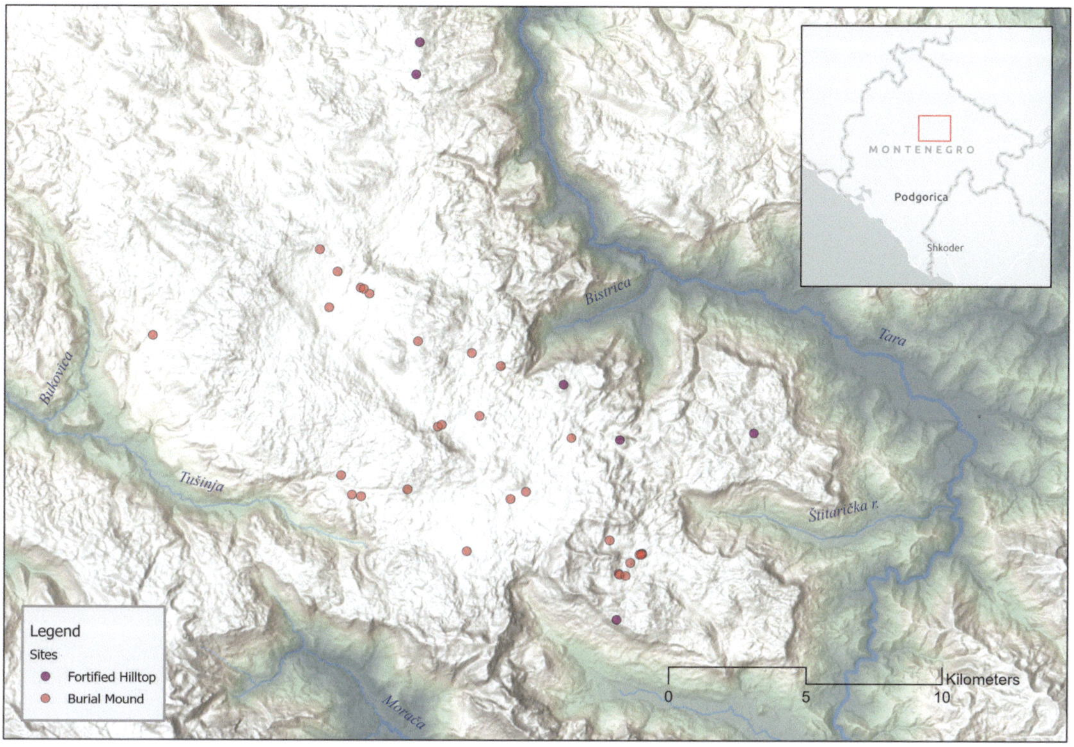

Fig. 8.2 Spatial distribution of archaeological sites in the Sinjajevina's plain. (Map by Miloš B. Petričević)

passed the areas of mountains Sinjajevina and Durmitor. Fifty-nine prehistoric *tumuli* and ten fortified hilltops have been detected by employing remote sensing techniques and field campaigns (Fig. 8.2).

8.3 Fortified Hilltops

For now, the discovered fortified hilltops represent the biggest challenge as far as interpretation is concerned. The fact that there are, for the first time, concrete data collected via remote sensing about inaccessible areas of the High Karst Unit is thrilling but at the same time limiting because, based on the results obtained, there is not really anything else to say about the nature of these fortified locations. For the time being, these discoveries have no further analogies which would help in determining their nature or chronology, so leave their final attribution is still open.

8.3.1 Ckara Šanac (1,836 m a.s.l.)

A dry-stone construction with long and irregular form was discovered in close vicinity to Ckara's Peak by aerial photographic interpretation. The dry stone wall almost entirely surrounds the highest point of the hill. The diameter of its big axis in the northeast to southwest direction is 55 m, while the diameter of its small axis in the northeast to southwest direction is 20 m. The height of the preserved part of the dry-stone wall is approximately 30 to 40 cm.

The fortified space manifests only recent anthropogenic activities, i.e. conical piles of stones which represent stone cairns, i.e. trail markers. They are specific to hardly accessible in mountain areas, but can also be found in coastal regions (Mizin 2013:314).

The structure of the dry-stone wall is made of stones of different sizes and shapes which were collected and removed from the soil in the exact

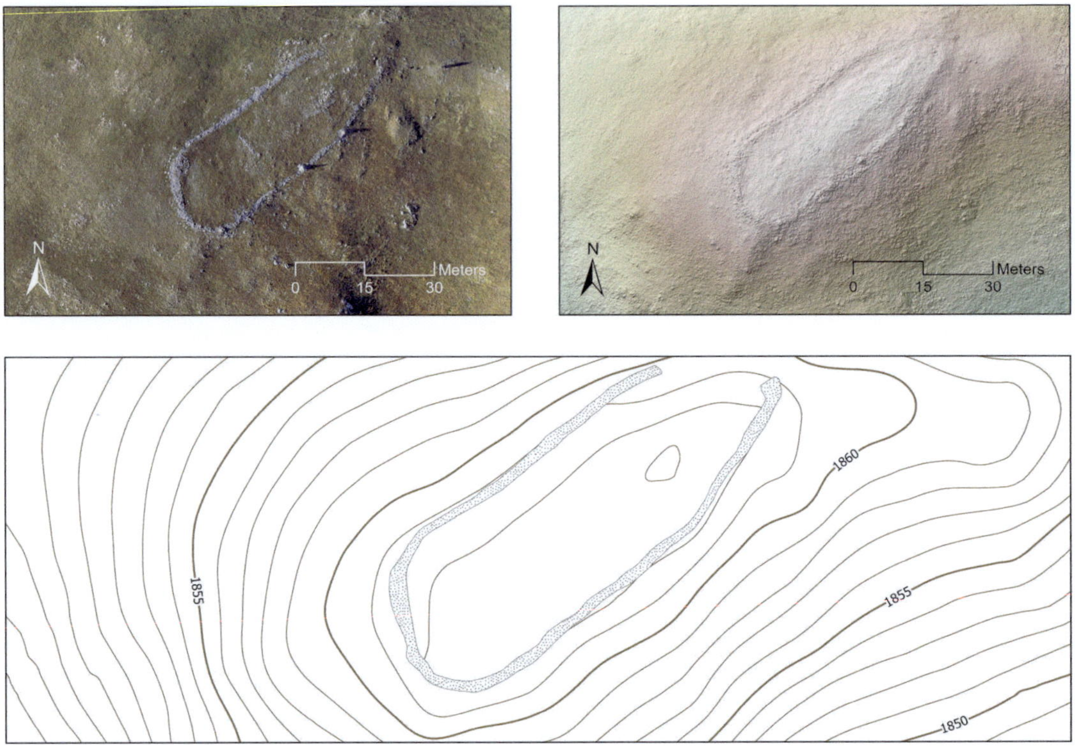

Fig. 8.3 Ckara Šanac: orthophoto (upper left corner), hill-shaded relief (upper right corner) and elevation map (at the bottom). (Generated by Miloš B. Petričević)

or close proximity to the fortified position. There is a noticeable abrasion to the stone material which most likely indicates that the material was actively transported by a glacier. Walls of the structure appear to be a mixture of glacial till, talus blocks, and possibly local rock fragments. No presence of archaeological remains was detected through the observation the stone mass. The visible surface distribution of the stones suggests the original stone construction was exposed for quite some time to the exogenic factors, and it represents a denudated mass of the original wall construction which was restored on several occasions (Fig. 8.3).

8.3.2 Mušovića Šanac (1,913 m a.s.l.)

The presence of a dry-stone wall was also detected above Katun Okrugljak on the location Mušovića Šanac. The hillfort plateau is walled between four highest points of the hill. The defensive walls connect the highest points of the hill by enclosing a polygonal space. In such manner, the distinctive plasticity of the terrain was used to achieve a more efficient and optimal construction. The inner space of the hillfort is slightly inclined towards the east which leaves the impression that the defended space is two-levelled.

As in the case of Ckara Šanac, the wall structure is made of stones of different sizes and shapes which were collected and removed from the soil in exact or close proximity to the fortified position. The wall structure is heterogenic and made mostly of glacial till, talus blocks, and possibly local rock fragments. No presence of archaeological remains was detected through the observation of the stone mass (Fig. 8.4).

8.4 Stone *Tumuli*

The discovered *tumuli* belong to the medium-sized formations of this kind. In most cases the

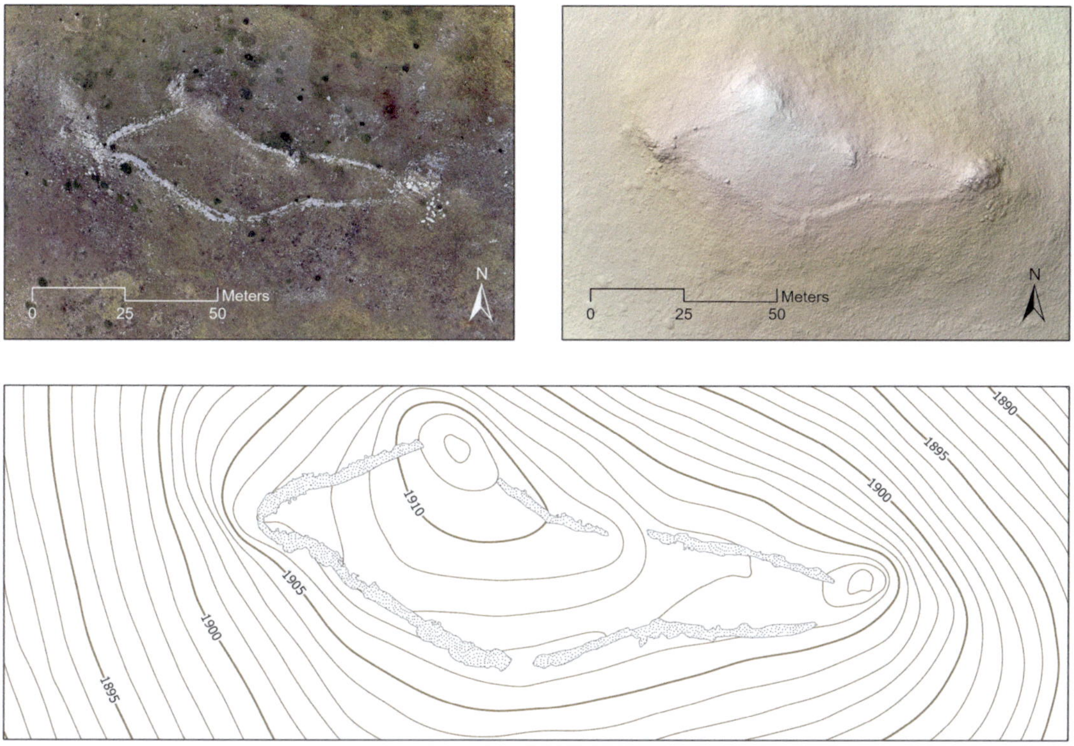

Fig. 8.4 Mušovića Šanac: orthophoto (upper left corner), hill-shaded relief (upper right corner) and elevation map (at the bottom). (Generated by Miloš B. Petričević)

corpuses of *tumuli* are made of stone piles and are built in the same pattern (Fig. 8.5). Certain discrepancies in the construction of a group of smaller *tumuli* do appear. Construction-wise, the visible parts of the stone pile mass reveal a heterogenic structure of the mound which is made of big and small-sized stones and earth.

The maximal elevation difference between the distributions of the prehistoric *tumuli* mentioned above is 617 m. The minimum elevation at which

Fig. 8.5 Real-Time Kinematic (RTK) mapping of typical stone tumulus in the Sinjajevina's plain. (Photo by Miloš B. Petričević)

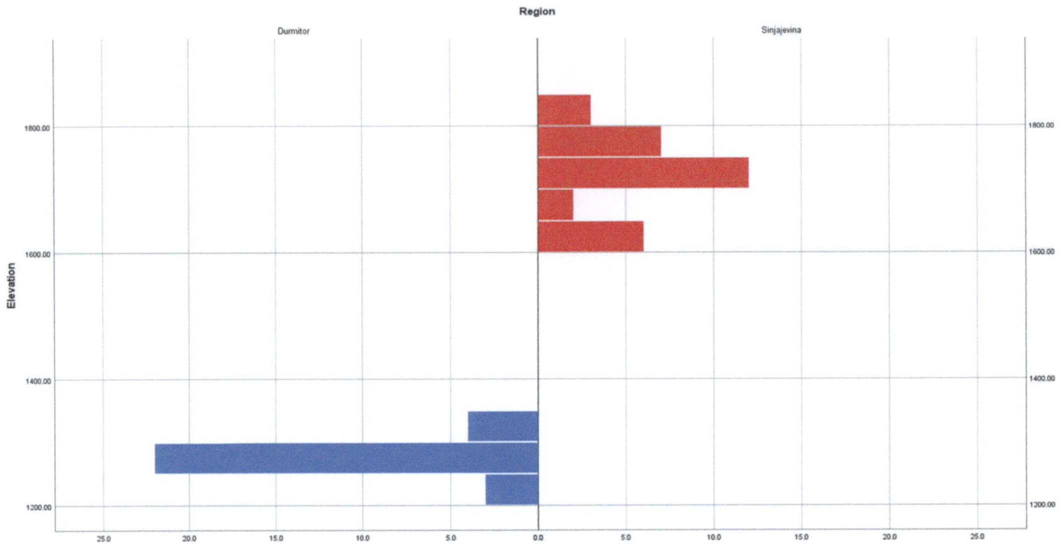

Fig. 8.6 Vertical distribution of burial mounds (*tumuli*) on Durmitor and Sinjajevina. (Graph by Miloš B. Petričević)

they are located is 1,220 m a.s.l. while the maximal elevation is at 1,837 m a.s.l. On average, the *tumuli* on Sinjevina are located on greater elevations above sea level than the ones located in the area of Durmitor and Drobnjačka jezera which can be attributed to the geomorphological features of the area in question (Fig. 8.6).

8.4.1 Stone *Tumulus* in Okrugljak Katun (1,793 m a.s.l.)

A tumulus with a 15 m diameter and a visible object slightly dislocated from its geometric centre was discovered west of Okrugljak Katun via aerial photographic interpretation. Subsequent ground truthing revealed that the object is a stone cairn which was, according to the herders, erected in the 1960s. The situation at site is rather unclear since the mound, which at first glance appears to be a stone tumulus, is missing. Clearly, the modest thickness of the stone mound has arose the question of the authenticity of the initial aerial photographic interpretation. However, orthophoto map interpretation and digital elevation model gave a better insight into the very nature of the discovered location. The distinctiveness of the vegetation cover has enabled to extend the analysis to a wider area of data visualization. The low plant density and relative uniformity of the vegetation cover made it easier to acquire more precise data via aerial imagery. In such way a precise digital surface model (DSM) was generated. For the display of data generated through the structure from motion (SfM), tools for the visualization of data obtained via aerial laser scanning (ALS) were used. The clarity/visibility of the tumulus' architectonic structure was increased by analysing the DSM with a specialized relief visualisation toolbox (RVT) and local relief model tool (LRM) (Kokalj et al. 2011; Hesse 2010). Peripheral parts of the mound have an apparent circular form and represent the original limits of the tumulus' base (Fig. 8.7).

8.4.2 *Tumuli* in Kolibština Katun (1,750 m a.s.l.)

The *tumuli* in the area of Kolibština belong to the medium-sized formations of this kind with an approximate 14 to 16 m diameter. They are built on the hills above a natural pathway to Staračko Polje to spatially emphasize a grave site. At first sight, the tumulus is made of stone piles mostly composed of glacial till, talus blocks and local rock fragments of undetermined size. A profile across of the devastated tumulus showed that

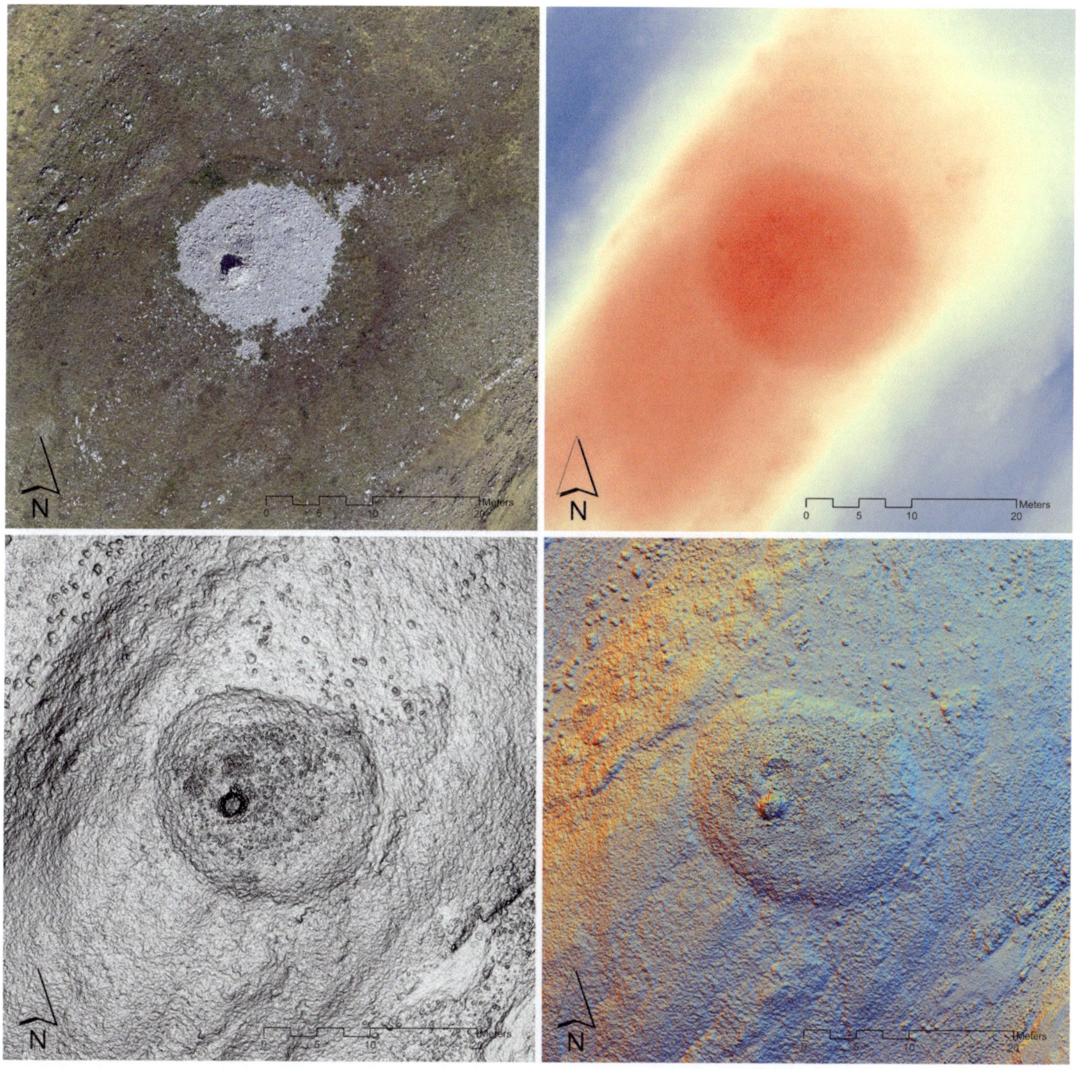

Fig. 8.7 Burial mound in Okrugljak Katun: orthophoto (upper left corner) and visualization of data using RVT and LRM tools. (Generated by Miloš B. Petričević)

the stone mound is not entirely made of stone (Fig. 8.8).

8.4.3 *Tumuli* on Margita (1,715 m a.s.l.)

Tumuli clusters on Margita represent the most dynamic part of the prehistoric landscape of Sinjajevina (Fig. 8.9). This tumuli cluster is located south-east from Srnjački Izvor on the very edge of a *katun* settlement which is grouped around Srnjačka Lokva. The dense necropolis is made of five *tumuli* of different sizes located on a relatively small distance from each other making it look like they are building a coherent sepulchral whole. This is for now the biggest registered cluster in the area of Sinjajevina. The recorded *tumuli* were not found in their original state. In addition, there are visible damages which occurred due to the erection of objects penetrating the *tumuli's* stone piles. These objects, with their size and shape, share similarities with herder's huts and stone corrals (Fig. 8.10).

Fig. 8.8 *Tumuli* in Kolibštine Katun: orthophoto (left) and hill-shaded relief. (Generated by Miloš B. Petričević)

Fig. 8.9 *Tumuli* on Margita with the Mužice top in the background. (Photo by Miloš B. Petričević)

8.5 Issues Related to Settlement Archaeology of High-Altitude Zones

The issues concerning the spatial distribution of prehistoric *tumuli* in the Central Balkans have been recognized as an important research component throughout Yugoslavian archaeological literature (Garašanin 1983; Lazić 1989). The discovery of prehistoric *tumuli* in the high mountain areas in Montenegro will most definitely set off new considerations among archaeolog-

Fig. 8.10 *Tumuli* cluster on Margita with visible remains of what are most likely to be herder's huts and stone corrals. (Photo by Miloš B. Petričević)

ical scientific circles. The discovered prehistoric *tumuli* in above 1,500 m a.s.l. zones indicate an efficient use of resources in high-altitude areas in prehistoric times. For now, we are able to define comparative frameworks in the analysis of the relationship between the mobility and settlement patterns in the prehistory of high-altitude areas of the Dinaric Alps only through comparative analyses of the customs of recent herders in the Southeastern Europe (Marković 1971; Gušić 1976; Lutovac 1976; Sterud 1978).

In support of transhumant pastoralism as a way of livelihood of the members of the Bronze Age communities, Milutin Garašanin (1983) attempted to answer the question why there were few, if any, permanent settlements in certain cultural groups of the Central Balkans. His answer consisted of a rather generalized argumentation that settlements had temporal, seasonal character (Garašanin 1983:714). However, it is necessary to point out that the emphasis in archaeological research of this area has been mainly put on necropolises, while the research of potential settlements has been ignored. Accordingly, the authors were limited by the quantity of useful information in determining the type of settlement and how its members earned for their livelihood. Our knowledge on the cultures of the Bronze Age in Montenegro is still exclusively founded on the explored necropolises under *tumuli* and scarce cave habitats (Primas and Della Casa 1996; Pušić 1999; Szymczak et al. 2010).

The fact is that at the moment we do not have a sufficient number of identified settlements which prevents us to define more precisely the settlement pattern. The assumption that necropolises under *tumuli* in high-altitude zones are a relict of transhumant activates of prehistoric communities has not been confirmed yet precisely because of the inability to detect settlements in the High Karst Unit and in the areas of the so-called winter quarters on lower heights above mean sea level.

On the other hand, monumental grave memorials like *tumuli* indicate an enviable level of community social organization. The concept of building an artificial hill of earth or stone over the remains of a dead is such a widespread occurrence that it can be defined as a phenomenon (Govedarica 2010). A tumulus, as a specific grave site, is of a greater significance than a mere grave memorial. It represents a personification of the power of a deceased, a place of cult and magic, a significant spatial join in direct connection with not only the territory but the time as well. Precisely the monumentality of a grave mound is an important indicator of social class differentiation, so it can be claimed that *tumuli* are resting places of tribal leaders or other dignitaries. Necropolises represent a very strong link in the chain of understanding cultural groups. Burial is a ritual which is very important for a community and hence is least susceptible to change and improvisation. The burial of community members on seasonal summer pastures was undoubtedly performed according to predefined burial rituals. However, the availability of materials needed for stone *tumuli* erection in the area of Sinjajevina was greatly limited. Therefore, summer pastures were used for the erection of *tumuli* precisely as a response to the specific environment. In such manner the necessary monumentality of *tumuli* was achieved by skilful morphological intertwining with smaller relief forms.

In most cases the discovered *tumuli* on Sinjajevina correlate to the very location of deserted medieval seasonal settlements—*katuni* or the surrounding area of the same. Logically, this brings up the question: does this occurrence imply a connection between the spatial distribution of grave sites and distribution of prehistoric seasonal settlements?

Another significant aspect may help to understand the High Karst Unit inhabitation pattern, which are freshwater springs. Herders had to store water by using or adapting topographic depressions in which snowmelt and rain runoff would accumulate. These surface water bodies, known in common parlance as *puddles* (mne. *lokva*) (Fig. 8.11), have been and still are the main source of livestock water. Several factors influence the location of seasonal settlements—*katuni*: terrain configuration, proximity and size of a pasture, water sources and available building material (Laković 2017:22). If we take into account that Sinjajevina is a waterless area with scarce surface water, then spring water should be regarded as an important factor in determining the location of the settlements. This assumption, theory-wise, enables the establishment of a prediction model for locating wider areas of prehistoric seasonal settlements.

The presumed settlement pattern provided by Milutin Garašanin (1983) could be amended in the sense of indirect defining of its dual character according to which the presence of permanent settlements should be confined to the lower heights above mean sea level while seasonal settlements—*katuni* to the upper High Karst Unit, which represent, at the same time, the utmost reach of herders' summer habitat. This brings up the question: why are the discovered *tumuli* located in the high-altitude zones of the southern Dinaric Alps? According to Gušić (1976), transhumant pastoralism corresponds to a low population density. Due to the increase in population density, herders shall be forced to find new and richer pastures in higher altitude zones (Gušić 1976:151). Thus, the exploitation of the High Karst Unit anecumene, as the ultimate upper stage

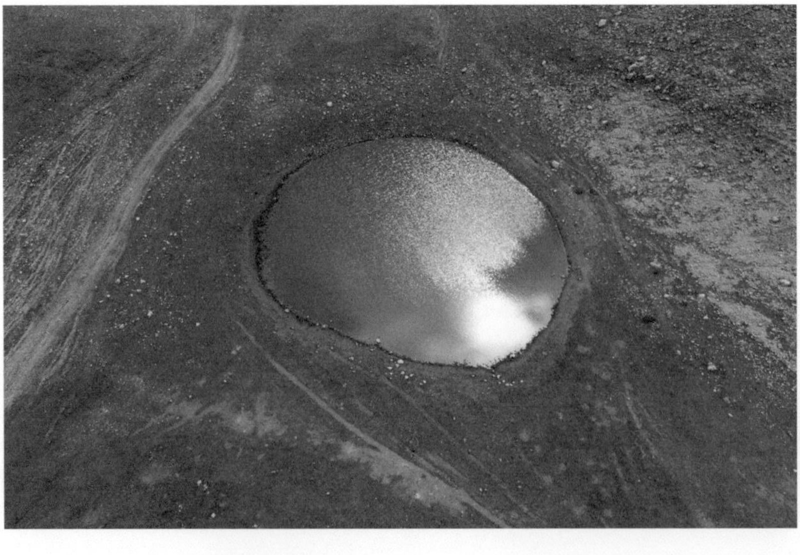

Fig. 8.11 Puddle near Okrugljak Katun. (Photo by Miloš B. Petričević)

in transhumance, was probably caused by population fluctuations as the conquered pastures of the highland of both Sinjajevina and Durmitor provided an unquestionable economic security which was essential to community survival.

8.6 Conclusion

For a long time, the mobility of transhumant herders has been a fundamental guarantee for the survival and progress of the population inhabiting the Dinaric Alps. When comparing the economy of the bearers of the Bronze Age cultures on the territory of Montenegro to the economy and mobility of recent herders, it can be concluded that the mechanism of seasonal movements of prehistoric herders was not that much different. By conquering the upper High Karst Unit, which is evidenced by tangible archaeological records in the area, herders provided bigger and better pastures for their herds. The discovery of prehistoric seasonal settlements—prehistoric *katuni* in the High Karst Unit would make an unfathomable contribution to the understanding of the mobility of prehistoric communities of the southern Dinaric Alps. Current research will most certainly provide new data crucial to the reconstruction of a transhumant matrix of the prehistoric communities from this area. Hopefully the publication of the same and the results obtained from the research conducted in the high-altitude areas of Montenegrin Hills will awaken interest of archaeologist community to carry out further archaeological surveys and research of this still less known area to Montenegrin archaeology.

Acknowledgements I wish to thank Filip Popović, an archaeological student on University of Primorska in Koper (Slovenia), who participated in survey activities on Sinjajevina. I also owe a debt of gratitude to Goran Rabrenović for the successful realization of these activities who provided necessary logistics. I would also like to express a special thanks to the lovers of Montenegrin mountains—Stevo Vasiljević and Neđeljko Jeknić for their selfless sharing of information and data on the records about the tumuli in the area of Sinjajevina and Durmitor.

References

Bešić Z, Garašanin D, Garašanin M, Kovačević J (1967) Istorija Crne Gore [A history of Montenegro]. Redakcija za istoriju Crne Gore, Titograd

Della Casa P, Walsh KJ (2007) Introduction: interpretation of sites and material culture from mid-high altitude mountain environments. Preistoria Alpina 42:5–8

Fuštić B, Đuretić G (2000) Zemljišta Crne Gore [Montenegro's soils]. Biotehnički institut, Podgorica

Garašanin M (1983) Grupa Belotić-Bela Crkva [Cultural group Belotić-Bela Crkva]. In: Benac A (ed) Praistorija jugoslavenskih zemalja IV: bronzano doba. Akademija nauka i umjetnosti Bosne i Hercegovine, Sarajevo, pp 705–718

Gluščević V (2004) Podgorica kao imigracioni centar ruralnog stanovništva Crne Gore [Podgorica as an immigration center of the rural population of Montenegro]. In: Zbornik radova: Selo u Crnoj Gori. CANU, Podgorica pp 353–383

Govedarica B (2010) Ideološki značaj grobnih tumula i sakralna simbolika kruga [Ideological meaning of grave tumuls and sacral circle symbolism]. Godišnjak Centra Za Balkanološka Ispitivanja 39:5–22

Gušić B (1976) Ekološki uslovi na prostoru transhumantnog stočarenja na Balkanu [Economic conditions related to the transhumant pastoralism on the Balkans]. In: Odredbe pozitivnog zakonodavstva i običajnog prava o sezonskim kretanjima stočara u jugoistočnoj Evropi kroz vekove: zbornik radova. Posebna izdanja Balkanološkog instituta, vol. 4. Balkanološki institut SANU, Beograd, pp 143–158

Hesse R (2010) LiDAR-derived local relief models—a new tool for archaeological prospection. Archaeol Prospect 17(2):67–72. https://doi.org/10.1002/arp.374

Kokalj Ž, Zakšek K, Oštir K (2011) Application of sky view factor for the visualization of historic landscape features in lidar-derived relief models. Antiquity 85(327):263–273. https://doi.org/10.1017/S0003598X00067594

Konjević V (2017) Kazivanje Ahmet-Bega Mušovića [The telling of Ahmet-Baig Mušović]. Matica 69:591–632

Kraus K (2011) Photogrammetry. De Gruyter, Berlin

Laković I (ed) (2017) Katuni Kučke planine [Katuni of the Kučka mountain]. Univerzitet Crne Gore, Podgorica

Lazić M (1989) Topografija i tipologija praistorijskih tumula u Srbiji i Crnoj Gori [The topography and typology of prehistoric barrows in Serbia and Montenegro]. Filozofski fakultet, Centar za arheološka istraživanja, Beograd

Lipovac NM (1987) Planina Sinjajevina: prilog poznavanju naših visokih planina [Mountain Sinjajevina: a contribution to the study of our high mountains]. Naučna knjiga, Beograd

Luhmann T, Robson S, Kyle S, Harley I (2011) Close range photogrammetry: principles, techniques and applications. Whittles Publishing, Dunbeath

Lutovac M (1975) Bjelasica. Zemlja i ljudi 25(17):11–22

Lutovac M (1976) Stočarska kretanja u planinama Crnogorskih brda i narodne regule pri iskorišćavanju pašnjaka [Pastoralism in the Montenegrin mountains and national regulations concerning/guidelines for pasture use]. In: Odredbe pozitivnog zakonodavstva i običajnog prava o sezonskim kretanjima stočara u jugoistočnoj Evropi kroz vekove: zbornik radova. Posebna izdanja Balkanološkog instituta, vol 4. Balkanološki institute SANU, Beograd, pp 23–40

Marković M (1971) Stočarska kretanja na dinarskim planinama [Pastoralism in the Dinaric Mountains]. In: Zbornik za narodni život i običaje južnih Slavena, vol 45. Jugoslavenska akademija znanosti i umjetnosti, Zagreb, pp 523–550

Mijović P (1992) Bronzanodopske gravire u Maji Popadiji na Prokletijama [Bronze Age engravings in Maja Popadija in Prokletije]. Glasnik Odjeljenja umjetnosti CANU 11:11–23

Milojević B (1935) Sinjajevina, Visitor i Zeletin: fizičko-geografska promatranja [Sinjajevina, Visitor and Zeletin: physical geographic observations]. Geografski Vestnik 11(1–4):76–87

Mizin V (2013) Stone cairns and simulacra: navigation, folklore, and tradition in the Arctic. Time Mind 6(3):313–329. https://doi.org/10.2752/175169713X13673499387082

Novaković P (2008) Arheologija prostora i arheologija krajolika [Spatial archaeology and landscape archaeology]. In: Olujić B (ed) Povijest u kršu: zbornik radova projekta: Naselja i komunikacije u kontekstu veza jadranskog priobalja i unutrašnjosti. Alpium Illyricarum Studia, vol 1. FF Press, Zagreb, pp 15–54

Petričević M, Vodolazhskaya L (2021) Bronze Age sundial from Prokletije (Montenegro). Archaeoastron Anc Technol 9(2):51–65. https://doi.org/10.48550/arXiv.2202.07923

Primas M, Della Casa P (1996) Velika Gruda. In Kommisssion bei R. Habelt, Bonn

Pušić I (1999) Pećina Vranjaj: preistorijsko stanište iznad Herceg-Novog [Vranjaj Cave: Prehistoric habitation above Herceg Novi]. CANU, Podgorica

Skok P (1974) Etimologijski rjecnik hrvatskoga ili srpskoga jezika I–IV [Etymological dictionary of the Croatian or Serbian language I–IV]. Jugoslavenska akademija znanosti i umjetnosti, Zagreb

Sterud E (1978) Prehistoric populations of Dinaric Alps: an investigation of interregional interaction. In: Redman C, Berman MJ, Curtin EV, Langhorne WT, Versaggi NM, Wanser JC (eds) Social archaeology: beyond subsistence and dating. Academic Press, New York, pp 381–408

Stojanović P (1976) Sezonska kretanja stočara i komunice u običajima i propisima Crne Gore [Seasonal movements of herders and communes in Montenegrin customs and regulations]. In: Odredbe pozitivnog zakonodavstva i običajnog prava o sezonskim kretanjima stočara u jugoistočnoj Evropi kroz vekove: zbornik radova. Posebna izdanja Balkanološkog instituta, vol 4. Balkanološki institut SANU, Beograd, pp 85–117

Szymczak K, Kot M, Petričević MB (2010) Preliminary report on the fieldworks in 2008 in Tamnica cave near Risan, Montenegro. Novensia 21:203–214

Tomičić Ž (2009) Suhopolje—Kliškovac: Od toponima do arheloške spoznaje! [Suhopolje—Kliškovac: from toponyms to archaeological discovery!]. Starohrvatska Prosvjeta 3(36):229–245

Vasić R (1972) Crteži na steni kod Gusinja [Rock art near Gusinje]. Starinar 23:131–135

Walsh K, Richer S, de Beaulieu JL (2006) Attitudes to altitude: changing meanings and perceptions within a “marginal” Alpine landscape—the integration of palaeoecological and archaeological data in a high-altitude landscape in the French Alps. World Archaeol 38(3):436–454. https://doi.org/10.1080/00438240600813392

9 Mountains as Medieval Man's Habitat in Montenegrin Territory

Miljan Gogić

Abstract

The formation and way of life of medieval communities was conditioned by social and natural factors, once isolated, and most often intertwined. Social factors can include anything that came out as a result of organised work of people in a community (rulers, i.e. their government with a developed network of organs). The natural factors that predominantly influenced the type of medieval communities were primarily relief and climate. These two factors were usually closely related and mutually conditioned. The types of communities and living conditions in them depended on the relief, since natural conditions provided opportunities for specific type of human communities to be characteristic of areas with certain geographical properties. Considering the natural living conditions, depending on the natural environment, three types of human habitats can be identified in the area of medieval Montenegro: coastal, continental lowland and mountainous. This classification is conditional and is based primarily on altitude and some features of the relief where the habitats existed. Each of these habitats was certainly socially shaped by the human community. Each of them was specific in its own way, very often determined by natural factors.

Keywords

Mountains · Cattle breeding · Glade · Middle Ages · Montenegro

9.1 Introduction

When studying any medieval topic related to the area of today's Montenegro, it must be borne in mind that Montenegro, when it comes to its current borders, is rather young country. The history of Montenegrin area in the Middle Ages must be observed through the history of historical–geographical units whose formation was influenced by specific historical circumstances, as well as economic conditions, social organisation and way of life. The first unit takes up the space of medieval Zeta. The following regions can be identified in its area: coastal region with its cities and districts, the Bay of Kotor, Skadar Lake with its surroundings and the Dinaric karst area. The second unit that bears more features of a geographical space than a historical area is Gornje Polimlje (Ćirković 1999:19–26). It included the upper stream of the Lim River. In the south, the border followed the ridges of the Prokletije mountains, which marked the watersheds of rivers Lim and Drim, while in the north it reached the valley

M. Gogić (✉)
Historical Institute, University of Montenegro, Podgorica, Montenegro
e-mail: miljangogic@yahoo.com

M. Zorn et al. (eds.), *Discourses on Mountains of Montenegro and Slovenia*, Historical Geography and Geosciences, https://doi.org/10.1007/978-3-031-99017-5_9

Fig. 9.1 Area of today's Montenegro with the areas that existed in the Middle Ages

of Lješnica, the Tivran gorge and the part of Mt. Bjelasica. The eastern border included sections of the Metohija region and Gornji Ibar, while in the west it went all the way down to the area of Kolasin, Kuca and Albania (Fig. 9.1) (Bakić 2005:10–15).

The area of Montenegro is mainly a hilly and mountainous land. Regarding the altitude range, the relief of Montenegro is dominated by mountainous land (Radojičić 2008).

The fact that there was only a small percentage of lowland in the Montenegrin area posed a major obstacle in the life of the medieval population. Economic activities and the possibility of providing livelihood influenced the fact that in the Middle Ages, life in this area took place in high altitude areas (Blagojević 1966; Istorija … 1970).

The living conditions of the population in the Middle Ages were determined by social and natural factors. Although we do not possess quantitative data on this, we can assume that the largest number of inhabitants in the continental area lived in the plains. Certainly, some activities required permanent or temporary residence in areas of higher altitudes, including mountains and mountainous areas (Blagojević 1966).

The time to which the first data refer, which speak to a greater extent about the life and population of people in the area of Zeta and Polimlje in the Middle Ages, is the period starting from the end of the twelfth century. The main sources that speak of the people's settlements in these areas, not only in mountainous areas, are church charters (Miklosich 1858; Jastrebov 1879; Novaković 1891, 1912; Grković 2004; Trifunović 2011a).

Mountains and mountainous areas were of great importance in the medieval monastic economy. When it comes to the mountains mentioned in the monastery charters, it must be borne in mind that the areas that were referred to as mountains in the Middle Ages cannot be equated with the modern notion of mountains. The term *mountain* in the Middle Ages implied an area at high altitude which necessarily includes pastures. Pasture areas covered only a part of a certain mountain massif, other parts were under forest or were bare and did not belong to mountain territories. When it comes to the altitude which mountain pastures were stretched across in the Middle Ages, a clear boundary cannot be determined. The monastery charters acknowledge mountains that can be determined to have been located at an altitude of less than 1,000 m above sea level, while it is known that mountain pastures extend above the belt of coniferous forests, on mountainous areas above 1,700 m (Jireček 1952:153; Blagojević 1966:10). The areas that were called mountains in the Middle Ages referred to parts of today's mountains of the same name. When it comes to the area of Montenegro, data on the population and generally connection of medieval man with mountainous areas are more extensive for the areas of Polimlje and Potar than for the southern parts of today's Montenegro.

The paper will discuss only those areas that were referred to as mountains in medieval documents.

9.2 Mountain Ownership

Being obliged to help the church, the rulers from the Nemanjić dynasty (Ćirković 2005) granted the church institutions landholdings along with dependent population, income and privileges. Until the middle of the fourteenth century, the area of Gornje Polimlje was almost entirely covered with the possessions of the most important and largest monastery seigniories. The situation was similar in neighbouring central Potarje. The area of medieval Zeta also almost entirely consisted of church estates, but it seems not as much as Gornje Polimlje. The reasons for this certainly lie in the fact that in Polimlje there were heritage estates of the Nemanjić dynasty, which they were able to assign to church institutions (Mišić 2003; Tomović and Pušica 2007).

It was almost obligatory that, during the formation, large monastery seigniories were given mountains or mountainous areas as part of the property necessary for life and operation. Certainly, the mountains that were assigned to monasteries were not the only mountainous areas that came into the possession of monastery seigniories. It is known that in the Middle Ages, villages usually had, except for possessions in the plains or lower areas, estates in the mountains or mountainous areas on which the agricultural district of the village (*atar*) rested (if geomorphological conditions allowed). These mountainous areas were considered the property of an entire village, and it was implied that the village had the right to use those mountainous areas, just as it had the right to other *atar* areas (Novaković 1965).

Some charters stated that an estate, usually a village, is granted and that it included, among other things, a mountain. If it was not directly stated that mountains or mountain areas were assigned, it is very often clear that it included mountain areas from the description of the villages' boundaries and their field reconnaissance (Blagojević 1966:15–17). The church authorities, as well as agricultural and livestock settlements in whose *atar* the mountains were located had the right to use the mountains and mountainous areas in the medieval Serbian state. Dušan's Code, by the way, divides all the mountains in the Serbian state into church, imperial and aristocratic (Bubalo 2010:91).

Vlach *katuni's* right to use the mountains posed another issue. The Vlachs represented the remnants of the Romanized indigenous population, which in the Middle Ages was a special category of the population that was mainly engaged in animal husbandry (Novaković 1965:29–53). Vlach *katuni's* (singular *katun*) are temporary settlements that are formed in areas where the population was engaged in livestock breeding, as a result of livestock movements or transhumance (Laković 2017); as Miloš Blagojević (1966:18–19) stated, with the exception of two Vlach *katuni* that belonged to the seigniory of St. Archangels near Prizren (Sinajinci and Dragoljevci) and the seigniories of St. Stefan in Banjska and St. Pantocrator in Dečani (all in preset-day Kosovo), which cannot be put in connection with the mountains. Data on Vlach *katuni* that were not part of the church estates are little known to us, but it can be reliably assumed that the Vlach population used to use the mountains for summer *katuni*. Since a significant part of the mountain areas with pastures was owned by monasteries, the Vlach population organized into *katuns* was forced to ask the monastery abbots for permission to use them and pay compensation. They also had to have permission in order to use the pasture areas in the lowlands. Similar case was with other livestock owners, the agricultural population, as well as the population that belonged to other monastery seigniories—they all had to pay a certain amount to use these areas. The fee that people used to pay for using monastic and village pastures, including those located in the mountains, was called *travnina*. It was usually paid in kind, by giving out a part of the cattle herd. When it comes to the monastery pastures, the abbot of the monastery was in charge of collecting fees (Blagojević 1966:84–89).

9.3 Mountains in the Area of Gornji Polimlje and Potarje (End of 12th till Middle of Fourteenth Century)

An overview of the mountains and mountainous areas that can be considered to be the habitats of medieval man in Montenegro will begin with a review of such areas that belonged to the diocesan and archdiocesan seigniories formed after 1217 (Janković 1985). The **Humska diocese** possessed, by a charter granted to it in the mid-thirteenth century by Serbian King Uroš (Komatina 2021), the mountain Smrječnik, which was located on the area of today's mountain Bjelasica. In the parish of Brskovo, the Hum seigniory was granted the village of Prošćenje, which stretched across the right and left banks of the river Tara in today's Mojkovac valley. This village had its own mountain Konj, acknowledged in the charter. It is today's Konjic mountain within the complex of the Mountains of Prošćenje. On the left bank of the Tara River, as a border of the village *atar*, the mountain Uborak is acknowledged, the existence of which is witnessed today by the Borova Glava *katun* on Sinjajevina. From the description of this village's borders in the charter and their reconnaissance in the field, it can be seen that it included the mountain areas of Sinjajevina on the left bank of Tara (1,600–1,800 m), as well as the mountain areas of its right bank (Novaković 1912:595, 601; Tomović 2005:40–41, 47). The life of Brskovo parish population, and also in the neighbouring Gornji Polimlje, was largely determined by economic activities in Brskovo. Mining production started here in it middle of the thirteenth century. Brskovo was the most important economic centre in the mainland of today's Montenegro during the second half of the 13th and in the fourteenth century, when trade and mining started flourishing (Gogić 2010).

As early as at the end of the twelfth century, during the founding of the **Hilandar seignory** (Hilandar monastery is in preset-day Greece), the Serbian Grand Prince Stefan Nemanja (Radujko 2016) granted the mountain Bogača located north-east of Plav to this monastery. Mountain Bogača is today's mountain Bogićevica, rich in pastures, whose one peak is called Bogdača today. A significant number of its peaks is located at over 2,000 m above sea level (Novaković 1912:384; Blagojević 1966:11; Bakić 2005:66–67). With the founding of the **Dečani seigniory**, this mountain was assigned to its monks (Ivić and Grković 1976:128, 260). An area of the mountain that belonged to the village of Dosuđa was also a part of the Hilandar's property in the mid-fourteenth century. This area was assigned to Hilandar by King Uroš in the mid-thirteenth century. According to the data from the charters, the mountain area that belonged to this village included the area of the Greben Mountain, where numerous *katuni* exist to this day (Novaković 1912:390, 391, 421, 431; Bakić

2005:74, 400; Tomović and Pušica 2007:33–34). Although it belongs to Albania today, it is necessary to state that Hilandar also had a mountain within the borders of the Vrmoša village, which is described in the sources as *Svetogorska mountain* (*Svetogorska* is an adjective derived from the toponym Month Athos in preset-day Greece, meaning "belonging, in relation or having properties of Mount Athos") (Ivić and Grković 1976:240; Tomović and Pušica 2007:59–60). The monastery also owned a mountainous area between Plav and Altin, which is now partly located in Albania (Tomović and Pušica 2007:37; Trifunović 2011a:102).

Large mountain areas in Gornji Polimlje region were given to the **Svetostefansko seigniory**, founded at the beginning of the fourteenth century (1314–1316) (Stanković and Erdeljan 2021; Stanković 2022). In the Goražde village's *atar*, in the Budimljanska parish, there was a mountain area, just as in the Kruševa village in Plav (Bakić 2005:391, 412; Tomović and Pušica 2007:40–41, 52–53: Trifunović 2011a:83, 84). The Gusino village covered a significant part of the Prokletije massif. This area includes today's mountains: Karanfili, Vezirova Brada, Zastan, Vojuša, Popadija, Ročica, Trojan and Konjski krš (Bakić 2005:62–63, 384–385; Tomović and Pušica 2007:42–43; Trifunović 2011a:85–86). Montenegrin mountain region located between the Plav area and Budimlje, also belonged to this seigniory. The whole region belongs to the Mojan mountain range (Lutovac 1933:72–73; Tomović and Pušica 2007:46–47; Trifunović 2011a:102–104). On the western outline of the Plav area, the Kolštica and Rikavac mountains were in possession of this seigniory (Tomović and Pušica 2007:48; Trifunović 2011a:97). The mountain area of the Hrastje village sat along these mountains, making a link between the Zeta and Polimlje estates of this monastery (Erdeljanović 1907:29, 91, 98–99; Trifunović 2011a:96–97; Trifunović 2011b:241, 224). North-east of Plav, the Banja, Gozbaba, Jelenak, Čakor and Viticrep mountains were granted to the Banjska monastery by the same charter (Bakić 2005:69–70; Tomović and Pušica 2007:49–50; Trifunović 2011a:99–102; Trifunović 2011b:235–237). A like Hilandar, the monastery of St. Stefan had a mountain complex in the area between Plav and Altin, that today belongs to neighbouring Albania (Tomović and Pušica 2007:43–44; Trifunović 2011a, 22). The Ulotino village in Plav that was assigned to this monastery in the mid-fourteenth century (1345), which until then had the status of a conditional good-*pronija*, had area that was included in the mountain (Novaković 1891:6; Blagojević 1966:32; Bakić 2005:413–414; Tomović and Pušica 2007:51). There are no data explicitly acknowledging that this monastery in Zeta included mountains, but the description of the boundaries of some villages (Helmica, Rujište) and their reconnaissance in the field shows that they included mountainous areas (Trifunović 2011a:87–91; Trifunović 2011b:239).

The **Gradac Monastery**, the endowment of Jelena Anžujska (Popović 2010) also possessed estates in Gornje Polimlje. Even though the charter of Gradac's estates has not been preserved, the charters of other monasteries show that the mountains in its possession were located in this area. For example, within the boundaries of the Dečani estates in the Budimlja parish, in the area of Komovi, the mountain estates of the Gradac monastery are mentioned (Ivić and Grković 1976:66, 244; Tomović and Pušica 2007:38). Within the Crne Gore range, between Plav and Budimlja, this monastery possessed mountain areas (Tomović and Pušica 2007:38–39; Trifunović 2011a:21). In addition, south-east of Plav, the Gradac monastery owned the mountain areas of Veliki Hrid and Mali Hrid mountains and the Rida Lake, in the mountain area between Plav and Altin (Bakić 2005:67–68; Tomović and Pušica 2007:39; Trifunović 2011a:22).

Monastery dedicated to Pantocrator in **Dečani** also owned land properties in Gornje Polimlje. Mountains are acknowledged in the data to be within the borders of the Komarani village (assigned to Dečani) near Plav, which shows that the village possessed mountainous areas in its *atar* (Ivić and Grković 1976:40, 234; Grković 2004:75; Tomović and Pušica 2007:56). Circumstances were similar for the Velika village, which partly leaned on the mountainous areas of Čakora, including the area of the mountains

Rzana and Lokva, which can still be identified in the field (Ivić and Grković 1976:40–41, 235; Grković 2004:75; Bakić 2005:404–406; Tomović and Pušica 2007:56). The Vrmoša village was also assigned to Dečani. In the description of its borders, which included mountain areas, there is a definition of what was considered a mountain in the Middle Ages, where it is stated that "*коудѣ се ни оре ни коси този все планина*". Grasslands that were not mowed and not intended for ploughing represented a mountain. Mowing or ploughing these areas would reduce pasture areas, which could jeopardize their basic purpose (Blagojević 1966:12; Ivić and Grković 1976:43, 240; Tomović and Pušica 2007:59–60). The monastery possessed a spacious mountain range that shared a border on its south side with the mountains belonging to seignory of St. Stefan monastery in Plav. The mountain range included: Hotina Gora, Štirnice, Dobra vodica i Sedmi bori, Treskavica, Košutica, Bela Vodica, Črn kamen, Hrid and Dol (Ivić and Grković 1976:56, 260; Grković 2004:75; Bakić 2005:65–66; Tomović and Pušica 2007:60–62). In the area of Budimlje, this monastery was assigned the Dobra Reka village, and within its border's mountains were acknowledged, i.e. the mountain areas that were a part of village's *atar* (Ivić and Grković 1976:45, 244; Grković 2004:75; Tomović and Pušica 2007:64). The importance of the mountains and their pastures for the economy of this seigniory is witnessed by the provision which states that the penalty for the unauthorized use of the monastery mountains and pastures, which was to be paid to the ruler, was 500 rams (Ivić and Grković 1976:56; Mihaljčić 2006:91, 165). This provision shows that the right of ownership over the mountains was granted by the ruler to the monastery to profit from grazing, while retaining control over their use and the right to impose fines (Tomović and Pušica 2007:62).

9.3.1 Life Traces in the Mountains of Gornje Polimlje

It is easy to understand that the mountains and mountainous areas of Gornji Polimlje that were assigned to monasteries were areas where human communities lived only during the summer months. This can be also confirmed by the above-mentioned charters. In some of these charters, specifically those referring to the seignory of the Dečani monastery, there are terms that directly refer to the existence of human communities in the mountains and mountainous areas. One of them is *glada* that appears in the charters issued to this monastery. According to the scientific interpretation, this term is the Proto-Slavic dialectism *golda* (plural *glade*), related to the Baltic *galda*, with the meaning of a barrier in the barn, an area fenced by wooden planks for domestic animals and poultry (Loma 2013:195). The term *glada* is mentioned repeatedly in these charters. In the borders of the Dečani village of Trepča, which shares its border with the Hilandar village of Dosuđe, in the highest parts of the Greben Mountain, *letske glade* are mentioned (Ivić and Grković 1976:41, 236; Tomović and Pušica 2007:34). Within the *atar* of Dečani village of Vrmoša, after the Svestosavska mountain, *Bnkećeve glade* and *Murićeve glade* are mentioned as its boundaries (Ivić and Grković1976:240). Within the Dečani village of Velika and the Ržana and Lokva mountains, *Salče glade* are mentioned as locality within in the mountainous area, where habitats existed (Ivić and Grković1976:235; Grković 2004: 75; Tomović and Pušica 2007:56). Salče glade is also mentioned in the boundaries of the Svetostefansko seigniory mountain complex, located northeast of Plav (Trifunović 2011a:101; Trifunović 2011b:235–236). It was the similar case with the Dobra Reka village's *atar*, where it is stated that the border of the *atar* went above *Halapine glade* (Ivić and Grković 1976:244; Grković 2004:75; Tomović and Pušica 2007:56). With the exception of *Letske glade*, etymological origin cannot be determined with certainty for toponyms that are related to the term *glade*. Regarding their origin, only conclusion we can draw is that they may be toponyms of Roman origin (Grković 1983:117–118). These toponyms are located at high altitudes that are classified as mountains or mountain areas. This term can be found only in the Dečani's charters—it is not acknowledged in any other documents of the

Middle Ages. Today, the term *glada* is used to denote objects in which one stays in summer habitats, in the area of Kučke mountains, while it is not used in lower, rural areas of Kuči (Erdeljanović 1907:235–237; Laković 2017:24–40). It can be stated with certainty that in the Middle Ages in the area of Gornji Polimlje the term *glade* denoted huts in the mountains and mountainous areas, where people stayed during a certain part of the year, when the weather conditions allowed it. The mountains of Gornje Polimlje, which were acknowledged in medieval charters as places where people lived during a certain part of the year, are the areas where summer *katuni* and local population's pastures still exist to this day. Population's way of life and income strategies when it comes to cattle breeding today are quite similar to those characteristic for Middle Ages (Lutovac 1933).

9.4 Mountains in the Area of Zeta (13th–Mid Fourteenth Century)

When it comes to the southern part of present-day Montenegro, i.e. the area of medieval Zeta, mountainous areas are also, although to a lesser extent, listed as areas assigned to churches and monasteries. Such a situation was also influenced by the fact that, among other factors, in a large part of medieval Zeta, the climate was different and milder in relation to Polimlje. In the area of Zeta, mountains used as summer pastures had a different role and significance than the ones in Polimlje. Flora was richer due to milder climate and the cattle breeders were able to stay longer during the year. Furthermore, the mild climate influenced the fact that the lowland areas in some regions (Zeta and Morača valleys, Skadar Lake area, coastal belt) could be used for grazing cattle during almost the whole year (Istorija … 1970).

During the formation of the **Žiča** archdiocesan **seigniory** in the area of medieval Zeta, this area was assigned the mountains north of Nikšić (medieval Onogošt)—Spora Nozdra, Javorje and Lukavica. These are today's parts of Duga and Golija, i.e. Javorje and Lukavica (Miklosich 1858:12; Tomović 2012:45).

The seigniory of the **Vranjine monastery** increased over time and included a larger number of villages. At the beginning of the fourteenth century, Serbian King Milutin granted the Selca mountain to this monastery. This is today's mountain Seoštik, which is located in the area between the Crnojevića river and Skadar Lake's confluence and the Budva Bay (Jastrebov 1879:224–226; Tomović 1987:39, 42). Certainly, the land incomes that this monastery received during the Middle Ages included other areas that were classified as mountainous areas (Škrivanić 1959:37–44).

In Zeta, the **Dečani monastery** was assigned the village of Kuševo, which reached the mountainous areas of Bukovik and Čemernik. This shows us that in addition to the plains, Kuševo also possessed mountainous areas where its population used to live and breed cattle during summer (Ivić and Grković 1976:246–247).

9.5 Mountains as Habitats in the Second Half of the Fourteenth and Fifteenth Centuries

Political circumstances in Zeta and Polimlje were changing during the second half of the fourteenth and fifteenth centuries. In Zeta, the Balšić dynasty started their rule, followed by the Crnojević dynasty, the last Montenegrin ruling dynasty. Brankovići and Lazarevići ruled over Polimlje during this period (Istorija … 1970). The mountains were certainly one of the most important habitats in these times, having in mind the general political and socio-economic circumstances of that period. Due to turbulent times and war, when some ways of income were often endangered, the mountains were recognized as areas where economy was stable, but not only as an area intended exclusively for cattle breeding. When Ivan Crnojević founded the **Cetinje monastery** at the end of the fifteenth century, he granted a mountain that was his hereditary estate to his seigniory.

According to the charter, that mountain, that is, the mountainous area, was located on the southern slopes of today's Lovćen, with Ivanova Korita as its centre. The usage of this property differs from the purpose and usage of other properties that are listed as mountains in other charters from the thirteenth and fourteenth centuries. Unlike mountains and mountainous areas in other regions that were, according to the charters from the earlier period, mainly used as cattle grazing areas during summer, the area of Lovćen was used as arable land. The charter states that the monastery can cultivate the property's land or lease it. In the case of leasing, a quarter of the obtained grain belonged to the monastery and monastery administration had to give consent to anybody wanting to cultivate the land (Šekularac 1987:239). There are no data on the use of this mountain in the preceding period, but it can be assumed with certainty that it was also used for agriculture. The karst area and the scarcity of arable land in this region affected the land cultivation at higher altitudes as well.

9.6 Features of Life in the Mountains

The way of living of the population who lived on monastic estates in the mountains and of those who came from agricultural villages in whose *atar* the mountainous areas entered, differed essentially only in the level of obligations they had, but we can safely state that their job description was the same. Cattle breeding in these areas initially included taking care of the cattle and milking, as well as preparing dairy products. Preserved sources indicate that mostly sheep, then goats, mountain cattle and to a lesser extent pigs were bred (Trifunovski 1963:32–33). The population was situated in the abovementioned and other similar buildings, probably built of stone and wood. Certainly, these facilities, i.e. the places where they were located, were at least relatively close to a water source. Although there is no data in the sources, we can assume that cattle in the mountains were often killed by wild animals, bears and wolves. Hunting wild animals, whose natural habitats were both mountains and highland areas, was one of the forms of entertainment of the medieval nobility and rulers, but also the responsibility of the ruled population (Mišić 1995).

The pasture quality in an area depended on several factors, among which the climatic conditions stand out. The harsh and cold climate that lasted during most of the year influenced the growing season in the mountains and mountainous areas to be shorter. This made a negative impact on the development and survival of vegetation, which also effected negatively nutritional value of grass. In mountainous areas, summers were short and winters long, so flora development was slow (Katić 1978:85). Such climatic conditions were certainly present in the area of Gornji Polimlje, whose mountainous areas are often characterized by cold springs, where conditions for grazing cattle were fulfilled as late as mid-June, while the vegetation for grazing would almost disappear as soon as in September. Medieval charters also acknowledge summer pastures that were used seasonally for grazing cattle. Analysing these charters, we can conclude that they were located at slightly lower altitudes compared to the areas that were called mountains (Blagojević 1966:39–42). While cattle grazed on the mountains and summer pastures during the summer; it was necessary to prepare access to winter pastures as well as a certain amount of hay for winter feeding. Monastic charters contain legal norms that speak of the dependent population's obligations regarding the food preparation for livestock (Blagojević 1966:68–84). Climatic conditions affected the fact that less food was needed to feed livestock during the winter in the Zeta than Gornji Polimlje, where flora for cattle graze would start growing later in year compared to the Zeta region. This difference in the start of growing season closely related to cattle grazing was even more drastic in the mountains and mountainous regions of Gornje Polimlje and parts of Zeta.

9.7 Conclusion

Montenegrin relief is characterized by large mountainous areas. During the Middle Ages, the largest part of these mountainous areas was inhabited only during a part of the year. During the formation of monastic seigniories since the end of the twelfth century and during the thirteenth and fifteenth centuries, they were assigned mountains in the areas of Gornji Polimlje and Zeta, as well as mountainous areas that entered the attached villages' *atari*. A large part of Gornji Polimlje, including its mountainous regions, was assigned to monasteries. According to the monastery charters, the term 'mountain' in the Middle Ages referred to areas that were not cultivated, mowed or ploughed in any way. Monastery seigniories, in addition to keeping their own cattle, leased these areas to Vlachs and professional cattle breeders, who had to pay a fee. This fee is called *travnina* in the sources. These areas were used for cattle breeding during summer months by the villagers. Villages' area included the neighboring mountainous areas, and agriculture was the main activity. The population that lived in the mountains was situated in the buildings most probably built of stone and wood. Such facilities are referred to in medieval charters as *glade*. In the area of Kučke mountains, the facilities for housing cattle breeders or *katuni* are called *glade* to this day. When it comes to the area of medieval Zeta, data from the end of the fifteenth century shows that in this area (Lovćen), mountain areas were used for agriculture, although they were predominantly cattle breeding areas.

References

Bakić R (2005) Gornje Polimlje, priroda, stanovništvo i naselja [Gornje Polimlje, nature, population and settlements]. Geografski institut Filozofskog fakulteta, Nikšić

Blagojević M (1966) Planine i pašnjaci u srednjovekovnoj Srbiji [Mountains and pastures in Medieval Serbia]. Istorijski Glasnik 19(2–3):3–95

Bubalo Đ (ed) (2010) Dušanov zakonik [Dušan's code] Službeni glasnik, Beograd

Ćirković S (1999) Osobenosti istoriografije o srednjovekovnom periodu Crne Gore [Historiography characteristics of Medieval Montenegro] In: Srednjovjekovna istorija Crne Gore kao polje istraživanja: zbornik radova Istorijski institut Republike Crne Gore, Podgorica, pp 19–26

Ćirković S (2005) Srbi u srednjem veku [Serbs in the Middle Ages]. Idea, Beograd

Erdeljanović J (1907) Kuči, Bratonožići, Piperi. Beograd

Gogić M (2010) Rudarska proizvodnja u srednjovjekovnom Brskovu [Mining production in the medieval Brskovo]. Istorijski Zapisi 83(1):195–212

Grković M (1983) Imena u Dečanskim hrisovuljama [Names in the Decani Monastery chrysobulls]. Institut za južnoslovenske jezike Filozofskog fakulteta, Novi Sad

Grković M (2004) Prva hrisovulja manastira Dečani [The first charter of the Dečani Monastery]. Centar za očuvanje nasleđa Kosova i Metohije, Beograd

Istorija Crne Gore [History of Montenegro], vols 2(1), 2(2) (1970) Redakcija za istoriju Crne Gore, Titograd

Ivić P, Grković M (1976) Dečanske hrisovulje [Decani Monastery chrysobulls] Institut za lingvistiku, Novi Sad

Janković M (1985) Episkopije i mitropolije Srpske crkve u srednjem veku [Dioceses and mitropolia of Medieval Serbia]. Istorijski institut, Beograd

Jastrebov I (1879) Prepis hrisovulja na Cetinju o manastiru Sv. Nikole na Vranjini [Transcript of chrysobulls from Cetinje about St. Nicholas' Monastery in Vranjina]. Glasnik Srpskog Učenog Društva 47:219–231

Jireček K (1952) Istorija Srba 2 [History of Serbs, vol 2]. Naučna knjiga, Beograd

Katić R (1978) Stočarstvo srednjovekovne Srbije [Cattle breeding in Medieval Serbia]. SANU, Beograd

Komatina I (2021) Kralj Stefan Uroš I Veliki i njegov vek [King Stefan Uroš the Great and his period]. Istorijski institut, Beograd

Laković I (ed) (2017) Katuni Kučke planine: monografija [Katunion Kučka Mountian: monograph]. Univerzitet Crne Gore, Podgorica

Loma A (2013) Toponimija Banjske hrisovulje [The toponymy of the Banjska Monastery's foundation charter]. SANU, Beograd

Lutovac M (1933) Stočarstvo na severoistočnim Prokletijama [Cattle breeding in the north-eastern Prokletije]. Geografsko društvo, Beograd

Mihaljčić R (2006) Zakoni u starim srpskim ispravama, pravni propisi, prevodi, uvodni tekstovi i objašnjenja [Laws in old Serbian documents, legal regulations, translations, introductory texts and explanations]. SANU, Beograd

Miklosich F (ed) (1858) Monumenta Serbica spectantia historiam Serbiae Bosnae Ragusii. Viennae

Mišić S (1995) Lov u srednjovekovnoj Srbiji [Hunting in Medieval Serbia]. Istorijski Glasnik 44(1–2):51–66

Mišić S (2003) Teritorijalno-upravna organizacija Polimlja (12–14 vek) [Territorial and administrative organization of Polimlje 12–14th century]. In: Kralj Vladislav

i Srbija XIII veka: zbornik radova. Istorijski institut, Beograd, pp 73–87
Novaković S (1891) Hrisovulja cara Stefana Dušana grobu mu majke kraljice Teodore [Tzar Dušan's chrysobulls on his mother Queen Teodora's tombstone]. Spomenik SKA, vol 9. Srpska kraljevska akademija, Beograd
Novaković S (1912) Zakonski spomenici srpskih država srednjeg veka [Legal monuments of Serbian states of the middle ages]. Beograd
Novaković S (1965) Selo [The village]. Srpska književna zadruga, Beograd
Popović M (2010) Srpska kraljica Jelena između rimokatoličanstva i pravoslavlja [Serbian queen Jelena between Roman Catholicism and Orthodoxy]. Pravoslavni bogoslovski fakultet Univerziteta, Institut za teološka istraživanja, Beograd
Radojičić B (2008) Geografija Crne Gore 1: prirodne osnove [Geography of Montenegro, vol 1: natural bases]. Dukljanska akademija nauka i umjetnosti, Podgorica
Radujko M (ed) (2016) Stefan Nemanja – prepodobni Simeon Mirotočivi: zbornik radova 1, 2 [Stefan Nemanja—Venerable Simeon the Myroblytes: collection of papers, vols 1, 2]. Institut za istoriju umetnosti Filozofskog fakulteta u Beogradu, Mitropolija Crnogorsko-primorska, Eparhija Budimljansko-nikšićka, Institut za srpski jezik SANU, Beograd, Cetinje
Stanković V, Erdeljan J (2021) Kralj Milutin i manastir Banjska: povodom sedam vekova od upokojenja Svetog kralja Milutina 1321–2021 [King Milutin and the monastery of Banjska: on the occasion of seven centuries since the repose of the Holy King Milutin 1321–2021]. Platoneum, Novi Sad
Stanković V (2022) Milutin sveti kralj [Milutin the Holy King]. Vukotić media, Beograd
Šekularac B (1987) Dukljansko- Zetske povelje [Doclea and Zeta charters]. Istorijski institut Crne Gore, Titograd
Škrivanić G (1959) Vlastelinstvo Sv. Nikole Vranjinskog [St. Nicholas Vranjski's seigniory]. Istorijski Zapisi 12(3–4):37–44
Tomović G (1987) Sopotski zabel Črmanj [Sopot's teritoty Črmanj]. Istorijski Časopis 34:37–59
Tomović G (2005) Kričani i Kričak [Kričani and Kričak]. Glasnik Zavičajnog Muzeja 4:39–48
Tomović G, Pušica S (2007) Srpski srednjovjekovni crkveni posedi u Gornjem Polimlju [The Serbian Medieval church estates in the Gornje Polimlje]. Mileševski Zapisi 7:31–67
Tomović G (2012) Žičko vlastelinstvo [The Žiča Archibishopric estate]. Naša Prošlost 13:9–52
Trifunović Đ (ed) (2011a) Povelja kralja Milutina manastiru Banjskoj 1 [Charter of King Milutin to the Banjska Monastery, vol 1]. Službeni glasnik, Beograd
Trifunović Đ (ed) (2011b) Povelja kralja Milutina manastiru Banjskoj 2 [Charter of King Milutin to the Banjska Monastery, vol 2]. Službeni glasnik, Beograd
Trifunovski J (1963) Geografske karakteristike srednjovjekovnih katuna [Geographical charateristics of Medieval katuni]. In: Simpozijum o srednjovjekovnom katunu. Naučno društvo SR Bosne i Hercegovine, Sarajevo, pp 19–39

Human Activity in the Slovene Alpine World in the Middle Ages

10

Gašper Oitzl

Abstract

At the time of the Slavic settlement, the flat periphery of the Alpine world was partly inhabited, and in the late tenth and eleventh centuries incorporated into the feudal social and economic system, with the territory being divided among several landlords. In the eleventh century, intensive colonisation at the lower altitudes began, culminating in the twelfth century and completed by the thirteenth century. At the same time, the settlement of the upland area started, which in some areas extended into the fifteenth century. Particularly in the Alpine valleys, animal husbandry and cheese-making were a principal source of livelihood. Arable farming was second to stock-raising; however, grain was cultivated throughout the region. Alpine streams and lakes were exploited for fishing. Deforestation intensified over time, initially to acquire more agrarian land, and from the fourteenth century onwards to meet the needs of charcoal production, which was the traditional fuel for the growing number of ironworks. Both the development of the iron industry and several transport routes via the Slovene Alps were central to the social and economic development of the region in the Late Middle Ages.

Keywords

Middle Ages · Medieval colonisation · Economy · Mountain grazing · Ironmaking · Alpine world · Slovenia

10.1 Introduction

In the Middle Ages, the relationship between man and the Alpine world was very different from today, determined primarily by the natural conditions in the high mountains and their immediate surroundings. In this paper, the Slovene Alpine world is defined by the upland areas near the mountain ranges of the Julian Alps, the Karavanks and the Kamnik-Savinja Alps. The study of human activity in the region focuses on the Alpine valleys and basins within the mountain ranges, including the periphery of the plains at the foot of the mountains (Fig. 10.1).

Until medieval colonisation, human activity in the Alps was somewhat limited. In the Roman Period, it was mainly the lowland area at the foot of mountain ranges that was settled, and the population was usually concentrated in countryside villas (*villae rusticae*) (Ciglenečki 2000:121; Vidrih Perko and Sagadin 2014:215–221). Archaeological research also points to the use of Alpine pastures, where cattle were grazed, and iron ore

G. Oitzl (✉)
Department of History and Applied Arts, National Museum of Slovenia, Prešernova 20, Ljubljana 1000, Slovenia
e-mail: gasper.oitzl@nms.si

M. Zorn et al. (eds.), *Discourses on Mountains of Montenegro and Slovenia*, Historical Geography and Geosciences, https://doi.org/10.1007/978-3-031-99017-5_10

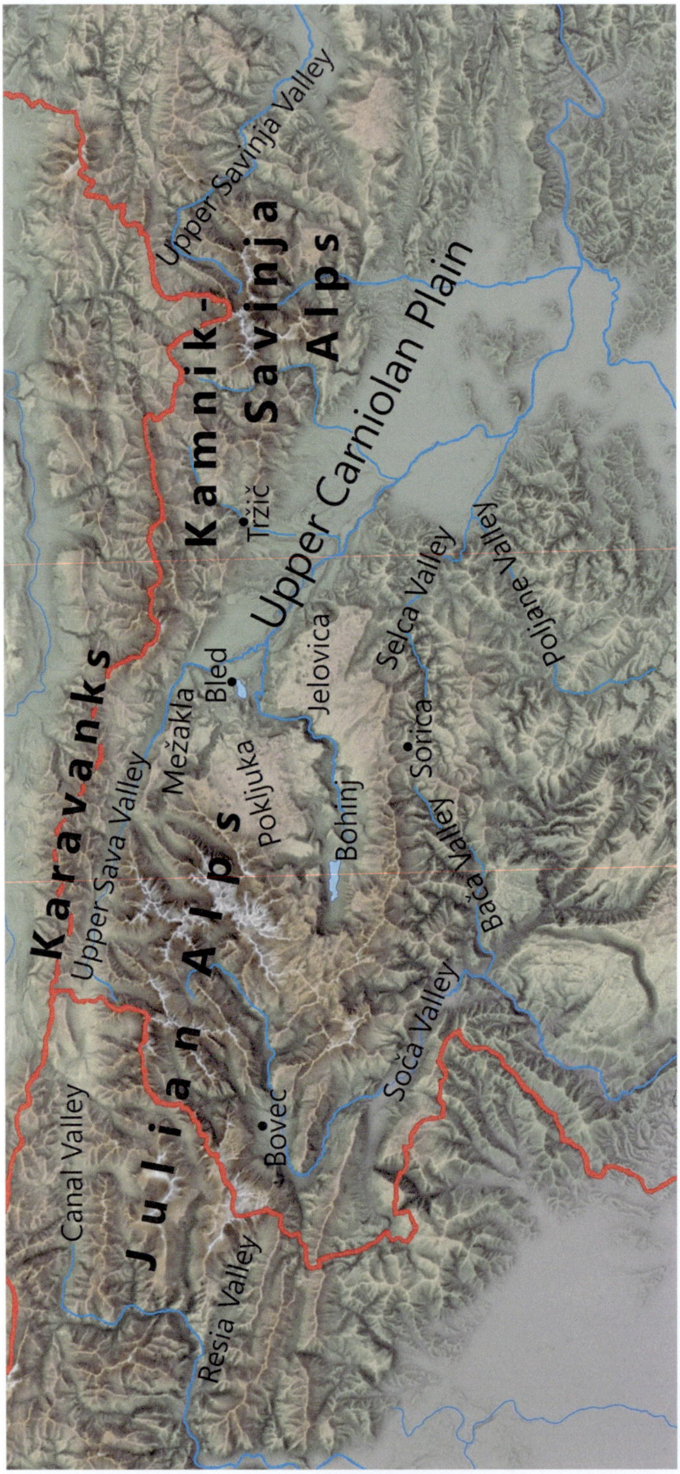

Fig. 10.1 Slovene Alpine world. (Map by Gašper Oitzl)

was mined back in antiquity (Cevc 2003; Horvat 2006; Ogrin 2006). In the turbulent period of Late Antiquity, the mountainous and remote land, far from the main transport routes, was a refuge from the military campaigns of Germanic tribes, the Huns, Avars and Slavs. Romanised indigenous peoples moved to fortified hilltop settlements and built temporary refuges and outposts at higher altitudes. In the fifth and sixth centuries, many such settlements emerged in the Slovene Alpine area, including those at Tonovcov grad near Kobarid, the Bled Castle Hill, Ajdna nad Potoki, Gradišče nad Bašljem, Sv. Lovrenc and Sv. Jakob above Preddvor, and Ter above the Upper Savinja Valley (Ciglenečki 2000).

The hilltop settlements were relatively small, with up to about twenty buildings and, at best, a population of just over one hundred. They strove to be as economically self-sufficient as possible, mainly engaging in animal husbandry, and ironmaking at Ajdna. Despite the stormy times, long-distance trade routes were still established, as testified by the pottery finds of the (North)African type at Ajdna. Remains of early Christian churches were discovered at several late antiquity hilltop settlements. The existence of a significant ecclesiastical centre at Tonovcov grad is suggested, which could have been a temporary episcopal seat or a bishop's refuge. Archaeological research reveals very different fates of said settlements. Ajdna nad Potoki saw a violent disruption in the second half of the sixth century, while the inhabitants of the settlement at the Tonovcov grad moved to the lowland in the early seventh century after peace had been restored. The research of burial site at the Bled Castle Hill might even suggest the cohabitation of the indigenous community with newly arrived Slavic settlers (Pleterski 2010; Modrijan 2011; Modrijan and Milavec 2011; Vidrih Perko and Sagadin 2014:219–221).

The valleys of the Julian Alps acted as a kind of buffer zone between the various polities in the Early Middle Ages. It was a broad swathe of no-man's-land, very sparsely populated or even uninhabited. Such was probably the case of the Upper Soča Valley and the Upper Sava Valley, which in the seventh and to some extent the eighth century constituted a somewhat a buffer zone between the Lombard Kingdom (to the west) and the Avar Khaganate (to the east). In the present-day Upper Carniola (*Gorenjska*), Slavic settlement began in the late sixth or early seventh century. The borderline areas of the early Slavic settlement were the plateaus of the Julian Alps (Jelovica, Pokljuka and Mežakla) and the southern foothills of the Karavanks and the Kamnik-Savinja Alps from Žirovnica to Komenda. Several settlements were located in the Blejski Kot area (Fig. 10.2), where an early medieval *župa* was established. It was a Slavic territorial community comprised of many settlements and governed by an elder head of community noted as *župan* (Pleterski 2013:10–12). However, based on archaeological research, these settlements were relatively small, with a few ten inhabitants. Similar conclusions could be drawn for other areas of early settlement. The Selca Valley was most likely populated up to present-day Železniki. In the Early Middle Ages, Bohinj and Dovje in the Upper Sava Valley were inhabited outside the continuous settlement. The Slavic settlement of the Upper Soča Valley began no later than the early eighth century, most probably stretching down to Koritnica in the Bača Valley and Bovec in the Upper Soča Valley (Blaznik et al. 1970:69–71; Kos 1971; Knific 2004; Pleterski 2013:158–163).

The course of settlement was, to a certain extent, dictated by mountain ranges. The Karavanks and Kamnik-Savinja Alps formed the dividing line between the Carinthian (*Koroška*) and Upper Carniolan settlement areas, the watershed between the Kamniška Bistrica and the Savinja Rivers bordered the settlement areas of Upper Carniola and Styria (*Štajerska*), and the central part of the Julian Alps with the Triglav mountain chain, Komna and the Bohinj-Tolmin Mountains were borderline area between the Littoral (*Primorska*) and the Upper Carniolan settlement areas (Blaznik et al. 1970:69–70). At the time of settlement, the Slavs came into contact with a Romanised indigenous population, and the contacts seem to have been particularly intense in the Upper Soča Valley, as indicated by the Roman and pre-Roman origin of many local toponyms (Snoj 2009). As to the economic

Fig. 10.2 Radovljica Plain with Blejski Kot area and Jelovica Plateau in the background. (Photo by Gašper Oitzl)

activities, the Slavs adopted from the indigenous population the mountain grazing. Such is indicated by archaeological finds that suggest a continuity of the use of mountain pastures in the Kamnik-Savinja Alps, as well as the origin of Slovene pastoral terminology, such as *tamar* (pen) and *vardevati* (to take care of), and some mountain names, such as Komna, Krn, Matajur and Porezen (Bezlaj 1967:31–32, 89–90; Cevc 2003).

10.2 Feudalisation of the Alpine World

Following Magyar incursions, which stormed Europe in the late ninth and mid-tenth centuries, the Slovene Alpine area witnessed gradual integration in the feudal social and economic system of the medieval Empire. The King of Romans (German king) had the vast majority of land in his possession, granting it to dioceses and high nobility, who established feudal seigniories. These had centres in castles on the periphery of the Alpine area, with the valleys, streams, plateaus and passes forming the borderline area between the various seigniories. As early as 973, the first royal grant for the area in question mentions the territory of Selca (*territorium Zelsah*) and the broader area of Ratitovec Mountain (*alpem Bosangam*) as the borderland of the grant. At that time, the King of Romans Otto II granted the Bavarian Diocese of Freising a large estate in the March of Carniola (Fig. 10.4), where the diocese established a seigniory with the seat at the Škofja Loka Castle. The seigniory included part of the Jelovica Plateau, the Selca Valley, the Sorica area and a large portion of the Upper Sava Valley (Schumi 1882:8; Blaznik 1973:11–18; Mlinar 2005:334–339).

In 1004, Henry II, King of Romans, granted to the Diocese of Brixen (Bressanone) in present-day South Tyrol (Italy), his estate of Bled (*predium quod dicitur Ueldes*), adding the castle and the land in the size of 30 royal farms (*mansus regalis,* Königshube) between the Sava Dolinka and the Sava Bohinjka Rivers seven years later (Fig. 10.4) (Schumi 1882:14, 15). The grant laid the foundations for the Diocese of Brixen seigniory with its seat at Bled Castle. Over the following decades, the diocesan estate, by way of new grants and land swaps, grew to include a large portion of Blejski Kot and Bohinj areas, some property at the Radovljica Plain and the foothills of the Karavanks from the Dobršnik Stream to the Tržiška Bistrica River, where the diocese was also granted the hunting rights (*wiltpannum*), and the forests; such also applied to the property on Pokljuka and Mežakla. In 1063, the diocese was vested with the eastern part of Jelovica (Schumi 1882:41; Bizjak 2006:47–61; Pleterski 2013:147; Štih 2014:45–48).

The documentation on the royal grants of the western Julian Alps is scarcer than that of the eastern counterpart. In 1007, King Henry II founded the Diocese of Bamberg in Franconia (*Franken*), vesting it with a large estate centred in Villach (present-day Austria), including Tarvisio (present-day Italy) and the Canal Valley down to the Pontebbana Stream (present-day Italy) on the border with Friuli (present-day Italy) (Moro 1957). The southwestern area of the Julian Alps, encompassing the Upper Soča Valley region, as part of Friuli, may have been granted to the Patriarchate of Aquileia in the second half of the tenth century. Such was definitely the case by the 1060s, when the Patriarch Rabenger (1063–1068) issued a charter in Tolmin (*in Tuimine*) (Schumi 1882:46; Kos 1948:12; Štih 2015:46). The centre of the extensive Tolmin seigniory (*Gastaldia*) was located at the Kozlov rob Castle above Tolmin, and the seigniory held in possession a large part of the Upper Soča Valley region and the Bača Valley up to the border with the March of Carniola along the peaks of the Julian Alps. The land property was more fragmented in Kobarid and particularly Bovec area, where it was held by many feudal lords in the twelfth and thirteenth centuries (Kos 1948:12–22; Oitzl 2016:56–57). In 1119, the Resia Valley (present-day Italy) became the property of the newly founded Benedictine monastery in Moggio (present-day Italy) (Fig. 10.4) (Härtel 1985:71, 77–80).

It is not entirely clear who were the first landlords of a significant part of the Radovljica Plain, the Lipnica Valley with part of Jelovica Plateau and the northwestern part of Bohinj area. The property was most likely included in the vast seigniory of the margraves of Carniola when the Counts of Sempt-Ebersberg still held this function in the late tenth and the first half of the eleventh centuries. During their period and that of the Counts of Weimar-Orlamünde, their successors in office, the property became hereditary. In the first half of the twelfth century, in a sequence of events not yet conclusively established, it came into the possession of the Counts of Ortenburg. They established a large seigniory, with centres at Waldenberg Castle below Jelovica and Kamen Castle below the Karavanks (Fig. 10.3), acquiring a large part of the Upper Sava Valley in the fourteenth century (Hauptmann 1953:271; Mlinar 2005:336–340; Bizjak 2006:47–49; Oitzl 2016:52–55).

The land property transfer in the eastern part of the Upper Carniola and the Kamnik-Savinja Alps is somewhat more straightforward. In the eleventh century, the seigniory of the margraves of Carniola, the territory left to the Sava River, eventually became hereditary property. Following the marriage of Sophie of Weimar-Orlamünde, the daughter and sole heiress of Poppo II, to Count Berthold II of Andechs, the estate passed as a dowry and inheritance into the hands of this Bavarian count family in the early or first half of the twelfth century. The seat of the extensive seigniory, which encompassed the Upper Carniolan Plain from the Kokra River in the west to the border with the later Styria province in the east, as well as a large part of the Kamnik-Savinja Alps, bordering Carinthia, was established at the Stari grad Castle above Kamnik (Fig. 10.4) (Štih 2000:11–13). The estate was soon fragmented. As early as the mid-twelfth century, the Cistercian monastery of Viktring in Carinthia acquired some property in the area of Preddvor, and the estate later even increased in size. On the slopes below Krvavec, the female Dominican monastery of Velesovo, founded in 1238 by the Andechs ministeriales, the brothers Gerloch, Weriand and Walter of Kamnik, held several farms, as well as land property in the Upper Carniola donated upon its establishment (Mlinarič 1999:70–77; Volčjak 2014:213–215, 225–227).

The first known landlord in the Upper Savinja Valley was Diepold Kager, possibly a member of a branch of Cham-Vohburg margraves from Bavaria. In 1140, he and his wife Truta granted most of their allodial property in the area to

Fig. 10.3 Kamen Castle on the late 17th-century copperplate. (Valvasor 1679:245, National and University Library archive)

the recently founded Benedictine monastery in Gornji Grad, and the rest to the Patriarchate of Aquileia (Fig. 10.4). The grant was the basis for establishing of an extensive monastical seigniory of Gornji Grad (Gestrin 1953:474; Ravnikar 2010:30–42; Štih 2015:47).

10.3 Colonisation

Once they acquired estates and established seigniories, feudal lords wanted to make the most of the land and organised the expansion of existing and building of new settlements, which was the task of seigniorial officials. More efficient methods of working the land facilitated colonisation, resulting in population growth. In certain areas, it started as early as the end of the tenth century and continued with considerable intensity throughout the High Middle Ages. Once the lowland had been occupied, the second, or upland stage saw the settlement of sun-bathed mountain slopes and forested parts of the valleys. The Alpine world was subject to second stage of colonisation from the first decades of the thirteenth to the fifteenth centuries, and, in some areas, as late as the sixteenth century. The new settlers usually enjoyed certain benefits—for the first few years after setting up the farm, they were exempt from paying dues and bounded labour (Blaznik et al. 1970:73–81).

Based on the research, colonisation on the periphery of the Alpine world seems to have culminated in the twelfth century. Such was the case in Blejski Kot and Bohinj, where it began as early as the eleventh century, with the first stage completed in the first half of the thirteenth century. According to the 1253 urbarium (rent-roll) of the Bled seigniory, nearly all settlements in the lowland area had been established by the mid-thirteenth century, and the number of farms did not change significantly over the following centuries (Bizjak 2006:47–54; Pleterski 2013:168–176).

Similar conclusions could be drawn for the Upper Carniolan Plain and the foothills of the Karavanks and Kamnik-Savinja Alps. In the High Middle Ages, some settlements at slightly higher altitudes emerged below Dobrča, between Tržiška Bistrica River and Preddvor, and between Kokra and Kamiška Bistrica Rivers. The settler-friendly parts of the Alpine valleys were inhabited relatively early, along the Tržiška Bistrica River down to Bistrica settlement and Stahovica in the fairly intensively populated Kamniška Bistrica Valley, with some individual farms above Preddvor in the Kokra Valley. The slopes above the transport route to Gornji Grad via Črnivec Pass were also settled. Following the foundation of the local monastery, colonisation of the Upper Savinja Valley was next to complete, at least in the surroundings of Luče and Solčava (Gestrin 1953:474–475; Kos 1971:7–9; Grafenauer 1985:15–16; Otorepec 1985:33–36).

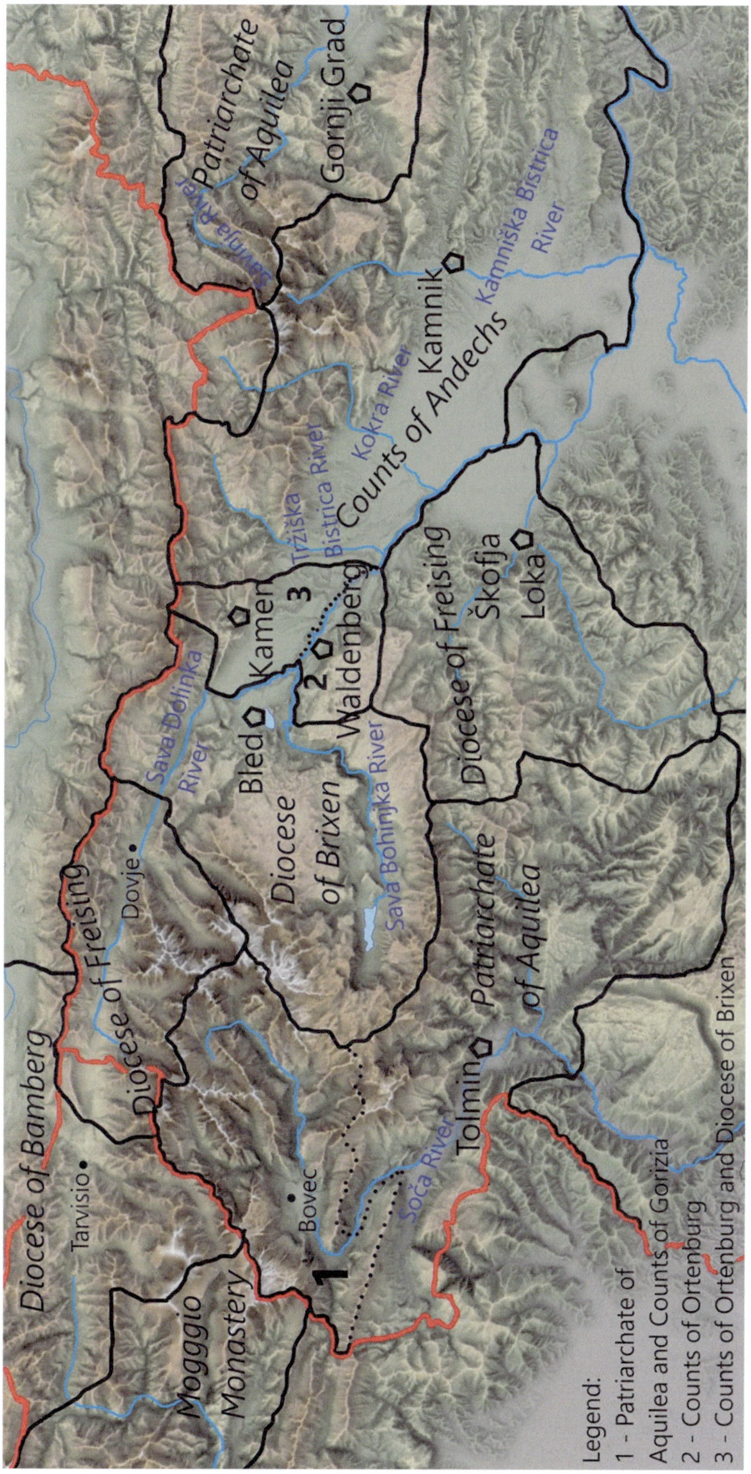

Fig. 10.4 Structure of seigniories in the Slovene Alpine world in the early thirteenth century. (*Source* Bajt and Vidic 2011:68–69; Map by Gašper Oitzl)

It is safe to assume that the Upper Soča Valley region was colonised early too. The oldest preserved urbarium of the Tolmin seigniory is dated 1377, when the upland settlement was nearly complete (Kos 1948). More records are available for Resia Valley, thanks to the preserved urbarium of monastical seigniory in Moggio (1237–1242). The valley was extensively colonised sometime between the mid-twelfth and the first decades of the thirteenth centuries. The farms increased from 23 in the early twelfth century to 49, and four new villages emerged. According to the 1459 urbarium, medieval colonisation in Resia was, to a great extent, completed in the first half of the thirteenth century (Kos 1954:298–306) (Fig. 10.5).

The colonisation of the Selca Valley (Fig. 10.1) took place somewhat later than in Blejski Kot and Resia, reaching its peak between the second half of the twelfth and the mid-thirteenth centuries, and its outcome is evidenced in the 1291 urbarium of the Škofja Loka seigniory. The farms in the Selca and Poljane Valleys grew from 153 around 1160 to about 750 at the end of the thirteenth century, i.e. nearly fivefold (Blaznik 1928, 1973:29). Similar growth can be observed in the Gornji Grad seigniory, where about 150 farms in 1140 increased to 745 in 1426 (Gestrin 1953:475). The outcome of lowland colonisation in the Upper Sava Valley is registered in the 1253 and 1291 urbaria of the Bled (Koroška Bela and Potoki) and Škofja Loka (Dovje) seigniories. In Dovje, four farms around 1160 went up twenty-eight by the end of the thirteenth century (Blaznik 1955:7–8; Bizjak 2006:187).

In terms of time, upland settlement varied even more from area to area. It began particularly early on the sunny slopes of the Kamnik-Savinja Alps east of Kokra River at an altitude of about 700 m, where the soil was fertile and could also be employed for grazing. As early as 1132, there were two manors (*curias*) on Štefanja Gora, and due to the mention of dues in cheese, Milko Kos assumed that the local settlements developed from cheese dairies. No later than the early or the first half of the thirteenth century, and possibly much earlier, Možjanca and Apno were established, and Stiška Vas and Šenturška Gora could also be the result of early upland colonisation (Kos 1960:57, 71–72; Kosi et al. 2021:20–21, 965).

Colonisation of the Tunjice Hills, Dobrave and the area near and above Bistričica River (Županje Njive, Okroglo and Bistričica) began no later than the thirteenth century. Somewhat later, the area above the Tunjice Hills, Zakal and Klemenčevo above Bistričica upstream, possibly also Ambrož pod Krvavcem (1079 m), were settled. In the thirteenth century, the remaining settlements below Dobrča were probably established, while some settlements in the hilly area between Tržič and Preddvor (Novaki, Gozd, Zalog and Srednja Vas above Goriče) emerged in the period of upland colonisation (Kos 1960:57; Grafenauer 1985:15–16; Otorepec 1985:33–37).

Upland colonisation in Blejski Kot (Laze and Kupljenik) and Bohinj (Nemški and Laški Rovt and Ribčev Laz) began in the first half of the thirteenth century, and by the first decades of the fourteenth century, it was mostly finished (Bizjak 2006:203–208). In the Early Modern Period, new settlements, i.e. Goreljek, Gorjuše, Koprivnik and Podjelje, emerged from pastures on the Pokljuka Plateau. The southern slopes of the Bohinj-Tolmin Mountains and the valley upstream of Bača River were also populated in the first half of the thirteenth century (Fig. 10.7). The Patriarch of Aquileia, Berthold of Andechs (1218–1251), brought to this area German-speaking colonists from his estates in the Puster Valley in Tyrol. They were organised in a special office called *rityria Rut*, had their magistrate, and 36 farms had been established by 1377 (Rutar 1882:42–43; Kos 1948:81–82). In the neighbouring Freising seigniory, the settlement of the Sorica area (Fig. 10.1) below Jelovica (Fig. 10.6) began in the 1280s, and the landlord also colonised the area with German-speaking settlers from the Puster Valley. The Sorica office is mentioned as early as 1286, 21 farms were established by 1291, 9 more by 1319, followed by a minor increase in their number (Blaznik 1928; Blaznik 1973:21–24, 38–39; Markelj 2019). In the Sorica area, a German dialect called *Zarzer Mundart* persisted until the early twentieth century (Markelj and Hoffman 2020).

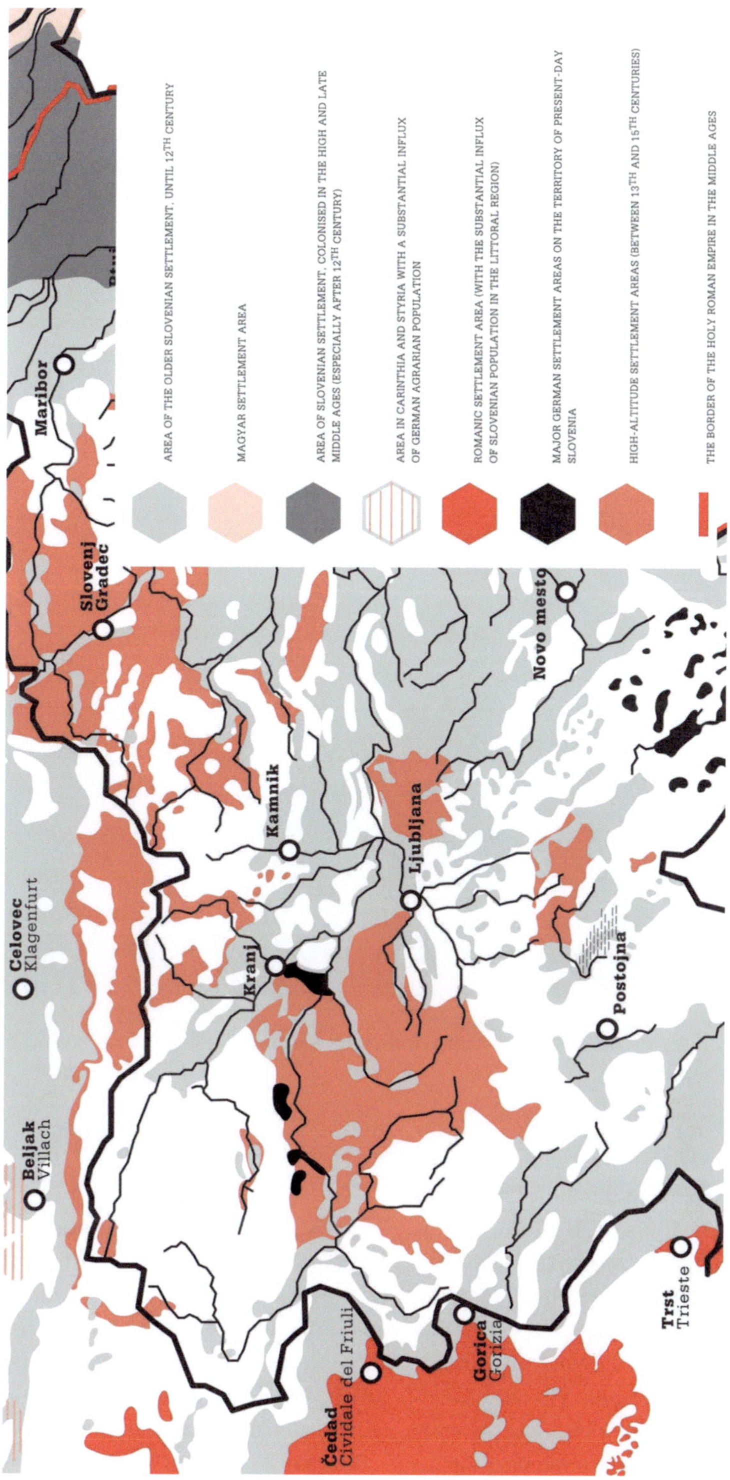

Fig. 10.5 Medieval colonisation process in the Slovenian territory. (*Source* Bajt and Vidic 2011:79; Map by Polonca Strman, authors archive)

Fig. 10.6 Settlement of the Sorica area began in 1283. The photo shows Spodnja Sorica. (Photo by Gašper Oitzl)

Fig. 10.7 The settlement of Stržišče was established during the colonisation of the Upper Bača Valley in the first half of the thirteenth century. (Photo by Gašper Oitzl)

Colonisation in the Upper Sava Valley (Fig. 10.1) also falls within a similar time frame; Kranjska Gora was founded around 1300, and Mojstrana in the early fourteenth century. It is also referred to in a source from 1362/63. No later than the second half of the fourteenth century, the settlement of *Hohentall* (probably today's Planina pod Golico) emerged, as well as Javorniški Rovt. In the fifteenth century, Srednji Vrh and Gozd Martuljek were probably settled, and Rateče and possibly Podkoren also significantly grew in size (Blaznik 1955:8; Otorepec 1995:17; Mlinar 2005:337–339; Bizjak 2006:197–198; Mlinar 2011:30–42; Mlinar 2018:71–79).

By the early fifteenth century, the upland colonisation of the Upper Savinja Valley (Fig. 10.1) was mainly finished, and it saw the establishment of numerous isolated farms, which also emerged in the Logar Valley and Matkov Kot. The outcome of colonisation is evident in the 1426 urbarium of the Gornji Grad seigniory, which lists 123 farms in the Luče and 50 farms in the Solčava offices, respectively (Gestrin 1953:475, 493–494).

In the fifteenth century, when upland colonisation was on the wane, sources reveal intensified deforestation to create assarts, serving as pastures or hay meadows, some of which were later permanently inhabited. Many assarts were cleared at the higher altitudes of the Radovljica seigniory, e.g. above Lake Bohinj, on Pokljuka, and Jelovica. A few were cleared in Radovna and the Upper Sava Valleys in the second half of the fifteenth century (Zwitter 2013:92–94). Assarts, mountain pastures, and isolated farms also emerged below Krvavec and in the Upper Kokra Valley from the fifteenth century onwards (Kos 1960:71–72).

Upland colonisation lacked the intensity of the lowland in terms of the number of settlers and farms, but it was much more intrusive in the Alpine area, especially concerning forests and their ecosystem. During colonisation of lowland and the expansion of farmland in the valleys, pastures moved to higher altitudes. However, deforestation increased with permanent settlement on the higher slopes of the Alps. Initially to prepare the terrain, and then to acquire additional agrarian land, primarily for pastures and hay meadows, while animal husbandry was the dominant economic activity in the Alpine region (Zwitter 2013:85–87).

10.4 Medieval Economy in the Alpine World

After settling permanently, a man had to find ways to survive in a seemingly inhospitable environment. However, the Alpine world holds "jewels" a man could not resist. These can be summarised in three points: (1) relatively favourable geographical and climatic conditions on the sunny side of mountain slopes; (2) geological composition that makes the hilly and mountainous areas rich in ores; and (3) its passable nature that facilitated and promoted the establishment of trade routes.

10.4.1 Animal Husbandry

In the Middle Ages, animal husbandry was the primary source of livelihood in the Alpine world, particularly at higher altitudes in the valleys. The less favourable the conditions for the development of agriculture, the more critical animal husbandry was for that area's economy. Such is evident from the numerous disputes over mountain pastures that can be traced in written sources. Breeding small ruminants was particularly relevant in Alpine valleys, such as Resia, the Upper Sava and the Upper Savinja. Goat breeding was typical of the mountain pastures at higher altitudes, e.g. in Bovec and Trenta, while sheep farming developed also on the fringes of the Alps. Other than livestock, the subjects paid their dues in cheese which was one of the leading export articles of the Alpine area. At the end of the sixteenth century, just under 13 tonnes of cheese were exported from Bovec to Friuli in a year. Nearly all farms in the Tolmin seigniory were claimed cheese by their landlords. In some places, there were cheese farms known locally as *švajge* (*swaig*). In the Selca Valley, their number amounted to 15 in 1500 (Novak 1970:380; Hassinger 1987:226; Oitzl 2018:69–70), and in the settlement of Kokra at the end of the fourteenth century, a cheese dairy farm is referenced to each of the four farms (*pey yeder hub ain swaig*) (Kosi et al. 2021:1763–1764). Cheese production was of vital importance in the Upper Savinja Valley, with a total of over 12,600 wheels of cow and sheep cheese paid as due to the seigniory per year (Gestrin 1953:504).

Other livestock farming activities included poultry farming, developed in much of the Alpine world, with the subjects paying their dues in chickens and eggs. Cattle breeding was most widespread in the Tolmin area (Kos 1948:31). Pigs were also kept at the lower altitudes, and as early as the second half of the tenth century there is mention of pig grazing in the Škofja Loka seigniory (Štih 2014:42), and later we find reference to them as a due paid by certain subjects to their landlords in the Selca Valley and Blejski Kot (Blaznik 1963:122; Bizjak 2006:176–177, 188). In the late fifteenth century, the burghers of Kamnik grazed pigs in the forest near Kamniška Bistrica River (Valenčič 1957:73).

Livestock was grazed on pastures near the village in spring and autumn and on higher mountain pastures during the summer. In order to use the pastures, the subjects had to obtain permission of the seigniory which imposed additional dues on them (Novak 1970:347–375). The first written records of mountain pastures date back to the twelfth century, the pasture (*alpes paschuales*) of the village of Mače below Mt. Storžič as early as 1156 (Kos 1960:71), and the *Stefan* pasture below Mt. Krn in 1178 (Bianchi 1861:266). In the mid-thirteenth century, pastures were fixtures of the village Nemški Rovt in Bohinj (Bizjak 2006:186). There are many references to pastures in the Upper Posočje area; in Trenta, the Trebiščina pasture was used as early as the fourteenth century (Kos 1948:13). In the fourteenth century, several mountain pastures in the Canal Valley can be traced in the written sources. In 1377, the Lipanca and Javornik pastures on Pokljuka were subject to sale, and the Radovljica urbarium from the late fifteenth century contains several records of pastures in the area of Bohinj and the Karavanks (AS 174 1498; Oitzl 2018:72–73). In the mid-fifteenth century, the pastures above Tržič (Javornik, Tegoška pasture and Kofce) are also recorded (Bizjak 2005:266–267), and in the late fifteenth century several pastures in the Kamnik-Savinja Alps (Kosi et al. 2021:17, 586, 665, 960, 972, 1257).

10.4.2 Agriculture

Agriculture was most developed in the flat fringes of the Alpine area, although many villages at higher altitudes also paid dues in grain. Wheat and oats were the main crops, and rye, barley, broad bean and groats were cultivated in some places. Flax frequently featured as a due paid in kind. Dues in grain were especially rarely imposed in the Upper Savinja Valley (Gestrin 1953:482–506; Mihelič 1997:82; Oitzl 2018:67–68).

10.4.3 Forestry

Much of the Alpine world was forested, and the subjects could, in principle, use the forest for their own needs (usually for construction and heating) with the permission of the landlord. At least as early as the fifteenth century, the extensive forests along the Kamniška Bistrica River could be used by the burghers of Kamnik (Valenčič 1957:72; Zwitter 2013:95–96). An official appointed for forest management (*vorsterius*) in the Selca office, is recorded as early as the late thirteenth century (Blaznik 1963:159). In the fifteenth century, the urbarial records listed many sawmills in the Julian Alps, while sawmills along the Kamniška Bistrica River are registered in the second half of the sixteenth century (Otorepec 1957:47; Oitzl 2018:76). More extensive deforestation only began towards the end of the Middle Ages, both due to clearing for pastures and hay meadows and mainly due to the development of the iron industry. Wood was used as a source for charcoal, the primary fuel in ironworks, until the nineteenth century (Mlinar 2014:187–190).

10.4.4 Hunting and Fishing

Hunting and fishing were usually a privilege of the landlords, who set up hunting offices administered by hunting officials. In the fifteenth century, they were recorded in the Radovljica and Škofja Loka seigniories (CKSL 1461; Blaznik 1963:347). The provincial prince appointed his own hunting official in the forests along the Kamniška Bistrica River (Valenčič 1957:72). The Weissenfels seigniory had a dedicated steward for fishing-related matters (Mlinar 2018:83) (Fig. 10.8). The Patriarch of Aquileia had his fishermen in the Tolmin seigniory, and he also leased the rights to fish. The monastery in Gornji Grad leased the fishing rights in the Dreta and Savinja Rivers in exchange for 100 smoked trouts (Umek 1970:505–506; Mihelič 1997:77).

Some Bled and Radovljica seigniories subjects who lived in the vicinity of the Sava Bohinjka River or Lake Bohinj had the right to fish for a specific time of the year, in return for which they were imposed dues in kind. In exchange for guarding border with the Republic of Venice in the mountains above Bohinj, subjects from Studor

Fig. 10.8 The 1498 urbarium of the Weissenfels seigniory contains a record of a dedicated steward who supervised fishing in the Laghi di Fusine (present-day Italy). (*Source* AS 1 1498:41)

and Srednja Vas in Bohinj were allowed to hunt chamois in high mountains. The subjects from the villages around Tolmin had the right to hunt certain species of birds (AS 174 1498:92; Kos 1948:31; Umek 1970:507; Bizjak 2006:185, 188, 199).

10.4.5 Mining and Metalworking

As to the non-agricultural industries, the mining of lead and zinc ore in Cave del Predil (present-day Italy) began at least in the middle of the fifteenth century (Wiessner 1951:185–187). The first records of clay mining in the Tunjice area for the pottery trade in Kamnik date back to the mid-fourteenth century. The local potters had to pay an annual fee for digging clay (Otorepec 1957:52). Ironworking was the most important non-agricultural industry in the Slovene Alps. They are not among the most mineral-bearing areas in Europe; however, they are relatively rich in iron ore. In Pokljuka and Jelovica, it is found at or just below the surface. In the Karavanks, the ore was mined underground (Mohorič 1969:38–41).

Intensive ironworking emerged in the Slovene Alps in the middle of the fourteenth century. The first ironworking centres were established in the Canal and Selca Valleys, where northern Italian ironmaking masters settled in the mid-fourteenth century and introduced technical innovations such as use of water power to run the ironworks. The following decades witnessed the development of a settlement named after the iron (Železniki, German: *Eisnern*). According to the 1501 urbarium, two ironworks (a smelting furnace with trip hammer) had a total of 14 master ironworkers (Blaznik 1963:322; Blaznik 1973:83–87).

By the end of the fourteenth century, two ironworking centres had been established in the Upper Sava Valley. In 1381, the landlord Count Frederick III of Ortenburg issued mining regulations which gave impetus to the development of ironmaking on the slopes of the Karavanks above Jesenice. The outcome is evident in the 1498 urbarium of the Weissenfels seigniory, reporting on seven smelting furnaces and three trip hammers in the Jesenice area (Lačen-Benedičič 2001; Mlinar 2016:51; Mlinar 2018:66, 72) (Fig. 10.9). Another ironworking centre in the valley was established at the end

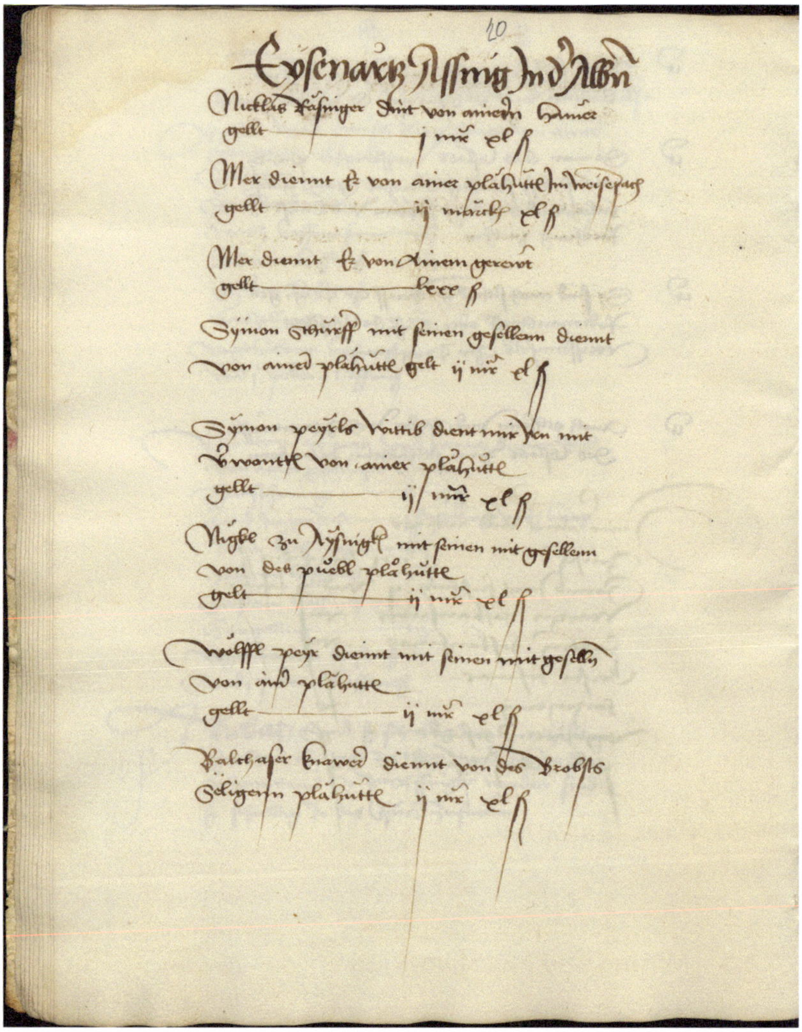

Fig. 10.9 Ironworks in the mountains above Jesenice, their owners and dues as per the 1498 Weissenfels urbarium. (*Source* AS 1 1498:20)

of the fourteenth century in Fusine in Valromana (Slovene: *Bela Peč*, German: *Weissenfels*; present-day Italy), in at the time uninhabited area below Rateče close to the provincial border with Carinthia. Unlike other centres, its principal activity was iron processing, with three trip hammers operating in Fusine at the end of the fifteenth century (Golec 2016:390–392; Mlinar 2018:82–83). Fusine was the only ironworking settlement in Carniola that developed into a market town. Referred to as burghers, its inhabitants are mentioned as early as the end of the fifteenth century, but its market-town status was not actually established until the 1540s (Golec 2016:390–395).

At least by the second half of the fifteenth century, three ironworking centres had been established along the north-eastern foothills of Jelovica, in Kolnica (today Spodnja Lipnica), Kamna Gorica and Kropa. At the end of the century, at least one trip hammer (most likely in a combination with a smelting furnace) operated in each of these settlements, while in Kropa there were two, and there was an additional smelting furnace below Kamna Gorica (AS 174 1498:104). At the end of the century, ironworks also operated at the foothills of the eastern Karavanks and Kamnik-Savinja Alps; however, the number of plants and the production scale were much smaller than in their Julian Alps counterparts. Three forges operated downstream the Tržiška Bistrica River, a blacksmith's trade was developed in Tržič, and an ironworks also operated at the Kamniška Bistrica River above Kamnik (Oitzl 2021:112–115).

The development of ironmaking was of vital importance for the Alpine world. As mentioned before, it had certain impact on forests, and ironworkers also had the right to use the surrounding mountain pastures. New settlements, including Železniki, Fusine and Kropa were built near the ironworks. The support industries of logging, charcoal production, mining, and packhorse transport were crucial to the operation of the ironworks, ensuring the supply of raw materials. These industries mainly employed a surplus population of the rural surroundings, who could earn additional income or even be able to make a living. In addition to local use, iron was exported mainly to the northern Italian lands, allowing the burghers of Škofja Loka, Ljubljana, Kamnik and Kranj to make considerable profit in iron and iron goods trade (Gestrin 1965:179–185; Mitterauer 1974:267–268).

Iron industry also contributed to the total income of the seigniories, as the owners of the works paid monetary dues for landlord's permission to operate. The share in total income was related to the size of the seigniory. In the Weissenfels seigniory, which was relatively small, the income from ironworks accounted for 24.2%, in the Radovljica seigniory 6.2% of the cash income, and in Škofja Loka only 0.6% of total income (Mlinar 2018:176; Oitzl 2021:293–298).

10.4.6 Transport

The last of the Alpine "jewels" is its passable nature, a combination of the Alpine valleys and passes that allowed the establishment of many transport routes, facilitating the movement of people, goods and ideas. The most prominent route in the region runs through the Canal Valley, providing easiest passage across the Eastern Alps and linking Carinthia with Friuli and the Adriatic ports. At certain times, it had an alternative via the Predel Pass and the Upper Soča Valley. Around 1400, the town of Cividale (present-day Italy) financed a large road-building project via the Predel Pass from Cividale to Tarvisio (present-day Italy). It involved the construction of roads, many bridges and protection against avalanches. Iron- and blacksmith's works were set up near Srpenica, making tools and construction material (Kosi 1998:191–192).

In the east, the passes across the Karavanks and Kamnik-Savinja Alps – in the Middle Ages especially Ljubelj and Jezersko—were crucial for connecting Carniola and Carinthia (Hassinger 1987:101; Kosi 1998:245–247, 253–257). Along the route over the Ljubelj Pass, a market town, named after the pass (*forum Lubellinum*), emerged in the valley of the Mošenik Stream by the mid-thirteenth century. This settlement was allegedly destroyed in a rock-fall, and by 1337 a new market town (literally named *Neumarktl* (German for "new market town")) was built a few kilometres further down, in the area of present-day Tržič (Zorn 2002:138–142; Golec 2020:441; Zorn and Mrak 2020).

Other transport routes ran mainly on the periphery of the Slovene Alpine area. On the southern edge of the Julian Alps, the Upper Carniola and Friuli were linked via the Selca Valley and Petrovo Brdo to the Bača Valley. It was an alternative to the main route running via the Poljane Valley and the Oselica Pass. The route across the Bohinj-Tolmin Mountains to the Bača Valley and on to Friuli was of great local importance, especially for the Bohinj area. In the Middle Ages, the Upper Sava Valley had a minor role in transport, although it provides passage both to the Canal Valley and Villach via the Koren Pass. The route over the Črnivec Pass, which connects Gornji Grad and Styria with Kamnik and Carniola, was also of secondary importance (Kosi 1998:247–257).

10.5 Conclusion

Since the tenth century, humans have intruded on the Alpine area, initially taking small steps. The first, lowland stage of colonisation encompassed mainly the flatland on the fringes of mountain ranges and the most favourable areas of settlement in the valleys. It was completed mainly in the thirteenth century. In some places, parallel to the lowland settlement, the second, upland stage, began in the thirteenth century. It was most intense in the thirteenth and fourteenth centuries, and most present-day settlements originate from medieval colonisation period. All the central valleys were inhabited, while the side valleys, the same as the mountain slopes, served mainly for grazing, haying and wood-cutting. Deforestation intensified in the Late Middle Ages. In addition to the growing need for agrarian land, the iron industry and related charcoal production prompted deforestation and human activity in the Alpine world in general. The intensive development of the iron industry from the middle of the fourteenth century onwards brought about a significant change in the social organisation and economy of the area. It also exerted a considerable impact on the natural environment, as the growing number and technical advancements of the ironworks needed more and more charcoal, which was used as fuel until the late nineteenth century.

References

AS 1 (1498) Urbar gospostva Bela Peč [Urbarium of the Weissenfels seigniory]. Vicedomski urad za Kranjsko, box nr 117. Arhiv Republike Slovenije, Ljubljana

AS 174 (1498) Urbar gospostva Radovljica [Urbarium of the Radovljica seigniory]. Terezijanski kataster za Kranjsko, box nr 246. Arhiv Republike Slovenije, Ljubljana

Bajt D, Vidic M (eds) (2011) Slovenski zgodovinski atlas [Slovenian historical atlas]. Nova revija, Ljubljana

Bezlaj F (1967) Eseji o slovenskem jeziku [The essays about Slovenian language]. Mladinska knjiga, Ljubljana

Bianchi G (1861) Documenta Historiae Forojuliensis Saeculi XIII. Ab Anno 1200 ad 1299. Summatim Regesta [Documents for Friulian history in the 13th century. From 1200 until 1299. Summarised Regesta]. Archiv Für Kunde Österreichischer Geschichtsquellen 26:229–302

Bizjak M (2005) Gutenberg in briksenska posest vzhodno od Tržiške Bistrice [Gutenberg and the Brixen estates east of the Tržiška Bistrica River]. In: Mihelič D (ed) Ad fontes: Otorepčev zbornik. Založba ZRC, Ljubljana, pp 255–269

Bizjak M (2006) Srednjeveški urbarji za Slovenijo 5: Urbarji briksenske škofije 1253–1464 [Medieval urbaria for Slovenia 5: Urbarium of the Brixen Diocese 1253–1464]. Založba ZRC, Ljubljana. https://doi.org/10.3986/9616568434

Blaznik P (1928) Kolonizacija Selške doline [Colonisation of the Selca valley]. PhD Thesis, University of Ljubljana, Ljubljana

Blaznik P (1955) Freisinška županija Dovje [Freising *županija* Dovje]. Zgodovinski Časopis 9(1–4):7–25

Blaznik P (1963) Srednjeveški urbarji za Slovenijo 4: Urbarji freisinške škofije [Medieval urbaria for Slovenia 4: Urbarium of the Freising Diocese]. SAZU, Ljubljana

Blaznik P (1973) Škofja Loka in loško gospostvo 973–1803 [Škofja Loka and Loka seigniory 973–1803]. Muzejsko društvo, Škofja Loka

Blaznik P, Grafenauer B, Kos M, Zwitter F (1970) Kolonizacija in populacija [Colonisation and population]. In: Blaznik P, Grafenauer B, Vilfan S (eds) Gospodarska in družbena zgodovina Slovencev. Zgodovina agrarnih panog 1: Agrarno gospodarstvo. SAZU, Ljubljana, pp 29–127

Cevc T (2003) Kontinuiteta antične pašne tradicije v planinah v Kamniških Alpah: Ob najdbi poznoantične in zgodnjesrednjeveške lončenine iz poznega 6.–9. stoletja v planini Krvavec [Continuity of the antique pasture tradition in the mountain pastures of the Kamnik Alps: on the find of the late-antique and early-medieval pottery from the late 6th–9th centuries on the Krvavec mountain pasture]. Traditiones 32(1):7–19

Ciglenečki S (2000) Archaeological investigations of the decline of antiquity in Slovenia. In: Bratož R (ed) Slovenija in sosednje dežele med antiko in karolinško dobo: Začetki slovenske etnogeneze = Slowenien und die Nachbarländer zwischen Antike und karolingischer Epoche: Anfänge der slowenischen Ethnogenese. NMS, SAZU, Ljubljana, pp 119–139

CKSL (1461) Centralna kartoteka srednjeveških listin za Slovenijo [Central record of medieval charters for Slovenia]. ZRC SAZU Zgodovinski inštitut Milka Kosa, Ljubljana

Gestrin F (1953) Gospodarska in socialna struktura gornjegrajske posesti [Economic and social structure of the Gornji Grad estate]. Zgodovinski Časopis 6–7:473–514

Gestrin F (1965) Trgovina slovenskega zaledja s primorskimi mesti od 13. do konca 16. stoletja [Trade of the Slovenian hinterland with the Littoral towns from the 13th until the end of the 16th centuries]. SAZU, Ljubljana

Golec B (2016) Posebnosti nastanka in razvoja fužinarskega trga Bela Peč [Features of the origin and development of the ironworking market town Weissenfels]. Kronika 64(3):389–412

Golec B (2020) Trške pravice in avtonomija Tržiča od nastanka trga do srede 19. stoletja [Tržič market rights and autonomy from the founding of the market town to the mid-19th century]. Kronika 68(3):439–468

Grafenauer B (1985) Srednjeveška kolonizacija in izgradnja fevdalne oblasti vzhodno od Kokre na Gorenjskem [Medieval colonisation and the building of the feudal authority east of the Kokra River in the Upper

Carniola]. In: Žontar J, Melik V, Gestrin F, Zupančič M, Stiplovšek M, Svetelj F (eds) Kamnik 1229–1979. Kulturna skupnost, Kamnik, pp 14–18
Härtel R (1985) Die älteren Urkunden des Klosters Moggio (bis 1250). VÖAW, Wien
Hassinger H (1987) Geschichte des Zollwesens, Handels und Verkehrs in den östlichen Alpenländern vom Spätmittelalter bis in die zweite Hälfte des 18. Jahrhunderts. F. Steiner, Stuttgart
Hauptmann L (1953) Razvoj družabnih razmer v Radovljiškem kotu do krize 15. stoletja [Development of the social conditions in the Radovljica area until the 15th century crisis]. Zgodovinski Časopis 6–7:270–284
Horvat J (2006) Arheološki sledovi v slovenskem visokogorju [Archaeological traces in the Slovenian mountains]. In: Cevc T (ed) Človek v Alpah: desetletje (1996–2006) raziskav o navzočnosti človeka v slovenskih Alpah. Založba ZRC, Ljubljana, pp 21–40
Knific T (2004) Arheološki sledovi blejskih prebivalcev iz pozne antike in zgodnjega srednjega veka [Archaeological traces of inhabitants of Bled from the late antiquity and early middle ages]. In: Dežman J (ed) Bled 1000 let. Didakta, Radovljica, pp 93–118
Kos M (1948) Srednjeveški urbarji za Slovenijo 2: Urbarji Slovenskega Primorja, del 1 [Medieval urbaria for Slovenia 2: Urbaria of the Slovenian Littoral area, Part 1]. SAZU, Ljubljana
Kos M (1954) Srednjeveški urbarji za Slovenijo 3: Urbarji Slovenskega Primorja, del 2 [Medieval urbaria for Slovenia 3: Urbaria of the Slovenian Littoral area, Part 2]. SAZU, Ljubljana
Kos M (1960) Starejša naselitev na Kranjski ravnini [Older settlement on the Kranj plain]. In: Žontar J (ed) 900 let Kranja. Občinski ljudski odbor, Kranj, pp 51–73
Kos M (1971) Naselitev Gorenjske v ranem srednjem veku [The settlement of the upper Carniola in the early middle ages]. Arheološki Vestnik 21–22:7–16
Kosi M (1998) Potujoči srednji vek: Cesta, popotnik in promet na Slovenskem med antiko in 16. stoletjem [Travelling middle ages: roads, travellers and traffic in Slovenian Territory between antiquity and the 16th century]. Založba ZRC, Ljubljana
Kosi M, Bizjak M, Seručnik M, Šilc J (2021) Historična topografija Kranjske (do 1500) [Historical topography of Carniola (until 1500)]. Založba ZRC, Ljubljana. https://doi.org/10.3986/9789612549749
Lačen-Benedičič I (2001) Ortenburški rudarski red [Ortenburg mining regulations]. Krajevna skupnost, Planina pod Golico
Markelj M (2019) Višinska kolonizacija zgornjega dela Selške doline: Primer razvoja posestne in populacijske strukture tirolske naselitve iz konca 13. stoletja [Colonization process in the upper part of the Selščica valley: example of the Tyrolean land unit and population structure analysis from the end of the 13th century] PhD Thesis, University of Primorska, Koper
Markelj M, Hoffman AD (2020) Soriški narečni govor/Zarzer(ische) Mundart (nem.): Nastanek in razvoj govora v sklopu preučevanja naselitvene slike nekdanjih nemških govornih območij na Škofjeloškem in v bližnji okolici [Sorica Dialect/Zarzer(ische) Mundart (German): The origin and development of speech as part of a study of the settlement of former German-speaking areas in and around Škofja Loka]. Loški Razgledi 67:131–147
Mihelič D (1997) Wirtschaftsgeschichte der Slowenischen Alpen. Histoire des Alpes 2:73–90
Mitterauer M (1974) Produktionsweise, Siedlungsstruktur und Sozialformen im österreichischen Montanwesen des Mittelalters und der frühen Neuzeit. In: Mitterauer M (ed) Österreichisches Montanwesen. Produktion, Verteilung, Sozialformen. Verlag für Geschichte und Politik, Wien, pp 234–315
Mlinar J (2005) Slepo črevo Kranjske. Prispevek h kolonizacijski zgodovini Zgornjesavske doline [The caecum of Carniola. A contribution to the history of colonisation of the upper sava valley]. Zgodovinski Časopis 59(3–4):333–340
Mlinar J (2011) Gornjesavska dolina v poznem srednjem veku [The upper sava valley in the late middle ages]. In: Mlinar J (ed) Med Julijci in Karavankami: Zgodovinske in naravne podobe Gornjesavske doline. Gornjesavski muzej, Jesenice, pp 25–42
Mlinar J (2014) Das Eisenhüttenwesen und sein Einfluss auf Mensch und Natur in Spätmittelalter und Frühneuzeit: Beispiele aus dem westlichen Oberkrain. In: Štih P, Zwitter Ž (eds) Man, nature and environment between the Northern Adriatic and the Eastern Alps in premodern times. Znanstvena založba Filozofske fakultete, Ljubljana, pp 182–191
Mlinar J (2016) Povednost srednjeveških urbarjev: Primer belopeškega urbarja iz leta 1498 [Expressiveness of the medieval urbaria: example of the Weissenfels urbarium from 1498]. In: Žnidaršič Golec L, Bizjak M (eds) Urbarji na Slovenskem skozi stoletja. Arhiv Republike Slovenije, Ljubljana, pp 35–53
Mlinar J (2018) Urbarji belopeškega gospostva [Urbaria of the Weissenfels seigniory]. Novejši urbarji za Slovenijo 2. Založba ZRC, Ljubljana. https://doi.org/10.3986/9789610504184
Mlinarič J (1999) Posest vetrinjskega samostana v Preddvoru do 17. stoletja [The estates of the Viktring monastery in Preddvor until the 17th century]. In: Roblek T (ed) Preddvor v času in prostoru. Občina, Mladinska knjiga, Preddvor, Ljubljana, pp 69–81
Modrijan Z (2011) Continuity in late antiquity Slovenian fortified hilltop settlements. In: Heinrich-Tamaska O (ed) Keszthely-Fenékpuszta im Kontext spätantiker Kontinuitätsforschung zwischen Noricum und Moesia. M. Leidorf, Rahden, pp 151–171
Modrijan Z, Milavec T (2011) Poznoantična utrjena naselbina Tonovcov grad pri Kobaridu: Najdbe = Late antique fortified settlement Tonovcov Grad near Kobarid: finds. Založba ZRC, Ljubljana. https://doi.org/10.3986/9789612545871
Mohorič I (1969) Dva tisoč let železarstva na Gorenjskem [Two thousand years of the ironmaking in Upper Carniola]. Mladinska knjiga, Ljubljana
Moro G (1957) Zur Entstehung und Ausdehnung bambergischen Besitzes in Kärnten. Carinthia I 147:243–260

Novak, V (1970) Živinoreja [Animal husbandry]. In: Blaznik P, Grafenauer B, Vilfan S (eds) Gospodarska in družbena zgodovina Slovencev. Zgodovina agrarnih panog 1: Agrarno gospodarstvo. SAZU, Ljubljana, pp 343–394

Ogrin M (2006) Arheološke raziskave v Julijskih Alpah. Bohinj in Blejski kot [Archaeological research in the Julian Alps. Bohinj and Blejski Kot]. In: Cevc T (ed) Človek v Alpah: desetletje (1996–2006) raziskav o navzočnosti človeka v slovenskih Alpah. Založba ZRC, Ljubljana, pp 96–110

Oitzl G (2016) Prostor—narava—človek v Julijskih Alpah: Zgodovina nekega prostora med letoma 500 in 1500 [Area—nature—human in the Julian Alps: history of an area between 500 and 1500 AD]. MSc Thesis, University of Ljubljana, Ljubljana

Oitzl G (2018) Gospodarstvo v prostoru Julijskih Alp v srednjem veku [Medieval economy in the area of the Julian Alps]. Zgodovinski Časopis 72(1–2):62–103

Oitzl G (2021) Družbene in gospodarske posledice železarjenja na Slovenskem v poznem srednjem veku [Social and economic effects of ironworking in the Slovenian territory in the late middle ages]. PhD Thesis, University of Ljubljana, Ljubljana

Otorepec B (1957) Donesek h gospodarski zgodovini Kamnika do 16. stoletja [Contribution to the economic history of Kamnik until 16th century]. Kamniški Zbornik 3:43–61

Otorepec B (1985) Iz najstarejše zgodovine krajev v kamniški občini [From the earliest history of settlements in the municipality of Kamnik]. In: Žontar J, Melik V, Gestrin F, Zupančič M, Stiplovšek M, Svetelj F (eds) Kamnik 1229–1979. Kulturna skupnost, Kamnik, pp 33–41

Otorepec B (1995) Gradivo za zgodovino Radovljice v srednjem veku [Material for the history of Radovljica in the middle ages]. In: Sinobad J (ed) Radovljiški zbornik 1995. Občina, Radovljica, pp 10–38

Pleterski A (2010) Zgodnjesrednjeveška naselbina na blejski Pristavi: Tafonomija, predmeti in čas = Frühmittelalterliche Siedlung Pristava in Bled: Taphonomie, Fundgegenstände und zeitliche Einordnung. Založba ZRC, Ljubljana. https://doi.org/10.3986/9789612545666

Pleterski A (2013) The invisible Slavs: Župa Bled in the prehistoric early middle ages. Založba ZRC, Ljubljana. https://doi.org/10.3986/9789612546083

Ravnikar T (2010) V primežu medplemiških prerivanj [In the grip of the nobles' quarrels]. Pedagoški inštitut, Ljubljana

Rutar S (1882) Zgodovina Tolminskega, to je: zgodovinski dogodki sodnijskih okrajev Tolmin, Bolec in Cerkno ž njih prirodoznanskim in statističnim opisom [A history of the Tolmin area: historical events in court districts Tolmin, Bovec and Cerkno with their natural historic and statistic description]. Devetak, Gorica

Schumi F (1882) Urkunden- und Regestenbuch des Herzogtums Krain I [Book of charters and regesta of the Duchy of Carniola I]. Narodna tiskarna, Ljubljana

Snoj M (2009) Etimološki slovar slovenskih zemljepisnih imen [Etymological dictionary of Slovenian geographic names]. Modrijan, Založba ZRC, Ljubljana

Štih P (2000) Kranjska v času grofov Andeških = Krain in der Zeit der Grafen von Andechs. In: Eržen A, Aigner T (eds) Grofje Andeško-Meranski: Prispevki k zgodovini Evrope v visokem srednjem veku = Die Andechs-Meranier. Beiträge zur Geschichte Europas im Hochmittelalter. Zveza kulturnih organizacij, Kamnik, pp 11–37

Štih P (2014) Mensch und Wald in den Ostalpen (bis zur großer Kolonisation). In: Štih P, Zwitter Ž (eds) Man, nature and environment between the Northern Adriatic and the Eastern Alps in premodern times. Znanstvena založba Filozofske fakultete, Ljubljana, pp 36–51

Štih P (2015) Izvor in začetki škofijske posesti na današnjem slovenskem ozemlju [Origin and the beginnings of the diocesan estates in the present-day Slovenian territory]. In: Bizjak M (ed) Blaznikov zbornik. Založba ZRC, Muzejsko društvo, Ljubljana, Škofja Loka, pp 35–48. https://doi.org/10.3986/9616500805

Umek E (1970) Ribištvo in ribolov [Fishing]. In: Blaznik P, Grafenauer B, Vilfan S (eds) Gospodarska in družbena zgodovina Slovencev. Zgodovina agrarnih panog 1: Agrarno gospodarstvo. SAZU, Ljubljana, pp 495–520

Valenčič V (1957) Bistriški gozd in kamniški meščani [Bistrica forest and burghers of Kamnik]. Kamniški Zbornik 3:69–103

Valvasor JW (1679) Topographia Ducatus Carnioliae Modernae [Topography of the contemporary Duchy of Carniola]. Wagensperg

Vidrih Perko V, Sagadin M (2014) Gorenjska v antiki [Upper Carniola in antiquity]. Kamniški Zbornik 17:207–223

Volčjak J (2014) Das Dominikanerinnenkloster Velesovo/Michelstetten im Laufe seines Bestehens. In: Specht H, Andraschek-Holzer R (eds) Bettelorden in Mitteleuropa: Geschichte, Kunst, Spiritualität. DASp, St. Pölten, pp 211–232

Wiessner H (1951) Geschichte des Kärntner Bergbaues II: Geschichte des Kärntner Buntmetallbergbaues mit besonderer Berücksichtigung des Blei-und Zinkbergbaues. Verlag des Geschichtsvereines für Kärnten, Klagenfurt

Zorn M (2002) Rockfalls in Slovene Alps. Geografski Zbornik 42:124–160

Zorn M, Mrak I (2020) Skalni podor z Velikega vrha na Košuti—ljudsko izročilo in zgodovinski dogodek [Rock avalanche from Mount Veliki Vrh in the Košuta Massif—intangible cultural heritage and a historical event]. Kronika 68(3):349–360

Zwitter Ž (2013) Krčenje in agrarna raba gozdov na Gorenjskem v času višinske kolonizacije [Deforestation and agrarian use of forests in the Upper Carniola during upland colonisation]. In: Rogelj B, Potočnik Slavič I, Mrak I (ed) Gorenjska v obdobju glokalizacije. Znanstvena založba Filozofske fakultete, Ljubljana, pp 85–98

Part III
The Downfall of Cultural Landscape

11 Abandoning Traditional Identity—Depopulation of the Mountain Areas of Montenegro Throughout the Second Half of the Twentieth Century

Ivan Laković

Abstract

Traditionally perceived as a mountainous country, Montenegro is facing trends that, having accelerated since the mid-twentieth century, are genuinely changing its character in this regard. Reorientation from extensive agriculture practices, dominantly cattle rearing, to the activities and lifestyle adjusted to the industrialized urban centers determined the gradual but constant depopulation of the mountain areas, leaving there ever fewer and mostly elder households. Such trends have their roots in the development of the Montenegrin urban settlements since the late nineteenth century; however, the balance was drastically broken only after WWII when the new social and economic paradigm pressed on forced industrialization and urbanization as the comprehensive political, social, and economic strategy. Generational experience of living with and in the mountains was quickly replaced by a completely different context, depriving the new generations not only of the intention and reason to live there but also of the knowledge needed to organize and lead a life in the rural parts of the country. Within only half a century, the socio-demographical picture of Montenegro was so thoroughly changed, those mountainous areas, once the cradle of its traditional identity, nowadays often resemble deserted areas with sparsely scattered inhabitants.

Keywords

Mountain areas · Depopulation · Local knowledge · Montenegro

11.1 Introduction

In the centuries following the falling under Ottoman rule, the only Christian communities on the territory of contemporary Montenegro that maintained some degree of autonomy were the ones inhabiting the mountainous parts of the area. These areas became the nuclei of resistance and future establishment of an independent state, traditionally consisting of two parts, *Crna Gora* and *Brda*, both reflecting the mountainous context in their names. Crna Gora, whose Italian translation *Montenegro* remained an internationally recognized country's name, literally means *The Black Mountain*, while the Brda, the area in its central and north-eastern part, means the *Hills* or *Highlands* (Fig. 11.1). In all its history of surviving, cultivating, enlarging, and defending its geographical and social mountain realm, Montenegro finally became a part of a joint South Slavic country (Kingdom of Serbs, Croats and Slovenes/Yugoslavia) after WWI. The

I. Laković (✉)
Historical Institute, University of Montenegro, Podgorica, Montenegro
e-mail: ivan.lakovic@gmail.com

M. Zorn et al. (eds.), *Discourses on Mountains of Montenegro and Slovenia*, Historical Geography and Geosciences, https://doi.org/10.1007/978-3-031-99017-5_11

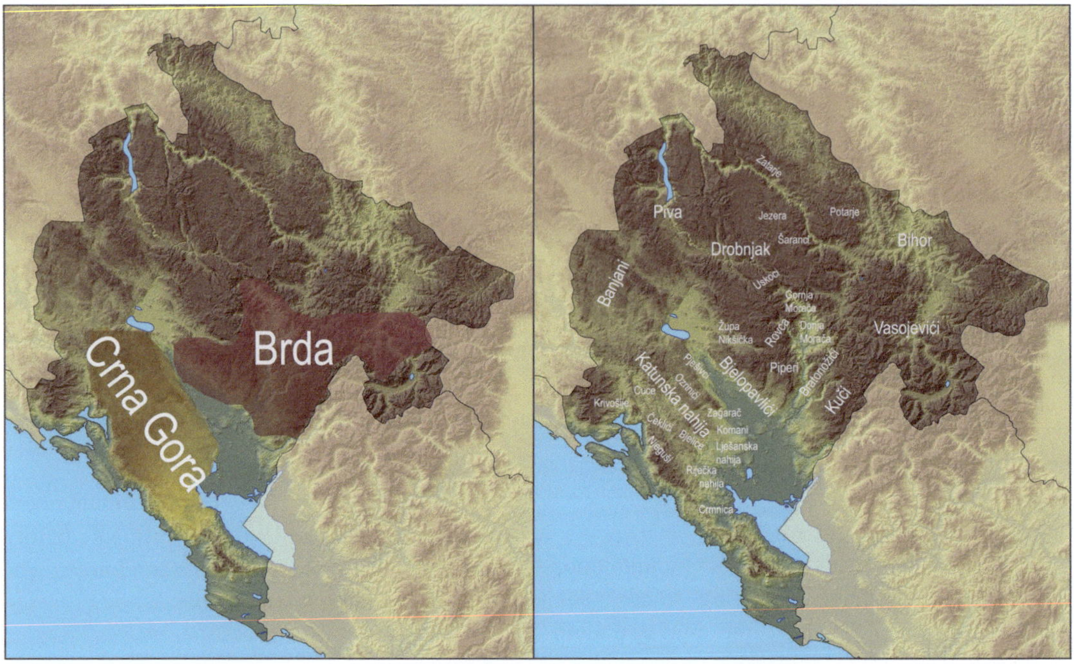

Fig. 11.1 Maps of traditional tribal division in Montenegro. (*Source* Relief … 2012)

everyday life of its people, throughout these times, was predominantly rural and agriculturally oriented, without towns except the Cetinje and newly established Danilovgrad, which began to obtain the urban outlines only in the second half of the nineteenth century (Marović 1998).

Such life was almost totally determined by geographical and climate conditions. The lack of arable land implied the full orientation toward cattle rearing. Extensive agriculture, limited by both the quantity and the quality of available soil, supplemented it where and as much as possible. Throughout this period, a solution to the given problem was sought in territorial enlargement toward the more fruitful, flatter, and hospitable areas in the valleys of Zeta and Morača rivers, by the north shores of Skadar Lake, as well as in the higher, but still agriculturally suitable areas of Grahovo and Nikšić poljes. Since these efforts successfully ended only in the nineteenth century, a way to survive in the meanwhile had to be found. The herds and flocks of cows, sheep, and goats, as well as the whole domain of their rearing, protecting, housing, providing food and water, use and trade of their products, thus became the central axis of the life, shaping all the other strategic political, social and economic tendencies. The climate conditions, characterized by hot and dry summers, additionally imposed the problem of summer grazing, where the food for cattle could not be provided in needed amounts within the traditional domain of villages. For this reason, a system of transhumance was established, where the population was spending the late fall, winter, and early spring in their villages, while the late springs, summers, and early falls were spent in seasonal settlements called *katuns* (Laković 2017; Fig. 11.2, built around the pastures in the mountains).

Organization of life within the *katuns* had many common characteristics across the country. The possession rights were being acquired and maintained through military clashes with previous owners, with examples of buying out the areas for *katun* foundation being extremely rare. While the number of *katuns*, especially those in the vicinity of related villages, were the private property of their owners, the majority belonged to the common one. It was, mostly, related to clans and families within the tribes, while only rarely to

Fig. 11.2 Look of the traditional Montenegrin mountain *katuns*. (Photo by Ivan Laković)

villages as the community. However, *katun* was related to the locality where the housing cottages and auxiliary objects were being constructed, with strictly determined pastureland and forest surface that belonged to it. Every member of a given community had a right to the cottage and the immediate surface around it, which could not be sold to any outsider. All the aspects and categories of such defined seasonal stay were being determined by relevant tribal agreements and enforced by their authorities (Rašović 1963; Radonjić 2004).

11.2 Consolidation of the Mountain Life in Montenegro and Highlands

This cycle included the vast majority of the population, crystalizing the comprehensive corpus of rules determining rights, obligations, and hierarchy in both establishing *katuns* and their use, as well as cattle grazing at the surrounding pastures. While a number of Montenegrin and Highlands' tribes occupied the mountains in central and northern parts of the country, the ones more southerly located turned to the Orjen-Lovćen-Rumija mountain chain. Since these mountains are not rich in grasslands, many households from adjacent tribes were renting their animals to those with *katuns* in more pasture-abundant regions. They were paid in a certain amount of dairy products, which provided both the income and enabled an otherwise impossible survival of their animals during the summer. In such constellation, survival or sustainable life of the village was preconditioned by having the *katun* or connection with those owning one (Laković 2017:4). That is why a significant portion of the warfare history of the region was the struggle to occupy and preserve the mountain areas for establishing *katuns*, with most of them resulting from successful military campaigns against previous owners, regardless of their ethnic or religious affiliation.

On the other side, tribes establishing their rule over certain mountain regions from the sixteenth to the eighteenth century used the newly obtained areas to increase their economic, social and political potential. Their herds and flocks, enlarged due to copious grazing during the summer in the mountains, needed adequate continuity in the winter months, inevitably pushing them to gradually occupy the bordering areas of southern lowlands in the Zeta valley, expanding the tribal territory and founding new villages. That is why the understanding of Montenegrin and Highlands

Fig. 11.3 Abandoned *katun* at Kuči Mountain. (Photo by Ivan Laković)

Fig. 11.4 Abandoned *katun* at Sinjajevina Mountain. (Photo by Ivan Laković)

tribes' growth starts with recognizing the role of mountain pastures and associated *katuns*, since their enlargement and strengthening were not the precondition to further venture toward mountains, but the consequence of consolidating possession over the mountain resources (Čubrilović 1963:329–330; Fig. 11.3). By the end of the nineteenth century, all the Montenegrin and Highlands' tribes had their mountain territory determined and defined, with the last significant change coming from Prince Nikola's allowing the parts of the Bjelopavlići tribe to establish *katuns* on Sinjajevina Mountain after its takeover and inclusion in Montenegro's territory (Radonjić 2004:446; Fig. 11.4).

The general economic circumstances improved during the first half of the twentieth century, and modernization of the society attached more significance to the urban areas acquired either following the wars of 1876–1878 (Nikšić, Podgorica, Kolašin, Bar and Ulcinj) or the First Balkan war (Pljevlja, Bijelo Polje, Mojkovac, Berane, Plav, Peć, Đakovica, and Prizren) (Pejović 1955:213–214; Blagojević 1991:84). Nevertheless, the main determinants of the societal structure remained almost unchanged, with the greatest percentage of the population still pursuing the traditional agricultural, transhumance-defined lifestyle (Marović 2018:336). The mountain *katuns* saw their prime in this period since the circumstances finally

Fig. 11.5 Ruins of *katuns* on Kuči Mountain, having reached their primes in the middle of twentieth century. (Photo by Ivan Laković)

settled down, and *katuns'* lifestyle got relieved of, at least partly, of its warfare side. Furthermore, all technological innovation available (and they were becoming ever more available) was improving everyday life significantly, both in terms of accommodation comfort and food production. According to our research during the KATUN project (No. 01–648) in 2015–2017, the period between the two world wars was the heyday of the *katuns'* life, since the greatest number of *katuns* in the Montenegrin mountains was active in these years (Laković 2017:4; Fig. 11.5).

11.3 Social-Economic Circumstances After WWII

The situation started to rapidly change after WWII. As the forces, led by the Communist Party of Yugoslavia (CPY), secured their unquestionable authority over the country, the comprehensive reform based on communist ideology was carried out within all the spheres of political, social, and economic domains. This concept assumed economic growth through rapid industrialization and change of the social and demographic structure. Since the majority of the population in interwar Yugoslavia lived in rural, agricultural communities, the general strategy of the new elite was to increase the number of industrial plants in urban areas to enable economic growth, but also to strengthen the electoral base of its power—the working class. Based on simplified interpretations of Marxist social theories, deriving from the historical experience of highly industrialized societies, where the main class confrontation raged between the huge and impoverished working class and their "bourgeois" counterparts, this concept was not fully applicable to the domain of agricultural, rural population of the Balkan country. The experience of the Soviet revolution and subsequent civil war 1917–1923 transformed the parole "power to the workers" into "power to the workers and peasants" due to the sheer fact that the dominant majority of the impoverished Russian population, seen as the supportive base for the new order, overwhelmingly lived in rural communities. Yugoslav communists used that model since there was no way to secure massive support, but ideological mistrust in relatively autonomous, free peasantry remained (Obradović 1999:138–139; Milošević 2015:104–105).

Traditional leftists' prejudices assumed that free peasantry, especially if it achieved a certain degree of economic self-sustainability, would be more inclined toward the conservative side of the political spectrum. In such conceptions, a villager who secured his private, economic independence based on the cultivation of his property and accustomed to "transferring" such independence to the sphere of political affinities would have a little sympathy for delegating his freedom of decisions and actions to any authority that he, personally, disagreed with. The native democracy of developed peasantry was considered one of the main threats to the new communist organization in rural communities, so the new authorities faced even riots against forceful collectivization according to the Soviet model. Here industrialization and urbanization were supposed to have a deeper and longer-lasting impact—aside from economic growth, the demographic picture was to be changed in favor of a more easily controlled and influenced population. The factory worker, living on the 6th floor, would depend not only on his monthly salary to provide the food and clothing for his family or pay the bills but also on a plethora of communal services, rules, and duties out of his reach or control. Formally free, he would find himself repeatedly conditioned to "stay on the right track" and obey the suggestions and commands from above, so he was not only needed as the part of the production process but also as the preferred model of the citizen in the new society. In this comparison, a substantially more independent peasant versus the worker incorporated into a spider web of the urban context, the former could not count on faring well (Šljukić 2011).

In Montenegro, this picture was among the most drastic in the whole of Yugoslavia (Vukčević 1974:18). As mentioned before, the urban population constituted a significant minority in the interwar period, 6,5% in 1921 (Ivanović 1976). During wartime, it suffered great losses, especially in the biggest city—Podgorica. It was one of the most bombed cities in WWII, being the target of 64 raids by both Italian at the beginning and Allied air forces in the later stages of the war. The city, which had had app. 14,000 inhabitants before, saw the end of the operations with less than 4,000 of them alive (Kovačević 2003). Even if not so heavily destroyed, other urban centers lost a significant portion of the population, which was even more visible given the initially low population numbers. In the following decades, all of them witnessed almost tenfold growth when the new development strategy, formulated according to Soviet-inspired 5-years plans, was implemented through the construction of numerous industrial plants, housing, transportation, medical, educational, and commercial facilities for the workers and their families, as well as for those employed in the relevant governmental bodies. In a relatively short period, the demographic trends turned in favor of the urban population growth, not based on increased birth rate, but at the expense of depopulation of rural areas, predominantly those in more remote regions. According to data cited by Kerckhof et al. (2016), about 75% Montenegrins working in agriculture in 1948, fell back to 5% in 1981.

In the period 1946–1980, the main flow of migration of the rural population from all parts of Montenegro was toward the capital Podgorica, while the regional centers, such as Nikšić, Pljevlja, Berane, Bijelo Polje, and Cetinje mainly attracted the influx of people from surrounding areas. A significant number also moved to the coastal cities (Herceg Novi, Kotor, Tivat, Budva, Bar, and, to a smaller extent, Ulcinj), especially since the tourism expansion created numerous economic opportunities there. Motivation for this drive was easily found in the perspective of a much easier, modern, and secure life, where the governmentally sponsored job meant not only the regular and, in average, satisfactory salary, high probability of solving housing questions, more choices for the children's education and easier access to medical care facilities. Compared to the previous routine, which consisted of hard work from dawn till dusk in all weather conditions, living in the old and often less than appropriate houses without decent road, electricity, and water supply infrastructure, this was a huge, civilizational step forward (Gluščević 2008:111–117).

As the society evolved, the transition from the previous lifestyle had the biggest impact on the first generation of newcomers, who had to directly adjust to the new circumstances. They kept the connection with their villages and *katuns*, maintaining it as much as it was possible, but those efforts had already become those of the outsiders, who lived somewhere else and did something else for a living. They became the visitors, almost tourists in their native soil, where they were born, they would

Fig. 11.6 Current look of pulp and paper mill in Berane. (Photo by Ivan Laković)

return to after retirement in the best case. On the other side, that connection faded even more for their children. Born in the cities, raised in an urban environment, and visiting the parents' native home more or less frequently, they were establishing a relationship with it as part of family heritage and a source of self-definition in a tribal sense. But, while their parents sometimes perceived the area through a dilemma of whether to come back or not, it did not exist for them. Unlike their elders, they already were on their ground, so regarding the possible future organization of the life, they didn't face the question of eventual coming back, but of going somewhere else. Regardless of all the stories, visits, family relations, weddings, funerals, money, and efforts invested in the maintenance of family houses or properties, even in buying or building new ones, the reality was that dealing with them was the pastime, not the real life.

11.4 Industrialization Boost

That real-life was based on the fact that, since the end of WWII, Montenegro witnessed the previously unimaginable level of industrialization. The coal mines were opened in Pljevlja (lignite) and Berane (brown coal, Fig. 11.6) in which also a large pulp and paper mill became the biggest and most important regional employer. That was the case also with ironworks, furniture plant, brewery and bauxite mines in Nikšić, aluminum and furniture plant, construction machine factory and food processing enterprise in Podgorica, or household appliance factory in Cetinje. All these enterprises were gigantic, employing thousands of workers in the case of several biggest ones, meaning that they needed a considerable workforce that had to be brought from somewhere. Surrounding rural areas, in the highlands and mountains, were the immediate and readily available source (Kalezić 1976:14–16; Radović and Vukotić 1983:51–53; Glusčević 2008:97–100; Marović 2018:561–566). It is a certain historical irony that almost all of those industrial facilities bankrupted and got abandoned after the Yugoslav collapse in the early nineties.

The first incoming flow came with the construction workers who were employed to build the future urban, traffic, and industrial infrastructure. Living within several hours' walking distance from towns, they would set off early, before the dawn, forming lines descending from their villages. After the work, they were going back the same way. Children attending elementary and high schools would be joining along, benefiting from the protection of elders during this hike to and from towns. Since they were too many, while the road and transportation infrastructure remained insufficient for a long time, such a "hike to work" lasted throughout the 1950s. In any case, the main characteristic of that trend was that occupation of the people living in villages changed; they were spending a whole day working away, dedicating to their traditional activities only their free time when available. The next step was moving to towns permanently.

That and the previous phases were overlapping for a long time. After improving the transportation facilities to most villages during the 1970s, and especially after a certain number of people could afford a car, a part of the rural population opted to commute to work without moving to the towns. Problems in obtaining accommodation, family obligations, insufficient qualification for better-paid jobs, and many other issues influenced such a decision. But, in the end, the said category soon became a clear minority, the one between those who stayed in the countryside and those who left permanently. Such a trend still exists, so it can even be said that it has the function of reversing the trend of rural areas' depopulation by offering a model of sustainably getting the best of both worlds.

Still, the main tendency, having lasted for the previous seven decades, was the permanent dislocation of people and their activities to the urban areas. New jobs, apartments, houses, education, habits, community, everything that defines the new, modern, and comparably better life than the previous one, marked the context of societal progress and improvement of general well-being in Montenegro. People were driven by the wish to change their lifestyle based on better, governmental and ideological propaganda, incomparable differences in the pace of growth between the rural and urban areas, as well as deep disbelief in longer-term sustainability and perspective of staying in the countryside. This way the state got the expectedly obedient mass of urban population, who, for its part, got rid of hard, poor, uncertain, and "outdated" lifestyle. Nobody seemed to care about the repercussions it had on both rural and mountain contexts, which found themselves in the spiral of ever stronger depopulation and abandonment (Vujović 2012:321–322). Nevertheless, it would be irrational to claim that the new social development strategy was purposely intended to destroy the rural domain. It was understood, even taken for granted, as the inexhaustible fountain of human resources, which will only benefit from common growth and progress. It was expected that the modernization of the country, envisaged through its urbanization and industrialization, will not significantly harm the villages and mountain settlements, but relieve them from the pressure of the surplus of population they had no resources to support.

In this sense, enormous efforts were made to build the roads to remote villages and provide electricity there. Since many of them were located in the mountainous areas without direct and reasonably short access to the local or regional urban centers, these roads had the direct intention of supporting the survival of rural communities, as the civilizational achievement that they had every right to. All the relevant government services participated in these endeavors, from state-owned construction companies, public utility enterprises to regular military units, who can be credited for the construction of roads to numerous mountain settlements (Marović 2018:592–599).

11.5 Infrastructural Context

This activity blended with the delivery of regular peacetime military tasks, providing a huge service to the local population. Since the system of border control and protection was the responsibility of the Army, a chain of border military watchtowers was established along the state border with Albania. The roads, connecting those watchtowers in the mountain parts with existing communications, were passing through the inhabited but remote *katun* areas, becoming the precious connection with their villages. Army units and services also had the duty of maintaining these roads, without placing a burden on the local government bodies. Aside from roads in border areas that were built at mutual benefit, Army units were engaged in

Fig. 11.7 Path carved through Mrtvica canyon. (Photo by Ivan Laković)

the construction of the roads in the central, mountainous part of Montenegro, where the inaccessibility of terrain would impose unbearable costs to the local authorities if such projects were pursued. For instance, General Danilo Jauković organized and headed the construction of the path through the Mrtvica canyon in 1973 (Fig. 11.7) to allow direct access to the villages in Rovca, as well as the road that connected Gornja Morača and Uskoci with the Adriatic main road (Simonović 2015).

The new roads that enabled easier and faster access to many *katuns*, also had a positive impact on the quality of life there. Traditional cottages were built from the materials readily available on the construction site, which imposed the solutions often compromising the level of their comfort, security, resilience, and longevity. The roads enabled the transportation of different types of construction materials, which despite its impact on the traditional landscape, made the life of the local population way much easier. What required a full hiking day for generations back became accessible by car in a few hours (Boljević et al. 1986:48–98).

For a couple of decades, from the 1950s through the 1970s, it seemed that everything was getting better. The middle-aged and elder population that, for the most part, retained their rural lifestyle habits was still a large category, with strong influence on the younger family members to dedicate the resources and energy to spend some time in villages and mountains. Using the social policy allowing relatively early retirement, they could get back to traditional cattle rearing practices while still in good health and strong, supplementing their pensions with added income. Better housing facilities and easier availability and transport of the products from the general markets made their life in villages and *katuns* much more comfortable, with the latter benefiting from their season's coinciding with school summer vacancies. That enabled the school children to spend summers with their parents or (ever more) grandparents shepherding their herds and flocks, giving the perspective of necessary generational continuity and leaving the impression that the new, sustainable cycle was created. Having become the choice instead of necessity, it seemed that the traditional lifestyle found a way to fit into contemporaneity.

There is no exact data on how many new cottages were built in the *katuns* on Montenegrin mountains in this period, but according to our research on Kuči Mountains from 2015–2017, at least one-fourth of existing structures originated in the post-war decades. They had a more solid construction with advanced techniques and better materials used, showing the widespread trend of renewing and modernizing the mountain domains, adapting it to contemporary standards. It is worth mentioning that almost all of the cottages, locally called *gladas*, built until the mid-1970s, were built with a traditional, agricultural purpose in mind, with those made for leisure and tourism coming much later (Laković 2017:79–267). Aside from areas that earlier became known as the part of tourism offer (Durmitor), this should apply to every mountain in Montenegro with developed *katuns'* life. That was translated into a strong communal drive toward retaining and improving their generational activity, using the opportunities offered by new times.

11.6 (Un)expected Response

Although it could appear that things unfolded favorably for the development of mountain areas in these times of overall societal growth, this was precisely when the foundation for their decline was laid. The bright side of the picture concerns those who still have not adjusted their mental, social and professional set-ups to new parameters. These positive tendencies were only the accompanying effects of the current that has already been rushing in a different direction, only initially influencing this domain in a way that would indicate a brighter future. Despite all technological, infrastructural, and economic progress, mountain areas suffered because the general social strategy shifted from agriculture to industrialization, physically displacing the context of dominant everyday life from rural and mountainous to urban environments. Collateral benefits of such development could not remedy the consequences of the comprehensive shift that drove the professional engagement of masses elsewhere, transforming the mountains into domains where only and the ever-decreasing number of die-hards stuck to their traditional lifestyle. The others, wanting to maintain some connection with both the mountain and related family heritage, found the solution in building cottages for relaxation and recreation. But put together, these were the reactions of the already defeated—either of those stubbornly keeping to their own without aspirations to make any wider impact, or those accepting the change and fitting into what was at that time possible and, it is fair to say, preferable.

For the generations, mountain life was not an option but an indispensable and unavoidable context without a substitute. Once the situation changed, the forces driving it disappeared or substantially lost power. When survival depended on suffering the weather, lack of housing comfort, hardship of dealing with mountains' tempers to take advantage of their pastures and forests, there was no alternative but to endure it all. As long as circumstances persisted in such constellation, the development of mountain life was the societal answer to the imposed historical challenge, and all its capacities were engaged within such an effort. Being a long-lasting chronological category, transhumance in direction village-*katun* and back, with all the accompanying aspects, has always been an outcome of the agricultural paradigm, whenever it was occurring and whoever the players were. In this regard, nothing has substantially changed since the Middle Ages (Vojvodić 2021:73–74).

The second half of the twentieth century brought the very change of the said paradigm, with agriculture being replaced as the dominant activity and the source of livelihood. Its importance was recognized, a significant deal of the strategic investments was directed toward it, and it continued to play an important role within all the conceptions of contemporary life. The difference was made by the fact that it was no longer either only or the most important strategical economic and social context. An ever-growing number of people started earning for living outside of agriculture, and while still staying in the countryside, they were not forced to perform a full transhumance cycle. They could retain from it what they wanted or were allowed by the changed reality, but the level of commitment within such a decision varied. Even more, during the period of three generations, it decreased several times, since the rhythm of contemporary life and the related sum of obligations, expectations, and projections have ever less included dealing with the mountains outside of tourism, sports, and relaxation. Since these determinants have never been decisive in historical orientation toward mountain areas, they could not prevent their abandonment when new life priorities took over.

It appeared that it did not matter much that the logistics were improved or that the activities became less stressful because they did not so decidedly determine the family's nutritional prospects. While struggling to reach their mountain *katuns* by narrow paths through rugged middles of nowhere, people hated it but kept on coming up since that was the only life they knew and were accustomed to. Having no viable alternative, they hoped for better roads, motorized transportation, and housing comfort on par with no more than a lower contemporary average. But when those roads reached their cottages—the clear majority of *katuns* were indeed covered by roads until the end of the twentieth century, they were mainly used to pack the movable household items and move down. In the context where the villages struggle not to get deserted with dubious results, the related mountain *katuns* face even harder challenges. Traditionally functioning as an agricultural supplement to the villages, they were left without enough people who still found them needed for that purpose. Completely defined by such a particular lifestyle, in times when they lost their importance, the *katuns* could not find the other type of users who would prolong their relevance. Transformation into tourism settlements is ongoing in attractive areas accessible by good road infrastructure, which devastates their authentic nature and landscape. Unfortunately, it is an expected consequence of the destruction of the previous, rules-based system, which relied on the generational wisdom and knowledge of how to live with the mountains, not to fight against them. Current construction pursuits, aside from notable exceptions, more than often demonstrate a different perception—ignorance backed with the sense of fake power, through building inappropriately designed structures at inappropriate sites and using inappropriate materials (Laković et al. 2020:251–252, Fig. 11.8).

11.7 Conclusion

The generation of the population, who stubbornly insisted on retaining transhumance as the way of life passed, has been replaced by the next which, although connected with that domain, had to take an additional effort to include it into the current lifestyle. Their results varied, but when the dusk of life came upon them, the perspective of continuation of mountain activity passed to their inheritors, who had a completely different attitude, vision, and conception. Within those three cycles, a gradual alienation through the remain—come back—visit hierarchy was visible, indicating that in the fourth iteration, it would be hard to expect even retaining the wish to visit the areas that once were the main source of life for their ancestors. As a result, the current picture of Montenegrin mountain areas uncovers the societal failure to provide a successful and sustainable response to changed

Fig. 11.8 Traditionally and newly built houses in Kuči Mountain. (Photo by Ivan Laković)

historical, economic, and social circumstances, which emerged after WWII. Despite the overall economic growth, improved standard of living, and general well-being of people, there appears that the domain of rural, especially mountain communities has become their collateral damage. Abandonment of the settlements and properties, as well as large-scale depopulation of the whole mountain regions, threaten to irreparably disrupt the connection that inhabitants of Montenegro still have with their mountain environment, alienating them from a substantial part of their historical, cultural, and geographical heritage. Reinverting mentioned trends is impossible without comprehensive and strategic planning of development of mountain areas in the country, especially one assuming that traditional and generationally proven practices need and deserve societal and governmental support. If approached with targeting their sustainability potentials and recognizing the sum of resources they contain for overall progress, it would be possible to expect the evolution and modernization of tradition-based and inspired lifestyle and economy. In small societies, as the Montenegrin is, positively engaging such niche-like domains can have visible impact on general development, while preserving important segments of its heritage through adequate and thought-out valorization.

References

Blagojević O (1991) Ekonomsko stanje seljaštva u Crnoj Gori 30-ih godina ovog vijeka [The economic situation of the peasantry in Montenegro in the 1930s]. Istorijski Zapisi 64(1–2):77–91

Boljević Lj, Mališić M, Ražnatović S, Petrović D, Slavić G, Radunović R, Šestović V, Keković S, Borović S (1986) Razvoj putne mreže u periodu od 1945 do 1985 godine [Development of the road network in the period from 1945 to 1985]. In: Putevi u SR Crnoj Gori 1945–1985. Društvo za puteve SR Crne Gore, Titograd, pp 48–98

Čubrilović V (1963) Postanak plemena Kuča [Genesis of the Kuči tribe]. Zbornik Filozofskog Fakulteta 7(1):313–335

Glušćević VM (2008) Sociološko-demografska analiza kretanja stanovništva u podgoričkoj opštini 1945–1991 [Sociological-demographic analysis of population movements in the municipality of Podgorica 1945–1991]. Istorijski institut Crne Gore, Podgorica

Ivanović Z (1976) Urbanizacija SR Crne Gore [The urbanization of SR Montenegro]. Geografski Glasnik 38(1):117–128

Kalezić Ž (1976) Strukturalne promjene na crnogorskom selu u XX vijeku [Structural changes in the Montenegrin village in the 20th century]. Pobjeda, Titograd

Kerckhof A, Spalevic V, Van Eetvelde V, Nyssen J (2016) Factors of land abandonment in mountainous Mediterranean areas: the case of Montenegrin settlements. SpringerPlus 5. https://doi.org/10.1186/s40064-016-2079-7

Kovačević B (2003) Savezničko bombardovanje Crne Gore 1943–1944: građa [Allied bombardment of Montenegro 1943–1944, documents]. Pobjeda, Podgorica

Laković I (ed) (2017) Katuni Kučke planine: monografija [The katuns of the Kuči Mountain: monograph]. Univerzitet Crne Gore, Podgorica

Laković I, Kapetanović A, Pelcer-Vujačić O, Koprivica T (2020) Endangered Mediterranean heritage—case study of *katuns* at the Kuči Mountains in Montenegro. Land 9(8). https://doi.org/10.3390/land9080248

Marović B (1998) Stočarstvo Crne Gore 1860–1953 [The cattle breeding in Montenegro 1860–1953]. Istorijski institut Crne Gore, Podgorica

Marović B (2018) Ekonomska istorija Crne Gore [Economic history of Montenegro]. Privredna komora Crne Gore, Podgorica

Milošević S (2015) Komunistička partija Jugoslavije o seljačkom i agrarnom pitanju u periodu između dva svetska rata [The communist party of Yugoslavia on the peasant and agrarian question in the period between the two world wars]. Tokovi Istorije 23(2):101–127

Obradović M (1999) Promena svojine i ekonomske strukture u istočnoevropskim zemljama posle Drugog svetskog rata (1941–1952) [Changes in property and economic structure in Eastern European countries after the Second World War (1941–1952)]. Istorijski Zapisi 72(3–4):137–148

Pejović Đ (1955) Agrarna politika crnogorske vlade u krajevima oslobođenim u balkanskom ratu 1912. godine [The agrarian policy of the Montenegrin government in the areas liberated in the Balkan war of 1912]. Istorijski Zapisi 8(1–2):215–242

Radonjić P (2004) Katuni Bjelopavlića na planini Sinjavini [The *katuns* of Bjelopavlići at the Sinjavina Mountain]. In: Vlahović P (ed) Selo u Crnoj Gori. CANU, Podgorica, pp 445–460

Radović Lj, Vukotić V (1983) Stanovništvo Crne Gore do 2000 godine [The Montenegrin population until the year 2000]. Institut za društvenoekonomska istraživanja, Titograd

Rašović M (1963) Pleme Kuči, etnografsko istorijski pregled [The Kuči tribe, ethnographical-historical overview]. Zavod za izradu novčanica Narodne banke, Beograd

Relief map of Montenegro (2012). https://commons.wikimedia.org/wiki/File:Relief_map_of_Montenegro.png. Accessed 14 May 2023

Simonović B (2015) Danilov put kroz kanjon Mrtvice [Danilo's path through Mrtvica Canyon]. Dan, 14 Nov 2015. Podgorica

Šljukić S (2011) Seljaštvo između ideologije i teorije [Peasantry between ideology and theory]. Kultura Polisa 8(15):319–334

Vojvodić U (2021) Transhumance in medieval Serbia—examples from the Pešter plateau and Northwestern slopes of the Prokletije mountains. In: Dixon P, Theune C (eds) Seasonal settlement in medieval and early modern countryside. Sidestone Press, Leiden, pp 69–80

Vukčević R (1974) Razvoj privrede Crne Gore [Development of the Montenegrin economy]. Ekonomski fakultet, Titograd

Vujović T (2012) Promjene u rasprostranjenosti stanovništva u Crnoj Gori (koncentracija u gradovima) i socio-patološke pojave [Changes in the distribution of the population in Montenegro (concentration in cities) and socio-pathological phenomena]. In: Stanovništvo i razvoj. Centar za ekonomska istraživanja Instituta društvenih nauka, Beograd, pp 321–330

Mountain Pastures in the Slovenian Alps: Their Role in Shaping and Sustaining the Cultural Landscape

12

Mimi Urbanc, Špela Ledinek Lozej, and Mateja Šmid Hribar

Abstract

This chapter focuses on mountain pastures and their role in shaping and maintaining landscapes on the southeastern fringe of the European Alps. Mountainous cultural landscapes witness centuries of interactions between people and their environment, resulting in multi-layered, multi-structured, and multi-functional entities. Although mountain pastures seem to be a strong pillar of stability of Alpine landscapes, they are nevertheless susceptible to constant change within the dynamic interplay of economic, social, and political factors. The question arises how to address current challenges and megatrends, such as the general decline of agriculture, especially small-scale farming, the depopulation of mountain areas, and booming tourism. The overall objective of this chapter is to highlight some essential landscape-related aspects of mountain pasture farming from historical and contemporary perspectives: landscape change, drivers of change, current challenges, landscape as a form and as a process, actors and stakeholders, and perspectives. We have explored these issues through three case studies from different parts of the Slovenian Alps. The paper concludes with a reflection on mountain landscapes in general and specifically on the users' perspectives on mountain pasture management in these landscapes.

Keywords

Cultural landscape · Mountain pastures · Pastoralism · Agrarian community · Slovenian Alps

12.1 Introduction

The first thing Drago, a long-time shepherd on the Tegošče Pasture, does every morning is to step outside the hut and take a few deep breaths. As he watches the grazing animals and listens to the cowbells, his eyes take in the splendour of colours from grass green to sky blue and the scents of the grassland, trees, and herbs fill every breath. Drago is one of a few dozen shepherds that move to a mountain pasture every summer, tend their herds, and continue the long tradition of grassland management in the Slovenian Alps.

M. Urbanc (✉) · M. Šmid Hribar
ZRC SAZU Anton Melik Geographical Institute, Ljubljana, Slovenia
e-mail: mimi.urbanc@zrc-sazu.si

M. Šmid Hribar
e-mail: mateja.smid@zrc-sazu.si

Š. Ledinek Lozej
ZRC SAZU Institute of Slovenian Ethnology, Ljubljana, Slovenia
e-mail: spela.ledinek@zrc-sazu.si

M. Zorn et al. (eds.), *Discourses on Mountains of Montenegro and Slovenia*, Historical Geography and Geosciences, https://doi.org/10.1007/978-3-031-99017-5_12

The Alps stretch across the centre of Europe, which, along with their numerous resources, makes them an economic, social, and civilisational heart of the continent. The Alps reflect a long-lasting relationship between nature and humans, characterised by the management of farmland and forests (Barker et al. 1991; Petek 2005; Carrer 2015). Seasonal grazing plays a prominent role, and its results, i.e. green mountain pastures, have become the iconic postcard motifs of the Alps (Krauß and Olwig 2018) and central for tourism (Wanner et al. 2021).

The key aspect of humanity's relationship with nature is the human adaptations to the natural conditions in the Alps and the development of appropriate living strategies. We understand the cultural landscape as the result of human activities that have produced a variety of material and intangible elements. The mountain pasture landscape is therefore composed of numerous elements, which—either individually but mostly as a complex unit—constitute not only a life resource but also a pillar of identity and belonging.

Alpine pastures and their associated cultural landscapes are vulnerable to natural disasters (Deléglise et al. 2019) and socioeconomic shifts (Potthoff et al. 2020; Bevione et al. 2022). They are the result of continuous human effort and are affected by myriad factors ranging from the agricultural situation to social demographics. The key question is how to manage mountain pastures and, indirectly, cultural landscapes; how to strike a balance between their declining function as a resource for smallholder livelihoods and their increasing value as cultural heritage within the booming tourism industry.

This chapter is based on research conducted between January and November 2021. The empirical part involved several rounds of in-depth interviews with representatives of individual mountain pastures, more specifically with herders, managers, and/or heads of agrarian communities (ACs; ownership-based groups) and pastoral communities (interest-based groups). The co-authors conducted the interviews individually between July and September after having discussed the interview questions at length. The criteria for selecting mountain pastures were activity, location (wide geographical coverage), and differences among them in most aspects. Regarding these differences, we relied on our prior findings (Ledinek Lozej 2002, 2020, 2022; Petek and Urbanc 2007; Ledinek Lozej and Roškar 2018; Potthoff et al. 2020; Urbanc and Šmid Hribar 2021). The cases studied here are the Matajur Pasture in the southwestern part of the Julian Alps, the Krstenica Pasture in the eastern part of the Julian Alps, and the Tegošče Pasture in the Karawanks (Fig. 12.1).

12.2 General Overview of Agriculture in Slovenia

Although Slovenia shares many similarities with other Alpine countries, it also differs from them in some details. Before becoming independent, Slovenia's distinctions stemmed from its political, economic, social, and cultural context (Potthoff et al. 2020). Turbulent efforts at democratisation and national emancipation in the 1980s and the early 1990s led to the formation of a sovereign state in 1991. The resulting far-reaching changes led to a multi-party system, democracy and market capitalism (Drozg 2007), and finally to EU accession in 2004 and the introduction of the euro in 2007.

The general situation in agriculture reflects the legacy of communist Yugoslavia: neglect of agriculture (Klemenčič and Genorio 1993), a 10-hectare land maximum, limited dialogue between decision-makers and the public, and imposition of decisions from above (Streifeneder 2010). All these factors are essential for understanding the current situation. The peculiarities are still noticeable: fragmented ownership, the prevalence of part-time farming (Udovč et al. 2006; Razpotnik Visković and Seručnik 2013), the need to supplement agricultural income with other sources (Bojnec and Latruffe 2013), the absence of more profit-oriented agriculture (Knežević Hočevar 2015), low productivity (Bojnec and Latruffe 2013), and the unfavourable age and education structure of landowners (Knežević Hočevar and Černič Istenič 2014).

Fig. 12.1 Location of three case studies: the Matajur, Krstenica, and Tegošče pastures

In addition to domain-specific characteristics, agriculture also reflects the regional and national economic, social, and spatial situation: (1) reluctance to participate in decision-making processes and distrust of politics (Streifeneder 2010), (2) the social and spatial immobility of the rural population (Juvančič and Erjavec 2005; Cirman 2006; Drozg 2012) a low degree of urbanity, and (3) polycentric regional development (Streifeneder 2010). This development encouraged the establishment of non-agricultural activities, local off-farm employment, and income-generating opportunities in smaller rural towns from the 1970s onwards (Nared et al. 2017), allowing rural settlements to remain vibrant.

12.3 Alpine Pasture Farming

Alpine pastoralism is a particular form of transhumance, a movement of people and their livestock between permanent settlements in valleys or lowlands and temporary summer settlements in the Alpine and subalpine areas. It effectively complements stable farming and cultivation of fields and meadows close to the village (Netting 1981; Viazzo 1989; Orland 2004; Ledinek Lozej 2022). This use, documented in archaeological sources for the Bronze Age (2200–800 BC) and written sources for the Middle Ages (Gilck and Poschlod 2019; Turk et al. 2020), has produced agricultural landscapes consisting of grazed ecosystems and a variety of collectively owned and managed land types (Petek and Urbanc 2007; Potthoff et al. 2020).

A diverse array of values is ascribed to these landscapes by individuals, communities, and societies (van Zanten et al. 2016). Their functions, and therefore their values, have changed over time. In the past, they were an essential pillar of highland farming, and communities were often economically dependent on them (Burns 1963; Cole and Wolf 1974; Netting 1981; Petek and Urbanc 2007; Mazzocchi and Sali 2019). In modern times, aspects other than economic ones are coming to the fore, in particular biodiversity, ecosystem services, aesthetic values, tourist appeal (Scolozzi et al. 2014; Schirpke et al. 2019; Pagot et al. 2025), and cultural heritage conservation (Kianicka et al. 2010; von Glasenapp and Thornton 2011), which benefit both locals and tourists (Deléglise et al. 2019).

Not only have the functions of mountain pastures changed in recent decades, but also the intensity of their use. Their former overuse and high economic importance (Baur and Nax 2018) have been replaced by underuse, or land abandonment (Gabrovec et al. 2020). The reasons for this are manifold: recent economic and social developments (see Price 2015; Egarter Vigl et al. 2017; Mazzocchi et al. 2019; Carrer et al. 2020; Gabrovec et al. 2020; Potthoff et al. 2020) as well as pressures from tourism and recreational activities and other non-agricultural uses such as nature conservation (Bätzing 2021). These processes have led to a loss of resilience, ecological sustainability, and ecological knowledge (Schirpke et al. 2013; Hernandez-Morcillo et al. 2014; Gretter et al. 2018; Pecher et al. 2018).

12.4 Mountain Pastures in Slovenia: Governance Issues

Alpine pasture farming reflects general agricultural features but is aggravated by physical marginality and harsh production conditions. One factor, namely the ownership of mountain pastures, deserves special attention. The special long-term feature of collective ownership was interrupted by the political and economic situation in Yugoslavia after the Second World War (see also Petek and Urbanc 2007; Potthoff et al. 2020). More than 1,000 collective property structures, namely ACs were affected by land expropriations (Petek and Urbanc 2007; Cerar et al. 2011). Nevertheless, some of them survived the political upheavals and maintained their traditional agricultural practices by transforming themselves into associations that were still allowed to use the property they had owned. After 1991, in the wake of political changes and the introduction of a market economy, ACs were re-established. Three acts (the Denationalization Act in 1991; the Act on the Reestablishment of Agrarian Communities and Restitution of Their Property and Rights in 1994; and the Agricultural Communities Act in 2015) addressed ACs and their common land. These acts are not without problems (see Šmid Hribar et al. 2018; Potthoff et al. 2020). Premrl (2013) reports that by March 2013, 638 ACs had completed the procedures to acquire a total of 77,486.47 hectares, representing 3.67% of Slovenia's territory. However, the latest data show that only a fraction of former mountain pastures (Jordan 1945; Melik 1950; Vojvoda 1967; Schlamberger 1995; Register … 2020) are still active as is obvious in the case of the Julian Alps (Figs. 12.2 and 12.3).

12.5 Case Studies

Despite some general similarities, presented cases are quite diverse (see Table 12.1).

12.5.1 Tegošče Pasture in the Karawanks Mountains

Tegošče Pasture (Sln. *Tegoška planina*; a *planina* is a mountain pasture: a complex of grazing areas, forests, and built infrastructure) is located in the eastern part of the eleven-kilometre-long Košuta Ridge on the border between Slovenia and Austria in the Karawanks Mountains below Mount Tegošče (Sln. *Tegoška gora* (2,044 m a.s.l.), Germ. *Höhe Spitze*). It is one of five still-active mountain pastures that line the sunny southern slope of the mountain. The Košuta Ridge is generally steep,

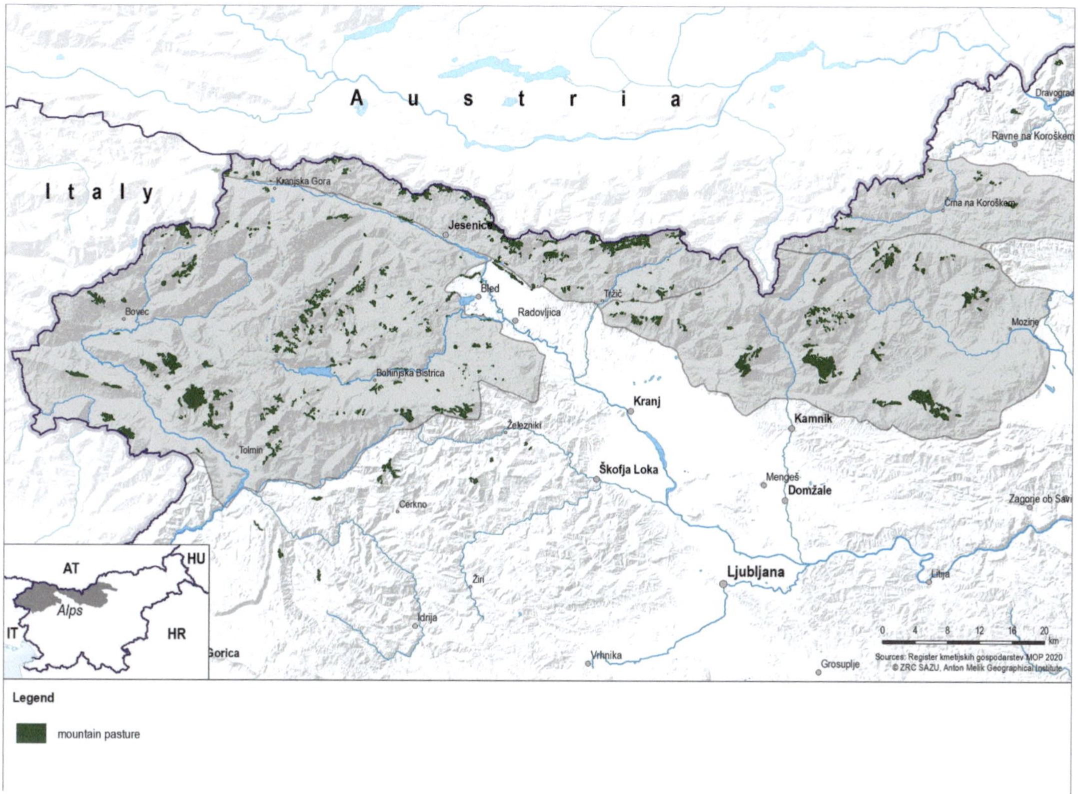

Fig. 12.2 Mountain pastures in the Alpine part of Slovenia that are still active (as of 2020)

and the mountain pasture took full advantage of the locally less steep slope (Fig. 12.4). The lower part of the pasture is covered by grassland and the steepness increases with altitude (covered by forest). Glacial and karst processes transformed the original grassy slope into hummocky grassland full of small-scale surface erosion elements that provide landscape diversity (Fig. 12.5, top right).

The Tegošče Pasture covers about 131 hectares (93 ha of grazing areas and 37 ha of forest) and lies at an altitude of 1,301 to 1,626 m. Several gullies cut through the slope (Fig. 12.5, top left). The mountain pasture is located on contact between limestone and moraine, which provides for water sources. Due to a thin layer of soil, the grassland is prone to summer rainfall deficiencies and a resulting lack of grass. The grazing season lasts roughly from mid-June to mid-September. Due to the high altitude, mid-September is already quite risky, due to possible snowfall. The hut is serviced outside of the grazing season, except in winter.

The tradition of grazing at the Tegošče Pasture dates back to 1444 (Bizjak 2020), when the Križe (nearby local centre) Imperial District Office owned it. Two noblemen prevented the farmers from exercising the right to use the resources (grass and wood). Information about the pasture in the following centuries is sparse. One piece of information comes from the end of the nineteenth century, when a new large landlord named Born became a co-owner (Gabrovšek 1998). After the Second World War, the Tegošče Pasture shared the fate of other pastures that were in common ownership: nationalisation. Between 1948 and 1993, the Tegošče Pasture was *de iure* managed by the Križe Cooperative, but the former co-owners were allowed to continue their established practices.

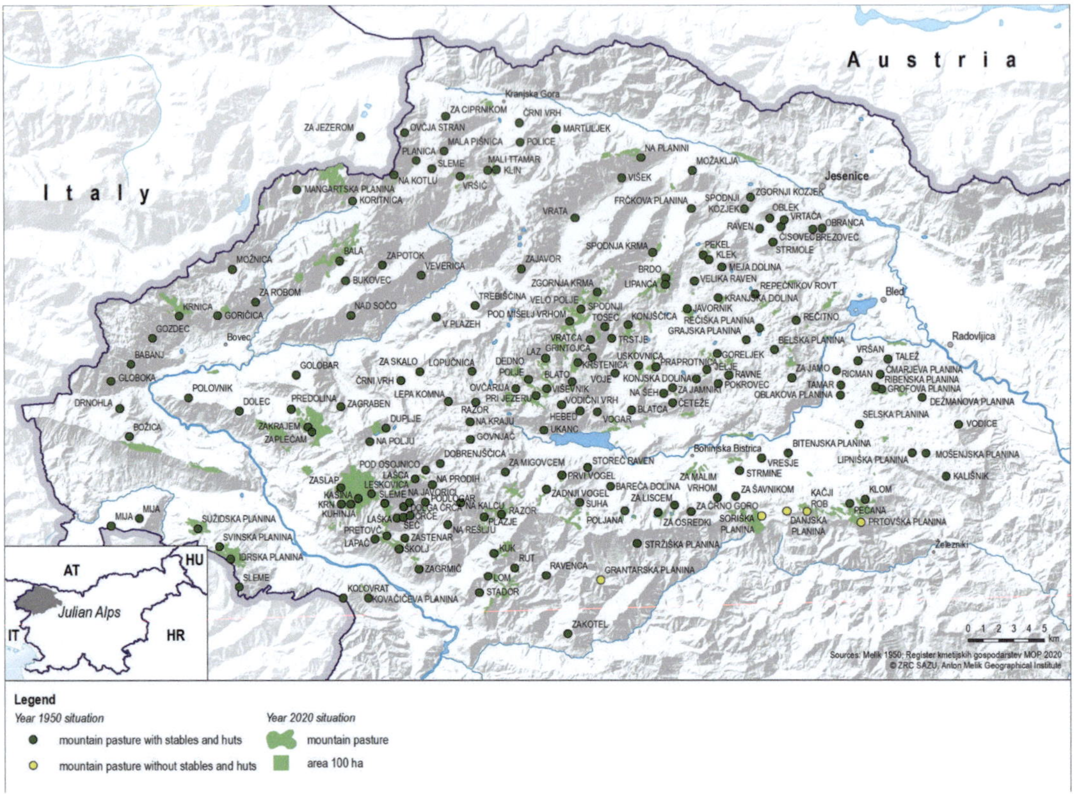

Fig. 12.3 Mountain pastures in the Julian Alps that were active in the 1940s (Melik 1950)

With a brief interruption of a few years in the 1950s, the Tegošče Pasture was active all the time. After Slovenian independence, the Tegošče AC exercised its right to reclaim its common land, but the case has been pending ever since (Gabrovšek 1998; Aljančič 2021; Oranič 2021).

People have adapted to the difficult living conditions in the mountains by setting the buildings on the least steep slope. The structures follow the local architectural style, characterised by wooden constructions, small windows with shutters, and steep roofs (Fig. 12.5, top centre). Avalanches are quite common on the slopes of the Košuta Ridge (Volk Bahun and Zorn 2020). People have passed down a story about an avalanche that swept away the old hut, probably in the winter of 1951–1952. The present building, a combination of a hut and a stable, was renovated after a fire in 1984. Recently, a small, separate chalet was built near the main building for grazing beneficiaries to stay overnight. Due to their lack of interest, it has been rented out to tourists instead.

In addition to abundant pasture and water, access is an important factor for operating the pasture. In the 1970s, the Yugoslav army built a gravel road (Oranič 2021). Because of its line of contact between permeable and less permeable rock, the pasture has two water sources with sufficient runoff. Nobody knows when the first reservoir was arranged. The second was built more recently and has made it possible to divide the pasture into two parts (with the widest gully serving as the boundary) and thus also to divide the livestock according to whether the presence of a bull is appropriate or not (Aljančič 2021; Oranič 2021).

Table 12.1 Overview of characteristics of the Krstenica, Tegošče, and Matajur pastures (data based on interviews; refers to the 2021 pasture season)

	Matajur pasture	Krstenica pasture	Tegošče pasture
General			
Total area (ha)	150	403	131
Grazing area (ha)	94	34	94
Altitude (m)	From 1,250 (common stables and a dairy) to 1,295 (upper part with the Konjarska Guta emergency shepherd's hut)	From 1,550 (Pod Ogradmi) to 1,945 (Jezerski Preval); dairy hut 1,670	From 1,301 to 1,626; hut 1,450
Owner(s)	Livek AC, Municipality of Kobarid	Bohinjska Češnjica, Jereka, Podjelje, Koprivnik AC	Tegošče AC
Manager(s)	Matajur Grazing Community	Bohinjska Češnjica, Jereka, Podjelje, Koprivnik AC	Tegošče AC
Livestock, no. and structure	113 (cattle mixed), 3 horses	30 (cattle mixed)	113 (cattle mixed)
No. of co-owners	96+ Municipality of Kobarid	Around 160	36
No. of users*	9	6 (5 co-owners and 1 non-co-owner)	19 (3 co-owners and 16 non-co-owners
Infrastructure			
Road access	Yes, gravel road, mountain pasture is accessible by car	With limitations; poor quality field road passable only by offroad vehicles	Yes, gravel road, mountain pasture is accessible by car
Electricity	Yes, a generator	Yes, photovoltaic and a generator	Yes, photovoltaic and a generator
Cellular network coverage	Not disruption free	Solid	Not disruption free
Economic aspects			
Dairy production	Yes	Yes	Yes
Accommodation facilities	No	No	Yes
Day tourist catering	Moderate; cheese sales	Moderate; cheese and beverage sales	Moderate; beverages and meal sales

* Holdings whose livestock graze on the mountain pasture irrespective of ownership status

12.5.2 The Matajur Pasture in the Western Julian Alps

The Matajur Pasture (Sln. *Planina Matajur*), also known as the Idrsko Pasture (Sln. *Idrska planina*; named after the settlement of Idrsko), extends from 1,250 m to 1,295 m above sea level on the eastern side of the ridge of Mount Matajur (Ital. *Montemaggiore*, 1,642 m) in the Julian Alps on the Slovenian-Italian border. The predominant water-bearing rocks allowed the establishment of several mountain pastures on both sides. Unlike on the Italian side, where mountain pastures declined after the 1960s and especially after the 1976 earthquake, three have survived on the southern (Slovenian) side, but milk production takes place only on the Matajur Pasture. Almost two-thirds (about 93 ha) of it is Alpine grassland; the rest (45–50 ha), mostly steeper slopes, are covered with forest (Fig. 12.6).

The Matajur Pasture boasts a long settlement history. The village of Idrsko is said to have had two mountain pastures as early as the end of the sixteenth century, but before 1874 it was mainly sheep and goats that were kept there (Stres 1994). Following the state-supported

Fig. 12.4 Tegošče Pasture is located at the point where the slope of Mount Košuta becomes less steep. (Photo by Mimi Urbanc)

Fig. 12.5 Cultural landscapes of mountain pastures in Slovenia are characterised by a challenging natural environment, good adaptation to that environment, and a high level of naturalness. (Photos by Mimi Urbanc (top), Mateja Šmid Hribar (centre), Špela Ledinek Lozej (bottom centre), Aleš Zdešar (bottom left and bottom right), authors archive)

Fig. 12.6 Matajur Pasture with a large stable in the lower part. (Photo by Mateja Šmid Hribar)

Fig. 12.7 Central part of the Krstenica Pasture. (Photo by Aleš Zdešar, authors archive)

promotion of cattle breeding and modern dairying introduced by the Swiss cheesemaker Thomas Hitz (Ledinek Lozej 2020), the number of cattle increased, and a stable was built. Later, around 1922, the farmers built seven stables at the new location (under the present dairy and stable). In 1923 a shepherd's hut was built and sheep grazing was banned (Stres 1994). This development reflects the spread of livestock farming in Idrsko. Each year, the farmers selected seven shepherds, each of whom herded up to 16 dairy cows in a stable. The cheese was processed in a common dairy.

The period following the Second World War began with the nationalisation of the Matajur Pasture. In 1951, when a common stable for 120 dairy cows and a cheese dairy were built (Fig. 12.6), and the water supply was arranged, the pasture moved to its present, third location. The following year, the newly founded Planika cooperative trading company became the owner of the pasture. However, uninspiring agricultural policies and accelerated industrialisation, which led to a decline in livestock in the valley, delayed full-scale operations. In the 1960s, Planika began to cooperate with the Josip Kraš food company in

Zagreb, which needed high-quality Alpine milk to process milk powder and make chocolates. The need for milk and the reconstruction after the 1976 earthquake provided the impetus for the revival of the pasture. In 1978 the stables were renovated, and the Idrsko Pasture merged with the former Livek Pasture (50–60 ha), because the latter had no milk processing, but the farmers were interested in cheese production (Stres 2021).

After Slovenia became independent, the denationalisation of the mountain pasture was delayed because the owner, the Kraš factory, was now a foreign (Croatian) company. In the meantime, the Livek AC was established, but the Idrsko AC was not, because the co-owners had ceded their property to the Municipality of Kobarid in exchange for free grazing rights. In 2008, the Livek AC acquired ownership of their part of the mountain pasture, but the administrative procedure was not completed until 2018 due to bureaucratic obstacles (Dreščček 2021). The Matajur Pasture is therefore now owned by the Livek AC and the Municipality of Kobarid.

In 2004, a stable was converted into a kitchen so that they now have a separate cheese dairy and a room for curd production according to the regulations. In 2013, they modernised the milking parlour, which enables nine cows to be milked at a time. This changed not only the way they milk but also the way they graze. The cows are now outside all the time, even at night, and they are only herded into the pen in the morning and evening to wait for milking (Stres 2021). In the past, only locals from the Idrsko settlement grazed on the Matajur Pasture, whereas in recent years, due to a shortage of cattle, owners from surrounding villages have also been admitted. Undergrazing causes the grasslands to become overgrown quickly. In 2021, there were two herders and one cheesemaker, 45 dairy cows, 68 other cattle, and three horses on the Matajur Pasture, belonging to nine owners from six villages. They graze for 105 to 110 days, usually coming up the mountain on the last weekend in May and staying at the dairy until mid-September. Barren cattle can graze for another two to three weeks, depending on the weather.

The Matajur Pasture is connected to the valley by a road that was built by the Italians in 1915/1916 during the First World War, but it has been in poor condition in recent years. The Matajur Pasture receives electricity from a generator purchased in the 1980s, when an amateur radio station was also established. Telephone and cable connections are not available, but there is cellular coverage. The Matajur Pasture does not (yet) play a significant role in tourism. There is no tourist accommodation or catering facilities, and only cheese can be purchased there (Fig. 12.5, centre right). The potential is there, but the lack of a workforce hinders the development of tourist services.

Overall, in addition to the recent partial decline in grazing, which has led mainly to the overgrowth of lower-lying pastures (Fig. 12.5, centre), drought has also been a challenge. Urbanc and Šmid Hribar (2021) reported that climate change is already being felt in nearby areas, affecting, in particular, the distribution of precipitation and the decrease in the number of days with snow cover (Hrvatin and Zorn 2017).

12.5.3 The Krstenica Pasture in the Eastern Julian Alps

The Krstenica Pasture (Sln. *planina Krstenica*), with grazing areas between 1,550 and 1,945 m above sea level, is the second-highest mountain pasture in the Julian Alps that is still active. It is situated below the ridge that runs from Mount Krsteniški Stog (1,879 m a.s.l.) to Mount Hribarice (2,388 m a.s.l.). The dairy and the huts took advantage of the open, flat ground (Fig. 12.5, bottom left and right) made possible by less permeable rocks (Melik 1951). Despite its location in the heart of the Julian Alps with Mount Triglav (2,864 m a.s.l.), the highest and most iconic peak in Slovenia, and within Triglav National Park, the Krstenica Pasture is off the beaten track.

Nowadays, it comprises 34 hectares of pasture and about 370 hectares of scrub. Grassland areas have fluctuated greatly over the last century: between 403 hectares in 1923, 38 hectares between 1963 and 1982, 80 hectares in 1995 (Jović 2016:13), and 70 hectares in 2014. The Krstenica

Pasture is owned by the Bohinjska Češnjica, Jereka, Podjelje, and Koprivnik AC, which owns several other mountain pastures at various altitudes. The grazing season on the Krstenica Pasture lasts from the end of June to the beginning of September. Unlike in the past, when grazing rules determined starting and ending dates (until the mid-twentieth century), today weather conditions play a decisive role.

The history of the Krstenica Pasture has been dynamic, and especially its location and ownership have been subject to many changes. Until the beginning of the twentieth century, it was situated higher, but then it moved to the water source below the present settlement. It was not until the 1930s that the cheese dairy and huts settled in their present location. Ownership went hand-in-hand with general social development. Starting in the Middle Ages, the Krstenica Pasture was most likely owned by the Bled Dominion, and the Češnjica community had servitude rights. After the emancipation of serfs in 1848, the peasants of Češnjica redeemed and regulated their serfdom rights. Each shareholder had the right (and duty) to pastureland and to build themselves a hut. Herding and milk processing were carried out individually; this changed when professional cheesemakers took over the common milk processing in the 1870s. This practice lasted until the Second World War (Ledinek Lozej 2002).

After the Second World War, the Krstenica Pasture was nationalised and transferred to the Radovljica Municipal Land Fund, and the AC was abolished. However, the management regime remained almost unchanged; that is, individual grazing and collective (cooperative) dairying. In the 1970s, after the centralisation of milk processing at the industrial dairy plant in the nearby local centre of Srednja Vas, the Krstenica Pasture was abandoned for a few years (Ledinek Lozej 2002). Nevertheless, pasture farming recovered thanks to the efforts of some individual herders. At first, they cared only for the cattle of family members and close relatives, but gradually the circle expanded almost to the entire former AC and occasionally beyond.

The restitution of property and rights to the (members of) the Bohinjska Češnjica, Jereka, Podjelje, and Koprivnik AC in 1993 did not have a significant impact on the management model, introduced by the above-mentioned individual herders, who mobilised young helpers for daily chores, either from their own relatives or from the families of the cattle owners. After their retirement, two younger relatives took their place in management for two seasons (Ledinek Lozej 2002; Ledinek Lozej and Roškar 2018). Since 2011, the AC has shown greater management ambitions; a vital step was taken in 2019, when an external cheesemaker and herders were recruited.

Improvements to infrastructure, particularly water supply solutions, dairy structure, and pasture accessibility, have been ongoing, either by the AC as a whole or by individual users. Running water was first provided from the spring on Jezerca in 1988. Until then, ponds were created with rainwater, rainwater was harvested, and spring water was carried from a spring water tank (arranged in 1934), a ten-minute walk away. Nowadays, water is piped to all the huts.

The dairy was built in 1935 by the Commission for Agrarian Operations to provide modern buildings on the mountain pastures. In 1954, it was equipped with a built-in cauldron and in 1989 with running water. In 1994, a screw/auger was installed to move the fire to and from the cauldron mechanically (Fig. 12.5, bottom centre). The dairy building was renovated in 1998, and various other improvements were made in the twenty-first century.

Infrastructural improvements have been made since the late 1980s. A generator was installed for electricity, and a smaller solar power plant was installed in 1999 and upgraded in subsequent years. Since the mid-1980s, when the first caterpillar tractor arrived on the Krstenica Pasture, grazing beneficiaries have made significant efforts to keep the road passable for tractors and better off-road vehicles. Before the introduction of cellular phones, the Krstenica Pasture established a wireless radio station with the village. Since 1999, GSM phones have offered solid coverage.

Although the AC cannot register catering and accommodation services, these activities are carried out on a small scale. Conversion of the pasture huts into holiday cabins is also not allowed, but it still takes place: Four cabins are rented out on a long-term basis, and one cabin is available for short-term rental, although this is against the Triglav National Park Act. The rest of the huts are used by locals.

12.5.4 Synthesis

12.5.5 Landscape Change

Landscapes are not only a legacy of the past, but also a locus of ongoing processes. Constant change is inherent in them by nature. Alpine pasture landscapes are no exception. Given their complex structural and multifunctional nature, it is hardly surprising that—like all mountain landscapes—they are likely to be a site of negotiation and even conflict (Scolozzi et al. 2014). The ACs have some power to bring about change and development, but they use their potential to safeguard well-established patterns (albeit not always successfully). On the Krstenica Pasture, for example, spontaneous overgrowth is constantly lamented. Comparable complaints about overgrowth, the 'green desert', as Vranješ (2008) reports, are also typical of other Alpine areas such as the Trenta Valley (in western Julian Alps).

The cultural landscape is not at the forefront of the interviewees' thoughts or actions. They focus on their everyday management practices: maintaining pastureland, fences, roads, and so on. Still, they are aware of the value of their activities in sustaining the landscape. The landscape is a byproduct of land-based activities and, at the same time, a representation of respondents' perceptions, values, and expectations, and also of the wider iconic Alpine imaginaries.

Three factors contribute to maintaining the cultural landscape. The first is related to the management priorities of farms, which are not geared towards rapid development and the introduction of high-intensity livestock farming. This is mainly due to small and fragmented ownership that does not allow for large-scale investments. Moreover, the AC representatives do not welcome innovations in the way they farm. Second, the fragile, harsh environment and formal framework do not provide a good platform for rapid change. Third, subsidies play a crucial role in preventing land abandonment and maintaining open space, and so—contrary to some claims (Favry and Pfefferkorn 2005)—they directly support the maintenance of the Alpine pasture landscapes.

Most of the changes in the last generation or two have been in infrastructure, particularly technical equipment and access. Regarding equipment, the introduction of milking machines and electricity generators is worth mentioning. As for access, the construction and maintenance of roads are crucial for the functioning of the mountain pasture, but challenging given the length of the roads and the harsh natural environment. On all three mountain pastures, some building improvements have been made to improve the staff living conditions.

Spontaneous overgrowth by forest and scrub is one of the biggest challenges for cultural landscapes in Slovenia (Gabrovec and Kumer 2019). Natural succession is a nationwide problem that is particularly pronounced in less-favoured areas (Ciglič et al. 2012). Not surprisingly, mountain pastures are among the most affected sites. This problem is widespread on the Krstenica Pasture and somewhat less pronounced on the Matajur Pasture, whereas the Tegošče Pasture is not affected (Table 12.2). The problem of overgrowth stems from the decline in livestock, which leads to the abandonment of areas with difficult grazing conditions (steeper slopes and difficult access). On the Krstenica Pasture, spontaneous afforestation and scrub encroachment have been addressed since the mid-1980s (Mlakar 2021; Sodja 2021), albeit with varying degrees of intensity. Recently, clearing has been running out of steam; regular efforts have been reduced to sporadic campaigns. On the Matajur Pasture, overgrowth is a greater problem at lower elevations than at higher elevations. The snow prevents spontaneous overgrowth at the upper edge of the pasture.

The reverse process—that is, the spread of grazing areas over forest land—is not present. All in all, the balance between grazing areas and forest areas is most stable on the Tegošče Pasture, whereas the Krstenica and Matajur pastures struggle with it with varying success.

Table 12.2 Tangible and visible changes of selected landscape elements

	Forest and scrub overgrowth					Abandoning less-favoured areas					Grazing areas replace forest					Landslides, avalanches, erosion					Construction of locally atypical buildings				
Krstenica Pasture	x						x								x				x						x
Matajur Pasture		x							x						x				x						x
Tegošče Pasture					x					x				x				x							x

Legend:	Very important	Important	Neutral	Less important	Unimportant

Natural hazards are not an issue for the three pastures, or at least they are not perceived as an issue, but they are not entirely absent either. Severe weather events such as thunderstorms (damaging roads on the Matajur Pasture) and lack of rainfall (less grazing on the Tegošče Pasture) occur occasionally.

12.5.6 Drivers of Change

Although mountain pasture landscapes are perceived as relatively stable entities, they do change at a slow pace of transformation. There are many external drivers (Table 12.3). Some of these (policy frameworks) are imposed from above and are directed towards specific goals, others (technological advancement) are part of the overall technological change; despite being adopted voluntarily, they are inevitable. Climate change is recognised as a problem, but water shortage is felt only on the Matajur Pasture. All three mountain pastures occasionally suffer from restricted grazing due to sporadic lack of precipitation. However, climate change does not steer management in general, but influences day-to-day practices such as water use efficiency. The strength of the drivers is quite consistent across the three mountain pastures.

Table 12.3 External drivers of change

	Technological advancement															Political framework																				Miscellaneous				
	Accessibility					Electrification					Communications					Policies (EU, national, local) in general					Subsidies					Protected areas (parks, Natura 2000)					General changes in the agricultural sector					Climate change				
Krstenica Pasture	x					x					x						x					x								x	x									x
Matajur Pasture		x				x						x				x					x									x		x						x		
Tegošče Pasture	x					x						x					x				x							x			x								x	

Legend:	Very important	Important	Neutral	Less important	Unimportant

Table 12.4 Internal drivers of changes

	Personal values, beliefs, and attitudes										Farm-related issues										Additional income				
	Co-owners' expectations and requirements					Co-owners' lack of interest					Farm succession and inheritance problems					Farm orientation					Tourism as an additional activity				
Krstenica Pasture			x			x					x							x							x
Matajur Pasture			x				x					x								x				x	
Tegošče Pasture				x				x					x				x						x		

Legend:	Very important	Important	Neutral	Less important	Unimportant

Modern mountain agriculture is influenced by a myriad of general political, social, technical, and environmental factors. The importance of drivers related to technology and policy (except protected area instruments) is deemed far more important than climate change (see Table 12.3). The interviewees were far more interested in more direct and tangible problems, such as accessibility, electrification, and subsidies than climate change and nature protection.

Among the external drivers (Table 12.3), one group stood out, not specifically in the Likert scale questions, but in the free conversation, namely the general changes in the agricultural sector affecting farm orientation and farm succession/inheritance. These issues came to the fore in different contexts. In general, all three mountain pastures remain active, yet to a lesser extent on the Krstenica Pasture than in the past. The greater decline of livestock in Krstenica is attributed partially to the management issues in the last years, and to a greater extent to the reduced role of agriculture in the economy of the Bohinj region, where tourism and forestry have gained the upper hand. Maintenance of the pasture is mainly due to the resilience of the respective communities and the endeavours of individuals. They are persistent in the face of change. The general fear is that the number of small family farms with livestock will continue to drop, due to changes in farm orientation and the abandonment of farming.

The internal drivers (Table 12.4) of change are generally not so clearly positioned on the 'very important' part of the scale spanning from 'unimportant' to 'very important'. Further on, our interviewees had some difficulties assessing the first two drivers (owner expectations/requirements and lack of interest) because they are related to personal values, beliefs, and attitudes. The co-owners' lack of interest seems crucial; when this is solved, other internal problems are not as pronounced. The reliability and trustworthiness of the ACs and the pasturing community on the one hand, and a critical mass of farmers interested in using the opportunity on the other, are paramount. In this respect, the Tegošče and Matajur Pastures show a considerable degree of vitality, but the Krstenica Pasture less so.

12.5.7 Contemporary Challenges

Although the challenges in mountain farming are many, one looms large for all: how to attract enough livestock to keep the mountain pastures active. The essential subsidies—the EU support to farmers under the Common Agricultural Policy—are only justified if a mountain pasture complies with requirements, and stocking density per hectare is an important factor. A pool of interested livestock farms nearby is the key point to close the gap. In addition, the future of cheese production on the Krstenica Pasture is currently very uncertain due to recruitment difficulties (a wider range of skills and flexibility is required, but wages are low).

There is also a legal/formal problem related to administrative procedures. Only the Matajur Pasture has no pending reprivatisation proceedings and is registered under the latest law (passed in 2015). This lack of appropriate administrative arrangement increases the vulnerability of the Krstenica and Tegošče Pastures. The legal uncertainty does not affect the pasture management practices, but it is perceived as an imminent threat. Moreover, in contrast to other problems that all mountain pastures proactively face, this challenge is viewed with resignation. The co-owners feel like completely powerless onlookers.

12.5.8 Landscape as a Form or a Function?

The question of whether a form or a function is (more) important becomes highly relevant when it comes to landscape conservation (Table 12.5). In the context of Alpine pastures, the question is whether grazing should be supported as the main function or the formal outcome of that function, such as cultivated open grassland and buildings, to name but a few. The answer is clear and unequivocal: both aspects are crucial. Mountain pastures are a great achievement of human adaptation to the environment and can only survive if the appropriate conditions are present. If we want to preserve their form, maintaining the activity that shaped them is the only way.

Although the respondents attribute pastoralism exclusively to agriculture, they are not indifferent to its integration in the tourism industry. They are aware of the importance of their farming activities for sustaining the cultural landscape and other cultural assets such as traditional knowledge, which are linked to a sense of belonging and therefore have the potential to be instrumentalised for community building and heritage-making.

12.5.9 Actors and Stakeholders

Even though the groups directly associated with the selected pastures are small and fairly closed, they are not coherent. They are composed of co-owners (individual members and the municipalities of Tržič and Kobarid in the case of the Tegošče and Matajur Pastures, respectively) and users interested in pasturing. Internal differences in the views and expectations of individual members arise from the way they farm, if they farm at all, and the size of their farms. Both municipalities refrain from taking an active role.

The ACs are fragile entities for several reasons. First, membership is not clearly defined, because not all the reprivatisation cases in the Krstenica and Tegošče ACs have been completed. Second, even if the co-owners are known, they may be either physically or mentally absent. They can be seen as 'detached owners' following Kumer and Štrumbelj (2017), who studied small forest owners in Slovenia. The detached group far outnumbers the other, opposite group, which is the active inner core, or 'engaged group'. However, this group is

Table 12.5 Importance of mountain pastures for selected elements of the cultural landscape

	Farm economics	Maintaining the cultural landscape	Tradition (in general)	Maintaining traditional knowledge	Sense of belonging	Community building	Quality of life of the local population	Quality of life of the non-local population
Krstenica Pasture	Neutral	Important	Important	Important	Important	Important	Important	Very important
Matajur Pasture	Neutral	Very important	Very important	Very important	Important	Important	Neutral	Important
Tegošče Pasture	Important	Very important	Important	Very important	Important	Important	Neutral	Important

Legend:	Very important	Important	Neutral	Less important	Unimportant

proactive and vital enough to keep the mountain pasture alive.

As things progress, both groups face their own challenges. The engaged group is threatened by the generation gap, changing management priorities and the general decline of small-scale and part-time farmers. Respondents did not express fear of the potential problems associated with inheritance and succession, but rather the unpredictability thereof. This uncertainty applies not only to grazing and managing mountain pastures, but to farming in general. Similarly, the detached group is susceptible to impending taxation, which is likely to affect their attitude to common property. As long as there is no cost to the common land, detached co-owners do not hinder or impede the management.

The drive to achieve the required cattle density has brought in non-proprietary grazing users that now outnumber the Tegošče Pasture co-owners. However, the Tegošče Pasture seems to have reached a stable pool of interested users that have been loyal for two decades. Their rights and obligations differ from those of co-owners only in minor details: they have to pay fees for livestock insurance and have no right to sit on management bodies.

The willingness to participate in management goes hand in hand with the direct use of the mountain pasture (consumption of grazing rights). The cessation of grazing usually leads to an owner's disengagement from management. There was an exception to this rule in the Bohinjska Češnjica, Jereka, Podjelje, Koprivnik ACs; one of the former presidents had no cattle. The lack of participation has forced engaged co-owners to introduce novel governance principles, although they tend to retain long-established principles. This is particularly true with the Tegošče Pasture, where the board has tried to varying degrees to encourage inactive members to contribute to the decision-making process.

Stakeholders in grazing have a very pragmatic attitude towards pasturing. Although they do not see its monetary value, they see other practical benefits: less effort required in the home stall during the pasture season, the possibility of obtaining dairy products and, most importantly, health benefits for the livestock. In addition, respondents are aware of the importance and relevance of social, economic, technological, and political realities and their impact on the functioning of the mountain pasture.

All three mountain pastures function and are managed according to established and long-standing general rules, which are followed by most grazing stakeholders without any problems. However, some internal tensions related to recent changes in management were identified at the Krstenica Pasture. In general, most tensions are felt in the process of coordinating, aligning, and harmonising the particular interests and objectives of all stakeholders; that is AC (landowners), breeders (cattle owners), and staff (labour). Roles and interests may overlap, co-exist or clash, change over time, or form different alliances due to the purely personal circumstances specific to all small communities. It is challenging to outline priorities and management directions. Despite all this, operational management runs smoothly in all the case studies, but the inherent tensions could affect future activities. The current way of organising and managing is generally considered adequate, but the individualism that has permeated the modern world is on the rise here as well.

12.5.10 Landscape Perspectives

The interviewees are well aware of the volatile line between a well-functioning and an abandoned mountain pasture. The Krstenica Pasture is in a delicate phase, whereas the Tegošče and Matajur Pastures seem to be sufficiently stable. However, in the long term, the situation is not sound. There are several reasons for this: (1) the generation gap and related uncertainty, (2) ownership and orientation of the farm, (3) recruitment of shepherds and cheesemakers (Krstenica and Matajur Pastures). Recruitment is a burning issue. Both professions require specific knowledge, practical skills, and experience that are difficult to obtain. The pool of potential employees is therefore limited, and co-owners tend to recruit among themselves.

Table 12.6 Perspectives on landscape development

	Forest and scrub overgrowth will prevail					Less-favoured areas will be abandoned					Forest will be replaced by grazing areas					The pasture will suffer from landslides/avalanches/erosion					Management will be difficult due lack of water					Locally atypical buildings will be constructed					Tourism will prevail; cattle only to keep the form				
Krstenica Pasture		x					x								x				x					x					x			x			
Matajur Pasture		x					x							x					x				x							x					x
Tegošče Pasture				x					x						x				x					x					x					x	

Legend:	Very much possible	Possible	Neutral	Hardly possible	Impossible

Agriculture functions within a dynamic and unpredictable political, economic, technological, and social environment. The interviewees are well aware of this. Their experience is that modern farming practices, a decline in small and medium-sized non-industrial family farms, changing social relations, and family structures do not support the preservation of pasture farming. The most likely scenario is that less-favoured areas will be abandoned, which will lead to grassland being overgrown by forest and scrub. The reverse process is highly unlikely. Natural disasters and water supply issues are not expected. Further on, neither local atypical construction nor prevalence of tourism are foreseen (Table 12.6).

The mountain pastures will survive if a sufficient number of farms continue to engage in mountain pasture farming. This depends on finding the right balance between modern interventions, especially technological advances, and traditional practices. Due to the limited importance of grazing for the farm economy, subsidies are crucial.

12.6 Conclusion

The history of mountain pastures is a history of agriculture and its influence on shaping the landscape, but also (and above all) a history of the survival, perseverance, and determination of the people associated with mountain farming. Life on the land and its management have shaped the cultural landscape, which is not intentional but a side effect of human activity. In other words, Alpine landscapes, for example, are a byproduct of farming practices in disadvantaged areas. However, the main agents—farmers—live, work and act in the context of the general and sector-specific situation projected into the policies. The fact that the landscape is exposed to different sectoral policies and partial interests is often the cause of management challenges.

All three mountain pastures have successfully maintained their primary agricultural role despite the various social, economic, and political upheavals that have appeared over the last generation or two. Changes occurred in all areas: technical equipment (electricity, milking), infrastructure (roads), social (deagrarisation, changed family structure) and farm orientation. The most significant change is connected with people; participation in mountain pasture underwent a significant transformation. Many co-owners are no longer active, and some new breeders have taken their place. A new stakeholder group emerged, which plays an important role but has no decision-making power. Expanding and reshaping decision-making bodies are one way to make a community more robust, active, and thus more resilient in the face of unpredictable changes. The balance between maintenance and natural succession is fragile and requires considerable human effort.

Uncertainty concerning humans is perceived as an impending threat, mainly because it depends on a broader dynamic and unpredictable social and political developments. In the last two decades, national and EU policies have played a crucial role, not only through direct payments for the preservation of grazing, but also by shaping an overall agricultural situation that can provide a sustainable environment for smallholder agriculture. Promoting agricultural activities to prevent land abandonment and overgrowth, especially in marginal areas with a mountain grazing tradition in the Alps, is crucial for maintaining cultural landscapes. In the long term, targeted measures encouraging farmers to integrate mountain pasture management into their business would be beneficial. This preserves the landscape in the most natural way and promotes other benefits such as biodiversity and tourism potential.

The importance of mountain pastures in the Alps goes far beyond the local level; on the contrary, it is recognised nationally and worldwide. Significantly, mountain pasturing and dairying was inscribed in the Register of the Intangible Cultural Heritage of the Republic of Slovenia in 2020. Transhumance, the seasonal driving of livestock along migratory routes in the Mediterranean and the Alps was inscribed on the UNESCO Representative List of the Intangible Cultural Heritage of Humanity in 2019, and the FAO has declared 2026 as the International Year of Rangelands and Pastoralists. Combining national and international initiatives and developing cross-cutting, targeted interventions could enable Drago's descendants to listen to cowbells, smell the grass and wild herbs, and enjoy the splendour of a blue sky.

Acknowledgements This study was conducted as part of the research programs The Geography of Slovenia (P6-0101) and Heritage on the Margins: New Perspectives on Heritage and Identity Within and Beyond National (P5-0408), financed by the Slovenian Research and Innovation Agency.

References

Aljančič J (2021) Interviews carried out (in person) in August and September 2021

Barker G, Grant A, Beavitt P, Christie N, Giorgi J, Hoare P, Leggio T, Migliavacca M (1991) Ancient and modern pastoralism in central Italy: an interdisciplinary study in the Cicolano mountains. Papers Br School Rome 49:15–88. https://doi.org/10.1017/S0068246200009673

Bätzing W (2021) Alm- und Alpwirtschaft im Alpenraum: Eine interdisziplinäre und internationale Bibliographie. Marktgemeinde Bad Hindelang, Augsburg, Nürnberg

Baur I, Nax HH (2018) Adapting governance incentives to avoid common pool resource underuse: the case of Swiss summer pastures. Sustainability 10. https://doi.org/10.3390/su10113988

Bevione M, Courtonne J-Y, Buclet N, Longaretti P-Y, Desvaux Q (2022) Analysing the vulnerabilities and capabilities of wealth creation activities in the Maurienne valley in the French Alps. Reg Environ Change 22. https://doi.org/10.1007/s10113-022-01908-0

Bizjak M (2020) Razvoj zemljiških gospostev na ozemlju Tržiča v srednjem veku [Development of seigniories in the territory of medieval Tržič]. Kronika 68(3):361–374

Bojnec Š, Latruffe L (2013) Farm size, agricultural subsidies and farm performance in Slovenia. Land Use Policy 32:207–217. https://doi.org/10.1016/j.landusepol.2012.09.016

Burns KR (1963) The circum–Alpine culture area: a preliminary view. Anthropol Q 36(3):130–155. https://doi.org/10.2307/3316628

Carrer F (2015) Herding strategies, dairy economy and seasonal sites in the southern Alps: ethnoarchaeological inferences and archaeological implications. J Mediterr Archaeol 28(1):3–22. https://doi.org/10.1558/jmea.v28i1.27499

Carrer F, Walsh K, Mocci F (2020) Ecology, economy, and upland landscapes: socio-ecological dynamics in the Alps during the transition to modernity. Hum Ecol 48:69–84. https://doi.org/10.1007/s10745-020-00130-y

Cerar G, Kliner P, Papež M (2011) Prihodnost agrarnih skupnosti [The future of agrarian communities]. Zelena Dežela 102:7–10

Ciglič R, Hrvatin M, Komac B, Perko D (2012) Karst as a criterion for defining areas less suitable for agriculture. Acta Geogr Slov 52(1):61–98. https://doi.org/10.3986/AGS52103

Cirman A (2006) Housing tenure preferences in the post-privatisation period: the case of Slovenia. Hous Stud 21(1):113–134. https://doi.org/10.1080/02673030500391213

Cole JW, Wolf ER (1974) The hidden frontier: ecology and ethnicity in an Alpine valley. University of California, Berkeley

Deléglise C, Dodier H, Garde L, François H, Arpin I, Nettier B (2019) A method for diagnosing summer mountain pastures' vulnerability to climate change, developed in the French Alps. Mt Res Dev 39(2):27–41. https://doi.org/10.1659/MRD-JOURNAL-D-18-00077.1

Dreščček B (2021) Interviews carried out (in person) in July 2021

Drozg V (2007) Prispevek h gospodarski geografiji Slovenije [A contribution to the economic geography of Slovenia]. Revija Za Geografijo 1(2):67–88

Drozg V (2012) Mobility and the lifestyle of the Slovene population. Geografski Vestnik 84(1):163–170

Egarter Vigl L, Tasser E, Schirpke U, Tappeiner U (2017) Using land use/land cover trajectories to uncover ecosystem service patterns across the Alps. Reg Environ Change 17:2237–2250. https://doi.org/10.1007/s10113-017-1132-6

Favry E, Pfefferkorn W (2005) The Alps over the past 30 years: changes in spatial structures and cultural landscapes. In: Pfefferkorn W, Egli HR, Massarutto A (eds) Regional development and cultural landscape change in the Alps—The challenge of polarisation. University of Bern, Bern, pp 21–60

Gabrovec M, Kumer P (2019) Land-use changes in Slovenia from the Franciscean Cadaster until today. Acta Geogr Slov 59(1):64–88. https://doi.org/10.3986/AGS.4892

Gabrovec M, Kumer P, Ribeiro D, Šmid Hribar M (2020) Land use in Slovenia. In: Perko D, Ciglič R, Zorn M (eds) The geography of Slovenia: small but diverse. Springer, Cham, pp 279–290. https://doi.org/10.1007/978-3-030-14066-3_18

Gabrovšek I (1998) Tržiške Alpe – problem samotnih kmetij in planinskega pašništva [The Tržič Alps – The problem of solitary farms and mountain pastures]. BSc Thesis, University of Ljubljana, Ljubljana

Gilck F, Poschlod P (2019) The origin of Alpine farming: a review of archaeological, linguistic and archaeobotanical studies in the Alps. The Holocene 29(9):1503–1511. https://doi.org/10.1177/0959683619854511

Gretter A, Ciolli M, Scolozzi R (2018) Governing mountain landscapes collectively: local responses to emerging challenges within a systems thinking perspective. Landsc Res 43(8):1117–1130. https://doi.org/10.1080/01426397.2018.1503239

Hernandez-Morcillo M, Hoberg J, Oteros-Rozas E, Plieninger T, Gomez-Baggethun E, Reyes-Garcia V (2014) Traditional ecological knowledge in Europe: Status quo and insights for the environmental policy agenda. Environ Sci Policy Sustain Dev 56(1):3–17. https://doi.org/10.1080/00139157.2014.861673

Hrvatin M, Zorn M (2017) Trendi pretokov rek v slovenskih Alpah med letoma 1961 in 2010 [River discharge trends in Slovenian Alps between 1961 and 2010]. Geografski Vestnik 89(2):9–35. https://doi.org/10.3986/GV89201

Jordan B (1945) Planine v Karavankah [Mountain pastures in the Karawank Mountains]. Geografski Vestnik 17:49–104

Jović D (2016) Paša in planšarstvo na bohinjskih planinah – pogled v prihodnost [Grazing and Alpine farming in Bohinj mountain pastures – future perspective]. PhD Thesis, University of Ljubljana, Ljubljana

Juvančič L, Erjavec E (2005) Intertemporal analysis of employment decisions on agricultural holdings in Slovenia. Agric Econ 33(2):153–161. https://doi.org/10.1111/j.1574-0864.2005.00378.x

Kianicka S, Knab L, Buchecker M (2010) Maiensäss – Swiss Alpine summer farms – an element of cultural heritage between conservation and further development: a qualitative case study. Int J Herit Stud 16(6):486–507. https://doi.org/10.1080/13527258.2010.505045

Klemenčič V, Genorio R (1993) The new state of Slovenia and its function within the frame of Europe. GeoJournal 3:323–335. https://doi.org/10.1007/BF00806724

Knežević Hočevar D (2015) Kdor ne tvega, ni kmet! K antropologiji tveganja in negotovosti [Who does not take a risk is not a farmer! Towards the anthropology of risk and uncertainty]. Glasnik Slovenskega Etnološkega Društva 55(3–4):61–69

Knežević Hočevar D, Černič Istenič M (2014) In pursuit of knowledge-based Slovenia: is knowledge transfer to agriculture stuck in faculties. Anthropol Notebook 20(3):103–112

Krauß W, Olwig KR (2018) Special issue on pastoral landscapes caught between abandonment, rewilding and agro-environmental management. Is there an alternative future? Landsc Res 43:1015–1020. https://doi.org/10.1080/01426397.2018.1503844

Kumer P, Štrumbelj E (2017) Clustering-based typology and analysis of private small-scale forest owners in Slovenia. Forest Policy Econ 80:116–124. https://doi.org/10.1016/j.forpol.2017.03.014

Ledinek Lozej Š (2002) Pričevanja o nekdanji ureditvi in življenju v planini Krstenica [Testimonies of past life on the Krstenica dairy mountain]. Traditiones 33(1):69–90

Ledinek Lozej Š (2020) Znamčenje tolminskega sira [Branding Tolmin cheese]. Traditiones 49(3):53–80. https://doi.org/10.3986/Traditio2020490304

Ledinek Lozej Š (2022) Mountain pasture in Friuli: past and present. In: Bindi L (ed) Grazing communities: Pastoralism in the move and biocultural heritage frictions. Berghan, New York, Oxford, pp 222–240. https://doi.org/10.2307/j.ctv2vr8tp1.15

Ledinek Lozej Š, Roškar S (2018) Alps of the Julian Alps. In: Jaćimović M (ed) Održivi razvoj planinskih područja: Iskustva, izazovi i perspektive. Naučni skupovi, Odjeljenje prirodnih nauka 148(20). Crnogorska akademija nauka i umjetnosti, Podgorica, pp 121–140

Mazzocchi C, Sali G (2019) Assessing the value of pastoral farming in the Alps using choice experiments: evidence for public policies and management. J Environ Plann Manag 62(4):552–567. https://doi.org/10.1080/09640568.2018.1430557

Mazzocchi C, Sali G, Ruggeri G (2019) Tourists' preferences for Alpine pastures maintenance. Landsc Online 68:1–18. https://doi.org/10.3097/LO.201968

Melik A (1950) Planine v Julijskih Alpah [Mountain pastures in the Julian Alps]. SAZU, Ljubljana

Mlakar C (2021) An interview carried out (in person) in September 2021

Nared J, Bole D, Breg Valjavec M, Ciglič R, Goluža M, Kozina J, Razpotnik Visković N, Repolusk P, Rus P, Tiran J, Černič Istenič M (2017) Central settlements in

Slovenia in 2016. Acta Geogr Slov 57(2):7–32. https://doi.org/10.3986/AGS.4606

Netting RM (1981) Balancing on an alp: ecological change and continuity in a Swiss mountain community. Cambridge University Press, Cambridge

Oranič D (2021) Interviews carried out (in person) in August and September 2021

Orland B (2004) Alpine milk: dairy farming as a pre-modern strategy of land use. Environ Hist 10(3):327–364. https://doi.org/10.3197/0967340041794286

Pagot G, Šmid Hribar M, Rail LF, Walters G, Hymas O, Liechti K, Haller T, Urbanc M, Dalla Torre C, Joye J-F (2025) Territories of commons: a review of common land organizations and institutions in the European Alps. Environ Res Lett 20(6). https://doi.org/10.1088/1748-9326/add1f4

Pecher C, Bacher M, Tasser E, Tappeiner U (2018) Agricultural landscapes between intensification and abandonment: the expectations of the public in a Central-Alpine cross-border region. Landsc Res 43(3):428–442. https://doi.org/10.1080/01426397.2017.1315062

Petek F (2005) Spremembe rabe tal v slovenskem alpskem svetu [Land use changes in Slovenian alpine regions]. Založba ZRC, Ljubljana. https://doi.org/10.3986/978 9612545123

Petek F, Urbanc M (2007) Skupna zemljišča v Sloveniji [Common land in Slovenia]. Geografski Vestnik 79(2):41–62

Potthoff K, Smrekar A, Šmid Hribar M, Urbanc M (2020) The past and perspective development of pasturing and tourism in the mountains: Insights from Norway and Slovenia. Geografski Vestnik 92(1):81–111. https://doi.org/10.3986/GV92105

Premrl T (2013) Analiza stanja agrarnih skupnosti v Sloveniji na podlagi podatkov upravnih enot [Analysis of the situation of agrarian communities in Slovenia on the basis of administrative data]. Gozdarski inštitut Slovenije, Ljubljana

Price M (2015) Mountains: a very short introduction. Oxford University Press, Oxford. https://doi.org/10.1093/actrade/9780199695881.001.0001

Razpotnik Visković N, Seručnik M (2013) Ugled kmečkega poklica in polkmetov v slovenski družbi po drugi svetovni vojni [Respectability of the farming profession and part-time farmers in Slovene society after World War II]. Geografski Vestnik 85(1):25–36

Register kmetijskih gospodarstev [Register of agricultural holdings] (2020) Ministrstvo za kmetijstvo, gozdarstvo in prehrano, Ljubljana

Schirpke U, Altzinger A, Leitinger G, Tasser E (2019) Change from agricultural to touristic use: effects on the aesthetic value of landscapes over the last 150 years. Landsc Urban Plan 187:23–35. https://doi.org/10.1016/j.landurbplan.2019.03.004

Schirpke U, Leitinger G, Tasser E, Schermer M, Steinbacher M, Tappeiner U (2013) Multiple ecosystem services of a changing Alpine landscape: past, present and future. Int J Biodiv Sci Ecosyst Serv Manag 9(2):123–135. https://doi.org/10.1080/215 13732.2012.751936

Schlamberger V (1995) Planine in skupni pašniki v Sloveniji [Mountain and common pastures in Slovenia]. Ministrstvo za kmetijstvo, gozdarstvo in prehrano, Ljubljana

Scolozzi R, Soane ID, Gretter A (2014) Multiple-level governance is needed in the social-ecological system of alpine cultural landscapes. In: Padt P, Opdam N, Polman C (eds) Scale-sensitive governance of the environment. Wiley, New York, pp 90–103. https://doi.org/10.1002/9781118567135.ch6

Sodja A (2021) An interview carried out (in person) in July and August 2021

Stres J (2021) An interview carried out (in person) in July 2021

Stres P (1994) Idrsko in Mlinsko: oris preteklosti vasi pod Matajurjem, od davnine do 15. septembra 1947: ob 910-letnici prve omembe vasi Idrsko in Mlinsko [Idrsko and Mlinsko: a historical outline of the villages below Matajur, from antiquity to 15 September 1947: on the 910th anniversary of the first mention of the villages of Idrsko and Mlinsko]. Branko, Nova Gorica

Šmid Hribar M, Kozina J, Bole D, Urbanc M (2018) Javno dobro, skupni viri in skupno: vpliv zgodovinske zapuščine na sodobno dojemanje v Sloveniji kot tranzicijski družbi [Public goods, common-pool resources, and the commons: the influence of historical legacy on modern perceptions in Slovenia as a transitional society]. Urbani Izziv 29(1):96–109

Streifeneder T (2010) Die Agrarstrukturen in den Alpen und ihre Entwicklung unter Berücksichtigung. Utz, München

Turk M, Velušček, A, Pavlin P, Tecco Hvala S, Grahek L, Horvat J, Šašel Kos M, Modrijan Z, Pleterski A (2020) The history of Slovenia: archaeological evidence from Prehistory to the Slavs. In: Perko D, Ciglič R, Zorn M (eds) The geography of Slovenia: small but diverse. Springer, Cham, pp 127–141. https://doi.org/10.1007/978-3-030-14066-3_8

Udovč A, Kovačič M, Kramarič F (2006) Socialno-ekonomski tipi kmetij po podatkih popisa kmetijskih gospodarstev v letu 2000 [Socio-economic types of farms according to the census of agricultural holdings in 2000]. In: Kavčič S (ed) Slovenija v EU - izzivi za kmetijstvo, živilstvo in podeželje: Zbornik 3. konference Društva agrarnih ekonomistov Slovenije. Društvo agrarnih ekonomistov Slovenije, Ljubljana, pp 71–79

Urbanc M, Šmid Hribar M (2021) Livek: A mountainous border area's transformation from a ski paradise to a resilient community. In: Martinez G (ed) Culture and climate resilience: perspectives from Europe. Palgrave Macmillan, Cham, pp 45–68. https://doi.org/10.1007/978-3-030-58403-0_4

van Zanten BT, van Berkel DB, Meentemeyer RK, Smith JW, Tieskens KF, Verburg PH (2016) Continental-scale quantification of landscape values using social media data. Proc Natl Acad Sci USA 113(46):12974–12979. https://doi.org/10.1073/pnas.1614158113

Viazzo PP (1989) Upland communities: environment, population and social structure in the Alps since the sixteenth century. Cambridge University Press, Cambridge

Vojvoda M (1967) Spreminjanje gospodarskih funkcij in fiziognomije sezonsko poseljenega pasu v slovenskih Alpah [Changing economic functions and physiognomy of the seasonally populated zone in the Slovenian Alps]. Unpublished material. Inštitut za geografijo Univerze v Ljubljani, Ljubljana

Volk Bahun M, Zorn M (2020) Snežni plazovi v Občini Tržič [Avalanches in the municipality of Tržič]. Kronika 68(3):945–958

von Glasenapp M, Thornton TF (2011) Traditional ecological knowledge of Swiss alpine farmers and their resilience to socioecological change. Hum Ecol 39:769–781. https://doi.org/10.1007/s10745-011-9427-6

Vranješ M (2008) Prostor, teritorij, kraj: produkcije lokalnosti v Trenti in na Soči [Space, territory, place: productions of locality in Trenta and on the Soča river]. Založba Annales, Koper

Wanner A, Pröbstl-Haider U, Feilhammer M (2021) The future of Alpine pastures – agricultural or tourism development? Experiences from the German Alps. J Outdoor Recreat Tour 35. https://doi.org/10.1016/j.jort.2021.100405

Population Migrations from Poddurmitor Region (Consequences and Alternative Development Paths)

13

Rade Šarović and Goran Ćeranić

Abstract

The municipalities of Žabljak, Šavnik and Plužine in the Durmitor National Park in northern Montenegro have experienced a steady population decline since the Second World War, after which the inhabitants of the rural areas migrated uncontrollably to the larger urban centres due to the rapid industrialization process. The migration of rural population and deagrarization led to almost complete deruralization of the north of Montenegro. The paper discusses issues related to out-migration, but also presents an alternative development model.

Keywords

Population · Migration · Deagrarization · Industrialization · Žabljak · Šavnik · Plužine · Durmitor

13.1 Introduction

Population migrations from one place to another, life style changes are not solely characteristics of Montenegro and its mountain areas. A problem of deserting rural, usually mountainous areas of some territories are almost a common problem for all former Yugoslav republics (Horvat 1976). However, comparing to Montenegro, in all other societies migration process was directed by economic development policy (mostly having rural to urban direction) and aligned with the development of a wholesome economy of a republic. This was not the case in Montenegro, where the migrations from northern, mountainous parts of the country, including Durmitor area, were so intense, that they resulted in demographic, social, economic and naturally spatial changes. Emphasizing industry and urban settlements throughout many years (especially Podgorica, Nikšić and some coastal municipalities) and neglecting agriculture, villages and small towns like Žabljak, Šavnik and Plužine, resulted significant structural changes that cannot contribute to a modern development of the municipalities. Unfortunately, these municipalities did not only lose inhabitants of some villages. Along with having deserted villages, towns had no development strategy or an idea of how to stop further migrations (Šarović 2012).

13.2 Study Subject Matter and Objective

Transformation in economy activities and in population size of Montenegrin mountainous area, represent a complete village deagrarization, population exodus and partial devastation of towns. In the core of these social processes are trends

R. Šarović (✉) · G. Ćeranić
Faculty of Philosophy, University of Montenegro, Nikšić, Montenegro
e-mail: rades@ucg.ac.me

G. Ćeranić
e-mail: gorance@ac.me

M. Zorn et al. (eds.), *Discourses on Mountains of Montenegro and Slovenia*, Historical Geography and Geosciences, https://doi.org/10.1007/978-3-031-99017-5_13

and changes which consequently resulted in the development of coastal region towns and the devastation of villages and towns of the north of the country. These processes as permanent facts changed the image of Montenegrin society in all. The defining characteristics of the stated changes and processes (comparing to all other European countries and former Yugoslav republics) are the promptness of the process and a mass number of people leaving their homes. However, all sudden social changes cause other numerous social changes, in this case affecting the core of Montenegrin society, its production, consumption, and naturally the wholesome lives of its people (Šarović 2012).

Subject matter of this paper is migrations on the territory of Montenegro, specifically from municipalities of Poddurmitor region, Žabljak, Šavnik and Plužine.

This paper objective is of a futuristic nature, focusing on alternative paths and possible development models for the devastated areas of northern, mountainous parts of our country, specifically Poddurmitor region.

13.3 Deserting Mountains of Montenegro (Historical Context)

Population migrations and society deagrarization are manifested in many different ways and have different consequences for the society itself. However, changes in Montenegro, seen in the economy structure, caused the development of industry and its sectors on one hand, and agriculture declension on the other. This represented a trigger for and the most distinctive manner of mountain area devastation, i.e. Montenegrin coastal and central area development. In the socialist Yugoslavia (Montenegro), rural area inhabitants out migrated without control, due to prompt industrialization process, hard lives of overburdened peasants and an inadequate social and political treatment (their inclusion). While direct deagrarization was initially dominant, i.e. direct employment of unqualified and semi-qualified workers from villages into towns, indirect deagrarization was more dominant in a later stage—transfer of educated and qualified workers into non-agricultural activities. Agriculture, rural and mountain areas of former Yugoslavia, were an automatic source of the initial capital as well as the main resource of workers for constructing the industrial country. Development models were ensured through an unequal exchange between agriculture and industry, always favouring industry, and causing a wholesome disintegration of rural society in social, economic, cultural, political and all other senses (Zupančić 2002).

Migrations of rural population and deagrarization are the processes that automatically caused a complete deruralization of the north of Montenegro (Šarović 2012). Migrations occurred mostly in the Northern region where the agriculture is extensive, and villages small and undeveloped (Fig. 13.1). Three municipalities that represent a subject of this paper are a typical example of this process on the north of Montenegro.

Deruralization of Montenegrin mountain area has started as the Second World War finished. Its process was slowed down in the last decade of twentieth century during the crisis of Montenegrin society on a general level. This period was a time when, in some way, villages were the only somewhat safe life and work surrounding. At the time of prominent economic and political collapse, in the 1990s, peasants did not fall on a social and economic scale as other social classes did—labourers, clerks, office workers. Finally, they were least dependent on the wholesome (adverse) social trends that were highly manifested at the time. This automatically made an influence on small towns as well. Therefore, this period brought a somewhat demographic stability to Poddurmitor region (Šarović 2012). Deagrarization processes happened too fast and without control causing many negative consequences for Poddurmitor towns and villages. Although a proportion of agriculture population, i.e. their farms, in the total population is not small it is certainly disproportional to the level of general social development and agriculture development in particular. On the other hand, quality components of agriculture population are not encouraging at all. This

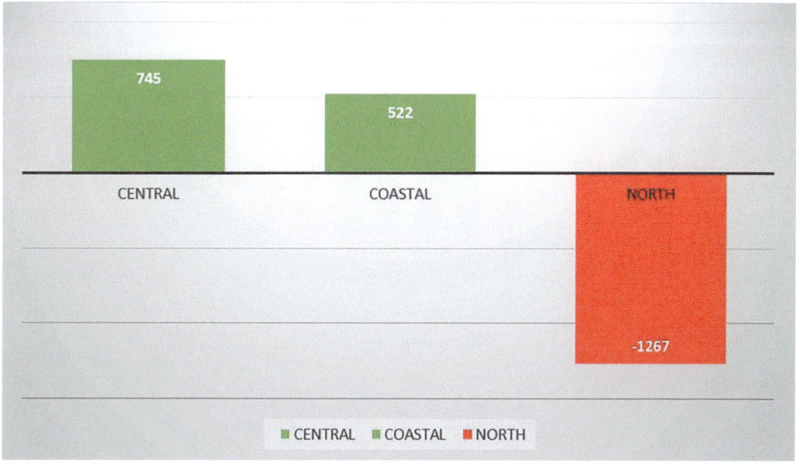

Fig. 13.1 Migration balance by regions of Montenegro in 2021 (number of persons who immigrated or emigrated from the region) (Monstat 2022a)

population is rapidly growing old, resulting to have a high percent of so called elderly households, as well as households without heirs (Raonić 2009).

13.4 Exodus of People of Poddurmitor Region

Poddurmitor municipalities in the current state of agro-technological underdevelopment, even with some favourable conditions for agriculture development (good natural conditions for agriculture and area that has been partially preserved from devastation), especially farming, are facing the problem with population (regarding its extremely unfavourable age structure) which represents extremely restrictive factor for agriculture development. However, not only villages of Durmitor mountain area are seen as a problem today. Figure 13.2 shows that migration process deeply shook urban centres of this part of Montenegro. After socialism breakdown, turbulent post socialist period caused migrations towards central and southern parts of the country. Reasons are numerous and vary from hostile climate and undeveloped infrastructure to unfavourable economic situation of underdeveloped north, where investments are obtained only if they are absolutely necessary (Vlahović 2004:81).

Regarding Poddurmitor villages, there is a highly noticed fact that most of them experience a deep demographic crisis, especially in the past few decades. Over 80% of these villages have a declining rate of natural population growth, and about 70% of them are more or less depopulated. Villages that almost have no inhabitants and those that are completely depopulated are not rare either (Vlahović 2004:22).

Out of total number of villages in Žabljak and Šavnik, almost 85% of them do not exist any longer, having less than 20 inhabitants and extremely unfavourable age structure (Raonič 2009). "*In this unequal exchange with towns, these villages lost a significant number of people that, throughout different life circumstances, developed committed life aspirations. It was not possible to realize these aspirations while living in a village, and this was also a common belief*" (Hodžić 2006:196).

Some farms stopped with their production and some villages ceased to exist in the northern part of Montenegro, causing a constant negative tendency of natural population growth in the second half of twentieth century. In the period between 1966 to 1970 birth rate was 13,4%, falling to –1.2% until 2004, and after having a negative record of –7.9% in 2018. Table 13.1 shows identical processes for the municipalities that are part of this research (Bakić and Mijanović 2006:19–27; Monstat 2019). The most extreme ratio of birth rate and death rate in Montenegro is present in the three municipalities of Poddurmitor region (according to the most

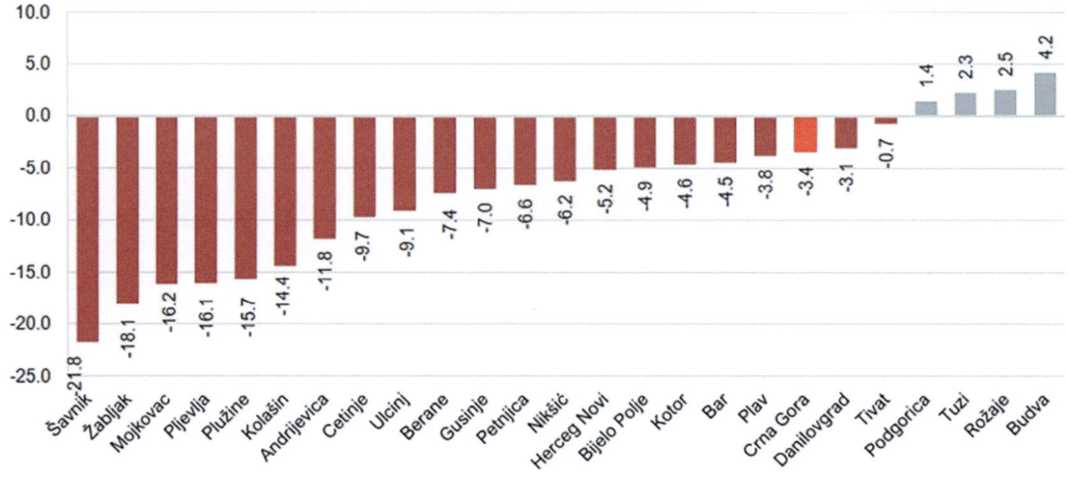

Fig. 13.2 Natural increase rate by municipalities of Montenegro in 2021 (Monstat 2022b)

recent data, average natural population growth is –7.4) (Monstat 2019).

Population migrations from rural to urban areas have caused huge spatial inequalities in the whole Montenegrin territory. The best indicators of this claim are seen in the fact that only four municipalities of Central and Northern region together (Podgorica, Danilovgrad, Gusinje, Petnica) have a positive rate of natural growth, comparing to Coastal region which experiences an ongoing growth (Monstat 2019). However, if a comparison is made on a level of the whole country, we can see a dramatic decline in natural growth (Table 13.1) and a threat of the problem of so called *Birth Dearth* which our neighbours (Serbia, Croatia, Bosnia), bigger in population and territorially, are already facing (Mitrović 1998).

Negative rate of natural growth, 80% of elderly households in villages, exodus of the young, social unrest, are only some of phenomena which caused total transformation (devastation) of rural and urban Durmitor area. However, today when far much more people live in towns than in villages, we increasingly feel a need for peasant's work and products of their work. A need for food, healthy surroundings, peaceful place to rest, are only some of the specificities that should be a base for the plans and strategies for the revitalization of Montenegrin rural heritage, principally Durmitor and its towns and villages.

13.5 Alternative Development Models of the Municipalities Studied

There are many examples of how to revitalize isolated and deserted areas of a country. There are many cities and towns all over the world, especially in Scandinavian countries, which used

Table 13.1 Natural population growth of Montenegrin regions and Poddurmitor municipalities from 1966 to 2018 (Monstat 2019; Bakić and Mijanović 2006; Šarović 2012)

	1966–1970	1976–1980	2000–2004	2011	2018
Northern region	13.4	8.1	−1.2	−7.4	−7.9
Šavnik	10.7	3.3	–5.3	–8.7	–12.5
Žabljak	11.5	7.5	–2.7	–8.4	–1.9
Plužine	12.2	6.5	–2.3	–10.8	–9.3
Central region	11.5	9.6	2.5	1.0	1.0
Coastal region	8.2	8.7	2.5	2.0	2.5
Montenegro	15.5	10.5	5.5	2.2	1.2

good ideas and strategies to develop successfully (Mihanović 2019). Some of these ideas would easily be implemented, some would have to be modified, but most of them are applicable for the cases in Montenegro area. In this paper, we emphasize some of these ideas.

Municipalities of Poddurmitor area are places where traditional lifestyle is still present, and globalization did not make distinctive traces of society modernization. This is why these municipalities are of a special interest from the aspect of pure nature, sustainable agriculture and healthy products, which are a dream of each tourist who appreciates ecology. In this respect, Poddurmitor towns have a solid base to become small green oases in a world of global rush and race for money. While researching such issues, Jeffrey Sachs asks a question: what can make a city both green and sustainable place to live? According to his opinion, sustainable cities are economically, socially, inclusively and politically, ecologically sustainable. In other words, they have to encourage efficient economic activities and enable benefits out of those activities for their citizens, but while performing them, biodiversity, clean air, physical health and safety of their citizens must be maintained, especially at this time of climate changes and increasing extreme weather conditions (Sachs 2014:10–15).

Many of the tourists that visit Montenegro are much more interested in mountain areas and traditional lifestyle than in modern functioning of urban centres (Vlahović 2004:82). Today's priorities are to find something original, communicate with locals, discover places and authentic culture with the help of stage acts and modern mass media. This is not only a call to create tourist attractions but to open job opportunities in the north of the country, not only in tourism but in many other industries. This is the way to stimulate poverty stroked areas (principally Poddurmitor region and its villages) and to expand tourist offer ensuring longer stays of guests.

Operationalization of the given ideas can be realized through a movement which idea is just perfect for small towns of Montenegro, such as Žabljak, Plužine and Šavnik. We'll emphasize one of lines of actions, which can be related to the future of these urban settlements, and which started in Italy at the end of the last century and today is a serious, global tourist brand (Cittaslow 2021).

Carlo Pertini is one of the founders of the idea of *Slow Life* and city Orvieto was one of the first members of *Slow City* movement. Today this city represents a mirror and the international centre of the movement. This movement already lives in 30 countries and 275 cities around the world, but none of the former Yugoslav republics are represented (Cittaslow 2021).

Speaking of architecture of these towns, first to notice are small family restaurants (trattorias) which serve local specialties. *Slow Cities* are fully oriented towards local architecture preservation and their philosophy supports the idea of renovation of old facades and run–down houses and buildings. In our case, a model for new facades and roofs would be of a characteristic wooden house (Fig. 13.3). In this way, the new constructions would have a local, traditional look, and would fit the mountain town image. *Slow Cities* seek to reduce traffic, enlarge green areas and pedestrian zones, and finally remove traffic areas from city centres. Philosophy of life of slow city people is based on encouraging locals and their guests to focus on a healthy lifestyle connected with nature. In this way, each *Slow City* tends to emphasize its characteristics, which make them unique and memorable, while their inhabitants are aware and proud of them.

This philosophy of *Slow City* is perfect for small (quiet) towns of Poddurmitor urban settlements. Each inhabitant would have a good life respecting the environment and a chance to enjoy development solutions and services which enable cosy and easy life in their cities. The life in slow city would have to be less vivid and quick, and certainly more emphatic towards people, nature and future generations (Cittaslow 2021). Life in a slow city gives the best and best produced food. Traditional knowledge and skills are used, inhabitants commonly help each other, thus making the feeling of unity while passing knowledge from old to young generations. The goal of living in a *Slow City* is to preserve the environment, participate in community life which respects traditional values and is open towards new cultures.

Fig. 13.3 Characteristic wooden house in Durmitor. (authors archive)

Now if we imagine a person from one hectic city, for example London, a word *slow* before *city* must sound truly tempting and attractive for him/her. However, this idea does not promote any technophobia or laziness, but a tendency to make life more humane and to eliminate speed and stress from lives. *Slow City* would not jeopardize business activities, but in a long-term sense encourage them, as a better natural ambience, more humane life and a happy person benefit business industry. This is not an image of the world without airplanes and the internet. This is about environment preservation, termination of illegal exploitation of natural resources and violation of balance of nature. The point is to determine a pace of life on personal and community level. Whether it will be hectic or slow, it depends on us.

Slow Cities set a new philosophy of living. Domestics and visitors of *Slow City* are "obliged" to spend more time on reflection, talk more with each other and in that way create a sense of unity and hospitality. All in all, this philosophy resists familiar and negative aspects and side effects of homogenization, unification, globalization, but does not refuse any values that improve quality of life.

Montenegro is an "ecological state" for over than 30 years. The declaration was adopted in 1991 (EPA 2021) on Durmitor which represents a solid base for revitalization of the three towns that are facing a historical turning point. Montenegro as an "ecological state" has to harmonize economic and technological development, needs for employment (job creation and a better life quality of employees), general social development and environment preservation as a characteristic living area of clean soil, water, air, nature and food, peace and quiet and hospitality.

13.6 Conclusion

As the Second World War finished, demographic filling of cities for meeting the needs of industry, was executed in an inadequate and harsh way for a democratic society, by pressuring peasants. Individual farming stopped having any particular importance on a social level, which strongly affected a wholesome integrity of village, village life and life on farms. Comparing with the uprising power of cities, farms experience stagnation and become marginalized society part, without any real influence in a social chain of events (Vlahović 2004:39). The changes mentioned above lead to changes not only in Poddurmitor area, but in the whole country, that made a genuine agrarian into partially developed urban society, which until today has not changed.

With these socio-cultural and historic facts intervened, it is essential to recognize the right time and create a new plan, i.e. new development strategy. First of all, the three municipalities discussed are to create a common vision, common goal, strategy, action plan and their implementation. The strategy (for example *Slow City*) has to rely on a clearly defined vision of the future, based on the existing natural resources, motivation, abilities, intellectual resources and perception of the place.

However, this process will not be easy at all. It is necessary to create a unique brand which will consequentially influence urban planning, use of business premises, education and training of specialists' personnel, nature and culture events protection, etc. The branding process of Poddurmitor municipalities, apart from creating connections within communities, must also influence relations and connections with other institutions in the country and abroad, with other places and personal level connections.

A successful process of a city (town) branding, *Slow City* brand, also brings numerous benefits for local inhabitants. It becomes much easier to attract investors and tourists, cultural activities flourish, and most importantly, people are encouraged to move and settle in these places, the outflow of young people decreases, education is enhanced, public revenue and standard are increased, all in all cumulative changes are made and the city becomes a better place to live (Mihanović 2019).

This is not just a mare theory but on the contrary, good examples of practice around the world confirm (Mihević 2015) the thesis of the importance of branding and good development strategy and positive effects that result out of them. Successfully branded cities, towns and regions improve an image of a country, naturally vice versa—a well branded country makes a branding process for its cities and regions much easier (Mihević 2015). This is one more reason why branding cities is very important, positive effects are not only local, they do not affect only the branded city but the whole country (Cittaslow 2021). These are solid grounds for finding a solution for the situation as bad as it is, in order to achieve a revitalization of the north of Montenegro. This process needs to be done according to the modules which developed western countries used to create agribusiness, village tourism, recreational tourism, etc. (Vlahović 2004:523). Considering the given theoretical framework and a chance to achieve the *Slow City* brand for the three Poddurmitor towns, there is no single reason why Montenegro with its incomparably better climate and altimetry conditions shouldn't apply a part of this story. Especially in Poddurmitor region.

References

Bakić R, Mijanović D (2006) Demografska kretanja u Crnoj Gori tokom druge polovine XX vijeka [Demographic trends in Montenegro during the second half of the 20th century]. Geografski institut Filozofskog fakulteta, Niksić

Cittaslow: international network of cities where living is good (2021). https://www.cittaslow.org/. Accessed 12 Aug 2021

EPA (2021). Agencija za zaštitu životne sredine, Podgorica. https://www.epa.org.me/deklaracija-o-ekoloskoj-drzavi/. Accessed 12 Aug 2021

Hodžić A (2006) Selo kao izbor? [Village as a choice?]. Institut za društvena istraživanja, Zagreb

Horvat B (1976) Jugoslovenska agrarna teorija i politika u poslijeratnom razdoblju [Yugoslav agrarian theory and politics in the post-war period]. Pregled, Sarajevo

Mihanović E (2019) Značaj koncepta sporog grada za lokalni razvoj [Significance of the slow city concept for local development]. Univerzitet u Splitu, Ekonomski fakultet, Split

Mihević Z (2015) Brendiranje gradova kao ekonomska nužnost [City branding as economic necessity]. Tehnički Glasnik 9(2):198–201

Mitrović M (1998) Sociologija sela [Rural sociology]. SDS, Beograd

Monstat (2019) Procjene stanovništva i osnovni demografski pokazatelji u 2018 godini [Population assessments and basic demographic indicators in 2018]. Uprava za statistiku, Podgorica. https://www.monstat.org/cg/page.php?id=53&pageid=53. Accessed 10 Aug 2021

Monstat (2022a) Migracije unutar Crne Gore u 2021 godini [Migrations within Montenegro in 2021]. Saopštenje 35. Uprava za statistiku, Podgorica. https://www.monstat.org/uploads/files/demografija/migracije/2021/Migracije%20unutar%20Crne%20Gore%20u%202021.%20godini.pdf. Accessed 26 May 2022

Monstat (2022b) Procjene stanovništva i osnovni demografski pokazatelji u 2021 godini [Population estimates and basic demographic indicators in 2021]. Saopštenje 60. https://www.monstat.org/uploads/files/demografija/procjene/2021/procjene%20stanovnistva%20i%20osnovni%20demografski%20indkatori.pdf. Accessed 1 Jun 2022

Raonić S (2009) Durmitorsko područje i njegovo selo [Durmitor area and its villages]. Sociološka Luča 3(1):79–88

Sachs J (2014) Doba održivog razvoja [The age of sustainable development]. Centar za međunarodnu saradnju i održivi razvoj, Beograd

Šarović R (2012) Perspektive seoskog domaćinstva i razvoj ruralnog područja u Crnoj Gori [Perspectives of rural household and development of rural area in Montenegro]. In: Vlahović P (ed) Demografski i populacioni problemi u Crnoj Gori. CANU, Podgorica, pp 380–403

Vlahović P (ed) (2004) Selo u Crnoj Gori: zbornik radova [Village in Montenegro: proceedings]. CANU, Podgorica

Zupančić M (2002) Vitalni poljoprivrednici i drugi akteri u razvoju ruralnih područja [Vital farmers and other actors in the development of rural areas]. In: Seferagić D (ed) Selo: izbor ili usud. Institut za društvena istraživanja, Zagreb, pp 35–71

Part IV
Identity and the Mountains

14 The Archetypal Discourse of the Mountain Rumija as a Source of Montenegrin Identity and the Significance of Its Existence Through the Character and Work of Prince Vladimir

Miomirka Rakonjac, Predrag Živković, and Vladimir Bakrač

Abstract

The significance of the mountain Rumija is reflected in its geographical location, the fact that it represents the original place of the emergence of the Montenegrin statehood and spirituality, as well as its multidimensional mythical and symbolic feature. The symbolic meaning of this mountain and its historical, social, and cultural components are involved in the context of historical science, even today. The analysis of our paper is focused on the identity conception related to the precursor of Montenegrin state-building sublimated by the mountain Rumija, the personality of Prince Vladimir who ruled the first Montenegrin state, as well as the pilgrimage held every year on the mountain Rumija and dedicated to carrying the cross on which Prince Vladimir was killed. The pilgrimage to the mountain Rumija represents not only a collective memory of the ruler who ruled the first Montenegrin state and suffered a martyr's death, but it also unites people of different religious affiliations (Orthodox, Catholic, and Muslim), which guarantees multi-confessional harmony as an important identity principle in modern Montenegro.

Keywords

Mountain Rumija · State-building · Identity · Hierophany · Pilgrimage · Multi-confessional harmony · Cross of Prince Vladimir

14.1 Introduction

In the paper, we try to present the geoidentity of Rumija with all the accompanying geographical, mythological, cultural, and, above all, sociological elements. The paper goes in the direction of discovering the enigmatic currents that are formed in mythological picture and time and that make Rumija and its accompanying spatial environment distinctive. Although we have pretentiously announced a theoretical step forward in terms of its temporal recognition, it seems that such an act is necessary if we take into account the constructive vitality of the space we explore, the space that integrates "the grimaces" of history, the pride of mythology, the introversion of space and those endemic

M. Rakonjac (✉) · P. Živković · V. Bakrač
Faculty of Philosophy, University of Montenegro, Nikšić, Montenegro
e-mail: miomirka@ucg.ac.me

P. Živković
e-mail: predragz@ucg.ac.me

V. Bakrač
e-mail: vladimir.b@ucg.ac.me

M. Zorn et al. (eds.), *Discourses on Mountains of Montenegro and Slovenia*, Historical Geography and Geosciences, https://doi.org/10.1007/978-3-031-99017-5_14

characteristics that coalesce with local people. We build scientific authenticity, in addition to the necessary geographical determinants, on the sociological valorisation of Rumija as a kind of *cultural and religious pilgrimage* (Simić 2019), a place that witnesses the multicultural harmony and dialogue, and above all, as an aggregate of psychogeographical factors that affect the mindset of the inhabitants of surrounding towns. Here we endeavour, if it is allowed at least in the imaginative act, to emphasize "the cultural turmoil" that Rumija has for its worshipers. Therefore, it is possible to encounter in the paper the sociological and political meanings of Montenegrin statehood and the omnipresent religious authority that forms a particular culture of remembrance, which is a sufficient challenge to try, if not provide an adequate response to the hypothetical categories.

From the abovementioned chronological array of introductions to the given topic of the research, the contents that indicate the geographical but primarily historical and cultural significance of Rumija for the Montenegrin society are presented. The exclusive reference to its geographical peculiarity would represent a reduction of its historical and cultural potential but also a detachment from the framework of identity affiliation and recognition that belong to its symbolic area.

14.2 The Theoretical Review of Identity Determination of the Mountain Rumija

The key concepts for our observation are the concepts of culture and identity. Referring to cultural theorist Stuart Hall, the important question concerning the analysis of identity as an element inherent to both the individual and the community is the question of *who am I and where do I come from* (Hall 1996:4). If we invert the question to follow the logical order of phenomena and give primacy to the question of *where do I come from*, we would clearly see the importance of the socio-spatial dimension that indicates the significance and meaning of identity for determining each individual. When analysing the identity of Rumija, we will take into account the cultural and symbolic dimension of space that points to *the meaning* that the space of Rumija has for the identity of Montenegrin society and culture. *The meaning* that the individual or community gives to the different spatial and cultural attributes within it is reflected in the formation of consciousness about the importance of space for the community and the individual. This hypothesis leads us to think that a human is the only one who can "*understand the symbolic reality that is both semantic and normative, upon which life in the community is based*" (Kalanj 2003:50), which confirms that the human activity is referred to creation, transmission and adoption of the symbols that "*deeply permeate the organization of the psyche and enable its integration into the collective order*" (Kalanj 2003:50). The story of identity that is emphasized in the previous lines has relevance for our work precisely because the mountain Rumija represents a place that integrates the spiritual and material heritage, but it is also the precursor of Montenegrin state-building, and therefore it is given a specially valuable and semantic connotation. Rumija is precisely the point where cultural, historical, and geographical bases of particular importance for Montenegrin society intersect. A permanent return to this symbolic tonality represents a deed of remembrance and an active mode of factual existence, not only in the meaning of state-building but also in religious and spiritual meaning, thus defying each attack of impersonality and historical baselessness.

In the following lines, we will present some historical facts that will indicate the temporal and historical context from which the first Montenegrin state *Doclea* emerged but also the importance of the mountain Rumija for that context. "*Montenegro is partly a mountainous country, and its inhabitants were destined for the mountains. Two mountains have special significance because they represent the axes of Montenegrin state-building, the mountain Rumija, and since the fifteenth century, the mountain Lovćen*" (Radojević 1997:29). Rumija is the southernmost Montenegrin mountain located between Lake Skadar in the northeast and the Adriatic Sea in the southwest. At the same time, its position near the Mediterranean made it suitable for defence and the formation of the core of the first Montenegrin state—*Doclea*. From the top of Rumija, we can see the slopes on which the towns and palaces of the Doclean rulers were located. We can also see the Oblik hill on which Prince Vladimir with his army resisted the army of the Macedonian Emperor Samuil in the fortified town in 997 to protect Doclea (Radojević 1997).

In order to understand the significance of the historical context, it is important to point out that the area inhabited by the ancestors of the Montenegrins was Doclea at the beginning of the seventh century. In his work *De administrando imperio*, the Byzantine Emperor Constantine Porphyrogenitus (905–959) calls the ancestors of the Montenegrins the Docleans, explaining they got this name because they inhabited the area around the Roman town Doclea. At the beginning of the ninth century, the Docleans formed a principality that was under the Byzantine administration, as evidenced by the seal of the Doclean Prince Petar (father of Prince Vladimir), which has the inscription "Petar, Archon of Diokleia" (Andrijašević 2017:148). The centre of state formation of the Doclean Slavs, ancestors of Montenegrins, was located at the foot of the mountain Rumija, where the source of Montenegrin spirituality is, i.e. the place of origin of the Montenegrin medieval tradition (Jovović 2017:197). According to the historian Živko Andrijašević, Doclea was attacked during the rule of Prince Vladimir by the Macedonian Emperor Samuil in 997, and the first conflict took place near the river Bojana. In that conflict, the Doclean Prince Vladimir was captured and brought into captivity in Prespa. Vladimir spent some time in captivity, after which, thanks to love and marriage with Samuil's daughter Kosara, he was released and returned to Doclea as governor (Andrijašević 2017). According to the historian Sreten Perović, who used "*The Chronicle of the Priest of Doclea*", Prince Vladimir was young, beautiful, spiritual, and educated. During his captivity in Prespa, Kosara, daughter of Emperor Samuil, often visited him. She eventually fell in love and married him with her father's permission. Kosara returned with him to Doclea and was his loyal wife until his death in 1016 (Perović 2016). Marriage approval was not a sentimental issue according to historical sources, but the Macedonian Emperor Samuil approved the marriage for state reasons since Macedonia, like Doclea, was under the constant threat of war from Byzantium, i.e. the Eastern Roman Empire. Therefore, Samuil put his son-in-law Vladimir again on the throne in Doclea, giving him the entire area of Durres in today's Albania (Perović 2016:20).

Other sources, such as "*The Slavic Life*", point to the fact that the love between Prince Vladimir and Samuil's daughter Kosara represents the embodiment of true lovež. "*Samuil had a daughter named Kosara; she had great mercy for the poor and the slaves, and she often visited dungeons and comforted the slaves. Blossomed with youth, kind and prudent, she saw Vladimir and fell in love with him. Kosara then asked her father to give her his slave for her husband. The father could not refuse his daughter; so he got Vladimir out of the dungeon, gave him Kosara as his wife, returned the whole state to him, and sent him away with honour and many gifts*" (Slovensko žitje 2016:58).

Vladimir showed loyalty to Emperor Samuil until he died in 1014. The same kind of loyalty Vladimir showed to Samuil's successor, the Macedonian Emperor Radomir (1014–1015). However, Radomir's successor Vladislav decided to kill the Doclean Prince Vladimir for fear that Vladimir, as Samuil's son-in-law, had political pretensions towards Macedonian areas. He did so by inviting him to Prespa. As Vladimir did not trust the Macedonian Emperor Vladislav, he asked him to send a wooden cross as a symbol of security and guarantee. When he received the cross, Vladimir headed towards Prespa. Despite the guarantee he had sent, Emperor Vladislav broke his promise and killed Vladimir in 1016 (Andrijašević 2017:149). The key reasons for his canonization given in the literature are the following: after his death "*many miracles began to happen on his grave, such as the healing of the sick, the appearance of mystical light,* etc.*, which attracted the attention of many pilgrims*" (Vujičić 2017:38), the martyr's death by which he redeemed his deep Love and deep Belief (Perović 2016:20), but also the tragic end of earthly life that also marked the beginning of his saintly cult (Jovović 2017). His canonization is the product of the sacrifice he made for his people, which is confirmed by the following sentences he uttered during his surrender to Emperor Samuil: "*It seems to me, dear brothers, that it is time for me to fulfil the words from the Gospel: The good shepherd lays down his life for the sheep. It is better, I think, brothers, that I lay down my soul for all of you, and gladly surrender my body to be killed, than that you perish from hunger or sword*" (Velimirović 2002:55–56). In the elements of sacrifice, faithful devotion, and obeying the canons of ethics, which have followed the life of Prince Jovan Vladimir, we find sociological varieties that relate to patterns of the culture of remembrance (more in Kuljić 2006), as well as Christian memory and doctrine, i.e. its "sacred history and eschatology", i.e. "the persons of the Holy Trinity", one of which is interpreted as "the sacrificial character of the Son (kenosis)" (more in Dugin 2011:16–17; Živković 2018:42).

The cult of St. Vladimir soon spread beyond Doclea. "*He was recognized as blessed in Dalmatia and Albania, and in Macedonia he was recognized as a miracle worker, while he is mentioned in frescoes and icons in Bulgaria as the Bulgarian Emperor. It is believed that he was never officially canonized*" (Vujičić 2017:38).

Thus, the focus of identity interpretation shifts from the personal to the collective level, i.e. to the socio-cultural identity. In this way "*personal and collective identity, an individual and culture are brought into connection, and the indispensable process that accompanies this connection is the interiorization of cultural attributes*" (Lučić 2012:31).

14.3 The Cultural and Symbolic Plane of the Mountain Rumija

In terms of finding the origin of spatial identity and its inherent culture, we refer to *the culture of remembrance* that constantly reminds us of the significance of the space that is the subject of our analysis. Collective memory is an important process for the survival of the culture of remembrance because the absence of collective memory means oblivion (Kuljić 2006). On the path to preserving the territorial recognition in the time of powerful social fluctuations to which the area of Rumija and its surroundings was exposed in the past, the memory through formal or informal manifestations, customs, and monuments represents an important segment for maintaining the continuity within the identity discourse. The identity discourse of the mountain Rumija and its surroundings is reflected in the ritual of traditional cultural pilgrimage to the mountain, where the *cross* of Prince Vladimir (which for centuries has been the foundation of multi-ethnic and multi-religious harmony on which the identity of Montenegro is based) plays the significant role and Prečista Krajinska as a symbol of the beginning of the cultural and spiritual identity of Montenegro.

14.3.1 The Mountain Rumija as a Place of Pilgrimage

Each society has its holy places and individuals that cannot be freed from beliefs that have religious content or religious connotation. One of the religious manifestations, which takes place on the mountain Rumija, and is at the centre of our consideration, refers to the pilgrimage—going to a holy place. The sociologist Emil Durkheim (1982) unites the sacred and the secular in his definition of religion, and via the process of pilgrimage, which takes place every year on the mountain Rumija, these two moments (sacred and secular) come to the fore, so we will use his definition of religion as relevant to this context. Namely, there was a church on the top of the mountain dedicated to the Holy Trinity, built by Prince Jovan Vladimir, which made this mountain a place of pilgrimage for the faithful and thus gained the glory of a holy place. According to some historians, the temple was built in the Byzantine era, although historical sources date back to the thirteenth century, or even earlier (Markuš 2018).

The fact that a pilgrimage is organized every year on the day of the Holy Trinity when the cross of St. Jovan Vladimir is carried, shows that it represents a holy place for the inhabitants of the region of Bar (Bar is a town in Montenegro located at the foot of the mountain Rumija). It is not only an Orthodox rite, but members of other confessions in Montenegro, Roman Catholics, and Muslims also participate in the pilgrimage (Šistek 2017). All this supports the statement that the mountain Rumija has earned the epithet of one of the identity symbols of this region.

From a historical perspective, it is important to emphasize that the relics of St. Vladimir rested in the church of Bogorodica Krajinska until 1215, but it is believed that after the fall of the Doclean state under the domination of Raška at the end of the twelfth century Vladimir's relics and all other spiritual strongholds of Montenegrin ancestors Docleans were unprotected. Therefore, after the creation of the Despotate of Epirus in 1215 on the territory of today's Albania, the Epirians brought the relics of St. Vladimir to Durres. However, *the cross* he held in his hands when he was executed, after leaving the church of Achilles in Prespa, remained to be preserved in Prečista Krajinska (Radojević 1997:32). After the conquest of Shkodra and Krajina in 1479, Prečista Krajinska fell into the hands of the Turks, and the cross of St. Jovan Vladimir as a significant relic took the priest from Kom Monastery on Lake Skadar and before his death (fifteenth century) handed it to his daughter who got married in the village Mikulići near Bar, on the other side of the mountain Rumija. Thus, from the fifteenth century, the pilgrimage of carrying the cross started from the southern slopes of Rumija (village Mikulići) instead of Prečista Krajinska Monastery (Radojević 1997:32–33). By carrying the cross to the mountain Rumija in a pilgrimage procession, the believer pays homage to the holy place, which is also one of the oldest traditional religious rites in Christianity. Holy places are of great importance for believers since they have not only religious but also historical and cultural values (Tartalo 2010). Thus, the pilgrimage is a path where the sacred place and time, the pursuit of one's own catharsis, and the integration of believers are connected, regardless of their religious or ethnic affiliation, as well as identification with a sacred symbol. In the context of explaining the pilgrimage to Rumija, the two elements: sacred space and sacred time are identified. "*The concept of a sacred space implies the idea of repeating the original hierophany, (since this space was sanctified during the rule of Prince Vladimir and the sanctity of the same space has lasted in the collective memory of Montenegrin citizens until today), giving it a special meaning, i.e. hierophany separated it from the surrounding secular space*" (Elijade 2011:432).

Why is a mountain defined as a sacred space? Not by accident. If we go back to the period of ancient Greece, we can clearly see that the mountain had a strong symbolic feature. Thus, Mount Olympus had a special semantic and symbolic role in terms of establishing the connection between Heaven and Earth, and people who would "conquer" its peak felt closer to God. This is why Mircea Eliade uses the metaphorical determinant "sacred place" for the mountain (Elijade 2011). If we once again refresh the memory of the meaning of the very peak of the mountain as a sacred place, we will see that in this specific case the peak of Rumija has preserved the tradition and the original initiation of the sacred place, which is related to the discoveries of the Sacred Cross and pilgrimage to the Sacred Land that have occurred since the fourth century (more in Grčić 2010:159). The very path to the centre of the world, or the sacred mountain, represents an ascent and attempt to make a human fictitiously rise above the things that disturb the social and individual harmony, and to experience the spiritual catharsis on that path as well as to approach his/her ideal form, i.e. God by reaching the peak. The cross dedicated to *the name of this great ruler and originator of state-building of Montenegrin people in the Vojislavljević dynasty* is carried to the highest peak of Rumija every year, for centuries, on Holy Trinity Day (22nd May according to the old calendar, and 4th June according to the new calendar) in honour of preserving the tradition, character, and work of Prince Jovan Vladimir, who has left the mark on Montenegrin identity discourse by his existence. The cross of St. Vladimir in Montenegro was declared a national cultural monument in 1953 (Šistek 2017). Along with the relic of the cross, a social practice that provides collective memory of the mountain Rumija is a ritual of religious pilgrimage with the aim of restoring the spiritual centre of the Church of the Holy Trinity on Rumija, which was destroyed in the fourteenth century after the Turkish invasion, and built by the Doclean Prince Vladimir. Every year, people of different religious confessions carried votive stones during the religious pilgrimage believing that the church would be rebuilt when enough stones were collected. Here, we will give a brief description of the pilgrimage "*And here they are all on top of Rumija before the sun, where all three faiths—Orthodox, Catholics, and Muslims light votive candles, and then divide: Orthodox and Catholics go with a cross around a pile of stones three times, and Muslims stand piously. The church rite is performed, the end of which is announced by shooting with rifles—celebration*" (Vučinić 1889:240–242). After the rite of carrying the cross on the top of Rumija, "*the gathered people went down to the nearby plateau, where the feast that united different religious confessions began. The celebration ended with a prayer in the Church of St. Nikola, and those who did not have the strength to climb to the very top of Rumija also participated in it*" (Radojević 1997:35).

By analysing the symbolic determinant of the pilgrimage on the mountain Rumija, we single out three important segments, the first is reflected in the restoration of the common good of the community through carrying the stones in order to restore the church from the past. The second important symbolically oriented element of the pilgrimage organized on the mountain Rumija concerns the carrying of the cross of the Holy Prince Jovan Vladimir, which unites the sacred and the secular. The third segment is reflected in the continuous duration of the pilgrimage to the mountain Rumija, immediately after the death of Prince Vladimir in 1016, which unites people of different confessions and nationalities, and which indicates the perspective of the development of Montenegrin society through coexistence and multi-confessional integrity in the modern identity concept.

This multi-confessional harmony is confirmed by many towns in Montenegro but also by the town of Bar (the seat of the Archdiocese of Doclea; it was settled by Slavs in the seventh century, but until the middle of the eleventh century remained under the rule of Byzantium in the region Dirahion (Prevalis); it was the seat of the diocese until the middle of the eighth century meaning that it was a very important spiritual, cultural, and political centre) (Bar Istorija 2022), positioned at the foot of Rumija since it represents a harmonious example of monotheistic religions, especially Christianity (within it Orthodoxy with a prestigious and magnificent temple dedicated to St. Prince Jovan Vladimir and Catholicism with the historical matrix of the Archdiocese of Bar), but also Islam, which is represented by the most famous Omerbašić mosque built at the end of the seventeenth and the beginning of the eighteenth century (Kasalica 2008). Three different religious confessions have existed for centuries and continue the tradition of harmonious life at the foot of Rumija both in its monumentalism and prestigious architectural and artistic edifice, and in the political and state format that is recognized in multiculturalism as an essential feature of Montenegrin identity. The common, collective value of the three religious communities is certainly piety towards the cult of St. Jovan Vladimir, who integrated and pacified with his sacrifice each hint that points to confessional diversity. We will prove this in the continuation of the analysis through Christian doctrine and teaching about kenosis for which the act of sacrifice of St. Jovan Vladimir, as well as the very peak of Rumija, has sacred and esoteric significance inherent to those societies that rely on the significance of traditional and esoteric narratives, because: "*Tradition …, for it, is an eternal form of supernatural knowledge, a superior form of being that originates from an absolute being. These principles are not learned through rationalist epistemology but through exercise, ontology, spiritual discipline, the asceticism of a knight, self-transcendence*" (Đurković 2019:242; Živković and Bakrač 2020). If we start a more detailed search and listen to esoteric accents, we should note that Rumija has an esoteric significance that is reflected in the "*sacred acquaintance of an individual with divine forces*" (Stojanović 1997;

Biedermann 2004; Elijade 2011; Evola 2020). The town of Bar, located at the foot of Rumija, stands out in esoteric terms. The esoteric tradition of the town of Bar is identified with the olive tree, which, if we recall Hans Biedermann's *Dictionary of Symbolism*, radiates with polyvalence of meaning. "*Images of the Gods were carved from its wood; the sacred grove in Olympia consisted of olive trees, and at tournaments olive branches were presented to the winners ... In the Bible the dove that Noah sends out from the ark after the flood returns with an olive branch in its beak as a sign of the restoration of peace with God; 'pouring oil on troubled waters' is not merely a figure of speech, and (olive) oil also softens, cleanses, serves as fuel and food. It is also used for the anointing of kings, priests, and the sick ... In the Christian tradition pure olive oil is mixed with balsam and aromatic spices to make the 'chrism' used in baptism, confirmation, holy orders, and the anointing of the sick*" (Biedermann 2004:227).

Epistemological results suggest the symbolic value of those toponyms that classify Rumija and its surroundings as cultural and identity impressive mountain massifs that are considered close to European esoteric heritage due to their authenticity (whether it is mythological ornamentation or literary and historiographical) (Cf. Stojanović 1997; Simić 2019; Živković 2021).

14.3.2 Prečista Krajinska as a Symbol of the Beginning of the Cultural and Spiritual Identity of Montenegro

In the monastery Prečista Krajinska Prince Vladimir was buried until 1215 when the despot of Epirus Mihailo conquered Krajina and ordered the transfer of Saint Vladimir's relics to Durres (present-day Albania). By transferring his relics to Durres, this Doclean saint was proclaimed the heavenly protector of the town of Durres (Perović 2016:21).

An analytical approach to this topic would not have the dimension of expediency if we excluded the centre of the original Montenegrin spirituality from the analysis, namely the Church of the Assumption of the Virgin or the Church of St. Mary in Krajina. There are various historical sources about its origin. According to some statements (Gagović 2012), the church dedicated to the Assumption of the Virgin or the Church of St. Mary was built by Petrislav Vojislavljević (father of Jovan Vladimir) as part of the monastery complex Prečista Krajinska in Krajina, as a family endowment in the middle of the tenth century, where he was buried. The church dedicated to the Assumption of the Virgin was a triconch building with a slightly deeper aisle of the altar, and a narthex on the west side. It was built of cut rubble, which is characteristic of coastal construction. Apparently, the older Paleo-Byzantine triconch on Topolica served as a model for this building; the triconch was damaged by Saracen robberies at the time, but the event, related to the Diocletianic Persecution and the persecution of the last Archbishop of Bar by Emperor Samuel in 926 or 927, could decide its fate. (Gagović 2012:3). When Prince Jovan Vladimir came to power, the dome of the church was built as well as a rectangular building—residence, thus the church was given a new look and it became the oldest monastery centre on Lake Skadar (Gagović 2012). According to other historical sources (Lukić 1975; Ukgjini 2017), the church was built by Prince Vladimir. The body of Prince Jovan Vladimir rested in this monastery complex. His wife Kosara transferred the body from Prespa to the church in Krajina after his murder. To confirm her immense love for Prince Jovan Vladimir and to symbolically indicate that death is not the end of their love, but that they exist as one being in the transcendental sense, Kosara became a nun, took the name Theodora, and established a convent *Collegium scarum Virginium* at Krajina Monastery on the slopes of the mountain Kozjak. Two years after Vladimir's death, Kosara died and she was also buried in St. Mary's Church (Gagović 2012). From this monastery, "*the students of Kliment Ohridski spread literacy in this narrow region*" (Lukić 1975:166).

The endowment of Prince Vladimir and his wife Kosara has become "*a hotbed of culture and later the seat of an unrestrained and violent element carried by the aspiration towards national liberation. According to many studies, the Krajina monastery was ready for action, believing that constant rebellion and mass movements would exact the obedience of the hesitant and the fearful in the interest of state independence*" (Lukić 1975:163). This monastery complex is the largest church building in the area of the former Doclea.

The beginnings of literary works from the middle of the ninth century are related to the Prečista Krajinska monastery since the representative works "*Life of Prince Vladimir of Zeta*" by an anonymous author from Krajina and "*The Kingdom of the Slavs*" i.e. "*Regnum Sclavorum*" by the Doclean Pope were created there. These are also the first written traces of the history of Montenegro (Prečista Krajinska 1985).

The remains of this monastery have been preserved in Ostros (near Bar) to this day. Due to the negligence of the relevant state institutions, this Sanctuary of Loyalty and Love is not conserved and protected, and therefore it is inaccessible to travellers and visitors.

14.4 Conclusion

In the paper, we pointed to the significance and meaning of the mountain Rumija and its surroundings for Montenegrin identity, taking into account the various aspects of its history, mythology, and culture that have been revived for centuries. All aspects mentioned in the paper have great importance for preserving the identity characteristics of Montenegrin society and confirm that this area is a source of Montenegrin state-building and spirituality. It is also a source of a high level of development of multi-ethnic and multi-confessional harmony that is maintained to this day through the practice of religious pilgrimage to the "holy peak" of the mountain Rumija, in which people of different ethnic and religious affiliations participate, in honour of the cross on which Prince Vladimir was killed. Therefore, his suffering in the name of protecting his people, whose symbol is a preserved *cross*, reconciles the differences that abound in today's Montenegrin society. An indispensable value aspect of this space is the monastery complex Prečista Krajinska located below the mountain range of Rumija. The significance of the monastery complex was especially emphasized during the life of Prince Vladimir and Princess Kosara, which was reflected in the multiple historical and cultural contexts and confirmed by the facts: that it was the endowment of Prince Vladimir (the first prince who was declared a saint), that Prince Vladimir's wife was buried in this monastery, and that already at the turn of the ninth and tenth centuries, and further on, it was a hotbed of culture and literacy in the area of Doclea, the precursor of Montenegrin state-building.

Although this complex should have a special significance for the Montenegrin society and culture not only from the historical aspect but also in terms of its preservation, today it is located in a very dilapidated edifice and it is not possible to assess much of the significance it has had in the past based on its current appearance.

References

Andrijašević Ž (2017) Crnogorsko nasljeđe ranog srednjeg vijeka (država, društvo, kultura) [Montenegrin heritage of the early Middle Age (state, society, culture)]. In: Borozan Đ (eds) 1000 godišnjica svetog Jovana Vladimira. CANU, Podgorica, pp 147–160

Bar Istorija (Bar history) (2022). https://me.visit-montenegro.com/main-cities/bar/bar-history/. Assessed 10 Jun 2022

Biedermann H (2004) Rečnik simbola [Dictionary of simbolism]. Plato, Beograd

Dugin A (2011) Metafizika blage vesti [Metaphysics of the good news]. Logos, Beograd

Đurković M (2019) Desna misao u XX veku [The right thought in the 20th century]. IK Zorana Stojanovića, Sremski Karlovci, Novi Sad

Elijade M (2011) Rasprava o istoriji religija [Discussion on the history of religions]. Akademska knjiga, Novi Sad

Evola J (2020) Razmišljanje o vrhovima: planiranje kao metafora za duhovnu potragu [Meditations on the peaks: mountain climbing as metaphor for the spiritual quest]. Ukronija, Beograd

Gagović Z (2012) Manastir Prečista Krajinska [Monastery Prečista Krajinska]. Cetinje. https://www.academia.edu/4891198/Zdravko_Gagovic_MANASTIR_PRE%C4%8CISTA_KRAJINSKA. Accessed 11 Jun 2021

Grčić M (2010) Geografija religija [Geography of religions]. Geografski fakultet Univerziteta u Beogradu, Beograd

Hall S (1996) Who needs 'identity'? In: Hall S, du Gay P (eds) Questions of cultural identity. Sage, London, pp 1–17

Jovović I (2017) Sveti knez Vladimir Dukljanski—milenijumska ličnost Balkana [Saint Prince Vladimir Dioclean—the Millenium personality on the Balkans]. In: Borozan Đ (eds) 1000 godišnjica svetog Jovana Vladimira. CANU, Podgorica, pp 195–206

Kalanj R (2003) Zov identiteta kao prijeporno znanstveno pitanje [The call of identity as a disputable scientific issue]. Socijalna Ekologija 12(1–2):47–68

Kasalica V (2008) Kulturna baština Crne Gore [Cultural heritage of Montenegro]. Kulturni centar, Bar

Kuljić T (2006) Kultura sećanja [Culture of memory]. Čigoja štampa, Beograd

Lukić V (1975) Prečista Krajinska i njen osnivač Jovan Vladimir u svjetolosti istorijskih činjenica [Prečista Krajinska and its founder Jovan Vladimir in the light of historical facts]. Starine Crne Gore: Godišnjak Republičkog Zavoda Za Zaštitu Spomenika Kulture Crne Gore 5:161–179

Lučić M (2012) Neposrednost identitetskih spekulacija [Immediacy of identity speculations]. Sociološka Luča 6(1):31–38

Markuš: Brojni su dokazi da je crkva na Rumiji postojala (2018) Barski Portal, 27 Oct 2018. http://www.barskiportal.com/lat/dru%C5%A1tvo/marku%C5%A1-brojni-su-dokazi-da-je-crkva-na-rumiji-postojala.php. Accessed 13 Jun 2021

Perović S (2016) Dukljanski knez Vladimir i makedonska princeza Kosara u legendi i istoriji [The Doclean Prince Vladimir and the Macedonian Princess Kosara in legend and history]. In: Perović S (ed) Dukljanski knez Sveti Vladimir 970–1016: 1000 godina crnogorske državotvornosti. Fondacija "Sveti Petar Cetinjski", Cetinje

Prečista Krajinska (1985). https://archive.org/details/precista-krajinska. Assessed 10 Jun 2022

Radojević M (1997) Rumija i kult Sv. Vladimira [Rumija and the cult of St. Vladimir]. In: Crnogorske planine: putopisi i zapisi. Obod, Unirex, Podgorica, Cetinje

Simić Ž (2019) Satanski tango u prokletoj avliji: ka sociologiji saznanja srpske kulturne tradicije [Satanic tango in the damned yard: to the sociology of knowledge of the Serbian cultural tradition]. Prometej, Novi Sad

Slovensko žitije [The Slavic Life] (2016). In: Čitanka stvaranja o Sv. Jovanu Vladimiru. Obod, Cetinje

Stojanović T (1997) Balkanski svetovi: prva i poslednja Evropa [Balkan worlds: the first and last Europe]. Equilibrium, Beograd

Šistek F (2017) Čija je Rumija? [To whom does Rumija belong]. Matica 18(70):59–100

Tartalo P (2010) Bliže Bogu: poklonička putovanja na sveta mesta hrišćanstva [Closer to god: pilgrimages to the holy places of Christianity]. Pravoslavna reč, Novi Sad

Ukgjini N (2017) Barska nadbiskupija i Manastir Svetog Jovana Vladimira u Krajni [The Archdiocese of Antivari and the Monasteries of St. Vladimir in Kraje]. In: Borozan Đ (eds) 1000 godišnjica svetog Jovana Vladimira. CANU, Podgorica, pp 171–193

Velimirović N (2002) Sveti kralj Jovan Vladimir: čitanka [Holy King Jovan Vladimir: a reading book]. Srpska književna zadruga, Hemijsko-prehrambeno-tehnološka škola, Beograd

Vučinić NJ (1889) Zavjet mrkojevićki: izlet na Rumiju [The Mrkojevic vow: trip to Rumija]. Nova Zeta 1(6):240–242

Vujičić R (2017) Sveti Vladimir Dukljanski [Saint Vladimir Doclean]. In: Perović S (ed) Dukljanski knez Sveti Vladimir 970–1016: 1000 godina crnogorske državotvornosti. Fondacija "Sveti Petar Cetinjski", Cetinje

Živković P (2018) Mit o zapadnoj imepriji [The myth of the western empire]. Sociološka Luča 12(1):39–47

Živković P (2021) Drvo divlje kruške ili drvo pobožnih strasti [A wild pear tree or a tree of pious passions]. In: Đukić N (ed) Film i identitet: Istraživanje društveno-političke upotrebe filma. Univerzitet u Banja Luci, Fakultet političkih nauka, Banja Luka

Živković P, Bakrač V (2020) Konzervativizam kao prolegomena razvoja duha multipolarnost [Conservatism as a prolegomenon of the development of the spirit of multipolarity]. Sociološka Luča 14(1):30–39

Slovenian Mountain Attachment as a Hallmark of a Nation

15

Matej Ogrin and Peter Mikša

Abstract

Like many other places in the Alps, Slovenian mountains were inhabited as early as in prehistoric times. Since then, mountains have been not only an environment for hunting and gathering, but also a refuge and in some areas sacred places. Over the centuries, Slovenes have developed a unique relationship with the mountains, especially since the nineteenth century, and mountains have become an integral part of the national identity. Slovenia's highest mountain, Triglav, has been adopted as the most important national symbol. Mountains have thus acquired a unique meaning in the consciousness of every Slovene. Mountains feature importantly in Slovenian literature and in culture more generally, in winter and summer sports, mountaineering, leisure, nature protection, and historical tradition. This paper discusses the importance of the mountain tradition for the Slovenian nation. It turns out that the perception of the mountains is very intense and positive, even though the territory of Slovenia does not particularly stand out in terms of high elevations, and the vast majority of the population lives in the lowlands.

Keywords

Mountain geography · History of mountains · Identity · Slovenian Alps

15.1 Introduction

Today, the image of mountains in Slovenia is inextricably linked with representations of the idea of Slovenian identity. Triglav, the highest summit in the territory of Slovenia, is depicted on two official symbols of the Republic of Slovenia: on the official national coat of arms and thus also on the flag (see Chap. 19 in this volume). The mountain motif can also be found on the first Slovenian stamp from 1919, where it is background to the "chainbreaker" figure symbolically breaking the shackles of the Austro-Hungarian "prison of nations" and embarking on the path of freedom and a bright future. In 1934, Triglav was placed in the coat of arms by Slovenia's most famous architect, Jože Plečnik (Mikša 2021) as a symbol of the Slovenes. During the Second World War, many Slovene members of the armed units of the partisan resistance movement wore hats in the stylized form of Triglav and its three peaks. In the process of "socialist" nation-building

M. Ogrin (✉)
Department of Geography, Faculty of Arts, University of Ljubljana, Ljubljana, Slovenia
e-mail: matej.ogrin@ff.uni-lj.si

P. Mikša
Department of History, Faculty of Arts, University of Ljubljana, Ljubljana, Slovenia
e-mail: peter.miksa@ff.uni-lj.si

M. Zorn et al. (eds.), *Discourses on Mountains of Montenegro and Slovenia*, Historical Geography and Geosciences, https://doi.org/10.1007/978-3-031-99017-5_15

during the second half of the twentieth century, Slovene collective and individual identifications, as well as ideological and cultural categorizations of Slovenian identity, became irrevocably associated with representations of the mountain world and with Alpine and mountaineering leisure activities. In 1945, Triglav was given a central place in the Slovenian coat of arms, thus indirectly becoming one of symbols of Yugoslavia. Winter sports and other mountain-related sports acquired the character of Slovenian national sports, and the achievements of Slovene competitors became a source of national pride: many Slovene public enterprises donated funds for mountaineering expeditions to the Himalayas (Mikša 2022); when the first major successes of Slovenian skiers in Alpine skiing began in the 1980s, production lines and classes in schools were halted as workers and schoolchildren crowded in front of televisions in factories and schools. Mountain tourism flourished, while in other Yugoslav republics the Slovenes were often categorized, often pejoratively, as a "nation of skiers." Nationalized mountain iconography found its place in cultural and mass production: for example, in Oberkrainer music (the Avsenik Brothers Ensemble) and film (the Slovenian version of *Bergfilme,* for example the *Kekec* films). After the disintegration of the multinational state and the establishment of the Slovene nation-state in 1991, the appropriation of the mountain world as a fundamental building block of Slovenian national identity became self-evident and ubiquitous in Slovenian high and mass culture as well as in the public space. With the introduction of the euro, Triglav also found its place on the national side of the 50-cent coin, and in this way continues, similarly as in the socialist era, to symbolize the (mountain) identity of Slovenes in a multinational European connection (Mikša and Zorn 2021).

Paradoxically, however, the symbolic connection of the Slovene identity to the Alpine world cannot really be explained by statistical data and geographical indicators. The high mountain topography of Slovenia can hardly be compared with that of Austria or Switzerland. In fact, Slovenia's territory encompasses relatively little area that could be characterized as a high mountain environment and thus there are also very few truly Alpine inhabitants. So why is the image of "Sloveneness," of Slovene national identity and also actual patterns of national identification, so centered around mountains and the Alpine world, when this cannot be justified by objective geographical arguments? Why are mountains depicted in every tourist brochure and why do Slovenes like to talk at family lunches or Sunday picnics about how every Slovene must climb the highest Slovenian mountain, Triglav, at least once in their lifetime?

15.2 The Presence of Humans in Slovenian Mountains Dates Back to Prehistoric Times

Archaeological excavations in recent decades (Cevc 1996, 1998, 2006a, 2006b; Bremšak 2000; Horvat 2002; Bizjak 2004; Gleirscher 2006) have provided evidence for the presence of humans in Slovenian mountains since prehistoric times. It is a mistake to think that in prehistory people did not venture into the mountains. They were driven there by both curiosity and necessity. However, there are far fewer traces of human activity in the mountains from these times, as there are no written records from those periods, and many material traces have been erased by time.

A prehistoric visitor to the mountains was certainly a hunter, and later also a miner and a shepherd—most likely in this order. But humans also sought a connection with the spiritual world by going into the mountains (Lovšin 1944).

So far, the oldest traces of human activity in the Slovenian mountains have been dated to the Lower Paleolithic (100,000–500,000 years BC), and specific finds under Mt. Pršivec are believed to be older than 200,000 years. Finds on Kriška mountain pasture below Mt. Krvavec in the Kamnik-Savinja Alps also belong to the same period, but the Paleolithic presence of humans in the Slovenian mountains has not yet been confirmed (Stele 2007). There are many more traces of human activity from the period 30,000–40,000 years BC. The most important finds from this period are located in Potočka zijalka Cave and Mokrica Cave as well as on some mountain pastures in the Bohinj area (V Lazu, Ovčarija, Vodični Vrh, Ogradi) (Odar 2006; Stele 2007). Slovenian mountains were also visited by humans in the Metal Ages: Studies have confirmed the presence of humans from the Bronze Age on (Ogrin 2006; Stele 2007).

In addition to mining, grazing was also an important motive for going into the mountains (Fig. 15.1). Alpine inhabitants were driven to the mountains due to the lack of grazing areas in the valley. In the vicinity of Vinschgau in the western part of South Tyrol (Italy), some traces are interpreted as evidence of grazing activity as early as the nineth and tenth centuries BC, and in the Dachstein mountains (Austria) even as far back as in the fourteenth century BC (Cevc 2006b). We have not yet found material evidence of prehistoric grazing in the Slovenian Alps, but it can be inferred from intangible heritage. Thus, there is a preserved description of the expedition to thc mountain pasture Za Liscem. Based on folklore known to date, true Alpine farming in Slovenian mountains took place from the eighth to the nineth century onward, when feudal relations also came into being (Cevc 2006b).

Along with mining, hunting, and grazing, the mountains were also a place of spirituality. Humans sought the nearness of God in the mountains. From the summits there was not only a clear view of the landscape below, but a person was also "closest" to God. Hunting, mining, and grazing in the mountains were often outside their reach, and so they often needed "help from above." Many sacred places, such as outdoor altars, have been discovered in the mountains, and this is also evidenced by the names of some peaks, referred to as holy mountains (Cevc 2006a). Even today, in some parts of the Alps,

Fig. 15.1 A view from Zajamniki Alp. Bohinj mountain pastures are today livestock grazing areas and tourist destinations. Archaeological finds indicate that this area was used by miners before the arrival of herders. (Photo by Aleš Zdešar, authors archive)

there is a tradition of placing a cross on the summits, an indication of the modern connection of mountains with the divine.

15.3 The First Explorers and the First Documented Approaches in Slovenian Mountains

Among the first to describe and write down for the first time the earliest known approaches to the lower-lying Slovenian mountains were Johann Weikhard von Valvasor, in his massive work *The Glory of the Duchy of Carniola*, published in 1689. In writing about mountains Valvasor (2017) divides them into *snežniki*—the highest mountains capped by snow all year round—and *gore* (mountains). As a remnant of times past and as evidence of viewing mountains as the abodes of gods, demons, and other mythical creatures, Valvasor mentions an interesting superstition according to which a storm could easily be conjured at the top of the high mountain Krma (Orožen 1903). In the eighteenth century the Slovenian Alps began attracting great interest from some naturalists, particularly foreign ones, this was also the century, when Triglav was depicted on the map for the first time (Fig. 15.2). South Tyrolean Giovanni Antonio Scopoli and Breton Balthasar Hacquet were serving in Idrija at the time. Both were doctors for mercury mine workers and both were very interested in botany, which was their reason for going to the mountains. Between 1755 and 1766, Scopoli traveled through a large part of Carniola and was the first person documented to set foot on Mt. Storžič (2,132 m) in 1758, and in 1759 on Mt. Grintovec (2,558 m) in the Kamnik-Savinja Alps (Mikša 2013).

Fig. 15.2 In 1744, Triglav first appears on the map of *Ducatus Carnioliae Tabula Chorographica* (Chorographic Map of the Duchy of Carniola) by Janez Dizma Florjančič de Grienfeld, stating that it is the highest summit in this area (Gašperič et al. 2020). (GIAM ZRC SAZU archive)

At the end of the eighteenth century, Hacquet focused on Triglav. He successfully climbed it in 1779 and 1782 (Mikša and Zorn 2016). Along with other foreign explorers of Slovenian mountains (Franz Xaver von Wulfen, Henrik Freyer), representatives of the local nobility such as Žiga Zois and his brother Karel, Count Franc Jožef Hanibal Hohenwart, and Count Rihard Ursini Blagaj should also be mentioned (Mikša 2013). The above-mentioned educated people and representatives of the nobility of Carniola (*Kranjska*), the first explorers of the Slovenian mountains, were mostly interested in natural science for scientific reasons (studying geology, minerals, botany, etc.).

At that time, the mountains were also of increasing interest to Slovenian priests, for whom an important motive was romantic admiration for the mountains. One of these was Valentin Vodnik, who is well known from literary history, and it is also worth mentioning the brothers Jakob and Ivan Dežman, chaplains in Bohinj, and Valentin Stanič. Stanič and Vodnik described their ascents in detail and wrote poems about the mountains (Lovšin 1944). Local mountain guides can be included in a special category of visitors to the Slovenian mountains in this period: without them, the intellectuals mentioned would not have risked a dangerous mountain trek. The reasons for their ascents were not scientific or romantic, but, like Alpine dairy farmers, miners, and hunters, their interest in the high mountain environment was economic (Zorn 2005).

15.4 Mountains as a Place of National Awakening and a Center of Struggle Against the Germans

The first mountaineering club to be established was the Alpine Club, based in London and founded in 1857 by the British. The first mountaineering club in the Alps was Austrian, founded in 1862 (Mikša and Ajlec 2015). The first attempt to establish a Slovenian mountaineering club was made in 1872, but the application was rejected and the idea was put on hold for the next twenty years (Lovšin 1944). In 1874, the Carniolan Section of the German-Austrian Alpine Club (DÖAV) (Mikša and Ajlec 2022) was founded in Ljubljana, and thus the Germans were the first in Slovenia to have their own mountaineering club. The Germans, who also had a strong economic influence in the Slovenian lands, were able to quickly obtain sufficient funding for marking trails, erecting German signs (Fig. 15.3) and inscriptions, and building huts so that visitors to the mountains were given the impression that they were on German speaking territory (Lindtner 1933). It was, of course, about marking the space as German. They were assisted in this by another German mountaineering organization—the Austrian Tourist Club. The first Slovenian mountaineering club was founded in 1893 in Ljubljana, and Slovenian mountaineers began to build, maintain, and mark mountain trails and build Slovenian huts (Mikša and Ajlec 2015). In the second year of operation, in 1894, the Slovenian Alpine Club (SPD) built the first Slovenian mountain hut, Orožen Hut on the mountain Črna Prst (Mikša and Ajlec 2015). Through building huts, using Slovenian names of peaks, creating routes and marking them with Slovenian inscriptions and signposts, especially in the vicinity of Triglav, Slovenian mountaineers wanted to prevent foreigners—Germans—from appropriating the Slovenian mountain world as much as possible. The tension in the relations between the SPD and the German-Austrian Alpine Club is also reflected by fact that there were two different summit registers on many peaks, one in Slovene and one in German (Mikša 2017).

15.5 The Battle for Triglav and the Birth of the Symbol

During the first years of the SPD's operation, Jakob Aljaž (Fig. 15.4), a parish priest in Dovje since 1889, acquired an important role through his activities in the club and the struggle to lay claim to the Slovenian mountains. For some time, Aljaž nurtured the idea of building a small tower on the top of Triglav—as a mountain shelter, but also as a defender against the ubiquitous German

Fig. 15.3 Signboards denoting Triglav-Seen-Hütte (the present-day Triglav lakes hut), Deschmannhaus (the present-day Valentin Stanič hut) and Maria Theresia-Hütte (the present-day Planika hut). (Slovenian Alpine Museum archive; Photo by Matej Ogrin)

influence (Aljaž 1922). He prepared a plan for the tower and then in 1895 he bought the "peak" of Veliki Triglav, an area of 16 square meters, from the municipalities of Dovje and Mojstrana for one florin. This was not the only important act by Aljaž in the Triglav range. In 1896, Aljaž built two more mountain huts—the Triglav hut on Kredarica, which is still the highest standing building in Slovenia, and Aljaž hut in the Vrata Valley (Mikša 2015).

This period of the German-Slovenian national struggle for the mountains and the construction of mountain huts can be described as positive in at least one respect. It is a fact that the national struggle encouraged the construction of huts that would otherwise never have been built in such numbers. As it was necessary to build mountain trails to reach the huts, the national struggle is also important for the rapid development of this type of infrastructure.

15.6 The Role of Mountains in the Formation of Slovenian National Identity

Through this national struggle, Slovenian mountains became firmly entrenched in national identity and a visible symbol of recognition and representation for many individuals, as reflected in the multitude of postcards with the motif of Slovenian mountains and the inscription "Alpine greetings" or the inscription "Greetings from Triglav"

Fig. 15.4 Jakob Aljaž (1845–1927)—initiator, owner and builder of Aljaž Tower on the top of Triglav (Lovšin 1944). (Authors archive)

with the image of Aljaž Tower. The erection of the tower in particular was the act that resonated most with the public in the following years and indicated the triumph of the Slovenes over the Germans. Through Aljaž, his purchase of the summit and the construction of Slovenian huts below Triglav, the Slovenes felt as if they were wresting their land back from German hands into their own. Triglav thus became an increasingly powerful symbol of Slovenia. After the First World War, the border between Italy and the Kingdom of Serbs, Croats and Slovenes, later renamed the Kingdom of Yugoslavia, ran through its summit, and Aljaž Tower stood inside Yugoslavia a little more than two meters from the border. Triglav became the highest mountain of the new state (Poljanec and Mikša 2017; Fig. 15.5).

Triglav was first depicted as a symbol of the Slovenes by the architect Jože Plečnik in the plan for the garden wayside shrine now in front of the Church of St. Martin in Bled. Three coats of arms of the then Yugoslav nations (Serbian, Slovene and Croatian) are depicted under the image of the double-headed royal eagle, with Plečnik using Triglav for the first time in the Slovenian coat of arms, above which he painted a six-pointed star of the Counts of Celje. During the Second World War, Triglav appeared as a symbol of the Liberation Front at the end of 1941, drawn by architects Marjan Tepina and Edo Ravnikar. In 1947, Triglav officially appeared as a national symbol in the new Slovenian constitution, when it was depicted in the coat of arms. It remained there until Slovenia's independence. In 1991, it was included as the main element in the new coat of arms of the Republic of Slovenia, and through the coat of arms also in the flag of the Republic of Slovenia (Fig. 15.6) (Mikša 2017). It is the only such example in Europe besides Slovakia where the figure of a mountain is given a place in the national flag.

Like other Alpine nations, Slovenes forged a large part of their national mythology in the mountains. But Slovenes did not stop there. They also successfully transferred mountain mythology to the Slovenian cultural space, which is instilled into every Slovene, whether through the universally appealing adventures of a young mountain hero Vandot's *Kekec* (Fig. 15.7), with their clear message of how good thinking and courage can overcome many obstacles, or perhaps the tale of Goldhorn (*Zlatorog*) with its clear message of how greed leads people and the world around them to ruin. The legend of King Matjaž was set in a mountain (Peca), and the mountain of Bogatin (*bogat* meaning rich) was named after the treasures to be found there. The first Slovenian feature film *In the Kingdom of Goldhorn* (orig. *V kraljestvu Zlatoroga*) (1931) also brings to the world the image of Slovenian mountains and the adventures of the pioneers of Slovenian mountaineering. Of all the professional journals published in Slovenia, *Planinski Vestnik*, published by the Alpine Association of Slovenia, enjoys the reputation of being the oldest still published journal in Slovenia, as it

Fig. 15.5 King Peter II on Triglav. Triglav had symbolic importance for the entire Yugoslavia, as seen in the ascent of Yugoslav King Peter II (in the middle, writing in a notebook). The fact that the King of Yugoslavia, Karađorđević, and his entourage climbed to the top of Triglav in August 1939 is also a symbol of this monumental event (Poljanec and Mikša 2017). (Authors archive)

has been coming out since 1895. In absolute terms, it is ranked the fourth oldest in Slovenian periodical publication and in 2021 the magazine received the Order of Merit for significant cultural and professional value from the President of Slovenia (Red … 2022).

If you go to any of large Slovenian bookstores, you will quickly find that a lot of books are dedicated to mountains, from fiction, in-depth literature, stories and myths, children's, family, hiking, skiing and mountaineering guides, to books dominated by aesthetic photographs of mountains at different times of the year or aerial photographs, and many more. It seems that no other activity inspires Slovenes as much as activities related to mountains. If we take a moment to look at mountaineering as a sport—even among the sports literature we do not find as many books about any sport as about mountaineering and related feats. All this is just further evidence of how, in the search for Slovenian identity within multinational Yugoslavia, winter and other sports set in the mountains have become Slovenian national sports. Since 1968, when awards started to be given for Slovenian Athlete of the Year, practitioners in these types of sports (winter sports and mountaineering) have accounted for 43.5% of all awards.

Slovenes have identified themselves through mountain iconography in music, culture, and film. This is most evident in the *Oberkrainer* musical style that originated in the Slovenian Alpine region and has become a hallmark of Alpine folklore. Here, of course, we are referring to polka and the Avsenik Brothers Ensemble (Fig. 15.7) and many other groups. Popular melodies, with *Golica* at the forefront, have become a real hallmark not only of the *Gorenjska* region (Upper Carniola) but also of Slovenes more generally. Of course, mountain-related activities also include Alpine skiing. Although originally belonging to the Nordic disciplines, ski jumping at Planica certainly belongs in the mountain skiing context among Slovenes (Fig. 15.8). In recent decades, other types of skiing sports, such as snowboarding, cross-country skiing, ski cross and the like, have

Fig. 15.6 June 25, 1991, after the Slovenian Parliament adopted the Declaration of Independence, Slovenia became an independent state. The following evening, June 26, 1991, a ceremony was held at Republic Square in Ljubljana, at which the new Slovenian flag was raised for the first time. (National Museum of Contemporary History of Slovenia archive)

become increasingly popular. And again, the successes brought by sports on snow and/or in the mountains also occupy a special place in the Slovenian media space: one of the most watched sports broadcasts in Slovenia is the annual ski jumping contest in Planica. According to TV Slovenija, in March 2016, Peter Prevc's last jump was watched by as many as 731,300 viewers, or 93% of those who were watching television in Slovenia at the time (Planiški … 2016).

Regardless of sporting success or lack thereof, the best known and attended sports event in Slovenia is the ski jumping competition. Alpine skiing competitions such as the Vitranc Cup in Kranjska Gora (Fig. 15.9) and the Golden Fox Cup on Pohorje (1964–2024), which have been on the calendar of the World Cup of the International Ski Federation for more than 50 years, also have a special and long tradition.

Of course, we cannot ignore mountaineering and the associated successes that elevated Slovenes in the 1970s to one of the most notable mountaineering nations with many world-class achievements, leading to international renown and acclaim at home. For example, the ascent by what is probably still today the most difficult new route to the world's highest summit in 1979, the ascents of alpinist Tomaž Humar, the ski descent of alpinist Davo Karničar from the top of Everest and the fact that Slovenes have received seven *Piolets d'Or*, the most prestigious international award in mountaineering (Državna … 2016).

The Slovenian Olympic Committee, the most important national sports organization, has also chosen Triglav for its logo, which is no coincidence and is clearly a recognizable symbol of the Slovenes. Slovenes have also adopted the symbol of Triglav as a national symbol in sportswear and national team jerseys in many sports, and not just in winter or mountain sports. In addition to the established colors with a combination of blue, green, and white, hockey, football, basketball, cycling, and other national teams also have the emblem of Triglav. The promotional film used in the tourism campaign from 1986 is also very telling in terms of the symbolic value of mountains for Slovenes. This was the period when Slovenia began promoting tourism and a promotional video entitled "Slovenia, my country" (subtitled "the sunny side of the Alps") was created. The tourist campaign was designed exclusively for the promotion of Slovenian tourism, but soon triggered an unexpected wider emotional response from Slovenes, supported by the aspiration for an independent Slovenia. The video was immediately nationalized and is still considered to be by far the most successful and most recognizable promotional video in Slovenia (Mikša 2021). While it shows a variety of Slovenian macroregions, it is interesting that this video begins and ends in the Slovenian Alps and devotes most of its footage to Alpine landscapes.

Fig. 15.7 Nationalized mountain iconography found its place in cultural and mass production: for example, in Oberkrainer music (the Avsenik Brothers Ensemble; upper figure) and film (the Slovenian version of *Bergfilme*, for example the *Kekec* films; lower figure). (National Museum of Contemporary History of Slovenia archive)

With the introduction of the euro in Slovenia, Triglav was also featured on euro coins and thus still today marks the (mountain) identity of Slovenes in a multinational European union (Mikša and Zorn 2021). Thirty-five years later, the mountain landscape as a typical Slovenian backdrop also served to promote the Development Strategy of Slovenia (Strategija … 2022), although the Alps do not predominate in either the share of the territory or the population of the Alps in Slovenia. The online image of Slovenia is also interesting. On the 30th anniversary of the creation of the Republic of Slovenia, 25 June 2021, a web browser published a picture of the snow-capped peak of Triglav with the Slovenian flag, which was a reflection of one of the most iconic photographs taken in June 1991, the historic days of independence. At that time, a group of Slovenian alpinists climbed Triglav and hoisted the Slovenian flag on its summit as a sign of independence, and the picture was published on the eve of independence, thus sending a clear message to the world by marking the space with a new Slovene symbol.

Mountain symbolism, especially the figure of Triglav, is truly deeply rooted in Slovenes as a national symbol. Slovenia also marked the

Fig. 15.8 Planica ski jumping tournament. (National Museum of Contemporary History of Slovenia archive)

Fig. 15.9 Vitranc Cup with the famous finish slope, where the great stories of Slovenian skiing were written. The successes in Alpine skiing brought the Slovenians into the company of other Alpine nations and on the world map of sports. (Photo by Matej Ogrin)

most acclaimed international event in Slovenia since joining the European Union in 2004, i.e., the Slovenian presidency of the Council of the European Union in 2008 and 2021, with a logo that included the image of Triglav on both occasions.

The Google search engine also most often shows motifs from Alpine landscapes when the search term Slovenia is entered in various languages. The most frequent one is the iconic image of Bled with its Alpine lake, island, and mountains in the background.

15.7 Geographical Features of Slovenia as a Mountainous Country

It is certainly the case that in the absence of mountains and elevated topography, it is not possible to develop a mountain identity. However, it is not necessarily the case that the people of mountainous regions recognize this as an important part of their identity and include it in traditions of all kinds. In point of fact, Slovenia lies at a relatively low elevation. More than half of the country lies below an elevation of 500 m, and only 11% of the land lies above 1000 m. Only 2.5% of the territory of Slovenia lies above 1500 m, and less than 0.5% at an elevation above 2000 m, which represents a true high mountain environment. According to Wikipedia's list of countries by average elevation (List … 2017), Slovenia has an average elevation of 556,8 m. The same list shows that Slovenia ranks 69th of 163 countries in terms of elevation. Interestingly, of the Alpine countries, its average altitude is lower than that of Austria, Italy, Switzerland, and Liechtenstein, but higher than that of France and Germany.

Based on the distribution of the population according to elevation, the situation in Slovenia is even less "mountainous." In 1991 (Kladnik et al. 1995) as many as 88.7% of the population lived at an elevation below 500 m and only 0.3% of the population inhabited areas above 1000 m; above 1330 m there are no permanent settlements apart from some mountain huts. Recent studies also show the small population of Alpine mountains in Slovenia. The population in this region, which occupies 15.1% of the national territory (Perko and Ciglič (2020), has hardly changed in the period 1961–2011. In 2011, 89,581 people lived in the Alpine mountains, which is 40 less than 50 years ago, and the share of the population living in this region decreased from 5.6 to 4.4% (Perko and Ciglič 2020) We can see, therefore, that both in terms of elevation distribution and the location of the population, Slovenia is not a particularly high-elevation country, nor does it have much settlement at higher elevations.

15.8 Discussion

The attachment to the mountains and the mountain landscape is the result of many activities that were described in this paper. It is also the result of perceptions, experiences, values, value system, and positive collective experiences with the mountains, which serve as a backdrop or space for experiencing collective success (alpinism, mountaineering, cross-country skiing, Alpine skiing). Over the centuries, Slovenians have lived not only in the mountains, but also with the mountains. We have formed the character of the nation in the mountains, we have used the mountains as an instrument of self-awareness, we have also used them to mark the space and for cultural struggle during the construction of identity both within the Austro-Hungarian monarchy and later in the Kingdom of Yugoslavia against its German and Italian neighbors (Šaver 2011).

Positive perceptions and experiences were also reinforced later during a more politically favorable time for Slovenes. After Second World War, positive perceptions can be associated with activities related to mountains, such as mountaineering, which became a mass manifestation of a specific and positive attitude toward mountains (Mikša and Ajlec 2022). Alpine and Nordic sports, which have become a very important symbol of Slovenianism in recent decades, are also associated with the mountains. The attachment to the mountains is also a consequence of the described cultural tradition, in which the mountain landscape, the mountain people with their special character, and the mythological beings form an important cultural environment, which is also embedded in the Slovenian symbolic space through other symbolism. The perception of mountains as a mythological space is by no means only a Slovenian specialty; this phenomenon is also found in many other mountain areas (Cigale 2017). There are far fewer nations that have adopted the symbols of mountains as national symbols, e.g., Ecuador, Kyrgyzstan, Slovakia, and also Slovenia (Mikša et al. 2017). What is striking about Slovenians is the transfer of mountain symbolism to the level of national symbolism, which is also accepted by Slovenians without prejudice, even though more than 90% of the population lives outside the Alpine mountains. Another important strategic function of mountain landscapes is related to mountain topography. Mountains are home to many indigenous people and cultures (Byers et al. 2013) and closed mountain topography often served as a refuge from invaders for many people (Mikša et al. 2017), and this was no different for the Slovenes.

15.9 Conclusion

In the formation of mountain identity, the intertwining of mountains with aspects of identity that the nation incorporates into both its symbolic and real space of self-awareness is of greater importance than the physical predominance of mountain landscapes in the territory of a nation. Mountain landscapes have offered people refuge from invaders, and knowing the mountains and living with them has also shaped a symbolic space for many nations. This is evident in the case of Slovenes as well as of other Alpine nations, as the scaling of mountains was also recognized as a national struggle and as a means of asserting themselves before other nations. This occurred initially within the Austro-Hungarian monarchy, when Slovenes were not yet recognized as a nation. Later, in the family of Yugoslav nations, they represented not only the physical but also the symbolic bond between Yugoslavia and the Alpine countries, and at the same time, an Alpine identity, which was a uniquely distinguishing feature for Slovenes within Yugoslavia. With the creation of the new state, Alpine and mountain identity only gained in significance among Slovenes, as mountain symbolism in the fields of culture, sports, and politics was further strengthened. The online image of Slovenia is also becoming increasingly important, sending images of mountain landscapes, with Bled as the most frequently depicted image, to the world as the most important photographic impressions.

References

Aljaž J (1922) Planinski spomini [Mountain memories]. Planinski Vestnik 22(3–4):33–44

Bizjak J (2004) Oskrbovalno zaledje železnodobnega in srednjeveškega Bleda v visokogorju Julijskih Alp [The supply hinterland of Iron Age and Medieval Bled in the highlands of the Julian Alps]. In: Dežman J (ed) Bled tisoč let: Blejski zbornik 2004. Didatka, Radovljica, pp 131–134

Bremšak M (2000) Življenje v planinah pred tisočletji [Life in the mountains thousands of years ago]. Planinski Vestnik 100(4):150–154

Byers AC, Price LW, Price MF (2013) An introduction to mountains. In: Price MF, Byers AC, Friend DA, Kohler T, Price LW (eds) Mountain geography: physical and human dimension. University of California Press, Berkely, pp 1–10

Cevc T (1996) Nova spoznanja o planšarstvu na Veliki planini v Kamniških Alpah [New findings on alpine dairy—farming on Velika Planina in the Kamnik Alps]. Traditiones 25:69–79

Cevc T (1998) Planina na Stanu (1450 m) pod Kamniškim sedlom (1884 m) v antiki in srednjem veku [The Alp Na Stanu (1450 m) bellow Kamniško sedlo (1884 m) in antiquity and the middle ages]. Traditiones 27(1):9–23

Cevc T (2006a) Verovanja v Alpah [Beliefs in the Alps]. In: Cevc T (ed) Človek v Alpah: desetletje (1996–2006) raziskav o navzočnosti človeka v slovenskih Alpah. Založba ZRC, Ljubljana, pp 125–137

Cevc T (2006b) Arheološki dokazi o pašništvu v alpskem visokogorju [Archaeological evidence of pastoralism in the Alpine highlands]. In: Cevc T (ed) Človek v Alpah: desetletje (1996–2006) raziskav o navzočnosti človeka v slovenskih Alpah. Založba ZRC, Ljubljana, pp 113–124

Cigale D (2017) Tradicionalni in sodobni odnosi človeka do gora na primeru Andov [Traditional and contemporary attitudes of man toward mountains: the case of Andes]. Geografski Obzornik 64(3–4):52–63

Državna odlikovanja za dosežke slovenskih alpinistov, plezalcev in planincev [State awards for the achievements of Slovenian alpinists, climbers and mountaineers]. http://www.friko.si/friko-info/drzavna-odlikovanja-za-dosezke-slovenskih-alpinistov-plezal cev-planincev. Accessed 11 Aug 2021

Gašperič P, Šolar R, Zorn M (2020) Kartografski zakladi slovenskega ozemlja = Cartographic treasures of Slovenian territory. Založba ZRC, Ljubljana. https://doi.org/10.3986/9789610504276

Gleirscher P (2006) Zum Nachweis römischer Almhütten am Dachsteinplateau und in den Steiner Alpen (Kamniške Alpe). In: Mandl (ed) Alpen: Archäologie, Almwirtschaftsgeschichte, Altwegeforschung. ANISA, Haus im Ennstall, pp 23–30

Horvat J (2002) Arheološki sledovi v slovenskem visokogorju [Archaeological evidence of pastoralism in the Alpine highlands]. Kamniški Zbornik 16:193–202

Kladnik D, Orožen Adamič M, Perko D (eds) (1995) Mali atlas Slovenije [Small atlas of Slovenia]. DZS, Ljubljana

Lindtner H (1933) Planinski piparji [Mountain society Piparji]. Planinski Vestnik 39(6–7):167–175

List of countries by average elevation. https://en.wikipedia.org/wiki/List_of_countries_by_average_elevation. Accessed 6 Aug 2021

Lovšin E (1944) V Triglavu in v njegovi soseščini: planinske študije in doživetja [In Triglav and its neighborhood: mountain studies and experiences]. Slovensko planinsko društvo, Ljubljana

Mikša P (2013) Prvi raziskovalci slovenskih gora in prvi dokumentirani pristopi nanje [The first researchers of the Slovene mountains and the first documented climbs on them]. Zgodovinski Časopis 67(3–4):390–405

Mikša P (2015) "Da je Triglav ostal v slovenskih rokah, je največ moja zasluga": Jakob Aljaž in njegovo planinsko delovanje v Triglavskem pogorju ["It is largely owing to me that Triglav remained in Slovene hands:" Jakob Aljaž and his mountaineering activities in the Triglav mountain range]. Zgodovinski Časopis 67(1–2):112–123

Mikša P (2017) Triglav in Jakob Aljaž [Mount Triglav and Jakob Aljaž]. Viharnik, Ljubljana

Mikša P (2021) Od gore do simbola: Triglav in njegova vloga pri Slovencih [From a mountain to a symbol: Mount Triglav and its role among Slovenes]. Geografija v Šoli 29(2):6–14

Mikša P (2022) The mountains that wrote them: Slovenians climbing and writing about eight-thousand-metre peaks. Slavica Tergestina 28(1):152–174. https://doi.org/10.13137/2283-5482/33711

Mikša P, Ajlec K (2015) Slovensko planinstvo = Slovene mountaineering. Planinska zveza Slovenije, Ljubljana

Mikša P, Ajlec K (2022) Slovensko planinstvo skozi čas [Slovenian mountaineering through time]. Planinska zveza Slovenije, Ljubljana

Mikša P, Zorn M (2016) The beginnings of the research of Slovenian Alps. Geografski Vestnik 88(2):103–131. https://doi.org/10.3986/GV88206

Mikša P, Ogrin M, Glojek K (2017) Od kod gorska identiteta Slovencev? [Where does the Slovenian mountain identity come from?]. Geografski Obzornik 64(3–4):10–28

Mikša P, Zorn M (2021) Nacionalni boj za gore in slovenska identiteta [The national battle for the mountains and the Slovenian identity]. Časopis za Zgodovino in Narodopisje 92(1):67–88

Odar B (2006) Ledenodobni lovci v visokogorju [The ice age hunters in the highlands]. In: Cevc T (ed) Človek v Alpah: desetletje (1996–2006) raziskav o navzočnosti človeka v slovenskih Alpah. Založba ZRC, Ljubljana, pp 59–70

Ogrin M (2006) Arheološke raziskave v Julijskih Alpah [Archaeological research in the Julian Alps]. In: Cevc T (ed) Človek v Alpah. Založba ZRC, Ljubljana, pp 96–110

Orožen F (1903) Kaj pripoveduje Valvasor o Krmi (Triglavu) [What Valvasor tells about Krma (Triglav)]. Planinski Vestnik 9(12):201–202

Perko D, Ciglič R (2020) Slovenia's landscapes. In: Perko D, Ciglič R, Zorn M (eds) The geography of Slovenia: small but diverse. Springer, Cham, pp 211–225. https://doi.org/10.1007/978-3-030-14066-3_14

Planiški rekordi tudi pri televizijski gledanosti [Planica records also in terms of television viewership] (2016) MMC RTV Slovenija. https://www.rtvslo.si/sport/zimski-sporti/planica/planiski-rekordi-tudi-pri-televizijski-gledanosti/388753. Accessed 6 Aug 2017

Poljanec N, Mikša P (2017) Kralj Peter v slovenskih Alpah [King Peter in the Slovenian Alps]. Ljudski muzej, Rogaška Slatina

Red za zasluge najstarejši slovenski reviji [Order of Merit to the oldest Slovenian magazine] (2021) MMC RTV Slovenija. https://www.rtvslo.si/zabava-in-slog/ture-avanture/red-za-zasluge-najstarejsi-slovenski-reviji-planinski-vestnik/592268. Accessed 20 Dec 2021

Šaver B (2011) Nazaj v planinski raj [Back to the mountain paradise]. Založba FDV, Ljubljana

Stele F (2007) Ljudje v Alpah [People in the Alps]. F. Stele, Komenda

Strategija razvoja Slovenije 2030 [Development strategy of Slovenia 2030] (2022). https://www.gov.si/assets/vladne-sluzbe/SVRK/Strategija-razvoja-Slovenije-2030/Strategija_razvoja_Slovenije_2030.pdf. Accessed 30 Nov 2022

Valvasor JV (2017) Slava Vojvodine Kranjske: izbrana poglavja [Glory of the Duchy of Carniola: selected chapters]. Mladinska knjiga, Ljubljana

Zorn M (2005) Fremde und einheimische Naturforscher und Geistliche—die ersten Besucher der slowenischen Berge (Ende des 18. Jahrhunderts bis Anfang des 19. Jahrhunderts). In: Mathieu J, Boscani Leoni S (eds) Die Alpen!: Zur europäischen Wahrnehmungsgeschichte seit der Renaissance. Peter Lang, Bern, pp 223–235

16 Spatial–Temporal Landscape of Durmitor and Drobnjak: Its Role in Creation of Highland Community's Local Identity and Its Toponyms

Obrad Samardžić, Rade Šarović, and Olga Radonjić

Abstract

Durmitor mountain consists of a very specific relief, which significantly influenced the creation and preservation of a unique identity of Durmitor people. In a semantic field a mythoetic origin of the word Durmitores, similar to Olympus, means the habitat of the gods, and in the field of geography it's primarily described as the mountain rich in waters in that way emphasizing its defining landscape, natural beauties and wealth all featuring this area. In this paper we primarily want to research spatio-temporal landscape of Durmitor and Drobnjak and its role in creating the identity of this mountainous area of Montenegro. While researching this topic we encountered numerous toponyms which indicate various interferences of natural surroundings, religious beliefs, ideological and cultural influences in shaping a specific identity, characterized by a socio-cultural ethos of this area. The starting point of our research of the role that Durmitor toponyms have in locals' "semantics of life" is the combination of sociological research, historical sources, ethnological and anthropological material and, of course, pioneering attempts by individuals to depict this specific natural and cultural landscape of Durmitor and Drobnjak.

Keywords

Identity · Toponyms · Culture · Religion · Ideology · Time · Space · Drobnjak · Durmitor

16.1 Introduction

In our attempt to present spatio-temporal landscape of Durmitor and Drobnjak as well as its role in forming the inhabitants' identity, we start from the idea that our understanding of time and space, as pointed out by sociologists Aron Gurević (1977:226) and Émile Durkheim (Dirkem 1982:12–13), is an important component of social awareness, and the manner of perception and apperception of time and space reveals numerous basic needs and society values, its groups and individuals. Moreover, as these authors emphasize, spatial–temporal understanding is one of important "character parameters" of a society.

Durmitor and Drobnjak represent a marvelous mosaic of natural beauty creating vivid mountain landscape with lakes, canyons, rivers, plateaus and valleys. Talking about Durmitor massif and

O. Samardžić · R. Šarović (✉) · O. Radonjić
Faculty of Philosophy, University of Montenegro, Nikšić, Montenegro
e-mail: rades@ucg.ac.me

O. Samardžić
e-mail: obrads@ucg.ac.me

O. Radonjić
e-mail: olga.radonjic999@gmail.com

M. Zorn et al. (eds.), *Discourses on Mountains of Montenegro and Slovenia*, Historical Geography and Geosciences, https://doi.org/10.1007/978-3-031-99017-5_16

its influence on the character of the inhabitants, goes along with talking about Drobnjak and its people—three Montenegrin municipalities are located in Durmitor mountain area, Žabljak, Šavnik and Plužine. Therefore, Durmitor people are not only people of Drobnjak. Area of Piva under Durmitor mountain is nothing less beautiful, attractive, significant or important. However, this paper will focus only on one special locality—Drobnjak, due to more precise thematic focus and analytical research.

The name of Durmitor initially relates to one of the biggest and most beautiful Southern Balkan mountains (2,522 m) (Fig. 16.1), while Drobnjak has a broader meaning and is related to geographical locality but also to people who live in Durmitor mountain area. Therefore, Drobnjak is both geographical and ethnological category. As historian and geographer Andrija Luburić (1930) points out, borders of Drobnjak were set in 1860, after the war that Montenegro and Herzegovina had with Turks, Drobnjak became the part of Montenegro according to Berlin congress (1878) regulations. It is naturally divided on Drobnjacka vala (*vala*—gorge), Jezerska plateau, undulations of Saranci area in the very north and karst Uskoci, and toward Morača in the east (Ostojić 2010:37). Drobnjak river bed or Niski Drobnjak (lower Drobnjak area) or, as some would call it Župa-Komarnica (Fig. 16.2), is a spacious karst depression carved with slopes, several river beds which flow into Bukovica river which flows into Piva. Mount Vojnik and plateau Krnovo separate it from Nikšićko Polje, and Mount Javorje and Lola from Morača areas. This part of Drobnjak area includes Šavnik town and Duži, Previš, Godijelja, Dobra Sela, Petnjica, and Komarnica as some of larger villages. Southern from Šavnik, there are four villages which are also part of Drobnjak: Mokro, Miloševići, Gornja Bijela, and Donja Bijela—the last two lean on Krnovo Glavica slopes, while Mokro together with a small village Gradac are located on a small plateau above Šavnik and Pridvorica river, and under Vojnik mountain. Medium

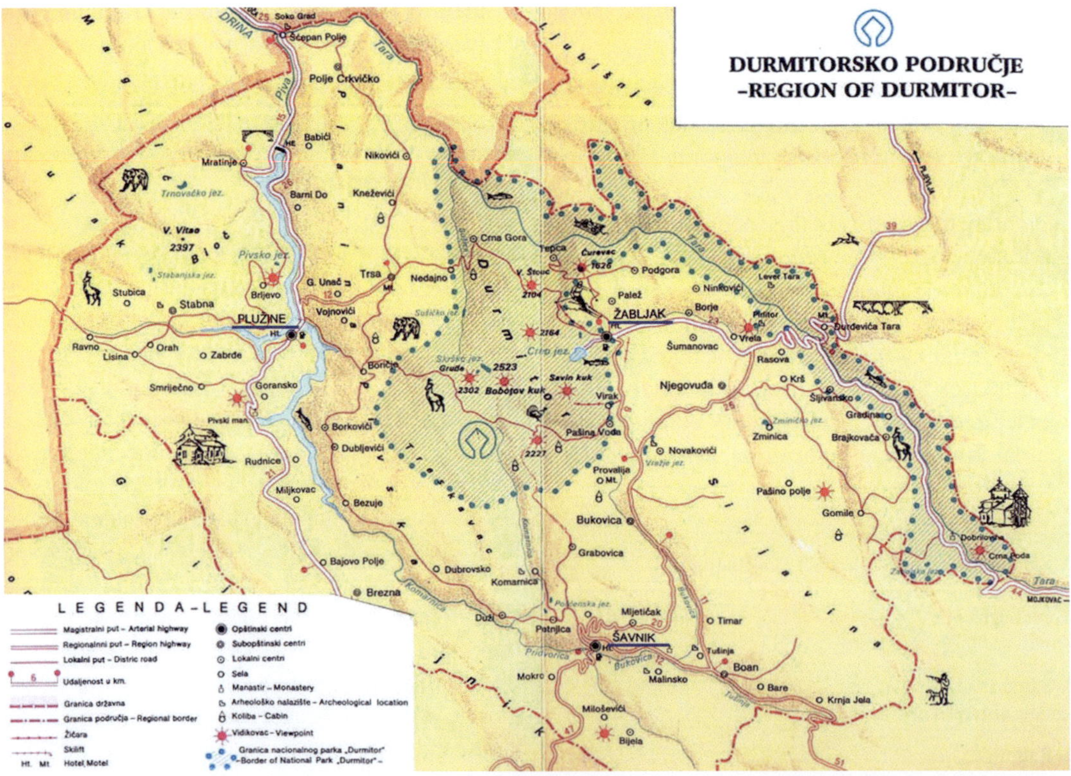

Fig. 16.1 Region of Durmitor (*Source* Turismo … 2022)

Fig. 16.2 Župa-Komarnica—central place of Drobnjak region. (Authors archive)

altitude plateau (1,500 m), called Jezera, encompasses north-east part of Drobnjak tribe and the area between massifs of Durmitor, Sinjajevina, Štuoc and Ivica, and Tara river in the north. In Jezera Žabljak town is located, and very near is a famous Crno jezero (*Black Lake*), a lake many times depicted in legends. A part of the plateau, called Šaranci, spread over the northern part of Sinjajevina, bordering with Tara river on one and Jezerska plateau on the other side. Finally, Uskoci, as a subgroup of Drobnjak (some would also claim it to be another tribe), mostly covers the area east of Savnik toward Morača area, and southern Bukovica river basin where Boan, Timar, Bare and Krnja Jela are placed, while it's divided from Gornja Morača by a mountain pass called Semolj.

Names and features of localities and villages near and at the foot of Durmitor mountain are commonly related to relief shapes (*oronyms*). Many of the toponyms are semantically related to history and human activities (*anthroponyms*), while *phitonyms* connect flora with some toponyms and *zoonyms* relate to fauna. In the semantic genesis of toponyms formation of this mountainous area, as pointed out by ethnographer Đorđe Ostojić (2010:22), names were developed via associations, each had some reason of its own. For example, locality Vrela associate to a couple dozen of small and large wellsprings (*vrelo*); Borje relates to some black pine trees (*bor*); Javorje got its name for numerous maple trees (*javor*); Nadgora, Podgora, Šumanovac, etc., are villages which names are phytonyms.

However, some of the localities received a special attention where their names' choice had a special symbolic for the people of Durmitor area. This paper is mostly interested in those localities which names were formed in accordance with Durmitor mountainous mentality so that it can be preserved. There are more than few toponyms relating to honor, heroism, battle for freedom, and free spirit of these highlanders. The wholesome Durmitor history witnesses this, history which is mostly orally passed from generations to generations through popular epic poems. According to the words of historian Nićifor Dučić, Drobnjak history is actually a one heroic and "bloodshed epic" (Karadžić and Šibalić 1997:6). Epic poems passed along with gusle sound, transferred into stories thanks to a storyteller's imagination, left a print in the space as well. A common characteristic of all Durmitor epic poems and legends, as well as of toponyms here mentioned, is knighthood, humanity, and bravery with high ethical principles which all characterize patriarchal morality of Durmitor inhabitants.

The goal of our paper is to determine cultural, historical, religious, and ideological influences on creating toponyms of Durmitor and Drobnjak area on one hand, and researching a reflexive connection between the sense of belongingness and the identity of the people who live in Durmitor.

16.2 Cultural-Historical Development of Drobnjak and "the Prints" of Cultural-Religious Heritage in Its Toponymy

A great number of philosophers, sociologists, anthropologists, geographers, ethnologists, saw geographical conditions as direct and indirect influence on the identity of people, society, and culture. The most prominent names related to this theory would be geographer and philosopher Charles Montesquieu and geographer Friedrich Ratzel (1909). They, like many other researchers, supported theoretical views which see society as deeply colored by geographical features. This approach does not follow these logics of geographical determinism, but it's based on the interference of geographical and cultural-historical factors as the constructors of a mental map and identity of people and community where they live.

For sociology, geography, anthropology of the area, are of a great importance indirect influence of geographical factors on people and mutual influence of great historical events, mentioned cultural zones and migrations on preservation of the culture and identity of local population. Geographical conditions, together with the influence of the mentioned social "indirect factors" determine, what geographer Jovan Cvijić names as "*the main ethnographic and sociological facts*," (spatial display of Balkan people) typical ways of their economy, land possession relations, types of village and town settlements, items of material culture (houses and other additional village units, some utensils, and tools), but also spiritual contributions such as mental characteristics, cultural patterns, moral and religious understandings, artistic expressions, and other individual and group socio-cultural characteristics (Cvijić 2013:8).

Chronologically seen, there are many examples of how social elements have left trace both on Durmitor people and the area which they inhabited. Even the ancient Romans left the everlasting trace by giving the name *Durmitores* to this mountain they admired. *Durmitores* meaning mountain—inhabitance of gods. Luburić and Tomić point out that this area was inhabited by the Illyrians and their tribes, before the Slavs. They left behind their cemeteries, *stećci* (medieval tombstones) and medieval walls, best preserved in Tepca village at the foot of Durmitor mountain. However, this period does not leave any "deeper trace" in the toponyms of the area, even though there is an assumption that *Tara*, name of the river and its canyon, originates from this period. According to historical records and tombstones inscriptions, it is certain that many of old cemeteries belonged to the Celts, Romans, Bogumils, Greeks, etc. *Stećci* and other tombstones from that period witness their presence in this area. These old cemeteries "*are records of life, its ending, as well as the customs and sorts of the culture of saints of the peoples who inhabited this area from the ancient times. In many respects they represented some sort of the acropolis or mausoleum as monuments of a culture which vanished a long time ago from Drobnjak area. Ancient shrine Mitrium should also be mentioned here; located on the left canyon side of the Tara (Lever Tara) originates from 2nd or 3rd century AD. It was erected to glorify eastern deity Mitra*" (Karadžić and Šibalić 1997:144–146).

Drobnjak as a name was first mentioned in thirteenth century as a family name (Bran Drobnjak, 1285 archive) and represents one tribe, and later on a geographical locality. In the dictionary of Serbian Academy of Sciences "*Drobnjak as the term has many meanings; one is related to a height, how tall these people were, which possibly changed over time. Another opinion is related to a type of limestone typical for this area, etc.*" (Karadžić and Šibalić 1997:129). According to some other opinions, *Drobnjak* term originates from "*a Greek word dromos which means a road, relating to an old Roman road which was leading from Dubrovnik, over Risan, Drobnjak area, Pljevlja, Prijepolje, Skoplje to Constantinople*" (Karadžić and Šibalić 1997:129). The name *Drobnjak* from Turkish Ottoman records in 1477 was not mentioned as the name of the tribe, but in accordance with its territorial determinants, this area was called *Nahia Komarska*. However, according to one traditional story Drobnjak got its name as people of Drobnjak slayed and tore apart intestines (*drob* therefore *Drobnjak*) of a Turkish pasha Husret beg, Bosnian regent from 1541, in a village Mokro near Savnik. Also, as Karadžić and Šibalić (1997:130) note "*there's a legend among Bosnia and Herzegovina Muslims that Husret beg was killed in Drobnjak, after which his body was taken and buried in Sarajevo Mosque, but his intestines were buried at the place where he was killed. It was after these intestines that Drobnjak got its name.*"

As well as a greater part of today's Herzegovina, Drobnjak had different rulers over the centuries; it was included in the territories of some stronger states which ruled these areas and went through all the stages from occupation, liberation, independence, occupation again, etc. However, fifteenth century represents a turning point when we talk about demographic and cultural development of the area. Throughout new migrations families from Banjani and Nikšićke rudine (Herzegovina) inhabited Drobnjak and set new basis for religion and ideology within tribe communities who live in this area. That is also a period when the tribe organization of this community reached its full form and strength. Many tribes which inhabit this area were mentioned from thirteenth and fourteenth century, which shows that a strong tribe awareness characterized the period, and a certain mild form of tribal organization (Karadžić and Šibalić 1997:29–35). In the broadest sense, the tribe represented a community of the inhabitants who live and work in a specific territory, a community which is "*organised on the principles of specific territorial and ethnical bounds, with its people having some significant common elements seen in their customs, religion, language, common ancestors, tradition and interests which influence a tighter connection of the people in the tribe*" (Karadžić and Šibalić 1997:29). Since that time, we have a more visible footage in the names of localities in the area, which were *katuns* (summer pasture villages) and later became permanent settlements. In Montenegro and Herzegovina, a common way of naming localities is via connection between the name of a tribe and places which tribe members inhabit. As Karadžić and Šibalić (1997:29, 129) state the name of a specific tribe was almost always identified with a geographical name, and therefore involved the territory where tribe inhabitants live—tribe *Drobnjak*, inhabitants *Drobnjaci*.

Forming and preserving the population identity and naming specific localities due to the influence of cultural, religious and ideological facts is surely immanent for Drobnjak and Durmitor area. Throughout a Turkish rule a great struggle of Durmitor population was inevitable in order to preserve the identity and inherited tradition. From 1482 to 1857, the period of Turkish rule in Balkan, the people of Durmitor constantly fought against this great empery in spite of cruel campaigns of the Turks. Five centuries were a period more than enough for some localities to define their position against the conquerors (Srdanović 2007; Ostojić 2010). However, their names and meanings were actually some kind of indications for future generations, as they constantly glorify courage, heroism, and honor of this people, as well as their battle for freedom and identity preservation.

We will look back on a few distinguished examples which show how history, fight for freedom and patriarchal moral was transferred into toponymic ground. This is how village *Bosača*, 1,600 m, a village with the highest altitude on Balkan, got its name. It was named after a girl who was defending her honor from some Turks and ran barefoot (*bosa*) toward Black Lake, losing her life in a cold Durmitor winter (Ostojić 2010). A locality *Boan* used to be a town which represented a center of a freedom fighting spirit of Uskoci tribe. The name comes from a Turkish word *buhan*, which means *this inn* (Ostojić 2010). The people of Uskoci proudly carried this name as throughout the history very few soldiers lingered here.

While following historical events at this area, it is inevitable to spot numerous geo-historical determinants woven into names of certain localities, which were mostly related to campaigns of Turkish conquerors and heroic resistance of local, native population. For example, each spring where Turkish army stayed for a period, was named *Pashinac* (belonging to pasha). On the other hand, a name of another locality tells about how one Turkish pasha was killed and how a brave girl from Drobnjak fought against a brute Turkish conqueror. The legend tells about a wife of a local voivode Ilija Kosorić, Anđa Kosorić, who beheaded pasha Selimović to revenge her husband. The name of this village according to Ostojić (2010) is originally from eighteenth century, when this pasha who was in charge of Pljevlja area "*came to Jezera in 1705 and settled his camp near a spring of water. He tricked and murdered Ilija Kosorić, then ordered his soldiers to bring him Ilija's wife. Anđelija, Anđa managed to enter his tent with a hidden knife, killed pasha and took to her village Kosorić both heads of her husband and her victimizer*" (Ostojić 2010:34). This is how a village and this spring got the name *Pašina Voda* (Pasha's water).

A locality *Ramovo Ždrijelo* is one of mountain passes which joins Jezera plateau and Njegovuđa. It got its name after the battle of Vuk Lopušina and Turkish hero Ramo Prekić (Krstajić 2010). Vuk managed to tie up Ramo and take him that way to Nikšić, so the name of this place stands till today as a warning and a symbol of a defeated Ramo Prekić. Together with the mentioned names, there are many localities which even today remind us on Turkish ruling in Montenegro. *Smailov Pod, Omerova Jama, Pashina Gomila, Pashino Polje, Anovi, Kadino Brdo* are only some of those which testify about five centuries of Ottoman presence in Balkan area.

16.3 Reflection of Literal Creativity on Toponyms at Durmitor

One of the most prominent examples of a human connection with the environment is seen in folk tales, poems, and lementations. This is specifically emphasized in this area where, for centuries long struggle with the conquerors was common, conquerors who strove to defeat soul and spirit of a highland man. In such circumstances the main characters of folk poems and lamentations were heroes, extraordinary characters, courageous, strong, righteous, etc. Understanding these heroes through folk poems, seeing them as colossi and comparing them to mountain prominence and steadiness is very common. This is seen in one folk lament which describes severe days and events of warfare in Saranci and Jezera area, where a mother grieves her son, well known for his height and beauty. In her lament she compares "*his wide shoulders with a tall Stožer mount where he lost his life, and with shadows of a forest growing on it*" through the words (Karadžić and Šibalić 1997:87):

"Milivoje, sunce moje, Kad ujutru sunce grane, Od Stožera krvavog, I udari sjen od gore Misli majka pleći tvoje."	"Milivoje, oh my dear son, When the morning sun strikes From a blooded Stožer mount Spreads a shadow from the forest, Oh, a thought comes of your shoulders."

These hard times jeopardized not only the people's existence, but also their identity, as well as religious beliefs, as the Turks strove to impose Islamic religion and identity. However, despite a tendency to impose different religious views and beliefs, orthodox people of this area fought and succeeded to preserve their Orthodox Christian religion. Devotion to orthodox faith was deeply wooven into identity and consciousness of these people. Consequently, there are many toponyms which names associate to religious belongness and identity. One of the most typical examples of this kind is related to village *Pridvorica*, located on a river Bukovica bank. On the right bank of Bukovica river there is a ruin of a castle and a church erracted by one of prominent members of Nemanjić dynasty, who was canonized by Serbian Orthodox church. From this period originates a folk poem and the name of the village itself:

"Bože mili, čuda velikoga, Car Simeun gradi Studenicu, a sluga mu crkvu Pridvoricu."	"God in Heaven, the stupendous wonder, Tzar Simeun builds Studenica and his servant a church Pridvorica."

The view of a central and most beautiful part of Durmitor mountain, also testifies our claims. *Savin Kuk, Savina Voda,* and *Savina Pećina* (Sava's Peak, Water, and Cave) are localities of Durmitor mountain with the most numerous epic stories, and descriptions, which attract many visitors with their breathtaking beauty. The mentioned localities were named after Rastko Nemanjić (Saint Sava) who often visited Durmitor leaving to its people education and enlightenment as a heritage together with numerous legends related to his name. According to a legend, *Crno Jezero* (Black Lake) and *Savina Voda* (Sava's Water) on *Savin Kuk* peak are related to one of Sava's visits. As the legend says, Sava miraculously made a spring of water on the mountain peak so that his student follower could quench his thirst. He made a cross sign with his stick and a water spring appeared, which is active today throughout the whole year. According to another legend a mountain top saves hoofprints of a winged divine horse carved into a stone (Karadžić and Šibalić 1997:84). Therefore, we can see that legends tell stories about Durmitor and describe it as "*a place where gods rest*" and "*a place where ferries live and compete with each other during the night on the mountain and its lakes*" while "*the swans fly over so called Gorske Oči* [mountain eyes]—*lakes of this secretive mountain*" (Karadžić and Šibalić 1997:84).

One more village, *Bogomolja*, has a name which testifies about the resistance of Drobnjak people against assimilation and oppression which was unsuccessfully present throughout five centuries. The name is from the period when the Turks burnt own a monastery of Saint Archangel Michael (*arhangel Mihailo*) in Tara area in 1875, and its inhabitants could not practice their religion as they didn't have a temple any more. However, this didn't stop them and they exercised their religious ceremonies while facing the monastery ruins, on the cliffs, which thus were named *Bogomolja* (meaning a place of worship).

Most of the toponyms in this area got their names in the period between fifteenth and nineteenth century. After World War I the area became a part of the Kingdom of Serbs, Croats, and Slovenes (later Kingdom of Yugoslavia) and after World War II it became a part of socialist Yugoslavia. However, this period did not leave any big changes in names of Drobnjak localities. When we talk about the period of Yugoslavia Kingdom, there are two names that are noticeable: *Gospodareva glavica* (Master's hill) and *Kraljeva Stolica* (King's chair) that are toponyms serving as reminders of visits of Montenegrin king Nikola Petrović (in 1903) and his grandson Aleksandar Karađorđević (in 1932), king of Yugoslavia. These two grandees both loved and respected Durmitor mountain as well as the identity of Drobnjak people. Facts to testify this statement are their several visits to these localities and some investments which enabled development chances for this area (Šarović and Kasalica 2022).

The final dismemberment of Federal Republic Yugoslavia (in 1991) did not bring any significant changes in the matter of toponyms of Durmitor area. Only a few localities witness the great victim Durmitor people suffered during the World War II. *Durmitor Republic* as this part of free Montenegro was called, left behind localities such as *Titova Pećina* (Tito's Cave) near Black Lake, *Savina Koševina* (Sava's mowing), *Veljkova Greda* (Veljko's cliff), *Danilov Grob* or *Mramor* (Danilo's grave or marble stone) are some of the toponyms, which without any great influence witness this part of history.

Finally, we have to emphasize that spatial identity of Durmitor people and location names were not only born within heroic poems and legends. Drobnjak identity involves stories as well which in a romantic (often tragical) way depicted life of the young who lived under this mountain. There are so many stories and legends tangled into Drobnjak mentality and identity which became eternal as the part of Drobnjak landscape. Some of locations which names are related to (unfortunate) love affairs *Draga, Djevic Kamen, Todorov Do, Ružica, Okamenjena djevojka, Kamenovani ljubavnici* (Darling, Virgin's rock, Todor's valley, Rose, A Lady Patrified, Lapidated Lovers) have an admirable lyrical context, that even worldwide known poems can look upon.

16.4 The Dinaric Type of Man and His Relation to Time and Space

One of the most prominent geographers, anthropologists, ethnologists of the Balkan area, who also researched northern part of Montenegro (late nineteenth–early twentieth century), is Jovan Cvijić (1865–1927). According to Cvijić this area is populated by Dinaric race (the Dinaric type), who not only inhabit Dinaric area but also all its neighboring areas where migrations from Dinaric area occurred. Most of Drobnjak and Durmitor territory belong to a variation of Dinaric race, named *crnogorsko-plemenski* (Montenegrin-tribal). Cvijić, whose ancestors also belong to this Dinaric type, saw it as ideal type and wrote: "*People of Dinaric area are of a lively spirit and subtle intelligence. Endowed with vivid and various sensitivity, they more often than not follow their vivid and rich imagination, as well as their first impulse of a thrill or a wrath. Their actions are normally inspired by moral or spiritual incentives; material interests are of a secondary importance. To provoke their greatest strength, one should dare their emotions and sensitivity, individual and national pride; question their honor or the ideal of justice and freedom. Those are the main reasons which motivate passion of the Dinaric people and cause conflicts between them making their life events fortunate or unfortunate more than any kind of selfishness or greediness would make. A Dinaric man does not believe in obstacles that cannot be surmounted. His faith is unwavering and confidence doesn't see limits.* [...] *He never abandoned his people's and social ideals, even when current circumstances and events deceived him and everyone turned their backs on him. This persistence and the absolute faith in the national ideal are the main truth of his history.* [...] *This is the reason why a sense of sacrifice and self-renunciation almost sees no limits: in the course of history the Dinaric people while sacrificing themselves caused the exhaustion of their whole people. The masses suffered peacefully the unspeakable difficulties all for the sake of social and national ideal*" (Cvijić 2013:188).

16.5 Influence of Everyday Life on the Creation of Toponyms

Toponyms are tightly related to practical activities and people's lives. Their origin is based on active experience and the inherited tradition, as well as on associations on specific landscape where people live. Deeply connected with the nature, people here carefully organized their lives as to fit the demands of natural events. Day and night, sunrise and sunset, moonrise and moonset, the change of seasons all dictated the pace of life of the traditional rural communities. Therefore, people here experience time and space with a huge effect of "nature's signs." While living in the heart of a snowy mountain and doing livestock farming, people of Drobnjak, taught by tradition, carefully followed natural phenomena so that they could foresee some heavy snow or rain precipitation, which ensured them to find shelter on time and prepare for different extreme weather conditions. Following the "sings from the sky" and behavior of domestic animals and using them as a guidance shaped through earlier experience, people often predicted flawlessly nice weather or frost or strong winds brining rain or snow. Watching mountain peaks if they are "*light and bright before the sunrise, if the moon is clear and white,*" if "*the stars of the milky way*" are clear, if the "*west sky is red on sunset and the east is glowing*" (Karadžić and Šibalić 1997:102) ensured some weather forecast. People of this area have a belief that evening lightning from all directions of the sky announces some long period of sunny days. Years were remembered by heavy snows, frost, hunger, illnesses, and wars. Therefore, weather conditions had a significant symbolic meaning and are connected to various natural disturbances or social events which distorted usual life of this mountain communities. These people "*remember how summer rains, storms, hailstorms and early snow, destroyed all their hopes in good harvest season. Drought would bring hunger, poverty and misery, and also cause people to leave their homes in search for better life conditions*" (Karadžić and Šibalić 1997:102). A saying from Durmitor area: "*it's better to meet a hungry wolf in February than a man wearing light clothes*" (Karadžić and Šibalić 1997:102), depicts how weather conditions are interpreted. On the other side, knowing the inner life of *katuni* (summer pasture settlements) and watching how domestic animals react and behave, were also some sings used for the weather forecast. So, if cows rush early in the evening to their shelter, and sheep restlessly bleat or an anxious sound of birds is heard, some bad weather is expected. Sociologist and historian Lewis Mumford (Mamford 2009:29) notes that shepherds measure time from the time their sheep lamb, while farmers measures time via days of sowing and harvesting. This is why people from this area, while living with their cattle empirically realized how sheep, cow or horse inarticulately react on their inner urges, needs and wishes, pain and oppression. A shepherd was never caught by dark in the mountain pasture and he never had a watch to know the time. Sheep would tell him the time. If their pupils are constricted, the shepherd knows the sun is high. If their pupils are dilated, and sheep are moving, the sun is close to setting (Vlahović 2018:168).

This kind of perception of time via numerous "signs of nature" has its own analogy in experience of people of this area and in naming the locations. As the matter of fact, Durmitor area, as the ethnologist Petar Vlahović (2018:168) points out, is a home of numerous shapes and landscapes of a fantastic beauty: "*Shepherds gave names to each of those shapes, and the names related to their color, content, appearance or something else specific. Those names are melodic, soft, sharp, dim, gloomy and then light again – to man's eye and ear, as well as the character of these tribes is.*"

There are many researchers (as Karadžić, Šibalić, Tomić and others) who pointed out creative imagination and originality of Drobnjak people when naming hill tops, valleys, rivers, etc. Therefore, they named a stream *Skakavac* (Hopper), while as the matter of fact "*it is a wellspring, short and quick fountain, where water 'jumps from one rock to another', 'jumping its way' to the Tara river. Another strong and fierce wellspring was named Ljutica* [ljut—angry, fierce] *as it angrily and fiercely rushes into waters of Tara, slitting its flow and calming the rapids of Tara*" (Karadžić and Šibalić 1997:133). On the other hand, areas which are wild, rocky and mountainous or distinct thanks to some visually specific features were named after either some symbol, which describes their distinctiveness, for example *Prijeka Stijena* (*prijeka*—savage) or visual appearance, for example *Žuta Stijena, Crvena Stijena* or *Đevojačka Stijena* (yellow, red and Maiden's cliff). There are also some specific locations and peaks which visually resemble to human elbow, and therefore the names: *Savin Lakat, Veliki Lakat, Donji Lakat* or *Tepački Lakat* (*lakat*—elbow).

Typical example of how people of Drobnjak define their homeland is in a name *Dobra sela* (*dobra*—good). Even today the answer to the question "*Where is Dobra sela?*" is "*There, where the good people are.*" However, according to geographer Svetozar Tomić (1901), name of this village was not created because of its position, its fruitfulness nor any other good quality. The legend related to this toponym, which Tomić points to, talks about different, nice and good quality life of this place. Aspirations toward good and goodness, rightfulness and honesty and effort to preserve those values, surely made a huge influence when naming permanent and temporary settlements where people of Durmitor area lived. These reasons made us do the research and identify that in the whole territory of today's Montenegro there less locations with a prefix "*dobrog*" or "*dobra*" in their names, than just in Drobnjak area, which is only one of several tribes (Vasojevici, Kuci, Bjelopavlici, Banjani, Uskoci, etc.) in Montenegro. What most certainly influenced this phenomenon is an effort to convince the descendants in a good quality of life of a place, but also to preserve identity of a tribe and a family. For example, one of the most populated and beautiful villages in Drobnjak is named *Pošćenje*. A story about its name says their inhabitants were very proud of their honesty (*poštenje*) and bravery (*čojstvo*), and these two words combined gave the name to this village. More toward Mojkovac, in the south, village named *Dobrilovina* is hidden close to Tara canyon. Donja Dobrilovina, as a part of Dobrilovina village, is in the territory of Drobnjak area, and its name may be related to good life conditions and the pleasure its inhabitants showed because they lived there.

As the geographical location of Drobnjak represented a crossroad of Roman imperial routes from Dubrovnik and other merchant centers toward Constantinople, name of one of the villages here reminds of time and role this area used to have at the crossroad of the routes. Thus, a village *Dubrovsko*, which spreads over a spacious, dry and waterless area and which soil is less fertile than anywhere else in Drobnjak, was named after a great merchant town of Dubrovnik (modern-day Croatia). It was thought (and wished that others share the same idea) that Dubrovsko was as comfortable and beautiful to live in, as Dubrovnik (Dubrovsko traded wool, meat and milk with Dubrovnik). This village was also called *Little Dubrovnik*, and therefore today's name, Dubrovsko. Numerous villages, hamlets, and localities witness a specific relation of the identity of Durmitor people with their habitat. Names such as *Dobri Nugo, Dobri do, Časni do, Dobrojevići*, *Dobro polje,* and many others that didn't find a place on this list, witness the claim.

A subjective approach may be recognized in this paper, as well as, similar to Cvijić's, idealization of the Dinaric type of man from Drobnjak. According to Andrija Luburić "*Cvijić himself was saying he is from Drobnjak. Before his death he gave me some data about the past of his family, with a request to include it in this work*" (Luburić 1930:16). However, neither did Cvijić, nor have we approached with any kind of non-critical romanticism, but on the contrary—using an ideal type of methodological process. Hence, this mountain which is described in Vuk Karadžić's dictionary as a column which upholds the sky (Karadžić 2006) and where according to legends gods rest, mountain which beauty made Romans, Greeks, and Kriči stand still, mountain which both saints and sinners, nobles and the poor, scientists and laymen bow down to, a place where even Hitler's army didn't stay longer than a few months when 2448 gave their lives for Durmitor Partizan Republic (Jokić 1978:72), mountain which means everything to the highlanders who live here and whose identity is woven into it, carrying the name *Durmitorac* proudly, surely deserves this kind of opinion and objectivity of this paper work.

16.6 Epilogue

In a deep "semantics of life" landscape of Durmitor and Drobnjak and its role in the identity of people of this highland area show that man's relation with time and connection with nature represent a significant feature of his social awareness. Specific character features of a man of this mountain area, as an individual and as a member of cultural historical community (family, brotherhood, tribe, the people), as well as specific geographical and relief characteristics of this mountain landscape are all tightly woven into condensed social forms, of people's identity as well as of the meaning of the space they live in. Through the prism of a "long duration" (Fernand Braudel; Brodel 1992) many of conquerors and armies (Ottomans, Romans, and Germans) marched through Durmitor area, leaving the devastation behind (destroyed settlements and villages). Different invaders and their influence did not bypass Dutmitor area. While defending themselves and fearing how to preserve their identity, people here left indelible and deep imprint on the space. Their affection to homeland, struggle, heroism, and faith were shaped into the space and landscape, forming the names of villages, hamlets, peaks, hills, valleys, preserving themselves eternally in that way for the future generations. Throughout all of its history people of this area never related to their surrounding and nature using the "subject-object" principle, but the principle of inner mutual well-being. A close and intimate unit and mutual interpenetration of humans and nature was developed; man and the nature are connected organically and intertwined mystically. Durmitor was not only a shield and shelter for warriors, nor just a fertile soil feeding farmers and workers. In fact, this area is not plain and secular for its inhabitants, but a place of mystical and holly power depicted in a constant dilemma: whether the people guarded Drobnjak and Durmitor with swards and guns, or this colossus guarded the people with its abysses and ridges? It appears that, throughout the time, the principle of "mutual guardians" best describes the relation of people who live here with their natural habitat.

References

Brodel F (1992) Spisi o istoriji [Writings on history]. Srpska književna zadruga, Beograd

Cvijić J (2013) Balkansko poluostrvo i južnoslovenske zemlje [Balkan Peninsula and South Slavic countries]. Narodna biblioteka Srbije, Beograd

Dirkem E (1982) Elementarni oblici religijskog života [Elementary forms of religious life]. Prosveta, Beograd

Gurevič AJ (1977) Vreme kao problem kulturne istorije [Time as a problem of cultural history]. Kultura 36–37:226–242

Jokić G (1978) Durmitorska partizanska republika [Durmitor partisan republic]. Industrija import, Titograd

Karadžić S, Šibalić V (1997) Drobnjak: Porodice u Drobnjaku i njihovo porijeklo (Drobnjak, Jezera, Uskoci i Šaranci) [Drobnjak: Families in Drobnjak and their origin (Drobnjak, Jezera, Uskoci and Saranci)]. IŠ "Stručna knjiga", Beograd

Karadžić V (2006) Srpski rječnik [Serbian dictionary]. Feniks Libris, Beograd

Krstajić B (2010) Mitovi i legende durmitorskog kraja [Myths and legends of the Durmitor region]. Grafokarton, Bijelo Polje

Luburić A (1930) Drobnjaci, pleme u Hercegovini [Drobnjaci, a tribe in Herzegovina]. MB Grafika, Beograd

Mamford L (2009) Tehnika i civilizacija [Technology and civilization]. Mediterran Publishing, Novi Sad

Ostojić ĐM (2010) Porijeklo naziva drobrnjačkih sela i zeseoka [The origin of the names of the Drobrnjak villages and settlements]. Skupština opština Šavnik, Pljevlja

Ratzel F (1909) Anthropogeographie, Teil 1: Grundzüge der Anwendung der Erdkunde auf die Geschichte. J. Engelhorn, Stuttgart

Srdanović TS (2007) Legende i predanja Durmitoraca [Legends and stories of Durmitor]. Grafo, Podgorica

Šarović R, Kasalica S (2022) Žabljak – 150 godina urbanog naselja; Crna Gora – 30 godina ekološke države [Žabljak – 150 years of urban settlement; Montenegro – 30 years of ecological state]. In: Društvena osnova: problemi i perspektive razvoja. Univerzitet Crne Gore, Filozofski fakultet, Niksić

Tomić S (1901) Drobnjak, antropogena istraživanja [Drobnjak, anthropogenic research]. Prosveta, Beograd

Turismo de Montenegro (2022). https://www.turismomontenegro.es/wp-content/uploads/2016/07/mapa_durmitor.jpg. Accessed 15 Jun 2022

Vlahović P (2018) Etnologija Crne Gore (ogledi i članci) [Ethnology of Montenegro (essays and articles)]. CID, Podgorica

17 Geographical Names and Place Identity in the Slovenian Alps

Matjaž Geršič, Primož Gašperič, and Matija Zorn

Abstract

Geographical names are part of intangible cultural heritage, which communities, groups, and individuals pass on from one generation to another and perpetually reproduce in response to their environment, nature, and history. In mountainous areas, geographical names were very important in the past mainly for three reasons. The first one was orientation in the area, especially for hunters, shepherds, and others living in alpine areas. The second reason is the expression of the national consciousness. Slovenians tried to fight against German influence through geographical names. The third reason is related to natural phenomena, which are mirrored through geographical names. These can be natural disasters, geographical features, and so on, which help in reconstructing landscapes from the past.

M. Geršič · P. Gašperič · M. Zorn (✉)
ZRC SAZU Anton Melik Geographical Institute, Ljubljana, Slovenia
e-mail: matija.zorn@zrc-sazu.si

M. Geršič
e-mail: matjaz.gersic@zrc-sazu.si

P. Gašperič
e-mail: primoz.gasperic@zrc-sazu.si

Keywords

Place name · Microtoponym · National identity · Mountains · Slovenia

17.1 Introduction

"Geographical names, or toponyms, are proper nouns that by definition are connected with a precisely defined geographical feature that they identify and individualize" (Kladnik et al. 2020:9). The first explanations of geographical names by Slovenian authors go back to the eighteenth century, when Johann Siegmund Valentin Popowitsch (Janez Žiga Valentin Popovič) discussed Slavic and non-Slavic toponyms in Europe from a comparative, typological, and etymological perspective in accordance with the linguistic methodology and orientation of that time. Around 1800, other works of this type were produced by Marko Pohlin, Jernej Kopitar, and Franc Serafin Metelko (Jakopin 1990:116; Šivic-Dular 2002:13). This period also saw the publication of the first work in which geographical names were broken down in detail by type of feature they refer to. This was a German study by Slovenian Urban Jarnik titled *Andeutungen über Kärntens Germanisierung* (An Outline of Carinthia's Germanization, 1826). Among the ethnic Slovenian provinces of the Habsburg Monarchy, Jarnik only covered Carinthia in his study, establishing

M. Zorn et al. (eds.), *Discourses on Mountains of Montenegro and Slovenia*, Historical Geography and Geosciences, https://doi.org/10.1007/978-3-031-99017-5_17

that many names in this province, for which the semantic motivation could be explained based on Slovenian common nouns, had been Germanized (Šivic-Dular 2002:13–14).

Onomastics began to develop as a science as part of Indo-European linguistics and other disciplines in the second half of the nineteenth century; the development of Slavic and Slovenian onomastics was primarily based on the following onomastic works published by Franz Miklosich (Franc Miklošič) in German: *Die Bildung der slavischen Personennamen* (The Formation of Slavic Personal Names, 1860), *Die Bildung der Ortsnamen aus Personennamen im Slavischen* (The Formation of Place Names from Personal Names in Slavic, 1864), and *Die slavischen Ortsnamen aus Apellativen I–II* (Slavic Place Names Derived from Common Nouns I–II, 1872–1874). His comparative grammars are also relevant for the study of Slovenian common and proper nouns (Jakopin 1990:117; Šivic-Dular 2002:15).

Around 1900, a substantial amount of onomastic material was also collected by non-linguists, especially the historians France and Milko Kos (rent-rolls and historical topography) and Pavle Blaznik (historical topography). An important mark on Slovenian onomastics was also left by the Croatian etymologist Petar Skok (Jakopin 1990:117). The reasons for increased interest in this topic at that time lie primarily in Miklosich's publications, as well the publication of the first Slovenian gazetteers (e.g., Peter Kosler's 1864 *Imenik mest, tergov in krajev zapopadenih v zemljovidu slovenske dežele* [Gazetteer of Towns, Market Towns, and Villages on the Map of Slovenian Territory]) and the publication of historical sources that included many names that were older than those in the works by Johann Weikhard von Valvasor (Janez Vajkard Valvasor) from the second half of the seventeenth century (Šivic-Dular 2002:14–15). A few years after Kosler's map, individual maps of *Atlant* (1869–1877) began to be issued. This was the first world atlas published in Slovenian, for which the geographical names were edited by Matej Cigale. This work is important especially from the perspective of handling Slovenianized foreign geographical names (Fridl et al. 2005; Urbanc et al. 2006; Kladnik and Geršič 2016). Slovenian gazetteers continued to be published at the beginning of the twentieth century, and an important contribution was made by Josip Brinar, who produced *Slovarček zemljepisnih imen, njih izvor in pomen* (A Pocket Dictionary of Geographical Names, Their Origin and Meaning, 1928).

In the early twentieth century, names were primarily studied and interpreted by Karel Štrekelj, Luka Pintar, Ivan Koštiál (surnames), Anton Breznik (surnames derived from the names of saints), and others (Jakopin 1990:17). Luka Pintar is considered the most productive researcher of that time. He wrote approximately fifty articles, in which he discussed names in the ethnic Slovenian provinces of the Habsburg Monarchy (i.e., Carniola, Carinthia, Styria, and the Littoral) from various angles. Karel Štrekelj wrote two historical and etymological studies of the originally Slovenian toponymy in the German- and Slovenian-speaking parts of Styria, and mention should also be made of Johann (Janez) Scheinigg, who focused on Carinthian place names (Šivic-Dular 2002:15). During that period, Fran Ramovš and France Bezlaj left a great mark on onomastics. Based on his knowledge of the history of the Slovenian language and dialectology, Ramovš created a solid basis for etymological research on geographical names, and, in reconstructing Slovenian (especially Slovenian phonetics) before AD 1000, he relied on the oldest written attestations of names in Slovenia. Bezlaj (1956, 1961) discussed Slovenian geographical names within a comparative etymological context, and he made an outstanding contribution to research on Slovenian hydronyms (*Slovenska vodna imena I in II* [Slovenian Hydronyms, vols. 1 and 2], 1956 and 1961; Jakopin 1990:17; Šivic-Dular 2002:15). Important onomastic works were also produced by Rajko Nahtigal, the Germanic studies specialist Jakob Kelemina, and the comparative linguist Karel Oštir (Jakopin 1990:17). In 1997, Bezlaj's etymological dictionary was joined by Marko Snoj's *Slovenski etimološki slovar* (Slovenian Etymological Dictionary; Snoj 1997). This was followed by *Etimološki slovar slovenskih zemljepisnih imen* (Etymological Dictionary of

Slovenian Geographical Names; Snoj 2009), which is vital for studying geographical names. It contains over four thousand Slovenian geographical names, which is only a modest percentage of the total, but it is nonetheless an indispensable tool for contemporary researchers of Slovenian geographical names.

In addition to Snoj, major contemporary etymological studies were also published by the linguists Alenka Šivic-Dular, Dušan Čop, Metka Furlan, Silvo Torkar, Matej Šekli, Maja Košmrlj-Levačič, and others. Geographical names have also been dealt with extensively by the geographers Julij Titl, Anton Sore, Ivan Gams, Drago Kladnik, Drago Perko, and Matjaž Geršič. Worth mentioning among the more recent works is Kladnik's detailed study of geographical names in Slovenian geography (Kladnik 2012).

This chapter focuses on the names of Slovenian mountains, or oronyms, which, depending on the scale, account for 7% to 40% of all geographical names in the Register of Geographical Names. The main part presents changes in oronyms in Slovenia using three examples and the specific importance oronyms have among proper nouns.

17.2 Formation of Slovenian Geographical Names

"Because of the geographical location of Slovenian ethnic territory and its history, a basic question in determining the etymology of Slovenian names is that of their origin—that is, whether a name is of Slavic, Germanic, Romance, or Hungarian origin. The answer to this question is sometimes known at first glance,... but often it is so hidden that it becomes clear only at the end of a successful etymological analysis" (Snoj 2002a:37).

The Slovenian name-formation process, or the formation and naming of geographical features in the territory inhabited by Slovenians, was most intense from the seventh to thirteenth centuries—that is, from the first wave of colonization after the settlement of the Slavs to the internal colonization a few centuries later. By the end of that period, there were almost more settlements in Slovenian territory than there are now, especially at higher elevations. In contrast to other Slavs, the Alpine Slavs began establishing permanent settlements relatively early: early historical sources mention around thirty place names of clearly Slavic origin even before or at least at the time of the arrival of the missionaries Cyril and Methodius in this area around AD 850 (Bezlaj 1965:113).

Geographical names in what is now Slovenia were also not immune to the turbulent historical developments after Slavic colonization (Jakopin 1990:116–117). Franz Miklosich was the first to study Slovenian geographical names from the etymological perspective. In line with the spirit of that time, the identification of the root morpheme sufficed for an etymological analysis. In turn, twentieth-century researchers, including Luka Pintar, Karel Štrekelj, Fran Ramovš, and France Bezlaj, also analyzed the linguistic traces indicated by the name's morphological characteristics, word or word-formational suffix, and declension, and the associated demonym and adjective (Snoj 2002a:38). Fran Ramovš—and, even more enthusiastically France Bezlaj and Dušan Čop, researchers from the second half of the twentieth century—advocated the idea that, in addition to taking into account morphological characteristics and conducting a critical analysis of medieval attestations, a precondition for successful etymological analysis was an analysis of dialectal forms. All too often, the standardization of names in Slovenia was in the hands of linguistically naive cartographers, which is why many standard or standardized name forms are misleading (Snoj 2002a:39).

Due to the complexity of linguistic and historical conditions, Slovenian names often contain pre-Slovenian elements (Illyrian, Celtic, Roman, and even pre-Indo-European), and they even more often reflect the later influences of German, Friulian, Italian, Hungarian, and Serbo-Croatian (Jakopin 1990:116–117). There is only a small percentage of pre-Indo-European substrate geographical names. Among them, the names of major rivers (e.g., the Sava) predominate, alongside the names of certain regions (e.g., *Koroška* 'Carinthia') and towns (e.g., *Kranj*). In some side valleys, even minor creeks (e.g.,

Mevlja) or towns (e.g., *Kozana*) have pre-Slavic names. However, there are practically no oronyms of substrate origin. The number of such name increases from east to west, which is quite informative about the oldest relationships between the Slavs and the indigenous inhabitants. However, most Slovenian geographical names are the result of Slovenian name-formation processes (Bezlaj 1965:113; Furlan 2002). To identify Slovenian names of pre-Slavic, substrate origin, it is vital to continuously identify the relationship between the phonetic and morphological characteristics of the inherited Slavic and the adstrate Romance, German (including superstratum), or Hungarian common noun and name material (Furlan 2002:29). Adstrate influences in Slovenian include the influences of neighboring languages from the beginning of Christianization (i.e., the ninth century) onward. They can be roughly divided into Romance, German, Hungarian, and South Slavic influences. Based on chronological and linguistic criteria, the Romance influences are divided into three layers: the non-differentiated Romance influences on the western ethnic border are difficult or even impossible to distinguish from the older substrate influences. These include names borrowed into Alpine Slavic, in which, on the one hand, a phonetic change typical of Alpine Slavic can be noticed, whereas, on the other hand, a specific linguistic feature may demonstrate that the name was borrowed several centuries after the Slavic settlement. A typical example is the oronym *Matajur* (Blaznik et al. 1970), which derives from the Romance name *Monte(m) maiōre(m)* 'higher mountain' (Snoj 2009). More recent adstrate influences in western Slovenia come from Friulian and Venetian Italian, and they can only be distinguished from one another if a name contains a special Friulian or Venetian Italian linguistic feature. Several toponyms of Romance origin can also be found in central Slovenia, but there they are already derived from Slovenian common nouns borrowed from Venetian Italian or Friulian (e.g., *Fužina*, *Fužine*). The Bavarian adstratum is the strongest in the north, especially in Carinthia, Styria, and northern Prekmurje. South of there, the Bavarian influences have more of a superstrate than adstrate character. Between the tenth and fifteenth centuries, over twenty large German enclaves developed in Slovenian ethnic territory due to colonization (Blaznik et al. 1970), of which only two survived until the twentieth century: the one in the Kočevje region (in southeast Slovenia) and the one in the Sorica region (in northwest Slovenia). The residents of other enclaves assimilated with the surrounding Slovenian population, but their dialects left deep traces in the Slovenian common noun and onomastic lexicon. Superstrate influences refer to influences of a language used by colonists that settled within compact Slovenian territory or frontally in the border areas of the former ethnic Slovenian territory, thus making it smaller. There are only a few Romance superstrate elements in Slovenian geographical names (e.g., the oeconym *Bergomaski* in the Laško area) because, unlike the Germanic people, the Romance people hardly ever colonized ethnic Slovenian territory and, except for the Canale Valley (today in northeast Italy), the Friulians did not colonize it at all. A rare example of Italian colonization is the arrival of settlers from the Bergamo area in Laško after 1544. The influence of a Church Latin superstratum can be observed in the names of saints (hagionyms) and the toponyms derived from them, which show typical features of (Church) Latin in which there are no recent Romance, Germanic, or Slavic changes. For example, influences of the Tyrolean superstratum can be seen in the toponym *Vinharje*, the name of a village above the Poljane Valley (*Poljanska dolina*), which is derived from the prepositional phrase **v Iniharje* 'to those from Innichen' (via Tyrolean *Innichaere*; Snoj 2002b).

Slovenia also has many toponyms that were formed from German anthroponyms, using Slovenian name-formation devices; for example, *Dekmanca* from the German anthroponym *Dietmar* with an added Slovenian suffix. Hungarian influences are mostly of a superstrate character and are historically determined by the tenth-century incursions of Hungarian nomadic tribes into the central Danube Basin and their later settlement of the area. Most of them can be found in the Rába Valley in Hungary, and to a somewhat lesser extent in Prekmurje, which belonged to the Hungarian part of Austria-Hungary until 1918, and even fewer in the area between the Drava and Mura rivers (in the Middle Ages and the early modern period, the latter was the border river between Styria and Hungary; Zorn and Mikša 2022). Very few of these influences can be found west of the Drava. Because the linguistic border between the Slovenian and Croatian dialects was never sharply defined (cf. Zajc 2006), South Slavic influences can only be defined as superstrate. This superstratum mainly derives from the Uskok refugees from the Balkans that settled the Habsburg Monarchy border area towards the Ottoman Empire in southeast Slovenia between the fifteenth and seventeenth centuries. Colonization of Croatian peoples is reflected in only a few toponyms in Slovenian Istria (e.g., *Hrvatini*; literally, 'Croatians'), Lower Carniola, and Carinthia. The superstrate influence of West Slavs is only demonstrable in the surnames of people that moved to Slovenian territory when it was part of Austria-Hungary (Snoj 2002b:41–45). Slovenian name forms are attested fairly late, usually only in the nineteenth century (Bezlaj 1969:10).

17.3 Oronyms

17.3.1 Origin

In addition to the names of mountains, Slovenian oronmys include the names of specific parts of mountainous regions and their special features, such as the names of mountain passes, mass movements, and so on (Čop 2002:93). In the mid-twentieth century, they were studied in great detail by Rudolf Badjura, who wrote several works on the topic, including *Gorski prehodi v luči ljudske geografije* (Mountain Passes in the Light of Folk Geography, 1951), *Ljudska geografija: terensko izrazoslovje* (Folk Geography: Field Terminology, 1953), and *Topografični oris 430 gorskih prehodov v Sloveniji* (Topographic Outline of 430 Mountain Passes in Slovenia, 1955). In addition to his concern for normative correctness, his efforts to promote the proper use of names reflect his concern for national identity because he believed that patriotism was also demonstrated through the use of Slovenian names of mountain passes, mountains, and other mountainous landscape elements, rather than German ones (Geršič et al. 2014).

Over the course of history, many peoples passed through what is now the Slovenian territory, which is reflected in the substrate character of oronyms: they are pre-Slavic names of Romance or Celtic origin. Foreign influences in the period after the Slavic settlement are reflected in the adstrate character of the names, most of which are of Romance and Germanic origin. Most oronyms are nonetheless of Slavic origin (Čop 2002:93–94). Their origin also reflects certain settlement characteristics of Slovenia, such as the German character of names of some elevations above the Sorica Pasture in the Julian Alps originating from the period following the area's German colonization in 1213. In addition, oronyms have not been immune to the development of tourism over the past three centuries because tourists and hikers have replaced certain well-established names with new ones (e.g., they replaced *Na Bavhi* with *Slemenova špica*, *Rutarško poldne* with *Visoka Martuljška Ponca*, and *Kuk* with *Kukova špica*; Čop 2002:96).

Oronyms may contain only one word or two or more words. Forms with a prefix or prepositional phrase are less common. One-word oronyms have one or two parts. Two-word or multi-word oronyms are often composed of common nouns preceded by an attributive adjective. Some mountains have several names, especially if they lie between two valleys or major settlements (Čop 2002:94). A good example is the mountain ridge between the Bohinj Valley to the north and the Bača Gorge to the south, which the locals from the Bohinj Valley call *Spodnje Bohinjske gore* 'Lower Bohinj Mountains' (in the southeastern Julian Alps; Figs. 17.1 and 17.8) and the locals from the area south of the ridge (i.e., the Soča Basin) refer to as *Peči* (Sect. 17.3.3.2).

There are various reasons why multiple names have become established for certain features. A typical example of a doublet is Mount Prisojnik/Prisank (2,547 m, Julian Alps; Fig. 17.2). Even though a linguistic analysis conducted by Šivic-Dular and Čop proved that the name *Prisojnik* is

Fig. 17.1 The Bohinj–Tolmin Mountains. (Photo by Matjaž Geršič, GIAM ZRC SAZU archive)

Fig. 17.2 Mount Prisojnik or Prisank (2,547 m) in the Julian Alps. (Photo by Matija Zorn, GIAM ZRC SAZU archive)

etymologically incorrect (albeit not in terms of naming) and their view was also confirmed by the Commission for the Standardization of Geographical Names (Klinar 1996:232), the commission nonetheless prioritized it over *Prisank* because it was already so widely used.

In general, the naming motivation for oronyms comes from personal names, geomorphology, weather, animals and plants, parts of the human body, lexemes related to objects, parts of buildings, foods, colors, whether the mountaintops have forest or grass, and mythological creatures (Čop 2002:94–96).

Slovenian linguistics examines oronyms (i.e., the names of mountain ranges, plateaus, mountains, and hills) at various levels. Usually, an individual oronym or a group of oronyms in a narrow area are studied. A comprehensive study like the one that Bezlaj (1956) conducted for hydronyms is not yet available for oronyms. Mention should be made of Henrik Tuma's work *Imenoslovje Julijskih Alp* (Place Names of the Julian Alps; Tuma 1929), which covers the oronyms in Slovenia's highest and most extensive mountain range. Tuma collected the material for this book on his hikes for over thirty years. He gathered the information on names from the locals, especially hunters, shepherds, and farmers, who were well acquainted with these names, which were passed down from past generations. In this way, he saved a large portion of Slovenian cultural heritage from oblivion, considering that geographical names are part of intangible cultural heritage, which is transferred from one generation to another, as a response to their environment, nature, and history (Klinar and Geršič 2014:419). Without Tuma, "we would have certainly felt a great gap in our place names today that no one else could ever fill, and many places or mountains would have had a foreign name!" (Čop 2000:119). Because Tuma was not a linguist, he also made many mistakes in his findings, especially when determining the origin of a name or its original form (Čop 2000:126).

17.3.2 Register of Geographical Names

Oronyms can be found on the oldest cartographic representations of Slovenian territory. Based on an analysis of fifty-eight medium-scale maps of Slovenia produced between the mid-sixteenth century and the end of the nineteenth century, Gašperič (2022) determined that oronyms were included on over 90% of maps, usually from the mid-seventeenth century onward. They were also provided on large-scale maps, such as the Josephinian military map (produced as part of the first Habsburg military survey at the end of the eighteenth century) and the Franciscan Cadaster (first half of the nineteenth century; Fig. 17.3). Many names of this type were later transferred to base topographic maps and other maps.

The Register of Geographical Names (REZI) is the basic database that can be used to conduct comprehensive analyses of the corpus of Slovenian geographical names (Perko et al. 2023). REZI is the official record kept by the Slovenian Surveying and Mapping Authority. Every geographical feature in the register is ascribed a unique endonym identifier. The register includes endonyms—that is, all geographical names, which are properly capitalized, taking into account the grammatical rules and any exceptions described in the latest edition of the Slovenian normative guide and Toponymic Guidelines for Slovenia (*Toponimska navodila za Slovenijo*; Radovan and Majdič 1995). REZI records names with a lasting chronologically, historically, ethnologically, or socially established identity from cartographic sources. It is set up at three levels of detail: at scales of 1:5,000 (REZI 5), 1:25,000 (REZI 25), and 1:250,000 (REZI 250). REZI 5 contains approximately 150,000 geographical names from the 1:5,000 and 1:10,000 base topographic maps. REZI 25 draws data from the 1:25,000 topographic map and the national topographic map of the same scale, and it includes around 60,000 geographical names. REZI 250 contains around eight thousand geographical names from the 1:250,000 map. In addition, REZI includes standardized geographical names from the 1:1,000,000 map.

Geographical names recorded at the 1:25,000 and 1:250,000 scales have been toponomastically reviewed (Perko et al. 2023). A toponomastic review comprises activities connected with collecting, recording, and processing toponyms in a specific area, including field collection and desk recording and examination of geographical names from various literature, contacts with the locals, and examination of the names' origin and their dialect and standard forms and pronunciation. A normative and grammatical review is also mandatory. Geographical names are represented graphically based on their occurrence, which means that every name or its part is represented as many times as it occurs in various sources or maps. It is represented in the form of an open polygon, which illustrates the location of the name or part of it in the source or on the map (Pravilnik … 2014). Geographical names in the register are divided into various types (Table 17.1).

Fig. 17.3 The oronym for **a** Mount Dobrča (1,634 m): **b** on the 1:28,800 map of the first Habsburg military survey from the end of the eighteenth century and **c** on the 1:2,880 Franciscan Cadaster from the first half of the nineteenth century. (Photo by Matija Zorn, all figures GIAM ZRC SAZU archive)

Table 17.1 The typology of geographical names in the Slovenian register of geographical names. The types and subtypes of geographical names related to mountains are marked in bold

Type code	Type	Subtype code	Subtype
1000	Place names	1101	Town
		1102	Non-urban settlement
		1103	Hamlet
		1104	Part of a settlement
		1105	Street
		1201	Farm
		1202	Religious structure
		1203	Castle
		1204	Isolated shelter
		1205	Important building or complex
2000	Hydronyms	2101	River, part of a river
		2102	Minor watercourse
		2103	Spring
		2104	Outfall
		2105	Waterfall
		2201	Standing water
		2202	Glacier
		2301	Sea
		2302	Part of a sea
		2303	Salt pans
3000	**Oronyms**	**3101**	**Mountain range, lowland, plain**
		3102	**Peak, elevation**
		3103	Depression, valley
		3104	**Part of an elevation or a valley**
		3105	**Saddle**
		3106	**Isolated rock**
		3107	Cave
		3201	Peninsula, island
		3202	Cape
4000	**Choronyms**	4101	Country
		4102	**Landscape**
		4201	**Landscape area**
5000	Urbanonyms	5101	Street
		5102	Rail track
		5103	Road
		5201	Intersection
		5202	Major traffic structure or section

Table 17.1 shows that oronyms also include certain types of names that are not necessarily connected with mountainous landscapes (e.g., the names of valleys). These are basically the names of various landforms. The question that arises here is how oronyms should be defined in the first place. For the purposes of this chapter, oronyms include all names coded 3102 (peak, elevation) and 3101, 3104, 3105, 3106, 4102, and 4201 (Tables 17.1 and 17.2) that are located in the Slovenian

Table 17.2 Basic statistics of oronyms in the Register of Geographical Names (*the names of peaks include three doublets)

Scale	1:1,000,000	1:250,000	1:25,000	1:5,000
Topographic objects: total (*n*)	479	4,221	51,239	117,889
Oronyms: total (*n*)	67	306	15,425	46,048
Percentage (%)	14.0	7.2	30.1	39.1
Code 3101 (*n*)	21	50	62	90
Percentage (%)	4.4	1.2	0.1	0.1
Code 3102 (*n*)	39	219*	8,408	10,591
Percentage (%)	8.1	5.2	16.4	9.0
Code 3104 (*n*)			1,540	1,269
Percentage (%)			3.0	1.1
Code 3105 (*n*)		10	210	138
Percentage (%)		0.2	0.4	0.1
Code 3106 (*n*)			11	26
Percentage (%)			0.02	0.02
Code 4102 (*n*)	7	24	41	83
Percentage (%)	1.5	0.6	0.1	0.1
Code 4201 (*n*)		3	5,153	33,851
Percentage (%)		0.07	10.1	28.7

mountainous areas as defined by Perko in Chap. 3 of this book. Because both the database of geographical names and the landscape typology are in vector form, the digital layer of geographical names was overlapped in GIS with the layer of the mountain area defined to demarcate the types of names mentioned above (except 3102, which was taken into account for the entire country).

17.3.3 Changes

Names are part of the corpus of a specific language. Language itself is a dynamic system, in which words are being created anew, some are disappearing, and the rules change over decades. Geographical names are also not immune to this process. The motives for changing them vary. With regard to place names and the names of streets, major political changes in particular provide ideological backgrounds for altering the names. These processes have been described in detail by Urbanc and Gabrovec (2005) in relation to place names and by Geršič and Kladnik (2016) for the names of streets. Oronyms undergo change as well. Three examples from different historical periods and with different naming motivations are presented below (Fig. 17.4).

17.3.3.1 The Kamnik–Savinja Alps

The standardized name for the mountain range extending between the Karawanks to the north and the Ljubljana Basin to the south was artificially created (Fig. 17.4). Geršič (2020) identified as many as seventeen different variants of the name on maps produced after 1900 (Fig. 17.5). Some appear only once, whereas others are more frequent. There are several reasons for this. Some forms may be the result of misprints or the use of foreign-language equivalents. The form with a hyphen with no spaces on either side, *Kamniško-Savinjske Alpe*, is the most common (it appears 310 times). It is semantically equivalent to the form with the conjunction *in* 'and' (i.e., *Kamniške in Savinjske Alpe* 'Kamnik and Savinja Alps'), which, however, is less frequent (it appears sixty-four times). Looking at the frequency of various name variants on a timeline (Fig. 17.6), it can be determined that up until around 1971 the two sections of the mountain range were referred to separately as *Kamniške planine* 'Kamnik Mountains' and *Savinjske Alpe* 'Savinja Alps', after

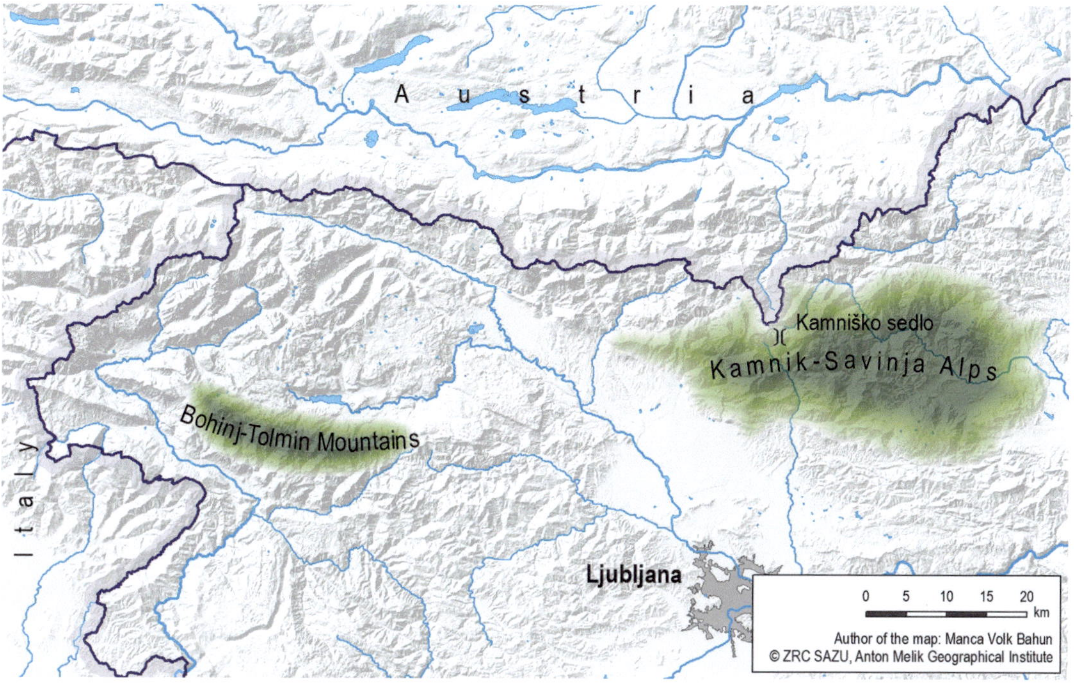

Fig. 17.4 Locations of the oronym case studies

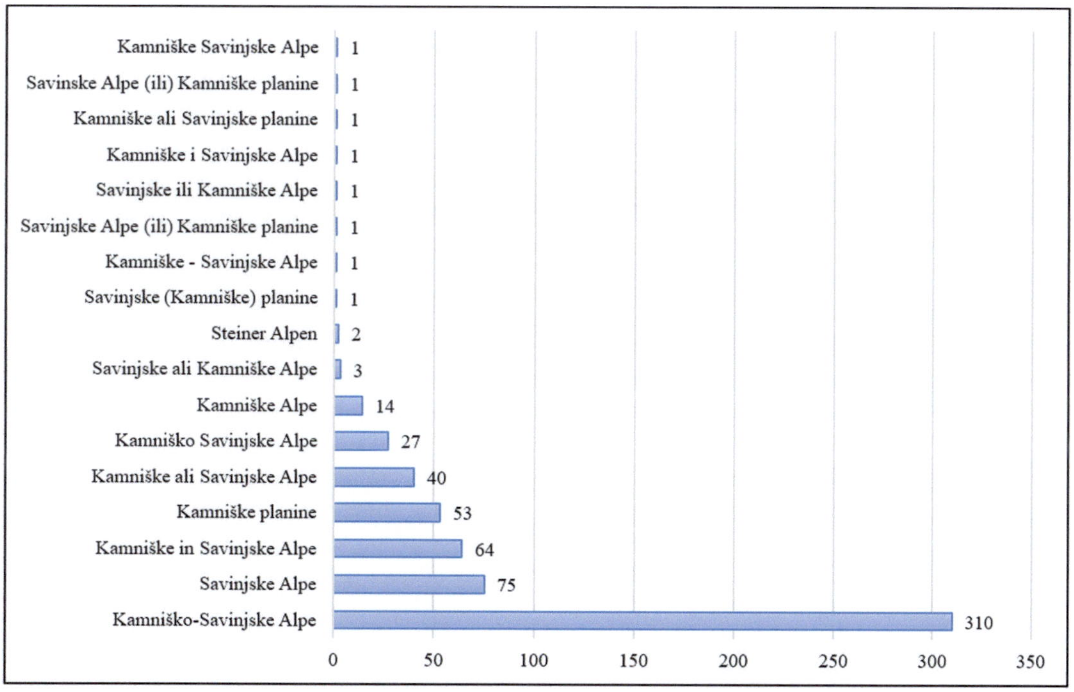

Fig. 17.5 Frequency of variant names for the Kamnik–Savinja Alps

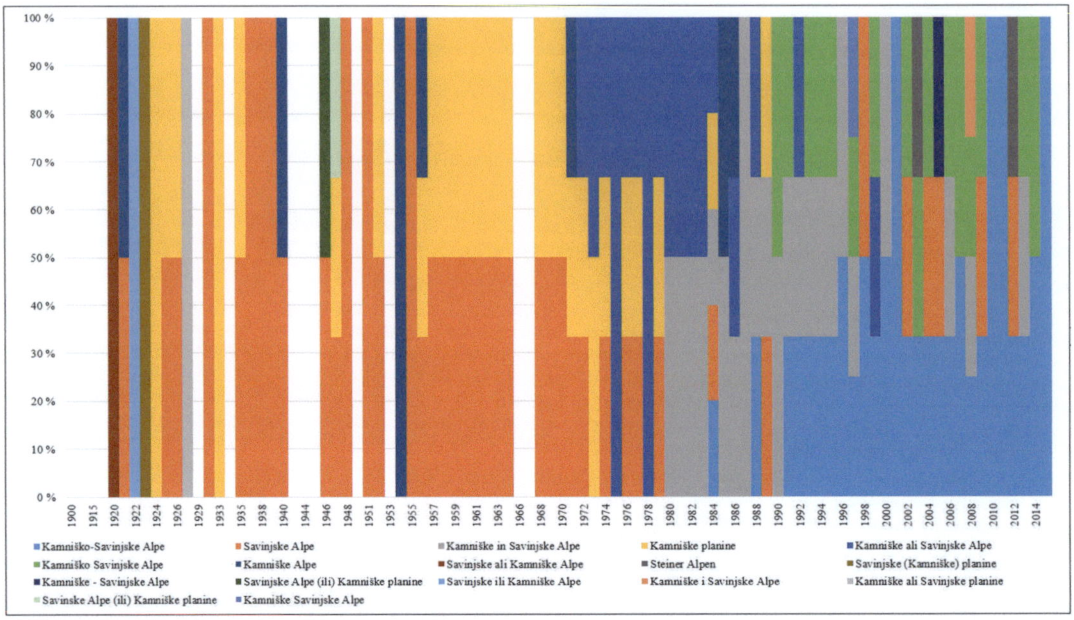

Fig. 17.6 Changes to names referring to the Kamnik–Savinja Alps over time

which the combined variant of the name began to be increasingly used. First, the form *Kamniške ali Savinjske Alpe* 'Kamnik or Savinja Alps' predominated until 1984 (and sporadically until 1999), followed by the form with the conjunction *in* 'and', which is still used today. The form with the hyphen first appeared in 1984 and it has become predominant. Later, the variant *Kamniške Savinjske Alpe* was also identified; it contains no conjunction or hyphen, which is most likely an error.

Historical sources show that the original name of this mountain range was *Snežniki* in Slovenian or *Schneegebirge* in German (literally, 'snow mountains'). It was used by Johann Weikhard von Valvasor in his *Slava vojvodine Kranjske* (The Glory of the Duchy of Carniola, 1689). Approximately a hundred years later, Belsazar Hacquet first referred to the same mountain range in Slovenian as *Ta Kamelshka Planina* 'Kamelk [Kamnik] Mountains' and later as *Steiner Alpen* 'Kamnik Alps' in German (Peršolja 1997). The form *Steiner Alben* first appeared on the maps of Carniola in 1744 (Florjančič pl. Grienfeld 1744; Gašperič et al. 2020). By the end of the nineteenth century, the variant *Steiner Alpen* became established in German. On the map of Carniola and Istria (*Krain – Istrien*), published by Bibliographisches Institut in Leipzig from 1885 to 1892, the variant *Sannthaler Alpen* 'Savinja Alps' appeared for the first time in reference to this area (Geršič 2020). In 1868, the area began to be studied by the Austrian mathematician and alpinist Johannes Frischauf, who adopted the name and also used it as the title for his 1877 guide (the second edition was published in 1881; Glonar 1926). The Slovenian equivalent of this name was advocated by the organizer and founder of organized mountain hiking in this area, Fran Kocbek. In his work *Savinjske Alpe* (The Savinja Alps; Kocbek 1926), he justified the use of this form based on the naming motivation for the Savinja River, which drains the eastern part of the mountain range (Peršolja 1997). The vernacular form *Grintovci* was widely used in Carniola or the lowlands of the Ljubljana Basin (south of the mountain range; Peršolja 1997). The word *grintavec* is a synonym for *dolomit* 'dolomite' (Slovar … 2014), which could also be the naming motivation for this variant. In seeking to explain the origin of this variant, the geographer Borut Peršolja (1997) draws parallels with the term *grintav*, which in his opinion could also mean 'rocky', but that is etymologically incorrect (Slovar … 2014). A possible explanation would be that the surface appears furrowed or resembles scabby skin, and the dialect word for 'scab' in Slovenian is *grinta* (Slovar … 2014).

Discussions on the appropriateness of the names used for this mountain range began after the Second World War. The geographer Anton Melik (1954) suggested that both elements of the name be retained with equal validity. A similar viewpoint was taken by the geographer Svetozar Ilešič (1972; Peršolja 1997), who also supported the use of both elements in the name—that is, the variants *Kamniške in Savinjske Alpe* or *Kamniške ali Savinjske Alpe*—but each could also be used separately when referring only to the easternmost or the Savinja section of the mountain range, or only to its western or Kamnik section (Peršolja 1997). In his work *Slovenske gore* (Slovenian Mountains; Gams 1982), the geographer Ivan Gams advocated the position that this was a single mountain range, which should also be reflected in its name. Therefore, he supported the variant *Kamniško-Savinjske Alpe*, which in his opinion showed both singularity and two starting points.

Table 17.3 Some examples of the names for the Bohinj–Tolmin Mountains used in non-cartographic sources

Name	Source
Krnovo obgorje (*Drežniške gore, Krnove gore, Kukovo/Razorovo pogorje*; from the Vrata to Suha Saddles) *Nemškorutarsko pogorje* (from the Suha Saddle to Mount Slatnik), continuing into the Jelovica Plateau	Rutar (1882)
Nemškorutarsko pogorje (from the Škrbina Pass to the top of Mount Slatnik)	Rutar (1892)
Bohinjski greben	Melik (1927)
Bohinjsko-italijansko pogorje *Goriško-Bohinjsko pogorje*	Tuma (1927a, b)
Bukovske planine in Komna	Tuma (1929)
Bohinjski greben *Spodnje Bohinjske gore* *spodnje Bohinjske gore* *V Pečeh* *Bohinjsko gorovje* *Bohinjsko pogorje* *bohinjske gore*	Melik (1954)
Bukovske gore	Gams (1994)

17.3.3.2 The Bohinj–Tolmin Mountains

Discussions have recently intensified regarding the name of the mountain range between the Upper Carniola Statistical Region (*Gorenjska statistična regija*; NUTS3) to the north and the Gorizia Statistical Region (*Goriška statistična regija*) to the south, or the Bohinj Valley to the north and the Bača Gorge to the south (Fig. 17.1). Since the end of the nineteenth century, the names have varied greatly across sources (Table 17.3).

In 2008, the Commission for the Standardization of Geographical Names standardized the form *Peči* 'Rocky Terrain'. Namely, it established that the form *s/Spodnje b/Bohinjske gore* 'Lower Bohinj Mountains' was relatively recent; it was only first mentioned in 1954 by Melik, and even then with different spellings. The name was artificially coined either as a common noun phrase with an attributive adjective or through conversion of an originally common noun phrase into a proper noun (*bohinjske gore* > *Bohinjske gore*) or conversion of its further structural expansions into a proper noun (*spodnje b/Bohinjske gore* → *gore na južni strani Bohinja* 'mountains on the southern side of the Bohinj Valley'). That this name is inappropriate is also evident from the use of the adjectival modifier *spodnji/zgornji* 'lower/upper', which in the Upper Carniolan dialect normally refers to the position along the course of the river (the upper position is closer to the source and the lower is farther from the source) or the position in the valley (the lower position is at the valley entrance and the upper is at the other end of the valley). However, in this case *lower* is used in the sense of *southern*, which disagrees with the practice that has developed in Slovenian local dialects over the centuries.

In 2006, it was established that the name *Spodnje bohinjske gore* 'Lower Bohinj Mountains' was not present in the local dialect. It turned out that this was a descriptive term for a mountain range in the southern part of the Lower Bohinj Valley.

The name *Peči*, which was also first mentioned by Melik, has greater name value compared to the name formed through the conversion of a homonymous common-noun phrase into a proper noun, but in general it is less informative in terms of location. Nonetheless, it is more appropriate from the linguistic perspective than the name *Spodnje bohinjske gore*.

The facts described and the absence of the name *Spodnje bohinjske gore* in the reference literature (e.g., *Atlas Slovenije* 1985, 1995) ultimately led to the standardization of the name *Peči*.

Later (in 2015) the commission decided that the suggested form *Bohinjsko-Tolminsko gorovje/greben* 'Bohinj–Tolmin Mountains/Ridge' was even more appropriate (Fig. 17.7) than *Peči* because it semantically refers to both regions that it separates. Therefore, the commission adopted a new standardized variant of the name and entered it in REZI. In 2021, the Municipality of Bohinj set up an information board at the Bohinjska Bistrica railway station on which it already used the new official name in a non-standardized form (*Bohinjsko-Tolminske gore* instead of *Bohinjsko-Tolminsko gorovje*; Fig. 17.8). That made some locals angry, and they requested that the previous name be restored. The local newspaper became their main medium, and they also turned to the Ministry of Culture and the Slovenian Academy of Sciences and Arts. To date, the commission has not changed its opinion.

17.3.3.3 The Kamnik Saddle

Kamniško sedlo 'Kamnik Saddle' (1,864 m; Fig. 17.9) is the name of a saddle in the Kamnik–Savinja Alps between Mounts Brana (2,253 m) and Planjava (2,392 m) or between the Logar Valley to the north and the Kamnik Bistrica Valley to the south.

This name was created in the mid-nineteenth century as a translation of the original German name *Steiner Sattel* (Fig. 17.10), which was used by Karel Dežman and Heinrich Hauffen in 1858 and later on by Johannes Frischauf between 1874

Fig. 17.7 The information board at the Uskovnica Pasture north of the Bohinj–Tolmin Mountains, using the new standardized name. (Photo by Matjaž Geršič, GIAM ZRC SAZU archive)

Fig. 17.8 The information board at the Bohinjska Bistrica railway station north of the Bohinj–Tolmin Mountains with the "new" name that has upset some of the locals. (Photo by Matjaž Geršič, GIAM ZRC SAZU archive)

Fig. 17.9 The Kamnik Saddle (1,864 m). (Photo by Matjaž Geršič, GIAM ZRC SAZU archive)

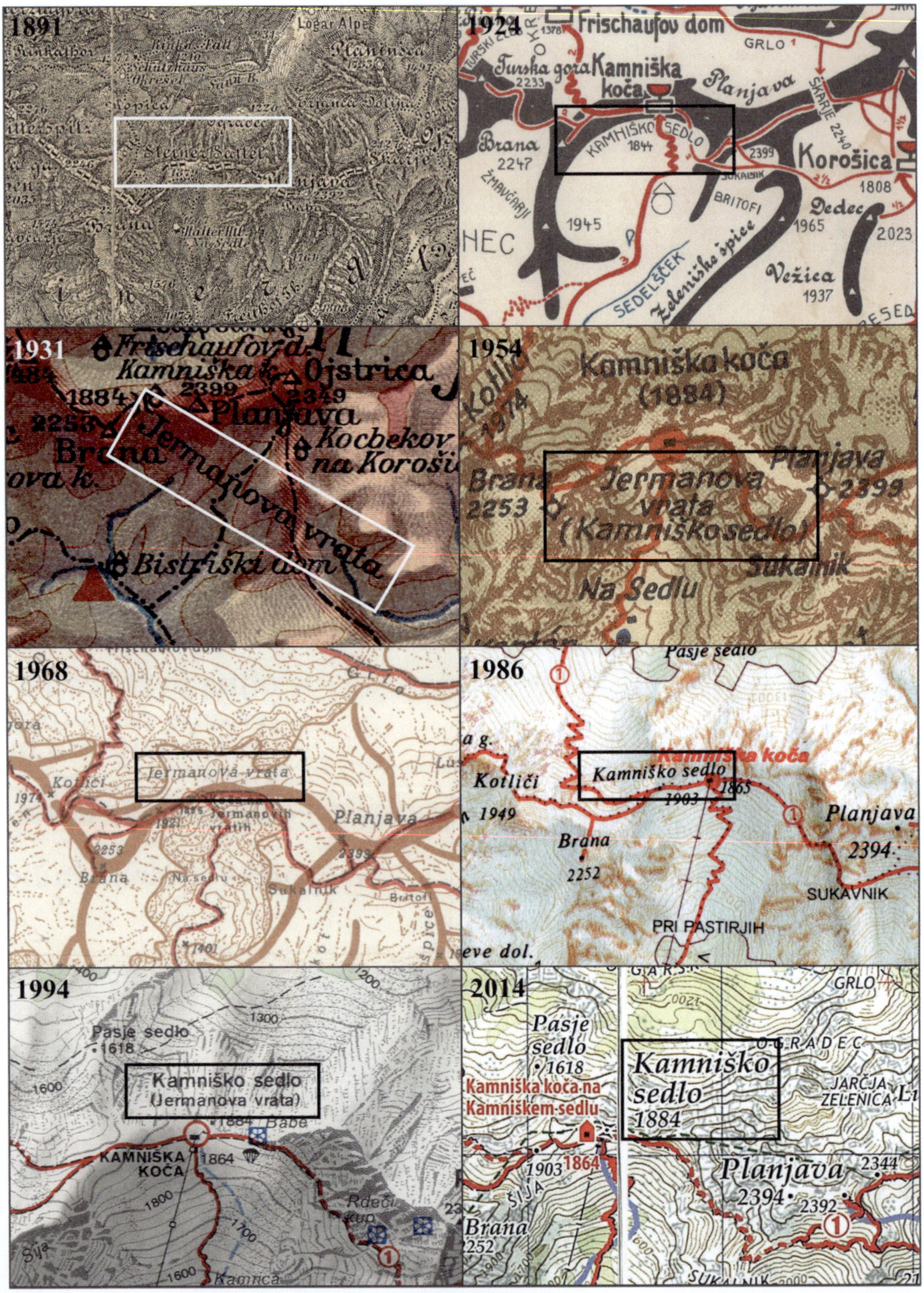

Fig. 17.10 Various names of the saddle between the Logar Valley to the north and the Kamnik Bistrica Valley to the south on selected maps from various periods. (GIAM ZRC SAZU archive)

Fig. 17.11 Ruins of the lodge above Dry Valley (*Suhi Dol*) that Frischauf built before 1876. (Photo by Matija Zorn, GIAM ZRC SAZU archive)

and 1877 (Deschmann 1856; Glonar 1926; Snoj 2009). Frischauf first visited the Kamnik–Savinja Alps in 1868 and he began to study them systematically. In 1874, he founded a society whose aim was to build mountain lodges and trails in the area. In 1876, he had three lodges built: one on the Korošica Pasture below Mount Ojstrica, one at the Okrešelj Cirque above the Logar Valley (called the Frischauf Lodge since 1908), and one above the Dry Valley (*Suhi Dol*; now in ruins; Fig. 17.11) near today's Zois Lodge on the Kokra Saddle (*Kokrško sedlo*) (Mikša and Ajlec 2015:31).

The credit for the name 'Kamnik Saddle' becoming established goes to Frischauf, even though that was not the original name of the saddle. One of the most important collectors of Slovenian folk names, Rudolf Badjura, drew attention to this in several of his works. He discussed this issue most extensively in his book *Gorski prehodi v luči ljudske geografije* (Mountain Passes in the Light of Folk Geography, 1951). What bothered him is that the term *sedlo* 'saddle' became used in Slavic terminology as a direct translation of German *Sattel*. Based on his own observations, he argued that traditionally this term had a completely different meaning in Slovenian than the German word. He published three works on this topic: the article *Slovensko in nemško geografsko sedlo* (The Geographical Term *Saddle* in Slovenian and German; Badjura 1950), followed by the already mentioned booklet *Gorski prehodi v luči ljudske geografije* (Mountain Passes in the Light of Folk Geography; Badjura 1951) a year later, and *Topografični oris 430 gorskih prehodov v Sloveniji* (A Topographic Outline of 430 Mountain Passes in Slovenia; Badjura 1955) a few years after that. In the first work, he explains that traditionally *sedlo* referred to an area on a mountain pasture where shepherds settled down with their herds and built cabins, or to an exposed place where they could sit and watch their animals. According to Badjura, *sedlo* can also refer to a selion in certain inhabited mountainous areas. He lists twenty-seven folk expressions that could be used for mountain passes instead. Based on a detailed analysis of historical sources from Valvasor (i.e., the second half of the seventeenth century) onward, he concludes that "saddles and other similar garbage that has now become nativized in Slovenia for mountain passes, all this has not always been part of our ethnic heritage!" and that "*sedlo* is not the right

Yugoslav term for a mountain pass and therefore has no native right to be used either by the South or North Slavs" (Badjura 1950:147–148).

In the topographic section of this article, Badjura (1950) first lists thirty-three toponyms containing the word *sedlo*, which he believes are inappropriate, and adds a more appropriate name for each. For example, he suggests *Jermanova vrata* 'Jerman Gate' instead of *Kamniško sedlo*, *Zdolška škrbina* 'Dolci Notch' instead of *Dolško sedlo* 'Dolci Saddle', and *Bašeljski vrh* 'Bašelj Peak' instead of *Bašeljsko sedlo* 'Bašelj Saddle'. He continues by listing thirty-eight toponyms in which, in his opinion, the word *sedlo* is appropriately used in both Slovenia and elsewhere. He concludes his article by commenting that the fact that some old Slavic terms were replaced with *sedlo* is "intolerable" and that "such chaos must be brought to an end as soon as possible!" He was convinced that the task of geography is "to define, maintain, and protect good native terms and names, rather than neglect and replace them with foreign and incorrect ones" (Badjura 1950:170).

A year later, Badjura (1951) expanded and further elaborated his previous study. The introduction is followed by a presentation of geographical categorizations of mountain passes and an onomastic typology of terms for mountains passes, and a concluding chapter on the topographic names of mountain passes. However, Badjura's research on mountain passes did not end there. Four years later, he published a work (Badjura 1955) in which he sought to present a methodology for collecting material on mountain passes in all Slovenian regions to produce a gazetteer for the uniform use of names of mountain passes (Badjura 1955:5).

However, Badjura's persistent rejection of the term *sedlo* did not have a sound basis. Modern etymology has established that the word has its origin in Proto-Slavic and it is derived from the Proto-Indo-European root **sed-* 'to sit'. The Old High German word *satul*, from which the modern German word *Sattel* 'saddle' is derived, is also borrowed from Proto-Slavic (Snoj 1997:557). It can thus be concluded that exactly the opposite is true of what Badjura claimed. With regard to the meaning of the word *sedlo*, it is now understood as "a device used for sitting or carrying that is fitted onto an animal's back" or as "a wide, gentle passage across a mountain ridge from one valley to another" (Slovar … 1985:612). The *Standard Slovenian Dictionary* does not include any of the meanings that Badjura listed, but this does not mean that people did not know them in the past.

Even though Badjura was a fervent advocate of old names, he entitled his first work of this type (co-authored with Bogumil Brinšek) *Izlet na Kamniško sedlo* (A Trip to the Kamnik Saddle; Badiura and Brinšek 1908), and he only used the title *Jermanova vrata* (The Jerman Gate; Badjura 1961) for one of his last works. *Jermanova vrata* was definitely a well-established name for this topographic feature among the locals in the past, but Frischauf was evidently not familiar with it.

Legend has it that Jerman was a poacher that knew all the trails in the area. When the Ottomans attacked Klagenfurt, he led the Christian army, which the Ottomans had tricked and lured into Carniola, back to Carinthia using the shortest route, across the Kamnik Saddle. They arrived in Carinthia that same day. The Ottomans were defeated and Klagenfurt was saved, but Jerman fell in battle. In gratitude to him and to commemorate his efforts, they named the saddle after him (Cerar 2002).

17.3.4 Meaning

Geographical names perform a variety of onomastic functions. The Croatian onomastician Petar Šimunović (2009) distinguishes between naming, identifying, differentiating, connotative, social, emotive, descriptive, ideological, and other functions. One of the main functions of names is to designate a feature with that name. This function is denotative (or marking). The meaning must be known to both the creator of the name and the addressee. Without that, it means nothing (Šimunović 2009:29). All this also applies to oronyms. Their specific functions are not discussed here in detail, but a clarification is provided for some that are interesting from the perspective of oronyms in Slovenia.

In addition to providing orientation to the residents and visitors of mountainous regions, (especially) in the past oronyms were an important (national) identity factor. Before the Slovenian Alpine Club (*Slovensko planinsko društvo*, SPD) was established in 1893 (as well as later), mountain lodges in Slovenian territory were built by the German–Austrian Alpine Club (*Deutscher und Österreichischer Alpenverein*, DÖAV) in order to Germanize the Slovenian mountains (Mikša and Zorn 2021). Several of its sections were active in Slovenian territory, including the Carniolan Section (*Section Krain*) between 1874 and 1919 (Peternel 2023). The DÖAV used German names for its mountain lodges: it built the *Hubertushütte* in the Vrata Valley, the *Deschmannhütte* below Mount Triglav, and the *Vosshütte* at Vršič Pass. However, it also used German names for the peaks and other geographical features in the high mountains (e.g., passes and uncultivated land). Some ethnically conscious Slovenians fought against using German names in the Slovenian mountains. In the first half of the twentieth century, strong concern for these names was shown by two members of the Skala Alpine Club (*Turistovski klub Skala*): Henrik Tuma and Rudolf Badjura. They both came from families of wealthy town craftsmen, and they already became acquainted with mountains in their childhood. Schooling abroad made them even more aware of the importance of the Slovenian language, which is also evident from their works. Even though they were not trained linguists and made certain errors in their work, their legacy related to the conservation of oronyms in Slovenian is invaluable (Geršič et al. 2021). Their work, including the collection of Slovenian microtoponyms, continues (e.g., Klinar et al. 2012). Namely, the identity function of Slovenian geographical names is still important today. This is especially evident in the mountainous areas of neighboring Austria and Italy, where geographical names testify to the primary settlement of Slovenians. The outstanding cultural significance of exclusively Slovenian field and house names has also been confirmed by their inclusion in the UNESCO list of intangible cultural heritage in Austria (Piko-Rustia 2013).

In the mountainous regions, oronyms also have a semantic value. The geographical names preserved reveal the condition the landscape was in and its use when they were created, and they are a reliable indicator of a cultural landscape's transformation from its original condition to its present form (Peršolja 2002:364). They are thus key in studying land-use changes and in historical cartography.

Some names, especially those of non-Slavic origin, can be very old and testify to millennia-long use of an area. In South Tyrol (Italy), it was established that most geographical names in the mountainous regions originate from the High Middle Ages or later, which is connected with permanent settlement (Ortner et al. 2012:50). Geographical names also reflect natural phenomena, such as natural disasters. These are not the most frequent naming motivation for the geographical names in the Alps, but names connected with natural disasters are definitely among the oldest ones (Ortner 2014:73). For a phenomenon to enter the collective memory, it must be sufficiently intense and frequent (Ortner 2014:73). In the past, almost any natural phenomenon that "differed from its environment received its own name" (Flom 1924:205). When speaking of geographical names and natural disasters, what one has in mind is especially disasters with a localized character, which means they are limited to a specific area. These include rockfalls, landslides, avalanches, fires, and windfalls. Natural disasters on a larger scale, such as floods or earthquakes, are often named with common nouns rather than geographical names (Geršič and Zorn 2016). The importance of geographical names in studying natural disasters in the Slovenian Alps and producing hazard maps, especially for avalanches, has been highlighted by the geographer Miha Pavšek (2002). He reports that avalanches (Sln. *snežni plazovi*) are often named using noun phrases that contain the "word *plaz*, *plazi* or *plazovi* (*plazovje*)," either independently or with a modifier to the left or right. Pavšek also draws attention to names that contain the words *plazič*, *omel*, and *osip*, and summarizes Badjura, who mentions the terms *grundnek* and *gruntnik* borrowed from German, and *labinje* and *lavina*, which go back to Latin *labi* in the sense 'slide or skid downward'.

Several names of Slovenian settlements are "reminiscent" of landslides or avalanches, such as *Plaz* or *Na Plazu* in Podljubelj, *Plazne pri Izlakah*, *Plazovje pri Laškem*, and *Blasnitzen* (Sln. *Plaznica*) near Eisenkappel (Austria), in addition to several non-settlement names, such as *Vallone Blasic* (Sln. *Plazič*) south of the Raccolana Valley (Italy), and *Begunjski plaz*, *Šentanski plaz*, and *Birški plaz* in the Karawanks. For the Julian Alps, many geographical names with the motivational meaning 'slide' were already selected by Henrik Tuma. In turn, *Šraj peski* in the Kamnik–Savinja Alps or *Podrta gora* (2,061 m) in the Julian Alps are examples of names that refer to rockfalls. The reflection of natural disasters in geographical names in the Western Karawanks and western Kamnik–Savinja Alps were examined in detail by Geršič and Zorn (2016). Etymological research and the examination of the naming motivation for individual names in the mountainous regions provide interesting insights especially into certain natural characteristics of land (e.g., landforms, the structure and other characteristics of soil, and land orientation), the presence of water, weather phenomena, plants, animals, events in the natural environment, and human impacts, such as land use (Fig. 17.12) or cultivation. The semantic value of these names is also significant because

Fig. 17.12 **a** The name of the cleared part on the southern slope of Mount Dobrča is *Seče* (literally, 'mowing'), where farmers used to cut grass with a scythe to make hay, which they fed to their cattle in winter (Photo by Matjaž Geršič). **b** The Franciscan cadaster map of Zgornje Jezersko from the first half of the nineteenth century contains the microtoponym *na Poschkanische* (cf. Sln. *Požganica* 'burned-over land'), which may refer to the practice of burning to maintain the pastures on the northern slopes of the Kamnik–Savinja Alps. (GIAM ZRC SAZU archive)

many land-cultivation practices that were well-established in the past are now disappearing in many areas, and it is precisely the microtoponyms and other names that have been preserved in the mountainous regions that continue to indirectly testify to previous economic activities in these areas.

17.4 Conclusion

Geographical names in Slovenian mountainous regions share the fate of other geographical names. The oldest are substrate and pre-Slavic (Romance or Celtic) in origin, and most are the result of the Slovenian name-formation process. In this regard, mention should be made of the adstrate influences or the influences of neighboring languages from the beginning of Christianization, and the superstrate influences of languages used by the peoples that colonized what is now Slovenia. The corpus of Slovenian oronyms is extensive; in addition to the names of mountains and mountain ranges, it includes many other names, such as the names of mountain passes, valleys, uncultivated land, and so on. Doublets are an interesting feature: some elevations can have two names if they are located between two valleys or settlements.

No comprehensive onomastic analysis of Slovenian oronyms has been conducted to date, but some authors have studied them in specific areas since the early twentieth century. Oronyms can already be found on the oldest cartographic representations of Slovenian territory, usually from the seventeenth century onward, and later many of them were transferred to official military and other maps. Together with other name categories, oronyms are listed in the Register of Geographical Names. It was established that, depending on the scale of maps in the register, the oronyms as defined in this article account for 7% to 40% of all geographical names. The high share of oronyms taken from large-scale maps can mainly be attributed to microtoponyms. Excluding the names of mountains and other elevations, this share is between 0.1 and 4%.

Like certain other types of geographical names, oronyms also experience change. The motives for changing these names may vary, but some changes may lead to strong emotional responses in individuals. A good example is the change of the standardized name *Peči* (literally, 'mountains') into *Bohinjsko-Tolminsko gorovje* 'Bohinj–Tolmin Mountains'.

It is also established that oronyms perform a variety of functions. In the past, the (ethnic) identity function was important for Slovenians, but the importance of the semantic function is also highlighted. This may provide insight into certain past economic and other practices in mountainous regions, the occurrence of natural disasters, and so on. In some cases, geographical names may facilitate the reconstruction of a former landscape, which is why they are also an important source for historical geography, identifying environmental changes, and so on.

Acknowledgements This study was conducted as part of the research program The Geography of Slovenia (P6-0101) and the infrastructure program Natural and Cultural Heritage (I0-0031), financed by the Slovenian Research and Innovation Agency.

References

Atlas Slovenije [Atlas of Slovenia] (1985). Mladinska knjiga, Ljubljana
Atlas Slovenije [Atlas of Slovenia] (1995). Mladinska knjiga, Ljubljana
Badiura R, Brinšek B (1908) Izlet na Kamniško sedlo [A trip to the Kamnik Saddle]. Planinski Vestnik 14(1):12–14
Badjura R (1950) Slovensko in nemško geografsko sedlo [The geographical term *saddle* in Slovenian and German]. Geografski Vestnik 22:138–170
Badjura R (1951) Gorski prehodi v luči ljudske geografije [Mountain passes in the light of folk geography]. Državna založba Slovenije, Ljubljana
Badjura R (1955) Topografični oris 430 gorskih prehodov v Sloveniji [Topographic outline of 430 mountain passes in Slovenia]. Unpublished material, Ljubljana
Badjura R (1961) Jermanova vrata (1921 m) [Jerman Gate]. Planinski Vestnik 61(1):35–37
Bezlaj F (1956) Slovenska vodna imena I [Slovenian hydronyms, vol. 1]. SAZU, Ljubljana
Bezlaj F (1961) Slovenska vodna imena II [Slovenian hydronyms, vol. 2]. SAZU, Ljubljana
Bezlaj F (1965) Slovenski imenotvorni proces [Slovenian name formation process]. Jezik in Slovstvo 10(4–5):113–118

Bezlaj F (1969) Onomastika in leksikologija [Onomastics and lexicology]. Onomastica Jugoslavica 1:10–21
Blaznik P, Grafenauer B, Kos M, Zwitter F (1970) Kolonizacija in populacija [Colonisation and population]. In: Blaznik P, Grafenauer B, Vilfan S (eds) Gospodarska in družbena zgodovina Slovencev: Zgodovina agrarnih panog 1, Agrarno gospodarstvo. Državna založba Slovenije, Ljubljana, pp 29–127
Cerar I (2002) Jermanova vrata na Koroško [The Jerman gate to Carinthia]. Gea 12(2):68–69
Čop D (2000) Dr. Henrik Tuma in imenoslovje Julijskih Alp [Dr. Henrik Tuma and the place names of the Julian Alps]. In: Tuma H, Imenoslovje Julijskih Alp. Tuma, Ljubljana, pp 119–126
Čop D (2002) Gorska, terenska (ledinska) in vodna imena v Sloveniji [Mountain, terrain (field) and water names in Slovenia]. Jezikoslovni Zapiski 8(2):93–108
Deschmann C (ed) (1856) Jahresheft des Vereines des krainischen Landes-Museums. Vereines des krainischen Landes-Museums, Laibach
Flom GT (1924) Place-name tests of racial mixture in Northern England. Mod Lang Notes 39(4):203–212
Florjančič pl. Grienfeld JD (1744) Ducatus Carnioliae tabula chorographica [Chorographic map of the Duchy of Carniola]. Ljubljana
Fridl J, Kladnik D, Orožen Adamič M, Furlan M (2002) Predslovanska substratna imena v slovenščini [Pre-Slavic substratum names in Slovenian]. Jezikoslovni Zapiski 8(2):29–35
Fridl J, Kladnik D, Orožen Adamič M, Perko D, Urbanc M (eds) (2005) Atlant. Založba ZRC, Ljubljana. https://doi.org/10.3986/9789612545093
Furlan M (2002) Predslovanska substratna imena v slovenščini [Pre-Slavic substratum names in Slovenia]. Jezikoslovni Zapiski 8(2):29–35
Gams I (1982) Gore [Mountains]. In: Raposa K (ed) Slovenske gore. Cankarjeva založba, Ljubljana, pp 9–66
Gams I (1994) Termalni pas v Sloveniji [Thermal zone in Slovenia]. Geografski Vestnik 68:5–38
Gašperič P (2022) Zgodovinska kartografija ozemlja Slovenije [Histrorical cartography of Slovenian territory]. Geografija Slovenije, vol 37. Založba ZRC, Ljubljana. https://doi.org/10.3986/9789610505907
Gašperič P, Šolar R, Zorn M (2020) Kartografski zakladi slovenskega ozemlja = Cartographic treasures of Slovenian territory. Založba ZRC, Ljubljana. https://doi.org/10.3986/9789610504276
Geršič M (2020) Pokrajinska imena kot dejavnik identitete [The names of regions as an identity factor]. Geografija Slovenije, vol 36. Založba ZRC, Ljubljana. https://doi.org/10.3986/9789610504368
Geršič M, Batagelj B, Berčič H, Bokal L, Guček A, Kavar J, Kocutar S, Komac B, Marković Z, Mikša P, Torkar B (2014) Rudolf Badjura—življenje in delo [Rudolf Badjura's life and work]. Geografija Slovenije, vol 29. Založba ZRC, Ljubljana. https://doi.org/10.3986/978 9610503378
Geršič M, Gašperič P, Zorn M (2021) Henrik Tuma in Rudolf Badjura: imenoslovca med skalaši [Henrik Tuma and Rudolf Badjura: mountain-climbing onomasticians]. In: Mikša P, Vehar M, Lačen Benedičič I, Zorn M (eds) Kvišku kakor skala vodi naša pot: ob stoletnici nastanka turistovskega kluba Skala. Retrospektive, Ljubljana, pp 67–83
Geršič M, Kladnik D (2016) Street-name changes in Ljubljana. In: Jordan P, Woodman P (eds) Place-name changes: proceedings of the Symposion in Rome. Verlag Dr. Kovač, Hamburg, pp 249–274
Geršič M, Zorn M (2016) Odsev naravnih nesreč v zemljepisnih imenih—nekaj primerov iz Zahodnih Karavank in zahodnih Kamniško-Savinjskih Alp [Natural disasters reflected in geographical names—a few examples from the Western Karavanke Mountains and western Kamnik-Savinja Alps]. Kronika 64(3):575–582
Glonar J (1926) Frischauf, Johannes (1837–1924). In: Cankar I et al (eds) Slovenski bijografski leksikon, vol 2. Zadružna gospodarska banka, Ljubljana. https://www.slovenska-biografija.si/oseba/sbi193134/. Accessed 25 May 2023
Ilešič S (1972) The regions of Slovenia. Geografski Vestnik 44:20–31
Jakopin F (1990) Imenoslovje [Onomastics]. In: Javornik M (ed) Enciklopedija Slovenije, vol 4. Mladinska knjiga, Ljubljana, pp 116–117
Kladnik D (2012) Slovenian geography and geographical names. Geografski Vestnik 84(1):237–249
Kladnik D, Geršič M (2016) Cigale's world atlas and its influence on Slovenian exonyms. Forum Ifl 30:91–102
Kladnik D, Geršič M, Perko D (2020) Slovenian geographical names. Acta Geogr Slov 60(3):7–190. https://doi.org/10.3986/AGS.9394
Klinar K, Geršič M (2014) Traditional house names as part of cultural heritage. Acta Geogr Slov 54(2):411–420. https://doi.org/10.3986/AGS54409
Klinar K, Škofic J, Šekli M, Piko-Rustia M (2012) Metode zbiranja hišnih in ledinskih imen [Methods of collecting house and field names]. Gornjesavski muzej, Jesenice
Klinar S (1996) Prisank ali Prisojnik? [Mount Prisank or Mount Prisojnik?]. Planinski Vestnik 96(5):232
Kocbek F (1926) Savinjske Alpe [The Savinja Alps]. Goričar & Leskovšek, Celje
Melik A (1927) Morfologija in gospodarska izraba tal v Bohinju [Morphology and land use in the Bohinj Valley]. Geografski Vestnik 3(1–4):53–95
Melik A (1954) Slovenski alpski svet [Slovenian Alpine world]. Slovenska matica, Ljubljana
Mikša P, Ajlec K (2015) Slovensko planinstvo = Slovene mountaineering. Planinska zveza Slovenije, Ljubljana
Mikša P, Zorn M (2021) Nacionalni boj za gore in slovenska identiteta [The national battle for the mountains and the Slovenian identity]. Časopis za Zgodovino in Narodopisje 92(1):67–88
Ortner J (2014) Naturgefahren im Spiegel der Flurnamen. Der Schlern 88(10):61–75

Ortner L, Kaufmann R, Kathrein Y, Pidner J (2012) Die Landschaft und ihre Namen—Landwirtschaftliche Nutzung und Nutzungsänderungen im Spiegel der Flurnamen von Obergurgl und Vent (Ötztal). In: Koch E-M, Erschbamer B (eds) An den Grenzen des Waldes und der menschlichen Siedlung. Universität Innsbruck, Innsbruck, pp 39–73

Pavšek M (2002) Snežni plazovi v Sloveniji [Avalanches in Slovenia]. Geografija Slovenije, vol 6. Založba ZRC, Ljubljana. https://doi.org/10.3986/9789612545000

Perko D, Geršič M, Zorn M (2023) Standardization of geographical names on land and sea in Slovenia. In: O'Reilly G (ed) Place naming, identities and geography. Springer, Cham, pp 193–239. https://doi.org/10.1007/978-3-031-21510-0_9

Perko D, Urbanc M (eds) (2005) Atlant. Založba ZRC, Ljubljana. https://doi.org/10.3986/9789612545093

Peršolja B (1997) Geografski problemi imenoslovja Kamniško-Savinjskih Alp [Geographical problems of onomastics in the selected example of the Kamnik-Savinja Alps]. BSc Thesis, University of Ljubljana, Ljubljana

Peršolja B (2002) Zgoščeni imenik zemljepisnih imen [Concise gazetteer of Slovenia]. Dela 18:363–373. https://doi.org/10.4312/dela.18.363-373

Piko-Rustia M (2013) Slovenska ledinska in hišna imena na Koroškem na Unescovem seznamu [Slovenian field and house names in Carinthia in the UNESCO list]. Koledar za leto 2013:135–138

Pravilnik o podrobnejši vsebini registra zemljepisnih imen [Rules on the detailed content of the register of geographical names] (2014) Uradni list Republike Slovenije 54/2014. Ljubljana

Peternel MM (2023) "Ljubiteljem kranjskih Alp!": Kranjska podružnica Nemškega in avstrijskega planinskega društva ["To the lovers of the Carniola Alps!": Carniola branch of the German and Austrian Alpine Club]. Založba Univerze, Ljubljana. https://doi.org/10.4312/9789612971069

Radovan D, Majdič V (1995) Toponymic guidelines for Slovenia. Surveying and mapping authority of the Republic of Slovenia, Ljubljana

Rutar S (1882) Zgodovina Tolminskega, to je: zgodovinski dogodki sodnijskih okrajev Tolmin, Bolec in Cerkno ž njih prirodoznanskim in statističnim opisom [The history of the Tolmin region: historical events of the Tolmin, Bovec, and Cerkno judicial districts with their natural-history and statistical descriptions]. J. Devetak, Gorica

Rutar S (1892) Poknežena grofija Goriška in Gradiščanska [The Princely County of Gorizia and Gradisca]. Matica Slovenska, Ljubljana

Slovar slovenskega knjižnega jezika [Standard Slovenian dictionary] (1985). Državna založba Slovenije, Ljubljana

Slovar slovenskega knjižnega jezika [Standard Slovenian dictionary] (2014). Cankarjeva založba, Ljubljana

Snoj M (1997) Slovenski etimološki slovar [Slovenian etymological dictionary]. Mladinska knjiga, Ljubljana

Snoj M (2002a) O etimologiji osebnih in zemljepisnih imen na Slovenskem [On etymology of personal and place-names in Slovenia]. Jezikoslovni Zapiski 8(2):37–40

Snoj M (2002b) Adstratni in superstratni vplivi na slovensko imensko leksiko [Adstratum and superstratum influences in Slovenian onomastic lexis]. Jezikoslovni Zapiski 8(2):41–45

Snoj M (2009) Etimološki slovar slovenskih zemljepisnih imen [Etymological dictionary of Slovenian geographical names]. Modrijan, Založba ZRC, Ljubljana

Šimunović P (2009) Uvod u hrvatsko imenoslovje [Introduction to Croatian onomastics]. Golden marketing-Tehnička knjiga, Zagreb

Šivic-Dular A (2002) Zgodovina imenoslovja v Sloveniji [The history of onomastics in Slovenia]. Jezikoslovni Zapiski 8(2):7–28

Tuma H (1929) Imenoslovje Julijskih Alp [Place names of the Julian Alps]. Slovensko planinsko društvo, Ljubljana

Tuma H (1927a) Bohinjski kot. Planinski Vestnik 27(3):49–60

Tuma H (1927b) Bohinjski kot. Planinski Vestnik 27(7):148–152

Urbanc M, Gabrovec M (2005) Krajevna imena: poligon za dokazovanje moči in odraz lokalne identitete [Place names: a testing ground for demonstration of power and a reflection of local identity]. Geografski Vestnik 77(2):25–43

Urbanc M, Fridl J, Kladnik D, Perko D (2006) Atlant and Slovene national consciousness in the second half of the 19th century. Acta Geogr Slov 46(2):251–283. https://doi.org/10.3986/AGS46204

Zajc M (2006) Kje se slovensko neha in hrvaško začne: slovensko-hrvaška meja v 19. in na začetku 20. stoletja [Where Slovenian ends and Croatian begins: the Slovenian-Croatian border in the 19th and early 20th centuries]. Modrijan, Ljubljana

Zorn M, Mikša P (2022) Mejniki na štajersko-ogrski meji—nekaj primerov iz Občine Ljutomer in okolice [Boundary stones on the border between the Duchy of Styria and the Kingdom of Hungary. A few examples from the Municipality of Ljutomer and its surroundings]. Kronika 70(3):921–932. https://doi.org/10.56420/Kronika.70.3.18

18 Depiction of the Montenegrin Mountains as Seen in British, German and French Travelogues: Geosophy in Montenegro in the 19th and Beginning of the 20th Century

Božena Miljić and Danilo Kalezić

Abstract

Until the nineteenth century, Montenegro was almost unknown to the people in Western Europe, with whom it had no close diplomatic, cultural, or economic ties. However, since the nineteenth century, interest in this mystical, small Balkan country has been growing in Europe, which has resulted in the arrival of foreign envoys, diplomats, and scientists, coming on various missions. Even though they came to Montenegro with specific goals in mind, these foreigners left valuable travel notes about it. From these travelogues, as well as newspaper articles they published in their countries, the inhabitants of European cities were for the first time informed about Montenegro, its extraordinary nature, the highlanders who lived in these isolated areas, and the country's turbulent history and customs. Travelogues of British, German, and French travel writers (although, of course, there were travelogues from other countries, for this paper we will solely dwell on Western European sources) tell us about the perception that foreigners had about Montenegro, about which they knew almost nothing. It is understood that foreigners came to this little-known country with certain prejudices, sometimes even fears, but the fact is that they were all fascinated by the natural beauty of Montenegro. The mountain landscape, as the entire region after all has been mystified through these vivid travelogues. They extend factual geographical knowledge and deal primarily with the imagination of mentality, customs, and general life in mountains. Since travel writers bring specific philosophy of place, their published accounts became a way of claiming and understanding the environment and therefore significantly influenced the way mountains are perceived, evaluated, and utilized. Therefore, the principal aim of this paper is to single out the experiences foreign travel writers had when encountering Montenegrin mountains and the highlanders who lived on them–how they understood the fact that a Montenegrin man was a man of the mountain and that mountains influenced the characterology and the very identity of Montenegrins. By crossing it with historiographic analysis, the aim is to explore the ways nineteenth and early twentieth-century explorers from abroad understood the cultural identity of mountainous Montenegrin communities.

Keywords

Montenegro · Mountains · Travelogues · History · The Balkans

B. Miljić (✉)
University Donja Gorica, Podgorica, Montenegro
e-mail: bozenam@t-com.me

D. Kalezić
Historical Institute, University of Montenegro, Podgorica, Montenegro
e-mail: kalezic13@gmail.com

M. Zorn et al. (eds.), *Discourses on Mountains of Montenegro and Slovenia*, Historical Geography and Geosciences, https://doi.org/10.1007/978-3-031-99017-5_18

18.1 Introduction

"*When creating the world and arranging the plains and mountains on Earth, God carried the necessary rocks in a sack. After a while, the bag wore out and tore, and the rocks plummeted, making a huge pile, which was later called Montenegro.*" This legend was apparently interesting to travel writers, as it was recorded by almost everyone who heard it (Lainović 1949:69).

It is in human nature to be curious, to explore and learn, to discover the unknown–to travel. Since the earliest times remembered by written sources, it has been characteristic of humans to record what they have learnt and seen. Whether they went on war expeditions or relocated in search of better living conditions, people would leave traces of the new areas and civilizations they encountered. The whole world started getting closer and more familiar because wars led warriors from one end of the world to the other, and the discovery of new lands along with great geographical discoveries broadened people's horizons. It was the ancient epics that recorded the first travels, after which the travel literature evolved and became a very popular genre–it started with Renaissance travels and curious Baroque travellers, and continued evolving to Romantics and adventurers of the nineteenth century. Travelogues is an important genre in literature (Šačkova 2012), both from a literary and a cultural-historical point of view, because they pass on geographical knowledge, teach how people were observed, and give a subjective view of the world (Konstantinović 1986:623–624). Vladimir Gvozden (2012:185) writes that "*The travelogue as a story is based on an encounter*". The essence of travel writing is discovering and getting to know new unknown cultures, people and places, and it is the writer who colours with words what he has experienced. And that is the richness of travelogues, they bring us closer, more than any other literary genre, to certain country and its people–they are the first-hand experience, a primary source, which at the same time helps the readers to broaden their horizons, acquire new experiences and knowledge, break prejudices about the unknown. Of course, travelogues should be read and understood with a dose of criticism, because they are primarily a subjective experience of what is described, which depends on many factors–the writer may have gone to some unknown and mystical area with preconceived prejudices and already formed opinions, which may be even shaped by fear. We should also take into account that the writer's notes are written almost always in specific historical and political circumstances that can often change their perspective.

The purpose of this paper is not to correct historical or linguistic errors, even though the authors of this article are aware that there are many of them, given travelogue writers' limited knowledge of Montenegrin culture and language. Still, these errors do not, in any way, affect the essence of the description of natural beauties and landscapes. Except for the occasional mispronounced toponym, the descriptions of mountains and hills are a credible result of the writers' experiences.

Travelogues are especially important for discovering a country like Montenegro–hidden among the hills, geographically in Europe, but culture and civilisation-wise centuries away. The Montenegrins themselves were taught a lot about Montenegro, their own country, by occasional travellers who visited its most remote parts. A Montenegrin in Cetinje knew little about the beauties of Prokletije or Black Lake in Žabljak. For them, those places were just as far away as London, Paris, or Berlin. Parts of Montenegro were still under the Turkish rule, and the borders that bounded the territory of present-day Montenegro in the nineteenth century did not allow the people to move freely. Poverty, poor education, and insufficient general development of the country did not allow Montenegrins to rise culturally to the level at which most other European nations were in the nineteenth century. Hence, for a long time, there were no Montenegrins who were curious, inquisitive, artistic, and who visited natural beauties in search of inspiration, as Western European adventurers did. The harsh reality shaped by historical circumstances turned the Balkan highlanders to pragmatism where the preservation of the house and the hearth was a priority, nurturing the mentality of war and the need to protect, which gave no way to developing skills other than those essential for surviving.

Literature about Montenegro was scarce and inaccessible to everyone until the nineteenth century when geographical and political circumstances and the redistribution of forces on European fronts placed Montenegro and other Balkan states in the field of interest. From the middle of the nineteenth century, the area of the Balkan Peninsula gradually became known to Western European merchants and envoys, adventurers, and explorers who, as political circumstances changed, turned ever more towards the European part of Turkey. All of them left a precious written trace behind them from their travels (Lainović 1949:4). The wars against Napoleon Bonaparte changed the route of the so-called Grand Tours–educational trips, so instead of Italy and Western Europe, the English started discovering the forgotten areas of the Balkan hills (Todorova 2006:114). In search of noble ancient ideals, the philhellenes (lovers of Greek art and culture) set out for Greece, while at the same time, in his travels between 1809 and 1811, the English poet Lord Byron wrote precious verses about the struggle of Greek and other Balkan people for freedom (*Childe Harold's Pilgrimage*, 1812) and thus encouraged the arrival of others (Goldsworthy 2020:24). The nineteenth century was a time when travel writing came to life as a popular literary genre, but the expansion of travellers and their writing left the impression that the Balkans, and therefore Montenegro, had not been written about before, which is not the case. Representatives of diplomatic missions and various envoys, soldiers, or merchants who travelled to the Balkan countries on business and often visited several countries in one trip recorded their stays in this area. Such records were very accurate and were not intended for a wider audience, as is the case with romantic travelogues from the nineteenth century. We find important travelogues about the Balkans and Montenegro in the sixteenth century, but there are records of these areas in Western European historiography even from the Middle Ages when pilgrims and warriors who went to the Holy Land crossed the Balkan Mountains (Goldsworthy 2020:22).

Many people visited and remembered Montenegro, from famous Ottoman traveler Evliya Çelebi, who lived in the seventeenth century, to Italian scholar, botanist, and geographer Antonio Baldacci, at the end of the nineteenth and the beginning of the twentieth century. The literature from before the nineteenth century was quite inaccessible, which fortunately changed in the following century (Lainović 1949:4). The largest number of records about Montenegro were left by the Russians, who were the ones who visited the country the most, given the close diplomatic relations, but also because of the same affiliation to the Orthodox faith. These circumstances motivated Montenegrins to rely on the Russian Empire in many ways throughout history, so the presence of Russian envoys on Montenegrin territory was a common occurrence.

On the other hand, for a man on the western side of the continent, Montenegro remained unknown and challenging for a long time. Therefore, in this paper, we will focus on those travelogues left behind by the British, French, and Germans who came to Montenegro for different reasons in the nineteenth and early twentieth century. It was no longer just diplomats who came on missions, but also scientists, archaeologists, researchers, priests, pilgrims, and adventurers, and it is interesting that a significant number of travel writers were women. Although they wrote on various topics, this paper will be based on descriptions of mountains and their terrifying beauty the travel writers learned about when they first saw Montenegro–whether they travelled via Kotor or Skadar Lake and Albanian territory, the mountains welcomed them either way. With their steep cliffs and foggy peaks, they warned foreigners that they were entering the country and were soon to meet people who they knew nothing about, without knowing what to expect. All travel writers without exception wrote about the Montenegrin mountains with awe and admiration. They admired their wild beauty, while at the same time they prayed that nature is merciful to them as they passed through the mountain cliffs. What is common to all foreigners who stayed in Montenegro long enough to form a somewhat objective opinion about the country is that they connected the

Montenegrin way of life with the mountains from where Montenegrins originated. The French travel writer Vialla de Sommieres, on his way from Kotor to Cetinje, recorded the following: "*In any case, Montenegro seems sinister in every respect, with all roads that lead to it being difficult and dangerous... Everywhere, there is a bunch of either floating or alternately stacked rocks that show the splendor of nature, indulging in general disorder... This area was created for those who live here and only for them. The Creator has, no doubt, in the constant harmony of his decisions, guided by noble goals, adapted man to earth and earth to man.*" He went on: "*Like all highlanders, they irresistibly love freedom. They have not yet been reached by modern European civilization, so they live in an almost intact nature. Instead of living in towns or villages, each family prefers to stay in an isolated hut on their property*" (de Sommieres 1995:24, 269).

18.2 British Travelogues

The genre of travelogue was welcomed and popularized the most in Great Britain, and the nineteenth century was a time of frequent visits by the British to these parts of the world. There were different reasons for the visit, but as more and more pamphlets, articles, studies, and books about the Balkan countries were published in their homeland, the reasons for the arrival of the British and their perception gradually changed–the Balkan states were described as safe enough to visit–it was a "*brave, but not an impossible adventure*" (Goldsworthy 2020:23). Interestingly, throughout the nineteenth century, this area served to "*provide a reliable cure for both male spleen and female depression: an opportunity for passionate political engagement*" (Goldsworthy 2020:24), so some were even prescribed medical trips to this area for mental health recovery. In the middle of the nineteenth century, Romantics played a major role in awakening interest in the Balkans among Western Europeans, who, inspired by Byron's positive impressions, became interested in the Balkan mountains and fierce, proud highlanders (Albanians, Herzegovinians, Montenegrins): "*The Romantics also sparked an interest in Balkan folklore and superstitions, and began the tradition of partisanship for particular Balkan nations ... Thus, in the first three decades of the nineteenth century, the Balkans metamorphosed from a stopover point to a destination and, finally, a cause*" (Goldsworthy 2020:23).

For the British, Montenegro was interesting as part of the answer to the Eastern Question, in the Russian-British diplomatic games about domination in the Eastern Mediterranean. This was the period of uprisings and struggles for the final liberation of Montenegro from the Turks, which took place in 1878 at the Berlin Congress, and it was then that the largest number of British visits to Montenegro was recorded, although a significant number of visitors stayed in the capital Cetinje during the rule of bishops Peter I and Peter II Petrović Njegoš in the first half of the nineteenth century.

However, it was not exclusively the political momentum that inspired British visitors to write, because the impressions of the Ottoman and Greek heritage had already become uninteresting to them, as little remained to be discovered. On the other hand, they found new inspiration in observing the unusual people in the Balkans and the lives they led, which were far from what they were used to. Reverend William Denton, who left one of the most important travelogues about Montenegro called *Montenegro: Its people and their history* (London, 1877 (Table 18.1);

Table 18.1 British travelers to Montenegro in the nineteenth century mentioned in the text

Name	Title of travelogue (date of publishing)	Time of travel	Visited locations
William Denton	*Montenegro: Its people and their history* (1877)	1876	Bay of Kotor, Cetinje, Lovćen, Skadar lake
Mary Edit Durham	*Through the lands of the Serb* (1904)	1900–1903	Cetinje, Lovćen, Komovi, Kolašin, Tara Valley, Bjelasica, Morača canyon

Montenegro edition *Crna Gora, njen narod i njegova istorija*, CID, 1996) explained the fascination of travel writers with Montenegrin everyday life: "*Simple but heroic and proud mountaineers whose life could be depicted in Homeric terms: Their life is one of primitive, and but for the warfare to which they have been compelled, of almost Arcadian simplicity … Their history is one long epic, in which the deeds of heroism, wrought out in their mountain home, seem more fitted for the verse of the poet than for the sober pen of the historian*" (Goldsworthy 2020:20). Denton experienced particular enchantment when climbing Lovćen, which he perceived as a sacred mountain for Montenegrins, the place where their ruler Petar II Petrović Njegoš was buried: "*No matter how hard and tiring the climb to the top of the mountain is, and no matter how bare the top is, everything is forgotten at that unique and magnificent sight that bursts before the traveller's eyes while standing by the sepulchral chapel. The eye wanders over the peaks of the Montenegrin mountains without seeing the plains that stretch between its ridges, or a glimpse of the crater valleys that are found on almost all the heights. To the viewer, it seems as if looking down a stony sea, gray in its silent emptiness, through which, without hyperbole, waves as big as mountains are rolling*" (Denton 1996:31). Also, he recognized the highlander, Scottish mentality in the Montenegrins, stressing that the Scots had built roads through their harsh mountains, while the Montenegrins were still waiting for that civilizationally necessary step (Denton 1996:29). It was a characteristic of British travel writers to notice similarities in the lifestyle between Montenegrin and Scottish highlanders and their life in clans, while British politician William Gladstone regretted that "*Montenegro did not have its Walter Scott to sing the wonders of its history*" (Knežević 2001:55).

During the reign of Peter II, several British travelers stayed in Cetinje. They did not move much further from the surroundings of Cetinje (Fig. 18.1). However, even in Cetinje, visitors had something to admire–the astonishingly wild (Fig. 18.2), uncivilized, and frightening little country was ruled by a man, as Charles Lamb called him "Saul of his people" (Đurković-Jakšić 1963:81). The first British visitors to Montenegro mostly described the Bishop Petar II Petrović Njegoš, while natural beauties were secondary because, in essence, these travellers had not visited enough of the country, staying mainly in Cetinje area. Among them, we would like to single out the adventurer William Barnett, who was one of the first foreign visitors to Cetinje in 1829, during the reign of Peter I. Soon after him, in 1839, Edward Ledwich Mitford visited, followed by Sir Austen Henry Layard who wrote about Njegoš's Montenegro in his work called *Autobiography and*

Fig. 18.1 View of Cetinje Polje and Skadar Lake. (*Source* Đurković-Jakšić 1963:221)

Fig. 18.2 View form Krstac on Bay of Kotor. (*Source* Đurković-Jakšić 1963:219)

Letters, published only in 1903 in London. Those who followed were A. A. Paton, Clarence Paget and Henry Frederick Winnington Ingram, and finally Gardner Wilkinson, who saw Montenegro as "*mountain peaks dotted around high ridges that in some places looked like a sea with huge, petrified waves*" (Đurković-Jakšić 1963:161).

Wilkinson also listed the mountains that were part of Montenegro at the time: Lovćen, Štirovnik, Stavor, Garač, Pusti lisac, Dobrštak, Sutorman, and Kom, emphasizing that the latter no longer belonged to Montenegro (Đurković-Jakšić 1963:161).

Famous English female writers Georgina Muir Mackenzie and Adeline Paulina Irby also passed through Montenegro, publishing several books and studies from their travels, where they discussed Montenegrins, their costumes, customs, and way of life in great detail, and paid special attention to Montenegrin women whom they admired for their strength, while at the same time regretting the hard life they led. Although well known to the British public even then, Montenegrins don't know these writers much, because, unfortunately, their works *Christmas in Montenegro* (London, 1862) and *Travels in The Slavonic Provinces of Turkey and Europe* (London, 1867) were not published with translation in Montenegro. (Knežević 2001:40). The same fate befell the publications of Viscountess Strangford, the author of *The Eastern Shores of the Adriatic in 1863 with a Visit to Montenegro* (1864), just as most other travelogues that were not translated from English or published in Montenegro (Goldsworthy 2020:29). In her study, Goldsworthy comments on the number of British women who ventured on trips and their long stays in the Balkan region. She linked it to the favorable status that women travelers enjoyed in the Balkans–they even had an honorary status–rich women with good connections and reputation, they were often political advisors whose opinion was respected. Some of them, Lady Mary Edith Durham above all, were actively involved in political affairs in the Balkans, and therefore Montenegro (Goldsworthy 2020:30). Mary Edith Durham traveled across Montenegro from 1900 to 1903, writing her book, which she published in 1904 after returning to London. She was an observer, researcher, scientist and an active participant in the process in the Balkans at the very beginning of the twentieth century. Durham was highly respected in her home country;she was a member of two royal institutes, and she also engaged in scientific and political work. The translation of her book *Through the lands of the Serb* (London, Edward Arnold, 1904; Table 18.1) was presented to the audience in Montenegro and Serbia in 1997, published in Belgrade, under the title *Kroz srpske zemlje* (Durham 1997:3). In her writings, Durham focused on the people and the political and social situation. She gives descriptions of the regions she passed through more casually, and concentrates on the people she meets in those regions, customs and anecdotes, so that her writings cannot be viewed as a pure travelogue, but rather as a study of the Balkans. Nevertheless, she left several records of the Montenegrin landscape worth mentioning: "*The first sight of the country horrifies you. The horror of the desolation, the endless chain of bare mountain peaks, the barren wildness of the bare karst, magnificent in its grim solitude, speak at once of centuries of suffering. The next moment fills one with respect and admiration for a people who preferred freedom in this wilderness to slavery in a fertile plain*" (Durham 1997:6). On her way to getting to know the Balkan countries and their population, Durham passed through various remote areas, so the road also led to the obscure town of Kolašin in the valley of the Tara River, in the north of Montenegro, whose surroundings she described as "*Switzerland waiting to be explored—meadows overgrown with thick grass, a ring of wooded hills rich in fir and beech, and above them rise the blue mountain peaks*" (Durham 1997:212). Crossing Mount Kom, she described the sight that presented itself as "*magnificently unique—the gigantous jagged peak of Kom rose suddenly from the grassland not five hundred meters away, and stood all naked and lonely, all white against the angry purple sky, because the mist had frozen on it. Down below us, wisps of fog rolled and crawled like great dragons, and distant peaks rose one behind the other, distinctly blue in color. It was extremely magnificent, and dead quiet*" (Durham 1997:266).

It is also interesting to emphasize that in 1882, a famous archaeologist, Sir Arthur Evans, commented on the situation in Montenegro and its liberation struggle, while the poet Alfred Tennyson, inspired by Gladstone's descriptions of Montenegro, gifted it with a sonnet (*Montenegro: A sonnet*, published in the magazine *Nineteenth Century*, London, 1877) (Knežević 2001:162–163).

18.3 French Travelogues

The increased interest of the French in this area was caused by political circumstances. They followed the events on the Montenegrin and Balkan political scene where France played a very important role several times during the nineteenth century (Lainović 1949:5). When Napoleon's state reached the borders of the Bay of Kotor (which now forms part of Montenegro, even though in the nineteenth century was part of Austria and before briefly France) (de Sommieres 1995:5) after the fall of the Venetian Republic in 1797 and the new redistribution of forces, Vialla de Sommieres, a Colonel in the French army and the Commander of Herceg Novi during the French occupation between 1807 and 1813 (de Sommieres 1995:5) was one of the first foreigners to dedicate his time and research to writing about Montenegro (Fig. 18.3). His work *Voyage historique at politique au Montenegro* (Historical and political journey to Montenegro) (Paris, 1820; Table 18.2), the first work among travel writing literature dealing with Montenegrin history, culture, geography, and people, is also one of the most valuable and thorough works of that kind.

De Sommieres was inspired to write by a desire to do something useful, in honour of his homeland, and to explain to his compatriots why the Montenegrin people and their country, which were unknown to almost everyone in France, are important (de Sommieres 1995:7); just like its people's lives, the name of the country is connected with the mountains, says de Sommieres (1995:14): "*Montenegro was named after its position and appearance. The huge mountain massifs observed from all sides, once overgrown with fir trees, look very gloomy. That dark color, even more, pronounced because the mountain peaks protrude above the abundant layers of snow, must have determined the name of this country ruled by nature. They called it, in Illyrian, Cernogore or Crnogora, black hill or black mountain.*"

While describing the geographical position of the country, he says that Montenegro is made up of many hard-to-reach mountains, abysses, and powerful mountain streams, which together make its geographical position quite inconsistent, especially emphasized in the border areas with the Bay of Kotor–around Lovćen, Vrba, Cuca, Bukovica, and Bjelica (de Sommieres 1995:307). In his work, he paid great attention to the study of Montenegrin customs and lifestyle, connecting their destiny with the area in which they lived.

Fig. 18.3 Kotor, underneath Montenegro. (Drawing by Vialla de Sommieres) (*Source* Sommieres 1995:6)

During his stay in Montenegro, De Sommieres had the opportunity to visit the north of the country–the Hills, which, at the time, had been part of Montenegro for almost two decades, where he observed that the way of life did not differ from that in the center of Montenegro.

The aspirations of the French to become more actively involved in the events in Montenegro and to replace the established Russian influence with their own, aroused interest and encouraged the visits of numerous French scholars. In the second half of the nineteenth century, the Montenegrin prince Danilo Petrović Njegoš considered France ruled by Napoleon III to be an ally and, more importantly, gained hope that Montenegro would finally be granted complete independence from the Ottoman Empire. Therefore, it is not surprising that the

Table 18.2 French travelers to Montenegro in the nineteenth century mentioned in the text

Name	Original French title of travelogue (date of publishing)	English title translation	Time of travel	Visited locations
Vialla de Sommieres	*Voyage historique at politique au Montenegro* (1820)	Historical and political journey to Montenegro	1807–1813	Lovćen, Bay of Kotor, The Hills
Henri Delarue	*Histoire, description, moeurs, usages, lègislation, constitution politique, documents et pièces officielles—Avec une carte du Montènègro et des Pays adjoints* (1892)	History, description, tradition, customs, legislation, political constitution, documents and official documents—with a map of Montenegro and the associated countries	Second half of the 19. century	Skadar lake, Lovćen, Bay of Kotor
François Lenormant	*Turcs et Monténégrins* (1865*)*	Turcs and Montenegrins	Second half of the 19. century	Skadar lake, Lovćen, Bay of Kotor
Frédéric Ferrière	*Monténégro: Journal d'un médecin Suisse*	Montenegro: The diary of a Swiss doctor	1875	Skadar lake, Lovćen, Zeta, Cetinje, Rumija, Sutorman, Vojnik, Župa
Henri Avelot and Joseph de La Nézière	*Monténégro, Bosnie, Herzégovine* (1895)	Montenegro, Bosnia, Herzegovina	1895	Lovćen, Orjen, Njeguši
Alice Nolte	*Essais sur le Monténégro* (1907)	Esseys on Montenegro	1907	Zeta, The Hills

French envoys were present at the prince's court. One of the prince's close people was Henri Delarue, his secretary and French teacher, but also the author of the memorandum that Prince Danilo sent to the great powers at the Congress of Paris, where the position of Montenegro was discussed. The articles he published about Montenegro in the French press helped the local society to better understand Montenegro's position on the European map (Lainović 1949:8). In addition to records of political events, Delarue left behind travelogues published in 1892–an article in the newspaper "Road to Montenegro" and the book *Histoire, description, moeurs, usages, lègislation, constitution politique, documents et pièces officielles—Avec Une carte du Montènègro et des Pays adjoints* (History, description, tradition, customs, legislation, political constitution, documents and official documents–With a map of Montenegro and the associated countries; Table 18.2), which provides valuable information on Montenegrin circumstances, people and custom (Lainović 1949:10). Delarue also noticed how the hills are woven into the essence of the Montenegrin man: "*A traveller, who is approaching Montenegro from the Turkish territory, sees a new and unexpected sight. But he sees, despite the barren land, signs of work and diligence. Even the smallest plots of land, at the foot or cracks of rocks, have been cultivated and protected from every calamity with so much diligence and attention that it is a pleasure to watch them. It is obvious that Montenegrins love their country. Conspicuously dressed, with a weapon on their belt or at their fingertips, they cheerfully drive a team of small oxen, or they dig the rocky land with a hoe*" (Lainović 1949:67).

In these words, it was documented that no matter how harsh and barren the Montenegrin mountains were, life took place on them and people's lives were tied to that karst, no matter how difficult the conditions were.

A few years after Delarue's stay, the book *Turcs et Monténégrins* (Turks and Montenegrins; Table 18.2) was published in 1865 in Paris by François Lenormant, a famous French scientist and archaeologist, which was based on his memories from Montenegro. He wrote (Lenormant 2002:5): "*It seems that nature prepared the soil of Montenegro for the fight. Surrounded by cliffs from all sides, in the heart of the area subjugated by the Ottomans, it represents an invincible fortress for the defenders of freedom.*"

Lenormant's work begins by comparing his experience of the first encounter with the Montenegrin hills with the records of his predecessors, so he remembers Delarue's descriptions of the rising, suddenly petrified sea, which waited for foreigners entering the Bay of Kotor, going from Kotor over the Mount Lovćen to get to Montenegro. Lenormant then discusses the characteristics of the country, and its geographical position, listing the highest peaks and mountains, their heights and specifics, the rivers that intersect them, the forests that surround them, along with a natural border Skadar Lake. He also makes the connection between the Montenegrin character and the land (Lenormant 2002:10): "*Montenegrins have, indeed, nature and characteristics as strong as their mountains. Tall, stocky, of strong health- the warrior life they lead from childhood develops their extraordinary strength.*"

From the travelogues about Montenegro printed in French, in this paper we will mention the records of Frédéric Ferrière, called *Monténégro: Journal d'un médecin suisse* (Montenegro: The diary of a Swiss doctor; Table 18.2), who stayed there for 70 days as a member of the mission of the International Committee of the Red Cross to help organize the Montenegrin Red Cross for the care of refugees and prisoners hurt in the Herzegovinian uprising of 1875 (Ferrière 2007:7). Travelling from Cetinje to Danilovgrad, through a rocky karst area with fertile land and without water where he spent most of his time treating people, Ferrière (2007:55) wrote: "*All around us is a bare rock where one feels uncomfortable, because nowhere in sight is there a bush, not even lichen ... Soon we will reach the top of the mountain: On the left side, the rocky sea ends with a snow-covered peak Lovćen; on our right are the rocks, while behind us there is Skadar Lake.*"

Leaving Danilovgrad and Župa, where he was stationed, Ferrière describes this wooded and fertile area as the most beautiful, so much so that he wanted to make a map of Župa for future travellers since the localities of that part of Montenegro were unknown to foreign visitors. Going further, he met the mountain Vojnik, which he called the "*Mont Blanc of this area*" (Ferrière 2007:120). Ferrière also left several records about the Montenegrin area towards the Albanian border with Skadar Lake (Ferrière 2007:135, 138): "*which is surrounded by mountains that look magnificent – on the left, bare and steep rocks that sink right into the lake, and on the right, the hills are lower and overgrown with bushes and forests, while the shores of the lake are covered with water lilies and sedges. On the other side of the lake from a distance, Ferrier enjoyed the view of the snow-capped high slopes of Rumija, one of the largest peaks of this mountain range, together with Lovćen and Sutorman.*"

At the very end of the nineteenth century, the book called *Monténégro, Bosnie, Herzégovine* (Montenegro, Bosnia and Herzegovina; Table 18.2) by Henri Avelot and Joseph de La Nézière was published in 1895, dedicated to the Montenegrin prince Nikola I Petrović Njegoš, who hosted the authors on one occasion. The travelogue was additionally enriched with illustrations–as many as 200 drawings (Fig. 18.4) and four watercolors with motifs of Montenegrin people and natural beauties. The impressions of these two Frenchmen did not differ from those of other visitors to Montenegro–when sailing through the Bay of Kotor, they first saw the slopes of the Montenegrin mountains Lovćen and Orjen, which hid Montenegro "*in an incredible decor and separated the country from the civilized world. The huge mountain massif of Montenegro, the part facing the sea, is so vertical that you can clearly see the rocks, both at the top and at the foot of the mountains*" (Avelot and de La Nézière 1996:24).

Passing further through the slopes of Lovćen on their way to Cetinje, which was their main destination, just like for almost all other visitors to Montenegro, the travelers spent the night near the village of Njeguši, from where the ruling dynasty originated, enjoying the most wonderful views imaginable, because to them, under the moonlight, it seemed as if the mountain itself were a

Fig. 18.4 Kotor and Lovćen at the end of the nineteenth century. (Drawing by Henri Avelot and Joseph de La Nézière). (*Source* Avelot and de La Nézière 1996:26)

hotbed that scattered the light, making the mountain peaks look like towers and castles (Avelot and de La Nézière 1996:37). Avelot and de La Nézière (1996:130) wrote with sympathy about Montenegro and its wild inhabitants—the proud highlanders that their country carefully hides in its cracks.

Among the French travelers who wrote down their memories of Montenegro were two women, Juliette Adam and Alice Nolte, who wrote in her *Essais sur le Monténégro* (Essays on Montenegro; Table 18.2) that Montenegro is divided into two parts–the massif of Montenegro, which is in the true sense of the words Black Mountain, and the Hills massif, which was separated by the fertile Zeta valley. For Nolte, who poetically expressed her feelings for the hard, isolated life in this area, "*the tragedy of nature in Montenegro was deeply touching, a strange corner of the globe that survives in the divine harmony of silence, desolation, and death*" (Nolte 1996:17–23).

18.4 German Travelogues

Scientists from the German-speaking countries have left a particularly rich and significant written trace about Montenegrin geographical characteristics. After the Berlin Congress and the significant territorial expansion of Montenegro, new, still undiscovered mountain areas opened up to foreign explorers and travelers. Among the first, Bernard Schwartz wrote about Montenegrin peculiarities in 1883, and the Austrian geographer

Oscar Baumann was the first explorer to climb Bobotov kuk, the highest peak of the Durmitor mountain (Hassert 1995:9–10). Kurt Hassert, the first geographer to take a comprehensive look at Montenegro (Hassert 1995), whose work on Montenegrin geographical features titled *Reise durch Montenegro nebst Bemerkungen uber Land und Leute* (Journey through Montenegro with remarks about the country and its people; Vienna, Prague and Leipzig, 1893; Table 18.3) in the form of a travelogue is one of the most important works in the field, climbed to the same peak. Unlike most other travel writers, Hassert visited much more of Montenegrin territory, giving detailed depictions of the Montenegrin mountains. The valuable descriptions that stand out are those of Durmitor in the north of Montenegro, which, along with the Piva Mountains formed the border with Bosnia. Hassert visited Montenegro as many as four times: in 1891, 1892, 1897 and 1900, and his notes were later supplemented by Kurt Kaiser between 1930 and 1934. On his first trip, Hassert toured the Montenegrin mountains for five months, which was an unprecedented endeavor. His choice was to publish his extensive research in the form of a travelogue, in order to bring Montenegro closer to the German citizens in a more popular and interesting way. He published articles and studies about Montenegro in the German press, but also in the newspaper *Glas Crnogorca* (Montenegrin Voice) in Cetinje. In addition to the travelogue, it is important to mention his article from 1894 called *The Climb to Prutaš on Durmitor*, and especially *Contributions to the Physical Geography of Montenegro with special reference to karst* (1895). On his next travels, he visited Rumija, located between Skadar Lake and the Adriatic Sea, he returned to Durmitor, Bjelasica, Komovi and Kučke mountains, and finished his expeditions on Lovćen, "*the most glorious peak of the Black Mountains*" (Hassert 1995:10–18).

From what is today's tourist attraction Jezera, Hassert saw a sad plateau, while Sinjajevina, the mountain opposite Durmitor, revealed its ancient image only in some places where it was covered with forests and houses. In front of him stood "*an inhospitable wasteland destroyed by ice storms and the negligence of the inhabitants. Since the cattle could no longer find enough food due to the harsh climate, the people did not even think of cultivating the land, so they had no choice but to leave the unbearable area … the barren rocky land was so impoverished with grass that the golden age of shepherd's life on Sinjajevina was long gone*" (Hassert 1995:189).

While touring the Durmitor area, Hassert wrote about the way the population lived back then with pity, the cattle would graze on barren land, the people would move to safer areas, finding shelter from the snow in their dilapidated houses in settlements where it was possible to survive winter. At that time, this geographer could not have known that a large number of inhabitants in those areas still live a similar life to the one he described with regret almost a century and a half earlier. He described the people with whom he toured the mountains, whom he met in inns and unexpected backwaters, making it clear from his travelogues that he nurtured sympathy for the Montenegrin people. Every Hassert's description of Montenegrin terrain shows that he was a man who knew his profession well. Visiting the mountain top of Crvena greda, he wrote (Hassert 1995:194): "*limestone slopes with spruce and beech stands, above which the majestic Crvena greda rose, owing its name to the reddish-brown decaying components of iron in limestone … The conspicuously circular shape of the hidden cauldron is related to the horizontality of the bottom and the position that is directly at the foot of the decaying and permeable rocks of the mountain.*"

With the shout of "*Hurray! The highest mountain in Montenegro is ours!*" he celebrated the ascent to Bobotov kuk, from where "*the view over Montenegro was unhindered, while the giant peaks of Durmitor rose menacingly*" (Hassert 1995:204–205).

Another geographer visited Montenegro in 1880 in order to do geographical research of the terrain. It was Berhard Schwarz, and his travelogue *Montenegro: Schilderung einer Reise durch das Innere nebst Entwurf einer Geographie des Landes* (Montenegro: Description of a journey through the interior together with a draft of geography of the country, 1888; Table 18.3), is considered to be one of the most detailed travelogues about

Table 18.3 German travelers to Montenegro in the nineteenth century mentioned in the text

Name	Original German title of travelogue (date of publishing)	English title translation	Time of travel	Visited locations
Kurt Hassert	*Reise durch Montenegro nebst Bemerkungen uber Land und Leute* (1893)	Journey through Montenegro with remarks about the country and its people	1891, 1892, 1897, 1900	Durmitor, Piva mountains, Rumija, Bjelasica, Komovi, Kučke mountains, Lovćen, Sinjajevina
Berhard Schwarz	*Montenegro: Schilderung einer Reise durch das Innere nebst Entwurf einer Geographie des Landes* (1888)	Montenegro: Description of a journey through the interior together with a draft of geography of the country	1880	Lovćen, Komovi, Vojnik, Durmitor, Kučke mountains
Hugo Grothe	*Durch Albanien und Montenegro: Zeitgemäße Betrachtungen zur Völkerkunde, Politik und Wirtschaftswelt der westlichen Balkanhalbinsel* (1913)	Through Montenegro and Albania: The ethnology, political circumstances and economy of the western Balkans	1912	Piva mountains, Rumija, Sutorman, Možura

Montenegro written in German (Schwarz 2014:426). Schwarz sympathized with the harsh destiny of the country, and found that the very mountains were to be blamed for that: "*the tragedy of Montenegro lies within the character of its mountains, that is the thing!*" (Schwarz 2014:337). He considered Montenegro to be one gigantic mountain that strived to the sky. The shape of the terrain in Montenegro could not be considered mountainous, but it all looked like a large high plateau Furthermore, Schwarz made a list of all toponyms with their characteristics, indicating that individual mountains could have a considerable length, but, as a rule, they rose from valleys and were small in width, while at greater or lesser heights they spread horizontally (Schwarz 2014:337).

As many his predecessors did, Schwarz started his journey in Kotor, heading over Njeguši and Lovćen to Cetinje. He registered Lovćen as the symbol of Montenegro "*the stone colossus*" that, although it has long since lost the reputation of the highest point on Montenegrin land, especially when such high mountains as Komovi or Durmitor have entered the borders of Montenegro, "*it still presented itself with its isolated position on the western edge of the Montenegrin plateau as one of the mountain peaks of the whole of this area that can be seen farthest. In Boka Kotorska, as well as on Lake Skadar, near Podgorica, as well as east of Nikšić, its characteristic shape stands out and imposes itself on our eyes in that dense multitude of mountain peaks*" (Schwarz 2014:66).

From the hill above Cetinje, he observed the panorama of Montenegro: "*there in the distance, on the horizon, a snow-white giant rises, an isolated boulder with three peaks, like some kind of crown. These are Komovi, a giant guard on the southeastern border of the Montenegrin country. To the left of it rise other peaks covered with snow, among them perhaps Prekornica, between the valleys of Zeta and Morača. There lies the area we want to go into. What joy and what fear overwhelms our hearts! To the right of Komovi, we can see the already familiar crown of the Upper Albanian Alps (Prokletije), also dressed in bright winter clothes. In front of them stretches the blue, endless surface of Lake Skadar, which softens the wilderness of this area with its appearance. From the place where we are, we have a view of the area that we could not see on the hill between Njeguš and Cetinje, namely the southeastern part from which Rumija (1,595 m) rises like a southern pillar of the 'black hills', which looks like a beautifully shaped pyramid placed on a wide base. The remains of winter are nowhere to be seen around it. The white snow ermine gives it such a royal look. But, in contrast to the vividness of the scenery in the distance, the immediate surroundings seem almost ugly. The true world of death and horror stretches all the way to Lake Skadar, it is the characteristic limestone barren of western Montenegro. There is not even a trace of grass or trees, everything is just bare rock … that's how our planet will look like once all the life on it has reached its goal, and then it will be floating around the universe like some kind of specter, so that eventually disintegrated into countless atoms*" (Schwarz 2014:91).

Passing through Kuči mountains, he compared its peaks (Stubica, Fundina and Orljevo) to the work of the Titans, who erected those steep, bare walls and thick, scattered, black rocks piled up and thrown up (Schwarz 2014:220).

Schwarz gave special attention to Vojnik mountain: "*Mountain Vojnik, 1,998 m high, a huge giant. From top to bottom it was covered with a royal mantle of snow, but at its broad base there*

Fig. 18.5 Entry from Bosnia to Montenegro at the beginning of the twentieth century. (Photo by Hugo Grothe) (*Source* Grothe 2005:19)

was a dark spot, a black forest, northern, German, the first we have seen on this journey" (Schwarz 2014:245). Finally, Schwarz described Durmitor, which was apparently the most exciting moment for him on this tour, as can be understood from his diary entries: "*Durmitor. It resonated in our hearts. So, we finally stand face to face, oh You proud ruler of the entire northwestern part of the Balkan Peninsula. I often spoke your name in awe until I could not even hope to approach your holiness. Thank you that today, when I see you for the first time, you didn't block my view with dense fog and keep that atmosphere for me in the future*" (Schwarz 2014:255). One should be reminded that by the end of the nineteenth century Durmitor was already famous for its scenery among the scientists and travellers, so Schwarz knew what kind of sights awaited him in this part of Montenegro.

Along with geographers, the nature of Montenegro was also interesting to botanists, who, although they focused on flora, wrote a few lines about the Montenegrin landscape. One of them, Erwin Janhen, went to the north all the way up to Komovi, where he wrote about peaks Kom Kučki and Kom Vasojevićki, saying they are 2,439 m and 2,460 m high, with alpine pastures, covered in rocks, cliffs and subalpine forests, while their peaks are made of pure limestone (Pulević and Vincek 1991:406).

The last travel writer whose work we will look at in this paper is Hugo Grothe, who visited the Balkans in 1912. His book*Durch Albanien und Montenegro: Zeitgemäße Betrachtungen zur Völkerkunde, Politik und Wirtschaftswelt der westlichen Balkanhalbinsel* (Through Montenegro and Albania: The ethnology, political circumstances and economy of the western Balkans, 1913; Table 18.3) represent his diary entries from war-torn Montenegro and Albania. He came to Montenegro via Bosnia (Fig. 18.5) and the first mountains he met were the Piva Mountains, covered in karst, desolate "*burdensome*

Fig. 18.6 Vojnik Mountain at the beginning of the twentieth century. (Photo by Hugo Grothe) (*Source* Grothe 2005:32)

Fig. 18.7 Mountains in Western Montenegro at the beginning of the twentieth century. (Photo by Hugo Grothe) (*Source* Grothe 2005:52)

monotonous, where he saw the true face of the land of the black mountains" (Grothe 2005:21).

Grothe had another image of the rest of Montenegro, "*the tamed Zeta with gurgling streams and slightly rounded mountain domes*" (Grothe 2005:51), mountain ranges of 1,600 m above the coastal town of Bar "*that are evenly distributed in rows that are, when looking from distance, more distracting than limiting*" (Grothe 2005:57).

The mountain ranges of Rumija, Sutorman and Možura seemed to Grothe like huge mountain scenery and pictures of fjords. Those pictures (Figs. 18.6 and 18.7) escorted Grothe from Montenegro to Albania, where he continued his journey (Grothe 2005:68).

18.5 Conclusion

This paper dealt with travel literature from nineteenth and early twentieth century that focused on Montenegrin mountains, a "signature sign" of this country, and of its people. Mountains gave Montenegro not only dramatic scenery, but their ruthlessness decided on the way of life the people living on them lead. Secluded and mysterious, those mountains attracted the curiosity of European travellers, but, as Montenegro became more and more open and known to the European public, they attracted different researchers who embarked on scientific journeys to this part of the Balkan Peninsula. Given the great number of travelogues written about Montenegro, the selection of such travel writings was made for this paper. The authors focused primarily on travelogues written in Western Europe, mainly in English, French and German and on those travelogues that were translated into Montenegrin, and in that way became known and close to Montenegrin public. Thanks to these travelogues, Montenegro was no longer a terra incognita, but a country that attracted the attention of the Europeans, stirred interest in its history, and admiration for the nature of those who visited it. The majority of travelogues on Montenegro that we mentioned in this paper were published in local magazines or anthologies abroad and for a long time remained largely unknown to Montenegrins. A great number of travelogues have never been translated into Montenegrin, which has made their use for scientific purposes even more difficult. Thanks to published translations in the last 20 years, at least some of these extremely valuable publications are now available to readers in Montenegro.

Historical circumstances had a decisive impact on travels to Montenegro, as is understood from these travelogues. Only when Montenegro started to open and break from the political (and cultural) isolation in the second half of the nineteenth century did the travellers start to trespass its borders. In that first period of foreign visits to Montenegro, visitors were restricted to areas under Montenegrin rule. That meant that the first travellers didn't have the opportunity to visit northern mountains, like Durmitor, Vojnik, Sinjajevina and Komovi, as those were still outside the Montenegrin borders. That is why most remarks were made about Lovćen, the mountain that overlooked Montenegrin capital Cetinje. Not only that the main path to Montenegro went from Kotor across Lovćen, but it was inside the Montenegrin borders. By the end of the nineteenth century, those borders expanded, and along with them the travels to the north of the country. Some writers focused on nature, come on customs, while others paid attention to Montenegrin people, and the politics in the Balkans. Whatever the reason for their visits may have been, or the primary interest of the travelogues these enthusiasts had left, they objectively described Montenegro and gave a realistic picture and valuable information about the country at a given time.

References

Avelot H, de La Nézière J (1996) Crna Gora i Hercegovina [Montenegro and Herzegovina]. CID, Podgorica

de Sommieres V (1995) Istorijsko i političko putovanje u Crnu Goru [Historical and political journey to Montenegro]. Izdavački centar “Cetinje”, Cetinje

Denton W (1996) Crna Gora: njen narod i njegova istorija [Montenegro: its people and their history] CID, Podgorica

Durham ME (1997) Kroz srpske zemlje [Through the lands of the Serb]. Srpska Evropa, Belgrade

Đurković-Jakšić L (1963) Englezi o Njegošu i Crnoj Gori [The English on Njegoš and Montenegro]. Grafički zavod, Titograd

Ferrière F (2007) Crna Gora: Dnevnik švajcarskog ljekara [Montenegro: a journal of a Swiss doctor]. CID, Podgorica

Goldsworthy V (2020) The Balkans in the nineteenth-century British travel writing. In: Youngs T (ed) Travel writing in the nineteenth century: filling the blank spaces. Anthem Press, Cambridge, pp 19–36. https://doi.org/10.7135/UPO9781843317692.002

Grothe H (2005) Kroz Crnu Goru i Albaniju: etnologija, političke prilike i privreda zapadnog Balkanskog poluostrva. [Through Montenegro and Albania: ethnology, political circumstances and the economy of the western Balkans]. CID, Podgorica

Gvozden V (2012) Ima li putopis svoju teoriju? [Does a travelogue have a theory?]. Putopis: časopis za putopisnu književnost 1(1–2):183–194

Hassert K (1995) Crna Gora 1: putopisi [Montenegro 1: Travelogues]. CID, Podgorica

Knežević S (2001) Crna Gora i Velika Britanija: odabrani radovi [Montenegro and Great Britain: selected papers]. Istorijski institut Crne Gore, Podgorica

Konstantinović Z (1986) Putopisna književnost [Travel literature]. In: Živković D (ed) Rečnik književnih termina. Nolit, Beograd, pp 623–624

Lainović A (1949) Tri Francuza o Crnoj Gori [Three Frenchmen on Montenegro]. Istorijski institut NR Crne Gore, Cetinje

Lenormant F (2002) Turci i Crnogorci [Turks and Montenegrins]. CID, Podgorica

Nolte A (1996) Esej o Crnoj Gori [Essay on Montenegro]. Narodna biblioteka "Radosav Ljumović", Podgorica

Pulević V, Vincek D (eds) (1991) Crna Gora vrata Balkana: putopisi i zapisi evropskih botaničara [Montenegro the gateway to the Balkans: travelogues and records of European botanists]. Obod, Cetinje

Šačkova VA (2012) Putopis kao žanr umetničke književnosti: pitanja teorije [Travelogue as a genre of the art literature: questions of theory]. Putopis: časopis za putopisnu književnost 1(1–2):195–202

Schwarz B (2014) Crna Gora, [Montenegro]. CID, Podgorica

Todorova M (2006) Imaginarni Balkan [Imagining the Balkans]. Biblioteka XX vek, Belgrade

Mount Triglav and Its Role Among Slovenes

19

Peter Mikša and Kornelija Ajlec

Abstract

Mount Triglav, the highest mountain in the Julian Alps, is also the highest peak of the Republic of Slovenia. It is depicted in the national coat of arms and, consequently, in the flag of the country. Thus, it is the most prominent Slovenian national symbol. This status is rooted in the historical processes and events that linked Mount Triglav with Slovenes. The name Triglav appeared on a map for the first time in 1744, and its first depiction dates back to 1778. With the development and organization of mountaineering and alpinism in the second half of the nineteenth century, Slovenian mountains, and in particular the Triglav mountain range, aroused the interest of German mountaineering clubs, which began to build huts and lay out trails to these huts and to nearby summits in order to mark this area as German. This general sentiment and purely patriotic motives led Jakob Aljaž, who worked as a parish priest beneath Triglav, to purchase the summit of Triglav in 1895 and erect a tower there. In the following years, the public increasingly took notice of his action, a visible indicator of the victory of Slovenes over Germans. Gradually, Triglav became an ever-growing symbol of Slovenian identity. The architect Jože Plečnik first depicted the mountain as a symbol of Slovenes in 1934. At the end of 1941, during the occupation of Slovenia in the Second World War, Triglav was used as a symbol of the Liberation Front. In 1947, it was officially used as a national symbol in the new Slovenian Constitution, when it was depicted in the coat of arms. It remained in every Yugoslav and Slovenian constitution until Slovenia's independence in 1991and afterwards.

Keywords

Triglav · Alps · Battle for the mountains · Symbolism · Coat of arms · Slovenia

19.1 Introduction

Today mountain images are inseparably linked to an understanding of the concept of Slovenian identity. Mount Triglav (literally 'three-head'—that is, 'three peaks'), the highest peak on the territory of present-day Slovenia, and before that of Yugoslavia (Fig. 19.14), with an elevation of 2864 m, is depicted in two official symbols of the Republic of Slovenia: the official national coat of arms and the flag (Fig. 19.1). Triglav is also more generally omnipresent in the Slovenian space. With its three peaks, it is an artistically powerful

P. Mikša (✉) · K. Ajlec
Department of History, Faculty of Arts, University of Ljubljana, Ljubljana, Slovenia
e-mail: peter.miksa@ff.uni-lj.si

K. Ajlec
e-mail: kornelija.ajlec@ff.uni-lj.si

M. Zorn et al. (eds.), *Discourses on Mountains of Montenegro and Slovenia*, Historical Geography and Geosciences, https://doi.org/10.1007/978-3-031-99017-5_19

Fig. 19.1 Mount Triglav appears in two out of three national symbols: in the coat of arms of the Republic of Slovenia and through it in the flag of the Republic of Slovenia (in the centre). In the photo, the Slovenian flag is flying alongside the flags of the Radovljica Alpine Club and the Alpine Association of Slovenia (PZS; on the right); the photo is taken in front of Roblek Mountain Hut and Mount Triglav can be seen in the background. (Photo by Peter Mikša)

symbol with a unique and dominating presence. Its form is unique in the Slovenian geographical space. Its image is used by virtually all political groupings, many associations, institutions, music groups and small businesses, it can be seen at rallies, in graffiti, etc. Triglav is so ubiquitous and taken for granted in Slovenian society that residents no longer notice all its depictions (Kozole 2021). Slovenes are as proud of the mountain as if they had built it themselves (Velikonja 2018).

Why and how did Triglav and mountains in general become a symbol of Slovenian national identity? The prominent place of mountains, and especially the highest mountain, among symbols of Slovenian identity has its roots in past historical processes and events that have given mountains a special place in the hearts of Slovenes. Particularly during the rise of Slovenian national consciousness in the nineteenth century, the symbolic meaning of the mountain landscape and Alpine culture acquired a central places in the "shaping" of Slovenes in the "struggle" with German national consciousness. At that time, efforts for the national independence of Slovenes spread from towns, newspapers, camps and reading rooms to the mountains. But the mountains, including Triglav, had to be "discovered" first.

19.2 Discovery of Mountains

People who lived below the mountains, such as gatherers of herbs, hunters following wounded game, and shepherds searching for lost livestock, have always ventured into them (Mikša 2013a). From the Enlightenment on, the attitude of local residents towards mountains began to change as a result of the increasingly frequent visits to them by foreigners. In return for payment, they began guiding and accompanying occasional visitors to the mountains, carrying their loads and ensuring their safety. The first tourists were mountain explorers who were professionally closely connected with the outdoors and the mountain environment due to their activities (Mikša and Ajlec 2015). These individuals considered their visits to the mountains to be research work, as they studied and described the mountain environment during their treks and were the first to record their ascents of Slovenian peaks. Natural historian and polymath Johann Weikhard Freiherr von Valvasor (1641–1693) was among the earliest to describe the first known approaches to lower-lying Slovenian mountains and the first to write them down, in his great work *The Glory of the Duchy of Carniola*, an extensive scientific survey of the Carniola region dating from 1689 (Mikša and Zorn 2016). The names of hills, valleys and rivers appear earlier, for example as boundary markers of estates in various documents, such as gift deeds (Mikša 2013a:392).

The first cartographic depictions appeared a few decades later. Among the most famous early maps of Slovenian mountains was the Chorographic Map of the Duchy of Carniola (*Ducatus Carnioliae Tabula Chorographica*) by Janez Dizma Florjančič de Grienfeld from 1744 (Gašperič 2007:269; Gašperič et al. 2020:82–85). On it, the name Triglav appears for the first time–*Mons Terglou Carnioliae Altissimus,* with an elevation of 1399 Parisian fathoms recorded for it. Although the measured elevation is inaccurate, it is still of great importance, as it indicates that Triglav was already considered the highest mountain in Carniola (*Kranjska*, present-day central, southern and north-western Slovenia) (Fridl and Šolar 2011:214).

At the end of the eighteenth century, Balthasar Hacquet was among the most important early mountain explorers. Triglav was a particular favourite for him. He tried to climb it for the first time in 1777, making the first documented successful approach to the summit, but he did not succeed in reaching it, achieving only the top of Mali Triglav (Little Mount Triglav). He certainly climbed it in 1782 (Mikša 2014). He also devoted attention to Triglav in his work *Oryctographia Carniolica oder physikalische Beschreibung des Herzogthums Krain, Istrien und zum Theil der benachbarten Länder* (Oryctography of Carniola or a Physical Geography of the Duchy of Carniola, Istria, and Parts of Neighbouring Lands; Hacquet 1778–1789). Among other things, in 1778, in the first part of this work, he published the first pictorial representation of the highest mountain in Carniola, which he calls Veliki Terglau (Hacquet 1778) (Fig. 19.2).

Hacquet's visits to mountain peaks were supported financially and in other ways by the Carniola patron and naturalist Sigismund (Žiga) Zois (Gspan-Prašelj 1991), who was an important mineralogist on a European scale at the time (Lovšin 1944). This leading Slovenian Enlightenment intellectual, naturalist and patron and initiator of cultural life in his time encouraged and supported Hacquet mainly because of his interest in geology, especially in minerals. In fact, it was Zois who encouraged and financially supported the first ascents of Slovenian peaks. Hacquet's fellow climbers also came from Zois's circle (Munda 1986).

19.3 Discovery and Ascent of Triglav

The first documented ascent to the top of Triglav was made in 1778 by four locals from Bohinj, who later became known as "the four brave men". They were hunter Štefan Rožič, miners Matevž Kos and Luka Korošec, and physician Lovrenc Willomitzer. The ascent took place as part of the research expedition of Baron Sigismund Zois, who, after Hacquet's failed attempt in 1777, promised a reward of six ducats to whomever reached the summit of Triglav (Mikša 2013a; Mikša and Ajlec 2015), an amount equivalent to about 90 daily mining wages (Deržaj 1980). The journey of the "four brave men" on the first approach to the summit took three days (Richter 1821). During the descent, they marked the path with small stones so that they would know how to return along the same route the next time, which in all likelihood represented the first documented marking of a path in the Slovenian mountains.

The ascent of Triglav thus took place eight years before the first ascent of Mont Blanc, 22 years before the first ascent of Grossglockner and 87 years before the first ascent of the Matterhorn. Slovenes are truly proud of being the first, as visitors to Lake Bohinj can see for themselves while admiring the monument to the "four brave men" unveiled in 1979 (Fig. 19.3).

Fig. 19.2 Copperplate engraving of Triglav in Hacquet's Oryctographia Carniolica, which is the first pictorial depiction of Triglav (1778). (Authors archive)

19.4 "Struggle for the Mountains" and for Triglav

19.4.1 Up Until the First World War

With the development of industrial society in the nineteenth century, the concept of leisure time starting taking shape in towns. People made use of it in various ways. Among other things, mountain tourism began to develop, attracting an ever-increasing number of people of different nationalities in a multilingual state. As a result, a national struggle between Germans and Slovenes also began in this part of the Alps, carried over from other spheres of social life at the time, especially from larger towns such as Ljubljana, Maribor, Celje, and others. The "struggle" was manifested in a race to be the first to reach mountain summits, in the establishment of mountain trails, and in the construction of huts and other essential infrastructure in the high mountains. The entire activity can be characterized as a struggle for marking and dominating mountain areas by one ethnic group (Mikša et al. 2017).

In the initial phase, the Germans were considerably more organized than the Slovenes. Their activities took place within the framework of German mountaineering organizations, which, in

Fig. 19.3 Monument to the first climbers to reach the summit of Triglav, erected in 1978 on the 200th anniversary of its first ascent. It is also called the Monument to the Four Brave Men and stands in the park in Ribčev Laz in Bohinj, near the church of St. John the Baptist. One of the men is pointing to the top of Triglav. (Photo by Mitja Sodja, authors archive)

Fig. 19.4 Dežman Hut or Deschmannhaus below Triglav was a hut owned by DÖAV. (Authors archive)

accordance with the aforementioned activities, began to consolidate the presence of Germans in the "Slovenian" mountains. Marking of mountain trails was carried out exclusively using German signposts, only German was spoken in the mountain huts, and German guests were given priority over Slovenian guests when it came to lodging. Thus, at the opening of the *Triglav-Hütte* (today named after Valentin Stanič) in 1887 (Mikša 2015), a representative of the German and Austrian Alpine Club (Deutsches und Österreichisches Alpenverein—DÖAV) from Berlin described Triglav as the German king of the Alps. The president of the Carniola branch of DÖAV, Karel Dežman (Deschmann), after whom the hut was soon renamed *Deschmannhaus* (Fig. 19.4), characterized it as part of "German land" (Debelak-Deržaj 1949:46).

At that time, Slovenes did not have their own mountaineering organization that could counterbalance the German pressure. The Slovenian Mountaineering Society (*Slovensko planinsko društvo*, SPD) was founded only six years later, on 27 February 1893. The SPD worked mainly to defend the Slovenian mountain environment against foreign influence, as a bastion of the Slovene language, as it strove to use Slovene in marking routes and to use Slovene names of mountains. It operated under the slogan: "Let's preserve the Slovenian face of Slovenian mountains" (Mikša and Ajlec 2022:111).

SPD member Janko Mlakar wrote on the 60th anniversary of the founding of SPD that the society was founded "*not so much out of enthusiasm for the mountains, but primarily as a defence against the surge of German visitors*", since "*a foreigner felt at home in our mountains, but we felt like strangers on our home soil*" (Mlakar 1953:213).

A year after its foundation, SPD had already erected the first two mountain huts: Orožen Hut on the mountain Črna Prst (Julian Alps), named after Fran Orožen, head of the main society in Ljubljana, and Kocbek Hut on Molička mountain pasture beneath Ojstrica (Kamnik-Savinja Alps) (Raposa 1992).

The members of the society celebrated every opening of a Slovenian hut as a national holiday at a time when the German language dominated in the mountains. The local population also participated, preparing a big celebration in areas closest to the huts, which awaited SPD members and honoured guests. Upon arrival, they were greeted by the celebratory firing of mortars and triumphal arches bearing uplifting inscriptions such as "*We salute you, sons of the mountains, who work for the glory of the homeland*!" or "*The first mountain hut opens, the whole Bohinj valley rejoices*!" (Ravnihar et al. 2009:7). The opening of individual huts was often marked by the issuing of postcards, which were popular and numerous at the time, as they were a modern way of informal mass communication. On postcards marking these openings, national flags can be seen flying from or next to the huts (Ravnihar et al. 2009). The depictions of Slovenian huts on these postcards were thus proof of the Slovenian presence in the Alps.

In addition to the huts, SPD members intensively marked the routes, equipping them with Slovene signposts and inscriptions with the Slovene names of the mountains. They worked particularly intensively in the Triglav range, where the Germans were also heavily present (Mikša and Zorn 2018a). Since the Slovenian mountaineers considered that the Germans were trying to appropriate the Slovenian mountain environment, they wanted to neutralize the activities of the DÖAV, especially in the vicinity of Triglav. Of course, these acts did not go unnoticed. Thus, the Germans coined the term *trutzhütte* (Ravnihar et al. 2009:7), or defiant huts, for Slovenian huts, because they were built in defiance of German actions, whereas they referred to theirs as *schutzhütte*, or shelter huts (Mikša 2017).

In 1895, Jakob Aljaž, a priest in Dovje, a village near Triglav, and a member of the SPD, purchased a piece of land on the summit of Triglav from the Dovje municipality for the sum of one florin, which in that period sufficed for 50 eggs, 10 L of milk or 4 pigeons (Mikša and Ajlec 2015:39) and erected a steel tower on the top, just under two metres high and a bit over one metre in diameter, made of thick galvanized sheets and iron columns fused with concrete (Mikša 2015). The tower, which quickly acquired the name "Aljaž Tower" (Fig. 19.6), became a national symbol for most Slovenes in the following decades, and at the time of its creation it served as a "lightning rod" for the struggle for the mountains between Germans and Slovenes.

Aljaž played an extremely important role in the development of Slovenian mountaineering. Besides the tower, he also initiated and built a number of structures in the mountains around Triglav: Aljaž Hut in the Vrata Valley (1896; Mikša 2015), Triglav Hut on Kredarica (1896; Mikša and Vehar 2016), protected climbing routes between Mount Mali ("Little") Triglav (2,725 m a.s.l.) and the top of Mount Triglav (2,864 m a.s.l.) (1895 and 1897), Tominšek Route from Vrata to Kredarica (1903) and Stanič Shelter (1895) just below the peak of Triglav (Mikša 2015, 2017). Jakob Aljaž was even sued by the Germans for erecting the tower and building huts, who claimed that, for example, he had destroyed the first order triangulation point on Triglav, and interfered with constructions on foreign property. Aljaž succeeded in rebuffing the charges, and the tower and huts remained (Mikša 2015).

Fig. 19.5 Opening of the first hut on Kredarica in 1896 (Lovšin 1944) and the present-day building. (Upper figure: authors archive; bottom figure: photo by Matija Zorn, authors archive)

The construction of the Triglav Hut on Kredarica or "Kredar'ca" (Fig. 19.5), as Slovenians generally call it today, was stimulated by an event in the German Deschmannhaus in August 1895, during the erection of Aljaž Tower (Aljaž 1923). As Aljaž recalls: "*At that time, we spent the night in the old, small Dežman hut, because Slovenes did not yet have any huts there. There was heavy fog. That's why I didn't go to the top of Triglav in the morning, but stayed in the hut, from where I could hear the individual pieces of the Triglav tower being hammered together. I talked with the caretaker of Dežman hut* [...]. *He said to me kindly and confidentially: 'It's lucky for you that there were no Germans here tonight; otherwise there would be no room for you.'* [...] *German tourists, members of the 'Alpen-Verein' – of course – have priority over Slovenes. At that moment, I resolved to build a hut on Triglav, just for myself and a few friends. But if the Slovenian Mountaineering Society wakes up and responds to my call, we will build a large hut, where exactly I will look into next week; if I can't find a place elsewhere and no one wants to help, I'll put my hut right alongside Dežman hut, 10 paces away...:—that's how offended my national pride was.*" (Aljaž 1922:84–85). The hut was ceremonially opened on 10 August of the

Fig. 19.6 Aljaž Tower was erected on the summit of Triglav by Jakob Aljaž in 1895. Between the wars the Italian-Yugoslav border ran alongside it (the Rapallo border): the stone marking the border stood beside the tower, which was just two metres inside Yugoslav territory. (Authors archive)

following year (Mikša and Zorn 2018a:24). Today, the greatly expanded building, which in 1909 was renamed Triglav Hut on Kredarica (Mikša and Vehar 2016), remains the highest-elevation building on the territory of Slovenia to this day and, precisely because of the background of its construction, a true symbol of Slovenian identity (Fig. 19.5). In 1933 and 1934, when the Italian-Yugoslav border passed through the Julian Alps, the Italians built the Triglav barracks of Viktor Emanuel III on the opposite side of the head of Triglav (Italian *Caserma Rifugio Vittorio Emanuele III di Monte Tricorno*) (Žorž 2014, 2015). After the Second World War, when this side of Triglav also became Yugoslav, the proposal was floated to turn the barracks into a mountain hut (it would have been called Kugy Hut). However, since it stood at an elevation of 2,520 m, the Italian building would have stood five metres higher than the "holy Kredar'ca", which represented a national and consequently also a political issue, so the realization was held up by the highest authorities and never carried out. Quite a few other Italian military buildings, erected in what are today Slovenian mountains have been converted into mountain huts or refuges (e.g., Klement Jug Hut in Lepena Valley, the hut on Razor mountain pasture, and so on).

By erecting a hut on Kredarica (Fig. 19.5), and above all by purchasing the summit of Triglav and erecting a tower (Fig. 19.6), Aljaž prevented the Germans from appropriating the peak of Triglav, and proclaimed the peak a national mountain, a symbol of Slovenian identity, and at the same time called for Slovenes to gather around it, regardless of the provincial borders that administratively divided the Slovenian national territory at the time. This call met with a good response, as Slovenes, at a time when they, as a minority nation in a predominantly German-speaking country, were fighting for greater rights, experienced such developments as part of strengthening patriotism and raising national consciousness. The erection of the tower was an act that resonated with the public even then, and especially in the following years, as it demonstrated the ascendancy of the Slovenes over the Germans in "Slovenian" mountains. In 1895, after the construction of Aljaž Tower, the journal Planinski Vestnik wrote: "*We announce to all Slovenes the very happy news that we have regained our old ownership of the giant of the Slovenian mountains, the magnificent Triglav. This year, we put Aljaž Tower, already famous throughout Europe, on top of it.*" The inscription on the tower (Fig. 19.6) was also the first Slovene inscription on Triglav and its environs, since prior to that all inscriptions were in German (Mikša 2015). Aljaž then donated the tower and the land to the SPD (Mikša 2013b). From 1895 up to the time before the Second World War, the SPD took care of Aljaž Tower, and since the Second World War it has been taken care of by the Ljubljana Matica Alpine Club. Aljaž Tower was declared a cultural monument of national importance in

October 1999, and state property in November of the same year (Mikša 2013b:161).

The sheet metal tower represents Slovenian patriotism, pride, national identity and the fight for Slovenian land and the mother tongue in a tangible form. There are countless records, articles, songs and poems, works of art, souvenirs, badges, books, brochures, as well as relatively tasteful advertising and commercial materials inspired by the tower and all that it represents (Kozole 2021).

19.4.2 The Interwar Period

While it seemed that the fight for the mountains had ended here and Triglav was safely ensconced in thr Slovenian national territory, this feeling of stability was soon upended by the restructuring of international relations and the redrawing of national borders after the First World War. After the collapse of the Habsburg Monarchy and the signing of the Treaty of Rapallo (1920) between the newly established Kingdom of Serbs, Croats and Slovenes and the Kingdom of Italy, the border between the two countries was set deep inside the Slovenian ethnic territory (Mikša and Zorn 2018b). The new border crossed the Julian Alps along its central ridge, and the national border also ran along the top of Triglav (Mikša and Zorn 2018b) (Fig. 19.6). This opened up a new front in the "battle for the mountains", this time between Slovenes and Italians. The Western Julian Alps increasingly took on an Italian character, e.g. with Italian geographical names and symbols and new buildings that associations and the state built in the high mountains. The Slovenes resisted, and as a result mountaineers from both sides competed to see who would visit the peak of Triglav, or Monte Tricorno as the Italians called it, in greater numbers, as well as other mountains in the Julian Alps (Škodič 2018). In addition, graffiti with a nationalist theme often appeared on Aljaž Tower, the surrounding rocks and mountain huts. Moreover, a "painting war" broke out between Italians and Slovenes, in the course of which Aljaž Tower was repainted several times in the colours of the Italian tricolour flag and then again in the colours of the Yugoslav tricolour (Mikša and Zorn 2021) (Fig. 19.7). The demarcation of the border on the ground (and the described "battle" at the summit of Triglav) continued until 1925, when it became clear that Aljaž Tower would stand two and a half metres inside the Kingdom of Serbs, Croats and Slovenes, later called the Kingdom of Yugoslavia (Mikša and Zorn 2021).

The delineation of the new border, which the Slovenes considered unjust, and aggressive Italian expansionist policy were contributing factors in ensuring that Triglav became even more firmly entrenched as a symbol of the Slovenes.

Fig. 19.7 Aljaž Tower in the 1920s, painted in the tricolour of Yugoslavia. (Authors archive)

Fig. 19.8 Triglav was used as a marking of their territory for: **Italians** (upper left) on propaganda postcards (*Saluti dalla Guardia alla frontiera*) featuring boundary stone no. 10, which stood on the highest peak of the Rapallo border (Triglav; the drawing does not show an accurate image of the landscape); **Yugoslavs** (upper right)–members of the Yugoslav nationalist organization ORJUNA standing in front of Aljaž Tower painted in the colours of the Yugoslav tricolour (in the period 1920–1925, the border on the ground had not yet been definitively determined and it was not clear on which side of the national border Aljaž Tower would stand at the top of Triglav); **Germans** (lower right) put Triglav on a postage stamp during the Second World War; **Slovenes** (lower left)–Slovenian Partisans reached the top of Triglav in 1944, displayed the Slovenian national tricolour flag there and tore down the boundary stone of the Rapallo border while shouting: "*No borders here ever again!*". (Authors archive)

19.4.3 During the Second World War

During the Second World War, the part of the Rapallo border that ran through the Julian Alps served as the state border between Italy and Germany. The Germans considered the Slovenian mountains to be the southern border of the Third Reich, so at the end of 1941, they also secured

Fig. 19.9 Triglav as a symbol for the Liberation Front. (National Museum of Contemporary History of Slovenia archive)

this border with minefields, wire fencing, watchtowers, pillboxes, etc. (Mikša and Zorn 2021). These were also placed near high mountain passes in the Julian Alps, regardless of the difficult access. In that period, the Slovenian "battle" for the mountains became part of a broader struggle against the occupying forces. The two struggles merged, which was evident in among other things such as the caps of the Partisan resistance movement, which were also worn in the Slovenian lowlands–the "Triglavka" caps had the stylized shape of Triglav with its three peaks (Fig. 19.9). The connection between the two struggles was manifested particularly in the adoption of Triglav as a symbol of the Liberation Front of the Slovene Nation (OF) in 1941, featuring in its illegal press, on leaflets and posters, Partisan money, in the names of its subgroups and newsletters, on stamps, etc. (Mikša 2018). In 1944, Triglav gained additional symbolic meaning, as the Partisans hung the Slovenian flag on the summit and uprooted the boundary stone of the Rapallo border under the motto "*No borders here ever again!*" (Čurin 2007) (Fig. 19.8). Immediately after the end of the war, when the area west of the Rapallo border came under Yugoslav administration and Triglav was no longer an unjust border for the Slovenes, mountaineers set out for the mountains in large numbers, where they demolished boundary markers and removed wire barriers. Special work brigades were organized for this purpose (Svetina 1980).

19.4.4 After the Second World War

With the end of the Second World War, the national struggle for the mountains also ended. The names of peaks, first ascents, huts and other national markings in the mountains became exclusively Slovene. From being a Slovenian symbol, Triglav also became a symbol of Yugoslavia, which, at the same time, in a multinational country subconsciously always denoted Slovenes (Mikša et al. 2017). In other Yugoslav republics, Slovenes were categorized, often pejoratively, as "*narod skijašev i planinarov*", i.e. a nation of skiers and mountaineers. For more on the influence of Triglav and of Slovenian mountains more generally on the construction of Slovenian (mountain) identity, see Chap. 15 in this volume.

19.5 Triglav in Pictures, Poetry and Stamps

The first depiction of the Triglav range (not Triglav on its own) dates back to 1701. It appears in a drawing of a panorama of the Triglav range by the parish priest Matija Ločnikar (Mikša 2017). More than four decades later, the work by Florjančič mentioned above, a map of the Carniola area, was published, where Triglav is marked with the words "*Mons Terglau Carniolae altissimus*" This was followed by the depiction by Hacquet from 1778, also noted above (Fig. 19.2). In the 19th and early twentieth century, the mountain landscape and Triglav were depicted pictorially by early mountain landscape painters, for example Anton Karinger and Marko Pernhart, later also by Ladislav Benesch, Valentin Hodnik, impressionists Ivan Grohar, Rihard Jakopič, Matija Jama and others (Mikša 2021). In 1904, the painter and illustrator Maksim Gaspari, as part of his proselytizing activities aimed at awakening national consciousness, produced the postcard "From the nation for the nation", in which he portrayed the Slovenian territory with three key spatial identifiers: Triglav, Ljubljana Castle and the Adriatic Sea (Kučan 1998).

During this time, Triglav was also a means of evoking Slovenian identity in poetry and prose. France Prešeren, the most prominent Slovenian poet, in his Introduction to *The Baptism on the Savica* (written in 1835), placed the Slovenes under the protection of the "grey chief of Carniola" and outlined the creation of a symbol with which Slovenian national (spatial) identity would later be most strongly associated (Strojin 1980, 2009; Kučan 1998). In his poem *Duma*, written in 1904–1908, the poet Oton Župančič referred to his homeland with Triglav and the Karawanks range as the main toponyms (Župančič 1908). The top parts of the tombstone for his three literary companions from Slovenian modernism (Ivan Cankar, Dragotin Kette and Josip Murno), which was designed by the architect Dušan Grabrijan in 1924 in Ljubljana and erected in the central city cemetery of Žale in 1925, is designed so as to symbolically portray Triglav (Mikša 2021) (Fig. 19.10).

In 1919, Triglav was portrayed on the first Slovenian stamps, the "Verigar" series from 1919. The draft designs for them were made by the painter Ivan Vavpotič, and Triglav was shown in the background beneath the legs of a slave breaking his chains (Mikša 2021) (Fig. 19.10). Triglav also appears on the first postage stamps of newly independent Slovenia in 1991, on which the national coat of arms is depicted (Fig. 19.11). Triglav has a main role in the first Slovenian feature-length (silent) film *In the Kingdom of the Goldhorn* from 1931, and the second film–*The Slopes of Triglav* from 1932–even has Triglav in its name.

Triglav was first depicted as a symbol of the Slovenes in 1934 by Slovenia's most famed architect Jože Plečnik in the plan for the garden wayside shrine, called *Šverljuga's Sign of Mary*, which now stands in front of the parish church in Bled. He placed it on the stone mantle of the statue of Mary, Mother of God. On the wings of the double-headed royal eagle, he depicted the three coats of arms of the then Yugoslav tribes, as the nations living there (Serbian, Slovenian and Croatian) were renamed in the unitary Kingdom of Yugoslavia. The Slovenian coat of arms is shaped like Triglav, above which shines the six-pointed star of the Counts of Celje, which also appears in the present-day Slovenian coat of arms (Mikša 2021) (Fig. 19.10). Plečnik also used Triglav as a Slovenian symbol at the entrance of the Ljubljana University Library. This became the Slovenian National Library in 1945. When the legal status of the Slovenian National Library was recognized in 1945 and the University Library became the National and University Library in Ljubljana, a new inscription was placed above the entrance and the Slovenian coat of arms was placed on the balcony balustrade until 1948. It was probably drawn by Peter Trpin under Plečnik's supervision (Raposa 1997).

19.6 From Symbol to Coat of Arms

However, these were not the only depictions of Triglav. It is the most familiar: three cones, the middle of which is higher, and the side ones are equal. This is how Triglav was described in the first Slovenian constitution as a description of the Slovenian republic's coat of arms from 1947. The stylization comes from the time of the Second World War, when it was taken over by the resistance side immediately after the occupation of Slovenia. It became the symbol of the Liberation Front of the Slovenian Nation (OF) (Fig. 19.9). We first encounter this image at the height of the national liberation struggle in late 1941. It

Fig. 19.10 Triglav appears as a mountain symbol of Slovenia as early as in 1919, on the first Slovenian stamp, the "*Verigar*" (Chainbreaker) (left; authors archive), then in 1934 Slovenia's most important architect Jože Plečnik placed it in the coat of arms as a symbol of the Slovenes (centre; photo by Peter Mikša). The tombstone of the three greatest modern-era Slovenian literary figures (Ivan Cankar, Dragotin Kette, Josip Murn; right; photo by Peter Mikša) is also designed in the shape of Triglav

was created by the architects Marjan Tepina and Edvard Ravnikar and proposed at a meeting of the central committee of the Architects' Liberation Front. Reportedly they decided to adopt this symbol because it could be drawn quickly and easily (Kos 1983). As such, it is said to have been seen for the first time on the leaflets that OF members spread on the streets of Ljubljana on 28 October 1941, on the eve of the celebration on 29 October, the anniversary of the founding of the State of Slovenes, Croats and Serbs in 1918 (Kos 1983). In November, the Triglav symbol was already widely established as a symbol of the Liberation Front. It also appeared on the cover of the illegal newspaper *Slovenski Poročevalec* (Slovenian Reporter), for the first time on 8 November 1941 (Slovenski poročevalec 1941). In the communiqué of the Executive Committee of

Fig. 19.11 Triglav also appears on the first postage stamps of the newly independent country. (National Museum of Contemporary History of Slovenia archive)

the Liberation Front published in the newspaper on 5 December 1941, it says that for the holiday of 1 December 1941, when the Liberation Front successfully held the second plebiscite demonstration of its members, the streets of Ljubljana were strewn with Slovenian flags, and the "symbols of the Liberation Front—Triglav with the letters OF—shone" from the walls. All these early symbols of the Liberation Front had the letters OF written in addition to the cones of Triglav (Kos 1983). At first, this acronym was written in various places, but finally it came to settle between the cones. Towards the end of the war, the letters OF were replaced by a (red) star.

After the adoption of the decree on the annexation of the Slovenian Littoral to a united Slovenia and Yugoslavia in mid-September 1943, and after the meeting of the Assembly of Delegates of the Slovene Nation at the beginning of October of the same year in Kočevje, a new part was added to the sign based on an idea of a writer France Bevk: wavy lines at the foot of the figure of Triglav, a symbol of the Slovenian sea. The first drawings of the completed symbol were made by painter Božidar Jakac. His draft for a postage stamp, which was dated 16 October 1943, has all the previously developed parts of the future national coat of arms: the image of Triglav, the sea, a star and the abbreviation OF, which was soon dropped (Mikša 2021).

The stylized symbolic role of Triglav can also be found on Partisan means of payment–bonds, receipts, certificates and vouchers, both those issued by the Economic and Financial Committee of the Liberation Front, and later those issued by the Monetary Institute of Slovenia under the presidency of the Slovenian National Liberation Council. In 1944, a new stage of development of the coat of arms can be seen on Partisan money. Triglav is drawn on vouchers with wavy lines at its foot, representing the sea. Triglav is surrounded by a kind of wreath of rays that enclose a five-pointed star above the highest cone. For a contemporary testimony about the design of this Partisan money, see Simčič (1969). This is already the image of the republic's coat of arms, which featured linden branches (Kos1983) only when the constitution was enacted (Fig. 19.12).

19.7 Triglav in the Republic Coat of Arms

At its first meeting on 26 April 1945, the presidency of the Slovenian National Liberation Council (SNOS) adopted a draft of the Slovenian republic coat of arms. The decision of the presidency was the first official decision on the image of the future republic coat of arms. The question of the final image of the coat of arms arose only during the preparation of the draft constitution of the People's Republic of Slovenia, i.e. after the second session of SNOS. It was launched by the Presidency of the National Government of Slovenia. From the letter of the organizational secretary of the SNOS presidency, Josip Globevnik, dated 19 October 1946, we learn that the presidency called on some architects and

Fig. 19.12 Triglav depicted on Partisan money, on a Slovenian payment voucher for five lira (first series, draft created by Branko Simčič) from 1944. (National Museum of Contemporary History of Slovenia archive)

professors to make drafts of the republic's coat of arms (Kos 1983). Architects Boris Kobe, Edvard Ravnikar, Branko Simčič, painter Anton Gojmir Kos and engineer Herman Hus submitted their drafts. The drafts submitted were largely similar in content and design, but they differed in artistic execution. They were reviewed by the President of the Presidency of the Slovenian National Liberation Council Josip Vidmar, Prime Minister Miha Marinko and Deputy Prime Minister Marjan Brecelj, who unanimously decided on the draft of architect Branko Simčič. The presidency sent Simčič's draft in two copies to the presidency of the national government, which sent one to Belgrade, where they wanted to harmonize the images of the republics' coats of arms. The Slovenian draft in Belgrade apparently did not undergo any changes, as it was published in the daily newspaper in unchanged form on 15 December 1946 (Osnutek … 1946).

On 8 January 1947, the Constitutional Committee adopted the text of the third article of the draft constitution, which refers to the republic's coat of arms. The article stated: "*The national coat of arms of the People's Republic of Slovenia is a field surrounded by ears of grain. The ears are tied at the bottom with a ribbon and interlaced with linden leaves. Between the two tops of the ears is a five-pointed star. In the lower part of the field, there are three wavy lines at the bottom, representing the sea. Above them rise three cones, of which the middle one is higher, and the lateral ones are equal; the cones represent Triglav.*" (Kos 1983:13). This text was adopted as the third article of the constitution of the People's Republic of Slovenia at the last session of the Constituent Assembly on 16 January 1947 (Ustava Ljudske … 1947). In the ceremonial edition of the constitution from 1947, the coat of arms is painted in gold with a red border, while the star is red and has a golden border (Kristan 1989). This coat of arms differs in details from the last published draft. Triglav is a bit higher, the ears of grain are full and the spikelets reach right up to the star, and the linden leaves are also more evenly spaced. We consider this depiction of the coat of arms to be the original. The Slovenian coat of arms is also described in the Constitution of the Federal People's Republic of Yugoslavia from 1948: "*The national coat of arms of the People's Republic of Slovenia is a field surrounded by ears of grain. The ears are tied at the bottom with a ribbon and interlaced with linden leaves. Between the two ears is a five-pointed star. In the lower field, there are three wavy lines at the bottom, representing the sea. Above them rise three cones, of which the middle one is higher, and the lateral ones are equal; the cones represent Triglav.*" (Ustava FLRJ … 1948).

The description of the coat of arms in the 1963 constitution (Ustava Socialistične … 1963) underwent only stylistic changes, and was also included in unchanged form in the 1974 constitution (Ustava Socialistične … 1974) (Fig. 19.13).

Fig. 19.13 Triglav in the Slovenian coat of arms at the session of the Constituent Assembly in 1946. (National Museum of Contemporary History of Slovenia archive)

Fig. 19.14 The first author of the article next to the republican coat of arms, which hung (from 1947 till 1961) in the People's Assembly of the People's Republic of Slovenia. (Authors archive)

19.8 Triglav's Inclusion in the State Coat of Arms

In the process of Slovenia's independence from Yugoslavia, it was decided and planned that the new country of Slovenia would have its own coat of arms and flag. In the Draft Constitution of the Republic of Slovenia in 1990, there is no mention of the coat of arms and the flag (Poročevalec … 1990). On 8 April 1991, the Assembly of the Republic of Slovenia announced a public competition for the design of the coat of arms of the Republic of Slovenia (Skupščina … 1991) in the daily newspaper, and the proposals received were evaluated by the relevant commission on 23 and 27 May 1991. They were divided into three groups, and four finalists were selected from the 87 proposals received. The prize was not awarded, but the works of authors Branko Vovko, Marko Pogačnik, Bojan Haberl and Dane Petek were purchased. In the end, the proposal of sculptor Marko Pogačnik was selected (Mikša 2021). This coat of arms is designed in the form of a shield, in the middle of which there is a silhouette of Triglav in white on a blue background, with two blue lines representing the Slovenian sea and rivers, and above it three golden stars arranged in the form of an inverted triangle, reminiscent of the gold colour in the historical flag of Carniola. The three colours of the Slovenian tricolour are completed by the colour red used in the border of the shield on the left and right sides (Fig. 19.11). The official record of the new Slovenian coat of arms in the 1991 constitution reads: "*The coat-of-arms of Slovenia has the form of a shield. In the middle of the shield, on a blue background, is a representation of Mount Triglav in white, under which there are two undulating blue lines symbolising the sea and rivers and above which there are three golden, six-pointed stars forming a downward-pointing triangle. The shield is bordered in red. The coat-of-arms is designed in accordance with a set standard of geometry and colour. The flag of Slovenia is the white-blue-red Slovene national flag with the coat-of-arms of Slovenia. The ratio of the width of the flag to the length thereof is one to two. The colours of the flag are in the following order: white, blue and red. Each colour occupies a horizontal band covering one third of the area of the flag. The coat-of-arms is positioned in the upper left portion of the flag such that it lies with one half in the white field and the other in the blue field* (Ustava … 1991).

19.9 Conclusion

Today, Slovenes are strongly affiliated with mountains. The highest mountain Triglav (2,864 m) is the country's premier national symbol. Upon the independence of Slovenia from Yugoslavia (whose highest peak was also Triglav) in 1991, a stylized Triglav became the central figure in the coat of arms of the Republic of Slovenia, and through the coat of arms was also placed on the flag of the Republic of Slovenia. This placement has its reasons in past historical actions and events that give Triglav a special status in the eyes of Slovenes. Triglav first appears on a map in 1744, and the first pictorial representation can be seen in 1778. With the development and organization of mountaineering and alpinism in the second half of the nineteenth century, Slovenian mountains, and especially the Triglav range, began to attract the interest of German mountaineering societies, who started building huts in Slovenian mountain areas and trails that led to the huts as well as to mountain summits. This was also a way of marking the territory as German. It was in this general atmosphere that Jakob Aljaž, a priest serving in the valleys below Triglav, bought the peak of Triglav for one florin in 1895 for purely patriotic motives, and erected a tower there. This was an act that resonated more and more with the public in the following years and served as a harbinger of the victory of the Slovenes over the Germans, with Triglav becoming an ever stronger symbol of Slovenian identity.

Triglav as a symbol of Slovenes was depicted for the first time (1934) by the architect Jože Plečnik on the garden wayside shrine, called *Šverljuga's Sign of Mary*, which today stands in front of the Church of St. Martin in Bled. During the Second World War, Triglav appeared as a symbol of the Liberation Front of the Slovenian nation at the end of 1941, drawn by architects Marjan Tepina and Edo Ravnikar. In 1947, Triglav officially appeared in the form of a coat of arms as the state or at that time the republican symbol (Slovenia was then one of the six republics of the Federative People's Republic of Yugoslavia). Despite changes to the Yugoslav and Slovenian constitutions, it remained there until the independence of Slovenia in 1991, when it was included as the main component in the new coat of arms of the Republic of Slovenia. The figure of Triglav as a mountain, which has found a place in such important national symbols, is a unique example in Europe, and very rare throughout the world.

More than 245 years have passed since the first ascent of Slovenia's highest peak. In 1778, three

Fig. 19.15 Triglav, Slovenia's highest mountain and symbol of Slovenian identity. (Photo by Aleš Zdešar, authors archive)

people from Bohinj climbed the remote mountain: miners Luka Korošec and Matevž Kos, hunter Štefan Rožič, and the organizer of the ascent Lovrenc Willomitzer. Today, Triglav is relatively easily accessible along several protected hiking and climbing routes, and there is a saying in Slovene: "If you have not been to the top of Triglav, you are not a true Slovene". This is also why today it is by far the most visited peak in the Slovenian mountains, summited by more than 3,000 hikers on fine summer days, with well over 50,000 visitors per season.

Slovenes are as proud of Triglav as if they had built it themselves, but in truth Triglav built Slovenes over the course of nineteenth and twentieth century historical processes (Fig. 19.15).

References

Aljaž J (1922) Planinski spomini [Mountain memoirs]. Planinski Vestnik 22(3–4):84–85

Aljaž J (1923) Oris mojega življenja [Chronicle of my life]. Planinski Vestnik 23(10):65–67

Čurin J (2007) Na Triglavu nikoli več meje! [No borders on Triglav ever again!]. Slovenske Novice 17(213):16–17

Debelak-Deržaj M (1949) Kronika Triglava [Triglav chronicles]. Planinski Vestnik 49(2):45–55

Deržaj M (1980) Zgodovina meritev višine Triglava [A history of measuring Triglav's elevation]. In: Strojin T (ed) Triglav, gora naših gora. Obzorja, Maribor, pp 141–163

Fridl J, Šolar R (2011) Vpliv razvoja kartografskih tehnik na podobe zemljevidov slovenskega ozemlja od 16. do 19. stoletja [The influence of the development of cartographic techniques on the images of maps of Slovenian territory: from the 16th to the 19th century]. Zgodovina knjige in bralne kulture na Slovenskem 55(4):205–221

Gašperič P (2007) Cartographic images of Slovenia through time. Acta Geogr Slov 47(2):245–273. https://doi.org/10.3986/AGS47205

Gašperič P, Šolar R, Zorn M (2020) Kartografski zakladi slovenskega ozemlja = Cartographic treasures of Slovenian territory. Založba ZRC, Ljubljana. https://doi.org/10.3986/9789610504276

Gspan-Prašelj N (1991) Žiga Zois. In: Munda J (ed) Slovenski biografski leksikon, vol 15. Slovenska akademija znanosti in umetnosti, Ljubljana, pp 832–846

Hacquet B (1778–1789) Oryctographia Carniolica, oder Physikalische Erdbeschreibung des Herzogthums Krain, Istrien, und zum Theil der benachbarten Länder, vols 1–4. Johann Gottlob Immanuel Breitkopf, Leipzig

Kos J (1983) Grb Socialistične republike Slovenije [Coat of arms of the socialist Republic of Slovenia]. Arhivi 6(1/2):7–14

Kozole E (2021) Arhiv Triglav: Triglav v 300 slovenskih znakih [Triglav archives: Triglav in 300 Slovene signs]. Studio Ljudje, Ljubljana

Kristan I (1989) Grb Socialistične republike Slovenije [Coat of arms of Socialist Republic of Slovenia]. In: Javornik M (ed) Enciklopedija Slovenije, vol 3. Mladinska knjiga, Ljubljana, pp 234–235

Kučan A (1998) Krajina kot nacionalni simbol [The landscape as a national symbol]. Znanstveno in publicistično središče, Ljubljana

Lovšin E (1944) V Triglavu in v njegovi soseščini: planinske študije in doživetja [Triglav and its environs: mountain studies and experiences]. Slovensko planinsko društvo, Ljubljana

Mikša P (2013a) Prvi raziskovalci slovenskih gora in prvi dokumentirani pristopi nanje [The first researchers of the Slovene mountains and the first documented climbs on them]. Zgodovinski Časopis 67(3–4):390–405

Mikša P (2013b) Matica planinstva: Ob 120 letnici Planinskega društva Ljubljana-Matica [The mother of mountaineering: on the occasion of the 120th anniversary of the Ljubljana Matica Alpine Club]. Planinsko društvo Ljubljana-Matica, Ljubljana

Mikša P (2014) Exploring the mountains—Triglav at the end of the 18th century. In: Štih P, Zwitter Ž (eds) Man, nature and environment between the Northern Adriatic and the Eastern Alps in Premodern Times. Znanstvena založba Filozofske fakultete, Ljubljana, pp 202–215

Mikša P (2015) "Da je Triglav ostal v slovenskih rokah, je največ moja zasluga": Jakob Aljaž in njegovo planinsko delovanje v Triglavskem pogorju ["It is largely owing to me that Triglav remained in Slovene hands:" Jakob Aljaž and his mountaineering activities in the Triglav mountain range]. Zgodovinski Časopis 69(1–2):112–123

Mikša P (2017), Triglav in Jakob Aljaž [Mount Triglav and Jakob Aljaž]. Viharnik, Ljubljana

Mikša P (2018) Kako je Triglav pristal v grbu Republike Slovenije [How Triglav came to be in the coat of arms of the Republic of Slovenia]. In: Zorn M, Mikša P, Lačen Benedičič I, Ogrin M, Kunstelj AM (eds) Triglav 240. Založba ZRC, Ljubljana, pp 15–28. https://doi.org/10.3986/9789610500841

Mikša P (2021) Od gore do simbola: Triglav in njegova vloga pri Slovencih [From a mountain to a symbol]. Geografija v Šoli 29(2):6–14

Mikša P, Ajlec K (2015) Slovensko planinstvo = Slovene mountaineering. Planinska zveza Slovenije, Ljubljana

Mikša P, Ajlec K (2022) Slovensko planinstvo skozi čas [Slovenian mountaineering through time]. Planinska zveza Slovenije, Ljubljana

Mikša P, Ogrin M, Glojek K (2017) Od kod gorska identiteta Slovencev? [Where does the mountain identity of Slovenes come from?]. Geografski Obzornik 64(3–4):10–28

Mikša P, Vehar M (2016) Kredar'ca: ob 120-letnici postavitve prve koče pod Triglavom [Kredar'ca: on the 120th anniversary of the construction of the first hut beneath Triglav]. Planinsko društvo Ljubljana-Matica, Ljubljana

Mikša P, Zorn M (2016) The beginnings of the research of Slovenian Alps. Geografski Vestnik 88(2):103–131. https://doi.org/10.3986/GV88206

Mikša P, Zorn M (2018a) National "marking" of Slovenian mountains before World War I. Igra ustvarjalnosti 6:22–29. https://doi.org/10.15292/IU-CG.2018.06.022-029

Mikša P, Zorn M (2018b) Rapalska meja: četrt stoletja obstoja in stoletje dediščine [The Rapallo border: a quarter century of existence and a century of heritage]. In: Ajlec K, Balkovec B, Repe B (eds) Nečakov zbornik: procesi, teme in dogodki iz 19. in 20. stoletja. Znanstvena založba Filozofske fakultete, Ljubljana, pp 605–642

Mikša P, Zorn M (2021) Nacionalni boj za gore in slovenska identiteta [The national battle for the mountains and the Slovenian identity]. Časopis za Zgodovino in Narodopisje 92(1):67–88

Mlakar J (1953) 60 let slovenskega planinstva [60 years of Slovenian mountaineering]. Planinski Vestnik 53(5):209–221

Munda J (1986) Lovrenc Willomitzer. In: Munda J (ed) Slovenski biografski leksikon, vol 14. Slovenska akademija znanosti in umetnosti, Ljubljana, pp 698–699

Osnutek ustave Ljudske republike Slovenije [Draft constitution of the People's Republic of Slovenia] (1946). Slovenski poročevalec 7(294, priloga), 15 Dec 1946

Poročevalec Skupščine Republike Slovenije in Skupščine SFR Jugoslavije za delegacije in delegate [Reporter of the assembly of the Republic of Slovenia and the assembly of the SFR Yugoslavia for delegations and delegates] 16(17), 19 Oct 1990

Raposa K (ed) (1992) Stoletje v gorah [A century in the mountains]. Cankarjeva založba, Ljubljana

Raposa K (1997) Prenova vhodnih vrat Plečnikove palače Narodne in univerzitetne knjižnice v Ljubljani [Renovation of the entrance doors of the Plečnik Palace of the National and University Library in Ljubljana]. Narodna in univerzitetna knjižnica, Ljubljana

Ravnihar M, Dolar N, Dolar M (2009) Pozdrav z vrhov: slovensko planinstvo na starih razglednicah [Greetings from the peaks: Slovenian mountaineering on old postcards]. Medium, Žirovnica

Richter FX (1821) Die Wochein. Illyrisches Blatt 16:61–63; 17:65–66; 18:69–70; 19:75–76; 20:77–80; 21:82–83

Simčič B (1969) Kako smo risali partizanski denar [How we drew Partisan money]. Borec 20(5):20–25

Škodič D (2018) Določitev meje na Triglavu in vojašnica Morbegna [Determining the border on Triglav and the Morbegna barracks]. In: Zorn M, Mikša P, Lačen Benedičič I, Ogrin M, Kunstelj AM (eds) Triglav 240. Založba ZRC, Ljubljana, pp 205–212. https://doi.org/10.3986/9789610500841

Skupščina Republike Slovenije, 6100 Ljubljana, Šubičeva 4 razpisuje javni natečaj za oblikovanje grba in idejnih aplikacij grba Republike Slovenije [The Assembly of the Republic of Slovenia, 6100 Ljubljana, Šubičeva 4, announces a public competition for the design of the coat of arms and conceptual applications of the coat of arms of the Republic of Slovenia] (1991). Delo 33(81):1, 8 Apr 1991

Slovenski poročevalec 2(24):1, 8 Nov 1941

Strojin T (ed) (1980) Triglav, gora naših gora [Triglav, the mountain of our mountains]. Obzorja, Maribor

Strojin T (2009) Zgodovina slovenskega planinstva: Slovenska planinska organizacija: SPD-PZS: 1893–1948–2003 [History of Slovenian mountaineering: the

Slovenian mountaineering society: SPD-PZS: 1893–1948–2003]. Didakta, Radovljica
Svetina T (1980) Triglav—gora naših gora [Triglav, the mountain of our mountains]. In: Strojin T (ed) Triglav, gora naših gora. Obzorja, Maribor, pp 16–32
Ustava Ljudske Republike Slovenije [Constitution of the People's Republic of Slovenia] (1947). Ustavodajna skupščina Ljudske Republike Slovenije, Ljubljana
Ustava FLRJ in ustave ljudskih republik [Constitution of the Federal People's Republic of Yugoslavia and and the constitutions of the people's republics] (1948) Uradni list FLRJ. Beograd
Ustava Socialistične republike Slovenije [Constitution of the Socialist Republic of Slovenia] (1963, 1964) Uradni list SR Slovenije. Ljubljana
Ustava Socialistične republike Slovenije [Constitution of the Socialist Republic of Slovenia] (1974) Uradni list SR Slovenije. Ljubljana
Ustava Republike Slovenije in Ustavni zakon za izvedbo ustave Republike Slovenije [Constitution of the Republic of Slovenia and the Constitutional act implementing the Constitution of the Republic of Slovenia] (1991). Mavrica, Celje
Velikonja M (2018) Narava kot politika: motiv Triglava v grafitih in street artu [Nature as politics: the motive of Mount Triglav in graffiti and street art]. In: Zorn M, Mikša P, Lačen Benedičič I, Ogrin M, Kunstelj AM (eds) Triglav 240. Založba ZRC, Ljubljana, pp 225–235. https://doi.org/10.3986/9789610500841
Žorž G (2014) Sedemdeset let smo jo napačno imenovali [For seventy years we used the wrong name]. Planinski Vestnik 114(10):4–9
Žorž G (2015) Bila je zgolj začasni, zasilni objekt [It was only a temporary, emergency building]. Planinski Vestnik 115(7–8):12–14
Župančič O (1908) Samogovori [Monologues]. Katoliško tiskovno društvo, Ljubljana

20 "The Issue of Mt. Lovćen" in World War I and Italy

Slavko Burzanović

Abstract

The paper discusses the military and political significance of the Montenegro's mountain Lovćen, which dominated the Austro-Hungarian naval base in the Bay of Kotor. The ambitions of Austro-Hungary to occupy this mountain through diplomatic or military means are indicated. Also pointed out are the steps of the Italian diplomacy to foil that attempt or failing this, to provide territorial compensation for their country. The Austro-Hungarian conquest of Mt. Lovćen in January 1916 was an overture to the occupation of Montenegro. Opponents of the Montenegrin King Nikola Petrović Njegoš and the Montenegrin government fabricated news on the existence of a secret agreement between Montenegro and Austria-Hungary with reference to surrendering Mt. Lovćen to the enemy without adequate defense. This had severe foreign-political and internal-political consequences for Montenegro and contributed to the vanishing of the Montenegrin state from the political map of Europe at the end of World War I. In this context, the significance of Mt. Lovćen for the identity for the Montenegrin people is analyzed, as well as its symbolic or propaganda significance for the Austro-Hungarian conqueror.

Keywords

Mt. Lovćen · Bay of Kotor · Austria-Hungary · Italy · Montenegro

20.1 Introduction

The inferiority of the Italian navy in relation to the Austro-Hungarian fleet on the Adriatic was partly conditioned by the jagged nature of the eastern Adriatic coast mostly occupied by Austro-Hungary and in part by the developed naval infrastructure in which the naval bases in Pula and the Bay of Kotor stood out. The military-strategic importance of the Montenegro-based Mt. Lovćen that dominates the Bay of Kotor and from where the base in Kotor could be threatened by artillery failed to go unnoticed by the Italian political and military forces. Thus, Italian military experts visited Lovćen in 1902 in strict confidentiality due to the alliance between Vienna and Rome to consider the possibility of equipping the mountain camp with artillery. In 1904, Italy allegedly sold and in fact donated to Montenegro several artillery weapons, including some large-caliber ones intended for positions on Mt. Lovćen (ASDMAE 1906).

The military-strategic importance of Mt. Lovćen was the reason why on several occasions

S. Burzanović (✉)
Historical Institute, University of Montenegro, Podgorica, Montenegro
e-mail: sburzanovic@ucg.ac.me

M. Zorn et al. (eds.), *Discourses on Mountains of Montenegro and Slovenia*,
Historical Geography and Geosciences, https://doi.org/10.1007/978-3-031-99017-5_20

in the period from 1912 to 1916 it was the focus of Italian, Austria-Hungarian and Montenegrin diplomacy, but also that of other countries interested in the Balkan and Adriatic area. During the First Balkan War, Italian envoys had been alarmed by the information that the implication of the surrender of part of Mt. Lovćen to Austro-Hungary was the foundation of the Austro-Hungarian consent to the Montenegrin occupation of Shkodra (modern-day Albania). That was the key reason for Italian Foreign Minister Antonio Paterno Castello di San Giuliano to renounce his initial support for Montenegro's aspiration to annex Shkodra. Somewhat later during the Shkodra crisis in the spring of 1913, one of the reasons Italy failed to consent to Austro-Hungary's unilateral action against Montenegro was the fear that Vienna might secure control over Mt. Lovćen (Burzanović 2014:284–290).

The Italian interest in the Mt. Lovćen fortifications also came to the attention of Austro-Hungary. As early as May of 1914, the Austro-Hungarian General Staff and the Ministry of Foreign Affairs were intrigued by the information that the Montenegrin government intended to carry out extensive fortification works on Mt. Lovćen, who were meant to get assistance from the Italian government in the form of a number of cannon and some 6,000,000 lire. The position of the Austro-Hungarian Supreme Command was that if Montenegro performed the mentioned works unaided, Vienna could not object for reasons of principle but there would be a disloyalty of allies in the case of Italy's involvement, thus making any promises by Italy worthless. Austro-Hungarian Ambassador to Rome Kajetan von Mérey and Cetinje envoy Eduard Otto considered the information to be unfounded. According to Otto's report dated July 18, 1914, certain fortification works were carried out on Mt. Lovćen and several artillery pieces seized in the Balkan War were set up on the mountain, all of which was carried out from the modest budget of the Montenegrin state (Babović-Raspopović 1998:216–220).

20.2 The July Crisis of 1914

In the spring of 1914, Montenegro was the focus of Italian diplomacy, mainly in regards to the matter of a loan by which the Great Powers intended to compensate for the loss of Shkodra and enable economic and political stabilization within the country. Rumors about the possible unification of Serbia and Montenegro were also discussed, and the annexation of Mt. Lovćen to Austro-Hungary and the issue of compensation to Italy as a consequence of the application of Article 7 of the alliance agreement between Italy and Austro-Hungary were brought in connection with it. The danger of an outbreak of war after the Sarajevo assassination and later on, the onset of the war made the first topic pointless and lessened the relevance of the second matter, while the third topic, the issue of Mt. Lovćen, became even more relevant (Burzanović 2019:373–376).

In mid-April 1914, the foreign ministers of Italy and Austro-Hungary met in Opatija (modern-day Croatia). Minister Berthold emphasized the South Slavic threat to Austro-Hungarian interests that therefore would not allow for a unification of Serbia and Montenegro but which it would treat as *casus belli* (Ekmečić 1990:31). Minister San Giuliano was not unaware of the possibility of the South Slavic menace yet in principle he was not opposed to unification. San Giuliano conveyed to Hans von Flotow, the envoy of Germany, that nothing useful had been agreed in Opatija in this regard. In point of fact, Germany was interested in both Italy and Austro-Hungary reaching a new, more specific agreement on the Balkans.

The interest of von Flotow on how Italy would react if Vienna decided to hand over the Montenegrin coast to Albania and they in turn annexed Lovćen was an intimation of aggressive Austro-German policy. This scenario was in such a way unacceptable to the other Great Powers, primarily Russia, that it denoted that a European-wide war was possible as a consequence. As a matter of fact, he revisited this topic several times in his conversations with San Giuliano from April to July

1914. The first reaction of the Italian minister was rejection and resentment due to Vienna's covert effort to territorially join the Albanian coast, over which it already had a kind of hegemony, with its coastal possessions in Dalmatia. He claimed that the annexation of Mt. Lovćen could not even be discussed because it was absolutely unacceptable for the entire Italian nation, regardless of party affiliation, and that no Italian government could ignore it. However, he spoke about the need for an agreement that would make the Triple Alliance compact at a time when Austro-Hungary opposed Belgrade's plans to expand its territories to the Adriatic with the support of Russia, that is, to achieve Serbian-Montenegrin unification. As a precondition for such "*compactness*," San Giuliano, among other things, expected a change in Vienna's policy in Dalmatia, that is, a halt to the "*Slavicization of the Adriatic coast*." What he really had in mind were compensations. It was clear to the Austro-Hungarian ambassador to Rome, Kajetan von Mérey, that San Giuliano, by reaching a new inter-alliance agreement, wished to sanction Austro-Hungary's access to Mt. Lovćen, and in fact the Montenegrin coast, in order to obtain *terra irredente* in return. Austro-Hungarian Minister Berthold declined to engage in talks with the Italians that would lead in this direction. He avoided San Giuliano's attempt to raise the issue of compensation in Vienna in May 1914 through Ambassador Giuseppe Avarna in the context of Serbian-Montenegrin unification (Ferraioli 2007:795–798).

In the first days of July, the afore mentioned issue was brought up by the Austrian press, in which was voiced the opinion that Austro-Hungary should not oppose the unification of Serbia and Montenegro, but a condition for its recognition should be to request Mt. Lovćen's surrender and Serbia's obligation not to construct any Adriatic ports or a naval fleet. The assumption was made that the handover of Mt. Lovćen would not be agreeable to Italy, but rather that it should be swayed by diplomatic means that the security of the port of Kotor was desirable to Italy due to the possibility for the Austrian fleet swiftly joining the Italian navy in the Mediterranean. A strong Austro-Hungary was presented as a guarantor of peace (DDI 1964a:47). Given the existing relations among the Allies, such an interpretation could be qualified as being cynical. In connection with the eventual Austro-Hungarian occupation of Mt. Lovćen, there was also speculation that the compensation to Italy for such an act might be the occupation of Vlora (modern-day Albania) and the surrounding area (DDI 1964b:124).

During the July crisis caused by the Sarajevo assassination, Italian diplomacy maintained its earlier views regarding Mt. Lovćen, also considering Austro-Hungarian intentions and consequences of a possible conflict between Serbia and Austro-Hungary (DDI 1964c:217). Due to the alliance, German diplomacy continued to attempt to mediate between Rome and Vienna. In Berlin, the government considered it justified that Italy would receive appropriate compensation for the occupation of Mt. Lovćen or the territorial expansion of the Habsburg Monarchy in the Balkans (Giordano 2012:263). In talks with San Giuliano in Fiuggi on July 8, 1914, Ambassador von Flotow demanded to know the Italian conditions for the transaction. As the context of a possible Austro-Hungarian securing of Mt. Lovćen, von Flotow failed to mention the current crisis although he emphasized the danger that a possible unification of Serbia and Montenegro signified for Italy-Austro-Hungary relations. The Italian minister first announced that the occupation of Mt. Lovćen would cause the most severe crisis in relations between Italy and Austro-Hungary hitherto, and most probably a war (Caccamo 2008:17–18). Two days later, San Giuliano announced that the cost for accepting Austro-Hungary's occupation of Mt. Lovćen would be Trentino (DDI 1964d:298–299; Monzali 2004:274).

The expectations of German diplomacy that Italy would accept to join Austro-Hungary in the war against Serbia-Montenegro proved unrealistic. Without adequate compensation, Italy held the attitude that such assistance would be contrary to the principle of nationality upon which Italian unity was based and which they had always supported in favor of the development of the Balkan states (DDI 1964c:217). The positions of Vienna and Rome not only diverged in terms of the cost of the Italian war alliance, but also in terms of the benevolence of Italian neutrality. The Italian ambassador in Vienna, Avarna, assessed that in regards to the eventual occupation of Mt. Lovćen and the approval to take action against Serbia, the Austro-Hungarian government had no intention of making any concessions to Italy even regarding Vlore and its neighboring territories in Albania (DDI 1964e:254). San Giuliano and Avarna agreed that direct talks with Berthold on the issue of the unification of Serbia and Montenegro should be avoided. The minister considered that should there be an unfeasibility of unification, it was advisable for it to be achieved gradually so as not to give Austro-Hungary a reason to intervene (DDI 1964c:217; Caccamo 2008:18). The Italian representative in Cetinje, Gaetano Paterno, assessed that Mt. Lovćen did not hold the same significance for Serbia as it did for Montenegro. In fact, in the event of Serbian-Montenegrin unification when facing either war or a peaceful acceptance of Mt. Lovćen's annexation, Serbia would likely accept the latter as a *fait accompli* (DDI 1964f:163). Serbian diplomacy, aware of the importance that Italians attached to Mt. Lovćen, instrumentalized the issue. The Serbian ambassador in Cetinje expressed his concern to his Italian counterpart, Paterno, regarding the cordial stance of King Nikola and his government towards Austro-Hungary. Paterno, albeit not explicitly told so, concluded that Austro-Hungary had already made some offers to King Nikola (DDI 1964g:189). In Belgrade, the Italian charge d'affaires Giuliano Cora held talks with Nikola Pašić about Serbian-Montenegrin unification and asked for his opinion on the possibility of secretly ceding Mt. Lovćen to Austro-Hungary. Pašić's reply was that if King Nikola endorsed it, neither the Serbian nor the Montenegrin people would accept it. In fact, they would contest it fiercely. Unification was discussed in all its details as Mt. Lovćen held the same importance for Serbia as it did for Montenegro. According to Pašić's statement, without Mt. Lovćen, unification would be useless (DDI 1964h:281). Pašić tried to present Serbia as an interested and important partner of Italy in preventing Austro-Hungary from strengthening the significance of the Bay of Kotor's naval base and its position in the Adriatic. In a diplomatic game set up in this way, Montenegro was presented as an unreliable partner whose policy could harm Italian interests.

The issue of Mt. Lovćen appeared in reports from Belgrade and in connection with the war preparations. In point of fact, it was hinted that the Serbian government would send three batteries of artillery to reinforce positions on Mt. Lovćen (DDI 1964i:251). Montenegrin Foreign Minister Petar Plamenac announced to envoy Lazaro Negrotto Cambiaso some measures to protect Lovćen, although neither he nor the king believed that Austro-Hungary would try to seize it by a sudden military attack in the current atmosphere. The Minister bid Italy to contribute to the security of Mt. Lovćen, inducing Austro-Hungary to define its position on the issue (DDI 1964j:251–252; DDI 1964k:297).

San Giuliano recommended extreme caution to the Montenegrin government regarding the Austro-Serbian dispute so as not to provide a reason for an Austro-Hungarian attack on Mt. Lovćen (DDI 1964l:274). He also had in mind the possibility of a Montenegrin-Austro-Hungarian agreement on the ceding of Mt. Lovćen, and thus he ordered the Italian representatives in Berlin, St. Petersburg and Belgrade to present this issue to their governments. Vienna, on the other hand, merely awaited an opinion from Avarna regarding this issue (DDI 1964m:264). Ambassador Avarna had some doubts, pondering whether the ceding of Lovćen was viable given the current enmity between Serbia and Austro-Hungary. He was also unsure whether Austro-Hungary would be satisfied with just Lovćen as an adequate compensation in the current circumstances, i.e., if it were a sufficient guarantee against the dangers that the unification of Serbia and Montenegro could cause Austro-Hungary in the future (DDI 1964n:348).

20.3 Italian Neutrality and Lovćen

The Italian government declared neutrality immediately after the outbreak of war. Regarding the rumors about Austro-Hungary's plan for the capture of Mt. Lovćen, San Giuliano ordered envoys in Berlin and Vienna to caution these governments in a friendly manner that Austro-Hungarian overtures towards Mt. Lovćen would have a decisive influence on the entire Italian policy (DDI 1964o:401; DDI 1964p:422–423). It was also conveyed to the Entente governments that the serious consequences of the occupation of Mt. Lovćen would have severe ramifications for Italian-Austrian relations (DDI 1964q:442). Austro-Hungarian Minister Leopold Berthold and German Minister von Jagow reiterated assurances that this would not occur if Montenegro remained neutral (DDI 1964r:459; DDI 1964s:410). This was repeated during the talks between Italian envoy Avarna and Minister Berthold, as well as in the discussions between military attaché Giuseppe Alberico Albricci and Chief of the Austro-Hungarian General Staff Franz Conrad von Hötzendorf. The Italians were apprised that Austro-Hungarian troops had been ordered to eschew any action that might be seen as a provocation of hostility towards Montenegro. Both Minister Berthold and General Conrad believed that King Nikola was inclined to maintain neutrality but that those in the military and the general population were seeking war (DDI 1964t:457; DDI 1964u:486; DDI 1964v:487). The Italians had reason to doubt the sincerity of the Austro-Hungarian statements regarding Mt. Lovćen. Confirmation that the government in Vienna wanted to occupy Lovćen at all costs was obtained from members of the Austro-Hungarian embassy in London (DDI 1964w:429). Martin Franklin, the Italian consul in Budapest, reported that military circles claimed that not a single meter would be taken from Serbian territory but that Mt. Lovćen would undoubtedly be annexed in the event of Montenegro's participation in the war (DDI 1964x:340).

Italian Foreign Minister San Giuliano received from Cambiaso similar assessments from Cetinje regarding the mood of the government and people in Montenegro (DDI 1964y:479; DDI 1964z:483).

The Italian insistence on Montenegro's government to maintain neutrality was in vain as Montenegro declared war on Austro-Hungary on August 6, 1914. It entered the war exhausted by the previous Balkan wars, without adequate equipment and weapons and with its state coffers practically depleted (Rakočević 1969:47–50).

Even after the outbreak of the conflict, while waiting for the warring parties to offer Italy an adequate reward for entering the war, the government warned its formal Allies, Austro-Hungary and Germany, that one of the conditions for preserving Italian neutrality was Mt. Lovćen's inviolability as an important element of the strategic balance between the Adriatic countries. Italian diplomacy and the Supreme Command went a step further, demanding a formal announcement from Vienna that the Mt. Lovćen area would be excluded from the military offenses of the Austro-Hungarian army. Such demands were perceived by Vienna as being blackmail yet they were still met.

20.4 Deliberation on Italian Involvement in Defending Lovćen

In regards to its likely entry into war, the Italian Supreme Command also considered the possibility of landing troops in Montenegro, thereby seizing Mt. Lovćen and equipping it with sufficient artillery. The Italian generals agreed that the operation required the allocation of a significant army contingent as well as the weaponry, equipment and resources that the Italian army required in other areas (the operation plan being linked to the name of Admiral Thaon di Revel). Next, Officer Tito Montefinale was then sent on a special mission but the operation was dismissed as irrational, expensive and risky (Capelano and Gionfrida 2019:139–145).

The possibility of joint French-Montenegrin engagement in the Bay of Kotor and southern Dalmatia that would ensure that Mt. Lovćen was seized and the Austro-Hungarian naval base in the Bay of Kotor neutralized met with scant sympathy on the part of the Italians. The reason was the fear that the success of this engagement could have an adverse effect on Italy's territorial ambitions in Dalmatia. The information that a legion of Italian volunteers led by General Peppino Garibaldi might be involved in the Franco-Montenegrin operation caused concerns of altogether a different kind. The consequence of such an engagement would inevitably be the end of Italian neutrality, that is, the untimely entry of Italy into the war on the side of the Entente. For their own reasons, neither Serbia nor Russia supported this engagement and thus the French reduced their military cooperation with Montenegro by sending a smaller artillery unit to Mt. Lovćen, which was neutralized in a relatively short time by the operations of the far more powerful Austro-Hungarian artillery from land and sea (Burzanović 2019:385–387).

20.5 Italian-Montenegrin War Alliance

Montenegro enthusiastically welcomed Italy's entry into the war in May 1915, believing that Italy would contribute to ending the blockade of its ports and facilitate Montenegro's expansion along the Adriatic coast. With the secret Treaty of London, 26 April 1915, Italy, among other things, effectively accepted that in the event of an Allied victory, Boka Kotorska and Mt. Lovćen would be included in one country. Article 5 of the Treaty, envisaged military neutralization of the coast from Boka Kotorska to the mouth of Vojuša river, for which Italy was the most interested of all powers. The neutralization did not apply to the coast that Montenegro already owned, but Montenegro herself accepted the restrictions on fortification in 1909 (Burzanović 2019:398–399).

Albeit seeming paradoxical, Montenegro benefited more from Italian neutrality than from its wartime alliance. As already mentioned, in regards to Italy, Austro-Hungary gave up military operations on the Mt. Lovćen front until the spring of 1915. On the other hand, in order to encourage Italy's joining the Entente forces, the French-British fleet made significant efforts to show its might in the Adriatic waters by breaking the Austro-Hungarian naval blockade of the Montenegrin coast and enabling supplies for the Montenegrin army and the general population (Burzanović 2019:530).

After Italy entered the war, this task was carried out unsuccessfully by the Italian fleet, which tried to preserve its seacraft from the military operations by Austro-Hungarian ships and especially submarines.

From the summer of 1915, Italy not only avoided but also openly refused to execute this task, explaining this by the Montenegrin occupation of Shkodra, which was carried out despite the opposition of the other Allies. Although Italy, Serbia and Greece had violated the territorial integrity of Albania before Montenegro, only Montenegro suffered sanctions due to this, which in turn had a very unfavorable effect on its defensive capabilities (Burzanović 2019:399–401).

In effect, Montenegrin-Italian relations were at their lowest point thus far. The incident had its background in which the issue of Lovćen played a significant place. Namely, only ten days after entering the war, Montenegro became a victim of a campaign aimed at discrediting and isolating it from the Allies in order to prepare its annexation to the Kingdom of Serbia. According to his statement, the Serbian Assistant Minister of Foreign Affairs Jovan Jovanović misinformed the Italian representative in Belgrade, telling him that negotiations on the seizing of Mt. Lovćen in exchange for the consent of Austro-Hungary to the Montenegrin occupation of Shkodra were allegedly being negotiated between Cetinje and Vienna. In fact, Italian Ambassador Andrea Carlotti reported similar rumors from St. Petersburg. In Cetinje, British envoy John Francis Charles De Salis also contributed to their spread among the diplomatic corps. The Italian minister Cambiaso disputed them as unfounded but in the summer of 1915 he changed his position in the light of the Montenegrin occupation of Shkodra and the negotiations that had taken place between Montenegrin and Austro-Hungarian representatives to halt the air bombing of civilians in unprotected Montenegrin cities (Burzanović 2019:401–417).

On June 11, Sidney Sonnino informed Cambiaso that Montenegrin and Austrian envoys had previously held three clandestine meetings in one of Kotor's ramparts, and that on June 10, Prince Petar, the commander of the Mt. Lovćen front had also attended such a meeting. He perceived in the story a confirmation of the thesis that Montenegrin-Austrian relations had not been disrupted after the declaration of war (DDI 1973a:88). Informed that Vienna was accepting the occupation of Shkodra in exchange for Montenegro's inoperativeness on the front against Austro-Hungary, Sonnino asked for this information to be verified. According to a report by Italian military attaché Akame, Austro-Hungary wanted to transfer part of the artillery from the Montenegrin to the Italian front (DDI 1973b:107). After talks with his Serbian counterpart Ljuba Mihailović, Cambiaso's conviction in the Montenegrin-Austrian agreements was even stronger. In fact, he was told that the subject of the agreement between Cetinje and Vienna was also the Montenegrin part of Novi Pazar's Sandžak, which Montenegro would allow to be occupied by Austro-Hungary in order to prevent Serbia from absorbing it (DDI 1973c:183; DDI 1973d:194).

The Italians expressed new doubts about the commitment of the Montenegrin government and the king to the allied coalition regarding the French effort to help Montenegro when faced with the superior Austro-Hungarian forces that had previously defeated the resistance of Serbia. In a conversation with Ambassador Camille Barrer, Minister Sonnino wondered whether Montenegro had opted for a separate peace for which it would be rewarded with Shkodra while Austro-Hungary would rule Mt. Lovćen. The Italian government refused to accept the proposal of the French ambassador to Cetinje that the Allies grant Montenegro 40,000,000 francs for the fortification and defense of Mt. Lovćen. The new Italian envoy to Cetinje, Romano Avezzana, claimed that Mt. Lovćen was well established and that approving the loan and sending food was "*contrary to the interests of the Allies, especially Italy*," as Montenegro intended to conclude a separate peace (Živojinović 1998:169).

Realizing that Montenegrin King Nikola had lost faith in Italian support and the backing of Avezzana, he asked for at least one ship with food to be sent in order to improve the impression at court and among the people (Živojinović 1998:178).

The telegram of King Nikola to Russian Tsar Nikola II dated January 1, 1916 in which it was emphasized that only hunger could force Montenegro to stop its war operations reaped no rewards for Montenegro. Russia offered no help, and subsequently Montenegro unsuccessfully asked the allies for assistance. Furthermore, Montenegro was threatened with a disallowance of any support if it decided on a separate peace. The Russian ambassador to Rome, Girs, discussed the issue with Minister Sonnino on January 5, 1916, and they agreed on possible sanctions (DDI 1988a:191; DDI 1988b:210–211; Živojinović 1998:179).

Russia's emergency proposal for Montenegro failed due to Italian non-cooperation. Avezzana considered the proposal inapplicable as it imposed great obligations on the Allies, and especially on the Italians, and yet lacked any prospect for success. He suggested to Minister Sonnino to inform the Allies that Italy would partake in sending food to Montenegro "*within the limits allowed by our means and needs, insofar that our efforts and risks bring us only the most absurd and harsh criticism both from Montenegro and from Serbia*." Sonnino in actual fact did carry out something similar, with the remark that "*we should avoid expressions that bind us wholly*" (DDI 1988c:197–198; Živojinović 1998:180).

Neither envoy Avezzana nor Minister Sonnino believed that Montenegro was capable of resisting for longer or had the intention to do so. Avezzana believed that in the current circumstances, Montenegro had already decided upon a separate peace (DDI 1988c:197–198; DDI 1988d:203–204). In this period, Sonnino learned that the Honorary Consul of Montenegro to Milan was conducting secret negotiations on behalf of Crown Prince Danilo with German government envoy Count von Bernstorff. King Nikola rejected the offer but the distrust of the Allies towards him remained (Živojinović 1998:180).

20.6 The Fall of Lovćen

The expected Austro-Hungarian offensive began on January 5, 1916. Forty thousand Montenegrin soldiers defended a 500 km long front line facing a force better equipped and three and a half times its size. In the Battle of Mojkovac on January 6 and 7, the Montenegrin-Sandžak detachment stopped the Austro-Hungarian advance on its front, the position of which it managed to maintain until January 18, regardless of the loss of Bijelo Polje and Berane. The main Austro-Hungarian strike followed on the Mt. Lovćen front, which was defended by 8,000 soldiers (Figs. 20.1 and 20.2) against 50,000 enemy forces supported by the Boka navy. The ratio of artillery forces was 12:1 in favor of the attacking enemy. From January 8 to 11, the Austro-Hungarian army took control of strategic points on the Mt. Lovćen massif (Figs. 20.3 and 20.4), and on January 13, it entered an undefended Cetinje. Around 1,260 soldiers of the Austro-Hungarian forces were killed in the Mt. Lovćen operation (DDI 1988e:269; Andrijašević and Rastoder 2005:288). On January 9, King Nikola requested the Allied fleet to help defend Mt. Lovćen. The request was forwarded to the Duke of the Abruzzi, commander of the Allied forces in the southern Adriatic. However,

Fig. 20.1 Montenegrin artillery on Mt. Lovćen, 1916. (*Source* Martinović 2018)

Fig. 20.2 Montenegrin artillery on Mt. Lovćen, 1916. (*Source* Martinović 2018)

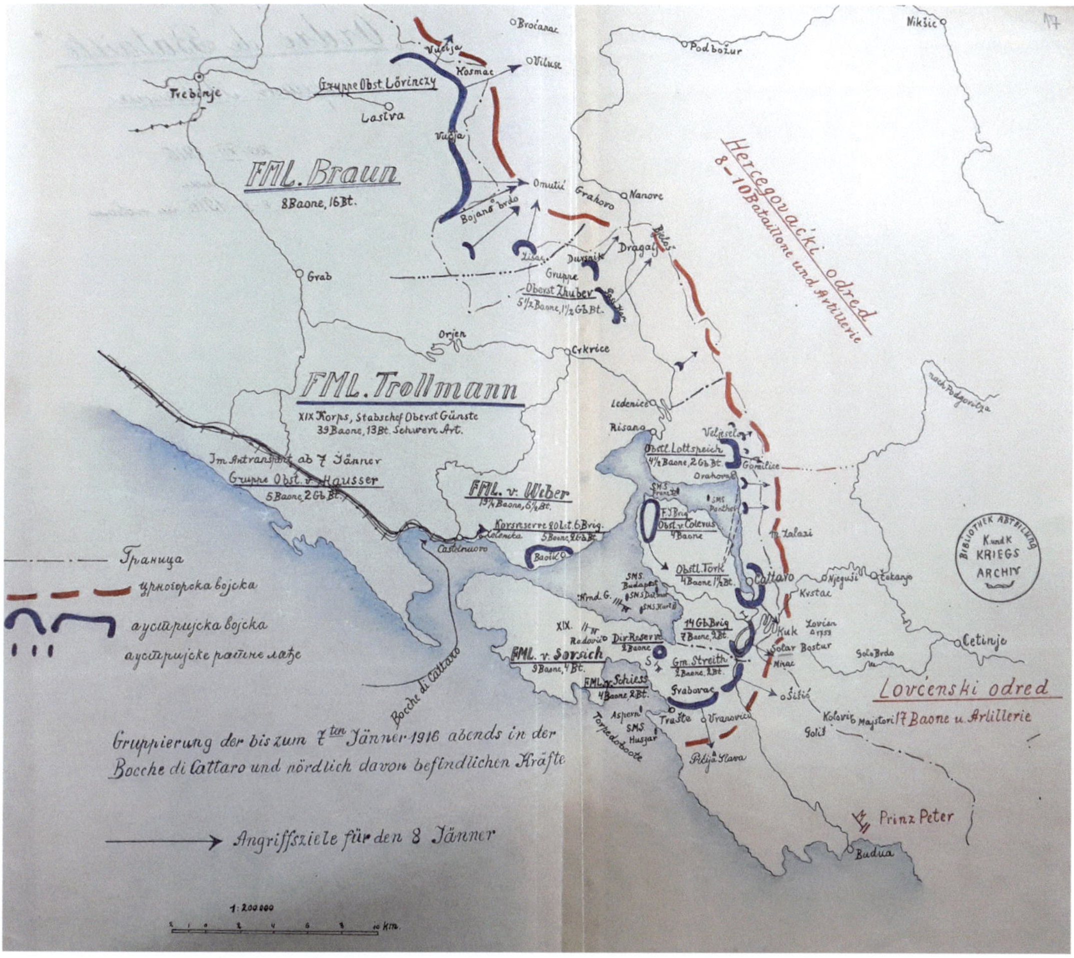

Fig. 20.3 The map of the Mt. Lovćen front on January 7th 1916. (University of Montenegro, Historical Institute archive)

the Italian fleet failed to intervene (Živojinović 1998:181).

The Montenegrin government withdrew to Podgorica with its court and diplomatic corps, demanding a truce. Vienna requested unconditional capitulation and set unacceptable conditions for the Montenegrin king and government (DDI 1988f:21; DDI 1988g:244–245). Avezzana informed Minister Sonnino that the King's statement on the continuation of the resistance was only an attempt to preserve military honor, although Montenegro was unable to continue the war (DDI 1988h:214–215). King Nikola moved to Brindisi with a part of his family and several officials by way of a ship made available to him by the Italian government. On January 18, 1916, the newly appointed chief of the Montenegrin Supreme Command, Serdar Janko Vukotić, ordered the retreat of the Montenegrin army and its concentration in the Podgorica region and on January 21, he issued an order for its disbanding. The *Order of Surrender* was signed on January 25, 1916 (DDI 1988g:244–245; Živojinović 1998: 185–186; Andrijašević and Rastoder 2005:289). It was an act of capitulation, although formally this provision was not an admission of defeat, especially after the statements by Prime Minister Lazar Mijušković and King Nikola that Montenegro is to continue the war with the Allies (DDI 1988i:290; DDI 1988j:598). King

Fig. 20.4 Austro-Hungarian soldiers on Mt. Lovćen, January 1916

Nikola found himself in exile without an army, which would significantly affect the fate of the state, the dynasty and him personally.

20.7 Reactions to the Capitulation of Montenegro and the Fall of Mt. Lovćen

Although it was clear that Montenegro would suffer the same fate as Serbia after its military collapse during January and February of 1916, the opinion emerged in Allied circles that it had succumbed too easily to Austro-Hungarian military force. For the Allies, Italy was the main culprit for its inadequate logistical and military support to Montenegro. In his diary on January 15, 1916, the Italian ambassador in London, Imperiali, noted the unease of diplomatic circles due to the fate of Mt. Lovćen and Montenegro. (Imperiali 2006:236). In this sense, the fall of Mt. Lovćen and its consequences in terms of the balance of forces of the warring parties in the Adriatic were especially emphasized due to the increase in the strategic value of the Austro-Hungarian naval base in the Bay of Kotor (DDI 1988k:240). The capitulation of Montenegro was also a blow to

Italian prestige in the Balkans. Serbian diplomacy accused the Italians in the presence of the other Allies of leaving Montenegro to its fate (Imperiali 2006:236). The Athens-based press wrote in much the same vein, not sparing the other Allied countries and claiming that any government that put faith in the Entente was doomed (DDI 1988l:235–236). Sonnino demanded from the Romanian Prime Minister Ion Brătianu to convince the Allies that Montenegro's capitulation was not of a great significance from a military point of view (Hrabak 1998:112).

The very act of the Austro-Hungarian occupation of Mt. Lovćen and the accusations of the Allies against Italy were disagreeable to the Italian public opinion. After the initial comments expressing sympathy and compassion for Montenegro, the Italian press reported on the fall of Mt. Lovćen and Montenegro as a consequence of clandestine agreements between Cetinje and Vienna (Vujović 1994:282; Živojinović 1998:187–191).

In political circles, in terms of their responsibility, opinions were initially divided. Foreign Minister Sidney Sonnino believed that Italian military leadership had done very little to prevent the fall of Mt. Lovćen and Montenegro. In their public appearances, Italian officials made correct assessments that the capitulation of Montenegro was a logical consequence to the general development of the Balkan battlefields. King Victor Emmanuel III was depressed by the news of the Montenegrin events but he predicted accurately that the entire story would fade in a week (Živojinović 1998:182).

The more the Allies' accusations irritated the Italians, the louder and more openly did Italian diplomats attack King Nikola for his alleged double game, and their self-critical remarks, ad absurdum, were articulated as regret that the Allies failed to prevent King Nikola from siding with the enemy. In the first reactions to the fall of Mt. Lovćen, Imperiali had words of criticism for Italy's attitude, assessing that it involved a vital Italian interest that was supposed to be secured independently of the Allies. He tried to relativize the responsibility the very next day by asking whether Montenegro had offered serious resistance. As the writing of the Italian and Allied press became increasingly uncomfortable due to the accusations against Italy as the main culprit, Imperiali, in a conversation with Edward Grey, tried to mitigate the negative effect of the Montenegrin events. He estimated that the fall of Montenegro was a natural consequence of the military collapse of Serbia, but he did not fail to comment on "the combinations of King Nikola." Minister Edward Grey assessed that Montenegro had not done much for Allied interests since the beginning of the war, and especially since the capture of Shkodra. He was under the impression that the Montenegrin resistance on Mt. Lovćen was lukewarm and that peace with Austro-Hungary might have been agreed for some time (DDI 1988m:239; Imperiali 2006:235). Minister Avezzana was also prone to suspicion of the Montenegrin government and King Nikola, but he emphasized the King's merits for saving the Serbian army and pointed out that Montenegro's short-term resistance had given the Italians time to decide whether to withdraw from or defend central Albania (DDI 1988e:277). Hence, the story of Mt. Lovćen in Italian political circles and public opinion would come to life again only after the end of the war.

The Austro-Hungarian government's attitude towards the "Lovćen issue", remained unchanged until the end of the war. At the end of 1916, considering a possible compromising peace with the Entante powers, Vienna always included possession of this mountain range as a precondition (Fischer 1967:310–311).Vienna's intention was to erect a monument on Štirovnik in honour of the Austro-Hungarian conquest of Lovćen known as "Adriatic Gibraltar". Due to the circumstances of the war, the project was not realized (Šistek 2012:112–113; Žaja Vrbica 2017:285–286).

20.8 Mt. Lovćen and the *Adriatic Issue*

The complete collapse of Austro-Hungary and its disappearance from the Balkans and the Adriatic resulted in Italy becoming the dominant power in the Adriatic. At the same time, a situation developed that Italy had once depicted to Vienna and Berlin as a *casus belli*: the Bay of Kotor and Mt. Lovćen found themselves in the same government building of the new Yugoslav state with which Italy was in a conflict of interest since its inception. In the process of negotiating and harmonizing the interests of the two states, Rome instrumentalized the *Montenegrin issue* by providing certain financial and diplomatic support to the Montenegrin government in exile. The issue of Mt. Lovćen reemerged in the public consciousness during the campaigns supporting the renewal of the Montenegrin state. Also, its survival within the independent Montenegrin state was declared as a vital Italian interest (Živojinović 2013:220–229). However, after the implementation of the *Rapallo Agreement* from 1920, the issue of Mt. Lovćen ceased to be interesting to Italian diplomatic and military circles although it was highlighted by numerous pro-Montenegrin committees formed in Italy and opposition politicians who accused the government of jeopardizing Italy's security by leaving Montenegro at the mercy of Belgrade (Burzanović 2019:507–511). These included Mussolini's Fascist Party, who after coming to power in 1922 changed their stance on this issue and put an end to all the activities by Montenegrin emigration in Italy more forcefully than their predecessors. In truth, in the first days after taking power, Mussolini, driven by the inertia of earlier opposition rhetoric, considered the possibility of renewing the Montenegrin state and dynasty and ordered a special investigation into the *Montenegrin issue* and within it the fate of Mt. Lovćen in World War I. However, political realism prevailed in relations with the Yugoslav kingdom and soon made this effort redundant (Burzanović 2019:523–526).

The conclusion is that, despite the great attention that Italian diplomacy and the public have paid to the issue of Mt. Lovćen for almost two decades, not much has been done by the Italians to preserve its inviolability, apart from a few diplomatic efforts. Moreover, no military effort was considered. In the relations between Rome and Vienna, and later on, Rome and Belgrade, the issue of Mt. Lovćen was used in negotiations to obtain other, more significant gains for Italy, i.e. compensations.

References

ASDMAE (1906) Archivio Segreto di gabineto, Cassete Verdi, cass. 20, Avarna a San Giuliano, Vienna, 29 gennaio 1906. Archivio Storico Diplomatico del Ministero degli Affari Esteri, Roma

Andrijašević Ž, Rastoder Š (2005) Istorija Crne Gore [The history of Montenegro]. Centar za iseljenike Crne Gore, Podgorica

Babović-Raspopović S (1998) Pitanje Lovćna u prepisci austrougarskih diplomata (maj-jul) 1914 [The Lovćen issue in the secret correspondence of Austro-Hungarian diplomats (May–July) 1914]. In: Raspopović R (ed) Crna Gora u Prvom svjetskom ratu: zbornik radova, vol. 2. Istorijski institut, Podgorica, pp 213–222

Burzanović S (2014) Montenegro in the Italian foreign policy during the Balkan Wars (1912–1913). In: Biagini A, Motta G (eds) Empires and nations from the 18th to the 20th century, vol 2. Cambridge Scholars Publishing, Cambridge, pp 284–290

Burzanović S (2019) Crna Gora u italijanskoj spoljnoj politici 1861–1923 [Montenegro in Italian foreign policy from 1861 until 1923]. CANU, Podgorica

Caccamo F (2008) Il Montenegro negli anni della prima guerra mondiale. Aracne, Roma

Capelano F, Gionfrida A (2019) Italy and Serbia in the first world war: the role of the Serbian army in Cadornas war. In: Pavlović V (ed) Serbia and Italy in the great war. Institute for Balkan Studies, Serbian Academy of Science and Arts, Belgrade, pp 137–160

DDI (1964a) No 74, Avarna a San Giuliano, Vienna, 4 luglio 1914. In: Torre A (ed) I Documenti diplomatici italiani, serie IV (1908–1914), vol. XII (28 giugno–2 agosto 1914). Ministero degli Affari Esteri, Istituto poligrafico dello stato, Roma, p 47

DDI (1964b) No 169, Bollati a San Giuliano, Berlino, 13 luglio 1914. In: Torre A (ed) I Documenti diplomatici italiani, serie IV (1908–1914), vol. XII (28 giugno–2 agosto 1914). Ministero degli Affari Esteri, Istituto poligrafico dello stato, Roma, p 124

DDI (1964c) No 330, Avarna a San Giuliano, Vienna, 18 luglio 1914. In: Torre A (ed) I Documenti diplomatici italiani, serie IV (1908–1914), vol. XII (28 giugno–2 agosto 1914). Ministero degli Affari Esteri, Istituto poligrafico dello stato, Roma, p 217

DDI (1964d) No 464, Bollati a San Giuliano, Berlino, 24 luglio 1914. In: Torre A (ed) I Documenti diplomatici italiani, serie IV (1908–1914), vol. XII (28 giugno–2 agosto 1914). Ministero degli Affari Esteri, Istituto poligrafico dello stato, Roma, pp 298–299

DDI (1964e) No 385, Avarna a San Giuliano, Vienna, 21 luglio 1914. In: Torre A (ed) I Documenti diplomatici italiani, serie IV (1908–1914), vol XII (28 giugno–2 agosto 1914). Ministero degli Affari Esteri, Istituto poligrafico dello stato, Roma, p 254

DDI (1964f) No 228, Paternò a San Giuliano, Cettigne, 14 luglio 1914. In: Torre A (ed) I Documenti diplomatici italiani, serie IV (1908–1914), vol XII (28 giugno–2 agosto 1914). Ministero degli Affari Esteri, Istituto poligrafico dello stato, Roma, p 163

DDI (1964g) No 274, Paternò a San Giuliano, Cettigne, 16 luglio 1914. In: Torre A (ed) I Documenti diplomatici italiani, serie IV (1908–1914), vol XII (28 giugno–2 agosto 1914). Ministero degli Affari Esteri, Istituto poligrafico dello stato, Roma, p 189

DDI (1964h) No 432, Cora a San Giuliano, Belgrado, 23 luglio 1914. In: Torre A (ed) I Documenti diplomatici italiani, serie IV (1908–1914), vol XII (28 giugno–2 agosto 1914). Ministero degli Affari Esteri, Istituto poligrafico dello stato, Roma, p 281

DDI (1964i) No 379, Cora a San Giuliano, Belgrado, 21 luglio 1914. In: Torre A (ed) I Documenti diplomatici italiani, serie IV (1908–1914), vol XII (28 giugno–2 agosto 1914). Ministero degli Affari Esteri, Istituto poligrafico dello stato, Roma, p 251

DDI (1964j) No 380, Negrotto Cambiaso a San Giuliano, Cettigne, 21 luglio 1914. In: Torre A (ed) I Documenti diplomatici italiani, serie IV (1908–1914), vol XII (28 giugno–2 agosto 1914). Ministero degli Affari Esteri, Istituto poligrafico dello stato, Roma, pp 251–252

DDI (1964k) No 462, Negrotto Cambiaso a San Giuliano, Cettigne, 24 luglio 1914. In: Torre A (ed) I Documenti diplomatici italiani, serie IV (1908–1914), vol. XII (28 giugno–2 agosto 1914). Ministero degli Affari Esteri, Istituto poligrafico dello stato, Roma, p 297

DDI (1964l) No 419, San Giuliano a Avarna, Roma, 22 luglio 1914. In: Torre A (ed) I Documenti diplomatici italiani, serie IV (1908–1914), vol. XII (28 giugno–2 agosto 1914). Ministero degli Affari Esteri, Istituto poligrafico dello stato, Roma, p 274

DDI (1964m) No 401, San Giuliano a Avarna, Bollati, Carlotti e Squitti, Roma, 22 luglio 1914. In: Torre A (ed) I Documenti diplomatici italiani, serie IV (1908–1914), vol. XII (28 giugno–2 agosto 1914). Ministero degli Affari Esteri, Istituto poligrafico dello stato, Roma, p 264

DDI (1964n) No 557, Avarna a San Giuliano, Vienna, 26 luglio 1914. In: Torre A (ed) I Documenti diplomatici italiani, serie IV (1908–1914), vol. XII (28 giugno–2 agosto 1914). Ministero degli Affari Esteri, Istituto poligrafico dello stato, Roma, p 348

DDI (1964o) No 652, San Giuliano a Bollati, Avarna, Carlotti e Ruspoli, Roma, 28 luglio 1914. In: Torre A (ed) I Documenti diplomatici italiani, serie IV (1908–1914), vol. XII (28 giugno–2 agosto 1914). Ministero degli Affari Esteri, Istituto poligrafico dello stato, Roma, p 401

DDI (1964p) No 686, San Giuliano a Bollati, Avarna, Carlotti e Negrotto Cambiaso, Roma, 29 luglio 1914. In: Torre A (ed) I Documenti diplomatici italiani, serie IV (1908–1914), vol XII (28 giugno–2 agosto 1914). Ministero degli Affari Esteri, Istituto poligrafico dello stato, Roma, pp 422–423

DDI (1964q) No 721, San Giuliano a Avarna, Bollati, Carlotti, De Facendis e Ruspoli, Roma, 29 luglio 1914. In: Torre A (ed) I Documenti diplomatici italiani, serie IV (1908–1914), vol XII (28 giugno–2 agosto 1914). Ministero degli Affari Esteri, Istituto poligrafico dello stato, Roma, p 442

DDI (1964r) No 753, San Giuliano a Imperiali, Avarna, Bollati, Carlotti e Ruspoli, Roma, 30 luglio 1914. In: Torre A (ed) I Documenti diplomatici italiani, serie IV (1908–1914), vol. XII (28 giugno–2 agosto 1914). Ministero degli Affari Esteri, Istituto poligrafico dello stato, Roma, p 459

DDI (1964s) No 668, San Giuliano a Imperiali, Avarna, Bollati, Carlotti, Negrotto Cambiaso, Fasciotti, Squitti e Ruspoli, Roma, 28 luglio 1914. In: Torre A (ed) I Documenti diplomatici italiani, serie IV (1908–1914), vol. XII (28 giugno–2 agosto 1914). Ministero degli Affari Esteri, Istituto poligrafico dello stato, Roma, p 410

DDI (1964s) No 750, Avarna a San Giuliano, Vienna, 30 luglio 1914. In: Torre A (ed) I Documenti diplomatici italiani, serie IV (1908–1914), vol XII (28 giugno–2 agosto 1914). Ministero degli Affari Esteri, Istituto poligrafico dello stato, Roma, p 457

DDI (1964u) No 801, Avarna a San Giuliano, Vienna, 31 luglio 1914. In: Torre A (ed) I Documenti diplomatici italiani, serie IV (1908–1914), vol XII (28 giugno–2 agosto 1914). Ministero degli Affari Esteri, Istituto poligrafico dello stato, Roma, p 486

DDI (1964v) No 804, Avarna a San Giuliano, Vienna, 31 luglio 1914. In: Torre A (ed) I Documenti diplomatici italiani, serie IV (1908–1914), vol XII (28 giugno–2 agosto 1914). Ministero degli Affari Esteri, Istituto poligrafico dello stato, Roma, p 487

DDI (1964w) No 697, Imperiali a San Giuliano, Londra, 29 luglio 1914. In: Torre A (ed) I Documenti diplomatici italiani, serie IV (1908–1914), vol XII (28 giugno–2 agosto 1914). Ministero degli Affari Esteri, Istituto poligrafico dello stato, Roma, p 429

DDI (1964x) No 541, Martin Franklin a San Giuliano, Budapest, 26 luglio 1914. In: Torre A (ed) I Documenti diplomatici italiani, serie IV (1908–1914), vol XII (28 giugno–2 agosto 1914). Ministero degli Affari Esteri, Istituto poligrafico dello stato, Roma, p 340

DDI (1964y) No 787, Negrotto Cambiaso a San Giuliano, Cettigne, 31 luglio 1914. In: Torre A (ed) I Documenti diplomatici italiani, serie IV (1908–1914), vol XII (28 giugno–2 agosto 1914). Ministero degli Affari Esteri, Istituto poligrafico dello stato, Roma, p 479

DDI (1964z) No 796, Negrotto Cambiaso a San Giuliano, Bari, 31 luglio 1914. In: Torre A (ed) I Documenti diplomatici italiani, serie IV (1908–1914), vol XII (28 giugno–2 agosto 1914). Ministero degli Affari Esteri, Istituto poligrafico dello stato, Roma, p 483

DDI (1973a) No 152, Negrotto Cambiaso a Sonnino, Cettigne, 11 giugno 1915. In: Anchieri E (ed) I Documenti diplomatici italiani, serie V (1914–1918), vol IV

(25 maggio–23 ottobre 1915). Ministero degli Affari Esteri, Istituto poligrafico dello stato, Roma, p 88
DDI (1973b) No 186, Sonnino a Negrotto Cambiaso, Roma 15 giugno 1915. In: Anchieri E (ed) I Documenti diplomatici italiani, serie V (1914–1918), vol IV (25 maggio–23 ottobre 1915). Ministero degli Affari Esteri, Istituto poligrafico dello stato, Roma, p 107
DDI (1973c) No 305, Negrotto Cambiaso a Sonnino, Scutari, 29 giugno 1915. In: Anchieri E (ed) I Documenti diplomatici italiani, serie V (1914–1918), vol IV (25 maggio–23 ottobre 1915). Ministero degli Affari Esteri, Istituto poligrafico dello stato, Roma, p 183
DDI (1973d) No 324, Negrotto Cambiaso a Sonnino, Cettigne, 1 luglio 1915. In: Anchieri E (ed) I Documenti diplomatici italiani, serie V (1914–1918), vol IV (25 maggio–23 ottobre 1915). Ministero degli Affari Esteri, Istituto poligrafico dello stato, Roma, p 194
DDI (1988a) No 275, Sonnino a Imperiali, Carlotti, Avezzana, Squitti e Ruspoli, Roma, 5 gennaio 1916. In: Curato F (ed) I Documenti diplomatici italiani, serie V (1914–1918), vol V (24 ottobre 1915–17 giugno 1916). Ministero degli Affari Esteri, Istituto poligrafico e zecca dello stato, Roma, p 191
DDI (1988b) No 300, Carlotti a Sonnino, Pietrogrado, 12 gennaio 1916. In: Curato F (ed) I Documenti diplomatici italiani, serie V (1914–1918), vol V (24 ottobre 1915–17 giugno 1916). Ministero degli Affari Esteri, Istituto poligrafico e zecca dello stato, Roma, pp 210–211
DDI (1988c) No 284, Avezzana a Sonnino, Cettigne, 8 gennaio 1916. In: Curato F (ed) I Documenti diplomatici italiani, serie V (1914–1918), vol V (24 ottobre 1915–17 giugno 1916). Ministero degli Affari Esteri, Istituto poligrafico e zecca dello stato, Roma, pp 197–198
DDI (1988d) No 293, Salandra a Sonnino, Roma, 10 gennaio 1916. In: Curato F (ed) I Documenti diplomatici italiani, serie V (1914–1918), vol V (24 ottobre 1915–17 giugno 1916). Ministero degli Affari Esteri, Istituto poligrafico e zecca dello stato, Roma, pp 203–204
DDI (1988e) No 379, Avezzana a Sonnino, Roma, 27 gennaio 1916. In: Curato F (ed) I Documenti diplomatici italiani, serie V (1914–1918), vol V (24 ottobre 1915–17 giugno 1916). Ministero degli Affari Esteri, Istituto poligrafico e zecca dello stato, Roma, p 269, 277
DDI (1988f) No 304, Squitti a Sonnino, Scutari, 12 gennaio 1916. In: Curato F (ed) I Documenti diplomatici italiani, serie V (1914–1918), vol V (24 ottobre 1915–17 giugno 1916). Ministero degli Affari Esteri, Istituto poligrafico e zecca dello stato, Roma, p 21
DDI (1988g) No 346, Sonnino a Imperiali, Tittoni e Carlotti, Roma, 20 gennaio 1916. In: Curato F (ed) I Documenti diplomatici italiani, serie V (1914–1918), vol V (24 ottobre 1915–17 giugno 1916). Ministero degli Affari Esteri, Istituto poligrafico e zecca dello stato, Roma, pp 244–245
DDI (1988h) No 308, Avezzana a Sonnino, Scutari, 13 gennaio 1916. In: Curato F (ed) I Documenti diplomatici italiani, serie V (1914–1918), vol. V (24 ottobre 1915–17 giugno 1916). Ministero degli Affari Esteri, Istituto poligrafico e zecca dello stato, Roma, pp 214–215
DDI (1988i) No 399, Tittoni a Sonnino, Parigi, 3 febbraio 1916. In: Curato F (ed) I Documenti diplomatici italiani, serie V (1914–1918), vol V (24 ottobre 1915–17 giugno 1916). Ministero degli Affari Esteri, Istituto poligrafico e zecca dello stato, Roma, p 290
DDI (1988j) No 808, Avezzana a Sonnino, Bordeaux, 13 maggio 1916. In: Curato F (ed) I Documenti diplomatici italiani, serie V (1914–1918), vol V (24 ottobre 1915–17 giugno 1916). Ministero degli Affari Esteri, Istituto poligrafico e zecca dello stato, Roma, p 598
DDI (1988k) No 339, No 340, Carlotti a Sonnino, Pietrogrado, 19 gennaio 1916. In: Curato F (ed) I Documenti diplomatici italiani, serie V (1914–1918), vol V (24 ottobre 1915–17 giugno 1916). Ministero degli Affari Esteri, Istituto poligrafico e zecca dello stato, Roma, p 240
DDI (1988l) No 331, De Bosdari a Sonnino, Athene, 17 gennaio 1916. In: Curato F (ed) I Documenti diplomatici italiani, serie V (1914–1918), vol V (24 ottobre 1915–17 giugno 1916). Ministero degli Affari Esteri, Istituto poligrafico e zecca dello stato, Roma, pp 235–236
DDI (1988m) No 336, Imperiali a Sonnino, Londra, 18 gennaio 1916. In: Curato F (ed) I Documenti diplomatici italiani, serie V (1914–1918), vol V (24 ottobre 1915–17 giugno 1916). Ministero degli Affari Esteri, Istituto poligrafico e zecca dello stato, Roma, p 239
Ekmečić M (1990) Ratni ciljevi Srbije 1914 [The war aims of Serbia 1914]. Srpska književna zadruga, Beograd
Ferraioli G (2007) Politica e diplomazia in Italia tra XIX e XX secolo: Vita di Antonio di San Giuliano (1852–1914). Rubbettino, Catanzaro
Fischer F (1967) Germany's aims in the First World War. W. W. Norton, New York
Giordano G (2012) Tra marsine e stiffelius: Venticinque anni di politica estera italiana 1900–1925. Nuova Cultura, Roma
Hrabak B (1998) Italijanske masonske vlade i Crna Gora [Italian masonic governments and Montenegro]. In: Raspopović R (ed) Crna Gora u Prvom svjetskom ratu: zbornik radova, vol 2. Istorijski institut, Podgorica, pp 103–128
Imperiali G (2006) Diario 1915–1919. Rubbettino, Catanzaro
Martinović S (2018), Crnogorska vojska [Montenegrin Army] 1854–1916. Fakultet za crnogorski jezik i književnost, Ministarstvo odbrane Crne Gore, Cetinje, Podgorica
Monzali L (2004) Italiani di Dalmazia: Dal Risorgimenro alla Grande Guerra. Le lettere, Firenze
Rakočević N (1969) Crna Gora u Prvom Svjetskom ratu [Montenegro in world war I]. Istorijski institut, Titograd

Šistek F (2012) Njegoševa grobnica na Lovćenu [The chapel of Njegoš on Mt. Lovćen]. Matica 51–52:105–140

Vujović D (1994) Ratna saradnja Crne Gore i Francuske 1914–1916 [French-Montenegrin war collaboration during WWI]. CANU, Podgorica

Žaja Vrbica S (2017) Slikar Marko Rašica u vojsci Austro-Ugarske Monarhije [Painter Marko Rašica in the Austro-Hungarian Army]. Ars Adriatica 7:281–290

Živojinović D (1998) Italija i Crna Gora 1914–1925: studija o izneverenom savezništvu [Italy and Montenegro 1914–1925: a study of a betrayed alliance]. Službeni list SRJ, Beograd

Živojinović D (2013) U potrazi za imperijom: Italija i Balkan početkom XX veka [Searching for the empire: Italy and Balkan at the beginning of 20th century]. Albatros plus, Beograd

Part V
Cultural Heritage

21 Examples of Immovable Cultural Heritage of Montenegrin Mountains

Aleksandra Kapetanović

Abstract

In Montenegro, which is predominantly a mountainous country, different segments of cultural heritage can be found in the mountainous areas. Especially important is the heritage of the most dominant element of Montenegrin mountains, temporary seasonal settlements–*katuns*, developed from a specific way of seasonal use of mountain pastures for cattle grazing. Although this segment of cultural heritage is still not formally protected, significant research activities and projects have been implemented in this field in the recent period. This paper focuses on the immovable cultural heritage of *katuns* which is observed as part of the overall cultural landscape that includes: huts, corrals, outbuildings, cisterns, ponds, religious buildings, cemeteries, roads and paths, as well as intangible practices. The analyses presented are focused on traditional architecture of *katuns*, especially different types of huts and their development.

Keywords

Immovable cultural heritage · Seasonal settlements–*katuns* · Traditional architecture · Huts · Transhumance · Cultural landscape · Montenegro

A. Kapetanović (✉)
EXPEDITIO Center for Sustainable Spatial Development, Kotor, Montenegro
e-mail: sandra@expeditio.org

21.1 Introduction

Although mountain areas are mainly perceived as zones with primarily natural values, they often possess significant segments of cultural heritage, created as a result of population's specific way of life. Cultural heritage, which can be movable, immovable and intangible, includes all aspects of the environment resulting from the interaction between people and places through time (Faro Convention 2015). The cultural landscape is recognised as one of the broadest and most comprehensive categories of cultural heritage, and it represents the combined works of nature and of man, illustrative of the evolution of human society and settlement over time (Operational … 1992). The immovable cultural heritage of the mountains is part of its overall cultural landscape.

In Montenegro, which is a predominantly mountainous country, an important–although not valued enough–segment of cultural heritage is found in the mountain areas. Although the majority of cultural heritage is concentrated in the lower regions with more favourable living conditions such as the coastal region and areas around lakes and river valleys, we can find various types and layers of cultural heritage that testify to the life

M. Zorn et al. (eds.), *Discourses on Mountains of Montenegro and Slovenia*,
Historical Geography and Geosciences, https://doi.org/10.1007/978-3-031-99017-5_21

during various historical periods in the mountain areas as well.

Protection of cultural heritage in Montenegro is regulated primarily by the *Protection of Cultural Property Act* (Zakon … 2010, 2019) which refers to cultural properties that are recognised as the validated part of the cultural heritage of general interest. Unfortunately, not all types and segments of cultural heritage are formally protected and have the status of a cultural property, including traditional and rural architecture (only two villages are protected and just a number of traditional architecture buildings) as well as cultural landscape (so far there are no protected cultural landscapes).

In this context, cultural heritage of Montenegrin mountains has not yet been systematically researched, sufficiently valued and protected. Out of all the cultural properties that are formally protected, very few are from the mountains, and these do not include traditional architecture, which is a part of the temporary, seasonal settlements—*katuns* (sing.*katun*; pl. Montenegrin *katuni,* Eng. *katuns*), and which represents the significant segment of the cultural heritage of the mountains (Laković 2017).

Data on *katuns* in Montenegro can be found in various sources albeit their comprehensive and systematic research as well as detailed research of their immovable cultural heritage has not been carried out so far. First mentions of *katuns* can be found in written sources from the 13th and 14th centuries (Kovijanić 1963; Crnić-Pejović 2000). Foreign researchers who visited Montenegro during the 19th and the beginning of the 20th century, provided valuable pieces of information related to *katuns* (e.g., Rovinski 2001; Baldači 2004). Highly significant are the anthropogeographical researches of certain areas of Montenegro, carried out by Yugoslav researchers, from the beginning of the 20th century, in which *katuns* and types of buildings within them are described in more detail (e.g., Erdeljanović 1907, 1926; Tomić 2000). Descriptions given by mountaineers from the 1930s (e.g., Urban 1938; Gušić 1964), and various researches during the 20th century represent an important source for the study of *katun* heritage (e.g., Lutovac 2000; Crnić-Pejović 2000, 2004).

However, in some Montenegrin areas, research activities and projects have been implemented in recent decades that represent significant progress in this field. Ethnographic field reconnaissance of *katuns* on Lovćen was conducted in 1996 by Tanja Vujačić, during the first phase of the Project on the Study and Protection of Montenegrin Traditional Architecture of the Republic Institute for Protection of Cultural Monuments in Cetinje (Vujačić 1997). Within the project carried out by the University of Montenegro titled "*Valorization of Montenegrin Katuni Through Sustainable Development of Agriculture and Tourism*—KATUN" implemented between 2015 and 2017, identification and mapping of all *katuns* in the area of the Kuči Mountain was conducted by the Historical Institute, as well as a detailed analysis of cultural heritage of *katuns* (Laković 2017; Kapetanović et al. 2017; Laković et al. 2020). This is the only project so far that has systematically and scientifically dealt with the topic of the cultural heritage of *katuns*. In 2018 and 2019, during the implementation of the "*Study of Traditional Architecture in the Municipalities of Berane, Petnjica, Andrijevica, Plav and Gusinje*", within the project of the Museum of Polimlje and the Regional Development Agency Bjelasica, Komovi and Prokletije, preliminary researches of the cultural heritage of *katuns* from this area were carried out (Kapetanović et al. 2019). The implementation of the project on *katuns* on the Sinjajevina mountain, led by the French National Center for Scientific Research (CNRS) within the project "IRIS—*Inspiring Rural Heritage: Sustainable Practices to Protect and Conserve Upland Landscapes and Memories*", was also conducted (IRIS 2022).

Although *katuns* are not protected as cultural properties, their value is recognized, especially within protected areas. Four out of the five National Parks of Montenegro (Lovćen, Durmitor, Biogradska gora and Prokletije) recognize *katuns* as a significant segment of traditional architecture (Nacionalni … 2022). Within certain Nature Parks, such as Orjen, *katuns* are described as a part of the cultural heritage (Nature … 2022).

Fig. 21.1 Carine katun, Kuči Mountain. (Photo by Aleksandra Kapetanović)

However, detailed research has not been done. In recent years *katuns* have been brought into light also through various projects. For example, since 2015 the Regional Development Agency for Bjelasica, Komovi and Prokletije implemented projects (e.g., Katuns As Cultural Heritage, Preserving Cultural Landscape of Albania and Montenegro, Katun Roads of Montenegro and Bosnia and Herzegovina) aimed at promoting the value of *katuns* as cultural heritage that contributes to rural and tourism development (Regional … 2022).

The analyses presented in this paper are based on data from existing documentation and sources as well as the results of the field research (Fig. 21.1). The author of this paper, in addition to participating in the most of the abovementioned research projects, researched *katuns* as part of the NGO EXPEDITIO's project conducted within the Prokletije area (Myteveli et al. 2012) and the Boka Kotorska Bay hinterland (EXPEDITIO 2022). In addition, the author has been visiting mountains in all parts of Montenegro and documenting *katuns*, while collecting extensive photo documentation as well as making architectural surveys since 2000.

21.2 Types of Cultural Heritage

Various types and layers of cultural heritage from different time periods can be found in Montenegrin mountains. Although the mountainous area of Montenegro has not yet been thoroughly archaeologically researched, there are traces (albeit modest) of material culture that indicate life in these areas since prehistoric times. In the Prokletije massif, at 1,700 m above sea level, there are engravings on a stone block dating back to the Bronze Age (Mijović 1992; Fig. 21.2a). A number of prehistoric burial mounds (*tumuli*) or forts (*gradine*) have been identified in other mountains as well (Marković 1994; Petričević 2019).

In the ancient and medieval periods, the roads that connected the Adriatic Sea with the inland of the Balkan Peninsula passed through the mountainous areas, so there are remains of medieval fortifications, such as Pirlitor on Durmitor, above the river Tara. Monolithic tombstones (*stećci*) represent a significant segment of the cultural heritage of the medieval period in the former Yugoslavia, and in Montenegro they can also be found in high mountain zones. On the Durmitor mountain at an altitude of about 1430 m, there

Fig. 21.2 **a** Bronze Age engravings on a stone slab, Prokletije; **b** Necropolis of *stećak* tombstones, Greek cemetery, Durmitor; **c** The church of the Holy Transfiguration, Ivanova Korita, Lovćen; **d** Kom fortress and a road from the Austro-Hungarian period, Crkvice, Orjen. (Photos by Aleksandra Kapetanović)

are two necropolises of *stećak* tombstones, built in the 14th and 15th centuries (UNESCO World Heritage … 2022a; Fig. 21.2b).

In the Montenegrin mountains we can find rare examples of sacral architecture, created mainly at the end of the 19th century (e.g., The Transfiguration Monastery, on Ivanova Korita on Lovćen (17th–19th century) (Fig. 21.2c); the church of St. Basil of Ostrog on the site of Ružica on Sinjajevina (1894); the church of St. Elijah on Krst od Prevlake on the Kuči Mountain (1900). At the places where the mountains, during different historical periods, were located within or near the border zones, different segments of defensive architecture were formed. In the coastal mountains there is a network of fortifications, roads and other infrastructure facilities from the Austro-Hungarian period (Pavićević 2019; Fig. 21.2d), from the 19th and the beginning of the 20th centuries. In the north-eastern part of the border zones towards Albania remains of watchtowers, observation posts and fortified combat positions, built for the needs of the Yugoslav People's Army in the 1950s and 1960s, have been preserved (Laković 2017). We can also find monuments and memorials, including those from the period of the World War II (Laković 2017). An exceptional example of the 20th century architecture is the mausoleum of Peter II Petrović Njegoš on Lovćen, built in 1974 based on the idea of sculptor Ivan Meštrović.

Along with all the abovementioned segments of immovable cultural heritage in the mountains, the most common part of the cultural heritage must be especially emphasized–traditional architecture, i.e., rural or vernacular architectural heritage.

This type of cultural heritage was created as a result of a specific way of life in the mountains, which is primarily connected to livestock breeding–the main economic activity for the most of the Montenegrin population from prehistoric times to the present day. Specific way of using the mountains is the seasonal use of pastures for cattle grazing, which was accompanied by the formation of temporary, seasonal settlements (*katuns*).

21.3 Katuns

Katuns are temporary, seasonal settlements that are formed in areas where the population was engaged in livestock breeding, as a result of livestock movements or *transhumance*. Due to problems with grazing around permanent villages in certain parts of the year, in winter or summer, it was necessary to relocate to areas with richer pastures. Depending on the position of the permanent settlement, climatic characteristics of the area, environmental characteristics, as well as political and historical circumstances, livestock movements were done in winter or summer, and these were either descents from higher to lower or rises from lower to higher mountain areas.

Bearing in mind that in Montenegro the largest part of the population was engaged in livestock breeding, it can be assumed that transhumance was present from the earliest periods, which can be confirmed by the remains of material culture but also by written sources, especially the ones from the beginning of the 13th century (Crnić-Pejović 2000). *Katuns* are mentioned in written documents in the 14th century. In the notarial books from 14th–16th centuries, which are kept in the Kotor archives, in 1331 the *katun* of *Kalođurđa Crnojeva* is mentioned. *Bostur*, a *katun* that still exists today on Lovćen, was first mentioned in notarial books in 1327. Although Bostur is not explicitly referred to in these documents as a *katun*, it was almost certainly the case even then (Kovijanić 1963). Evidence of a very developed practice through the following periods can be found in the *Report and Description of the Skadar Sandzak* from 1614 written by Marijan Bolica, a nobleman from Kotor, in which he stated that the people of Montenegro would bring their cattle to the pastures on Lovćen, starting from May, until October, when they would descend again to the coastal area (Bolica 2004). The significance of this practice is kept also in the historic name of this area called *Katunska nahija*, after the *katuns* that were located on Lovćen (from the term *Katuni* in 15th century, *Cattuni* in 1614, the name *Katunska nahija* (*nahija* ~ district) was created during the Turkish rule in Montenegro, during the 17th century; Erdeljanović 1926; Bolica 2004).

Regarding the mountains in the northern part of Montenegro, *katuns* have also been mentioned since the Middle Ages. Numerous pastures, meadows and *katuns* on Prokletije and other neighbouring mountains, Hajla, Čakor, Cmiljevica, Mokra and others, are also mentioned in the charters and other documents of medieval rulers. Turkish census defters from the second half of the 15th and 16th centuries also testify that there was seasonal cattle breeding in the area of Polimlje (Dašić 2000).

Throughout history there were different types of livestock movements in Montenegro, some of which lasted shorter, some longer, and some of them exist to this day (Cvijić 1987; Crnić-Pejović 2000; Guštin 2012).

During the winter, cattle breeders from the higher karst areas descended to the *zimovnik* (winter shelters) in the coast or fertile areas where conditions were good for winter grazing. Usually, shepherds would spend the nights and situate their herd with local land owners, on whose meadows or pastures they would let the herd graze. Some cattle breeders also built temporary stables, while others used caves to protect the herd from storms. This practice was common with cattle breeders from the hinterland of the Boka Kotorska Bay, who used winter grazing on the coastal sides of the bay, until World War II (Crnić-Pejović 2000; Dašić 2000).

Less frequently, there have been cases of livestock descents during the summer months from higher karst areas to lower areas with pastures. A rare example of this practise in Montenegro, which was preserved until the first decade of the 21st century, is the descent from the village in the slopes of the coastal mountain Rumija to the summer *katun* on the shores of Skadar Lake (Guštin 2012).

The most dominant type of livestock movement that has been taking place in Montenegro until today, primarily in the northern parts of the high mountains, is the ascent to the mountain pastures and mountain *katuns* during the summer months. During summer, the major part of the family would ascent, and stay in temporary seasonal settlements–*katuns*–formed for that purpose. These *katuns* are the subject of this paper.

Many remains of the *katuns* have been preserved in Montenegro in almost all the mountains, and some of them are still being used in the same way. Unfortunately, a larger number of them are abandoned or ruined. There are also examples of *katuns* having been transformed into permanent settlements (e.g., *katuns* in the north-western part of Sinjajevina; Lutovac 1976), and in the recent years into weekend resorts (e.g., some *katuns* on Kuči Mountain; Laković 2017).

Katuns in Montenegro can be found in all mountainous areas, from the coastal mountains Lovćen and Orjen, mountains in the central part including the Kuči Mountain, to the northern ones such as Durmitor, Sinjajevina, Bjelasica, Komovi, Prokletije and others (Fig. 21.3). They can also be found at various altitudes, from the Dugi doci katun in the hinterland of the Boka Korska Bay, at about 890 m to the Roman katun on Prokletije, located at 2,019 m above sea level (Fig. 21.4).

Fig. 21.3 Map of Montenegro with key mountains marked (*Source* Relief … 2012). (Map by Aleksandra Kapetanović)

Fig. 21.4 **a** Remains of Dugi doci katun (890 m), Boka Kotorska Bay hinterland; **b** Remains of Roman katun (2,019 m), Prokletije. (Photos by Aleksandra Kapetanović)

21.4 Cultural Landscape of *Katuns*

Katuns in mountainous areas were formed on naturally suitable terrains. The position of a *katun* is determined by the terrain's features that contribute to its formation: the configuration of the terrain, the proximity and size of pastures, water sources, the availability of building materials, etc. The existence of spacious pastures is crucial for their formation. In addition, an important factor for locating a *katun* is the proximity of water. *Katuns* are usually placed on mountain slopes, in the pasture zone, above the forest zone and below the zone of rocky mountain peaks, or along the edge of valleys and dolines. *Katuns* were formed as settlements of a compact type, in which units such as huts, with corrals and outbuildings, were grouped. However, some *katuns* were formed as areas with several grouped units or areas with scattered individual units (Fig. 21.5).

Katuns cannot be observed separately from the environment in which they are located, the pastures that belong to them, the water supply, but also all other natural factors. Available materials from which the huts and other constructions were built are what makes a *katun* unique. The basic elements of the architectural heritage of the *katuns*, which are part of the cultural landscape of the *katuns*, are: traditional mountain huts or cottages, corrals, outbuildings, cisterns or ponds, rare religious buildings, cemeteries, etc.; but also roads and paths that connect *katuns* and pastures, as well as intangible practices related to the culture of life on *katuns* (Fig. 21.6).

Fig. 21.5 **a** *Katun*'s position on the slope, Krivodoljski katun, Kuči Mountain; **b** *Katun* on the slope, Babino polje, Bogićevica; **c** *Katun* in the valley, Dobri do, Durmitor; **d** Remains of the *katun* in a karstic depression, Gropa Beljeva, Kuči Mountain. (Photos by Aleksandra Kapetanović)

21.5 Types of Katun Huts and Their Development

The essential element of the architectural heritage of *katuns* are huts–buildings for housing shepherds–which are the simplest form of housing. There are several terms for huts in different parts of Montenegro. In the central part of Montenegro, historically known as Old Montenegro, it is called *koliba* (Rovinski 2001:53), in the north-eastern Montenegro *stan*, or *dban* in the area of Kliment (Gušić 1964:60) and in the area of the Kuči Mountain *glada* (Cvijić 1987:290). All these terms are still being used.

Different types of huts can be identified, which have been built depending on the location of a *katun*, the available material, the purpose, social and historical circumstances and the period in which they were created. In addition to climatic characteristics, the lifestyle and the economic circumstances, the available construction materials have the greatest impact on the construction and building methods used for huts and on their shape and appearance. For the construction of huts, as well as all other facilities and buildings on the *katuns*, materials that were used were the ones that were available in that particular area, and that were at the same time suitable and of sufficient quality for construction. In such a mountainous area as Montenegro is, the basic building materials are stone in karst areas (e.g., Fig. 21.8) and wood in areas richer in forests (e.g., Fig. 21.9).

Fig. 21.6 **a** Cultural landscape of a *katun* of the southern karst mountains, Majstori, Lovćen; **b** *Katun* in the karst terrain, Momonjevo, Kuči Mountain; **c** *Katun* located nearby wooded area, Štavna, Komovi. (Photos by Aleksandra Kapetanović)

The simplest constructions that were built on *katuns* were firstly a type of shelter, as a shelter from rain and wind, and then the simplest types of huts. It is stated that the first shepherd shelter in Orjen was *guber*, a waterproof blanket woven from goat hair (Crnić-Pejović 2004:483). For the area of the Kuči Mountain, it is stated that the original huts were built of *pustovi* or *pustine*–cloth made of goat hair, i.e., *kostreti*. These huts looked like large nomadic tents and could be moved from place to place (Erdeljanović 1907:236). These simplest forms of construction have not been preserved in the Montenegrin mountains.

The oldest and simplest type of hut that was built had in the beginning only a roof that started immediately from the ground. It was constructed over a circular or rectangular base, with a simple construction of wooden rods and different types of covers–a roof (*krovina*) made of available materials. From these basic forms of huts, more complex types have developed, which had a dry-stone wall on which the roof structure and the roof are placed.

21.5.1 Huts with Circular Base—Dubirozi/Savardaci

One of the simplest and probably the oldest forms of huts, especially in wooded areas, were huts with a circular base, with a fireplace in the middle, covered with a cone-shaped roof, called *dubirog* (a Slavic term) or *savardak* (a Turkish term that comes from a word meaning a cone-shaped item). The simpler type of *dubirog* (pl. *dubirozi*) has only a cone-shaped roof starting from the ground, while the more advanced form has a lower circular dry-stone wall, on which a wooden construction for the cone-shaped roof is placed.

Pavel Apollonovich Rovinsky in 1880, describes the *savardak* as a type of temporary habitat–large huts made of long poles arranged in a cone shape and covered with straw, with a circular base and a fireplace in the middle (he states that they have a diameter of 4 and a height of 5 *hvata*–an archaic measure of length that amounts to around 1.8 m). He also mentions *busare*, which are of the same shape but covered with turf (Rovinski 2001). In 1898, the Italian botanist Antonio Baldacci also mentioned cone-shaped huts covered with bundles of rye straw in the area of Vasojević and on the Kuči Mountain (Baldači 2004). At the beginning of the 20th century, Svetozar Tomić described in detail the *savardak* in Drobnjaci (Tomić 2000) and Jovan Erdeljanović the ones in Kom (Erdeljanović 1907).

Dubirozi, as a type of hut on *katuns*, almost completely disappeared during the 20th century. During the field research conducted in several mountain zones (e.g., in the area of the Kuči Mountain by the Historical Institute of the University of Montenegro and in parts of the mountains Bjelasica, Komovi, Prokletije for the "*Study of Traditional Architecture in the Municipalities of Berane, Petnjica, Andrijevica, Plav and Gusinje*"; Laković 2017; Kapetanović et al. 2019), completely preserved authentic *dubirozi* were not found. Only circular stone walls remained of the authentic *dubirozi.*

During the field research in the area of the Kuči Mountain, carried out by the Historical Institute of the University of Montenegro within the project "*Valorization of Montenegrin Katuni Through Sustainable Development of Agriculture and Tourism*—KATUN" (Laković 2017), out of approximately 2,900 traditional huts identified, only about 20 are *dubirozi*. The walls of *dubirozi* are built of stone using the dry-stone wall technique, and have only a singular opening—the doors. The inner diameter of the preserved *dubirozi* is 4.50–5.60 m, and in some of them around 3.60 m. The thickness of the wall ranges from 60 to 140 cm, and the height is around 120 cm in most cases, up to 155 cm (Laković 2017) (Fig. 21.7).

21.5.2 Huts with Rectangular Base

In addition to huts with a circular base, rectangular base huts also appear very early. They were built out of different types of materials that were available on that particular location, primarily stone

Fig. 21.7 *Dubirog* remains with inner diameter of 5.50 m, Javorci katun, Kuči Mountain. (Photo by Aleksandra Kapetanović)

and wood. However, in some areas there are also *pletare* or *košare*, the walls of which are made of wicker, woven around upright sticks (Deroko 1968:14).

Describing the *katuns* in Vasojevići in 1880, P. A. Rovinski mentions *katafe*, small huts covered with branches (Rovinski 2001:58). In the 1930s, mountaineers who visited the area of Prokletije stated that the shepherds' huts in that area were built of wood or shrubs (Urban 1938:95). Starting as early as 1960s, such types of huts were an exception (Gušić 1964:67). Shrub-covered huts are also mentioned in the coastal mountains. It is stated that in Orjen the first habitats on *katuns* were small huts roughly built of stone or woven with shrubs, covered with leafy maple branches or ferns (Crnić-Pejović 2004:483). During field research conducted in the last two decades in several mountain zones, no authentic examples of this type of huts on *katuns* have been identified.

Today, the most typical and most common type of traditional huts preserved on the *katuns* of the Montenegrin mountains are huts with an elongated rectangular base, made of stone or wood (e.g., Fig. 21.6b–c).

21.5.3 Stone Huts

The first descriptions of stone huts on *katuns* can be found in Mariano Bolizza's text from 1614, in which he states that the huts on Lovćen are made of stone and covered with rye straw, with transverse beams, intertwined with dry branches (Bolica 2004:55). Pavel Apollonovich Rovinsky in a text from 1880 also describes the huts on the Lovćen katuni, for which he states are the same as the houses in which many people used to live in the villages, built of rough stone, of various sizes and shapes, with earthen floors and no windows (Rovinski 2001:51). At the beginning of the 20th century, Jovan Erdeljanović described the huts on the Kuči Mountain, made of dry-stone wall, with a wooden structure and a roof covered with wooden rods (*štice*) or rye straw (Erdeljanović 1907). In 1902, on the same mountain, Antonio Baldacci mentioned huts covered with straw and huts in higher zones that are smaller and covered with wooden planks (Baldači 2004).

The development of stone huts on *katuns* progressed just as the development of houses in villages—from simple, smaller, rough masonry huts to larger, better ones built with more quality. It is stated that on Orjen the first habitats on the *katuns* were small huts roughly built of stone, and after that *pozemljuše* or *potleušice* (single storey huts) were built to the height of a man, covered with straw or stone slabs (Crnić Pejović 2004:483).

The stone huts preserved to this day on the mountains from all zones, coastal, central or northern, are very similar: elongated rectangular bases, no windows, only with doors, dry-stone walls, a wooden roof structure and a gable or hipped roof, originally covered with straw (Fig. 21.8a) or wood, i.e., planks (*štica*; local, archaic term for a type of wooden plank) or *šindra* (roof shingle). The interior of the huts is almost always one room without any walls or barriers, with a fireplace in one part and a sleeping area in the other. On the inner walls of the huts, we can often find small dents—niches.

Originally, all stone constructions on *katuns* were made with the dry-stone technique, from blocks which are found locally and are roughly hewn or not hewn at all, without mortar as a binder material. The most typical type of stone that could be found in the area of a *katun* was limestone, and less frequent types include slate, marl or sandstone. This masonry technique was applied to huts, *dubirozi*, outbuildings, corrals and fences. The huts were built in such a way that they used the terrain to full advantage. Thus, there are many examples that in lower zones have rocks as part of the wall. Parts of the wall around the door and at the corners were stacked with more precision, from larger and finer blocks. During the 20th century, mortar began to be used subsequently on dry-stone wall structures, first lime and then cement mortar, in order to improve the insulation (Fig. 21.8b–c).

After field research of the *katuns* on the Kuči Mountain (Laković 2017), data were obtained that the internal width of the huts range mostly from

Fig. 21.8 **a** Hut covered with straw, Vuči do, Lovćen; **b** Limestone hut, Rikavac katun, Kuči Mountain; **c** Hut from marly limestone and sandstone, Vujadinovića katun, Kuči Mountain; **d** Hut on the Bižaljevac katun, Lovćen; **e** Plan, cross-section and elevation of the stone hut, katun Popovića, Carine, Kuči Mountain. (Photos and drawings by Aleksandra Kapetanović)

3.00–3.50 m (maximum 4.30 m), and length from 4.20–7.80 m, while the proportion of plans ranges from 1.25 up to 2.0 m (some examples that show the different dimensions and proportions of the plan: 2.30 × 5.70 m, 2.42 × 4.77 m, 3.66 × 6.65 m, 3.20 × 9.25 m, 4.34 × 8.58 m). The thickness of the wall usually ranges from 80–100 cm, although there are some that go up to 120–130 cm. Therefore, the external width of the huts can range from 4.80–6.00 m and length from 7.20–9.70 m. The height of the wall was usually 120–130 cm, up to 150 cm, rarely 170 cm. It is noted that in newer huts the walls are higher. The height of the roof in huts with masonry gable walls, which probably belong to the older type, ranges from 2.70–3.00 m, and roof inclination is approximately 30°. In huts that have flat walls with a gabled or hipped roof, the roof height can range from 3.70–4.20 m, and roof inclination is up to 50° (Laković 2017) (Fig. 21.8e).

In some *katuns*, the architecture followed the development of houses in villages, and the huts are more advanced. We can find such huts, for example, on *katuns* on Lovćen (Vujačić 1997). On the Bižaljevac katun we can find more recent huts with a higher-quality build, which have a window in addition to the door. In one of the recorded huts, the internal dimensions are 3.95 × 6.90 m, the wall height is 1.86 m, the height of the *listra* is 3.75 m, and the wall thickness ranges from 82 to 107 cm (Fig. 21.8d).

21.5.4 Wooden Huts

Wooden huts occur in mountainous areas rich in forests. Descriptions of wooden huts can be found in several authors. Pavel Apollonovich Rovinsky in 1880 described *stanovi* (huts) in the *katuns* as quadrangular elongated buildings made of thin logs, the roof with a slight slope is made of planks or tree bark, which are pressed with stones or covered with turf (Rovinski 2001:54). In 1902, Svetozar Tomić described in more detail the huts in Drobnjak, stating that they were built of logs, 1.20–1.50 m high, with a roof that is usually made of *lub*–tree bark bent over the roof ridge (Tomić 2000:76). Describing *katuns* in the Berane area, in the 1950s, Milisav Lutovac states that the huts are made of planks (*podnice*) placed between the sharpened wooden rods and covered with spruce bark (*lub*) or plank (*štica*) (Lutovac 2000).

In the area of north-eastern Montenegro, which includes the mountains Bjelasica, Komovi and Mokra, the most common type of huts are built entirely of wood, with elongated rectangular base, a low gabled roof and doors as the only opening. They were built of wooden planks of rectangular cross-section placed one above the other and connected at the corners by a notch. The planks are usually 20–30 cm wide, about 5.5–7 cm thick, while their length depends on the size of the hut, since the walls are always made of a single piece of plank. In some cases, instead of planks of rectangular cross-section, a type of semicircle was used. The first beams (*temeljače*) were laid directly on the ground or a stone was placed below.

The external dimensions of the width of the hut were usually 3.00–3.70 m, and the length ranged from 4.10 to 6.20 m (some examples that show the different dimensions and proportions of the plan: 2.70 × 5.70 m, 3.00 × 5.40 m, 3.08 × 4.10 m, 3.17 × 5.15 m, 3.22 × 6.20 m, 3.63 × 6.00 m). The huts are quite low and have a gabled roof. The height of the walls ranges from 1.10 m to 1.70 m, usually around 1.40 m. The height of the roof in the ridge ranges from 2.12 m to 3.40 m, usually around 2.40 m. The only opening on the hut is the door, which is always on the narrower side of the hut, in the very corner. The roof structure is wooden. Besides gable, there are also examples of hipped roof. The roof of the huts was covered with tree bark (*lub*) or wooden planks (*štice*). Currently, a number of huts is covered with *štice* and a more recent type of roofing made of wood (*šindra*), but there are no houses left that are covered with *lub* (Kapetanović et al. 2019) (Fig. 21.9a–d).

The hut interior functionally consisted of two parts, the front one right at the entrance where people would spend their days and sleep (this part had a fireplace), and the back part, *mljekar*–where milk was processed and milk products were kept. However, when there were more cattle, people would typically build two huts, one for a *mljekar* and the other for living and sleeping, or added *mljekar* to the hut in the form of annexes (Fig. 21.9e).

Along with the huts, there is another type of construction–*kućer*, placed on a sleigh, that was built for shepherds. It was being placed next to the corral so that it could be easily moved from one place to another. *Kućer* is of smaller dimension, around 1.50–1.80 m wide and 2.50 m long (Fig. 21.9f).

21.6 Other Elements of Architectural Heritage of *Katuns*

Besides huts, corrals are a crucial element of *katuns*. Corrals, as well as huts, can be made of stone or wood, depending on the characteristics of the location, whether it is a karst or forest area. The size and shape of the corrals vary depending on the configuration and morphology of the terrain and the size of the herd. Stone corrals were built, just like the walls of the hut, of dry-stone walls, but without a central infill made of smaller stones. Wooden corrals were built of wooden beams and

Fig. 21.9 **a** Typical wooden hut at Bjelasica, katun Plavac; **b** Hut with semicircle planks, covered with *štice*, Bogićevica; **c** Huts with hipped roof covered with *šindra*, Sinjajevina; **d** Plan, cross-section and elevation of the wooden hut, katun Varda, Komovi; **e** Grouping of wooden huts, Mokri do, Cmiljevica; **f** *Kućer*, Sinjajevina. (Photos and drawings by Aleksandra Kapetanović)

slats so that they can be easily moved if necessary (Picuno et al. 2017; Kapetanović et al. 2019) (Fig. 21.10a–b).

Regarding other elements of the architectural heritage of *katuns*, it is important to mention the water supply infrastructure. Depending on the natural features of the location, the following can be found on the *katuns*: natural springs (Fig. 21.10c), ponds or wells, cisterns, hollows and snow pits. Pond (*ubao*) enclosed by drystone is often built in karst areas, and specific construction with a corbelled dome can be found on mount Lovćen (Juvanec 2017) (Fig. 21.10d).

Fig. 21.10 **a** Stone corrals, Momonjevo katun, Kuči Mountain; **b** Wooden corrals, Okrugljak katun, Sinjajevina; **c** Velika Radeća spring, Kuči Mountain; **d** Pond (*ubao*) with a corbelled dome construction, Bižaljevac katun, Lovćen. (Photos by Aleksandra Kapetanović)

21.7 Conclusion

Based on the research conducted so far, it can be confirmed that in the mountains of Montenegro within the seasonal settlements (*katuns*), very important layers of immovable cultural heritage have been preserved, which is reflected in various types of traditional architecture, especially huts, corrals, etc. However, only an extensive study that should be conducted in all Montenegrin mountains could provide a complete picture of the overall scope of this important segment of the mountain cultural landscape.

Although this segment of cultural heritage has recently been to some extent recognized as valuable (e.g. within protected areas and especially through the various research and other projects dealing with rural and tourism development), it is still not formally protected, due to failure and inertia of the cultural heritage protection system, nor adequately treated within the planning and strategic documents. Thus there is a risk of the disappearance of *katuns*, due to decay, abandonment or uncontrolled transformation.

Having in mind all the abovementioned, in order to preserve the extremely valuable cultural heritage of *katuns*, it would be necessary to observe them as part of the overall cultural landscape of the mountains, and implement the following: comprehensive research of *katuns* in all Montenegrin mountains, which would include their identification, mapping and detailed analysis; formal protection of *katun's* cultural heritage at the national level; adequate treatment and protection of the cultural landscape of *katuns* within the planning and strategic documents; as well as sustainable management through the preservation of intangible practices and the involvement of the local community. Some successful examples of protecting this type of heritage in the other areas, such as the ones in the Eastern Alps, including Velika

Planina in Slovenia, can be considered suitable models for the segments of the proposed process (Cevc 1995). Besides that, designations at the international level in the last decades, from the inscription of The Causses and the Cévennes, Mediterranean agro-pastoral Cultural Landscape on the World Heritage List to the inscription of Transhumance, the seasonal droving of livestock on the Representative List of the Intangible Cultural Heritage of Humanity, can be considered as a significant reference and a way forward (UNESCO World Heritage … 2022b; UNESCO Intangible … 2022).

References

Baldači A (2004) Područje Cijevne – istraživačka putovanja u istočnu Crnu Goru i albanske planine (1900, 1901, 1902) [Cijevna area – research trips to Eastern Montenegro and the Albanian Mountains (1900, 1901, 1902)]. In: Pulević V, Vincek D (eds) Crna Gora vrata Balkana: putopisi i zapisi evropskih botaničara. Obod, Cetinje, pp 489–580

Bolica M (2004) Lovćen-Jezero-Duklja. In: Pulević V, Vincek D, Bušković V (eds) Crnogorske planine: putopisi i zapisi. Obod, Cetinje, pp 55–59

Cevc T (ed) (1995) Planšarske stavbe v Vzhodnih Alpah: stavbna tipologija in varovanje stavbne dediščine [Alpine buildings in the Eastern Alps: building typology and protection of the building heritage]. Znanstvenoraziskovalni center SAZU, Ljubljana

Crnić-Pejović M (2000) Zimska ispaša [Winter cattle grazing]. In: Tradicionalna narodna kultura u Crnoj Gori. Naučni skupovi, vol 54. Crnogorska akademija nauka i umjetnosti, Podgorica, pp 89–114

Crnić-Pejović M (2004) Ruralna naselja Sitnica i Vrbanj: katuni – stalna i vikend naselja [Rural settlements Sitnica and Vrbanj: Katuns – permanent and weekend settlements]. In: Selo u Crnoj Gori: naučni skup. Crnogorska akademija nauka i umjestnosti, Univerzitet Crne Gore, Podgorica, pp 461–498

Cvijić J (1987) Balkansko poluostrvo: sabrana dela, 2 [Balkan Peninsula: Collected works, vol 2]. Srpska akademija nauka i umetnosti, Beograd

Dašić M (2000) Stočarstvo i stočarska kretanja u sjevernoj i sjeveroistočnoj Crnoj Gori od XVI do početka XX vijeka [Livestock breeding and livestock breeding movements in Northern and North-Eastern Montenegro from the 16th to the beginning of the 20th century]. In: Tradicionalna narodna kultura u Crnoj Gori. Naučni skupovi, vol 54. Crnogorska akademija nauka i umjetnosti, Podgorica, pp 65–88

Deroko A (1968) Narodno neimarstvo I [Traditional architecture I]. Spomenik, vol 118. Srpska akademija nauka i umetnosti, Beograd

Erdeljanović J (1907) Kuči, pleme u Crnoj Gori [Kuči, Montenegrin tribe]. Srpski etnografski zbornik, vol 8. Srpska kraljevska akademija, Beograd

Erdeljanović J (1926) Stara Crna Gora: etnička prošlost i formiranje crnogorskih plemena [Old Montenegro: ethnic past and the formation of Montenegrin tribes]. Srpska kraljevska akademija, Beograd

EXPEDITIO (2022) Valorization of the vernacular heritage in Montenegro – a pilot area of Zalazi. https://www.expeditio.org/index.php?option=com_content&view=category&id=138&Itemid=544&lang=sr. Accessed 28 Feb 2022

Faro Convention (2015) Council of Europe Framework Convention on the Value of Cultural Heritage for Society. Council of Europe, Faro. https://rm.coe.int/1680083746. Accessed 28 Feb 2022

Gušić B (1964) Brada u Prokletijama [Brada in Prokletije]. Naše Planine 16(3–4):49–70

Guštin M (2012) On the slopes of Rumija: an owerview of the life by Lake Skadar and vernacular architecture. Littera picta, Ljubljana

IRIS (2022) Inspiring rural heritage: sustainable practices to protect and conserve upland landscapes and memories. https://iris-jpi.eu/. Accessed 28 Feb 2022

Juvanec B (2017) Crna Gora and Montenegro. AR: arhitektura, raziskave 17(1):82–93

Kapetanović A, Gligorić B, Rajić T (2019) Studija tradicionalna arhitektura na području opština Berane, Petnjica, Andrijevica, Plav i Gusinje [Study of traditional architecture in the areas of municipalities Berane, Petnjica, Andrijevica, Plav and Gusinje]. Polimski muzej Regionalna razvojna agencija Bjelasica, Komovi i Prokletije, Berane

Kapetanović A, Laković I, Koprivica T, Pelcer-Vujačić O (2017) Graditeljska baština katuna na Kučkoj planini [Architectural heritage of the katuns at the Kuči Mountain]. Istorijski Zapisi 90(1–2):91–107

Kovijanić R (1963) Pomeni crnogorskih plemena u kotorskim spomenicima (XIV–XVI vijek) I [References to Montenegrin tribes in Kotor Monuments (XIV–XVI Century) I]. Istorijski institut SR Crne Gore, Cetinje

Laković I (ed) (2017) Katuni Kučke planine: monografija [Katuns of the Kuči Mountain: monograph]. Univerzitet Crne Gore, Podgorica

Laković I, Kapetanović A, Pelcer-Vujačić O, Koprivica T (2020) Endangered Mediterranean mountain heritage–case study of katuns at the Kuči Mountain in Montenegro. Land 9(8). https://doi.org/10.3390/land9080248

Lutovac M (1976) Stočarska kretanja u planinama Crnogorskih Brda i narodne regule pri iskorišćenju pašnjaka [Livestock movements in the mountains of Montenegrin Brda and popular rules regarding the use of pastures]. In: Čubrilović V (ed) Odredbe pozitivnog zakonodavstva i običajnog prava o sezonskim kretanjima stočara u Jugoistočnoj Evropi kroz vekove. Srpska akademija nauka i umetnosti, Beograd

Lutovac M (2000) Ivangradska (Beranska) kotlina: regionalno-geografska ispitivanja [Ivangrad (Berane) Valley: regional-geographical studies]. Skupština opštine, Berane

Marković Č (1994) Tragovi materijalne kulture ranijih epoha na prostorima Nacionalnog parka „Lovćen" [Traces of material culture of early epochs in the "Lovćen" National Park]. In: Nacionalni park „Lovćen": prirodna i kulturna dobra. Crnogorska akademija nauka i umjetnosti, Podgorica, pp 239–246

Mijović P (1992) Bronzanodopske gravire u Maji Popadiji na Prokletijama [Bronze Age engravings in Maja Popadija on Prokletije]. Glasnik Odjeljenja Umjetnosti 11:11–23

Myteveli J, Rupa M, Kapetanović A (eds) (2012) Prokletije = Bjeshkët e Nëmuna: biodiversity and cultural heritage. SNV—Netherlands Development Organization, Kotor

Nacionalni parkovi Crne Gore [National Parks of Montenegro] (2022). https://nparkovi.me/. Accessed 28 Feb 2022

Nature Park Orjen (2022) Katuni. https://orjen.me/eng/cultural-heritage/katuni/. Accessed 28 Feb 2022

Operational guidelines for the implementation of the World Heritage Convention (1992). UNESCO, Paris. https://whc.unesco.org/archive/opguide12-en.pdf. Accessed 28 Feb 2022

Pavićević R (2019) Werk: austro-ugarske tvrđave u Crnoj Gori [Werk: Austro-Hungarian fortresses in Montenegro]. Nova Pobjeda, Podgorica

Petričević M (2019) Neznana Sinjajevina I [Unknown Sinjajevina I]. https://archaeoblogger.blogspot.com/2019/02/neznana-sinjajevina-i.html. Accessed 28 Feb 2022

Picuno CA, Laković I, Roubis D, Picuno P, Kapetanović A (2017) Analysis of the characteristics of traditional rural constructions for animal corrals in the Adriatic-Ionian Area. Sustainability 9(8). https://doi.org/10.3390/su9081441

Regional development agency for Bjelasica, Komovi and Prokletije (2022). http://www.bjelasica-komovi.me/en/project-management/. Accessed 28 Feb 2022

Relief map of Montenegro (2012). https://commons.wikimedia.org/wiki/File:Relief_Map_of_Montenegro.svg. Accessed 14 May 2023

Rovinski PA (2001) Kratak pregled Crne Gore [Montenegro – A brief overview]. In: Zapisi o Crnoj Gori. CID, Podgorica, pp 35–58

Tomić S (2000) Drobnjak: antropogeografska ispitivanja [Drobnjak: anthropogeographic research]. In: Drobnjak, Piva, Banjani. CID, Podgorica, pp 9–151

Vujačić T (1997) Lovćenski katuni [Katuni on Lovćen]. In: Seoski dani Sretena Vukosavljevića: devetnaesti simpozijum. Opštinska zajednica obrazovanja, Prijepolje, pp 47–62

UNESCO Intangible Cultural Heritage (2022) Transhumance, the seasonal droving of livestock. https://ich.unesco.org/en/RL/transhumance-the-seasonal-droving-of-livestock-01964#identification. Accessed 28 Feb 2022

UNESCO World Heritage Convention (2022a) Stećci medieval tombstone graveyards. https://whc.unesco.org/en/list/1504. Accessed 28 Feb 2022

UNESCO World Heritage Convention (2022b) The causses and the cévennes, Mediterranean agro-pastoral cultural landscape. http://whc.unesco.org/en/list/1153/. Accessed 28 Feb 2022

Urban S (1938) Na vrhovima Prokletija [On the peaks of Prokletije]. Hrvatski Planinar 34(3):93–96

Zakon o zaštiti kulturnih dobara [Protection of Cultural Property Act] (2010, 2019) Službeni list Crne Gore, No. 049/10, 040/11, 044/17, 018/19. Podgorica

Alpine Heritage in Slovenia: Between Authorisation and Imagination

22

Špela Ledinek Lozej, Primož Pipan, and Maja Topole

Abstract

The chapter analyses the official registers of heritage institutions in Slovenia. It localises the heritage units, i.e. natural values—sites, areas, and caves, immovable cultural heritage, museums, and intangible cultural heritage by macroregion—Alpine, Pannonian, Dinaric, and Mediterranean. It discusses possible correlations between historic contexts, national imaginaries, and the eventual outstanding position of the officially registered heritage in the Alpine macroregion. It concludes on the micro- and macroscale—on overlappings and discrepancies between national registers and local heritage-making, and on some supranational heritage initiatives enabling alternative heritage futures.

Š. Ledinek Lozej (✉)
ZRC SAZU Institute of Slovenian Ethnology, Ljubljana, Slovenia
e-mail: spela.ledinek@zrc-sazu.si

P. Pipan · M. Topole
ZRC SAZU Anton Melik Geographical Institute, Ljubljana, Slovenia
e-mail: primoz.pipan@zrc-sazu.si

M. Topole
e-mail: topole.maja@gmail.com

Keywords

Geography · Heritage · Museums · Heritage registers · National imaginaries · Mountain identity · Mountain landscape

22.1 Introduction

Slovenia lies at the crossroads of the Alps, the Pannonian Basin, the Dinaric Alps, and the Mediterranean. The Alps, following the landscape typology (Perko 1998; Perko et al. 2021), cover about two-fifths of the Slovenian territory. Therefore, Alpine imagery is largely embedded in the representations and identifications of the Slovenes. All children in Slovenia know the Alpine mythical creatures such as the Golden Horn (*Zlatorog*), *Krivopete* (a wild woman with her feet turned backwards), the Wild Hunter and *Pehta* (a female mythical being of an ambivalent nature), as well as the young folk hero *Kekec* (a trickster-like boy, whose cleverness defeats the big and wicked villain) (Kropej 2012). Lively popular folk music can be heard at village festivals and mass events. And there is a proverb that says that every "true Slovene" must climb the highest Slovenian mountain Triglav (2,864 m) in the Julian Alps. Its stylised depiction is the central element of the Slovenian coat of arms (Mikša 2018). Following various authors (Šaver 2005; Velikonja 2018), the ideological self-construction

M. Zorn et al. (eds.), *Discourses on Mountains of Montenegro and Slovenia*, Historical Geography and Geosciences, https://doi.org/10.1007/978-3-031-99017-5_22

and self-perception of Slovenes are believed to focus on the Alpineness.

In this context, the chapter assesses the share of the heritage in/of the Alps. It examines the authorised heritage as it can be found in the official registers of natural values and cultural, tangible and intangible heritage, and museums. It discusses possible connections between historic contexts, national imaginaries and the eventual outstanding position of the officially registered heritage in the Alpine macroregion. It concludes on the micro- and macroscale, on overlapping and discrepancies between national registers, on Alpine heritage and local heritage-making in Alpine communities, as well as some supranational initiatives.

22.2 Terminological and Methodological Prologue

The mountain landscape, village centres, hayracks, the Bovec dumplings (*krafi*), and the Slovene folk-pop music are Alpine elements, classified as heritage in the various heritage registers. Heritage is actually not something per se; landscape, objects, and practices have yet to be recognised, i.e. identified, defined, registered, or otherwise designated as heritage. To roughly summarise conventions, legislation, and various professional definitions (e.g. Smith 2006; Harrison 2013; Slavec Gradišnik 2014; Adell et al. 2015; Harrison et al. 2020; Heriskop 2020), heritage is any natural or cultural, movable or immovable, tangible or intangible element from an arbitrarily defined past, *recognised* as heritage. It is this recognition, this perception in the light of heritage, that revalues an element. Heritage is therefore a strategic set of elements from the past with implications for the present and future relations in different spheres. Due to the distinctions that individuals ascribe to the elements and, consequently, to the incredibly loose and inclusive definitions in conventions and legislation, the control over the defining of and evaluating what is and what is not heritage is in the hands of authorised institutions. These oversee heritage and heritagisation through administrative apparatuses and with the support of experts from different disciplines (Smith 2006). At the global level, such authorised institutions include UNESCO, ICOMOS, ICCROM, IUCN, Europa Nostra, and the Council of Europe. In Slovenia, there are authorised museums, archives, the Institute for the Protection of Cultural Heritage of Slovenia, and the Institute of the Republic of Slovenia for Nature Conservation. The latter two used to be one institution until 1999, but have since been operating under the auspices of two different ministries, the Ministry of Culture and the Ministry of the of Natural Resources and Spatial Planning (former Ministry of the Environment, Climate and Energy, and before Ministry of Environment and Spatial Planning), in seven regional branches. Various associations, regional development agencies, tourism operators and cultural workers are also involved in heritage-making. Initiatives and entries in the register are usually prepared by the professionals who manage the registers, although an element can be proposed as heritage by anyone (Zakon o varstvu kulturne … 2008). Heritage cannot "*exist independently of a process of categorising, ordering, listing, and subsequently conserving and/or archiving it*" (Harrison 2013:6). "*Categorisation is integral to the definition and management of heritage* […] *and we see this manifested in the many and varied modes of ordering that have been established to facilitate the collection, categorisation, preservation, and management of heritage objects, sites and practices in the modern world*" (Harrison 2013:33). The mainstream discourse is still influenced by authorised heritage discourse and conventional understandings in which heritage is thought of as a series of valued objects, places, or practices that are gathered together in a museum or on a list, register, or catalogue of some form.

In the analysis of the Alpine heritage in Slovenia, we focused on the "official" registers of heritage institutions (Fig. 22.1). Data were obtained from national registers. We included all natural values, namely, *sites* and *areas* from the Register of Natural Values, identified as a natural heritage in Slovenia under the Nature Conservation Act (Zakon o ohranjanju … 1999). Cultural heritage units were drawn from the Register of

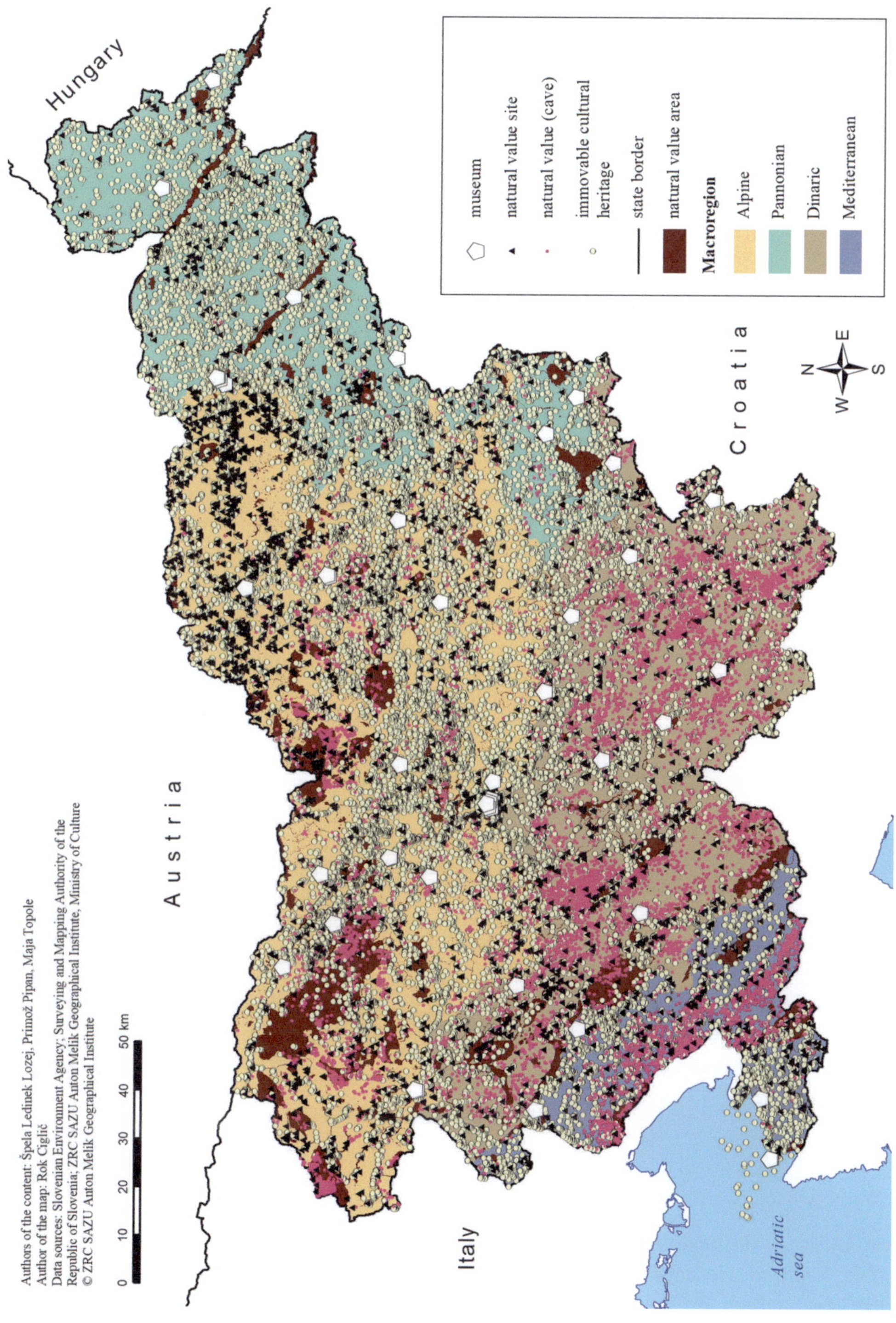

Fig. 22.1 Map of registered heritage in Slovenia (2021)

Table 22.1 Overview of data capture

Heritage types	Institution	Register	Description	Source (see References)
Natural values–sites; 2,660 units	Ministry of the Environment, Climate and Energy, Slovenian Environment Agency	Register of natural values	Under the Nature Conservation Act (Zakon o ohranjanju … 1999), a natural value is not only a rare, precious, or remarkable natural phenomenon, but any other valuable phenomenon, component, or part of living or non-living nature, a natural area or part of a natural area, an ecosystem, a landscape, or a designed natural feature. The following are considered to be sites and areas of natural value: geological phenomena, minerals and fossils and their deposits, surface and sub-surface karst phenomena, underground caves, gorges and ravines, and other geomorphological phenomena, glaciers and forms of glacial action, springs, waterfalls, rapids, lakes, marshes, streams, and rivers with their banks, the seashore, flora and fauna, their outstanding specimen and their habitats, ecosystems, landscapes, and designed natural features According to the Underground Cave Protection Act (Zakon o varstvu podzemnih … 2004), every cave automatically has the status of a natural value of national importance. Caves are the only natural value with this protection status in Slovenia. The location of the cave is the location of the entrance to the cave	Atlas okolja
Natural values–areas; 5,929 units				
Natural values–caves; 12,146 units				
Immovable cultural heritage; 30,442 units	Ministry of Culture, Cultural Heritage Directorate	Register of the immovable cultural heritage	Immovable cultural heritage is immovable property or parts of it with heritage value. It consists of archaeological sites, buildings, parks and gardens, memorial buildings and sites, other facilities and installations, settlements and parts thereof, and cultural landscapes	Kulturna dediščina
Movable cultural heritage; 57 units		List of Museums (for 2019)	A register of movable heritage is not yet in place, therefore we have considered the List of Museums. It includes museums that meet the space, financial, and staffing requirements for the preserving and protecting of the heritage. It includes museums that provide a national public service and other museums that ensure public access to the collections	Razvid muzejev
Intangible cultural heritage; 95 units		Register of the intangible cultural heritage	Intangible cultural heritage is "practices, representations, expressions, knowledge, skills … that communities, groups and, in some cases, individuals recognise as part of their cultural heritage" (Convention for Safeguarding … 2003). The Register of the Intangible Cultural Heritage has 95 units (June 2021). 28 units were excluded from the analysis because they cannot be localised; they refer either to the whole or several different areas in Slovenia, or they are listed in the Register by the location of the address of the association that registered them	Nesnovna dediščina

Immovable Cultural Heritage and the Register of Intangible Cultural Heritage. A register of movable heritage, which by law also includes legally defined categories of heritage such as "national treasures", has not yet been established. We have therefore considered the List of Museums instead. Archive and library material has not been taken into account in the present heritage analysis. The data capture is presented in more detail in Table 22.1.

In our analysis, we considered the geographical regionalisation, according to which there are four macroregions in Slovenia—Alpine, Pannonian, Dinaric, and Mediterranean, and within these nine landscape types—Alpine mountains, Alpine hills, Alpine plains, Pannonian low hills, Pannonian plains, Dinaric plateaus, Dinaric lowlands, Mediterranean low hills, and Mediterranean plateaus. The focus is on the Alpine macroregion, which covers 42.1% of Slovenia. Characteristic of it is the Alpine direction of ridges and valleys running from west to east. The Alps are divided into three landscape types: Alpine mountains (35.8%), Alpine hills (54.6%), and Alpine plains (9.6%). To the east they drop in elevation and gradually change into Pannonian landscapes, to the south they change into Dinaric landscapes, and in the Soča Valley to the southwest, they come close to Mediterranean landscapes (Perko et al. 2021). A statistical analysis of the heritage types was carried out. We determined the average absolute and relative values of each phenomenon (proportions within Slovenia, within macroregions and landscape types) and, to facilitate the comparison of individual types, we calculated their concentration indices, i.e. the ratio between the proportion in Slovenia and the proportion within each region or landscape type.

In continuation, we drew from historical, geographical, cultural, and ethnological literature on the role of mountains in the nation-building. We examined the relationship between the registered heritage in Alpine areas and Alpine imaginaries. We are aware that this is a preliminary attempt to contextualise the quantitative analysis of heritage registers, which does not (yet) provide comprehensive answers, but raises questions and opens directions for further study. We indicated it in the final discussion, which introduces and builds on some (unauthorised) grassroots heritage initiatives on the microscale, and on the participation of Slovenia in some Alpine-wide heritage initiatives and their international recognition and authorisation.

22.3 Examination of Official Heritage Registers

The following types of natural and cultural heritage were considered in the context of nationally registered heritage: 57 museums, 5,929 natural value areas, and 2,660 natural value sites (geological, geomorphological, hydrological, botanical, zoological, tree and ecosystem types, and designed natural features), 12,146 caves, 30,442 units of immovable cultural heritage (archaeological sites, cultural landscapes, settlements and their parts, parks and gardens, memorial sites and places, buildings, buildings with parks and gardens), and 67 units of intangible cultural heritage (knowledges and practices, oral traditions and expressions, performing arts, social practices, rituals, and festive events)—a total of 51,301 heritage units (see Tables 22.1 and 22.2 and Fig. 22.1).

Table 22.2 shows absolute and relative data on natural and cultural heritage types by the four macroregions of Slovenia and by the nine landscape types within them. They can be seen in relation to the surface area and population of each unit. The structure of individual heritage types can be better seen in Figs. 22.2 and 22.3 (within macroregions) and in Fig. 22.4 (within landscape types).

The Alpine landscapes, which cover 42.1% of Slovenia's surface area and are home to 47.1% of its population, are of above-average importance in Slovenia, especially in terms of the number of sites of natural value (56.7%), the number of museums (56.1%), and the surface area of sites of natural value (50.9%). The proportion of immovable and intangible cultural heritage is slightly above average in terms of area, and almost in line with the proportion of the population, which is 47.1% (47.8% in both cases). The Alps have the lowest proportion of caves (29.7%); in this area, they are understandably outnumbered by the Dinaric landscapes (52.8%), due to their greater degree of karstification (Fig. 22.3).

Table 22.2 Macroregions of Slovenia and their landscape types in numbers, including the associated heritage elements

Macroregion	Alpine			Pannonian		Dinaric		Mediterranean		Sea	Slovenia
Macroregion area (ha)	854,095			429,151		570,625		173,429		37	2,027,300
Macroregion area (%)	42.1			21.2		28.1		8.6		0.1	100.0
Landscape type	*Alpine mountains*	*Alpine hills*	*Alpine plains*	*Pannonian low hills*	*Pannonian plains*	*Dinaric plateaus*	*Dinaric lowlands*	*Mediterranean low hills*	*Mediterranean plateaus*	Sea	Slovenia
Landscape type - area (ha)	306,177	465,997	81,921	299,452	129,699	380,932	189,693	106,102	67,327		2,027,300
Landscape type - area (%)	15.1	23.0	4.0	14.8	6.4	18.8	9.4	5.2	3.3		100.0
Population (No. in 2017)	90,549	382,819	497,846	270,278	268,314	74,025	281,239	170,161	26,846		2,062,077
Population (% in 2017)	4.4	18.6	24.1	13.1	13.0	3.6	13.6	8.3	1.3		100.0
Museums (No. in 2019)	4	7	21	2	9	2	7	5	0		57
Museums (% in 2019)	7.0	12.3	36.8	3.5	15.8	3.5	12.3	8.8	0.0		100.0
Natural heritage - area (ha)	100,181.6	25,345.3	10,745.5	12,839.1	23,625.2	51,629.7	19,999.7	14,092.0	8,957.7		267,680.3
Natural heritage - area (%)	37.4	9.5	4.0	4.8	8.8	19.3	7.5	5.3	3.3		100.0
Natural heritage sites (No.)	477	874	156	226	84	358	264	143	78		2,660
Natural heritage sites (%)	17.9	32.9	5.9	8.5	3.2	13.5	9.9	5.4	2.9		100.0
Caves (No. in 2019)	2,779	719	113	92	2	4,447	1,963	187	1,846		12,148
Caves (% in 2019)	22.9	5.9	0.9	0.8	0.0	36.6	16.2	1.5	15.2		100.0
Immovable cultural heritage (No. in 2021)	2,978	6,829	4,736	4,046	2,400	2,607	3,802	2,199	807	37	30,441
Immovable cultural heritage (% in 2021)	9.8	22.4	15.6	13.3	7.9	8.6	12.5	7.2	2.7	0.1	100.0
Intangible cultural heritage (No. in 2021)	16	10	6	2	12	6	8	4	3		67
Intangible cultural heritage (% in 2021)	23.9	14.9	9.0	3.0	17.9	9.0	11.9	6.0	4.5		100.0

For illustrative purposes, the proportions of areas within Slovenia in each of the nine landscape types are shown in common graphs, which allow for a comparison between macroregions and landscape types according to each heritage type, based on the surface of each area and the number of inhabitants in that area in 2017 (Figs. 22.5, 22.6, and 22.7).

In Slovenia, there are 28 museums per 1 million inhabitants, with the largest number of them (33) in the Alps, 44 in the Alpine mountains, and 42 on the Alpine plains (Fig. 22.5a). There is 130 ha of protected area per 1000 inhabitants in Slovenia. The large surface of the protected area includes as many as 1106 ha in the Alpine mountains and 697 ha per 1000 inhabitants on the Dinaric Plateaus (Fig. 22.5b). There are 13 natural heritage sites per 10,000 inhabitants in Slovenia, 53 in the Alpine Mountains, and 48 on the Dinaric Plateaus. There are less than 10 per 10,000 inhabitants in the areas of plains and hills (Fig. 22.5c). Caves are the most frequent among the natural value sites. As they are associated with karstified areas, densities are highest in the case of the Mediterranean (688) and the Dinaric Plateaus (601). The Alpine mountains have 307 caves; Slovenia has 59 caves per 10,000 inhabitants (Fig. 22.5d). In terms of the number of immovable cultural heritage sites per 1000 inhabitants, the Dinaric Plateaus have the highest number

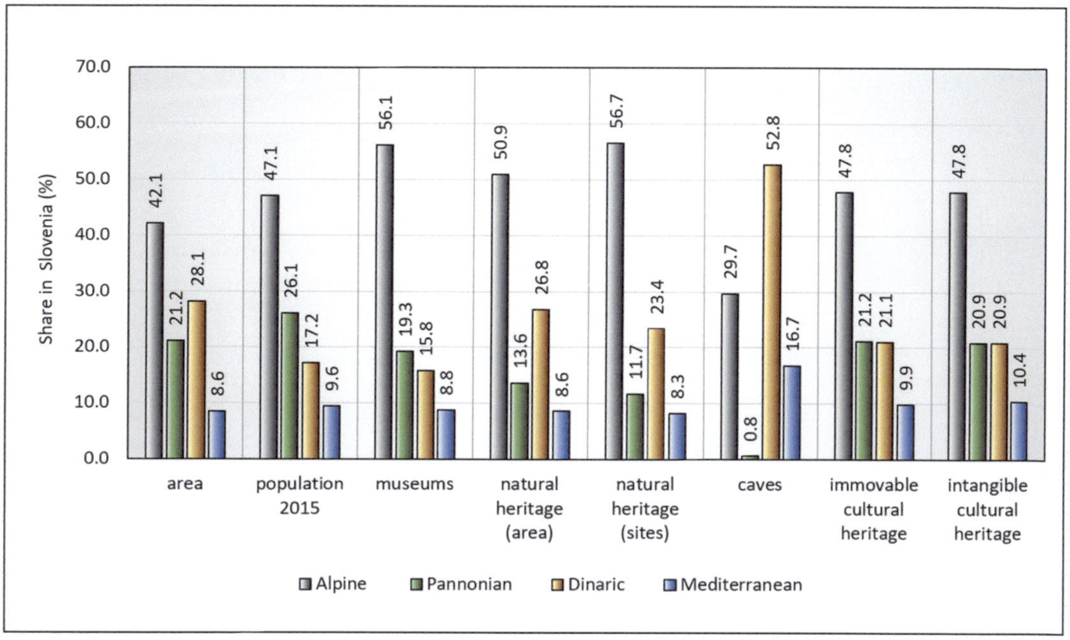

Fig. 22.2 Data from heritage registers in 2019 and macroregions in Slovenia

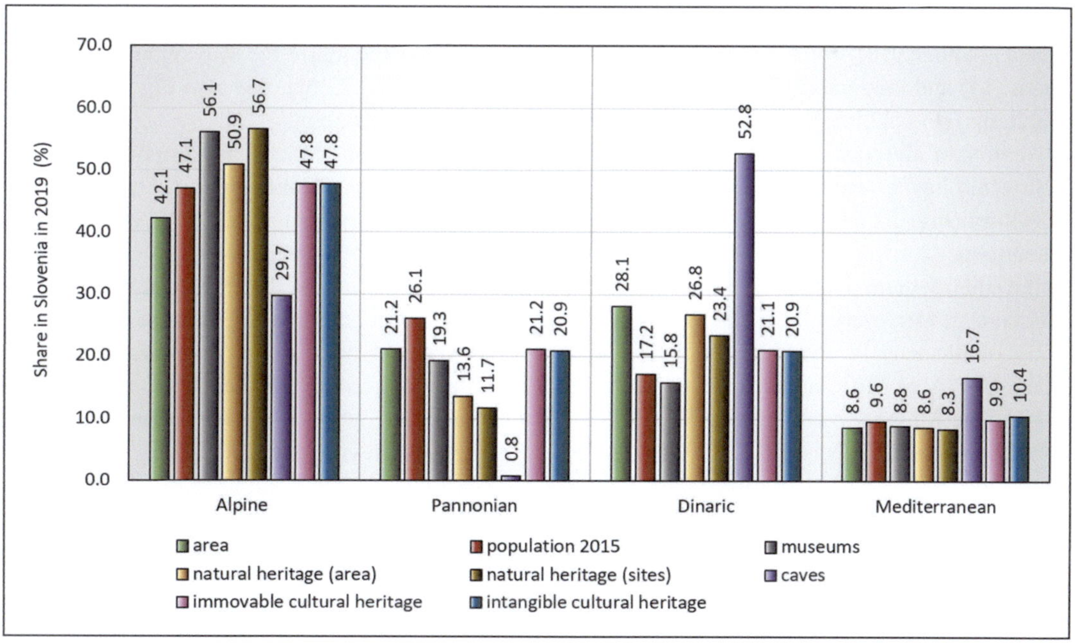

Fig. 22.3 Macroregions in Slovenia and heritage types in 2019

of them (40), followed by the Alpine Mountains (33) and the Mediterranean Plateaus (32). The figure for Slovenia is 15 immovable cultural heritage sites per 1000 inhabitants (Fig. 22.5e). As for Slovenian intangible heritage with merely 67 examples, there are only three units per 1000 inhabitants. The Alpine Mountains boast the

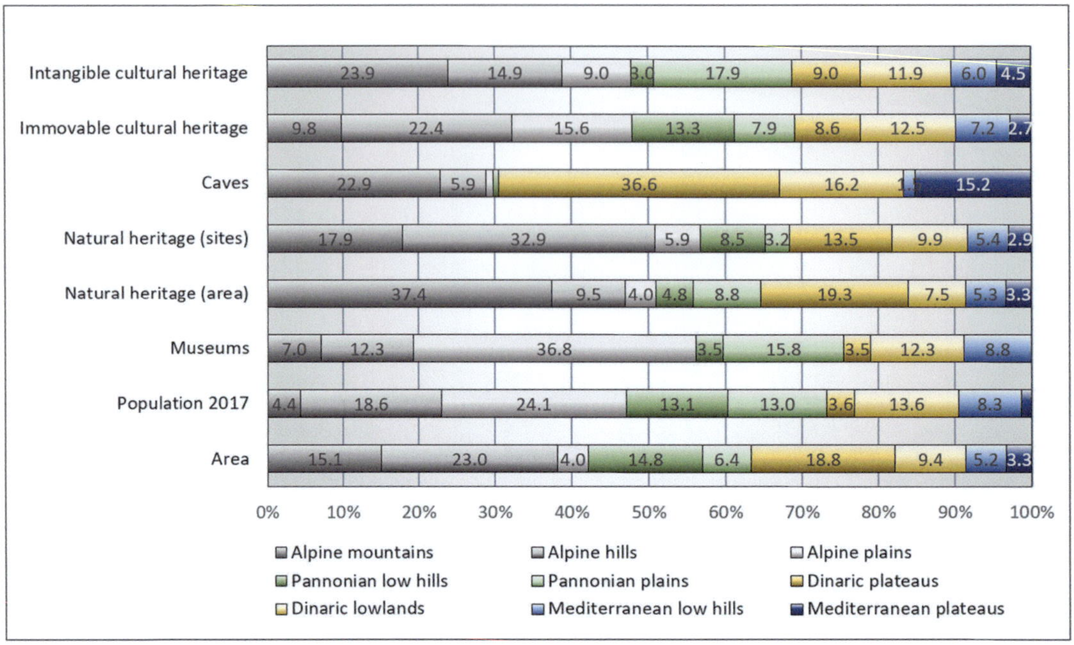

Fig. 22.4 Landscape types and heritage in Slovenia

highest number (18), followed by the Mediterranean (11) and the Dinaric Plateaus (8 per 1000 inhabitants) (Fig. 22.5f). See also Fig. 22.1 for the distribution of all types in Slovenia.

Slovenia has an average of 2.8 museums per 1000 km^2 (Fig. 22.6a). The highest number of museums per km^2 is in the plains—25.6 in the Alpine and 6.9 in the Pannonian Plains, while the number is lower everywhere else. In Slovenia, 13.2% of the land is protected, compared with 16% in the Alps. The largest area of protected land is in the Alpine Mountains (32.7%), followed by the Pannonian Plains with 18.2% (Fig. 22.6b). There are 13.1 natural value units per 1 ha in Slovenia. In the Alps, there are 17.6 (Alpine Plains 19, Alpine Hills 18.8, and Alpine Mountains 15.6 per ha) (Fig. 22.6c). In 2019, there were, on average, 60 known underground caves per km^2 in Slovenia. Mediterranean plateaus come first with 274 caves, followed by Dinaric Plateaus (117) and Alpine Mountains (91) (Fig. 22.6d). The density of immovable cultural heritage sites is highest (5.8 per km^2) on the Alpine Plains (Fig. 22.6e). Intangible cultural heritage is most frequently protected on the plains (Pannonian 9.3, Alpine 7.3), the figure for Slovenia being 3.3 per 1000 km^2 (Fig. 22.6f). See also Fig. 22.1 for the distribution of types.

Concentration indices are important indicators of the outlying phenomena. They show the ratio between the proportion of the area under consideration (macroregion or sub-macroregion) and the proportion of the region or country as a whole. The concentration index of 100 applies to Slovenia, a region with an index above 100 has an above-average concentration of a given phenomenon, while a region below 100 has a below-average concentration. Thus, the number of museums per surface unit in the Alpine Plains is nine times the Slovenian average (Fig. 22.7a). If the proportion of protected land is compared with the proportion of the population, the Alpine Mountains, the Dinaric Plateaus, and the Mediterranean Plateaus stand out (11, 7, and 3 times the average, respectively) (Fig. 22.7b). The same landscapes are also above average in terms of the number of protected sites of natural value (Alpine Mountains 4 times more, Dinaric Plateaus 3.7 times more, and Mediterranean Plateaus 2.2 times more than average) (Fig. 22.7c). The Mediterranean Plateaus

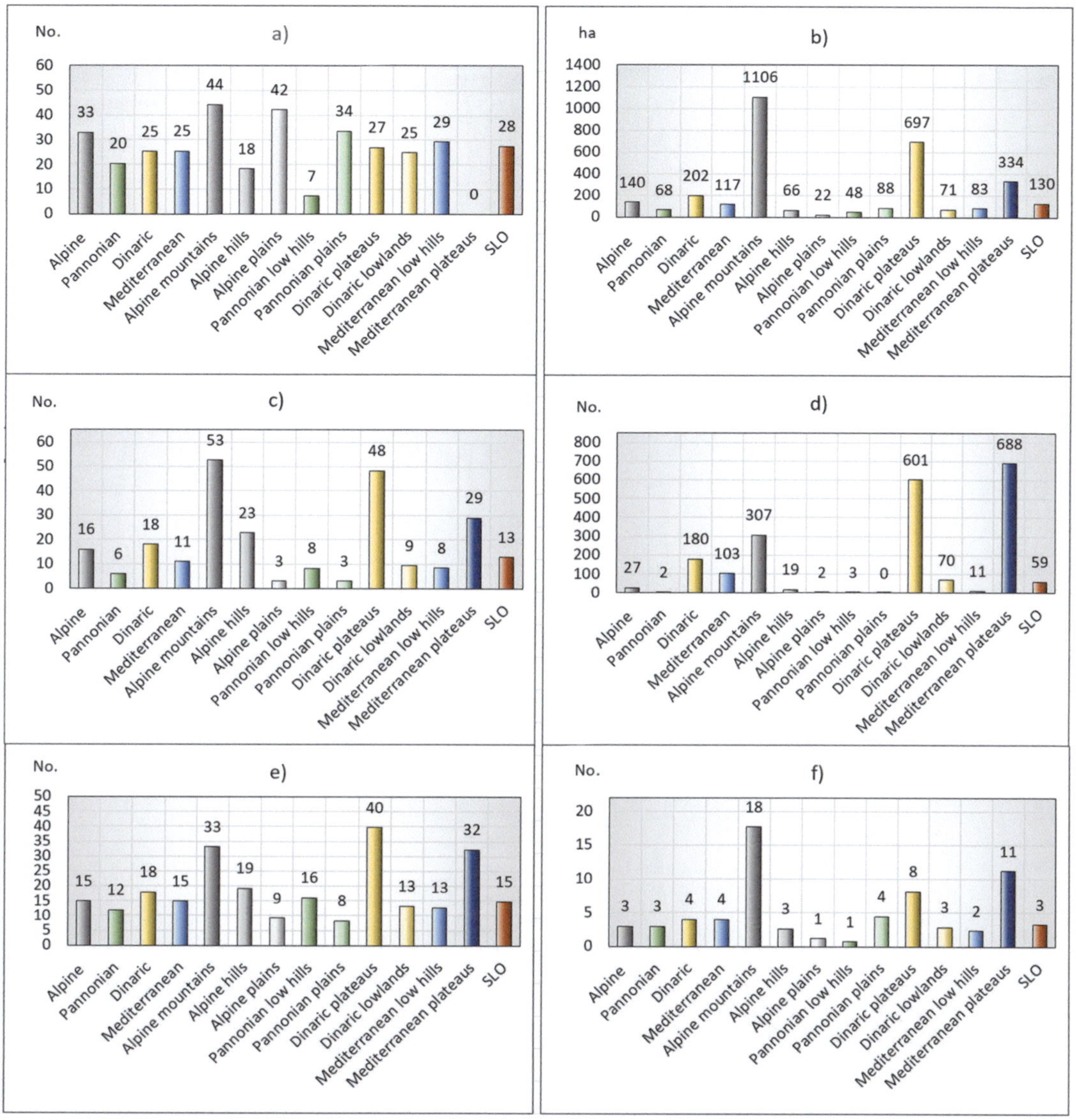

Fig. 22.5 Density of heritage units in relation to population: **a** the number of museums per million inhabitants, **b** natural heritage area per 1000 inhabitants, **c** the number of natural heritage sites per 10,000 inhabitants, **d** the number of caves per 10,000 inhabitants, **e** the number of immovable cultural heritage sites per 1000 inhabitants, **f** the number of intangible cultural heritage units per 100,000 inhabitants

have the highest cave density (11.7 times), followed by the Dinaric Plateaus (10.2 times) and the Alpine Mountains, 5.2 times more than the Slovenian average (Fig. 22.7d). In terms of the concentration of immovable cultural heritage within a given area, the Alpine Plains stand out (3.9 times), followed by the Dinaric Plateaus (2.7 times more), the Alpine Mountains, and the Mediterranean Plateaus (both 2.2 times more) compared to the number of inhabitants (Fig. 22.7e). In terms of intangible cultural heritage, the Alpine Mountains are a special case, with 5.4 times more protected intangible cultural heritage compared to the population (Figs. 22.7f).

As immovable cultural heritage is largely diversified, we have further developed the analysis of this type (Fig. 22.8, 22.9 and 22.10).

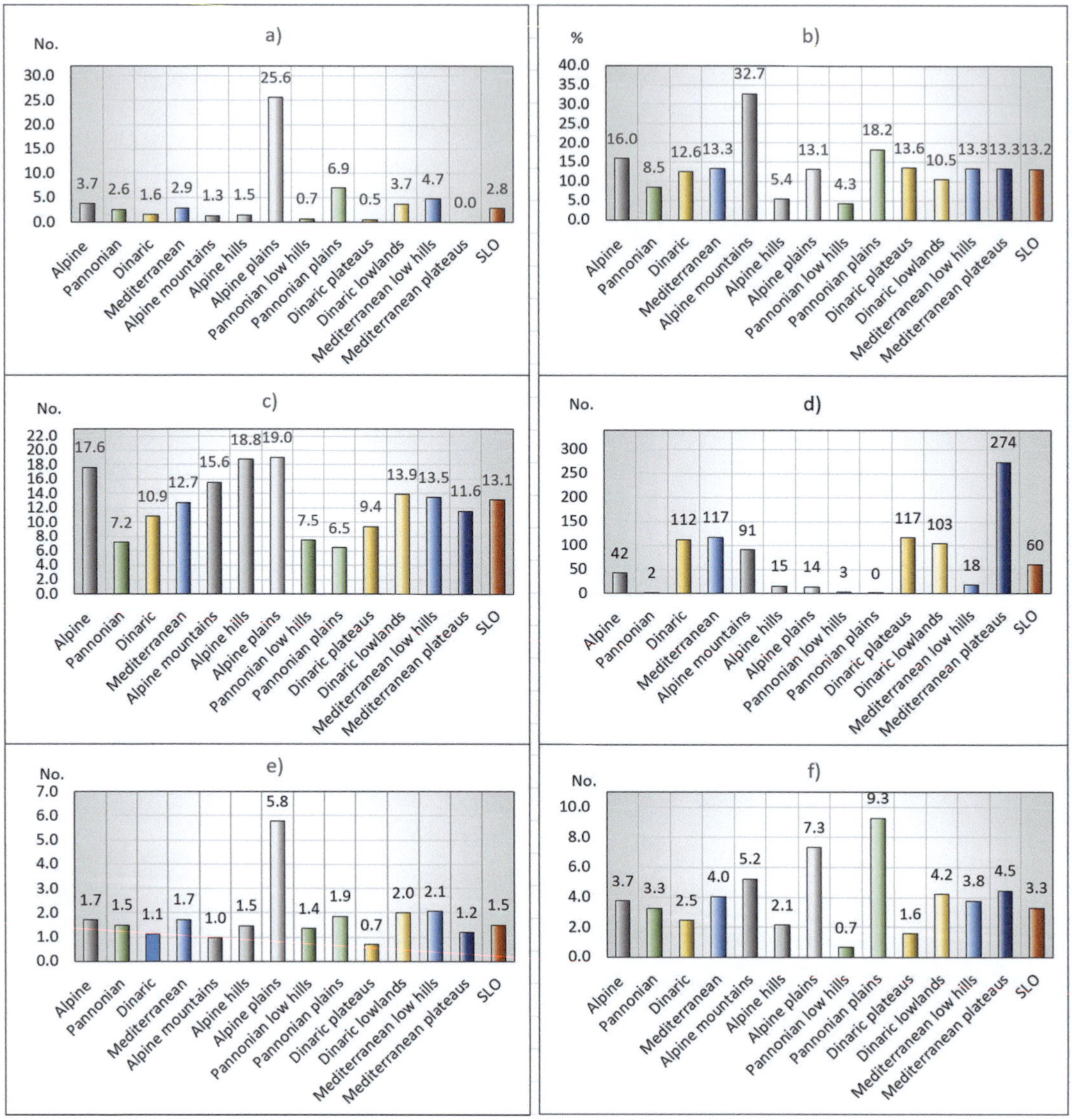

Fig. 22.6 Density of heritage units per area: **a** the number of museums per 1000 km^2, **b** natural heritage area (share/% of the whole territory), **c** the number of natural heritage sites per ha, **d** the number of caves per km^2, **e** the number of immovable cultural heritage units per km^2, **f** the number of intangible cultural heritage units per 1000 km^2

While Alpine landscape types have an absolute predominance in terms of the number of immovable cultural heritage sites, we should look at them with regard to their surface area and population. Concerning the surface area (Fig. 22.11), the concentration indices show that the Alpine Plains stand out, with an above-average representation of all types. They are followed by the Mediterranean Hills and Plateaus—both of which are largely characterised by settlements and parts of settlements, and finally the Dinaric Lowlands and Plains, and the Pannonian Plains. In the Alpine Mountains, only the cultural landscape is represented above average. In terms of population (Fig. 22.12), the Alpine Mountains and Mediterranean Plateaus stand out the most due to their lower population density. They are followed by the Dinaric Plateaus and the Mediterranean

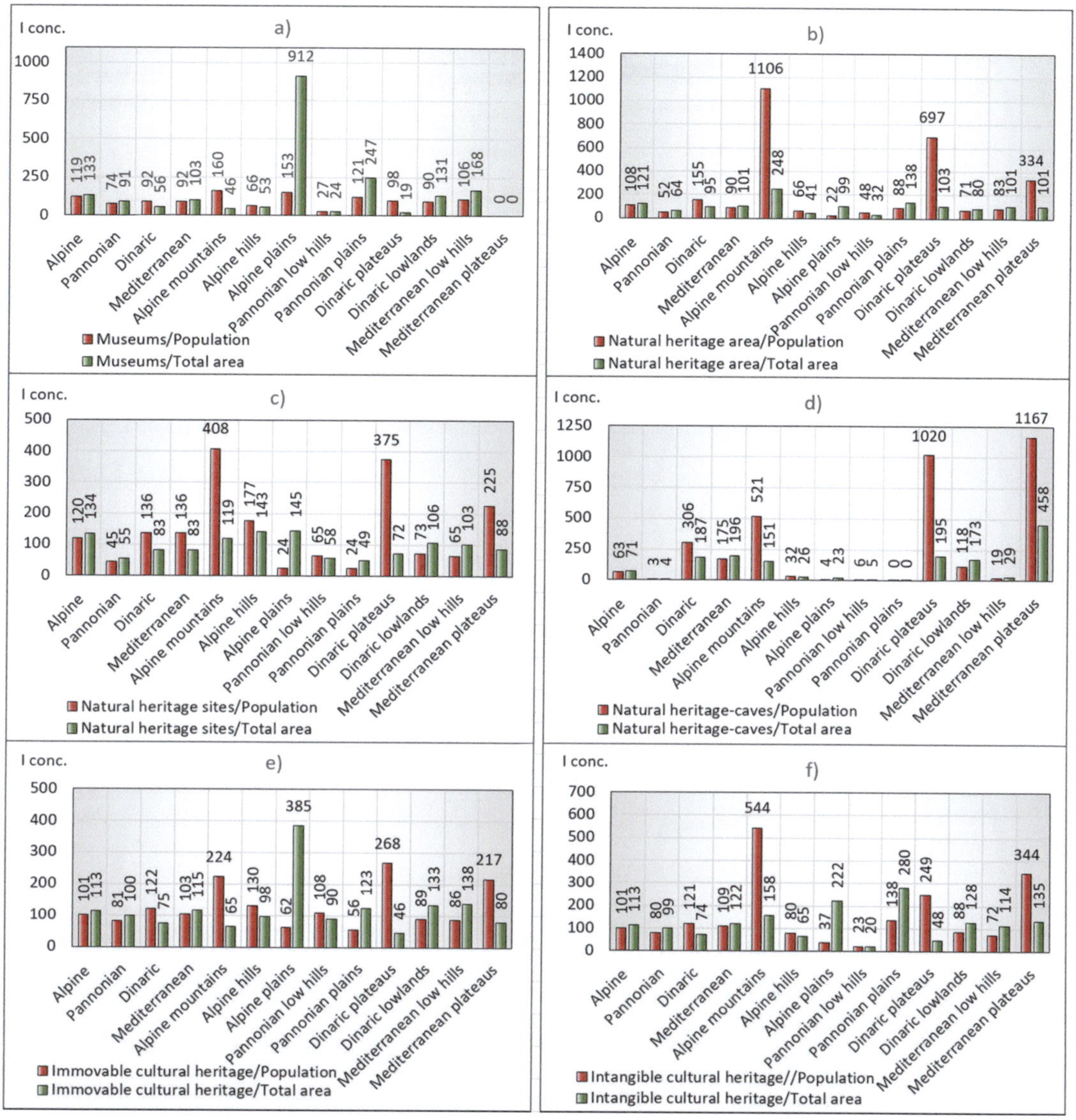

Fig. 22.7 Concentration of heritage units compared with the number of inhabitants and the area. Concentration indices for: **a** museums, **b** natural heritage areas, **c** natural heritage sites, **d** caves, **e** immovable cultural heritage, **f** intangible cultural heritage

Low Hills. Within these, two types in particular are worth mentioning: the settlements and parts of settlements are 8.8 times more common than average within the Mediterranean Plateaus, and the cultural landscape is 8 times more common than average in the Alpine Mountains.

22.4 Alpineness Between Registers and Imaginaries

The statistical analysis of the official heritage, which clearly illustrated a higher proportion of registered heritage units in the Alpine areas, was placed in the broader socio-historical context of the Alpine imaginaries. The analysis of visual representations of Slovenian landscapes showed

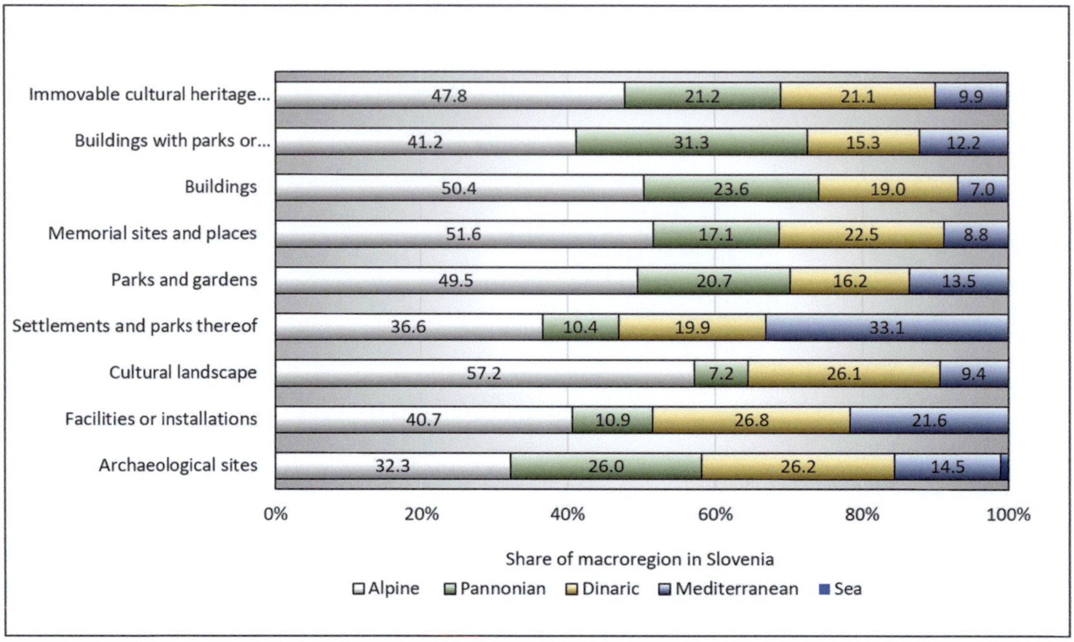

Fig. 22.8 Macroregions and immovable cultural heritage

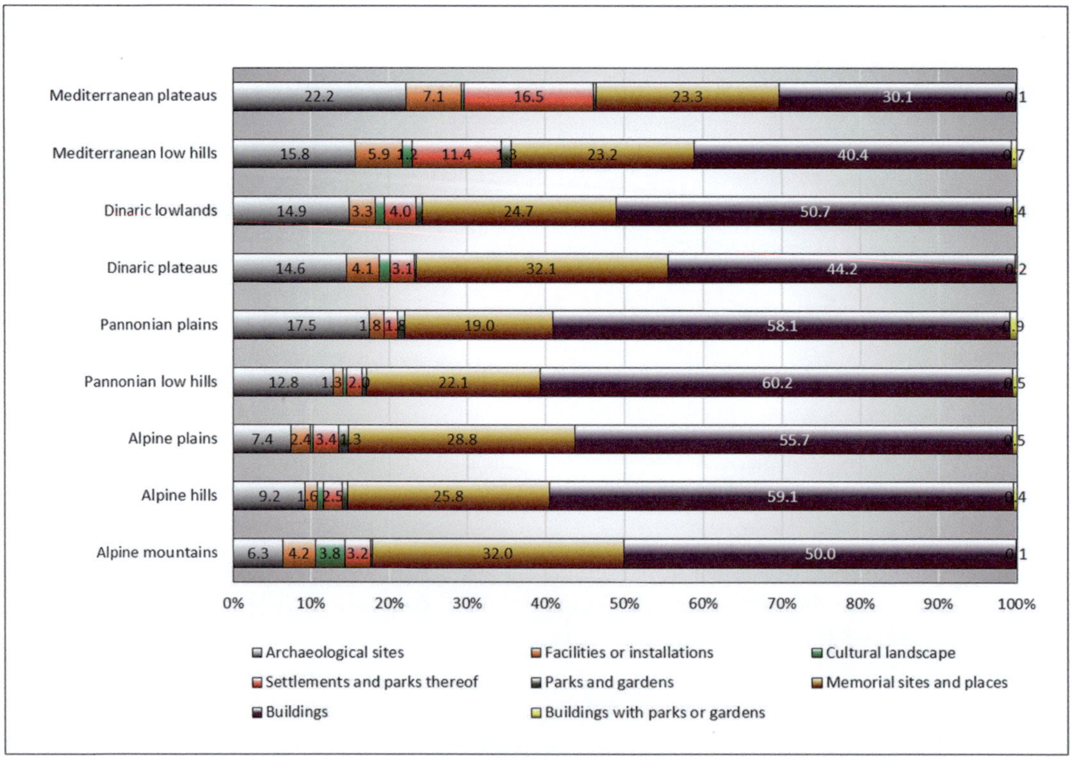

Fig. 22.9 Immovable cultural heritage per landscape types

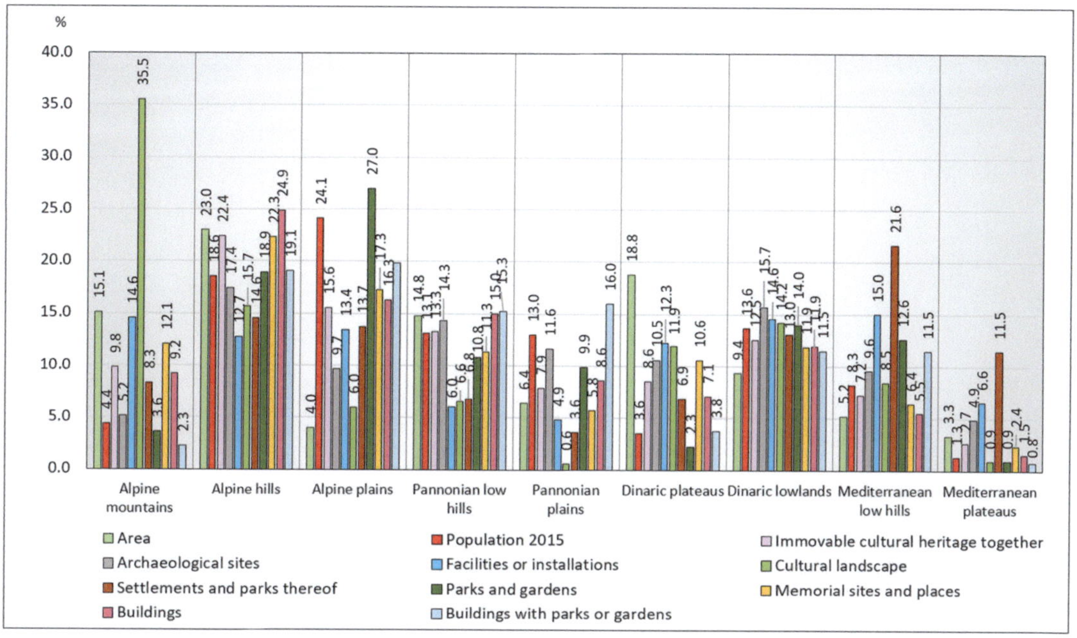

Fig. 22.10 Landscape types and immovable cultural heritage

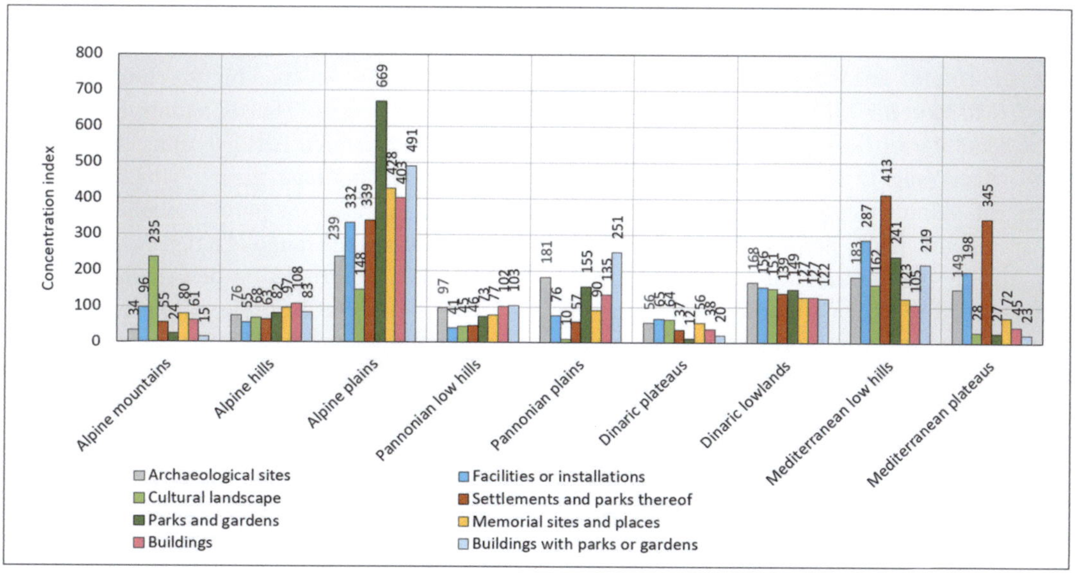

Fig. 22.11 Concentration of immovable cultural heritage per area of landscape types (Slovenia = 100)

an above-average representation of the Alpine landscape elements in the identifications of Slovenians (Kučan 1998). What is the relationship between the "official", i.e. registered heritage and the historically documented overlaps between Alpine and national imaginaries? The overlap has led many writers to designate mountains, especially Triglav, as a Slovenian national symbol (see, e.g. Strojin 1980; Šaver 2004, 2005; Mikša 2018), an inseparable element of Slovenian identity (Mikša et al. 2018), individual landscape elements as "symbols of national identity" (Kučan

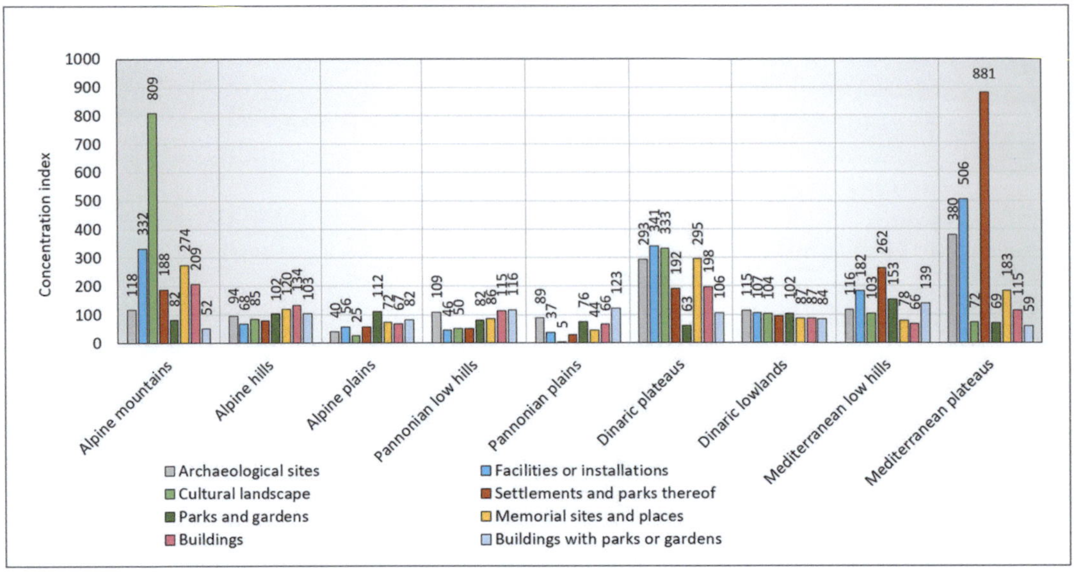

Fig. 22.12 Concentration of immovable cultural heritage per population of landscape types (Slovenia = 100)

1998:108), and their mutual interplay as "the Alpine culture of Slovenianness" (Šaver 2005), which constitutes "the mountain identity of Slovenians" (Mikša et al. 2018) or even "the pinnacle of the ideology of Slovene identity" (Velikonja 2018). According to Boštjan Šaver, the multiplicity of Alpine representations of national heritage stems from the fact that the Slovenian Alps are the most north-western Slovenian region and thus a geographical and symbolic bridge to the European world, in contrast to the Balkans to the south. According to Šaver, Slovenians "have always looked to our northern and western neighbours, and in the period of the awakening of nations, we formed similar (Alpine) national symbols" and a "historically petrified and insightful Alpine mythology" (Šaver 2005:271). In the twentieth century and the modern state, the latter has become the dominant Slovenian collective idea and ideology at the same time.

Authors who shed light on the naturalisation of the Alpineness, also co-create it. To what extent is heritigisation also involved in the naturalisation of the Alpineness? Can the fact that heritage in the Alpine region has more entries in the official registers than the one in the Mediterranean, Pannonian, and Dinaric regions be linked to the "historically petrified Alpine mythology" (Šaver 2005:271) and the "mountain identity of Slovenians" (Mikša et al. 2018:157)? An unambiguous answer cannot be given, of course. That is mainly because the number of registered heritage units depends not only on the "heritage view", which has been changing, but also on purely practical circumstances, such as the range of a regional unit of the Institute for the Protection of Cultural Heritage of Slovenia or the Institute of the Republic of Slovenia for Nature Conservation, the number of staff, their expertise and possible priorities.

In five points, we summarise the analysis of the registered heritage, outline the (re)construction and overlaps of the Alpine and national imaginaries, and indicate individual links and challenges. Instead of answers, we also raise separate new questions.

22.4.1 Establishing the View of the Mountains

The Alpine macroregion has the highest proportion of protected area—as much as 16% (Fig. 22.6b). The Alps stand out also in terms of the number of natural heritage sites per hectare

(Fig. 22.6c). The aestheticised and romantic view of "wild nature" and "pastoral bucolic" has been promoted and consolidated by Alpine explorers and mountaineers. Discovering and mapping the Alps and the mountain world became one of the central research interests of Enlightenment thinkers (Mikša and Zorn 2016). Later mountaineering, alpinism, and tourism, in turn, influenced the formation of a romanticised and aestheticised view of the mountain landscapes (Tschofen 1999; Fleming 2004; Hansen 2013; Debarbieux and Rudaz 2015; Smrekar et al. 2016). The juxtaposition of two mountain views—that of the *mountaineer-alpinist* and that of the *mountaineer-shepherd*—was vividly described by Josip Tominšek, the long-time editor of the *Planinski vestnik* alpine journal, in his memoirs of a conversation with a shepherd on the Korošica Alp in the Savinja Alps in 1889. When asked by the shepherd, why he keeps returning to the mountains, he replied, "The snow-capped mountains seem so close from our house; they invite me and tempt me to get to know them up close. They are, after all, the proudest part of Slovenian land". The shepherd replied, "You find these steep, white rocks beautiful; but my eyes look down on your green meadows, colourful fields, misty woods, glittering rivers, colourful villages; and all this is Slovenian too, is it not?" Tominšek concludes, "So we have two heritages and two longings!" (Tominšek 1949:8).

22.4.2 Protected Areas

The first heritage actions on the territory of present-day Slovenia are linked to Alpine areas. As early as the end of the nineteenth century, the Carniola Regional Assembly protected the Savica and Peričnik waterfalls by a special decree (Šaver 2005). In 1906, the seismologist and natural scientist Albin Belar, while taking an inventory of the natural monuments of Carniola, prepared the first proposal for the protection of the Triglav Lakes Valley and some other areas. In 1908 (the year of most references), he set about implementing it. The implementation was prevented, probably due to the simultaneous promotion of grazing in the mountains (Skoberne 2018) and the First World War. After the war, the protection efforts were resumed. In 1919, the Slovene Museum Society established the Department for the Protection of Nature and Natural Monuments, which in the following year submitted the memorandum *Spomenica* to the provincial Government of Slovenia. It specified the key requirements for nature protection in Slovenia, including the protection of the Triglav Lakes Valley in the Julian Alps. The protection regime and the Alpine Conservation Park in the Triglav Lakes Valley were established in 1924. Subsequently, there were proposals to extend and rename the conservation area. In 1961, the area was renamed Triglav National Park and in 1981 it was expanded to its present extent (Bajuk Senčar 2012; Zorn et al. 2015; Skoberne 2018; see also Chap. 26 in this volume).

22.4.3 Alpine Elements in National Iconography

In the 20th century, the images of Alpineness were reinforced in the wider national public through several visual media. The credit goes to landscape painters, the writers of the *Planinski vestnik* alpine journal, the first filmmakers, and postcard artists—photographic depictions of peaks, mountain outposts, mountain still-lifes, the national motifs of the painter Maksim Gaspari, and later wall calendars with mountain vedutas. Alpine iconography was also reinforced in other branches of art, in the cultural industry, and popular culture. *Krst pri Savici* (The Baptism on the Savica, 1835), an epic-lyric poem by France Prešeren (1800–1849) set in the heart of the Alps, challenged many artists in the visual arts, drama, ballet, and other artistic fields (Juvan 2015). The legend of the Golden Horn has also been interpreted in various ways in literature and popular culture (Kropej 2012). The mountain tales about the folk hero Kekec by Josip Vandot are well known, as are their film versions and popular depictions (Hladnik 1980). *Kekec* (1951) was the third Slovenian feature film, the first Slovenian children's film, and the first film to win an award at the

Venice Film Festival, in 1952. The sequel, *Srečno, Kekec!* (Good Luck, Kekec! 1963), was the first Slovenian colour film, followed in 1968 by the third film, *Kekčeve ukane* (Kekec's Tricks 1968). In addition to the Kekec trilogy, the film *Cvetje v jeseni* (Flowers in Autumn, 1973), based on the novel with the same title by Ivan Tavčar, plays an iconic role in the spreading and reinforcing of the Alpineness (Erjavec 1994; Kučan 1998; Zgonik 2002). An important role in the dissemination of Alpine imaginaries of Slovenianness in music was played by the popularisation of the accordion and the genre of popular folk music, which was introduced to the world public by the Avsenik Brothers Ensemble (Kovačič 2015).

The political and social dimensions of everyday life are strongly intertwined with the Triglav and the mountains. Triglav appears in the current state symbols, insurance, and food product names, on postcards, calendars, the name of a film production company; there is even the occasional fusion of Alpine and religious iconography (Šaver 2005; Mikša et al. 2018). However, mountains are often the centre of the world, the dwelling place of deities and beings with supernatural powers, the place of the sacred and spiritual in other mythologies, religions and ideologies (Eliade 1974; Šmitek 1998). "Alpine culture" (Šaver 2005) and "mountain identity" (Mikša et al. 2018) are therefore not nation-specific, since Alpine imaginaries can also be traced in Austrian and German lands during the nation-building in the nineteenth and twentieth centuries (Keller 2016) and even among non-Alpine nations; in addition to Austrian, German, and Slovenian mountaineers, the Czechs were also searching for "their own" Alps and mountain identity at the turn of the century. The Czech branch of the Slovenian Mountaineering Society built the Češka koča Mountain hut (the Czech hut) in the Kamnik-Savinja Alps at the end of the nineteenth century. In 2020, it was not only entered into the Slovenian National Register of the Intangible Cultural Heritage but also declared a cultural monument by the Jezersko Municipality (Roškar 2020).

22.4.4 Alpine Museum and Collections

As recorded in the *Planinski vestnik* alpine journal, the Czechs had already called for the collection of material for an "Alpine museum" in the Slovenian Alps as early as 1901 (Anon 1901). The Triglav Museum Collection was established in 1984, however, the initiative was realised in 2010 when the new-build Slovenian Alpine Museum opened. The analysis showed that the Alpine areas have the highest number of authorised museums, both in terms of surface area and population (Figs. 22.5a and 22.6a). The number is above-average, since the Alpine macroregion also includes the capital Ljubljana, where most of the national museums are located. The role of tourism in the opening of museums and the establishing of museum collections should also not be neglected. The Slovenian Alpine Museum was one of the last new buildings and major public investments in museum infrastructure in rural Slovenia. The museum, which is managed by the Jesenice Upper Sava Museum (*Gornjesavski muzej Jesenice*), is the result of many years of efforts by the local Alpine Association, the municipality of Kranjska Gora, and the national Alpine Association of Slovenia.

In addition to the museum and municipal collections and the collections of the Triglav National Park Public Institute, museumized monuments are also common in the Alpine macroregion—the birthplace of the poet France Prešeren in Vrba is the first museumized monument in the country (Roškar 2021). The Alpine region is also incredibly lively in terms of private collectors and museums—e.g. the Kobarid Museum, a museum of the First World War; the museum collection *Od planine do Planike* (From Mountain Pasture to Planika Dairy), depicting the development of the mountain economy and dairy farming in the Soča Valley. Private collecting is intense in the Soča Valley, the area where the Soča/Isonzo Front took place during the First World War. The heritagisation of the Soča/Isonzo Front started in 1990 with the establishment of the Kobarid Museum by local enthusiasts. It greatly developed in the following decades in connection with the establishment and development of the

Walk of Peace trail and brand by the eponymous foundation, the Walk of Peace in the Soča Region Foundation (Kumer et al. 2020; Ledinek Lozej and Pisk 2021). Today, there are 19 private collections in the Soča Valley related to the relics of the First World War, open to the public. However, it is estimated that there are between 50 and 100 more that have not been inventoried by any public institution (Kofol 2015). Sometimes they are linked to more or less institutionalised initiatives and networks (Ledinek Lozej 2020). These bottom-up private collections are perceived with ambiguity. Sometimes as the initiatives of the awareness-raising on the importance of the Soča/Isonzo Front heritage. Sometimes as problematic in terms of the Cultural Heritage Protection Act (Zakon o varstvu kulturne … 2008), which strictly forbids the collection of underground objects and the trade in these objects (Kofol 2015; Kravanja 2018).

22.4.5 International Nominations and the Alps

The Walks of Peace in the Soča Region Foundation initiated and prepared the nomination for The Walk of Peace from the Alps to the Adriatic–Heritage of the First World War, which was submitted to the UNESCO Tentative List in 2016, during the First World War centenary (2014–2018). At the moment, it came to a halt because UNESCO formed an expert body that shall re-consider the nominations of difficult heritage, based on wars and conflicts. Nevertheless, the following sites, that are located in the Alpine macroregion, form part of the supranational heritage registers:

- two out of five Slovenian sites on the UNESCO World Heritage List (WHL): (1) Heritage of Mercury—a transnational nomination of Almadén, Spain, and Idrija, Slovenia (the latter lying actually on the border between the Alpine and Dinaric macroregions); (2) the most recent one, the Works of the architect Jože Plečnik in Ljubljana (the multilateral nomination Prehistoric Pile Dwellings around the Alps might be added here; the Slovenian sites of this multinational nomination are located in the Ljubljana Marsh (*Ljubljansko barje*), hence already in the Dinaric macroregion);
- one out of five elements on the UNESCO Representative List of the Intangible Heritage of Humanity (ICH RL)—The Škofja Loka Passion Play;
- and two out of three European Heritage Label sites—the Franja Partisan Hospital and the Javorca Church.

With such a number of heritage elements on international registers, it is surprising that Slovenia was not included in the multilateral nomination of alpinism on the UNESCO ICH RL, especially because of the documented popularity of mountaineering and alpinism in Slovenia, which has been heritagised since the beginning of the twentieth century and culminated in the above-mentioned recent Slovenian Alpine Museum. This "tactical rescaling" (Debarbieux et al. 2023:12) of the nomination of alpinism was carried out by the amateur alpinists and professional mountain guides of the Mont Blanc region from France, Italy, and Switzerland. The UNESCO multinational instrument is driven by a mixture of idealism, realism, opportunism, political concerns, and balancing between a nationalist and diplomatic strategy (Debarbieux and Munz 2019; Debarbieux et al. 2023), being at the same time the tool and the aim of regional and transnational cooperation. This is at least attempted in the emergent multilateral proposal on Alpine Food Heritage that builds on multilevel community participation from the whole Alpine arc (Godina Golija and Ledinek Lozej 2018). Alpine Food Heritage is defined as "an integrated system of knowledges, skills, practices, and values that are interconnected and linked to a shared natural capital by Alpine people, from France to Slovenia" (Toward … 2021:11). Drawing on the Debarbieux et al. (2023:15) findings on the alpinism nomination case, we hope that it will enable a reflexive process on the Alpineness and build imaginaries that will go beyond the national.

22.5 Conclusion

In Slovenia, heritage has more entries in the official registers from the Alpine macroregion than from the Mediterranean, Pannonian, and Dinaric landscapes. The exploring of the mountain world in the Enlightenment and Romantic periods, and later mountaineering and alpinism, promoted an aestheticised view of the mountains. The first nature conservation proposals and the establishment of the Alpine Conservation Park, the forerunner of today's only national park in Slovenia, the Triglav National Park, are linked to the Alpine areas. The role of the mountains in the period of the nation-building and in contemporary everyday and ritual socio-political life is such that individual authors speak of the Alpine culture of Slovenianness and of the mountain identity of Slovenes. Based on the analysed material, we cannot deduce to what extent the authorised heritage discourse is a result and/or a co-creator of the naturalisation of the Alpinness—but we do point out some gaps that allow for alternative possibilities of heritage-making.

Acknowledgements This study was conducted as part of the research program Heritage on the Margins: New Perspectives on Heritage and Identity within and beyond the National (P5-0408), financed by the Slovenian Research and Innovation Agency. We would like to thank Rok Ciglič from the ZRC SAZU Anton Melik Geographical Institute for the spatial mapping of the heritage units, and Grega Žorž from the Directorate for Cultural Heritage, Ministry of Culture, for his guidance in obtaining the data.

References

Adell N, Bendix RF, Bortolotto C, Tauschek M (eds) (2015) Between imagined communities and communities of practice. Göttingen studies in cultural property, vol 8. Universitätsverlag, Göttingen

Anon (1901) Alpski muzej [Alpine Museum]. Planinski Vestnik 7(1):14

Atlas okoja [Atlas of the environment] (2021). Agencija Republike Slovenije za okolje, Ljubljana. https://gis.arso.gov.si/atlasokolja/. Accessed 19 Jun 2021

Bajuk Senčar T (2012) Natural and cultural heritage practices in Triglav National Park. Traditiones 41(2):53–64. https://doi.org/10.3986/Traditio2012410204

Convention for the safeguarding of the intangible cultural heritage (2003). https://ich.unesco.org/doc/src/15164-EN.pdf. Accessed 30 Sep 2021

Debarbieux B, Rudaz G (2015) The mountain a political history from the enlightenment to the present. University of Chicago Press, Chicago

Debarbieux B, Munz HA (2019) Scaling heritage: the construction of scales in the submission process of alpinism to UNESCO's intangible cultural heritage list. Int J Herit Stud 25(12):1248–1262. https://doi.org/10.1080/13527258.2019.1590445

Debarbieux B, Bortolotto C, Munz H, Raziano C (2023) Sharing heritage? Politics and territoriality in UNESCO's heritage lists. Territory, Politics, Governance 11(3):608–624. https://doi.org/10.1080/21622671.2020.1854112

Eliade M (1974) Patterns in comparative religion. New American Library, New York

Erjavec A (1994) Mountain photography and the constitution of National Identity. Filozofski Vestnik 15(2):211–235

Fleming F (2004) The Alps and the imagination. Ambio 33:51–55. https://doi.org/10.1007/0044-7447-33.sp13.51

Hansen HP (2013) The summits of modern man: mountaineering after the enlightenment. Harvard University Press, Cambridge

Godina Golija M, Ledinek Lozej Š (2018) Pomen in cilji projektov ohranjanja dediščine prehrane [The importance and objectives of projects aimed at safeguarding food heritage]. Etnolog 28:85–103

Harrison R (2013) Heritage: critical approaches. Routledge, Abingdon, New York

Harrison R, Desilvey C, Holtorf C, Macdonald S, Bartolini N, Breithoff E, Fredheim H, Lyons A, May S, Morgan J, Penrose S (2020) Heritage futures: comparative approaches to natural and cultural heritage practices. UCL Press, London

Heriskop (2020) Observing heritage-making on the margins. https://dediscina.zrc-sazu.si/en/home2/. Accessed 21 Mar 2022

Hladnik M (1980) Shema in značilnost Vandotove planinske pripovedke [The scheme and characteristics of Vandot's mountain tale]. Slavistična Revija 28(1–4):311–324

Juvan M (2015) The nation between the epic and the novel: France Prešeren's the Baptism on the Savica as a compromise "world text". Canadian Rev Comparat Lit 42(4):382–395. https://doi.org/10.1353/crc.2015.0039

Keller T (2016) Apostles of the Alps: mountaineering and nation building in Germany and Austria, 1860–1929. The University of North Carolina Press, Chapel Hill

Kofol K (2015) Privatne zbirke Gornjega Posočja ter njihovo sodelovanje s pooblaščenimi muzeji in pristojnimi institucijami [Private collections of the Upper Soča Valley and the cooperation of their owners with the authorised museums and competent institutions]. In: Dapit R, Ivančič Kutin B, Ledinek Lozej Š (eds) Le collezioni uniscono: Collezioni etnologiche, tradizione

orale e turismo culturale fra Alpi e il Carso. Università degli Studio di Udine, Udine, pp 271–281
Kovačič M (2015) V deželi harmonike – nacionalizacija harmonike v slovenskem kontekstu [In the land of the accordion – nationalization of the accordion in the Slovenian context]. In: Mlekuž J (ed) Venček domačih: predmeti, Slovencem sveti. Založba ZRC, Ljubljana, pp 87–116. https://doi.org/10.3986/9789612547714
Kravanja B (2018) Learning by collecting: amateur collectors and their shifting positions in the Isonzo Front heritagization and tourism adaptation. Folklore 73:95–116. https://doi.org/10.7592/FEJF2018.73.kravanja
Kropej M (2012) Supernatural beings from Slovenian myth and folktales. Založba ZRC, Ljubljana. https://doi.org/10.3986/9789610503385
Kučan A (1998) Krajina kot nacionalni simbol [Landscape as a national symbol]. Sophia, Ljubljana
Kulturna dediščina: odprti podatki [Cultural heritage: open data] (2021). Direktorat za kulturno dediščino, Ljubjana. https://data-mk-indok.opendata.arcgis.com/. Accessed 19 Jun 2021
Kumer P, Zorn M, Žorž G, Gašperič P (2020) From the Great War to interwar fortifications: changing narratives attached to the military landscape in Western Slovenia. In: O'Reilly G (ed) Places of memory and legacies in an age of insecurities and globalization. Springer, Cham, pp 247–263. https://doi.org/10.1007/978-3-030-60982-5_11
Ledinek Lozej Š (2020) Collaborative inventory–a participatory approach to cultural heritage collections. In: Nared J, Bole D (eds) Participatory research and planning in practice. Springer, Cham. https://doi.org/10.1007/978-3-030-28014-7_8
Nesnovna dediščina [Intangible heritage] (2021). Ministrstvo za kulturo, Ljubljana. https://www.gov.si/teme/nesnovna-dediscina/. Accessed 18 Jun 2021
Ledinek Lozej Š, Pisk M (2021) Collections, walks, and cultural landscapes: Slovenian-Italian cross-border heritage initiatives. Traditiones 50(1):79–10. https://doi.org/10.3986/Traditio2021500106
Mikša P (2018) Kako je Triglav pristal v grbu Republike Slovenije [How Triglav ended up on the coat of arms of the Republic of Slovenia]. In: Zorn M, Mikša P, Lačen Benedičič I, Ogrin M, Kunstelj AM (eds) Triglav 240. Založba ZRC, Ljubljana, pp 15–28. https://doi.org/10.3986/9789610500841
Mikša P, Zorn M (2016) Začetki raziskovanja slovenskih Alp [The beginnings of the research of Slovenian Alps]. Geografski Vestnik 88(2):103–131. https://doi.org/10.3986/GV88206
Mikša P, Ogrin M, Glojek K (2018) Gorska identiteta Slovencev [Mountain identity of Slovenes]. In: Zorn M, Mikša P, Lačen Benedičič I, Ogrin M, Kunstelj AM (eds) Triglav 240. Založba ZRC, Ljubljana, pp 157–165. https://doi.org/10.3986/9789610500841
Perko D (1998) The regionalization of Slovenia. Geografski Zbornik 38:12–57
Perko D, Ciglič R, Hrvatin M (2021) Landscape macrotypologies and microtypologies of Slovenia. Acta Geogr Slov 61(3):7–89. https://doi.org/10.3986/AGS.10384
Razvid muzejev [Records of museums] (2021). Ministrstvo za kulturo, Ljubljana. https://podatki.gov.si/dataset/razvid-muzejev. Accessed 19 Jun 2021
Roškar S (2020) Strokovne podlage za razglasitev za kulturni spomenik lokalnega pomena: Zgornje Jezersko – Češka koča (EŠD 30697) [Expertise for the declaration of a local cultural monument: Zgornje Jezersko – The Češka koča hut (no. 30697)]. Občina Jezersko, Jezersko. https://www.jezersko.si/files/other/news/169/27138803_Odlok_kulturni_spomenik_Ceska_koca.pdf. Accessed 10 Oct 2021
Roškar S (2021) Imamo spomenik – muzej. Kaj pa zdaj? Vzdrževanje in upravljanje muzejskih hiš [We have a monument – a museum. Now what? The maintenance and management of house museums]. In: Upravljanje kulturnom baštinom: od valorizacije do interpretacije. NU Zavod za zaštita na spomenicite na kulturata i muzej Bitola, Bitola, pp 277–293
Skoberne P (2018) Predlog Albina Belarja za zavarovanje Doline Triglavskih jezer (1908) [Albin Belar's proposal for protection of the Triglav Lakes Valley (1908)]. In: Zorn M, Mikša P, Lačen Benedičič I, Ogrin M, Kunstelj AM (eds) Triglav 240. Založba ZRC, Ljubljana, pp 167–177. https://doi.org/10.3986/9789610500841
Slavec Gradišnik I (2014) V objemih dediščine [Embraced by heritage]. In: Dolžan Eržen T, Slavec Gradišnik I, Valentinčič Furlan N (eds) Interpretacije dediščine. Slovensko etnološko društvo, Ljubljana, pp 8–24
Smith L (2006) Uses of heritage. Routledge, London
Smrekar A, Polajnar Horvat K, Erhartič B (2016) The beauty of landforms. Acta Geogr Slov 56(2):321–335. https://doi.org/10.3986/AGS.3039
Strojin T (ed) (1980) Triglav, gora naših gora [Triglav, the mountain of our mountains]. Obzorja, Maribor
Šaver B (2004) Nevidna alpska kultura slovenstva med "Balkanom" in "Evropo" [The invisible Alpine culture of Slovenianess between the "Balkans" and "Europe"]. Dialogi 40(11–12):110–121
Šaver B (2005) Nazaj v Planinski raj: Alpska kultura slovenstva in mitologija Triglava [Back to the mountain paradise: Alpine culture of Slovenianness and the mythology of Triglav]. Fakulteta za družbene vede, Ljubljana
Šmitek Z (1998) Kristalna gora: mitološko izročilo Slovencev [Crystal mountain: mythological tradition of Slovenes]. Forma, Ljubljana
Tominšek J (1949) Celjani, Savinjčani—turisti nekdaj in zdaj [People from Celje, Savinjska region—tourists in the past and now]. Celjski Tednik 1(9):8
Toward a multinational nomination of the element Alpine Food Heritage – Knowledge, skills, practices and values of Alpine communities – in the Representative List of intangible cultural heritage (2021). Lombardy Region, Milan
Tschofen B (1999) Berg – Kultur – Moderne: Volkskundliches aus den Alpen. Sonderzahl Verlaggesellschaft, Wien
Velikonja M (2018) Narava kot politika: motiv Triglava v grafitih in street artu [Nature as politics: the Mount

Triglav motive in graffiti and street art]. In: Zorn M, Mikša P, Lačen Benedičič I, Ogrin M, Kunstelj AM (eds) Triglav 240. Založba ZRC, Ljubljana, pp 225–235. https://doi.org/10.3986/9789610501138

Zakon o ohranjanju narave [Nature conservation act] (1999) Uradni list Republike Slovenije 56/1999. Ljubljana

Zakon o varstvu kulturne dediščine [Cultural heritage protection act] (2008) Uradni list Republike Slovenije 16/2008. Ljubljana

Zakon o varstvu podzemnih jam [Underground cave protection act] (2004) Uradni list Republike Slovenije 2/2004. Ljubljana

Zorn M, Smrekar A, Skoberne P, Šmuc A, Brancelj A, Dakskobler I, Poljanec A, Peršolja B, Erhartič B, Ferk M, Hrvatin M, Komac B, Ribeiro D (2015) Dolina Triglavskih jezer [The Triglav Lakes Valley]. Geografija Slovenije, vol 32. Založba ZRC, Ljubljana. https://doi.org/10.3986/9789610503620

Zgonik N (2002) Podobe slovenstva [Images of the Slovenianess]. Nova revija, Ljubljana

23 Recognition of Spatial Properties for the Location Assessment and Design of Bivouacs in the Slovenian Mountains

Kristijan Lavtižar, Alenka Fikfak, and Janez P. Grom

Abstract

Emergency mountain shelters, which are known also as bivouacs, offer visitors of the highlands a place to take refuge where there are no other options and offer an intermediate starting or resting point for strenuous climbing and hiking tours. The modern design of alpine bivouacs presents a challenge to architects, who must consider extreme natural conditions. The vulnerability of the natural environment requires restrained sizing with simple design solutions and construction details. The logic behind installing bivouacs in remote locations in a vulnerable natural space is to ensure that the bivouacs have as little accessibility and visibility as necessary so that the protection of the mountain environment as well as the safety of mountaineers can be assured. The study's emphasis was on identifying the spatial properties that define bivouacs with the parameters of placement, according to their basic rescue functions, and respecting the natural characteristics and values of the space. In the study cases, the characteristics of architecture, the characteristics of the space, and the influence of bivouacs on this space were questioned and quantified via GIS analysis. As part of the study, a survey was conducted with members of the Alpine Association of Slovenia. The results are followed with a discussion on the spatial guidelines for the placement and architectural design of bivouacs in a natural alpine environment.

Keywords

Bivouac · Mountain shelter · Sustainable use · Sustainable mountain tourism · Mountain architecture · Environmental and heritage conservation · Alps

23.1 Introduction

We climb the mountains to observe them in solitude in all their pristine natural beauty. The higher a person climbs, the more solitude and natural beauty they find. However, the higher a person goes, the more they are also exposed to the natural elements and the harsher environment they find (Flicking 2003). As mountain tourism evolved, we meet an increasing number of people and an increasing number of unnatural elements that were put there to ensure their safety. As such, the earth's undecaying monuments, as Nathaniel Hawthorne called the mountains (Hawthorne 2001), are at

K. Lavtižar (✉) · A. Fikfak · J. P. Grom
Faculty of Architecture, University of Ljubljana, Ljubljana, Slovenia
e-mail: kristijan.lavtizar@gmail.com

A. Fikfak
e-mail: alenka.fikfak@fa.uni-lj.si

J. P. Grom
e-mail: janez.grom@fa.uni-lj.si

M. Zorn et al. (eds.), *Discourses on Mountains of Montenegro and Slovenia*, Historical Geography and Geosciences, https://doi.org/10.1007/978-3-031-99017-5_23

risk due to unscrupulous and unsustainable human usage. In what manner we interact with the mountain environment and how we affect them is becoming a pressing issue of sustainable usage. As such, this study assessed the visual impact of three bivouacs in the Slovenian Alps. A methodological approach to evaluating architectural attributes and spatial characteristics was established. Subjective perception of the analysed bivouacs was verified with a widespread survey, conducted within a focus group. To better understand the role of the bivouacs, the word's definition becomes relevant as it distinguishes this typology from other built structures present in the Slovene Mountains. The built structures that are found in the Slovene Alps are predominantly small, modest, lone structures, such as mountain huts, bivouacs, and other buildings that are part of historical cultural heritage. All these architectural elements have a rich history, which form an integral part of the Slovene national identity connected to popular perception of the mountains.

The occasional visitor to the Slovene Mountains could come across some of the built elements of extensive military infrastructure. Most of these military structures are out of use and are just historical ruins. For example, close to the present border with Italy remains of the Soča/Isonzo Front from World War I can be found (Kumer et al. 2020). Then there are bunkers from the interwar period that helped protect the post-World War I Rapallo border and were part of the Italian "Alpine Wall" and the Yugoslavian "Rupnik Line". Few of these structures have been converted to tourist attractions and transformed into active use (Grom et al. 2021). Mountain huts (also known as mountain lodges) and the bivouacs (also known as mountain shelters) both presently hold an active role and both function in a different way. The role and the architectural articulation are somewhat connected to their alpine location. The mountain huts are placed on either the mountain foothills or the more open and accessible high mountain valleys and ridges. As the mountains became more accessible to a larger group of recreational visitors, so have the mountains developed and adapted with a greater option of services and accommodation (Ragelj 2015). Mountain huts are the structures that support these mass tourism requirements, which means these buildings provide a wider support programme, such as toilets, kitchens, restaurants, and accommodations. Their sizes and their spatial presence are thus much larger than that of a bivouac. The role of a bivouac is to ensure safety among the high mountains and offer overnight stays for one night while also completing a network of mountain outposts, where building proper huts is not possible (Kajzelj 2017). Historically, they were positioned in a location where they provided not only a shelter, but also offered an optimal starting point for alpine ascents. However, it is important to acknowledge that these stays are not meant for several days. They are intended as a starting point, a springboard in conquering the least accessible or most distant destinations in the mountains that go beyond the scope of a 1-day hike (Eržen 2017; Kajzelj 2017). Because of their alpine location, the scale of the bivouacs is exclusively of a more modest size and their services are therefore limited. The bivouacs are further built with the intent of forming a safety net—to reduce accidents, make the task of mountain rescuing easier, and to save lives. From this point of view, the Slovenian Mountains are almost too small and lay too low for this purpose to always be justified, yet the typology of bivouacs also finds its role as an element of national cultural heritage within the vast mountain hiking tradition (Kajzelj 2017).

One of the first historically widely recognized built elements in the Slovene Mountains was a small structure known as *Aljažev Stolp* (The Aljaž Tower). It is the single most iconic example of high mountain structures in the Slovenian Alps. It was built in 1895 on the summit of Mount Triglav in north-western Slovenia, as a tower, a storm shelter, and a triangulation point (Vidic 2000). Along with Triglav, it is a landmark of Slovenia and a symbol of Slovenehood. In the early 1890s, with the hopes of restraining the increased interest of foreigners in the Slovene Mountains, Jakob Aljaž, a priest in Dovje beneath Mount Triglav, used a piece of chalk and the floor of his room to draw the plans for a cylindrical tower with a flag on its top (Slovenian … 2014). In April of that year, he purchased the summit of Triglav for the sum of one

Austro-Hungarian gulden from the Municipality of Dovje, as how it was called at the time. Having done so, he secured himself the right to erect a building on his own real property (Mikša 2011). The tower was constructed from iron- and zinc-coated sheet steel by Anton Belec from Šentvid near Ljubljana. Aljaž, with the help of four workers, brought the parts of the tower to the summit of Triglav and put the tower together in only 5 h on 7 August 1895. Its opening ceremony and blessing took place in a small circle on 22 August that year (Belimezov 2007; Mikša 2015, 2017).

Within the general Slovene tradition, the bivouacs were initially required mainly because of the low network density of different outposts in Slovenian Mountains. This was when the network consisted of small huts with poor logistical connections and possibilities of visiting them (Kajzelj 1995). They were the only safe way of making safe overnight stays or sheltering when conquering the unknown mountain areas and greater mountain peaks with longer, multi-day excursions. Besides, mountaineering tourism was also not as popular as today and there were not as many tourists making as many trips, which meant not as many beds were needed. Only later in the late twentieth century, with more domestic and foreign visitors (Ogrin 2020), a greater economic incentive presented itself to help complete the patchwork network of mountain lodges (Ragelj 2015). With the economical exploitation of mountain tourism and the consequentially growing need to expand safety mechanisms and services to support these trends, the question of the impact on the delicate and vulnerable mountains became more pressing.

Besides the usual elements of pollution, visual pollution is a very pressing factors, as it tends to be permanent. Once an architectural element finds its way in the mountains, it is difficult to remove, and it becomes a permanent element of visual pollution. On the contrary, despite the mountain military constructs, such as bunkers, numbering into thousands, their visual impact is rather low (Grom and Štukovnik 2018). This originates from their conception, as they were designed to be unnoticeable and blend within the natural surroundings. On the other hand, the architectural reality of touristic mountain exploitation induces high visibility, which is incorporated by design.

Through our research we are trying to better understand the conflicting nature of a bivouac—it must be highly visible in order to carry out its primary role and grant safety to high mountain users and, in some respects, it must have as little impact as possible on its surroundings. It is a building inserted as an artificial element. Furthermore, the visual exposure by itself does not necessarily offer higher protection since sometimes it is merely a result of some architects wish to create attractive architecture, not to build the visibility of the bivouac on the premises of its basic role. Nonetheless, the impact is as such predominantly visual, as the bivouacs, as compared to mountain huts, offer no additional services and comforts.

23.2 Purpose and Objectives

The chapter focuses on the study of three bivouacs in the Slovenian Alps. The purpose of the three study cases was to outline the visibility aspects of the bivouacs and their distinctive architecture by defining their objective visual characteristics. In order to evaluate what it is that constitutes the appearance of this architecture, fundamental terminology of architectural design can be briefly explored. Within this study the relevant visual elements of architecture are synthesized to shape, colour, and texture. Depending on the priorities of a specific building, one or more elements can dominate to achieve a desired result.

We can use this articulation in architectural visual communication to determine the criteria that we adopt to assess the selected bivouacs for the case study and assess their inhibited attributes. Furthermore, these attributes speak to a certain level about the visual contrasting relationship between the bivouac and its natural surroundings. Moreover, understanding their spatial characteristics and properties of their placement within a location is (in the case of this study) more factual and inherently measurable. To understand their position and their visibility in the mountains, the distances between them, the network density, and their dependence on the network formed by adjacent bivouacs were analysed. Their sum is the result of the evaluation of attributes that gave us the numerical data—this was used to create a comparative table with quantifiable outputs. The article discusses the references presented and demonstrates that these three bivouacs behave in drastically different ways within the same mountain space.

Researching how and why the placement of a certain bivouac works or not was not our goal. Firstly, the purpose of the study was to understand how a specific bivouac functions in the mountain environment in its primary role of providing support in particular situations. Secondly, we try to understand how a singular bivouac contributes to strengthening the safety net of a network that is woven just densely enough to carry out its role and have as little impact on the environment as possible. The objective of the article is thus to answer the question of what prevails in the planning and construction of mountain bivouacs: their beauty or their visibility? The resolution is further provided with the help of a public opinion poll.

23.3 Tools and Methodology

Three different bivouacs were selected for this study. Their position is marked in Fig. 23.1. They are approximately 2 km apart from one another, covering a common area of the Slovenian Mountain range of the Kamnik–Savinja Alps. The area spans from Dovški Plaz on the west, sitting below the summit of Kočna (2,540 m), over Streha below Grintovec (2,558 m), towards Veliki and Mali

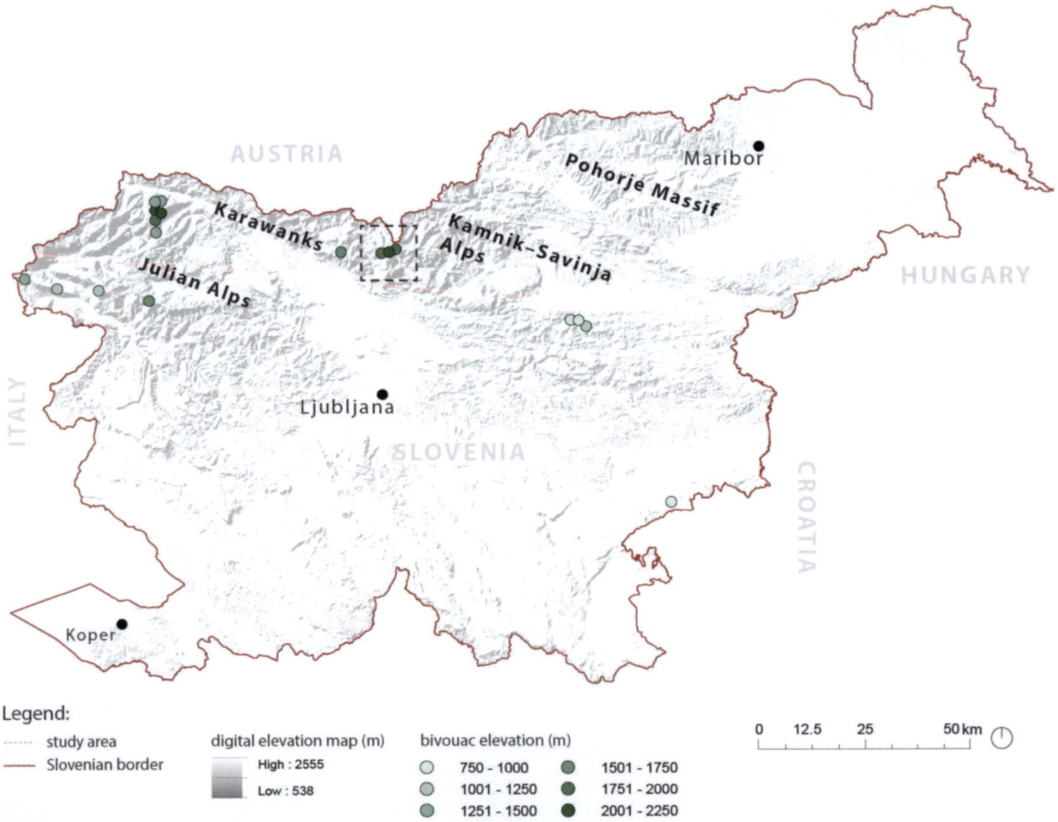

Fig. 23.1 Macro-location of mountain bivouacs in the Slovenian Mountains in 2021. Shown are mountain bivouacs (not including winter rooms or mountain lodge extensions), which are sorted by their elevation. Bivouacs discussed are marked in a square

Podi Plateaus underneath the peak of Turska Gora (2,251 m). This selection was made with the intention to distinguish between different architectural design approaches and periods of construction within the same space with the intent of making the analysis more comparable.

The spatial data for the case study analysis was assembled using GIS tools. Digital terrain model (DTM) data for each individual terrain datasheet covering the considered sample were processed with the tools *Make XY Event Layer*, *Feature to Point*, *Point to Raster*, thus obtaining a DTM. The obtained DTM in raster format.tiff was bound to the input data at a resolution of 5 × 5 m. Using *Resample*, we changed the resolution of the relief display to 1 × 1 m, which was necessary to produce precise layers. We used the *Contour tool* to make the lines, and the *Smooth Line* tool to process and better display them.

The analytical cluster for each example consists of a photograph of the bivouac in its latest condition in 2021, an aerial photo of the mountain site, and a GIS analysis of its visibility. The last part includes a comprehensive calculation of the area of terrain, from which any part of the bivouac's structure is visible at human height. This takes into the account the mountain terrain, its elevation, the bivouac size, and the average height of an observer (eyesight level) set at 1.75 m for each of the examples. The visibility of each cell centre was determined by comparing the altitude angle to the cell centre with the altitude angle to the local horizon. The local horizon was computed by considering the intervening terrain between the point of observation and the current cell centre. If the point lies above the local horizon, it was considered visible. The view range radius, however, is not set, as the maximum distance from the viewpoint captured in the area stands at 3 kms. The end "visible" output determines the raster surface locations visible to a set of observer features. This is then overlaid with a layer of marked mountain trails of varying difficulties.

All nearby mountain lodges are added to the map as well. The following result is a representation of the bivouacs' micro-location with regard to their purpose of ensuring their visibility, possibly a good view, and the safety of mountaineers. All spatial data, including the digital elevation map and the aerial photography, were taken from the website of the Surveying and Mapping Authority of the Republic of Slovenia (E-surveying ... 2021). Bivouac points and other points of interest were taken from the *maPZS* website of the Alpine Association of Slovenia (maPZS 2021), which is open source. The analytical cluster concludes with established criteria on visual quality along with their actual spatial placements in all three case studies. Table 23.1 includes a synthesis of all previously discussed architectural and spatial elements. The selected bivouacs were assessed by considering two separate sets of parameters. Within the first set we determined architectural attributes that define the appearance of the bivouacs at the architectural level. Again, this stems from the grounds of basic architectural theory, including

Table 23.1 The evaluation and sum of bivouac attributes

Example		Architecture			Position		Pollution
		Shape	Colour	Material	Visibility	Network	Visual impact
Bivouac	BI	+	–	+	+	–	+
Bivouac	BII	+	+	+	+	+	–
Bivouac	BIII	–	–	–	–	+	+

form, colour, and material. Within the second set spatial presence was determined through their visibility, i.e. how visible they are from any path of access, and the network placement, i.e. how well they are included and they complete the network with other adjacent bivouacs. The gathered objective and subjective observations are combined into the aforementioned table with simple positive and negative ratings for each example. As the perception of the visual characteristics of a certain element, bivouacs, in this case, is subjective, the gathered and analysed data were verified through a web-based survey.

The survey was organized through the Alpine Association of Slovenia and their members. They were selected because the organization includes the many different types of people who visit the Slovenian Mountains, varying in their age structure, sex, their level of mountaineering experience, their involvement with bivouacs, and the nature of their usual visit. It was distributed anonymously through their members via social media. The interview with the members of the Alpine Association of Slovenia was structured with the online tool *1KA* (1KA 2021). The survey was useful to collect a closed type of questions for the first 13 questions, circling the letter before the answer that applies. These were followed with an open-ended question concerning general opinions about Slovenian bivouacs.

23.4 Case Studies

Selected bivouacs have distinctly different architectural appearances. They retain their compact size, footprint, and proportions, yet they are made of differently textured materials, are of a different colour, and have different shapes. They also engage in a different composition dialogue with their surroundings. Each of the bivouacs has unique properties of visibility that relate to its architectural character. More importantly, for our scale of analysis, it relates to their position in space. Each caters differently to its own section of the trail and safety network.

23.4.1 Example I

The Kočna bivouac (Figs. 23.2 and 23.3) stands on the southwestern ridge of Kokrska Kočna at the elevation of 1,952 m as the first of the three selected, looking from west to east. A demanding

Fig. 23.2 Example BI, looking from west to east towards the Streha ridge. Bivouac name: Bivak v Kočni; year of construction: 1952 (restored in 2011); latitude/longitude: 46.350807/14.524442; altitude: 1,952 m; architect: unknown. (Photo by Kristijan Lavtižar, authors archive)

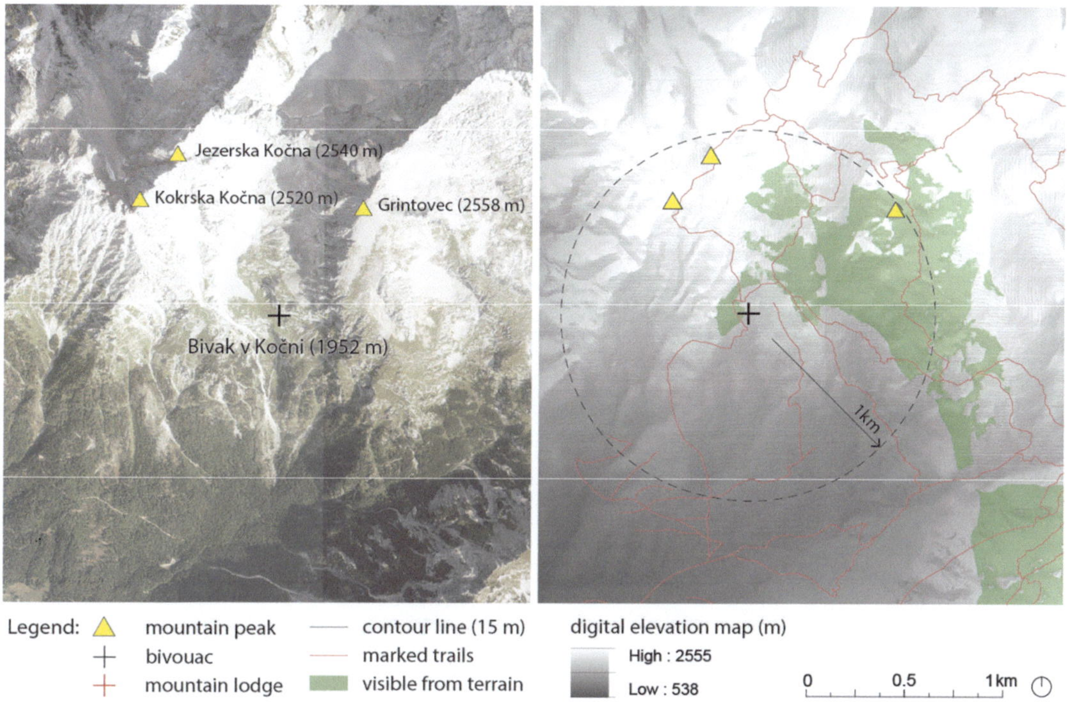

Fig. 23.3 Example BI–Bivak v Kočni. From left to right: aerial photo analysis, GIS visibility and network analysis

mountain trail runs nearby from the Suhadolnik farm above the Kokra Valley towards the top of Kokrska Kočna. Its construction dates back to 1952; it was coated originally in a dark, military-green colour and of a matte, dull texture (PZS 2021a). Its shape, that of a half-barrel, had an elongated, rounded hull constructed from sheet metal and a wooden lining. During restoration works in 2011 it was painted a metallic-grey colour of mate texturing (PZS 2021a). The building has a single window, which serves only to offer some daylight inside and some much-needed ventilation, as opposed to making it an observational window. Neither would it be positioned in the appropriate direction to present a compelling view. The limited glass surface limits effects on visibility by adding a glare effect on the façade. It has a capacity of 8 bunk beds. The overall design and construction of the building shell ensures an aerodynamic shape with a low-centred mass. This enables it to withstand the given climatic conditions and harsh winds. Mountain shelters in this shape and construction are typical of others built in that era in the Slovenian Alps (Kajzelj 2017). Its architectural articulation offers a clear contrast to the background of the surrounding natural space. Today, after the restoration works, it is even easier to spot from distances.

The structure of the bivouac was modelled at 3-m high in the GIS analysis (Fig. 23.3). The green-coloured area marks the terrain from which any part of the bivouac's structure is visible at average human height. The result showed that it is visible from a large part of the northeastern area, covering 23% visibility in a 1-km radius. This corresponds to approximately 2.5 km of all intersected marked trails inside the radius; however, most of these trails run high above the shelter which makes the bivouac considerably less accessible. It also stands as the only outpost in its near vicinity of 2 km. The bivouac's location on the distant mountain ridge is as such its biggest disadvantage, therefore clearly reducing its functionality. Due to selection for its location, the architectural attributes are not fully taken advantage of.

Fig. 23.4 Example BII, looking from the south direction towards Mlinarsko Sedlo saddle and Dolgi Hrbet ridge. Bivouac name: Bivak Pavla Kemperla; year of construction: 1973 (restored in 2009); latitude/longitude: 46.353364/14.547263; altitude: 2,104 m; architect: Miha Kajzelj. (Photo by Domen Avsec, authors archive)

23.4.2 Example II

The Pavle Kemperl bivouac (also known as the bivouac under Grintovec; (Figs. 23.4 and 23.5) was eponymously dedicated to the alpinist and mountain rescuer. It was standing at 2,104 m on a grassy hill below the eastern wall of Grintovec (2,558 m), which is one of Slovenia's most frequently visited mountains. It was effectively positioned on the crossroads of major trails in the area. It could accommodate up to a total of 8 people. The first bivouac was set up at the location in 1973 by Mountain Rescue team Kamnik (PZS 2021b). It was previously replaced due to wear in 2009 by a three-level structure that is built on a concrete base with minimal footprint that was anchored to the rock (Čelik 2010). Its dimensions were reduced on the level of its floor plan at 2 × 3 m but were more pronounced in its height at 4.5 m. Its walls were composed of an aluminium outer layer of a distinctive black-coloured finish and of matte texturing. It had a high ceiling, which enabled more comfortable temperatures at the top and easier ventilation (Lavtižar 2020). Long, vertical windows opened up towards the mountains and created a much brighter interior as opposed to the previous example but were still not reflective in the sense of adding to visibility with glass glare. The bivouac was removed from its location during our current research due to disagreements related to its maintenance. The process of its removal revealed deeper divides between the architects of the bivouac and its owner and caretaker—the local mountaineering association. The dispute between the two parties emerged due to its maintenance. Potential flaws related to the chosen construction details were revealed as the location of its placement was heavily exposed to the harsh mountain environment. Furthermore, doubts on the true intention of the location choice were raised as it was stated that the lead factor was its picturesque presence, with the convenience and functionality of its position being an afterthought (Stegnar 2021).

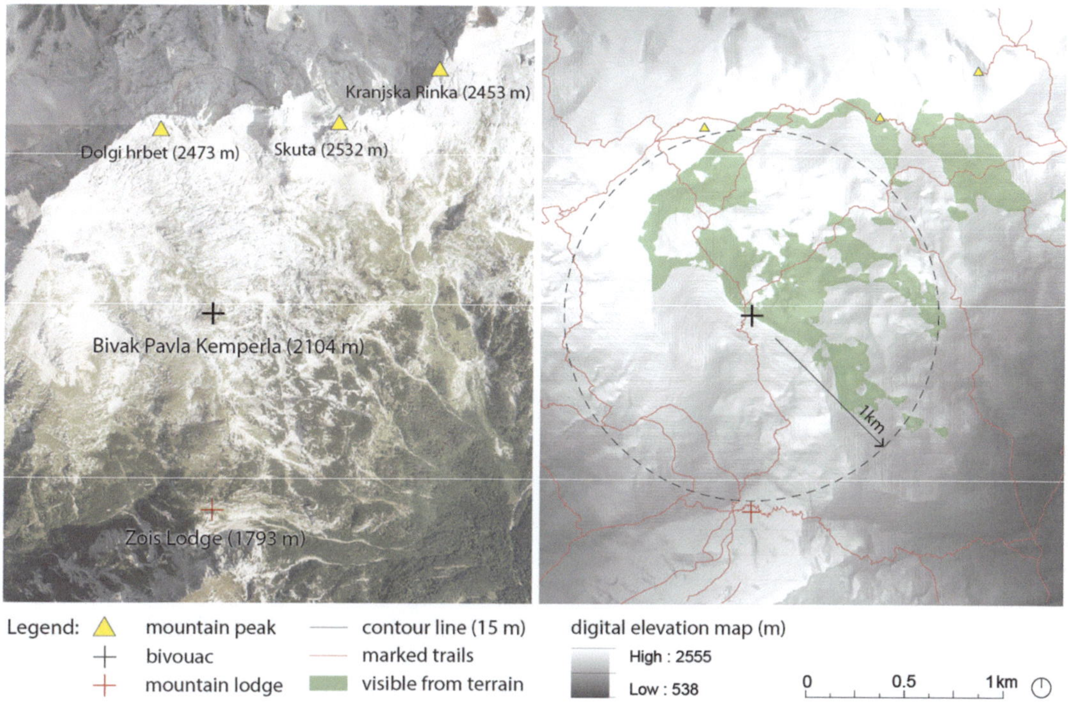

Fig. 23.5 Example BII–Pavel Kemperle bivouac. From left to right: aerial photo analysis, GIS visibility, and network analysis

The roof height of 4.5 m was used in the visibility analysis, which placed it first in terms of visibility in its immediate surroundings and with slightly better views from the nearby marked trails than its counterparts. The result shows the visible area covering 28% in a 1-km radius. This corresponds to approximately 1.9 km of all intersected marked trails inside the radius. Again, most of these intersected trails are elevated above the structure. Its closest mountain outpost was the Cojzova Koča Mountain Lodge, which is approximately 2 h away in the south direction over challenging terrain.

The black, contrasting colour of the facade made this bivouac stand out from its surroundings. Its rigorous geometry offered a clear contrast with the organic and fragmented surroundings. The vertical dark volume was visually distinctive from a distance in the light grey rock surroundings and especially in the white, snowy landscape, yet this made it more difficult to spot in darker hours of the day. Regardless, its strategic positioning enabled the architectural attributes to render this bivouac highly visible.

23.4.3 Example III

The third example is arguably the most famous bivouac in the Slovenian Mountains, due to its distinctive architecture and prominent position on the rocks. The bivouac Bivak pod Skuto (Figs. 23.6 and 23.7) stands on the edge of the Mali Podi Karst Plateau below Skuta (2,533 m) and Turska Gora (2,251 m), itself at 2,045 m (PZS 2021c). It was designed for another location in the Julian Alps, which was later rejected because the size of the bivouac was deemed too large (Čokl 2015) and a slope below the peak of Skuta was chosen as the new location. The only (unmarked) path to it leads from Konec in the Kamniška Bistrica Valley across Žmavčarje, which makes it quite difficult to access, yet it is often utilized by climbers that travel onwards up the mountains on unmarked routes. The outpost consists of three

Fig. 23.6 Example B III., looking from north to south towards the valley of the Kamniška Bistrica River. Bivouac name: Bivak pod Skuto; year of construction: 2016; latitude/longitude: 46.358956/14.567736; altitude: 2,045 m; architect: OFIS Architects and Harvard University Students. (Photo by Laura Kosi, authors archive)

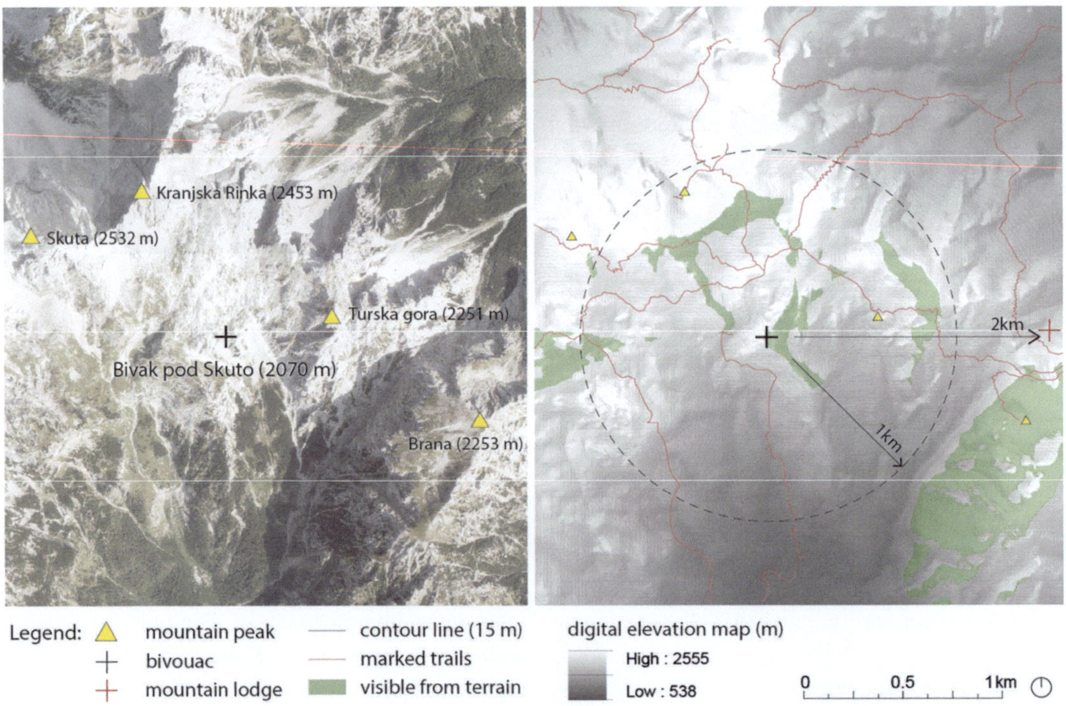

Fig. 23.7 Example B III–Bivak pod Skuto. From left to right: aerial photo analysis, GIS visibility and network analysis

modules, the first of which has an entrance and a storage area, the second a bedroom with a space for socializing and sleeping for 10, while the last room is intended exclusively for relaxation. Both side walls are glass and offer panoramic views of the valley towards Kamniška Bistrica Valley and Mount Skuta on the other side. Interestingly, this suggests other values and functionalities of the structure, besides offering the basic functionalities of a mountain shelter. The bivouac's colour and design coincide with the rocky surroundings, with its texturing mimicking the limestone appearance. It has a distinctive, stepped shape.

A height of 4 m was used as the input for the analysis of its visibility. The result showed a weak visibility on all sides surrounding the structure, measuring at less than 10% inside the 1-km radius. It has a weak intersection of marked routs of 340 m. This correlates to the reclusive micro-location in a somewhat depressed part of terrain. The closest neighbour would be the Kamniška Koča Mountain Lodge on Kamnik Saddle approximately 2 km air distance away to the east, which is considerably far away due to the rough terrain. The architectural attributes alone make this bivouac more difficult to spot, in particular, in poorer weather conditions. The shape, colour, and texture all tend to visually fuse the bivouac with its natural backdrop. Its architecture, positioning, and its functional components suggest this bivouac leans more towards offering a touristic attraction to this part of the mountains, rather than serving solely as a safety stop of a bare outpost.

23.4.4 Assessment of the Case Studies

The selected bivouacs were assessed by considering two separate sets of parameters (Table 23.1). Within the first set we determined architectural attributes that define the appearance of the bivouacs at the architectural level. Within the second set spatial presence was determined through their visibility, i.e. how much they are visible from access paths, and network placement, i.e. how well they are included in and upgrade the network with other adjacent bivouacs.

As can be observed, the first set (Table 23.1) mostly relates to subjective elements as they are conditioned by someone's assessment. But still the three determined attributes of shape, colour, and texture are observed through the comparison of the same attributes of the surrounding natural space. As such these elements are only evaluated through a perspective of contrast. Is there a contrast between the bivouac and the natural surrounding or is there not? If the answer to this question is yes, then there is contrast that gives the bivouac a clearly distinctive appearance in the space of its placement. Subsequently, the bivouac serves its purpose of a safety infrastructure. The second set is fractographic, since both determined criteria are measured. Its visibility is measured in percentages with the help of *ArcGIS*. The same was carried out for the assessment of a bivouac's contribution to developing the network.

The three considered bivouacs performed differently in contrast of expectations. The Pavel Kemperle bivouac (BII) performed the best, while the bivouac Bivak pod Skuto (BIII) underperformed. This result is unexpected as the worst performing bivouac is the most recent of the analysed ones to have been built. One's expectations would be that knowledge gained from the past would ensure a safer, better product. But this was not so as the oldest of the three bivouacs clearly performed the best. A closer look at the reasons why the BIII performed so badly reveals hidden flaws in its conception, as the original designers boasted about the exceptional views in their description of the project concept. As such it was designed and perceived as a tourist attraction and not as a safety mechanism.

It could be stated that assessing the characteristics of these bivouacs in the perspective of them carrying out the role of a safety net for mountain users is in distinct contrast to their role in the sense of visual pollutants of natural mountain areas. In this context, the BII was visually exposed, as it was dominating over the subtle nature as an artificial monolith. The BIII, on the other hand, is fused with the natural elements by design, as its architecture is trying to mimic them. From this point of view, the BIII is the least intrusive artificially inserted element in the natural space of the mountain, thus being the least visually polluting one. The BI's impact was evaluated as low despite the bivouac performing well in its primary role. The shape is without doubt a strong contrast as it is geometrical, but all other attributes are non-contrasting. This combined with its point of placement makes the BI a rather non-intrusive artificial element in the mountainscape with a reduced level of visual pollution.

23.5 Survey

The aim of the survey was to collect the opinions of those who hike in the mountains on the presence of the bivouac to help us answer the lingering question of the true importance of the visibility of bivouacs. The online poll was successful; it was open for 1 week from 24 until 31 May 2021 and gathered a total of 742 results from unique visitors. This returned 396 complete and valid responses, along with 346 incomplete or invalid responses. These include the respondents who completed the survey but had not had an encounter with a mountain bivouac yet, which is explained later in this chapter. The age group distribution was mostly even, leaning towards the younger half of the population, with 65% of the visitors younger than 40 years old.

To the question of "How often have you visited the mountains in the last year?" 23% had gone on 4–10 occasions, and 67% had gone more than 10 times in the last year. Half of all visitors expressed that they primarily use unmarked alpine trails and 66% stated that they travel throughout the year, regardless of the season. Hiking the popular mountains in Slovenia in the winter is especially demanding and dangerous during unfavourable weather conditions. From this we conclude that most responders to the survey can be considered experienced hikers who are also familiar with the mountains (specifically Alps) and their infrastructure. In the next question, it became clear that only 10% of respondents have never encountered or used a bivouac before. These were asked to not proceed with further questions, which include their opinion on the function and use of bivouacs. Of the remaining responders, 95% were recreational visitors, 3% were bivouac caretakers, and 2% were mountain rescuers. The questions on the nature of bivouacs are presented in Table 23.2.

The results in Table 23.2 show that the majority of respondents at 82% find the use of bivouacs for the purpose of tourism inappropriate (Q3). 93% of

Table 23.2 Selected questions (Q1–Q4) on the nature of bivouacs in the eyes of the public

Q1	What was the reason for visiting the bivouac? *multiple-choice question			Q3	Do you find the use of bivouacs for purpose of tourism appropriate?		
	Answer options	*Frequency*	*Percentage (%)*		*Answers*	*Frequency*	*Percentage (%)*
	Emergency sheltering	72	15		Yes	69	18
	Rest/overnight stay	221	46		No	321	82
	Short visit	141	29		Total	390	
	Other	45	9				
Q2	What do alpine bivouacs represent to you personally? *multiple-choice question			Q4	Do you consider mountains bivouacs as a visual disturbance in the natural mountain environment?		
	Answer options	*Frequency*	*Percentage (%)*		*Answers*	*Frequency*	*Percentage (%)*
	Architectural/tourist attraction	67	12		Yes	28	7
	Emergency shelter	353	61		No	369	93
	Necessary infrastructure	154	27		Total	388	
	Unnecessary infrastructure	3	1				

all respondents said they do not consider mountain bivouacs as a visual disturbance in the natural mountain environment (Q4). This might suggest there is a certain level of adjustment already in place for them to expect these structures already in this environment. It might stem from a heritage outlook, where this culture of mountaineering accepts them as they were developed and dotted the mountains already more than 50 years ago. At that time, this activity was still in its pioneering stages, when the mass mountaineering tourism that we know today did not yet exist. Consequently, the people who visit the Alps regularly have always felt their presence. Another outlook on this question could be that the necessity of this infrastructure and its importance prevails and overcomes any doubt about them or their presence. The question on their true meaning (Q2) shows 88% of answers fall under the scope of their utilitarian presence. Therefore, any possible visual impact is overlooked. Of course, it could also just mean that these bivouacs are so small and sparsely positioned in the mountains that their visual impact is almost disregarded and the natural mountain landscape unchallenged.

The following question "Which characteristics of the bivouac location are important in your opinion?" (Fig. 23.8) creates a clearer picture on the public opinion of the bivouacs' impact on the mountains. Most people chose inconspicuousness as the highest value, thus highlighting the visual pollution of bivouacs as an open issue—still leading the question of visibility and accessibility in close second.

The concluding query to the survey was an open-ended question: "Do you have any suggestions on improving the Slovenian bivouacs?" The response surprised us as it returned detailed responses from 27% of respondents, which reflects how many of them carry strong opinions on the matter and the mountain environment in general. Most comments were sceptical towards using the bivouacs as a tourist attraction or even an accommodation platform and warned of their primary objective—sheltering in emergencies and offering an exceptional stay, making multi-day hiking trips possible. Many of them report on the clear misuse of the bivouacs, as some people choose to spend the night there, rather than in the paid rooms offered in the nearby mountain lodges. Hygiene is another problem, as the bivouacs are not even equipped to offer such a service, lacking sanitation equipment and running water. Their remote location presents problems for their uninterrupted maintenance along with the number of visitors they can accommodate, which is why most of them are locked during the less frequent winter months.

Which characteristics of the bivouac location are important in your opinion? (1- insignificant, 2- less important, 3 - moderately important, 4 - important, 5 - very important).

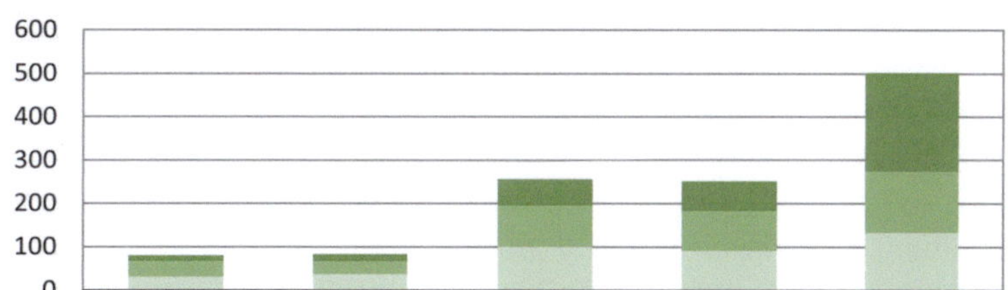

	1	2	3	4	5
Inconspicuousness / respecting the moutain landscape	13	17	62	70	229
Visibility	36	29	95	91	140
Accesibility	31	37	100	91	133

Fig. 23.8 The question on the characteristics of bivouacs regarding their location

The survey showed an interesting discrepancy between the public interpretation of the bivouac's visual impact on the mountainscape. The vast majority did not find the existing bivouacs a visual disturbance in the natural mountain environment (Q4 in Table 23.2). On the other hand, they expressed concern for the preservation of this landscape. This points to a dilemma on what the true limits on visual pollution are in improving bivouacs' role of ensuring safety in the mountains. The analysis clearly showed one of the bivouacs to be an architectural sculpture, while not carrying out its primary role as a safety tool, yet the public still accepts it as visually non-intrusive. In other words, at what point does a bivouac's visual presence become problematic? Perhaps all of the bivouac cases were somewhat successful in combining the properties of visibility and invisibility in the sense that they satisfied the practical and functional values of a bivouac. Their visual impact became second nature to the eyes of the mountaineering public because the pragmatic role was not critically overshadowed.

23.6 Conclusion

The popularization of mountain and highland tourism means that the number of visitors the environment receives increases each year, and the natural landscape cannot adapt by itself. This puts increasing pressure on the space of a formally protected, but still vulnerable natural environment. This is mainly due to the need to ensure the safety of visitors, for which it is necessary to continuously expand the network of bivouacs. As the bivouacs are an artificially introduced element into an unspoiled natural space, the question of proportionality arises between maintaining the visual integrity of the highlands and placing bivouacs in visible and easily accessible locations. In this study, we considered three cases of bivouacs to determine these relationships through the implementation of analyses and surveys. This included a set criterion of architectural and spatial attributes that define the appearance of the bivouacs and their visibility, which are innovatively woven within the analysis. The results showed a clear and problematic conflict between the function and purpose of bivouacs with the requirements for preserving the integrity of the natural highland environment. The result of the survey conveyed a similar conflicting message proving any compromise in this area challenging. New architecture outlooks for bivouacs in this space must adapt to the micro-specific properties of each new location and adjust accordingly. Therefore, simplistic guidelines to defining the criteria of visibility would not be of much use. Deciding what predominates, the logic of visibility or inconspicuousness, the necessities of each location should always be the deciding factor. This research lays the groundwork for further exploration, which should address this dilemma by establishing further criteria for where, how, in what form and in what numbers these bivouacs can be present in the highlands. Future research in this area should focus more on the properties of architectural concepts, different material choices, and how they affect visibility in the mountainscape. The inclusion of the public perception is necessary, as with the use of tools like our survey, as the question of visibility and its relation to visual pollution is always in part a subjective one.

References

1KA (2021). https://1ka.arnes.si/. Accessed 11 Aug 2021

Belimezov M (2007) Stolp ima 110 let [The tower is 110 years old]. https://www.gore-ljudje.si/kategorije/novosti/stolp-ima-110-let. Accessed 12 Aug 2021

Čelik M (2010) Bivak pod Grintovcem: pri gradnji v visokogorju je ključno razumevanje topografije [Bivouac under Grintovec: understanding topography is crucial when building in the highlands]. http://trajekt.org/2010/10/17/bivak-pod-grintovcem/. Accessed 12 Aug 2021

Čokl A (2015) Nov bivak pod Skuto [New bivouac under Skuto]. http://www.anzecokl.com/blog/2015/8/nov-bivak-pod-skuto. Accessed 9 Aug 2021

Eržen M (2017) Od nekoč do danes – izzivi obnov bivakov [From the past to the present – challenges of bivouac renovations]. Planinski Vestnik 117(4):7–11

E-surveying data (2021). https://egp.gu.gov.si/egp/?lang=en. Accessed 11 Aug 2021

Fickling D (2003) We knocked the bastard off. The Guardian, 13 Mar 2003. https://www.theguardian.com/world/2003/mar/13/everest.nepal. Accessed 18 Aug 2021

Grom JP, Mikša P, Fikfak A (2021) Pomen rapalske meje in vpliv na morfološki razvoj Idrije ter Žirov [The importance of the Rapallo border and the impact on the morphological development of Idria and Žiri]. Annales: Series historia et sociologia 30(1):117–134. https://doi.org/10.19233/ASHS.2021.08

Grom JP, Štukovnik P (2018) Sectoral division of Rapallo border defence systems and discovering the extent of the Rupnik line fortification system. Igra Ustvarjalnosti 6:30–40. https://doi.org/10.15292/IU-CG.2018.06.030-040

Hawthorne N (2001) Tales of the White Mountains. Durand Press, New Hampshire

Kajzelj M (1995) Alpinska arhitektura [Alpine architecture]. BSc Thesis, University of Ljubljana, Faculty of Architecture, Ljubljana

Kajzelj M (2017) Aktualni vidiki gradnje bivakov v Sloveniji [Current aspects of bivouac construction in Slovenia]. Planinski Vestnik 117(4):7–11

Kumer P, Zorn M, Žorž G, Gašperič P (2020) From the Great War to interwar fortifications: changing narratives attached to the military landscape in western Slovenia. In: O'Reilly G (ed) Places of memory and legacies in an age of insecurities and globalization. Springer, Cham, pp 247–263. https://doi.org/10.1007/978-3-030-60982-5_11

Lavtižar K (2020) Temelji naravnega prezračevanja stavb [Fundamentals of natural ventilation in buildings]. Igra ustvarjalnosti 8:20–27. https://doi.org/10.15292/IU-CG.2020.08.020-027

maPZS (2021) https://mapzs.pzs.si/home/trails. Accessed 11 Aug 2021

Mikša P (2011) Narodnostni boji v planinstvu na Slovenskem do 1. svetovne vojne [National conflicts among mountaineers in Slovenia before world war I]. Zgodovina za Vse 18(2):59–69

Mikša P (2015) Aljažev stolp in Jakob Aljaž – najvišji posestnik [Aljaž Tower and Jakob Aljaž – the highest landowner]. SLO: časi, kraji, ljudje 5(2):30–35

Mikša P (2017). Triglav in Jakob Aljaž [Mount Triglav and Jakob Aljaž]. Viharnik, Ljubljana

Ogrin M (2020). Turistični izzivi varnejšega obiskovanja slovenskih gora [Tourism challenges of safer visiting of Slovenian mountains]. https://www.slovenia.info/sl/novinarsko-sredisce/novice/11467-turisticni-izzivi-varnejsega-obiskovanja-slovenskih-gora. Accessed 11 Feb 2022

PZS (2021a) Seznam koč, zavetišč in bivakov: Bivak v Kočni (1952 m) [List of huts, shelters and bivouacs: Bivouac in Kočna]. https://www.pzs.si/koce.php?pid=107. Accessed 18 Aug 2021

PZS (2021b) Seznam koč, zavetišč in bivakov: Bivak Pavla Kemperla (2104 m) [List of huts, shelters and bivouacs: Pavel Kemperl's bivouac]. https://www.pzs.si/koce.php?pid=110. Accessed 18 Aug 2021

PZS (2021c) Seznam koč, zavetišč in bivakov: Bivak pod Skuto na Malih podih (2045 m) [List of huts, shelters and bivouacs: Bivouac under Skuta on Mali podih]. https://www.pzs.si/koce.php?pid=108. Accessed 18 Aug 2021

Ragelj T (2015) Bivaki nekoč in danes [Bivouacs then and now]. BSc Thesis, University of Ljubljana, Faculty of Sport, Ljubljana

Stegnar N (2021) Žalostna usoda bivaka na Kaninu: Kaj bo z njim, še ni znano [The sad fate of the bivouac on Kanin: What will happen to it is not yet known]. 24ur.com, 25 Dec 2021. https://www.24ur.com/novice/slovenija/bivak.html. Accessed 18 Jan 2022

Slovenian Alpine Museum (2014) Ali ste vedeli... [Did you know...]. https://www.planinskimuzej.si/en/mojstrana/what-makes-us-special/. Accessed 11 Aug 2021

Vidic M (ed) (2000) Ilustrirana zgodovina Slovencev [Illustrated history of Slovenes]. Mladinska knjiga, Ljubljana

The Chapel of Njegoš and the Mausoleum on Mt. Lovćen

24

Tatjana Koprivica

Abstract

The paper discusses the history of the construction of the chapel and mausoleum on Mt. Lovćen. Based on the concept by Peter II Petrović Njegoš, the chapel was built in 1845 and dedicated to his uncle Peter I Petrović Njegoš. The chapel was damaged during the 1915–1916 war operations. In the Kingdom of Serbs, Croats and Slovenes King Alexander Karađorđević had a new chapel built, inaugurated in 1925. The idea to build a mausoleum to Njegoš on Mt. Lovćen and to entrust its construction to Ivan Meštrović was first considered in 1924 and renewed in 1952. Meštrović accepted the invitation of the Montenegrin authorities in 1952 to design the mausoleum. Harold Bilinić, architect and Andrija Krstulović, sculptor, were engaged in this work. The construction of the mausoleum, which began in 1971, was completed in 1974. The paper also gives an overview of the reactions of the Yugoslav scientific and professional public to the building of the mausoleum on Mt. Lovćen.

Keywords

Mt. Lovćen · Chapel · Mausoleum · Ivan Meštrović · Petar II Petrović Njegoš

24.1 The Chapel of Njegoš on Mt. Lovćen from 1845

Mt. Lovćen, the "holy" Montenegrin Mountain, was a gathering place of fairies in folklore, as well as being celebrated by Kotor-based poets from the time of the Venetian Republic. Ljudevit Paskvalić (renaissance writer from Kotor, Montenegro) called it a pyramid in his epistles. Timotej Cizila (baroque priest and writer from Kotor, Montenegro) wrote in *Bove d'oro* that "*at certain times of the year a large number of witches and other malicious women from all over the world gather*" on Mt. Lovćen, but also that it holds "*the finest mountain crystal*" from which the reliquary of St. Tryphon was made (Čirgić 2018:183; Brajović 2019:42).

There are no known preserved archival documents of Montenegrin provenance about the origin of the chapel on Mt. Lovćen dating from the 1843 to 1846 period (Fig. 24.1). According to Austrian sources from the Zadar archives, Kotor District Chief Gabriel Ivančić submitted a report on May 18, 1845 to the Governor's Presidency in Zadar that he had been informed by his commissioners that Petar Petrović Njegoš (1813–1851, ruled 1830–1851) would build a church on Mt.

T. Koprivica (✉)
Historical Institute, University of Montenegro, Podgorica, Montenegro
e-mail: tkoprivica@ucg.ac.me

M. Zorn et al. (eds.), *Discourses on Mountains of Montenegro and Slovenia*, Historical Geography and Geosciences, https://doi.org/10.1007/978-3-031-99017-5_24

Fig. 24.1 Upper figure: Njegoš's chapel on Mt. Lovćen (a postcard from 1905 with a reproduction of a painting by Mihailo Milo Vrbica); bottom figure: Chapel on Mt. Lovćen, watercolor by Nikola Krasnov from 1923. (Authors archive)

Lovćen, dedicating it to St. Peter of Cetinje or St. Luke of Ljubotin (Milović 1984:527).

The construction of the chapel began on June 16, 1845 and it is believed that it was completed in the fall of 1845 (Milović 1984:528; Šistek 2012:109). Njegoš dedicated the chapel to his uncle and predecessor, Peter I Petrović Njegoš (1747–1830, ruled 1784–1830), whom he canonized in 1834. The chapel was also the first shrine dedicated to St. Peter of Cetinje, while the dedication of the chapel to Peter I represented a step further in strengthening the cult of the Petrović-Njegoš Dynasty.

The chapel, located in a prominent position, on Jezerski vrh (1,660 m) was meant to provide a visual symbol of Montenegrin freedom and independence. A simple building with a round structure measuring 5 × 3.5 m, built of carved stone, it was covered with stone slabs. The chapel pointed in the northeast–southwest direction and its entrance was from the southern, milder mountain slope (Radičević 1911:59; Mijović 1997:138–139). A stone altar was placed inside the chapel that Njegoš had built using state funds (Kilibarda 2004:84).

There are diverse observations in literature regarding the origins of Njegoš's project for the chapel, from local architectural solutions to role models in Italian art (Mijović 1997:117–120), which Njegoš knew well from literature and travels in Italy (Kilibarda 2014; Brajović 2015).

However, there are no preserved archival documents that speak of Njegoš's chapel being built for his tomb. Njegoš did not specify the place of burial in his will. Ljubomir Nenadović (Serbian writer and diplomat) attests about his intention to be buried on Mt. Lovćen (Nenadović 1889:163). The wish to be buried on Mt. Lovćen was allegedly recorded on October 30, 1851 in a letter sent by Vuk Popović (priest and writer from Risan, Montenegro) to Vuk Karadžić (philologist, anthropologist, and linguist) on November 22, 1851, stating that Njegoš had announced "*lay me to rest near the church*," which suggests that he did not build a chapel to contain his tomb (Šistek 2012:111).

Due to unfavorable weather conditions and the fear that Njegoš's Tomb may be desecrated, he was buried in the Cetinje Monastery on November 2, 1851. On September 8, 1855, in the presence of some 2,000 people, among which were senators and priests led by Prince Danilo and Princess Darinka Petrović-Njegoš, Njegoš's remains were entombed in the chapel on Mt. Lovćen (Klančić 1962:296–302; Batrićević 2018:90–96). Thus, the cult of Njegoš, the Petrović-Njegoš Dynasty, and Mt. Lovćen was reinforced, "*to which the Montenegrins directed their gaze like the Greeks towards Olympus*" (Pavićević 2007:310; Brajović 2019:42).

Many travel writers, visitors, and painters visited the chapel during the reign of Nikola I Petrović Njegoš. An important source for its study is the paintings of the chapel by such Montenegrin painters as Pero Poček and Mihailo Milo Vrbica (Fig. 24.1). As Jezerski vrh on Mt. Lovćen was within the area of Njeguši, Nikola Petrović Njegoš paid compensation to the inhabitants of Njeguši who performed minor upkeep works and secured the chapel (Kilibarda 2004:84). In 1879, the Montenegrin government hired craftsmen to repair the chapel and its lightning rod, as well as to maintain the path leading to the chapel (Radičević 1911:59–60).

24.2 Austro-Hungarian Plan to build a Monument on Mt. Lovćen

During World War I, Montenegrin positions on Mt. Lovćen represented the last line of defense for the capital of Cetinje and Montenegro itself. Although the Lovćen heights were an impressive natural fortress, the outcome of Operation Lovćen was determined by the vast superiority of the Austro-Hungarian forces both in manpower and in weaponry. In December 1915 and January 1916, after several days of shelling from 431 cannon from Austro-Hungarian ships and fortifications in the Bay of Kotor, Austro-Hungarian soldiers seized Mt. Lovćen and soon after the whole of Montenegro on January 11, 1916 (Andrijašević and Rastoder 2006:288).

During these military operations, Njegoš's chapel on Mt. Lovćen was damaged by at least one shell though there is no precise data on the nature of the damages. The preserved photographs show damage of about 1 m in diameter in the upper part of the chapel, between the door and the window (Mijović 1997:153–154). The Austro-Hungarian occupation authorities organized various construction works in Montenegro, including fortification works on Mt. Lovćen. However, there was no effort put in to repair the existing damage; on the contrary, there was an intention to erect the new monument on the highest peak of Lovćen-Štirovnik (1,749 m).

The governor of the occupying forces in Montenegro, General Victor Weber Edler von Webenau (1861–1932), at the beginning of July 1916 demanded from the Montenegrin Metropolitan Mitrofan Ban that Njegoš's remains be exhumed and transferred to Cetinje Monastery. Despite the protest of the citizens of Cetinje and the Metropolitan, Weber's order was carried out on the night of August 12/13, 1916. A commission made up of former Montenegrin Minister Ljubomir Bakić and intendant Mišo Popović carried out the order in the presence of two Austro-Hungarian officers. The commission stated that it "*found the church and the tomb in perfect order*" (Mijović 1997:154; Šištek 2012:112–113). The translation of Njegoš's remains to Blatište was done first manually and then by vehicle, so that the remains could finally be interred in Cetinje Monastery (Pejović 1963:500).

In July 1916, the Austro-Hungarian authorities announced a competition for the erection of a monument on Mt. Lovćen (Štirovnik) "*in honor of the occupation of this Adriatic Gibraltar, and for the descendants to value.*" Architects, engineers, technicians, and sculptors serving in the military forces from Njeguši to Dubrovnik had the right to participate in the competition (Rakočević 1960:640–641). The contest received 100 submissions from 17 applicants. General Weber opened an exhibition of the works that had arrived for the contest August 18, 1916, on the Emperor's Day, in the building of "Croatian Home" in Herceg-Novi (Rotković 2013:90; Žaja Vrbica 2017:285–286).

The design of the monument by Croatian artist Marko Rašica (1883–1963) was the chosen work. The plaster model of the monument, then exhibited, has not been preserved. The monument was to represent a genius figure of victory and glory. The planned height of the monument was to be 16 m to 32 m. The figure was depicted in a tunic with a helmet, arms crossed and resting on a sword, with a laurel wreath wrapped around its head, across the body and around the sword. The initials of Emperor Francis Joseph I were to be carved on the hilt of the sword and January

11, 1916, as the date of the fall of Mt. Lovćen was on the point. There was a three-sided tower at the back of the figure of victory. The monument was to be made of white stone, while the laurel wreath and the inscription were to be of red stone, "*so that every flag on the Adriatic could spot them and pay homage to the monogram as much as to the felicitous historical date.*" Due to the high altitude and the nature of the terrain, the monument was planned to be of "*a rugged construction.*" The figure of the genius of victory was conceived as a "*guard of the Adriatic and the eastern Adriatic coast*," and was to be turned in the direction of Italy (Vihtić 1916:1; Žaja Vrbica 2017:286). This depicted an emblematic animosity toward the former ally and current enemy, the Kingdom of Italy, which claimed Austro-Hungarian territories and dominance in the Adriatic. The Austro-Hungarian press reported on the results of the competition for the monument on Mt. Lovćen (Žaja Vrbica 2017:289). Due to the circumstances of war, the intended monument was never realized. In Vienna decided to repair the chapel on Jezerski vrh at the end of August 1918 (Anonymous 1918:4). The realization of this intention should have obviously had the effect of reducing the animosity of the Montenegrin population toward the occupying authorities. However, the intention of the occupying authorities to carry out restoration works on the chapel on Mt. Lovćen was not realized due to the Austro-Hungarian collapse and capitulation before the Entente countries that followed on November 3, 1918.

24.3 King Alexander Karađorđević's Endowment: New Chapel on Lovćen in 1925

After the damage caused by Austro-Hungarian shelling and years of exposure to the natural processes, Njegoš's chapel has moderate damage. At the end of October 1921, Đuro Pejović, the parish priest in Njeguši, sent a report on its state to Metropolitan Gavrilo Dožić. In the same year, the Construction Department from Cetinje led by engineer Milan Karlovac formed a project for the construction of a new chapel on Mt. Lovćen as well as for the road from Cetinje to Mt. Lovćen. These initiatives, however, were not implemented (Durković Jakšić 1971:154–157).

In 1921, architects employed by the Ministry of Construction of the Kingdom of Serbs, Croats, and Slovenes also made 25 draughts for the Lovćen chapel, and four of them by architect Sveta Lazić are renowned. In 1922, the Ministry of Religion selected the draught made by architect Viktor Lukomski, although this project failed to be realized as well (Durković Jakšić 1971:167–173).

In 1923, King Alexander Karađorđević formed a special committee for the chapel's restoration and the translations of Njegoš's remains that he financed personally. Architect Nikola Krasnov (1864–1939) was sent to Montenegro to chronicle the state of the chapel on September 24, 1923 (Acović et al. 2013:37–41) (Fig. 24.1). During 1924, the reparation of the old chapel and the construction of a new one were taken into consideration in Belgrade. A member of the Board, Svetozar Tomić, reported on the situation on Mt. Lovćen in 1924 (Durković Jakšić 1971:195, 211).

In the same year, King Alexander commissioned Ivan Meštrović (1883–1962) to build a monument on Mt. Lovćen (Kečkemet 2009:18). In the fall of 1924, the famous sculptor exhibited blueprints for Njegoš's monument at the Brooklyn Museum in New York. They were in turn published in the *Nova Evropa* journal in January 1925 (Ćurčin 1925:21) (Fig. 24.2). Meštrović's project was a square-shaped building surmounted by a dome with a pyramidal roof. The staircase at the entrance to the building was flanked with two sphinxes on each side. The figure of Njegoš in a semi-recumbent position with an eagle above his head was to be placed in the interior of the tomb. Meštrović's project was not formally accepted due to the exorbitant costs. In a letter from 1952, referring to the reasons for the refusal of the project, Meštrović quoted the opposition of church dignitaries as it was objectionable for Alexander Karađorđević's endowment to be designed by a Croat and a Catholic. Meštrović's model for Njegoš's sculpture served as a template for the monument the construction of

Fig. 24.2 Meštrović's plan of the Mausoleum on Mt. Lovćen from 1924. (Museums of Ivan Meštrović archive, Zagreb)

which was initiated in Belgrade in 1929 but which also failed to be realized (Malbaša 2013:284–285). Although King Alexander disapproved of Meštrović's project, it was Meštrović himself who created the equestrian statue of king in Cetinje, which was erected in 1940 but removed in 1941 (Koprivica 2009:193–196).

In 1925, King Alexander decided to restructure the chapel based on the project submitted by architect Krasnov from 1923. Architect Pero Popović was sent to Montenegro to inspect the state of the Mt. Lovćen chapel, and in May of the same year Krasnov submitted an Estimate for the Reconstruction of the Chapel on the Tomb of Bishop Peter on Mt. Lovćen. In the Zeta militarized zone, a Commission for the Construction of a Chapel on Mt. Lovćen was formed headed by officer Pantelija Đukić. The overseer of the construction works, which began on July 20, 1925, was engineer Veliša Popović (Durković Jakšić 1971:228–229).

Although the name of the renovation work was titled "reconstruction of the chapel," it involved the construction of a completely new edifice (Fig. 24.3). The old chapel was removed and the site was leveled. The foundations were

Fig. 24.3 The chapel from 1925 (left) and its interior (right). (Authors archive)

strengthened, and a massive stone fence made from larger blocks was built around the chapel. Only some of the material from the old chapel was used in the construction of the new sanctuary. Some 480 stones were from non-Lovćen quarries and 280 white stones were from the island Brač and were used in the building of the chapel floor. Compared to the design of the 1845 chapel, the entrance portal was more monumental, while some Byzantine arches were added to the windows and the chapel was completed with an attic cornice in the style of pseudo-classical Russian architecture that was counter to the tradition of Montenegrin architecture. Also, a cross crafted of Venčac marble was placed on the roof of the chapel (Mijović 1997:162–165; Komatina and Alihodžić Jašarović 2018:325).

The plastered interior was considerably unalike the starkness of the previous 1845 chapel which had a solitary stone altar. Serbian painter Uroš Predić had painted four icons of Montenegrin saints on marble slabs: St. Vladimir, St. Basil of Ostrog, St. Peter of Cetinje, and St. Stephen of Piperi. Predić stated that the site had been in total disarray, that none of the work had been completed, and that the icons could not be fixed into the frames prepared for them even though he had the exact measurements of them. Moreover, Predić painted the Christ Pantocrator in situ in rather dire circumstances. Also, he made a color mosaic with a depiction of Njegoš located above the Venčac marble sarcophagus, located on the right side of the entrance to the church (Fig. 24.3). Verses from Njegoš's *Gorski vijenac* (The Mountain Wreath) were inscribed on the first row of the vault on the plastered wall. To the left of the apse was a marble plaque with Njegoš's depiction donated by Njeguši from the Njeguški Committee for the Reconstruction of the Chapel which was dissolved under pressure from the authorities (Kilibarda 2004:81–83). The initials of King Alexander were engraved on the chandelier that was based on the one from the Kalenić monastery. A marble slab with the depiction of St. George, the work of sculptor Živojin Lukić, failed to fit into its intended niche above the entrance portal and thus it was shortened by 10 cm—with the horse's legs being cut off below the knee, while the dragon's upper parts and tail remained (Kilibarda 2004:82).

The new chapel was completed in less than 2 months and it was consecrated on September 10, 1925. Njegoš's remains were laid in the chapel on September 21, 1925 in the presence of King Alexander and Queen Natalie and other political and religious dignitaries. Furthermore, the ceremony had the status of a superior state event (Vujović 1994:167).

The fact that it was a new endowment by King Alexander and not merely the renovation of Njegoš's chapel is attested by the founder's inscription in Venčac marble above the entrance to the chapel (Vuksan 1926:16–17): "*In the name of the Father and of the Son and of the Holy Spirit, Amen. We, Alexander I, the King of the Serbs, Croats and Slovenes, the leader and initiator of the liberation of Serbia, grandson of Prince Alexander Karađorđević, and son of King Peter I, the great liberator and unifier of our people and Princess Zorka Petrović-Njegoš, have renovated this holy temple on Mt. Lovćen near Cetinje, built and chosen as the eternal abode by our ancestor Petar Petrović-Njegoš, the Bishop and Master of Montenegro, a worshiper of the hero of Topola Karađorđe, an immortal apostle and herald of the unity of our people, and which was destroyed in 1916 in the war for liberation and unification. May peace be upon the ashes of Peter II Petrović Njegoš and may glory be to his great name. May his holy blessing be served on our home and our people through all the ages. Amen. On Mt. Lovćen on September 21, 1925*" (Vuksan 1926:16–17).

King Alexander had the chapel built to honor the Karađorđević Dynasty (Folić 2011:445). Unlike Njegoš, who consecrated the chapel to St. Peter of Cetinje in 1845, the King dedicated the sanctuary to St. George. King Alexander announced the following to Metropolitan Dožić during the ceremony: "*I bequeath to you my endowment and the bones of the famous Njegoš, so that you can continue to preserve them*" (Vuksan 1926:30).

Some of the population considered the building of a new chapel on Mt. Lovćen as an expression of respect for the new state, the Kingdom of Serbs, Croats, and Slovenes and its ruler, for Montenegro and the Montenegrin traditions. For the others, the chapel was a symbol of the occupation and loss of the Montenegrin State in 1918, highlighting the domination of the Karađorđević Dynasty over the Petrović family and the destruction of the independence of the Montenegrin Church in 1920.

The 1925 chapel was damaged in April 1942 when it was shelled by fascist soldiers from a position above the Cetinje monastery. The people of Cetinje protested against the assault to the Italian fascist administration and Governor A. Pirzio Biroli. The Italians apologized, stating that it had not been their intention to bomb the chapel and assuring that the damage would be repaired (Anonymous 1942:1). The depiction of *Christ Pantocrator* in the apse was damaged (Kilibarda 2004:85). These damages were repaired in 1945 under the supervision of engineer Popović (Medaković et al. 1971:145).

24.4 Meštrović's Mausoleum on Mt. Lovćen

The idea of building Njegoš's monument on Mt. Lovćen dates from 1951, when the 100th anniversary of Njegoš's death was marked in Cetinje. The State Board for the celebration of this jubilee put forward to the Government of Montenegro the suggestion to build a notable monument to Njegoš in Cetinje. Furthermore, of all the solutions (from the artists Vanja Radauš, Lojze Dolinar, Vojin Bakić), received for the competition, none were accepted (Anonymous 1951:358, 447, 489, 531; Malbaša 2013:291–295; Jović and Kovačević 2013:203–204).

In April 1952, the Yugoslav ambassador to the USA, Vladimir Popović, contacted sculptor Ivan Meštrović at the initiative of the Montenegrin government. Meštrović received the invitation to build a monument to Njegoš with appreciation and delight. The Prime Minister of Montenegro, Blažo Jovanović, wrote to Meštrović in his letter dated May 21, 1952 the following: "*We imagine the monument to Njegoš to be in the form of a magnificent mausoleum … The mausoleum would be built atop Mt. Lovćen, where the small chapel is now located. We believe that such a monument, a great piece of art in honor of Njegoš, would be best fitted with verses by Njegoš and folk epics and not ecclesiastic motifs.*" Jovanović also stated that the monument should be based in the spirit of Njegoš's inspirations, justice, freedom and human dignity. What's more, Meštrović accepted to build the mausoleum without any reimbursement. He was not inspired, as he stated, by Bishop Rade, but "*by the poet Njegoš, who with his intellect and poetic gift brought together and creatively formed the 'wreath' of philosophy and spirituality of the people of Montenegro*" (Kilibarda 2004:8). Mate Meštrović acknowledged that his father had accepted the complex task for the following reason: "*To raise the awareness of the Montenegrin people and the significance of their identity,*" adding that he believed "*that those who oppose the mausoleum are against the independence of Montenegro!*" (Hudelista 2018).

After arranging the work specifications, the construction was to be completed in 1963 for the 150th anniversary of the birth of Njegoš. Meštrović worked on the mausoleum project for 8 months in a studio in Syracuse with the help of architect Harold Bilinić, who had presented the mausoleum plans in Cetinje in May 1953.

Meštrović's plaster scale model of Njegoš's sculpture was sent from the USA to Split in the same year. The monument was to be made of Jablanica marble, according to Meštrović's specification, and Njegoš's sculpture would be 3.74 m in height, over 3,10 m in length, 1,43 m in width and weigh 28 tons and carved by sculptor Andrija Krstulović (1912–1997) (Kilibarda 2004:9) (Fig. 26.4). The sculpture was transported from Split on a crane towed by the Proleter tugboat, and from Budva to Cetinje by a Faun truck. In Cetinje, the sculpture was placed in the courtyard of the State Museum (Fig. 24.4).

Fig. 24.4 Upper left: Andrija Krstulović in the studio in Split, carving a figure of Njegoš according to Meštrović's plaster model (*Source* Bezić Filipović 2015); upper right: Meštrović's sculpture of Njegoš in Cetinje (Museum of Yugoslavia archive, Belgrade); bottom: Josip Broz Tito with the Montenegrin leadership, touring Meštrović's sculpture of Njegoš in Cetinje (1959). (Museum of Yugoslavia archive, Belgrade)

Two caryatids (Fig. 24.5) made of Jablanica marble, more than 4,5 m high and weighing 8 tons each carved by Andrija Krstulović and his associates in Split according to Meštrović's specifications, arrived in Cetinje in June 1959. Disputing the importance of the project by Meštrović, the Serbian Orthodox Church (hereinafter: SOC) claimed in 1970 without justification that female figures were not seemly in a mausoleum, citing examples from Njegoš's verses referring to women ("*men suffer and women weep*," "*a woman will change a hundred faiths to do what she desires*," "*women's nature is a peculiar matter*") (Medaković et al. 1971:159). Meštrović appreciated Krstulović's work, believing that it was far more than mere reproductive art.

In 1956, the Horvat-Bilinić architectural office in Zagreb offered a solution and in 1958 Bilinić completed the construction plans for the mausoleum. Additionally, in 1960 he proposed plans related to the stonework. However, the preparation of the construction site was delayed due to problems with the access route. Meštrović died in January 1962 and in May 1962, according to a decision by the Executive Council of Montenegro, the construction of the mausoleum was suspended indefinitely due to a lack of resources (Kilibarda 2004:9).

From 1968, with the emergence of a new political and republic leadership led by the president of the Central Committee of the League of Communists of Montenegro Veselin Đuranović, more compelling work began on affirming of Montenegro's national policy. The issue of continuing with the construction was wholly political for Montenegro's government (Papović 2017:216). In December 1968, an initiative was launched to continue with the construction of the mausoleum, with the municipality of Cetinje in charge of continuing the works.

Although SOC, i.e., the Metropolitanate of Montenegro and the Littoral was against the construction of the mausoleum on Mt. Lovćen, it lacked sufficient political leverage to resist the works as Metropolitan Arsenije

Fig. 24.5 Andrija Krstulović in the studio in Split, carving out caryatids. (*Source* Bezić Filipović 2016)

Bradvarević had been incarcerated since 1954 for 11 years and 6 months due to political activity against the socialist system (Folić 2007:268). After the enthroning of Metropolitan Danilo Dajković (1895–1993) in 1961, SOC strongly protested against the construction of the mausoleum. Dajković also proposed compromise solutions—to either incorporate the chapel into the mausoleum or to erect the mausoleum on the plateau above Cetinje Monastery (Papović 2017:220). There were demonstrations in Cetinje on March 18, 1969 as a backlash to the stance taken by SOC that Njegoš's mausoleum had no place on Mt. Lovćen (Pušonjić 1969:6). Responding to the accusations of SOC, the municipality of Cetinje pointed out "*that the construction of the mausoleum was an expression of the wishes of the working people of Yugoslavia*," accusing the Church of being silent in 1925 when King Alexander had demolished Njegoš's chapel in order to build a new one (Batrićević 2018:243).

In April 1969, the Association of Orthodox Priests of Montenegro unanimously supported the construction of the mausoleum. The President of the Federal Commission for Religious Affairs, Milo Jovićević and Serbian Patriarch German held talks in Belgrade on May 5, 1969 regarding the stand of SOC on the issues of the Mt. Lovćen mausoleum. Jovićević warned Patriarch German that an assiduous stand of SOC against the construction of the mausoleum could lead to the renewal of the autocephalous Montenegrin church (Papović 2017:227–228).

The municipality of Cetinje responded on April 15 to the letter written by Lazar Trifunović (1929–1984, Serbian art historian, Professor at the Faculty of Philosophy, University of Belgrade) dated April 6, 1969 against the construction of the mausoleum and Meštrović's solution of Njegoš's sculpture. Trifunović was the head of the Commission for the Reorganization of the Museum in Cetinje which in February 1966 submitted an opinion to the Republic Secretariat for Education and Culture of Socialist Republic Montenegro against the construction of the mausoleum in Lovćen. Trifunović disputed Meštrović's aesthetic solutions, namely, the caryatids as a "German architectural element," also stating for the statue of an eagle above the figure of Njegoš in a sitting position that it was made in Bismarck's style. He considered the mausoleum to be secessionist (Medaković et al. 1971:59–60, 94–96). The municipality stated that the approach taken by Trifunović was discriminatory: of the whole Meštrović opus, especially important for Belgrade, the only thing that he believed to be wrong was his mausoleum on Mt. Lovćen and Njegoš's sculpture. The response also stated that there was something else in question, speculating that the monument had no right to be on Lovćen due to ideological-aesthetic and theoretical-artistic beliefs, how then could it be situated in Cetinje (Medaković et al. 1971:94–99) This ensued in both supporters and opponents of the construction of the mausoleum debating heatedly in the press during the following period.

In July 1969, a Committee for the construction of the mausoleum was formed, whose members were Veljko Milatović, Veselin Đuranović, Svetozar Vukmanović Tempo, Peko Dapčević, Jovo Kapičić and others, as well as Yugoslav intellectuals and artists Dušan Vukotić, Veljko Bulajić, Oskar Davičo, Branko Ćopić, Branislav Đurđev, Vladimir Dedijer, Ćamil Sijarić, Esad Mekuli, and others (Medaković et al. 1971:107–108). In July 1969, a group of artists and cultural figures sent an open letter against the construction of the Mt. Lovćen mausoleum. Among the opponents of the construction were Lazar Trifunović, Mihailo Lalić, Petar Lubarda, Risto Stijović, Gligor Stanojević, Miroslav Krleža, and others. During the following period, they led a strong campaign against Meštrović's project, also engaging cultural workers from abroad in the movement (Medaković et al. 1971). An explosive device and a threatening letter addressed to the Yugoslav Consulate General were found at a station in Hamburg, warning Yugoslav consulates that President Tito should "*order the chapel on Mt. Lovćen to be demolished within a month*" (Anonymous 1969:4).

In April 1970, the Metropolitanate of Montenegro and the Littoral submitted an appeal to the Constitutional Court of Yugoslavia in Belgrade against the Socialist Republic of Montenegro, considering the chapel its property. From 1925 to 1946, the chapel was the property of Cetinje municipality. From 1946 to 1956, it was owned by the Assembly of the Cetinje Municipality, and from 1956 it was under the protection of the Institute for the Protection of Cultural Monuments of Montenegro (Selhanović 2013:199). The Constitutional Court, presided over by Blažo Jovanović, decided in July 1970 that the court was not competent to judge the dispute, which was continued in Montenegro. In November 1970, the Constitutional Court of Montenegro rejected the appeal of the Metropolitanate.

Vladimir Dedijer (Dedijer 1970:6) noted that "*a part of the SOC hierarchy cannot abide to have a Croat weave a wreath on Mt. Lovćen to a Montenegrin, and for a Roman Catholic to build a monument to an Orthodox bishop.*" Also during the same year, SOC continued to campaign against the construction of the mausoleum. The Montenegrin leadership held talks with the leadership of Macedonia, whereupon there was an intense dispute with SOC over the status of the Macedonian Orthodox Church, as well as with the leaders of Serbia and Croatia, who provided the political and financial support for the construction of the mausoleum (Selhanović 2013:192, 201–219). A member of the Central Committee of the League of Communists, Predrag Ajtić, stated the following at a meeting held on June 26, 1970 in Titograd: "*According to our knowledge, all the Great Serbian, nationalistic forces to a great extent use the construction of Njegoš's mausoleum as a reason to make political statements, for making actions of course from unitarian positions, and lately they are starting to open a wide front. They are joined not only by aesthetic forces, but it also seems to us that there are informers and anarcho-liberal elements incarnated in that group at the Faculty of Philosophy. Personal, artistic motives, tastes,* etc., *in that entire gathering become secondary, and for us that issue is, above all, for us a political one*" (Selhanović 2013:202).

In the fall of 1971, the 1925 chapel was dismantled and relocated to the base of Mt. Lovćen. The Decision on the relocation of the chapel from Lovćenski vrh, made on September 3, 1969, states that the fittings will be temporarily laid out in Ivanova korita, until the decision on where the chapel will be erected is made (Medaković et al. 1971:199). The project for the construction of the chapel was made in 1987 but it was not realized (Žunjić 2013:212–213).

The construction of the mausoleum lasted from 1971 to 1974 (RTCG 1972). Supporters of the construction from the entire territory of Yugoslavia donated funds for its construction—some 1,028 companies and 2,678 individuals (Kilibarda 2004:159–186).

Complex works on the hard-to-reach terrain of Jezerski vrh began in the fall of 1971 (Fig. 24.6). Due to problems with his sight, Bilinić sought the help of Srđan Čulić, and Zadar engineers Niko Jakovčev, Slavko Šimatić, and Berislav Petravić also took part in the realization of the project (Milkota 2013). The Zadar-based *Donat* Company was engaged, and the *Engineering Design Institute* from Zagreb was appointed to build the access tunnel. The *Jadrankomerc*

Fig. 24.6 Mausoleum construction works. (*Source* Pobjeda, July 28, 1974)

Company from Split carried out the stone-cutting works (Piplović 1994:1, 5, 14–19). The *Titograd* construction company from Titograd and *Jugošped* from Belgrade also participated in the undertaking of the construction.

The mausoleum, being 37 m long, 15 m wide, and 10 m high, followed the outline of Lovćen Peak (Fig. 24.7). The main façade with the main entrance faced south. The interior of the mausoleum was simple, consisting of an atrium with a well at its heart, porticos on the left and right, and two monumental caryatids. The mausoleum is covered by an elliptical vaulted ceiling. Njegoš's earthly remains are interred in the crypt, with six side niches and one central alcove. In point of fact, the design of the mausoleum is a continuance of Meštrović's 1924 project (Kečkemet 2009:563–577; Komatina and Alihodžić Jašarović 2018:326–328). Facing the front door is a sculpture of Njegoš. The surfaces of the walls are covered with light gray marble slabs of Jablanica granite, while the granite floor is darker in color.

Fig. 24.7 The Mausoleum on Mt. Lovćen. (Photo by Lazar Pejović, authors archive)

Fig. 24.8 The opening of the mausoleum on Mt. Lovćen on July 28, 1974. (*Source* Pobjeda, August 1, 1974)

A road reaches the location, some 1,600 m above sea level, and the remaining 60 m of altitude is covered by a 135-m-long tunnel, in which there are 461 steps.

The mausoleum was inaugurated on July 28, 1974 (Fig. 24.8; RTCG 1974). In his speech, Veljko Milatović, President of the Presidency of Montenegro, emphasized that "*Njegoš's mausoleum is today becoming an inseparable part of Lovćen Mountain*" and pointed out that the mausoleum was a symbol of collective Yugoslav solidarity, and that the largest monument to Njegoš is "*a free Montenegro in a free community of equal Yugoslav peoples and nationalities*" (Batrićević 2018:247).

The complex on Mt. Lovćen was completed in 1978 by the construction of a restaurant—an Observation Deck designed by architect Marko Mušič. In the 1983 the interior of the restaurant was finished (Košir 2010:35–51; Komatina and Alihodžić Jašarović 2018:329; Stamatović-Vučković 2018:328).

24.5 Conclusion

The peak of the mountain Lovćen, Jezerski vrh, a symbol of Montenegrin independence, was marked by the construction of Njegoš's Chapel (1845), Alexander Karađorđević's Chapel

(1925) and Njegoš's Mausoleum (1974) by Ivan Meštrović. Construction works included the relocation and the re-translation of Njegoš's remains to Mt. Lovćen. The architectural solutions all incorporated elements of the times in which they were created, as well as the ideas of their founders and financiers. Furthermore, the artistic and aesthetic elements in the projects are intertwined with the political and the ideological.

References

Anonymous (1918) Cetinjske novine 212, 25 Oct 1918

Anonymous (1942) Glas Crnogorca 12, 14 Apr 1942:1

Anonymous (1951) Stvaranje 7–8

Anonymous (1969) Paket s eksplozivom i preteće pismo konzulu SFRJ u Hamburgu [A package of explosives and a threatening letter to the SFRY consul in Hamburg]. Politika, 21 Jul 1969:4

Ačović D, Gordić G, Kamilić V, Crvenković B (2013) Njegoševa kapela na Lovćenu i arhitekta Nikola Krasnov [Njegoš's chapel on Mt. Lovćen and architect Nikola Krasnov]. Etnografski muzej u Beogradu, Beograd

Andrijašević Ž, Rastoder Š (2006) Istorija Crne Gore od najstarijih vremena do 2003 [History of Montenegro from ancient times to 2003]. Centar za iseljenike Crne Gore, Podgorica

Batrićević B (2018) Bog našeg nacionalizma: recepcija, reinterpretiranje i korišćenje Njegoševa lika i djela u političkoj propagandi i diskursu vlasti u Crnoj Gori od 1851. do 2013. godine [The God of our nationalism: the reception, reinterpretation and the use of the character and works of Njegoš in the political propaganda and discourse of the authorities in Montenegro from 1851 to 2013]. Fakultet za crnogorski jezik i književnost, Nova Pobjeda, Cetinje, Podgorica

Bezić Filipović B (2015) Andrija Krstulović, kipar. https://youtu.be/btOQ5PcEJTo. Accessed 8 Aug 2021

Bezić Filipović B (2016) Omaž Andriji Krstulović [Homage to Andrija Krstulović]. Ministarstvo kulture Crne Gore, Cetinje

Brajović S (2015) Njegoševo veliko putovanje: meditacije o vizuelnoj kulturi Italije [Njegoš's Grand Tour: meditations on the visual culture of Italy]. CANU, Podgorica

Brajović S (2019) Portreti knjaza Danila I Petrovića Njegoša [The portraits of Duke Danilo I Petrović Njegoš]. Matica crnogorska, Podgorica

Ćurčin M (1925) Njegoševa grobnica na Lovćenu [Njegoš's tomb at Mt. Lovćen]. Nova Evropa 11(1):21

Čirgić A (2018) Natprirodna bića u tradicijskoj kulturi Crnogoraca [Supernatural beings in the traditional culture of the Montenegrins]. Fakultet za crnogorski jezik i književnost, Cetinje

Dedijer V (1970) Pismo Vladimira Dedijera [The letter of Vladimir Dedijer]. Borba 227, 20 Jul 1970:6.

Durković Jakšić Lj (1971) Njegoš i Lovćen [Njegoš and Mt. Lovćen]. Pravoslavlje, Beograd

Folić Z (2007) Država i vjerske zajednice u Crnoj Gori 1945–1965 [The state and religious communities in Montenegro 1945–1965]. Istorijski institut, Biblioteka "Đurđe Crnojević", Podgorica

Folić Z (2011) Vlada Kraljevine SHS i podizanje kapele na Lovćenu [Governement of the Kingdom of SCS and the chapel on Lovćen in 1925]. In: Zbornik radova Njegoševi dani III. Filozofski fakultet, Nikšić, pp 433–447

Hudelista D (2018) Bitka za Lovćen [The Battle for Lovćen]. Nedjeljnik, 15 May 2018

Jović T, Kovačević I (2013) Odbor za proslavu Njegoševe stogodišnjice [Committee for the Njegoš's centennial]. Glasnik Narodnog Muzeja Crne Gore 9:129–238

Kečkemet D (2009) Život Ivana Meštrovića (1883–1962–2002) II [The life of Ivan Meštrović (1883–1962–2002) II]. Školska knjiga, Zagreb

Kilibarda B (ed) (2004) Lovćen, Njegoš, Meštrović: projekt Njegoševa mauzoleja na Lovćenu i njegova realizacija (1952–1974.) [Mt. Lovćen, Njegoš, Meštrović: the project of the mausoleum of Njegoš on Mt. Lovćen and its realization (1952–1974)]. Nacionalna zajednica Crnogoraca Hrvatske, Matica crnogorska, Zagreb, Cetinje

Kilibarda V (2014) Njegoš i Italija [Njegoš and Italy]. Institut za crnogorski jezik i književnost, Podgorica

Klančić Lj (1962) Drugi Njegošev pogreb [The second funeral of Njegoš]. Istorijski Zapisi 15(2):291–302

Komatina D, Alihodžić Jašarović E (2018) Projekti Ivana Meštrovića i Harolda Bilinića na Lovćenu u tadašnjem političkom i ideološkom konceptu [Projects by Ivan Meštrović and Harold Bilinić on Lovćen in the political and ideological context of the period]. Prostor 26(2):320–331

Koprivica T (2009) Spomenici i pobjednici [Monuments and winners]. In: Zbornik radova sa naučnog skupa povodom 90. godišnjice Božićnjeg ustanka crnogorskog naroda. Fondacija "Sveti Petar Cetinjski", Cetinje, pp 189–196

Košir F (2010) Kolašin, Lovćen: od ljudskog ka izvanljudskom: arhitekta Marko Mušič [Kolašin, Lovćen: from human to trans-human: architect Marko Mušič]. CANU, SAZU, Podgorica, Ljubljana

Malbaša P (2013) Spomenici Petru II Petroviću Njegošu [Monuments to Petar II Petrović Njegoš]. Matica 55:255–314

Medaković D, Stojković Ž, Trifunović L (ed) (1971) Umetnost 27–28. Turistička štampa, Beograd

Mijović P (1997) Cetinje kao feniks [Cetinje as a phoenix]. Biblioteka "Đurđe Crnojević", Cetinje

Milkota A (2013) Pečat zadarske struke na Njegoševom mauzoleju [The seal of the Zadar profession on Njegoš's mausoleum]. http://057info.hr/kultura/2013-12-14/pecat-zadarske-struke-na-njegosevom-mauzoleju. Accessed 7 Aug 2021

Milović J (1984) O podizanju Njegoševe kapele na Lovćenu [On the erection of Njegoš's chapel on Mt. Lovćen]. In: Petar II Petrović Njegoš u svom vremenu. CANU, Titograd, pp 527–529

Nenedović Lj (1889) O Crnogorcima: pisma sa Cetinja 1878 [About Montenegrins: letters from Cetinje in 1878]. Novi Sad

Papović D (2017) Odnosi Crne Gore i SPC tokom izgradnje Njegoševog mauzoleja [Relations between Montenegro and the Serbian Orthodox Church during the construction of the Mausoleum of Njegoš]. Matica 69:209–270

Pavićević B (2007) Sazdanje crnogorske nacionalne države 1796–1878 V [The founding of the Montenegrin national state 1796–1878 V]. CID, Podgorica

Pejović Đ (1963) Zašto su Njegoševi posmrtni ostaci preneseni s Lovćena na Cetinje 1916. godine [Why Njegoš's remains were translated from Mt. Lovćen to Cetinje in 1916]. Istorijski Zapisi 16(3):498–501

Piplović S (1994) Harold Bilinić. Društvo arhitekata Splita, Split

Pušonjić B (1969) Cetinjski gimnazijalci demonstrirali pred mitropolitovom rezidencijom [Cetinje secondary school pupils demonstrate in front of the Metropolitan's residence]. Politika, 19 Mar 1969:6

Radičević F (1911) Pjesnikov grob na Lovćenu [The Poet's grave on Mt. Lovćen]. In: Boka, veliki ilustrovani kalendar za 1912. Kotor, pp 58–61

Rakočević N (1960) Prilozi istoriji autrougarske okupacije Crne Gore 1916–1918 [Contributions to the history of the Austro-Hungarian occupation of Montenegro 1916–1918]. Istorijski Zapisi 13(3):611–641

Rotković R (2013) Ko je, kada i zašto srušio Njegoševu kapelu [Who, when and why destoyed Njegoš's chapel]. Crnogorski Anali 1(2):87–93

RTCG (1972) Arhiv Zapisi o gradnji mauzoleja na Lovćenu [Archive records about constructing the mausoleum on Lovćen]. https://youtu.be/aYqOd15wPCo. Accessed 5 Aug 2021

RTCG (1974) Otvaranje Njegoševog mauzoleja na Lovćenu [Opening of the mausoleum on Lovćen]. https://www.youtube.com/watch?v=WgKkv8EPd2s. Accessed 7 Aug 2021

Selhanović J (2013) Mauzolej-partija-crkva [Mausoleum-party-church]. Arhivski Zapisi 20(1):191–219

Stamatović-Vučković S (2018) Arhitektonska komunikacija: objekti kulture u Crnoj Gori 1945–2000 [Architectural communication: cultural centers in Montenegro 1945–2000]. Univerzitet Crne Gore, Podgorica

Šistek F (2012) Njegoševa grobnica na Lovćenu [The chapel of Njegoš on Mt. Lovćen]. Matica 51–52:105–140

Vihtić B (1916) Spomenik na Lovćenu [The monument on Mt. Lovćen]. Prava Crvena Hrvatska 595, 16 Sep 1916:1

Vujović D (1994) Politički život u Cetinju i njegovoj okolini 1918–1941 [Political life in Cetinje and its surroundings 1918–1941]. In: Cetinje 1482–1982. CANU, Obod, Cetinje, pp 161–188

Vuksan D (1926) Spomenica Petra II Petrovića Njegoša—Vladike Rada 1813–1851–1925 [Memorial of Petar II Petrović Njegoš—Bishop Rade 1813–1851–1925]. Cetinje

Žaja Vrbica S (2017) Slikar Marko Rašica u vojsci Austro-Ugarske Monarhije [Painter Marko Rašica in the Austro-Hungarian army]. Ars Adriatica 7:281–290. https://doi.org/10.15291/ars.1395

Žunjić S (2013) Mauzolej na Lovćenu-istraživanje, tumačenje, zaštita [The Mausoleum on Mt. Lovćen—research, interpretations, security]. In: Prostori pamćenja, zbornik radova: vol 1, Arhitektura. Filozofski fakultet, Beograd, pp 200–213

Part VI

The Natural Environment

25 Mountain Forests in Montenegro: Usage and Public Perception

Jelena Lazarević, Olga Jokanović, and Ivan Laković

Abstract

Forests are an important natural resource of mountain areas in Montenegro, covering over 60% of the state territory. They provide protection, construction materials, firewood, food, and medicine. Collecting non-wood forest products, such as medicinal herbs, forest fruits, and mushrooms, is a source of income for the local communities and a considerable contribution to the national economy. Although the contemporary developed societies have been paying increasing attention to ecosystem services provided by forests and non-wood forest products, exploitation of forest goods here is still intensive, threatening important resources such as soil, water, and biodiversity. This paper points out the importance of Montenegrin forests and non-wood forest products, discussing also the negligence of relevant stakeholders toward these resources. Finally, it shows the results of a pilot survey conducted with citizens, teachers, and students on the issues related to forests, non-wood forest products, and forest-provided ecosystem services. Information on people's perception, attitudes, and behavior should serve as an important input to policymaking, management of forest resources, and strengthening public awareness on the matter.

Keywords

Forest management · Forest resources · Non-wood forest products · Public perception · Public awareness · Montenegro

25.1 Forests and Forestry Sector in Montenegro

Forests are one of the globally most valuable natural resources (Global … 2020a). Montenegro is a mountainous country with a pronounced relief, diverse climate, and geology. It is characterized by heterogeneous habitats and high biodiversity. According to the richness of species of flora, fungi, fauna, and diversity of ecosystems, it is among the leading countries in Europe (Stevanović et al. 1995; Lazarević 2011). According to the National Forest Inventory (2013), forests cover 59.9% of the area of Montenegro, with forest land (9.8%) and forestation (69.7%), creating the largest ecosystem in the country. The number of recorded tree species in forests is high (about 70 tree and about 90 shrub species), as well as 40 forest types (Anđelić et al. 2011).

J. Lazarević (✉) · O. Jokanović
Biotechnical Faculty, University of Montenegro, Podgorica, Montenegro
e-mail: enalazarevic@ucg.ac.me

I. Laković
Historical Institute, University of Montenegro, Podgorica, Montenegro

M. Zorn et al. (eds.), *Discourses on Mountains of Montenegro and Slovenia*, Historical Geography and Geosciences, https://doi.org/10.1007/978-3-031-99017-5_25

The total wood stock of all forests in Montenegro is 122 million cubic meters of wood, with an annual increment of 2.9 million m^3 (Global … 2020b). The average amount of wood per hectare is 159.6 m^3/ha for public forests in central and northern Montenegro (Dragović et al. 2017) that could be considered mountain forests, while the values are significantly lower in the forests of coastal and sub-Mediterranean part of Montenegro. Based on data on the average amount of wood produced per unit area, Dragović et al. (2017) conclude that the current state of forests, by the amount of wood, is the best compared to the countries in the region of Southeast Europe. On the other side, the condition of the forest fund in Montenegro is very unfavorable in relation to natural potentials. Over 50% of the total area under the forests is devastated or covered by forests of unfavorable growth, and below the average of South-East European countries (Global … 2005). Hence, it is usually emphasized that Montenegro is rich in poor forests (Martinović 1997; Global … 2005; Kohler et al. 2007; Nuhodžić and Ferlin 2010).

In terms of forest ownership, 52.3% are state forests managed by the national Forest Administration, while 47.7% are private. Although the division is not exclusive, the natural characteristics of state forests are better, while privately owned forests are mostly degraded, of coppice origin (36%), as well as shrubberies and maquis (47%) (Kohler et al. 2007; Nuhodžić and Ferlin 2010). Here should be mentioned that the coppice are degraded type of forests, characterized by considerably smaller volume and growth, weak assortment structure, and having less economic value compared to high forests. Ecologically, coppice forests are characterized by different habitat conditions in relation to high forests of the same species, inhabiting different associated plant and animal species. They are characterized by lower environmental efficiency in terms of CO_2 assimilation and carbon sequestration (Morison et al. 2012). Of the state policies and strategies, Montenegro adopted the concept of "ecological state" in 1991 (Deklaracija … 1991), proclaiming sustainable development policies. Further, the National Biodiversity Strategy and Action Plan for the period 2010–2015 (Nacionalna … 2010) and 2016–2020 (Nacionalna … 2015) were adopted. In the field of forestry, there are several strategic documents, policies, or strategies, as well as pieces of legislation and regulations that explicitly encourage sustainable forest management, operations, and use, such as the National Forest and Forest Land Administration Policy (Nacionalna … 2008), the Forest Law (Zakon o šumama 2010; Zakon o izmjenama … 2011; Zakon o izmjenama … 2015), and the National Strategy with the plan for the development of forests and forestry (Strategija … 2013). Unfortunately, they were neither accompanied by appropriate, more detailed legislation, nor with the implementation of existing ones in the field of both environmental protection and forestry. It is assumed that the low level of implementation of forestry strategies is caused by the fact that the goals have been missed or unrealistically set.

The care of forests in Montenegro is the responsibility of the Forest Administration of Montenegro on the basis of management programs and plans. However, management plans in the private sector are mandatory only for property owners up to 5 ha. The size of plots under forest in the private sector rarely (40%) exceeds 5 ha (Kaluđerović et al. 2008; Nuhodžić and Ferlin 2010), so these forests remain largely unregulated. According to Kaluđerović et al. (2008), 50.5% of interviewed forest owners do not manage their forests according to plan, while 24.9% do.

The number of employees in the forestry sector is relatively low in relation to needs, and in 2017 amounted to 395, with 94 forestry engineers and 163 skilled and highly skilled workers. It is characterized by an unfavorable age structure of employees, 55 on average. The number of employees in forest dependent industries in Montenegro was 1985, with the largest number of employees in the field of wood processing industry and production of wood, cork, and straw (Dragović et al. 2017). In addition to the small number of employees, especially engineers, the characteristic of this sector is that in managing positions (period from early 2000 to 2020) a majority of officials were outside of forestry professions (primarily economists and lawyers).

According to Global Forest Resources Assessment (2020b), the numbers are just slightly different: total employment in the forestry and wood processing industry is 2,373, while data about education level, age structure, and per gender were not presented.

Although rich in forest resources, wood production in Montenegro is limited to primary production due to the lack of developed secondary and tertiary economic cycles. As a result, wood is used mainly for local needs and exported as a raw material. Significant investment in technology, infrastructure, and processing capacities is envisaged in order to enable better economic development of the sector. The main forest product of Montenegro is wood (sales reach 89,645 m^3), and the value of goods on the market in 2015 reached 9,979,000 euros. The value of goods from the wood industry on the market is 157,000 euros and makes up only a small percentage of the total goods placed on the market, without the value of firewood (Dragović et al. 2017).

As of 2007, a concession model of forest exploitation was introduced in Montenegro, according to which state-owned forests were given for use to legal entities under concessions. In the period from 2007 to 2017, 68 concession agreements were concluded with the contracted felling in the gross amount of 3,617,700 m^3 of timber. The felling reached 2,747,860 m^3 of net wood mass, and the collected budget revenues, based on concession fees for the use of forests, amounted to 35,967,084 euros (Felin 2017). The expected budget revenues, which would enable sustainable forest management and the optimal volume of investments in forests, have never been achieved since the implementation of this forest use model. The concessionaires also failed to fulfill the purpose for which they received the concession—faster development of the wood industry by achieving a high degree of finalization of wood products. On the contrary, excessive export of wood raw materials to neighboring countries has emerged (Felin 2017). The weaknesses, in terms of legal aspects of the implementation of concessions, are incomplete implementation of the Law on Forests (Zakon o šumama 2010) and Concession Agreements, non-execution of the concessions in accordance with the agreed scope and dynamics, non-payment of concession fees, non-provision of guarantees for the concessions, etc. The key weakness in terms of forestry-technical aspects of the implementation of concessions was the failure to achieve a sufficient volume of planned investments in forests, especially in forest cultivation and protection, as well as forest infrastructure, as a prerequisite for sustainable forest management (Felin 2017).

The countries in the region did not use the concessions model in forestry, except for Slovenia that abandoned it in 2016. The forestry in EU countries is mainly managed by the state agencies or public forest management companies, with the participation of private companies in performing direct forest management tasks. That generally requires training and possession of licenses and certificates, which companies obtain from the state administration or forestry chambers of engineering (Felin 2017).

Illegal activities and corruption in the forestry sector are also perceived as a visible problem by forestry sector actors, who also distinguish between their causes, types, and consequences. Illegal logging was recognized as the main such category, but forestry sector actors differentiated the illegal deforestation caused by poverty and aimed at earning deficient income from the one pursued in quest for high profits. They have also differentiated between this activity performed by locals and by those supposedly engaged for preventing it or with more financial, political, or technological power to organize it on a higher level and at a larger scale—forestry management officials, rangers, wood processing, exporting lobbies, etc. For example, it was assumed that members of local communities were the key single actors performing illegal logging in the Montenegrin forests (41%). At the same time, their view was that illegal forestry activities causing the greatest damages to the natural ecosystems was enabled by misuse of the political influence and high-level corruption and were carried out in 25.6% cases by contractors in forestry and concessioners, 18.9% by forest rangers and police, 2.3% by management officials and employees in the forestry sector, 8.9%

by organized criminal groups, and 6.7% by wood processing and exporting enterprises. Together, they reached 62.4%, clearly dominating in the category (Bećirović 2020).

In Global Forest Resources Assessment (2020a), there is no data about forest disturbance by diseases and pests, severe weather events, or similar, while data about areas affected by forest fires is inconsistent and incomplete (Global … 2015, 2020a). Official data show that, on average, more than 100 fires are registered in Montenegro per year (Informacija … 2014). During 2012–2017, more than 3,500 forest fires were recorded and almost 9,700 in shrubberies and coppices (according to registered number of interventions by responsible services; Nacionalni … 2018). In 2012 alone, 7% of forests in Montenegro were burned. Fires occur in all regions of Montenegro, both in summer and winter. In the last 20 years, they have been recorded even in valuable coniferous forests in the north of Montenegro as well as in protected forest areas within national parks. Large-scale fires burned, among others, sizable portions of endemic pine forests in high mountain regions of Morača mountains, Kuči mountains, and Prekornica, causing total destruction of ecosystems (Lazarević 2017).

The values of forest ecosystem services, such as biodiversity conservation and recreational and protection functions, along with historical values, have not been estimated in Montenegro. They are not part of the calculation of forest benefits, although they are significant. For example, tourism is one of the key sectors of economic development in Montenegro, commonly including mountain tourism in forest areas. At least, for continuous and sustainable development of the tourism sector, it is important to take care of the preservation of all landscape characteristics, having in mind forest resources as such.

Scientific studies related to forest resources are still far from the satisfactory level. Despite the small number of engaged scientific researchers in Montenegro (3), there is a visible increase in both the quantity and quality of research, especially in relation to the health condition and pathogens of the mountain forests and ecosystems (Lazarević et al. 2012; Lazarević and Menkis 2018, 2020), and forest management (Curovic et al. 2020).

25.2 Resources of Non-wood Forest Products and Their Use in Montenegro

In addition to wood as raw material, forests provide other "products" that have a market value, today known in the literature as non-wood forest products (NWFP). NWFPs are considered important, especially for the Mediterranean area, where they play a role in sustainable forest management, forest protection, wildfire prevention, and rural economies, as well as traditional identity and knowledge, and recreation. Medicinal and aromatic herbs, forest fruits, berries, mushrooms, and truffles are collected, having a high market value and constantly in high demand. The latest report on the state of European forests estimates the total value of NWFPs in Europe at 2.27 billion euros, making more than 10% of the value associated with round wood, which is about 18 billion euros. The EU28 export of NWFP is over 2 billion euros worth annually, and import is over 3 billion euros (Mutke et al. 2019).

Resources of medicinal plants and edible mushrooms were recognized as a significant source of income for the population of mountainous areas in Montenegro, as well as for the national economy in the second half of the twentieth century. Yugoslavia was one of the main suppliers and exporters of medicinal and aromatic herbs and edible mushrooms to European countries. At that time, medicinal herbs were collected in an organized manner in Montenegro, and its primary processing was carried out domestically. After the disintegration of Yugoslavia, this activity has lost its economic significance. Although not in the same scope as before, it has recently gained in importance again. In Montenegro, large quantities of medicinal herbs and mushrooms are bought up every year. There is no exact data on how many people are engaged in collecting mushrooms for food, hobbies, and personal needs and how many are collecting for commercial purposes. According to estimates from 2006, the number of collectors of edible mushrooms and medicinal plants in Montenegro is about 30,000, and the income earned for them is an added value to agricultural income. In 2006, about 30 companies for the processing of medicinal herbs and mushrooms were registered (Lazarević and Bojović 2017). According to Global Forest Assessment (2015), the commercial value of NWFP goods was estimated at 1,227,000 euros from forest fruits and fresh mushrooms, 3,324,000 for forest honey, and about 180,000 euros from hunting. Based on the survey on rural communities, conducted on a representative sample of 1000 rural households, 19.3% of respondents collected forest fruits, medicinal plants, and mushrooms, of which 65.1% collected berries and medicinal herbs (2010). It is reported that 123 households collected about 38,000 kg of berries and 12,000 kg of tea, while 96 households collected about 14,500 kg of mushrooms. Only 5% of collected amounts went to the direct buyers, while the rest was sold to the repurchase companies. Researchers believe that this potential is still not used enough (2010). According to Kaluđerović et al.'s (2008) survey on 500 rural households, 38.6% of respondents do not collect NWFPs, while 29.8% collect them only for their own needs; 39.4% of respondents collect NWFPs to supplement the household budget, while 18.5% collect them as a family tradition. From the total number of those engaged in the collection of NWFPs, 56.9% collect mushrooms, 34.2% forest fruits, and 18.5% medicinal herbs. The average quantities of mushrooms sold in 2007 were 72.8 kg, forest fruits 80.0 kg, while the average quantity of medicinal plants sold in 2007 was 25.0 kg. Collectors sell NWFPs to purchasers (45.6%), directly to processors and consumers (10.4%), or do not sell (11.9%); 71.8% of them do not consider these sources of income sustainable. Only 3.1% of the total number of collectors took part in NWFP collection training programs, while others have never done so. In 2007, mushrooms were bought by 15 purchasers, forest fruits by 7, and medicinal plants by 6. The average annual quantity of mushroom bought-in in that year (per purchaser) was 805.3 kg, for forest fruits 934.2 kg, while the quantity of purchased medicinal plants was 103.3 kg.

According to Dragović et al. (2017), the annual income of the rural population derived from the NWFPs reaches about 5,000,000 euros, which is about half of the annual income from wood. According to Kaluđerović et al. (2008), the average household income from the sale of forest timber and firewood in 2007 amounted to 853.2 euros, while the sale of mushrooms amounted to 454 euros, from forest fruits 434.6 euros, and from medicinal plants 248.7 euros. Although the revenues from edible mushrooms and medicinal plants can be generated at different production levels, Montenegro's exports mainly consist of unprocessed and partially processed plant raw materials and mushrooms, thus resulting in lower earnings. On the other hand, various semi-finished products are imported, as well as the final products from medicinal herbs, which could be produced domestically in larger quantities and assortment. Montenegro lacks processing plants, as well as the staff that can support such development (Kaluđerović et al. 2008; Lazarević and Bojović 2017). This area is not regulated, although there were initiatives and interests from stakeholders and NGOs (Kaluđerović et al. 2008). For a long time, neither adequate data on the localities, quantities, and types of plants and fungi collected in Montenegro nor an accurate estimate of resources was available. There is no publicly available data on the quantities and types of plant material or products placed in the local market or exported. There is no control in any segment of the work, not even on the verification of the collected raw materials. It is recommended to collect only species estimated to be well represented in a location, which requires an expert assessment of resources (Hadžiablahović and Kasom 2005; Lazarević and Bojović 2017). Depending on the assessment of representation of species on a given locality or region, approximate quantities of material for collecting can be estimated, and annual collection plans should be developed. The collection should be done on the basis of buy-in plans, too. Organized buy-in of medicinal herbs and edible mushrooms needs to be carried out under professional supervision, what is not the case at the moment (Lazarević and Bojović 2017).

25.2.1 Example of Permits for Collection of Forest Goods/Non-wood Forest Products

Permits for the collection of medicinal plants and fungi used for commercial purposes are issued to legal bodies or registered individuals by the Environment Protection Agency of Montenegro (EPA) in accordance with the Rulebook (Pravilnik … 2010). Major negligence of the EPA and also of other decision-makers in the field toward NWFP, as a natural resource, is expressed and reflected in these permits, which we cite as an example. We performed an analysis of permits issued by the EPA for the years 2018, 2019, and 2020 (available at the EPA's website; https://epa.org.me/; Dozvole … 2018, 2019, 2020) on the basis of which we concluded the following:

- The collection is being done by 21–25 companies.
- 7 species of fungi (given at the genus level) and 64 plant species were collected. However, most companies collected 3 different types of mushrooms and up to 10 different plant species, while only a few companies sought and obtained permits for the other ones.
- The total quantities for which permits were issued were unrealistically large, both per company-purchaser and summarily. They amounted to, for example, 1,650,000 kg of mushrooms in 2020 (as well as on average in 2018, 2019, and 2020), 630,000 kg of chanterelles, 200,000 kg of morels; 1700,000 kg of blueberries, 1,000,000 kg of raspberries, 440,000 kg of blackberries; 72,000 kg of St. John's wort, 53,000 kg of ragweed, 34,000 kg of nettle, and 5,000 (and 34,000 later) kg of bitter gourd flowers per year.
- Companies reported hiring from several to about 130 collectors (on average about 10–12 people), indicating that one person was collecting hundreds of kilograms of mushrooms and hundreds of kilograms of plants at the same time in a short growing season, and it is obviously impossible for the registered persons to collect herbs and mushrooms in the stated quantities in the mentioned timeframes.
- Permits were issued for fresh, dried, frozen, or processed mushrooms, and this was often expressed in sum. These categories were not separated, and there was no regularity according to which certain categories were grouped in individual permits, meaning that the collected quantities from nature can be significant—up to 10 times higher—but they were not specified.
- In recent years, permits for the collection of truffles have been issued, although there were no officially published data on truffle habitats in Montenegro, and, despite that, the legal regulations for the collection of underground species in countries where they were developed are strict.
- Permitted quantities for medicinal herbs were mostly in the dry state (could also be fresh), while the forest fruits are predominantly fresh or frozen.

The permits issued in that manner did not result from the assessment of available resources but were affirmative responses to the requests of *collectors*, i.e., companies applying for permits, which are not mandatory, and it is not known whether the collectors were trained for the sustainable collection of plants and fungi from natural populations, and how the collection was conducted. It is commonly assumed that many coppice forests at lower altitudes were set on fire by the local population, in order to encourage the fruiting of mushrooms, and thus generate short-term income.

Here we also present, for comparison, an overview of purchased quantities of aboveground fungi in Slovenia, a country also rich in forest resources (Čater and Dijaci 2021) and with a tradition of picking and use of mushrooms. Similarly, as prescribed in Montenegro, it is prohibited to pick mushrooms in national parks and protected areas, collect protected species, deliberately destroy mushrooms, and use any preparations and tools that may damage the mycelium and the site. In contrast, it is prohibited to export fresh mushrooms across the state border. The quantity of wild mushrooms that can be individually collected and sold is limited to 2 kg per day. The sale of the permitted daily harvested quantities of mushrooms takes place at registered buyers (purchasers) who must keep a record of purchase (species and quantity of mushrooms, place, and date) and issue certificates of purchase. Data from the purchase records are forwarded to the Slovenian Environment Agency, which keeps purchase statistics from 1994 onwards. There are 20 species or groups of species in the purchase records (all the other species purchased in the 15-year period did not exceed 400,000 kg). Compared to the above Montenegrin data on the species for which permits were issued (but in fact, it is not known to what extent they were collected), Slovenia's total recorded collection of porcini in 2000 was 137,337 kg, chanterelles 5,196 kg, while the total quantity for the 20 registered mushroom species was 504,669 kg. In 2010, a total of 1,730 kg of porcini mushrooms and 205 kg of chanterelles were collected, with a total of 26,890 kg for all mushroom species (Medved et al. 2013). It is assumed that the larger quantities are being collected in Slovenia than we have shown here, since only the data on the quantities that purchase companies officially sell on the market is officially recorded and available.

25.3 Public Attitude Toward Forests

25.3.1 Methodology of Survey

People's perceptions, attitudes, and behavior are important for the conservation of forest resources. They represent the basis for developing a strategy for strengthening public awareness of the need to preserve this natural asset (Krajter Ostoić et al. 2017; Ranacher et al. 2020). A pilot survey was conducted with citizens (general population), teachers, and students on issues related to forests, non-wood forest products, and ecosystem services provided by the forest. We were interested in how the population perceives forests, their values and problems related to forests in Montenegro, how local communities cared about natural resources in the past, and what the current role of the local community is in relation to these issues. The second part of the questionnaire referred to non-wood forest products, their use, knowledge, and preservation.

During the time of the COVID crisis, we had limited opportunities for research in focus groups or in direct contact with respondents. Otherwise, the pandemic enabled greater focus and practice within the general population due to the use of Internet tools. The survey was conducted through an online *Google* questionnaire for the period from 30 June to 15 September 2020.

Respondents were divided into 3 groups, which were analyzed separately:

(1) All the citizens (general population), including the population over 18 years of age;
(2) Students of all faculties;
(3) Teachers, including kindergarten, elementary, and high schools teachers, as well as university teaching staff.

Teachers were selected as a group of respondents of particular importance as they were directly involved in the education process, working with the young population, and their opinions and attitudes were important for public perception.

Students were considered an essential social category of young and educated people, interested in participating in creating the future, already possessing certain knowledge and developed attitudes.

All the respondents were informed about the questionnaire through the different Internet tools and public networks, as well as in dialog with informal social groups. We repeated calls on 7 days' basis in the first month of the survey and 15 days later. Special attention was paid to providing information to teachers and students about the questionnaire.

The questionnaire was divided into 3 parts:

- The first part referred to the respondents' perception of forests and their general view on the state of forests, their quantity, and preservation in Montenegro. Some of the basic attitudes regarding the management of forest resources and the need for education were raised as a topic.
- In the second part, respondents were expected to identify and rank forest-related problems in Montenegro, express their views on forest management in the past compared to contemporary times, and the importance and participation of the local community in the management and care of resources, as well as its own participation in activities related to forests and nature protection in general.
- The third part of the survey refers to non-wood forest products: medicinal herbs, forest fruits, and mushrooms.

The questionnaire included 13 closed or semi-opened questions and statements related to the perception of non-wood forest products, satisfaction with their stewardship, and identified problems regarding NWFP, as well as knowledge level (number of known species).

The social data on respondents related to their gender, age, municipality and place of residence, level of education, and occupation were also collected. For students, we recorded the year of study (younger: I and II; or older: III and IV) and a group of faculties (natural, technical, agricultural, social, humanities), while for teachers, we were interested in which student population they worked with (younger or older), as well as groups of subjects in which they were engaged (natural or social).

25.3.2 Respondents in the Survey and General Issues

There were 394 respondents who participated in the survey. This was a representative sample, randomly based on the size of the total population in Montenegro and statistical requirements for generalizability (384 is set as minimal sample size) (Krajter Ostoić et al. 2017; online sample size calculator (Raosoft 2004)).

Among the "teachers" category, over 2/3 of the respondents were male. As for the age category, both younger and older teachers were equally represented; 43% of respondents were under 40 years. About 2/3 of teachers lived in Podgorica (capital), while the rest equally represented other cities and regions of Montenegro. The majority of teachers worked at university (51%), followed by primary (29%) and secondary school (17%), and 2 kindergarten teachers (3%). The response of teachers can be considered weak. Special attention was paid to this group of respondents during the survey. In particular, they were additionally informed about the existence of the survey and encouraged to respond. To enable a higher response rate to the questionnaire, the survey was set before the end of the school year and extended in the autumn period to better suit this group of respondents. 80% of the surveyed primary school teachers were engaged in work on subjects from the social-humanistic field; in high schools, the ratio was 50:50, while at university, the ratio was 3:2 for natural/mathematicians and technical–technological versus social-humanistic. However, their responses did not stand out in relation to others, showing significantly less interest in forest-related issues and this survey. This was especially true for primary school teachers.

Students showed an interest in the survey and responded quickly, and mostly within the first few weeks after being introduced to the questionnaire. About 70% of females responded; there was a slightly smaller number of respondents from junior years of study (about 40%). Then, students of natural, agricultural, and technical sciences had a higher response rate to the survey compared to their colleagues from social sciences and humanities. According to the place of residence, most students were from Podgorica and Nikšić, then from the coastal municipalities and from the north.

In the general population category, 2/3 of respondents were women. Respondents aged 25–65 were the most represented (25–40 and 40–65 equally). The answers of the older population were missing (only 2 persons older than 65), which was expected since Internet tools were used to conduct the survey. Around 2/3 of the respondents were university-educated, followed by the people with completed high school and about 15% of those with primary education. In terms of occupation, respondents from the social sciences and humanities, economics, law and administration, culture, arts, and public information, trade, and tourism (a total of 129 respondents) dominated. Respondents from the natural-mathematical field (40), agriculture, food production, and processing (25) were also represented as a group. The response rate of the employees in the technical occupations (construction, traffic, engineering) and even in forestry and wood processing and health care was low, although there were individual respondents in each of the categories set according to the official categorization of occupations.

Based on the above, the conducted research mainly covered the population of university-educated women (or those in the process of education), dominantly belonging to the group with social-humanistic education and occupations. This means that they were more interested in these problems, but also that they had a greater tendency to respond to surveys. The research did not cover the inhabitants of rural areas to the desired extent. About 55% of the total number of respondents lived in Podgorica, which was not bad considering the demographic structure of the country and the fact that 2/3 of the Montenegrin population lives there today.

Also, over 55% of respondents owned a forest somewhere, with 25% of teachers and students, as well as 33% of residents, using it for their own needs. It indicates their connection with rural and forest areas since the overwhelming majority of residents of Montenegrin cities (more than 95%) were either born or were the children and grandchildren of the people born in the villages. Also, based on the answers about the previous participation in forest-related activities, as well as on some other answers, there were people largely interested in environmental and nature protection issues and those often spending their time in nature.

25.3.3 Respondents' Answers

The majority of respondents mentioned that their favorite forest was at National Park Biogradska Gora, followed by those in Durmitor mountain. Others, located throughout the country, have been sporadically mentioned, depending on the personal connection of the respondents with a particular area.

Regarding the perception and condition of forests, most respondents emphasized that forests were personally important to them, with only three respondents being undecided. More than half of the respondents (53–57% in all 3 categories) owned the forest. 1/4–1/3 were using the forest for their own needs, with not more than 5% for sales.

Respondents agreed with the following statements: Montenegro is rich in forests (about 75% agree, 15% undecided, and 4% disagree); it is important to educate the rural population about the importance of the forest (77–80%) and that the children should be taught from an early age about the importance of forests (91–100%).

Further, respondents knew that the forests protect the land and water from degradation (>95%); forest fires are difficult to control and cause great material and environmental damage (>95% students and teachers, but also >85% of the general population). They (correctly) disagree with the statements: the forest can be felled indefinitely because it regenerates itself (>90%) and that burning plants to clear the grounds is a good practice (>60% firmly and some 20% in principle). Setting the forests on fire in order to collect mushrooms is not acceptable for 79–87%

of respondents, also showing a visible percentage of those considering such devastating practice as appropriate.

Regarding forest management, 69–79% of respondents did not agree with the statement that forests were well managed in Montenegro, and 82–90% thought that more inspectors were needed to implement the regulations related to forests. More than 75% of respondents believed that the forests were not a topic often present in the media and the public.

In the second part of the questionnaire, respondents identified and ranked forest-related problems. Almost all respondents (99.1–100%) believed that there were problems related to forests in Montenegro: excessive deforestation by concessionaires (79–87%), illegal cutting of forests by the local population (76–80%), forest fires (79%), inappropriate picking of forest fruits (45–55%), and waste disposal and garbage dumping (79–88%), as well as unplanned road drilling and irresponsible felling. Respondents confirmed that mini-power plants permanently endanger forests and forest communities (72–74%), and unplanned construction on forest land was a problem for about 80% of them. Respondents were undecided with regard to the issue of afforestation with inadequate or poor planting material. 88–93% of respondents believed that people who burn forests or cause forest fires were not punished appropriately. It was considered that the local population mostly (80%) leaves waste in the forests, and that waste was collected in about 50% by communal services, 28–39% by the local population, and about 10% by tourists.

More than 50% of respondents believed that human attitude toward natural resources was better in the past than today, with an additional 15–20% agreeing in principle. Respondents neither agreed nor disagreed with the statement that in the 1990s, the forests were managed better than today. The issue was elaborated on in 4 questions in more detail, but the answers were undecided for all.

More than 85% of respondents believed that representatives of the local community should participate in decision-making on the use of natural resources, as well as that the local community should benefit from the use of resources (67–80% of which 45–62% agreed strongly). Respondents also agreed that the local population must take care of their natural resources (93–97%).

Further, 70% of respondents from the student and general population categories, as well as 62% of teachers, thought that people set the forests on fire to pick mushrooms or strawberries. 98–100% are of the attitude that those, who knowingly burn the forests, should be reported to authorities (suspicion of 2% was expressed only by students), while 85–93% of respondents believed that those, who inadvertently set fire to the forest, should be held criminally accountable and persecuted.

In relation to participation in activities related to forests and nature protection in general, over half of the respondents (53–61%) participated in forest fire extinguishing actions, tree planting, or similar actions, and more than 98 expressed their readiness to participate in tree planting activities; about half of the respondents (50–60%) participated in garbage collection activities and about 90% expressed readiness to participate in such activities in the future.

In the part of the questionnaire about non-wood forest products, respondents answered as follows: they were undecided (33–40%) regarding whether forest fruits were collected in a way that endangered their long-term survival in nature. They were also undecided about whether only the local population should collect forest fruits (35–41%) or people from abroad should come and pick forest fruits in large quantities (34–56%).

Slightly less than half of the respondents were collecting medicinal herbs, mushrooms, or forest fruits (42–47%), of which over 93% collected them for their own needs. The largest number of respondents collected medicinal herbs and forest fruits, while most of them did not collect mushrooms at all.

The majority of respondents (59–65%) were not aware of regulations on the methods of collection or the quantities of medicinal plants, edible mushrooms, and forest fruits that can be collected in a certain area. Among them, 49–54%

thought that collectors poorly adhere to collection rules, with 2–7% thinking that these rules were followed. 83% (students) to 97% (teachers) thought that it would be useful to expand their knowledge about medicinal plants, mushrooms, and forest fruits. Respondents knew medicinal herbs and forest fruits well: about 50% knew more than 10 fruit species (e.g., strawberry, blueberry, roseberry, blacknerry, hazelnut, cherry), 30–40% of them knew up to 10 species of medicinal herbs (e.g., mint, thyme, sage, perforate St. John's-wort, yarrow, nettle). We did not receive data on revenues generated from the collection.

Teachers (70%) and students (87%), as well as the others, believed that forest conservation problems were not sufficiently represented in school curricula, while 2–4% in all categories had an opposite opinion.

25.4 Conclusions and Lessons Learned

The forests of Montenegro have a character of mountain forests and represent a valuable natural resource, providing important ecological and economic functions. Decision-makers, administrations, and agencies, formally responsible for dealing with forest and protection issues, manage this domain according to plans, but not entirely responsibly and in compliance with the proclaimed strategic documents. Among the numerous examples of such (mal)practices, the situation with permits for collections of mushrooms, forest fruits, and medicinal herbs was chosen as the one of the most obvious, depicting the general context of the discrepancy between the reality and proclaimed, obligatory by law and common sense principles. Unfortunately, Montenegrin forest resources are far from unlimited, where combined threats imposed by excessive felling, forest fires, lack of well-organized reforestation, and general negligence by relevant stakeholders seriously endanger the perspective of their preservation and future sustainable valorization.

On the other side, the public awareness on the matter, examined through the presented questionnaire, appears insufficiently strong to instigate any substantial changes. Aside from generally positive attitudes toward the existence and importance of forests, the need for their protection, and verbal support to the preservation and remedial activities, the Montenegrin population does not perceive the forest-related problems as those of societal priorities. Living in the urban areas, it has mostly lost the genuine contact and connection with forests, referring to them either from the standpoint of directly unrelated but positively tuned outsiders or, in the best cases, occasional and recreational users. Despite the clear support to positive practices and attitudes echoing from respondents' answers, they have also demonstrated a visible alienation of the community from the forests and the lack of awareness on services they provide. However, resulting from omnipresent tendencies of urbanization and abandonment of traditional backcountry areas covered with forests, it would have been surprising if the results had been different. Despite the technological progress and overall modernization, Montenegrin society is yet to find a balanced relation to one of its most valuable natural resources.

References

Anđelić M, Dees M, Pantić D, Borota D, Šljukić B, Čurović M (2011) Status of forest resources of Montenegro. Poljoprivreda i Šumarstvo 57(3):39–52

Bećirović D (2020) The survey on corruption and illegal activities in forestry: the report for the activity 3.2, regional action for combating forest crime and corruption. http://www.cnvp-eu.org/uploads/documents/197/Report.pdf. Accessed 15 Nov 2021

Čater M, Dijaci J (2021) Scientific support for close-to-nature forestry In: Čater M, Železnik P (eds) Forests and forestry in Slovenia. Slovenian Forestry Institute, Ljubljana. https://doi.org/10.20315/SFS.165

Curovic M, Spalevic V, Sestras P, Motta R, Dan C, Garbarino M, Vitali A, Urbinati C (2020) Structural and ecological characteristics of mixed broadleaved old-growth forest (Biogradska Gora - Montenegro). Turk J Agric for 44(4):428–438. https://doi.org/10.3906/tar-2003-103

Deklaracija o ekološkoj državi Crnoj Gori [Declaration on the ecological state of Montenegro] (1991). Skupština Republike Crne Gore, Podgorica. https://epa.org.me/deklaracija-o-ekoloskoj-drzavi/. Accessed 15 Nov 2021

Dees M, Anđelić M, Fetić A, Jokanović B, Tepavčević V, Boriota D, Vasiljević A, Weinreich A, Marković D, Terzić D (2013) The first national forest inventory in Montenegro: final report. Ministry of agriculture and rural development, Podgorica

Dozvole shodno zakonu o zaštiti prirode [Permits pursuant to the Law on Nature Protection] (2019). https://epa.org.me/wp-content/uploads/2019/01/dozvole_biljke_sajt.pdf. Accessed 15 Nov 2021

Dozvole shodno zakonu o zaštiti prirode [Permits pursuant to the Law on Nature Protection] (2020). https://epa.org.me/wp-content/uploads/2020/01/dozvole_biljke_sajt.pdf. Accessed 15 Nov 2021

Dozvole za sakupljanje, otkup i promet biljaka i životinja [Permissions for collecting, repurchase and trade of plants and animals] (2018). https://epa.org.me/wp-content/uploads/2018/03/dozvole_biljke_sajt.pdf. Accessed 15 Nov 2021

Dragović N, Ristić R, Pülzl H, Wolfslehner B (eds) (2017) Natural resource management in Southeast Europe: forest, soil and water. GIZ, Skopje

Felin (2017) Reorganizacija koncepta koncesionog korišćenja šuma [Reorganization of the concept of concessionary use of forests]. Ministarstvo poljoprivrede i ruralnog razvoja, Podgorica. https://wapi.gov.me/download-preview/821021b9-280b-4a22-80ad-49c58738bc5d?version=1.0. Accessed 15 Nov 2021

Global forest assessment: country reports Serbia and Montenegro (2005). FAO, Rome. https://www.fao.org/3/ai953e/ai953e00.pdf. Accessed 15 Nov 2021

Global forest assessment: report Montenegro (2015). FAO, Rome. https://www.fao.org/3/az279e/az279e.pdf. Accessed 15 Nov 2021

Global forest resources assessment 2020: Key findings (2020a). FAO. Rome. https://doi.org/10.4060/ca8753en. Accessed 15 Nov 2021

Global forest resources assessment: report Montenegro (2020b). FAO, Rome. https://www.fao.org/3/ca9825en/ca9825en.pdf. Accessed 15 Nov 2021

Hadžiablahović S, Kasom G (2005) Ljekovito bilje i samonikle jestive gljive u Srbiji i Crnoj Gori: monografija za sakupljače po principima organske proizvodnje [Medicinal herbs and wild edible fungi in Serbia and Montenegro: monograph for collectors according to principles of organic production]. GTZ, Podgorica

Informacija o šumskim požarima [Information on the forest fires] (2014). Ministarstvo poljoprivrede i ruralnog razvoja, Podgorica

Kaluđerović J, Daković M, Krsmanović A, Mirković M, Mulesković M, Golubović V, Glendza M, Milačić J (2008) Istraživanje o nedrvnim šumskim proizvodima: pod-sektorska analiza. Finalni izvještaj [Research on non-wood forest products: sub-sectoral analysis. Final report]. Institut za strateška istraţivanja i prognoze Podgorica i SNV Crna Gora, Podgorica

Kohler V, Janković R, Bakić M, Dukić G, Korać V, Radulović R (2007) Privatno šumarstvo [The private forestry]. Ministarstvo poljoprivrede, šumarstva i vodoprivrede, Podgorica

Krajter Ostoić S, van den Bosch CCK, Vuletić D, Stevanov M, Živojinović I, Mutabdžija-Bećirović S, Lazarević J, Stojanova B, Blagojević D, Stojanovska M, Nevenić R, Pezdevšek Malovrh Š (2017) Citizens' perception of and satisfaction with urban forests and green space: results from selected Southeast European cities. Urban Forest Urban Green 23:93–103. https://doi.org/10.1016/j.ufug.2017.02.005

Lazarević J (2011) Genetic resources in the forestry sector of Montenegro. Poljoprivreda i Šumarstvo 57(1):51–71

Lazarević J (2017) Šume i biodiverzitet Kučke planine [Forests and biodiversity of Kuči mountains]. In: Laković I (ed) Katuni Kučke planine: monografija. Univerzitet Crne Gore, Podgorica

Lazarević J, Bojović D (2017) Hrana i ljekovi iz prirode, priručnik o jestivim gljivama i ljekovitom bilju planinskih područja [The food and medicine from the nature, handbook on edible fungi and medicinal herbs of the mountain regions]. Univerzitet Crne Gore, Podgorica

Lazarević J, Keča N, Martinović A (2012) Mycorrhization of containerized *Pinus nigra* seedlings with *Suillus granulatus* under open field conditions. Forest Syst 21(3):498–507. https://doi.org/10.5424/fs/2012213-02895

Lazarević J, Menkis A (2018) Fungi inhabiting fine roots of *Pinus heldreichii* in the Montenegrin montane forests. Symbiosis 74:189–197. https://doi.org/10.1007/s13199-017-0504-5

Lazarević J, Menkis A (2020) Fungal diversity in the Phyllosphere of *Pinus heldreichii* H. Christ–an endemic and high-altitude pine of the Mediterranean region. Diversity 12(5). https://doi.org/10.3390/d12050172

Martinović V (1997) Neophodnost formiranja naučnoistraživačke organizacije u oblasti šumarstva Crne Gore [The necessity of forming scientific-research organization in forestry sector in Montenegro]. Poljoprivreda i Šumarstvo 43(1–2):105–122

Medved M, Bajc M, Božić G, Čas M, Čater M, Fereira A, Grebenc T, Kobal M, Kraigher H, Kutnar L, Mali B, Planinšek Š, Simončič P, Urbančič M, Vilhar U, Westergren M, Kranjc N, Kušar G, Levanič T, Poljanšek S, Jurc D, Jurc M, Ogris N, Klun J, Premrl T, Robek R, Železnik P, Gričar J, Piškur M (2013) Gospodarjenje z gozdom za lastnike gozdov [Forest management for forest owners]. Kmečki glas, Ljubljana

Morison J, Matthews R, Miller G, Perks M, Randle T, Vanguelova E, White M, Yamulki S (2012) Understanding the carbon and greenhouse gas balance of forests in Britain. Forestry Commission, Edinburgh

Mutke S, Bonet Lledos JA, Calado N, Calvo J, Taghouti I, Redondo C, Martínez de Arano I (2019) Innovation networks on Mediterranean non wood forest products. J Innov Sci Eng 3(1):1–10. https://doi.org/10.38088/jise.498559

Nacionalna strategija biodiverziteta sa akcionim planom za period 2016–2020 godina [National biodiversity strategy and action plan for period 2016–2020] (2015). Ministarstvo održivog razvoja i turizma, Podgorica. https://www.cbd.int/doc/world/me/me-nbsap-v2-me.pdf. Accessed 15 Nov 2021

Nacionalna strategija biodiverziteta sa akcionim planom za period 2010–2015. godine [National biodiversity strategy and action plan for period 2010–2015] (2010). Ministarstvo uređenja prostora i zaštite životne srdine, Podgorica. https://wapi.gov.me/download-preview/13a5960f-02cb-4ac2-bc5f-745851f6dee9?version=1.0. Accessed 15 Nov 2021

Nacionalna šumska politika Crne Gore [National forest and forest land administration policy] (2008). Ministarstvo poljoprivrede, šumarstva i vodoprivrede, Podgorica. http://www.upravazasume.me/razni_fajlovi/1316593609.pdf. Accessed 15 Nov 2021

Nacionalni plan zaštite i spašavanja od požara [National plan for protection and rescue from the fires] (2018). Ministarstvo unutrašnjih poslova, Podgorica. https://www.gov.me/dokumenta/eb0119d2-a238-4db0-97b9-fd48d3b63b7e. Accessed 15 Nov 2021

Nuhodžić M, Ferlin F (2010) Experiences in forming the associations of private forests' owners and their influence on development of forestry politics and legislative of Montenegro. In: Medved M (ed) Small scale forestry in a changing world: opportunities and challenges and the role of extension and technology transfer. Slovenian Forestry Institute, Ljubljana, pp 512–524

Pravilnik o detaljnim načinima i uslovima za sakupljanje, upotrebu i promet nezaštićenih divljih vrsta životinja, biljaka i gljiva koje se koriste u komercijalne svrhe [Rulebook on detailed manners and conditions for collecting, use and trade of unprotected wild species of animals, plants and fungi used for commercial purposes] (2010) Službeni list Crne Gore 62. Podgorica

Ranacher L, Sedmik A, Schwarzbauer P (2020) Public perceptions of forestry and the forest-based bioeconomy in the European Union. European Forest Institute, Joensuu. https://doi.org/10.36333/k2a03

Raosoft (2004) Sample size calculator. http://size.com http://www.raosoft.com/samplesize.html. Accessed 15 Nov 2021

Stevanović V, Jovanović S, Lakusić D, Niketić M (1995) Diverzitet vaskularne flore Jugoslavije sa pregledom vrsta od međunarodnog značaja [The diversity of vascular flora of Yugoslavia with overview of species of international importance]. In: Stevanović V, Vasić V (eds) Biodiverzitet Jugoslavije sa pregledom vrsta od međunarodnog značaja. Ecolibri, Biološki fakultet, Beograd, pp 183–217

Strategija sa planom razvoja šuma i šumarstva – nacionalna šumarska strategija [National strategy with plan for develoment forest and Forestry] (2013). Ministarstvo poljoprivrede, šumarstva i vodoprivrede, Podgorica. https://wapi.gov.me/download-preview/8c404309-33dc-4f9e-8bb9-ea05edc91ad9?version=1.0. Accessed 15 Nov 2021

Zakon o izmjenama i dopunama Zakona kojima su propisane novčane kazne za prekršaje [Law on amendments to the Law prescribing fines for offenses] (2011) Službeni list Crne Gore 40. Podgorica

Zakon o izmjenama i dopunama Zakona o šumama [Law on amendments to the Law on forests] (2015) Službeni list Crne Gore 47. Podgorica

Zakon o šumama [Forest law] (2010) Službeni list Crne Gore 74. Podgorica

Mountain Environment Protection and Protected Mountain Areas in Slovenia

26

Aleš Smrekar, Katarina Polajnar Horvat, and Matija Zorn

Abstract

The conservation of the natural environment in Slovenia goes back to the late nineteenth century. The first nature protection program in Slovenia was drawn up in 1920. The current Nature Conservation Act was adopted in 1999. Today there are one national park, three regional parks, and twenty-three landscape parks in the Slovenian mountains; this chapter presents one example of each. The beginnings of Slovenia's only national park (i.e., Triglav National Park) hark back to 1924. It is rich in typical alpine features and full of alpine flora and fauna. Its cultural landscape reflects the connection between people and nature. It covers 840 km^2 and contains thirty-three settlements with over 2,300 residents. Logar Valley Landscape Park was established in 1987. It covers 24 km^2 and has only around forty residents. The beauty of this glacial valley is reflected in numerous natural features. Farms, which have shaped the cultural landscape over the centuries, have also left their mark. Efforts to preserve the predominantly forested Pohorje Hills region date back to the 1920s. Pohorje Regional Park was established in 2024. It covers 59 km^2 and has no residents.

Keywords

Mountains · Natural heritage · Cultural heritage · Nature protection · Nature conservation · Triglav National Park · Logar Valley Landscape Park · Pohorje Regional Park

26.1 Introduction

26.1.1 Development of Protected Areas in Slovenia

Protecting areas with abundant natural resources and diverse landscapes has been Slovenia's priority for over a century (Bizjak et al. 2008). The first nature protection efforts were directed toward forestry, which was a highly important industry in the nineteenth century. The Auersperg noble family, which was among the largest landowners in what is now Slovenia, ordered their forest manager Leopold Hufnagl to monitor its forests more closely. In 1892, he prepared his first forest management plan, in which he added a small note recommending that a specific area be conserved as virgin

A. Smrekar · K. Polajnar Horvat · M. Zorn (✉)
ZRC SAZU Anton Melik Geographical Institute, Ljubljana, Slovenia
e-mail: matija.zorn@zrc-sazu.si

A. Smrekar
e-mail: ales.smrekar@zrc-sazu.si

K. Polajnar Horvat
e-mail: katarina.polajnar@zrc-sazu.si

M. Zorn et al. (eds.), *Discourses on Mountains of Montenegro and Slovenia*, Historical Geography and Geosciences, https://doi.org/10.1007/978-3-031-99017-5_26

forest. In this way, he was the first in Europe and among the first in the world to exclude a virgin forest from commercial use (Pachschwöll 2011). Hufnagl is thus considered the father of Slovenia's planned nature protection.

The importance of the natural environment and the need to conserve it because of its beauty and uniqueness were increasingly acknowledged in the early twentieth century (Berginc et al. 2007). In 1920, the Section for Nature and Natural Monument Protection of the Museum Society for Slovenia submitted a memorandum (Spomenica ... 1920)—which was effectively the first Slovenian nature protection program—to the Slovenian provincial government (Slovenia was part of the Kingdom of the Serbs, Croats, and Slovenes at that time); in terms of its concreteness and comprehensiveness, this program was among the most significant nature conservation documents of that period overall. It contained the initiative to establish alpine, subalpine, and marsh conservation parks, prohibit the destruction of rare and characteristic flora and fauna, protect caves with interesting flora and fauna, and raise public interest in nature protection. Even though at that time nature protection still lacked any legal or organizational structure, the authorities took the memorandum very seriously, which was also evident from the 1924 official protection of the Triglav Lakes Valley (*Dolina Triglavskih jezer*) in the Julian Alps (Erhartič 2012; Sect. 26.2). The nature protection cause evolved over the course of the entire twentieth century (cf. Piskernik 1963–1964; Polajnar Horvat et al. 2017; Komac and Zorn 2022).

An important step in the efforts to introduce a comprehensive nature conservation system was the 1999 adoption of the Nature Conservation Act (Zakon ... 1999), which provides a legal framework for establishing protected areas in Slovenia, defines measures to conserve biodiversity, and establishes a system for protecting Slovenia's natural values. This act provides for the most comprehensive environmental protection, which helped society recognize the benefits offered by the natural environment.

Slovenia is among the European countries with the best-preserved natural environment, great biodiversity, and very diverse landscapes (Ciglič and Perko 2013). Natura 2000 sites encompass a full 38% (7,675 km^2) of the country, and over 14% (2,925 km^2) of its territory is designated as protected areas (Fig. 26.1). These sites partially overlap, so that nature protection areas cover a total of 40.4% of Slovenia (Nature ... 2022).

26.1.2 Protected Areas in Slovenia's Mountainous Areas

The data on Slovenia's mountainous areas are summarized from Chap. 3 of this volume. In Slovenia's mountainous areas, which cover 11,531.1 km^2 (or 56.9% of the country), Natura 2000 sites comprise 4,904.3 km^2 (42.5% of the mountainous areas) and protected areas cover 2,241.8 km^2 (19.4%). Because these sites partially overlap, nature protection areas in the mountainous areas comprise a total of 5,173.2 km^2 (44.9%; Fig. 26.1).

Protected mountainous areas include one national park (the only one in Slovenia), three regional parks (out of four in Slovenia), twenty-three landscape parks (out of forty-seven in Slovenia), twenty-five nature reserves (out of sixty in Slovenia), and 184 natural monuments (330 in Slovenia; Table 26.1).

Three protected mountainous areas in Slovenia are presented below (Fig. 26.1; Table 26.2): Triglav National Park as the only national park in the country (Sect. 26.2), which encompasses nearly the entire Slovenian Julian Alps, Logar Valley Landscape Park (Sect. 26.3) as a large protected area within the Kamnik–Savinja Alps, and Pohorje Regional Park (Sect. 26.4) as an example of a protected area in the Alpine hills. The history and origin of each is presented, along with their distinctive natural features, cultural landscapes, the conflicts between land use and protection efforts, and why they are considered unique landscapes.

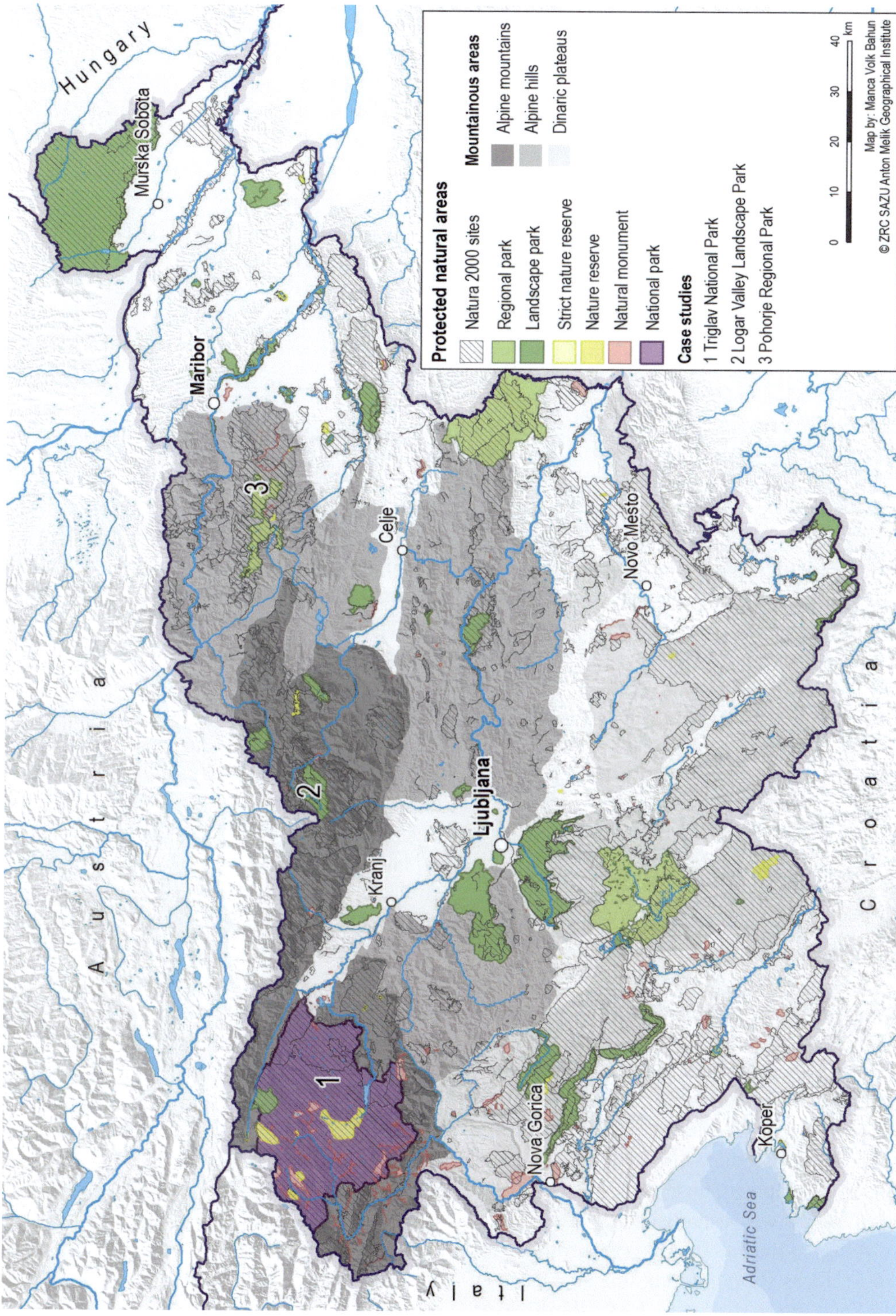

Fig. 26.1 Protected natural areas in Slovenia

Table 26.1 Protected areas in Slovenia. (*Source* Geoportal ... 2024)

Type	IUCN category	Mountainous areas (no.)	Mountainous areas (km^2)	Mountainous areas (%)	Slovenia (no.)	Slovenia (km^2)	Slovenia (%)
National park	II/V	1	839.82	7.28	1	839.82	4.14
Regional park	V	3	484.62	4.20	4	488.63	2.41
Landscape park	V	23	700.24	6.07	47	1,319.48	6.51
Nature reserve	I or IV	25	44.21	0.38	60	60.27	0.30
Natural monument	III	184	170.41	1.48	330	208.45	1.03
Natura 2000 sites			4,904.30	42.53		7,675.00	37.86

Table 26.2 The protected mountainous areas presented

Name (*original name*)	Type	IUCN category	Area (km^2)
Triglav National Park (*Triglavski narodni park*)	National park	II/V	839.82
Logar Valley Landscape Park (*Krajinski park Logarska dolina*)	Landscape park	V	24.31
Pohorje Regional Park (*Regijski park Pohorje*)	Regional park	V	59.11

26.2 Triglav National Park

26.2.1 History

The year 2024 marked one hundred years since the first site in what is now Triglav National Park was protected. At that time, when the area was part of the Kingdom of the Serbs, Croats, and Slovenes, Alpine Conservation Park (*Alpski varstveni park*) was established in the Triglav Lakes Valley in the Julian Alps (Piskernik 1962; Fig. 26.2). Ideas for protecting this area go even further back, to before the First World War, when the area belonged to Austria–Hungary. In 1907, the seismologist Albin Belar (1864–1939) produced a catalog of Carniola's natural monuments, in which he also presented reasons for protecting the Triglav Lakes Valley (Skoberne 2016a:32): "The area is exceptionally important in geotectonic terms (magnificent folds, full-breadth crevasses, grikes, and dolines like nowhere else in the Austrian Alps), and no less in geological and paleontological terms. … It is well known to landscape painters, whom the wild natural park offers magnificent views at every step (ancient trees, cirques, and lush flowers). The most appropriate area to protect would be the one around Double Lake by … the Triglav Lakes Lodge … where several hundred hectares of land could easily be protected because the area is infertile and owned by the Imperial and Royal Religious Fund. It should, by all means, be recommended that a protected area be established by the Seven Lakes, where any human intervention would be prohibited to preserve the last remnants of the exceptional high-mountain virgin forest, the habitat of ancient larches, for posterity." The preparation of this catalog was connected with the 1903 decree on collecting data on natural monuments and an initiative to prepare a monument protection and conservation law from two years earlier (Zorn et al. 2015; Skoberne 2016a).

In 1908, a year after the catalog was produced, Belar initiated a public debate to examine the conditions under which Komarča Natural Conservation Park could be established (*Komarča* is the name of an escarpment at the end of the Triglav Lakes Valley). The first

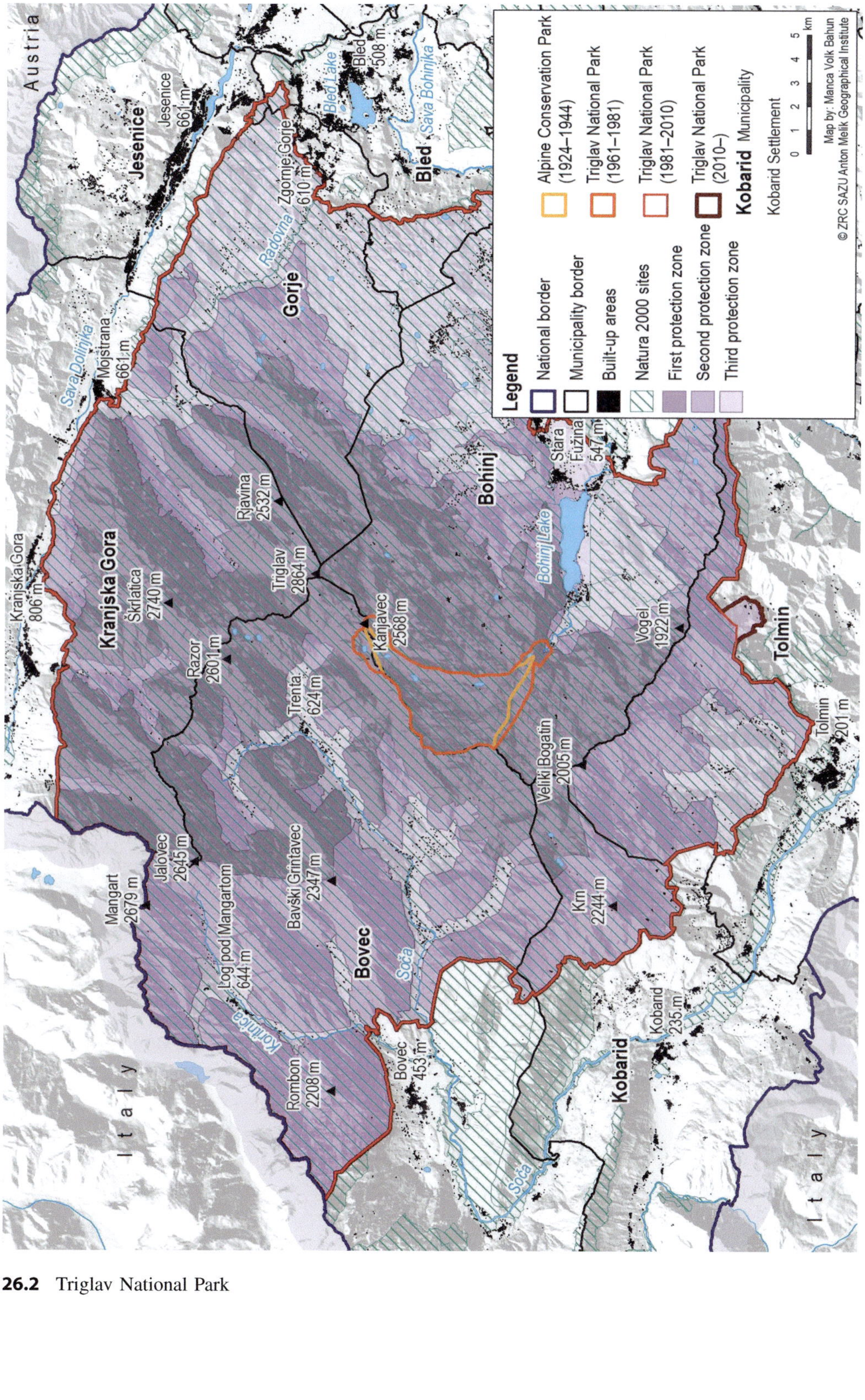

Fig. 26.2 Triglav National Park

meetings already showed significant differences in the views of the locals, who advocated grazing in the envisaged protected area (grazing rights were also protected by legislation; Bajuk Senčar 2013), and the park's proposers, who were strongly against it. The initial protection proposals were also regarded with suspicion by the members of Slovenian cultural circles because the Ljubljana branch of the German and Austrian Alpine Club was involved in the proposals, and they were afraid of the German dominance in Slovenian mountains. These initiatives were interrupted by the First World War. After the war and the formation of the new state (i.e., the Kingdom of the Serbs, Croats, and Slovenes), members of the Museum Society for Slovenia endorsed the initiative to develop a nature protection program introduced by the naturalist Ferdinand Seidl (1856–1942; Skoberne 2016a). The Section for Nature and Natural Monument Protection was established. In 1920, it produced the memorandum mentioned above, in which it also included the Seven Lakes Valley below Mount Triglav among the areas proposed for "high-mountain alpine parks," adding a note that this area was "highly important in geotectonic, geological, paleontological, zoological, and botanical terms. There the alpine flora and fauna have developed a wide range of special species and variants not represented elsewhere" (Spomenica ... 1920:71).

Because neither appropriate legislation nor a nature protection service were in place at the time, the requirements presented in the memorandum had very limited chances to be met. Hence, the area could only be protected based on an agreement between the landowner and the association, which wanted to set up a protection regime in this area. A twenty-year lease between the Directorate for Forest Protection, which managed the land that the Carniolan Religious Fund owned in the Triglav Lakes Valley, the Museum Society's Section for Nature and Natural Monument Protection, and the Slovenian Alpine Club, was signed on July 1, 1924. The lease prohibited all human activity in the Triglav Lakes Valley except hunting and tourist visits. Grazing was also forbidden, which left the fundamental dispute between the locals and the protected area's proposers unresolved (Zorn et al. 2015; Skoberne 2016a). The protected area covered approximately 1,400 hectares (Fig. 26.2; Piskernik 1962). The lease expired in 1944 (i.e., during the Second World War) and, due to changed ownership, it could not be extended after the war (Zorn et al. 2015; Skoberne 2016a).

After the Second World War, there were repeated attempts to protect the Triglav Lakes Valley again (Potočnik 1975), but the park's reestablishment was primarily hindered by grazing interests (Piskernik 1962) and lack of clarity about who had the authority to designate this area a park. A legal basis was only provided in 1959 through the adoption of the National Parks Act (Zakon ... 1959). The Triglav Lakes Valley was protected again as late as 1961 through the Ordinance Designating the Triglav Lakes Valley a National Park, referring to it as Triglav National Park, even though Mount Triglav was not part of it. The newly protected area was slightly larger than the initially protected site, covering approximately 2,000 hectares (Fig. 26.2; Piskernik 1962; Lukan Klavžar 2011; Zorn et al. 2015).

Mount Triglav (Fig. 26.3) only became part of the park in 1981 through the adoption of the Triglav National Park Act (Zakon ... 1981), which designated a major part of the Slovenian Julian Alps a protected area. The area protected covered 83,807 hectares or 4% of Slovenia (Fig. 26.2). The park was divided into two protection zones: a core zone (55,337 hectares) and an outer zone (28,475 hectares). A public institution was also established (*Javni zavod Triglavski narodni park*) to operate and manage the protected area (Bajuk Senčar 2013; Hazler 2013).

The park's current size (83,982 hectares; Table 26.2; Fig. 26.2) and protection zones were defined through the new (and still valid) Triglav National Park Act (Zakon ... 2010) adopted in 2010. The park is divided into three protection zones. The first protection zone (31,488 hectares or 37.5% of the park) is the core area with the strictest protection regime, the second zone (32,412 hectares or 38.6%) is still part of the core area, but with permitted traditional use of

Fig. 26.3 Triglav National Park is named after Mount Triglav (2,864 m), the highest peak in the Julian Alps and Slovenia, and an important Slovenian national symbol. (Photo by Matija Zorn)

resources. The third zone (20,082 hectares or 23.9%) has the most relaxed protection regime, which allows the promotion of sustainable development; it primarily includes permanently settled areas (Lukan Klavžar 2011; Načrt ... 2016).

Today "the primary aim is the protection of nature, conservation of outstanding nature and culture, protection of endemic, rare, and threatened plant species, natural ecosystems and elements of inanimate nature, as well as the conservation and maintenance of the cultural landscape" (Bajuk Senčar 2012).

26.2.2 Distinctive Natural Features

The Slovenian segment of the Julian Alps, which is protected as part of Triglav National Park, is predominantly composed of carbonate rocks, with a combination of karst and glacial landscapes. Karst features are especially prominent on plateaus, whereas valleys were reshaped by glaciers, either through glacial abrasion or accumulation (e.g., moraines). The park varies from 180 to 2,864 m in elevation. There is little surface running water, but there are many torrential streambeds and several glacial lakes (e.g., the ones in the Triglav Lakes Valley, the Krn Lakes, Lake Jezero na Planini pri Jezeru, Fig. 26.4, and Lake Bohinj), and peat bogs are a special feature. Rendzina is the predominant soil, followed by lithosol. The park is heavily wooded (forests account for 57% of its area), and it is characterized by great habitat diversity (Lukan Klavžar 2011; Načrt ... 2016). Beech forests predominate, and spruce is also common, along with larch at higher elevations, with alpine pine growing in the highest areas. Because of the past forest management practices (cutting down beech for charcoal and spruce afforestation at the end of the nineteenth and beginning of the twentieth century; Smolej 1984), spruce predominates in the park's central and northern parts. Forests are usually left to natural processes within the first protection zone and natural reserves, and sustainable and multifunctional forest management is common in the second and third protection zones (Načrt ... 2016).

At least 6,700 wildlife species live in the park (Fig. 26.5)—that is, at least 2,700 animal and 3,100 plant species, which include nineteen

Fig. 26.4 Lake Jezero na Planini pri Jezeru (1,453 m) covers 1.5 hectares and is up to 11 m deep (Dobravec and Šiško 2002). In 1951, fish were introduced to the lake, causing irreversible changes to its ecosystem. Today the lake is subject to eutrophication. At the end of summer, floating patches of filamentous algae spread along its shores. (Zorn et al. 2015; Photo by Matija Zorn)

Fig. 26.5 An abundance of bilingual signs raises visitors' awareness about the distinctive natural and cultural features of Triglav National Park. (Photo by Matija Zorn)

endemic plant species (e.g., Zois's bellflower), 163 bird species, thirty-six protected mammal species, and other protected species (e.g., the Soča trout; Načrt ... 2016; Nature ... 2018). Most importantly, though, the ecosystems in the park have been preserved contiguously from the valleys to the highest peaks (Skoberne 2016b).

The park also contains forty-six additionally protected areas, i.e., forty-three natural monuments and three nature reserves (which are also forest reserves), and over 1,070 natural values sites (including over seven hundred karst caves; Načrt ... 2016).

26.2.3 Cultural Landscape

The development of the cultural landscape in Triglav National Park is primarily connected with mountain pasturing, ironworking, and forestry, which created clearings within contiguous forests and hence contributed to the formation of a patchwork landscape. Iron ore mining in the area goes back to pre-Roman times. Due to a lack of economic viability, ironworking began to decline in the mid-nineteenth century but, because of its demand for charcoal, it led to extensive deforestation. Mountain pasturing is also associated with the first settlements in the area; many pastures were created (not all at the same time), especially where a water source was also available. These mountain pastures were primarily used for grazing cattle and horses, whereas sheep and goats grazed in more extensive higher and rockier areas. The demand for charcoal for the foundries led to planned forest exploitation. Deforestation caused the upper tree line to move from approximately 2,200 m down to approximately 1,800 m (Juvan 1975; Načrt ... 2016). Today, the process is reversed: the cultural landscape is being overgrown due to the abandonment of these activities (Sedej 1994; Načrt ... 2016; Fig. 26.6). This is changing the cultural landscape because the share of forest is increasing and grassland is disappearing. Changes in haymaking technology are causing traditional hayracks to disappear, and further changes in the cultural landscape are also resulting from the infrastructural needs of contemporary tourism (Skoberne 2016b). Nearly one-tenth of the park is agricultural land, of which four-fifths is meadows and pastures, whereas fields only account for 0.1% (Načrt ... 2016).

The cultural landscape is significantly shaped by immovable cultural heritage. There are nearly four hundred cultural heritage sites in the park, which are included in the national register of immovable cultural heritage (Lukan Klavžar 2011); these include fifteen sites classified as cultural monuments of national importance (Cultural ... 2024), over forty sites classified as cultural monuments of local importance, and several museum collections open to the public (Načrt ... 2016).

Immovable cultural heritage primarily includes architecture heritage. In this regard, stone construction in the southern part of the park can be distinguished from wooden construction in the northern part (Fig. 26.7). The herdsmen's settlements have different typological characteristics, based on which a distinction can be made between the mountain pastures in the Bohinj area, those in the Tolmin and Kobarid areas, and the sheep pastures above the Upper Soča Valley. In addition, the settlements and hamlets in the valleys also differ from one another, in which pronounced differences can be observed between the typical Bohinj, Tolmin, Bovec–Trenta, and Upper Sava houses (Zavarovana ... 2008). Active pastures are a significant part of settlement heritage in the cultural landscape; there are around fifty in the park (Hazler 2013).

An important part in shaping the cultural landscape is also played by numerous structures from the First World War (e.g., the remnants of barracks, dugouts, fire trenches, and cableways), when the Soča/Isonzo Front between Italy and Austria–Hungary ran across what is now Triglav National Park, and those from the interwar period, when various structures (e.g., barracks, fortresses, and boundary stones) were built on the border between the kingdoms of Italy and Yugoslavia, which also ran across the territory of today's national park (Kumer et al. 2020; Fig. 26.8). Consequently, many memorial structures (e.g., plaques, monuments, and military cemeteries) have been preserved in the park, along with various religious structures (e.g., churches, chapels, and shrines; Hazler 2013).

There are also numerous roads in the park (most of them are forest roads; national roads total 126.8 km), as well as hiking trails (826 km), mountain lodges (thirty-eight with approximately 2,800 beds), and ultimately the Aljaž Tower (Fig. 26.9), without which no Slovenian can imagine the summit of Mount Triglav (Lukan Klavžar 2011; Načrt ... 2016). There are thirty-three settlements within the park (twenty-one entirely and twelve partly) with a total population of over 2,300 (Triglav ... 2024). Compact clustered villages predominate in large valleys (e.g., the Soča Valley and the Bohinj area),

Fig. 26.6 Overgrown areas around Double Lake in the Triglav Lakes Valley in the 1920s (top; GIAM ZRC SAZU archive) and in 2014 (bottom). (Photo by Matija Zorn)

ribbon villages are less frequent (e.g., Strmec), scattered settlements are common in sunny areas at higher elevations (e.g., on the Pokljuka and Mežakla plateaus), and individual farms can be found in pocked valleys and on alluvial fans or higher terraces (Načrt ... 2016).

26.2.4 Conflicts

"To live in a national park means to accept and live with restrictions imposed by the Triglav National Park Act" (Fikfak and Bajuk Senčar 2015). A century ago, one of the main obstacles to protecting the area was the different views on the management of a potentially protected area: the locals were in favor of grazing, which together with milk production and processing provided a solid economic base, whereas the park's supporters advocated its prohibition because they believed it caused landscape degradation. The locals perceived the ban on grazing primarily as townsfolk's encroachment on their rights (Zorn et al. 2015; Skoberne 2016a).

Different views on how to manage this now protected area remain today; the locals in particular believe they need to deal with too many restrictions that have been imposed on them

Fig. 26.7 Architectural heritage in the mountain pastures of Triglav National Park: wooden construction predominates in the north (top: the V Lazu Pasture), and stone construction predominates in the south (bottom: the Prode Pasture). (Photo by Matija Zorn)

"from the outside." It needs to be added that, in the past decades, grazing has primarily declined as a result of economic changes (e.g., land use abandonment) rather than the park's restrictions (Bizjak 2001). Today, the traditional farming practices in the park (e.g., alpine pasturing) have heritage status (Bajuk Senčar 2013).

Every regulation that has been adopted in connection with the park's protection, however, reflects the interests of a specific period, including conflicting ones (Kmecl 2001). Hence, the 1981 law reflects the interests of tourism, agriculture, and hunting (Zorn et al. 2015). For a long time, hunting was the park's main source of finance, which is fundamentally at odds with the park's function (Kmecl 2001). Hunting was only prohibited in the park's core zone as late as 2000 (Bizjak 2001). Under the current regulation, hunting is not allowed within the first protection zone. It is allowed in the second and third protection zones, where it is planned according to the annual hunting management plans (Načrt ... 2016). Despite the different interests, the regulations have contributed to less dispersed development and fewer architecturally inappropriate structures in the park. Nonetheless, they have failed to prevent people from bending the rules on farming construction, such as converting stables into vacation houses or selling farmland and subsequently converting it into holiday villages (Zorn et al. 2015). It is estimated that there are over six hundred vacation units in the park, which are often associated with uneconomical use of space and other negative effects: inappropriate construction, demand for public infrastructure provision, a rise in property prices, seasonal pressures, and changes in the visual character of settlements and mountain

Fig. 26.8 The cultural landscape of Triglav National Park is shaped in part by First World War heritage (e.g., Holy Spirit Church on Javorca Hill, built by Austro-Hungarian soldiers in 1916 in memory of the soldiers fallen on the Soča/Isonzo Front; top) and the heritage from interwar period associated with the former national border between the kingdoms of Italy and Yugoslavia (e.g., the former Italian barracks in the Lepoč Valley; bottom). (Photo by Matija Zorn)

pastures, which often leads to loss of identity (Načrt ... 2016).

Illegal construction is a special issue. Development inside the park requires additional permits, especially nature conservation and cultural conservation permits, as well as permits from certain national authorities. However, acquiring these additional permits means that the entire process is more complex and longer, and it can be made additionally complex by potential lack of clarity among the stakeholders. It is estimated that there are between four hundred and six hundred illegally built structures in the park; most of them are found in the Bohinj area, which is also where most residents live (Bajuk Senčar 2013).

Fragmented land ownership is another challenge the park is dealing with, along with the fact that a great deal of land within the park is privately owned (i.e., over half, of which four-fifths of plots are smaller than five hectares; municipalities and the state each own less than a quarter of the land). The private owners, the municipalities, and the state have different views on where the public interest ends and where the private owners' rights begin (Bajuk Senčar 2013; Načrt ... 2016). Moreover, the park is divided among eight municipalities, and each has its own spatial plan and views on development (e.g., the development of tourism, which approximately a third of all business entities engage in; Načrt ... 2016; Žiberna 2016).

The role of the locals in creating and managing the protected area has gradually changed since the first idea of establishing a national park. Initially, they were excluded from the area through a lease agreement and a ban on grazing. During later expansions of the park,

Fig. 26.9 The Aljaž Tower at the summit of Mount Triglav is a cultural monument at the highest elevation in the country. (Photo by Matija Zorn)

they had more opportunities to participate in shaping the protection regulations, especially the park's management plan; the 1981 condition for expanding the park was prior consent from the locals, and the 2010 law provided for the local population's participation in the management plan (Bajuk Senčar 2013). Nonetheless, there has been historical mistrust among the stakeholders, and a feeling remains that the locals are in a subordinate position when it comes to shaping the conservation policy within the park and making decisions about its management. They are being heard, but not sufficiently considered. In the Bohinj area, which has the largest share of the population inside the park (more than half), there have been several (unsuccessful) initiatives so far to exclude settlements from the protected area. In this context, it is becoming increasingly clear that the fundamental problem is not their inclusion in Triglav National Park, but the lack of responsiveness from the relevant services and bodies. Consequently, the locals also resort to more extreme forms of drawing attention to issues (Zorn et al. 2015; Načrt ... 2016).

Conflicts are also connected with spatial disagreements between the interests and needs of individual recreational activities (Jeršič 2001). For example, in the late 1960s, winter sports (a project of building a ski resort with ski lifts around Mount Triglav; Mikša and Zorn 2017) was weighed against nature conservation. Efforts have also been made to tap into the sports potential of the valleys by developing the eastern shore of Lake Bohinj (Jeršič 2001).

Further challenges in the park have to do with inadequate spatial planning (especially prohibited spatial changes), inappropriate transport policy (especially constantly increasing car traffic; Fig. 26.10), increasing large-scale tourism in summer (e.g., an increasing number of daily visitors), a large number of people visiting specific points within the park (e.g., crowds flocking to Mount Triglav, which, at the peak of the hiking season, can be visited by up to three thousand hikers a day and around seventy thousand hikers a year), and soil and water pollution as a result of negligence (Bizjak 2001; Korsika 2012; Načrt ... 2016). The park is estimated to have over two million visitors per year and 1.3 million visitors in the summer season alone (Načrt ... 2016). The main pressures (Table 26.3) hindering nature conservation objectives are disturbing wildlife, a large number of hikers (Fig. 26.10), improper wastewater management, previous planned introduction of fish into isolated high-mountain lakes, abandonment of land use, afforestation of pastures or reduced landscape diversity, inappropriate spread of settlement, infrastructure development, and intensive forest management (Izhodišča ... 2013; Načrt ... 2016).

Table 26.3 Some pressures on and threats to the natural environment in Triglav National Park (Izhodišča ... 2013; Načrt ... 2016)

Pressure/threat	Description
Forestry	– Inappropriate mechanized logging (damage to the soil and vegetation) – Impact on animals – Pollution – Construction of a dense forest road network
Settlement	– Inappropriate land use – Settlements spreading beyond settlement areas – Illegal construction – Converting farm structures into vacation houses – Inadequate design and construction of new buildings – Changes in landscape and settlement patterns – Changes in local architectural heritage elements
Interventions in aquatic and riparian areas	– Water energy exploitation (small hydroelectric plants) – Changes for tourism and recreational purposes (clearing vegetation, reshaping banks, building new structures) – Water pollution
Abandonment of grazing	– Overgrowth, afforestation
Mechanized haymaking	– Tendencies to level specific landforms (e.g., hummocky meadows)
Disturbing animals	– Tourism and recreation outside intended areas – Uncontrolled pasturing – Air traffic – Riding snowmobiles and ATVs – Inappropriate feeding of animals
Gathering forest plants and berries	– Compromising and destroying habitats
Exploiting mineral resources	– Illegal extraction for use as construction material (sand and gravel)
Tourism and recreation	– Infrastructure development – Overtourism (a large number of people visiting certain mountain lodges, trails, and peaks) – Traffic – Hiking outside marked trails – Illegal overnight stays – Noise – Wastewater drainage and treatment – Light pollution – Shrinking habitats for animals
Waste	– Illegal dumps – Scattered waste along tourist routes
Invasive alien species	– Past intentional introduction (e.g., introducing fish into mountain lakes) – Planned spruce planting
Traffic	– Dense road network – Noise

(continued)

Table 26.3 (continued)

Pressure/threat	Description
Municipal infrastructure	– Not extensive enough, lacking in some places – Increasing demand for drinking water
Energy infrastructure	– Building small hydroelectric plants – Using electrical generators at mountain lodges (noise, emissions) – Fuel transport and storage
Public lighting and telecommunications	– Linear layout of mains changes landscape patterns – Light pollution – Installing mobile phone base stations
Natural disasters	– Damage to soil and vegetation – Threats to people, animals, and property – Damage

Fig. 26.10 At the peak of the tourist season, Triglav National Park is exposed to overtourism in certain areas: unregulated parking at the Vršič Pass (top; 1,611 m); overcrowded Mount Triglav (bottom). (Photo by Matija Zorn)

26.2.5 A Unique Landscape

Triglav National Park extends across the highest mountains of the Julian Alps, including their highest peak, Mount Triglav (2,864 m; Fig. 26.3)—which is also Slovenia's highest mountain and an important Slovenian national symbol—ridges, rockfaces, deeply cut glacial valleys, canyons, rushing streams, forests, mountain meadows, taluses, high-mountain (glacio)karst landscapes, and glacial lakes (Nature ... 2018). The park also contains Slovenia's highest lake (i.e., Upper Križ Lake (*Zgornje Kriško jezero*); elevation: 2,150 m; median area: 0.662 hectares; Dobravec and Šiško 2002), largest and deepest permanent lake (i.e., Lake Bohinj; area: 318 hectares; depth: up to 45 m), deepest cave (Čehi II Cave; depth: 1,505 m), longest cave system (the Tolminski Migovec Cave System; length: 43 km), deepest canyon (on the Koritnica River at the Kluže fortress, which is over 70 m deep in places), and largest natural arch (the Prisojnik Arch; Skoberne 2016b). This is where the Sava, Slovenia's longest river, has its source, and this is where Slovenia receives the most precipitation (over 3,000 mm per year, especially in the southern and southwestern parts of the park; Zupančič 1998) and where the lowest temperature in the country has been measured (on the Komna Plateau; −49.1 °C; Dovečar et al. 2009). The highest mountain lodge (the Triglav Lodge on Mount Kredarica, 2,515 m) can be found in the park, along with the highest chapel and meteorological station next to it (Fig. 26.11), the highest elevation cultural monument of national importance (the Aljaž Tower at the summit of Mount Triglav, 2,864 m; Fig. 26.9), the highest alpine road pass (the Vršič Pass, 1,611 m; Fig. 26.10), and the highest elevation road in the country (the road to the Mangart Saddle, 2,055 m; Skoberne 2016b).

The initial establishment of an alpine conservation park in the Triglav Lakes Valley was primarily the result of the area's beauty (Fig. 26.12); as was written then, the park was established in "what is the most beautiful part of our country" (Hafner 1925:62). "By creating this alpine conservation park, we have fulfilled our cultural duty because, only in this way, can we conserve this beautiful area, which, in its entirety as well as in all its living and nonliving parts, is a true natural monument for our posterity in all its beauty and attractiveness" (Hafner 1925:65). The valley left a strong impression on both previous and later visitors. In 1795, approximately 130 years before the conservation park's establishment, Count Franz Hochenwart wrote the following when laying his eyes on Double Lake: "in all the Carniolan mountains we cannot find a prettier and more enchanting view than this" (Hochenwart 1838:52). Of course, there is much more to the Triglav Lakes Valley and all of Triglav National Park than just their esthetics: the area has been a research challenge for many disciplines (Zorn et al. 2015).

Fig. 26.11 The Triglav Lodge on Mount Kredarica (2,515 m) is the highest mountain lodge in the country, standing below Mount Triglav, with Slovenia's highest meteorological station next to it. (Photo by Matija Zorn)

Fig. 26.12 Triglav National Park protects the unique mountain landscape of the Julian Alps as part of the Southeastern Alps: the Komna karst plateau covered in mountain pine (1,350–1,700 m; front), and the rocky limestone "desert" of the Krn Mountains with their highest peak, Mount Krn (2,244 m; back). (Photo by Matija Zorn)

Today, the park is part of the first UNESCO Man and the Biosphere Program (MAB) area designated in Slovenia and the Natura 2000 network, and it is a co-founder of Alparc, a network of protected areas in the Alps (Načrt ... 2016).

26.3 Logar Valley Landscape Park

26.3.1 History

The first ideas about safeguarding the natural environment in what is now the Logar Valley Landscape Park go back over a hundred years, when the Kamnik–Savinja Alps were proposed as an area suitable for conservation (Poličnik 2008) in the aforementioned 1920 nature conservation memorandum (Spomenica ... 1920). The memorandum suggested that the protected area extend "from Mount Lučka Baba near the Kamnik Saddle to the top of Mount Konj, and from there into the valley of Lučka Bela Creek, where it would include a small part of the high-elevation forest, then across Mounts Lastovec and Dleskovec to Mount Veliki vrh and from there along the foot of the northern steep slopes back to the Kamnik Saddle" (Spomenica ... 1920:70) (Fig. 26.13), but these conservation efforts failed. In 1932, the Savinja Branch of the Slovenian Alpine Club purchased the Okrešelj area above the head of the Logar Valley from the Ljubljana Archdiocese for nature conservation purposes (Kupna ... 1931). It is unknown who initiated the purchase of the Okrešelj mountain pasture in the Savinja Branch, but what is known is that the branch "intended to convert the Okrešelj area into a nature park" and that "the area extending from the tops of Mounts Ojstrica and Planjava down to Lučka Bela Creek would be very befitting for a nature park" (Orožen 1963:50). In 1934, the Okrešelj area began to be afforested to create a nature park (Orožen 1963). Just like with the protection of the Triglav Lakes Valley (Sect. 26.2), there were also problems in protecting the Okrešelj area and establishing a park because the Kingdom of Yugoslavia had no conservation-related legislation in place up until 1938, when the Decree on National Parks (Uredba ... 1938) was adopted. This decree stipulated that "in accordance with Article 100 of the 1938/1939 Financial Act and at the proposal of the Ministry of Forests and Mines, areas of extraordinary natural beauty, and research or historical importance, or those that allow people to enjoy nature, strengthen their physical and mental fitness, and serve the development of tourism may be selected and proposed for designation as national parks" (Uredba ... 1938). Hence in 1940, the Savinja Branch presented a proposal to establish Logar Valley–Okrešelj National Park (Meze and Ramovš 1978; Peterlin 1994). Everything suggested that the park would soon be established, and so the Savinja Branch had three signs made reading "Logar Valley–Okrešelj National Park, Protect

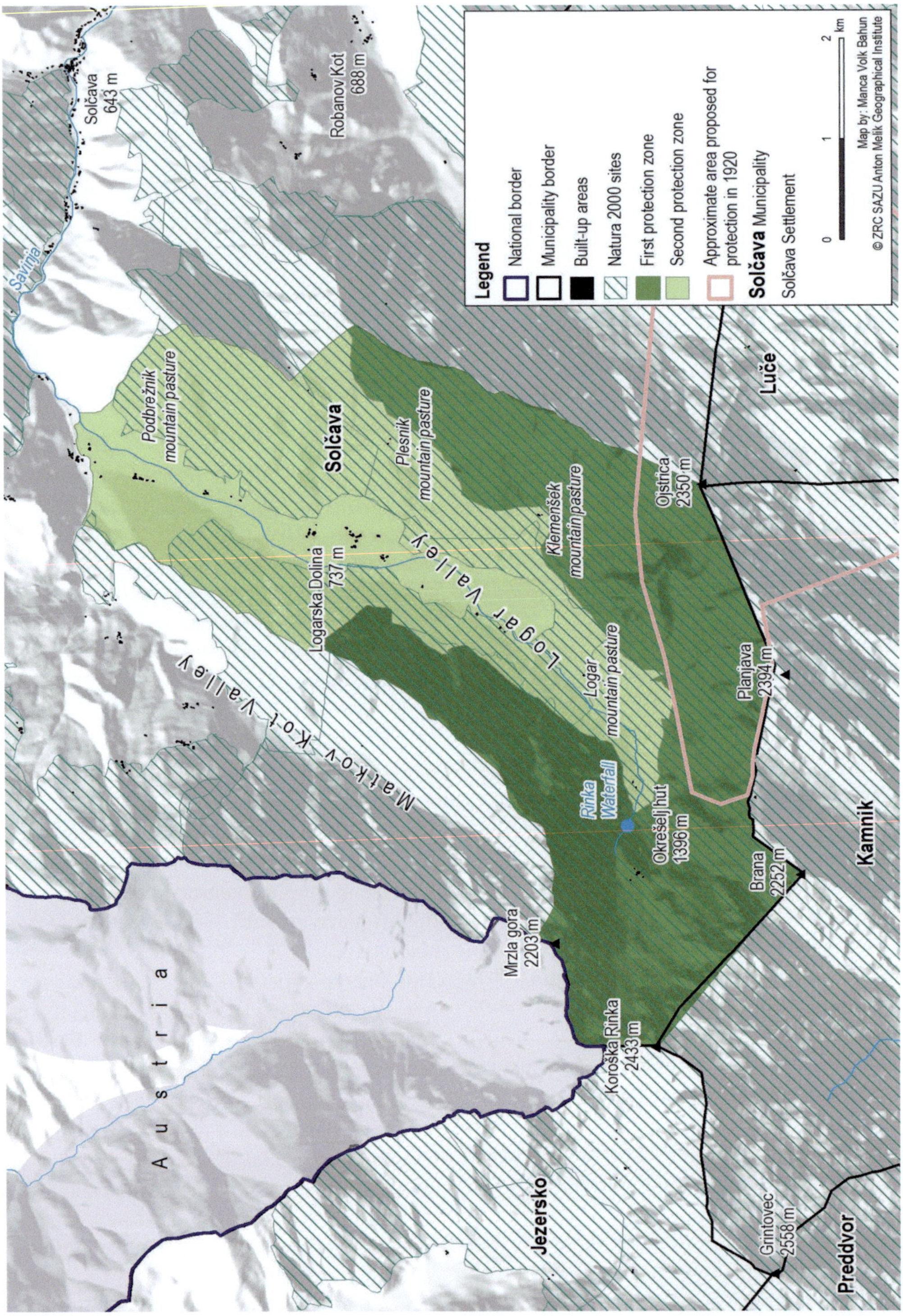

Fig. 26.13 Logar Valley Landscape Park

the Plants and Animals!" and put them up. Only a few months after it had submitted the proposal and received approval from the Provincial Nature Protection Committee, the Second World War began, preventing all further procedures (Peterlin 1994). After the war, there were no longer any perceivable efforts to formally protect the area, even though one of the signs mentioned above, which still stood at the entrance to the Logar Valley, continued to draw visitors' attention to these efforts for quite a while because the prewar signs had not been removed. The sign surely influenced the visitors and their environmentally friendly behavior because they did not think about whether the area was actually protected or not. Nature conservation became a common practice (Poličnik 2008). It was not until half a century later, in 1987, that the Logar Valley was declared a landscape park (Mencinger 2004; Fig. 26.13). The landscape park with an area of 2,431 hectares was established by the Municipality of Mozirje, but the municipal decree did not envisage that the municipality would also finance the park's operations. Therefore, the locals, mostly landowners within the park, agreed on development goals. In 1992, they founded the non-profit private company Logarska dolina d.o.o. and obtained a license to manage the park from the municipality. Due to the need to provide a systemic solution for managing the entire Logar Valley–Solčava Region tourist destination, in 2017 the Municipality of Solčava began managing Logar Valley Landscape Park through its Public Utility Unit (Prelesnik 2022).

26.3.2 Distinctive Natural Features

Logar Valley Landscape Park is rich in natural heritage (Perko and Orožen Adamič 1998). It contains over forty registered natural values sites, ranging from waterfalls, rock towers, caves, and dolines to glacial boulders, trees of exceptional dimensions (Fig. 26.14), flora, and fauna (Prelesnik 2022). The geological base, glacial activity, and other external factors have gradually shaped the valley into its unique present form. The park comprises a glacially transformed valley in the Savinja Basin 7 km long and up to 0.5 km wide (Meze 1976; Hrvatin 2010). The Logar Valley is the largest glacial valley in the Kamnik–Savinja Alps. It runs in a southwest–northeast direction between mountain ridges that formed along tectonic faults. During the Ice Age, the valley was transformed by the movement of large volumes of ice, which extensively filled it. The glacier from the Logar Valley merged with the glacier from the Matkov Kot Valley below Mount Klemenčja peč, and together they extended as far as the entrance to the Socka Gorge (Hrvatin 2010; Stojilković et al. 2013). Glacial boulders and moraines indicate the extent of former glaciation. Locals divide the valley into three parts. The wet, clayey, grassy landscape stretching across the lower part of the valley is called *Log*, the central, mostly wooded part is called *Plest*, and the mostly wooded and rubble-strewn upper part of the valley is called *Kot*. In the valley head below the Okrešelj Cirque, at an elevation of 1,280 m, the Savinja River rises, its short course taking it over 90 m Rinka Falls (Fig. 26.15). Its waters then quickly sink into the gravelly valley bed of Kotovec Creek and reappear in Log, a former glacial lake that has gradually become silted up. The locals refer to this section as *Črna* (literally, "black"), because the water here flows on lacustrine clay and therefore has a dark color. The river is named Savinja below the confluence with Jezera Creek, which flows from the Matkov Kot Valley. During heavy rain, the waters of the Palenk Creek descend in falls and cascades from below the Mount Logarska peč (1,536 m), but the most beautiful part of the falls is not visible from the valley. The waters of Palenk Creek are partially impounded for a water pipeline and a small hydroelectric power plant. The valley is surrounded by peaks rising over 2,000 m. To the south and west, the highest are Mounts Ojstrica (2,350 m; Fig. 26.16), Planjava (2,394 m; Fig. 26.16), and Brana (2,252 m; Figs. 26.16 and 26.17), Mount Štajerska Rinka (2,289 m; Figs. 26.16 and 26.17), Mount Koroška Rinka (2,433 m), and Mount Mrzla gora (2,203 m; Fig. 26.17). The mountain ridge between the

Fig. 26.14 Slovenia's largest larch, with a circumference of over 460 cm, a diameter of approximately 150 cm, and a height of 29 m, can be found in Logar Valley Landscape Park (near the Klemenča jama Lodge). (Photo by Matija Zorn)

Logar Valley and the Matkov Kot Valley to the north is slightly lower (Hrvatin 2010). There are extensive forests in the valley and on the slopes. Alpine beech forest with spruce predominates at lower elevations, and beech forest with larch predominates above 1,200 m. The Logar Valley is known for its beautiful meadows, rich in colorful flowers. This is the result of the traditional grazing practices in the region. The park is home to many plant and animal species, which have adapted to the Alpine environment. These include the chamois, the blackbird, the golden mountain flower, and the alpine lily (Logarska ... 2023).

26.3.3 Cultural Landscape

Logar Valley Landscape Park has a rich and diverse cultural heritage combined with the natural beauties of alpine landscapes (Fig. 26.16). The bottom of the Logar Valley was attractive for settlement and farming very early on, but at the same time it had a very unfavorable climate, and it was very difficult to access. The valley began to be settled in the fifteenth century, when forests were cleared for agriculture. The initial inhabitants obtained land for cultivation and pastures through controlled burning. In the past, there was

Fig. 26.15 Rinka Falls, with a vertical drop of 90 m, is located at the head of the Logar Valley. (Photo by Matija Zorn)

significantly more such farmland than there is today; people grew potatoes, buckwheat, wheat, garden vegetables, turnip, flax, and rye (Acman 2019). Today there are no longer any tilled fields in the valley, and grassland predominates (GERK 2023). Farmland can be found in the lower and central parts of the valley, where there are three farms (the Logar, Juvanija, and Plest farms), and on a few mountain pastures and small patches of grassland at higher elevations (e.g., the Plesnik and Logar mountain pastures). In the past, agricultural production was self-sufficient and extensive, and mountain pasturing was an important part of animal husbandry. Three mountain pastures high up on the slopes of Mount Strelovec (1,763 m), the Movznik Ridge, and Mount Krofička (2,083 m) were important in terms of pasturing (i.e., the Klemenšek, Plesnik, and Podbrežnik pastures), along with the Logar Pasture at Kot in the valley. These pastures were used for cows, sheep (there is the well-known Jezersko–Solčava sheep breed), and goats. Because of poor road connections, the Solčava region used to be better connected with Carinthia to the north than with Styria to the east because two roads were built toward the north due to intensive forest exploitation for charcoal burning. The valley's transport connectivity only improved with the 1894 construction of the road between Luče and Solčava downstream from the Logar Valley (Acman 2019).

The beginnings of tourism in the Logar Valley go back to the second half of the nineteenth century, when the first inn (i.e., the Plesnik Inn) opened in the valley. The first mountain lodge in the Okrešelj mountain pasture was built in 1876. Hiking became popular with the 1893 establishment of the Slovenian Alpine Club and its Savinja Branch (Poličnik 2008). With newly built roads, the valley became accessible to increasingly more travelers and hikers, which led to the construction of several mountain lodges (Meze 1976).

Settlements in the park can be divided into traditional, rural, and tourist types. The traditional type includes isolated farms with their archaic character and herdsmen's huts on mountain pastures, the tourist type comprises mountain lodges (Fig. 26.17), bed and breakfasts, and inns, and the rural type includes characteristic rural settlements at lower elevations. A peak in tourist arrivals was recorded in the first half of the 1970s, when many visitors arrived in the valley by bus. In the 1980s, organized tourist groups began to decline and private excursions with their own cars began to increase. Unregulated, large-scale parking and burning wood led to environmental degradation, which resulted

Fig. 26.16 The lower part of the Logar Valley is characterized by agricultural land use, especially meadows and pastures (top; the two prominent peaks in the background are Mount Ojstrica, 2,350 m, to the left and Mount Planjava, 2,394 m, to the right), and forests predominate in the upper part of the valley (bottom; the peaks in the background, from left to right: Mount Brana, 2,253 m, Mount Turska gora, 2,251 m, and Mount Štajerska Rinka, 2,374 m. (Photo by Matija Zorn)

in increasing awareness of the importance of the valley's value and the establishment of a protected area (Acman 2019).

Animal husbandry used to be a key economic activity in the park, but today forestry is coming increasingly to the fore, as is tourism. There are no farms on the eastern mountain slopes, but the farmers from the valley have converted some high-mountain level ground into pastures. In the past, forest was cleared for farmland in favorable areas, but now the process has been reversed: farms are being abandoned, and the steep pastures and meadows are being rapidly overgrown (Hrvatin 2010).

26.3.4 Conflicts

A great issue in Slovenian mountainous areas, including Logar Valley Landscape Park, is the abandonment of haymaking and pasturing, which leads to the afforestation of meadows and alpine pastures, and a consequent decline in landscape diversity. In some places, a trend of deforestation and meadow growth is present, but that is often merely the result of agricultural policy (i.e., direct payments or agricultural subsidies). On the other hand, such imprudent changes can disrupt the ratio between open land and forest, and hence lead to landscape degradation (Acman 2019).

Fig. 26.17 The Klemenšek Mountain Pasture with the Klemenča jama Lodge (1,208 m). In the background (from the left): Mount Brana, 2,253 m, Mount Turska gora, 2,251 m, Mount Štajerska Rinka, 2,374 m, and Mount Mrzla gora, 2,203 m. (Photo by Matija Zorn)

The negative trends of unplanned development of tourism facilities and converting agricultural and forestry structures into accommodation facilities after the Second World War primarily stopped, thanks to the establishment of a protected area. Similarly, the construction of individual vacation houses has never been permitted in the valley, and so the traditional settlement structure is well preserved despite some new tourism construction. Nonetheless, noncompliant and illegal development can be observed in places, along with development without required permits and approvals, failure to respect the characteristics of traditional architecture and cultural heritage, failure to observe the landscape configuration, inappropriate siting of facilities or activities, and so on. However, development is not the key environmental degradation factor in the park (Acman 2019).

One of the key degradation factors is most certainly tourism, and the associated traffic and infrastructure. On the one hand, it is an important source of income for the locals but, on the other, it has a significant impact on the environment and other activities in the valley, such as agriculture and forestry. Daily visitors predominate among the tourists, with long-stay tourists only accounting for about a tenth of all visitors. In terms of their reasons for coming, visitors can be divided into mountaineers, hikers, strollers, and transit or motorized visitors. Car arrivals in the Logar Valley have increased significantly since 1990. Because of unregulated traffic in the area, a need arose to create an entry point into the park and provide parking areas and traffic signs. In 1992, the park's manager introduced an entry fee for motor vehicles, achieving several effects: reducing the volume of motor traffic in the protected area, and informing, directing, and counting visitors. Nonetheless, during peak season the entry fee is an insufficient instrument to restrict entrance and regulate traffic; what is required is greater control, introducing stricter entry arrangements for motor vehicles or reducing their numbers, and building parking areas within and outside the park. There are still no alternative transport options (e.g., electric buses) available, even though they were already discussed in 1989 (Hazler Papič et al. 1989). It is vital to produce a comprehensive traffic regulation plan for the Logar Valley, select locations for parking areas, and ensure they are properly technically outfitted, taking into account the protected area's objectives (Acman 2019).

26.3.5 A Unique Landscape

Due to the valuable natural and cultural heritage it contains, Logar Valley Landscape Park is especially sensitive to various impacts of human activity. Its natural heritage is especially rich. The valley's size, shape, and landforms, and the presence of glacial traces rank it among the most beautiful Alpine valleys. The valley also features escarpments, waterfalls, creeks, and extensive forests and meadows, which contribute to the exceptional quality of the landscape. The Logar Valley is also known for its aesthetic appeal. Beautiful meadows, which abound in colorful flowers and offer picturesque views of the surrounding mountains, create spectacular scenes. This attracts many visitors and nature lovers to the park. The Logar Valley is home to a variety of plant and animal species, which have adapted to the Alpine environment. Its natural diversity is also reflected in a series of protection statutes; the park contains seven natural monuments and fifty-four natural values sites of national and local importance (Acman 2019). The natural monuments designated through government ordinances include the Savinja River, the source of the Črna River, Rinka Falls (Fig. 26.15), Palenk Falls, the Palenk Gorge, the Luknja natural arch, and the Matk natural arch (*Matkovo okno*) (Logarska ... 2023). The park is characterized by highly diverse flora and fauna, including the chamois, steinbock, peregrine falcon, golden eagle, and black grouse, and Jezersko–Solčava sheep in the mountain pastures. The typical flora includes yellow lady's slipper, Carniolan lily, auricula, Zois' bellflower, gentian, edelweiss, and many other species important for preserving biodiversity. In addition to natural heritage, the Logar Valley has a rich cultural heritage. Traditional livestock grazing practices, land cultivation methods, and preserved architectural heritage in the valley testify to past lifestyles and farming practices (Slapnik 2020).

26.4 Pohorje Regional Park

26.4.1 History

Nature conservation efforts in the Pohorje Hills also go back over a century; specifically, to 1920, when the aforementioned memorandum proposed that the virgin forest above Oplotnica near the village of Lukanja, which was owned by Count Windischgrätz, be designated a mountain or forest conservation park. It provided that this virgin forest should "continue to remain unexploited" (Spomenica ... 1920:71) and subsequently a conservation park. It no longer exists today; it is believed that it was cut down shortly after the Second World War, and so its precise location and size remain unknown (Jež et al. 2020).

In 1923, the idea emerged that "interesting parts of swamps should be protected as part of a marsh nature conservation park" (Kos 1923:64). Another call for protecting Pohorje's nature was made a few years later by Franc Dolšak (1877–1941), who, modeling on the protection efforts for the Triglav Lakes Valley, wrote that "certain parts of this mountain landscape should be protected, where flora and fauna appear before us in especially distinctive forms of their kind, which differ significantly from those below Mount Triglav and around the seven lakes [the Triglav Lakes Valley], where limestone and the rich flora of the southern limestone Alps predominate, and the rocks that form the mountain massif of all the Pohorje peaks are predominantly made of silicate" (Dolšak 1927:66). Thus, a well-grounded initiative was presented for protecting the Pohorje Hills, which could only be reproached for omitting the fauna. The next steps were only taken well after the Second World War. In 1987, the Monument Protection Institute produced the Expert Bases for Establishing Pohorje Nature Park, which were submitted to the responsible ministry a year later. The ministry requested certain modifications and additions. Hence in 1993, the report Pohorje

Nature Park: A Concept of Development with Guidelines for the Park's Establishment (Naravni ... 1993) was produced, followed by the formation of a committee consisting of representatives of all Pohorje municipalities, the Natural and Cultural Heritage Protection Institute, and the ministry, but the park was not yet established at that point. The third official proposal was produced in 1998 in the form of the Design Plan for the Future Park, but this also failed to result in the establishment of a protected area, primarily because of the disagreement between the municipalities and the ministry (Jež et al. 2020). In the process of joining the EU, by 2004 Slovenia adopted a series of measures to protect its natural environment, including in the Pohorje Hills (e.g., Natura 2000 sites cover around half of the Pohorje Hills; Fig. 26.1), but no new initiative was put forward to establish the park.

This was followed by several high-profile projects, which in 2017 led to the fourth initiative to establish Pohorje Regional Park presented by six Pohorje municipalities (Zreče, Vitanje, Mislinja, Ribnica na Pohorju, Lovrenc na Pohorju, and Slovenska Bistrica; Fig. 26.18); in 2021, the Municipality of Ruše also expressed its intent to join the initiative. In early 2024, the government adopted the Decree on Pohorje Regional Park (Uredba ... 2024), establishing a park covering 59.11 km^2 (Table 26.2).

The regional park is divided into two protection zones (Fig. 26.18) with protection regimes of various degrees. The first protection zone (11% of the park) is the core area with the strictest nature protection regime, and the second protection zone (89% of the park) is the area in which the traditional use of natural resources is permitted to facilitate sustainable farming, forestry, and wildlife management.

The explanation of the grounds for establishing the park provided that the highest parts of the Pohorje Hills (above 1,200 m) are characterized by an exceptional degree of natural conservation and landscape diversity, with frequent and exceptional natural features. The specific and diverse flora and fauna are the result of poorly permeable bedrock, rugged terrain, and an extensive network of surface waters (eUPRAVA 2022).

26.4.2 Distinctive Natural Features

The Pohorje Hills are the most distinct elevated landscape in northeastern Slovenia. They make up the southeasternmost part of the Eastern Alps and extend over 840 km^2 between the Drava Plain to the east, the Mislinja Valley to the west, the Drava Valley to the north, and the Vitanje Lowland to the south (Gams 1959). A distinction is usually made between the Eastern and Western Pohorje Hills (Žiberna and Zajc 2020).

The main surface landforms in the Pohorje Hills are ridges and valleys. Their ridge-like character, with rounded hill tops, comes most strongly to the fore in the Western Pohorje Hills, where the range is also the highest, with peaks extending above 1,500 m (the highest is Mount Črni vrh, which rises to 1,542 m). Toward the west, the dividing ridge descends in places down to the upper limit of former rural settlement, which was higher in the Western Pohorje Hills (extending up to 1,250 m) than in the Eastern Pohorje Hills (up to 1,150 m; Žiberna and Zajc 2020).

Within the Slovenian context, the Pohorje Hills have an unusual geological structure composed of igneous and metamorphic rocks. These create acidic soils and help shape ecosystems typical of the area. The poorly permeable bedrock led to the development of an extensive surface water network with many creeks and waterfalls (Nose Marolt and Gulič 2010). The terrain is wet and covered in many peatbogs with small lakes (Fig. 26.19). Ski resorts (Areh, Bellevue, and Tri Kralji) developed in the sections of the Pohorje Hills near Maribor and Slovenska Bistrica in the extreme east. In the Western Pohorje Hills, undulating terrain turns into a rounded ridge, which extends over 1,500 m but then starts descending to the northwest (Ilešič 1979). The main ridge descends so low in places that in the past it was possible to cross

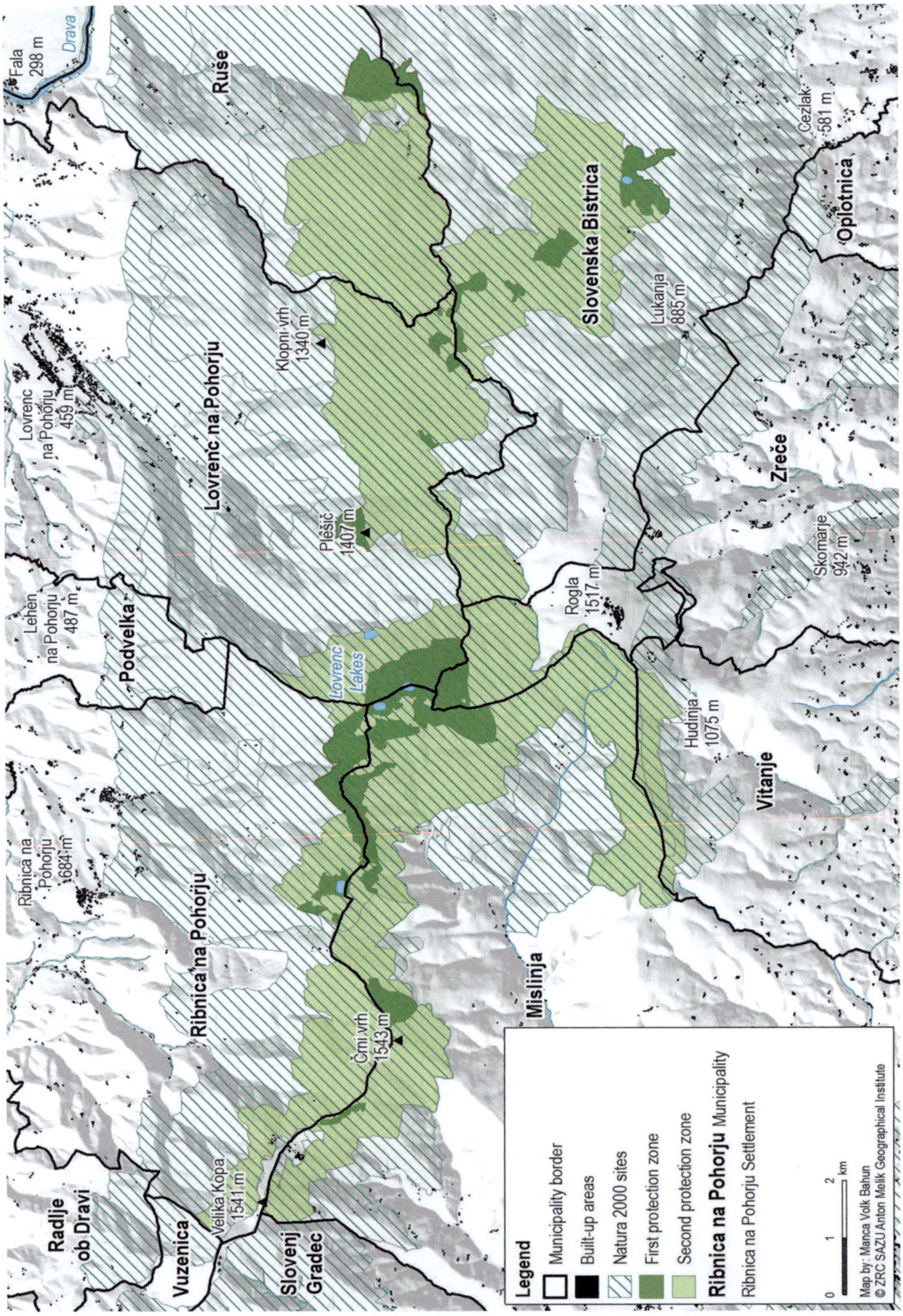

Fig. 26.18 Pohorje Regional Park

Fig. 26.19 Pohorje Regional Park has many peatbogs with small lakes, such as the Lovrenc Lakes. (Photo by Matija Zorn)

from the Drava Valley into the Mislinja Valley. Many gorges in the area were created by creeks, which carved them out in all directions, so that the water network there could be referred to as radial, with headwaters in the central part of the Pohorje Hills. Standing out the most is the 9 km long and 450 m deep Mislinja Gorge. The long ridges, which often extend from the top all the way down to the valley bottom, and the deep and steep gorges between them facilitated communication between a farm on the ridge and the valley more than between two farms on neighboring ridges (Gams 1959). That gave communications and life in general a radial character. The top part of the Pohorje Hills west of Mount Rogla (1,517 m) has a different character, with predominantly rounded peaks and more level terrain, which is dissected only by the channels of creeks with a gentle gradient. This area is the eastern Pohorje plateau, which extends between Mount Rogla and Mount Žigartov vrh (1,347 m), and whose northern and southern edges drop steeply into valleys (Žiberna and Zajc 2020).

The vegetation period at the foot of the Pohorje Hills is five to six months, in Šmartno na Pohorju, at an elevation of 775 m, it is still around five months, but it shortens to only three months on the highest peaks. Up to an elevation of 900 m, the blossoming of cherry trees, haymaking, and grain harvest occur only a week later than at elevations between 400 and 500 m. Precipitation increases with elevation, with the highest peaks receiving 1,300 to 1,600 mm. Because of the sub-Pannonian climate impacts, the eastern parts of the Pohorje Hills receive less precipitation than the western parts, despite a similar elevation. It snows a few weeks earlier on the ridges than in the rest of the hills and the lowlands. The peaks are covered in snow for up to 150 days or even longer, especially on the northern slopes (Žiberna and Zajc 2020).

The Pohorje Hills are, first and foremost, characterized by extensive forest cover. In 2019, forests covered over three-quarters of the area. The second most frequent land use type is meadows (over 17%), due to farms that predominantly engage in animal husbandry. The shares of orchards, fields, and vegetable gardens are negligible (below 2%). Meadows and cultivated land are common especially on the edges of the Pohorje Hills and in the Ribnica–Lovrenc Lowland; meadows are also typical around farms at high elevations and in the highest parts of the hills, which are used as pastures in summer and as ski areas in winter (Žiberna and Zajc 2020).

26.4.3 Cultural Landscape

Human activity has primarily shaped the landscape of today's Pohorje Hills from the late Middle Ages onward. The first isolated farms up to an elevation of 1,200 m were reportedly already established in the area in the mid-twelfth century. Large contiguous estates, not divided into fragments, were required to sustain families. A typical cultural landscape of extensive forest complexes formed during feudalism, with isolated farms and cultivated clearings in the middle (Cenčič 2020). It was probably already back then that cattle were grazed up to the highest peaks; initially they were probably grazed in forest pastures and later in cleared areas. From the eighteenth to the first half of the twentieth century, ironworking, glassmaking, and clearcutting resulted in great quantities of wood being cut. In the second half of the eighteenth century, many forest glassworks were established, which required great amounts of beech wood for melting quartz, as well as softwood for buildings, flumes, glassware packaging, and so on (Cenčič 2020). At the peak of their production, some glassworks had several hundred people working for them (Gulič 2020). During the interwar period, up to a third of the population in the Pohorje Hills worked in forestry and the sawmilling industry, and around three hundred sawmills operated in the area at that time. Hence, extensive cleared areas were created in the Pohorje Hills. Reforestation involved burning the remnants of trees in the spring, after which potatoes and turnips were planted, and rye, barley, or oats were sown the year after that. The grain seeds were mixed with spruce seeds, along with some pine and larch seeds. In this way, extensive spruce monoculture areas developed in the former habitats of fir–beech forests and high-mountain beech forests (Fig. 26.20). In some places, the clearings were not even reforested, which led to the formation of extensive grassland or marshy areas (Cenčič 2020). At the peak of the glassmaking and ironworking industries, which used enormous amounts of charcoal, charcoal burning became widespread and wood was used in large quantities. At the end of the eighteenth century, deciduous trees predominated above the settlement limit, especially beech. Conifers were scarce, unlike today, when they account for over two-thirds of all trees above 1,000 m (Žiberna and Zajc 2020).

In the nineteenth century, harvesting the eggs or pupae of brown ants was also very widespread

Fig. 26.20 Spruce predominates in the Pohorje Hills today. It was planted after the intense clearcutting of predominantly beech forests in the second half of the eighteenth century and in the nineteenth century to produce charcoal required by the ironworking and glassmaking industries. (Photo by Matija Zorn)

Fig. 26.21 Educational tourism is developed in and around Pohorje Regional Park: the Treetop Trail next to the Rogla ski resort. (Photo by Matija Zorn)

in the Pohorje Hills. They were sold as bird fodder and incense or for use in medicinal baths. Especially in spruce forests, brown ants built enormous anthills, up to 2 m high, which were almost completely removed (Gulič 2020).

More intense development of transport infrastructure also stimulated the development of tourism. From 1872 onward, a climatic health resort at Pension Büttner was gaining prominence in Lovrenc na Pohorju (Občina ... 2022). In 1901, the Drava Branch of the Slovenian Alpine Club was established in Ruše, which began to mark the hiking trails. This did not take place entirely according to plan because, for instance, a fine had to be paid to Count Thurn because a trail had been marked without his permission (Teržan 1967). Thanks to its gentle slopes and open views, the Pohorje Hills also attracted winter skiing and hiking, which provided the impetus for building mountain lodges in the 1930s. In the late 1950s, the first ski lifts were installed, providing the basis for the development of ski resorts in the following decades (Guček 2009). Contemporary tourism services (Fig. 26.21) have been developing in the sections of the Pohorje Hills near Maribor and Zreče, at the Kope ski resort, in Ribnica na Pohorju, and at the Trije Kralji ski resort. With the exception of tourist trails, the protected area contains no tourism infrastructure.

26.4.4 Conflicts

Logging, charcoal burning, and glassmaking were the main motives for settlement in the Pohorje Hills. Colonization that began in the medieval period spread to higher elevations, where individual farms surrounded by forests were established (Horvat et al. 2000). Thanks to these farms, extensive meadows developed on acidic soil over the centuries. The meadows (Fig. 26.22) are the result of traditional land use (grazing and haymaking), and their existence depends on human management. Today, they are endangered especially because of the abandonment of land and consequent afforestation on the one hand and excessive meadow intensification on the other. In both cases, the result is the same: the loss of former meadows and the subsequent loss of their patchwork structure. To avoid this, activities are underway to reestablish meadows rich in species by removing shrubs and trees and reestablishing the green cover. The highest quality meadows in terms of maintaining a high biodiversity level are

Fig. 26.22 Meadows in the Pohorje Hills reflect traditional land use (grazing and haymaking). Today they are threatened by afforestation. (Photo by Matija Zorn)

those with alternating land uses between years, in which their use alternates between pastures for grazing traditional livestock (i.e., cattle and sheep), meadows with late first-crop hay, and unused meadows. This forms a patchwork landscape, which makes it possible for organisms to move between meadows of various stages (Gulič et al. 2016). Thirteen farms that own meadows in the park participated in the project Life to Grasslands (Sln. *Življenje traviščem*) between 2015 and 2020 (Vintar Mally et al. 2020). The farm owners recognized the importance of their conservation.

In the eighteenth and nineteenth centuries, charcoal burning and glassmaking caused intensive clearcutting, which, along with later afforestation, led to the formation of spruce monoculture forests (Fig. 26.20), which are not as resilient against various disturbances (e.g., pests, wind, and snow; Jakša 2007).

Today, certain areas of the Pohorje Hills are dealing with overtourism, especially in and around the winter sports resorts (i.e., at the peak ski season in winter and the peak hiking season in summer; Fig. 26.23). Among other things, a large number of visitors accelerate trail erosion, and, if hikers stray from the marked and reinforced trails, they also cause soil erosion outside these trails (Fig. 26.24).

26.4.5 A Unique Landscape

With its ridges, which offer scenic views, abundant meadows, extensive forests (Fig. 26.20), numerous springs, bogs, and lakes (Fig. 26.19), and a picturesque landscape, the Pohorje Hills are a rare mountainous area comprised of igneous and metamorphic rock, which is why its ecological systems differ from those in the predominantly carbonate areas of the rest of Slovenia.

The central idea behind the vision Pohorje 2030 is as follows: "Our beautiful, green Pohorje Hills are just as they used to be in the past. … But they are nonetheless very different! Pohorje Nature Park carefully preserves the heritage of forests, local people, and nature, while creating new opportunities for landscape development and a better life for the locals. Modern family farms focus on ecological farming and, together with tourist resorts, on sustainable tourism. Our local natural resources and work provide the basis for our future. We promote environmentally friendly traffic and ensure that economic activities and their appertaining infrastructure have an appropriate place within the landscape" (Lešnik Štuhec 2011).

Fig. 26.23 At peak tourist seasons the Pohorje Hills also experience overtourism; top: "car tourism" on Mount Rogla (1,517 m); bottom: the resulting overuse of hiking trails in the park around Mount Rogla. (Photo by Matija Zorn)

Fig. 26.24 Because many visitors to the park fail to stick to marked and reinforced trails, soil erosion is also common outside these trails. (Photo by Matija Zorn)

26.5 Conclusion

Being among the European countries with the greatest landscape diversity (Ciglič and Perko 2013; Perko et al. 2020), Slovenia has a rich and diverse natural environment. Its exceptional patchwork landscape character is primarily reflected in a large share of Natura 2000 sites, which cover 38% of the country's land area (compared to the European average of just above 18%). This share is even larger in Slovenia's mountainous areas (i.e., 42.5%).

With their diversity and uniqueness, protected areas contribute to the conservation of biodiversity and geodiversity, and they provide shelter to endangered plant and animal species (Smrekar et al. 2020). These are special areas that, along with their great environmental value, also have important potentials that are key to the sustainable development of all of Slovenia (Lampič and Mrak 2008). In addition, they have a significant impact on the local community, tourism, and economy. Protected areas, such as national, regional, and landscape parks, facilitate the conservation of natural ecosystems and their sustainable management. Carefully controlled use of resources and the conservation of natural and social processes make it possible to preserve the diversity of natural and cultural heritage.

The state and local communities work together in developing and implementing effective policies and measures to conserve protected mountainous areas. Unfortunately, Slovenia is also facing challenges in nature protection, including inappropriate management, insufficient funding, changes to the environmental policy, and the impact of human activities, such as intensive farming and tourism. Therefore, it is key to continuously monitor and improve the management of nature protection areas. Despite all the challenges, the state of protected areas in Slovenia is fairly good compared to the rest of Europe (Smrekar et al. 2023).

Acknowledgements This study was conducted as part of the research program The Geography of Slovenia (P6-0101) financed by the Slovenian Research and Innovation Agency.

References

Acman T (2019) Strokovne podlage za prenovo akta o zavarovanju Krajinskega parka Logarska dolina [Expert bases for revising the Logar Valley Landscape Park Protection Act]. Območna enota Zavoda Republike Slovenije za varstvo narave Celje, Celje

Bajuk Senčar T (2012) Natural and cultural heritage practices in Triglav National Park. Traditiones 41(2):53–64. https://doi.org/10.3986/Traditio2012410204

Bajuk Senčar T (2013) Načrt upravljanja Triglavskega narodnega parka in kultura dediščinskih praks [The Triglav National Park Management Plan and the culture of heritage practice]. Traditiones 42(2):9–25. https://doi.org/10.3986/Traditio2013420201

Berginc M, Kremesec Jevšenak J, Vidic J (2007) Sistem varstva narave v Sloveniji [Slovenia's nature conservation system]. Ministrstvo za okolje in prostor, Ljubljana

Bizjak J, Vidic J, Berginc M, Hladnik J, Zupanc Hrastar S, Groznik Zeiler K, Kristanc J, Vičar A (eds) (2008) Zavarovana območja v Sloveniji [Slovenia's protected areas]. Ministrstvo za okolje in prostor, Ljubljana

Bizjak J (2001) Narodni park med vizijami, možnostmi in resničnostjo [A national park between visions, possibilities, and reality]. In: Bizjak J, Šolar M (eds) Dvajset let pozneje 1981–2001: 20 let Zakona o Triglavskem narodnem parku. Javni zavod Triglavski narodni park, Bled, pp 85–89

Cenčič L (2020) Gozdovi Pohorja [Pohorje forests]. Proteus 83(2–5):129–138

Ciglič R, Perko D (2013) Europe's landscape hotspots. Acta Geogr Slov 53(1):117–139. https://doi.org/10.3986/AGS53106

Cultural heritage (2024). Javni zavod Triglavski narodni park, Bled. https://www.tnp.si/en/. Accessed 1 Jul 2024

Dobravec J, Šiško M (2002) Geografska lega in opis jezer = Geographical location and description of lakes. In: Brancelj A (ed) Visokogorska jezera v vzhodnem delu Julijskih Alp = High-mountain lakes in the eastern part of the Julian Alps. Založba ZRC, Ljubljana, pp 49–76. https://doi.org/10.3986/9616358588

Dolšak F (1927) Za varstvo prirode na Pohorju [For protecting nature in the Pohorje Hills]. Glasnik Muzejskega društva za Slovenijo 7–8(1–4):65–69

Dovečar M, Sinjur I, Ogrin M, Vertačnik G (2009) Najnižja temperatura v Sloveniji [The lowest temperature in Slovenia]. Geografski Obzornik 56(1–2):27–31

Erhartič B (2012) Geomorfološka dediščina v Dolini Triglavskih jezer [Geomorphosites in the Triglav Lakes Valley]. Geografija Slovenije, vol 23. Založba ZRC, Ljubljana. https://doi.org/10.3986/9789612545895

eUPRAVA (2022) Uredba o regijskem parku Pohorje [Decree on Pohorje Regional Park]. https://e-uprava.gov.si/drzava-in-druzba/e-demokracija/predlogi-predpisov/predlog-predpisa.html?id=13635. Accessed 7 Jul 2023

Fikfak J, Bajuk Senčar T (2015) Triglavski narodni park: akterji, strategije, problemi, rešitve [Triglav National Park: actors, strategies, problems, solutions].

Traditiones 44(1):7–27. https://doi.org/10.3986/Traditio2015440101
Gams I (1959) Pohorsko Podravje: razvoj kulturne pokrajine [The Pohorsko Podravje Region: cultural landscape development]. SAZU, Ljubljana
Geoportal ARSO (2024). Agencija Republike Slovenije za okolje, Ljubljana. https://gis.arso.gov.si/geoportal/catalog/main/home.page. Accessed 1 Jul 2024
GERK (2023) Javni pregledovalnik grafičnih podatkov MKGP [Public graphic data viewer of the Ministry of Agriculture, Forestry, and Food]. Ministrstvo za kmetijstvo, gozdarstvo in prehrano, Ljubljana. https://rkg.gov.si/GERK/. Accessed 7 Jul 2023
Guček A (2009) Brez sape v hrib [Breathless up the hill]. Združenje žičničarjev Slovenije, Ljubljana
Gulič A (2020) Pohorje skozi čas [The Pohorje Hills through time]. Proteus 83(2–5):225–235
Gulič J, Štruc S, Debeljak Šabec N, Kamenik K (2016) Akcijski načrt za izboljšanje stanja in zagotavljanje dolgoročnega ohranjanja travišč za gozd in druga gozdna zemljišča pod gozdno masko ter zaraščajoče površine v območju s pojavljanjem habitatnega tipa 6230 [Action plan to improve and ensure the long-term conservation of grasslands for forests and other wooded land under forest cover and overgrown areas in the area where habitat type 6230 occurs]. Zavod Republike Slovenije za varstvo narave, Maribor
Hafner M (1925) Alpski varstveni park v Dolini sedmih jezer [Alpine conservation park in the Seven Lakes Valley]. Geografski Vestnik 1:62–65
Hazler V (2013) Varovanje dediščine v Triglavskem narodnem parku v razcepu med strokovnimi načeli, zakonodajo in vsakdanjo realnostjo [Heritage protection in Triglav National Park: caught between professional principles, legislation, and everyday realities]. Traditiones 42(2):27–48. https://doi.org/10.3986/Traditio2013420202
Hazler Papič M, Hazler V, Kolšek A, Lenarčič M, Strahovnik V (1989) Naravna in kulturna dediščina Logarske doline, njeno varovanje in razvoj [The natural and cultural heritage of the Logar Valley, its protection and development]. Zavod za varstvo naravne in kulturne dediščine, Območna enota Celje, Celje
Hochenwart Fv (1838) Beiträge zur Naturgeschichte, Landwirtschaft und Topographie des Herzogthums Krain, Heft 1. Laibach
Horvat A, Jenčič S, Kanop M, Kruder B, Pristovnik D, Štajnbaher S (2000) Bistriški vintgar [The Bistrica Gorge]. Občina Slovenska Bistrica, Slovenska Bistrica
Hrvatin M (2010) Logarska dolina [The Logar Valley]. In: DEDI - digitalna enciklopedija naravne in kulturne dediščine na Slovenskem. http://www.dedi.si/dediscina/431-logarska-dolina. Accessed 7 Jul 2023
Ilešič S (1979) Pogledi na geografijo [Perspectives on geography]. Partizanska knjiga, Ljubljana
Izhodišča za načrt upravljanja Triglavskega narodnega parka [Bases for the Triglav National Park Management Plan] (2013). Javni zavod Triglavski narodni park, Bled
Jakša J (2007) Naravne ujme v gozdovih Slovenije [Natural disasters in Slovenian forests]. Gozdarski Vestnik 65(3):161–176
Jeršič M (2001) Narodni park in prostorski razvoj [National park and spatial development]. In: Bizjak J, Šolar M (eds) Dvajset let pozneje 1981–2001: 20 let Zakona o Triglavskem narodnem parku. Javni zavod Triglavski narodni park, Bled, pp 51–58
Jež M, Kaligarič S, Gulič J (2020) Pohorje včeraj, danes, jutri [The Pohorje Hills yesterday, today, and tomorrow]. Proteus 83(2–5):56–62
Juvan J (1975) Človekova dejavnost na ozemlju Triglavskega narodnega parka v preteklosti: fužinarstvo, planšarstvo in gozdarstvo [Human activities in the Triglav National Park area in the past: blast furnace forging, Alpine pasturing, forestry]. In: Peterlin S (ed) 50 let Triglavskega narodnega parka. Prirodoslovno društvo Slovenije, Ljubljana, pp 73–77
Kmecl M (2001) Triglavski narodni park—včeraj, danes, jutri [Triglav National Park: yesterday, today, and tomorrow]. In: Bizjak J, Šolar M (eds) Dvajset let pozneje 1981–2001: 20 let Zakona o Triglavskem narodnem parku. Javni zavod Triglavski narodni park, Bled, pp 21–23
Komac B, Zorn M (2022) Hydropower landscape in the Slovenian Alps: pros and cons for the construction of hydropower plants in the Upper Soča Valley. Histoire des Alpes = Storia delle Alpi = Geschichte der Alpen 27:205–222
Korsika M (2012) Trume na Triglav, koliko pohodnikov lahko prenese gora? [Mount Triglav: how many hikers can the mountain support?]. Novice Svet 24, 9 Jul 2022. https://novice.svet24.si/clanek/novice/slovenija/62c7e5bdd1897/obisk. Accessed 11 Jun 2024
Kos F (1923) Varstvo prirode in prirodnih spomenikov: Odsek za varstvo prirode in prirodnih spomenikov [Protection of nature and natural monuments: the Section for Nature and Natural Monument Protection]. Glasnik Muzejskega društva za Slovenijo 2–3(1–4):64–67
Kumer P, Zorn M, Žorž G, Gašperič P (2020) From the Great War to interwar fortifications: changing narratives attached to the military landscape in western Slovenia. In: O'Reilly G (eds) Places of memory and legacies in an age of insecurities and globalization. Springer, Cham, pp 247–263. https://doi.org/10.1007/978-3-030-60982-5_11
Kupna pogodba št. 162/32 [Purchase contract no. 162/32] (1931). Nadškofijski arhiv v Ljubljani, Ljubljana
Lampič B, Mrak I (2008) Vrednote, vrednosti in razvojni potenciali območij varovanja [Valuables, values, and development potentials of protected areas]. Dela 29:145–159. https://doi.org/10.4312/dela.29.11.145-159
Lešnik Štuhec T (2011) Vizija trajnostnega razvoja 'zelene' ponudbe turističnega območja Pohorje 2030 [Vision for the sustainable development of "green" tourism in the Pohorje Hills until 2030]. Tatjana Lešnik Štuhec s.p., Šentilj

Logarska dolina (2023) Naturale values. https://www.logarska-dolina.si/en/attractions/nature/natural-values. Accessed 7 Jul 2023

Lukan Klavžar T (2011) Triglavski narodni park in kulturna dediščina [Triglav National Park and cultural heritage]. Kulturni in naravni spomeniki Slovenije, vol 214. Zavod za varstvo kulturne dediščine Slovenije, Ljubljana

Mencinger B (2004) Logarska dolina [The Logar Valley]. Naravni parki Slovenije. Ljubljana

Meze D, Ramovš A (1978) Logarska dolina [The Logar Valley]. Kulturni in naravni spomeniki Slovenije, vol 82. Obzorja, Maribor

Meze D (1976) Logarska Dolina [The Logar Valley]. In: Savnik R (ed) Krajevni leksikon Slovenije, vol 3. Državna založba Slovenije, Ljubljana, pp 225–227

Mikša P, Zorn M (2017) The Julian Alps (Slovenia): between protection and "modernization." Ekonomska i Ekohistorija 13:147–158

Načrt upravljanja Triglavskega narodnega parka 2016–2025 [2016–2025 Triglav National Park Management Plan] (2016). Javni zavod Triglavski narodni park, Bled

Naravni park Pohorje: koncept razvoja s smernicami za razglasitev [Pohorje Nature Park: development concept with designation guidelines] (1993). Gozdno gospodarstvo Maribor, Maribor

Nature areas under protection (2022). Agencija Republike Slovenije za okolje, Ljubljana. https://kazalci.arso.gov.si/en/content/nature-areas-under-protection-1. Accessed 21 May 2024

Nature parks Slovenia (2018). Slovenian Tourist Board, Ljubljana

Nose Marolt M, Gulič J (2010) NATREG - Kako upravljati (za)varovana območja narave, da bodo postala priložnost za trajnostni razvoj [NATREG: How to manage protected areas to turn them into sustainable development opportunities]. Regionalni center za okolje za vzhodno in srednjo Evropo Slovenija, Ljubljana

Občina Lovrenc na Pohorju (2022) Penzion Bitner. https://www.lovrenc.si/objava/150154. Accessed 7 Jul 2023

Orožen J (1963) Zgodovina Planinskega društva Celje [History of the Celje Alpine Club]. Unpublished material, Celje

Pachschwöll A (2011) Leopold Hufnagl (1857–1942): Ein ökologischer Vordenker. Unsere Heimat 82(1):30–33

Perko D, Orožen Adamič M (eds) (1998) Slovenija—pokrajine in ljudje. [Slovenia—landscapes and people]. Mladinska knjiga, Ljubljana

Perko D, Ciglič R, Zorn M (2020) Slovenia: a European landscape hotspot. In: Perko D, Ciglič R, Zorn M (eds) The geography of Slovenia: small but diverse. Springer, Cham, pp 1–20. https://doi.org/10.1007/978-3-030-14066-3_1

Peterlin S (1994) Narodni park Logarska dolina-Okrešelj: nekaj dokumentiranih drobcev iz let 1931–1932 [Logar Valley-Okrešelj National Park: selected documented fragments from 1931 and 1932]. Mohorjev Koledar 1994:43–45

Piskernik A (1962) Zgodovina prizadevanj za ustanovitev Triglavskega narodnega parka [History of the efforts to establish Triglav National Park]. In: Varstvo narave, vol 1. Zavod Republike Slovenije za varstvo naravne in kulturne dediščine, Ljubljana, pp 9–20

Piskernik A (1963–1964) Iz zgodovine slovenskega varstva narave [From the history of Slovenian nature conservation]. In: Varstvo narave, vol 2–3. Zavod Republike Slovenije za varstvo naravne in kulturne dediščine, Ljubljana, pp 59–74

Polajnar Horvat K, Smrekar A, Zorn M (2017) Environmental thought in Slovenia. In: Petrić H, Žebec Šilj I (eds) Environmentalism in Central and Southeastern Europe: historical perspectives. Lexington Books, Lanham, pp 21–33

Poličnik T (2008) Zgodovinsko ozadje odkupa Logarske doline-Okrešlja za narodni park [Historical background of repurchasing the Logar Valley–Okrešelj area for a national park]. BSc Thesis, University of Ljubljana, Ljubljana

Potočnik M (1975) Povojna prizadevanja za razglasitev današnjega (in jutrišnjega) Triglavskega narodnega parka [Postwar efforts to declare today's (and tomorrow's) Triglav National Park]. In: Peterlin S (ed) 50 let Triglavskega narodnega parka. Prirodoslovno društvo Slovenije, Ljubljana, pp 61–63

Prelesnik K (2022) Letno poročilo o delu v Krajinskem parku Logarska dolina 2022 [Annual report on the work of the Logar Valley Landscape Park 2022]. Občina Solčava, Solčava

Sedej D (1994) Kje so tiste stezice...: vrsta problemov v Triglavskem narodnem parku [Where are the trails of yore...: a series of problems in Triglav National Park]. Gorenjski glas, 3 Jun 1994, p 4

Skoberne P (2016a) Triglavski narodni park [Triglav National Park]. Mohorjev Koledar 2016:30–34

Skoberne P (2016b) Triglavski narodni park: kronski dragulj slovenske narave—dediščina za prihodnost [Triglav National Park: the crown jewel of Slovenian nature—heritage for the future]. Natl Geogr Slov 11(5):104–129

Slapnik M (2020) Zasnova Načrta interpretacije tematske Poti po Logarski dolini za podajanje predlogov [Design of the Logar Valley Thematic Trail Interpretation Plan for proposals]. Občina Solčava. Solčava

Smolej I (1984) Prispevek k zgodovini blejskih gozdov [Contribution to the history of the Bled forests]. Kronika 32(2–3):145–154

Smrekar A, Polajnar Horvat K, Ribeiro D (2020) Slovenia's protected areas. In: Perko D, Ciglič R, Zorn M (eds) The geography of Slovenia: small but diverse. Springer, Cham, pp 313–320. https://doi.org/10.1007/978-3-030-14066-3_21

Smrekar A, Polajnar Horvat K, Ribeiro D (2023) Institucionalni okvir ohranjanja narave v Sloveniji [Institutional framework for nature conservation in Slovenia]. Geografija v Šoli 31(2–3):9–21. https://doi.org/10.59132/geo/2023/2-3/9-21

Spomenica Odseka za varstvo prirode in prirodnih spomenikov [Memorandum by the Section for Nature

and Natural Monument Protection] (1920) Glasnik Muzejskega društva za Slovenijo 1(1–4):69–75
Stojilković B, Stepišnik U, Žebre M (2013) Pleistocenska poledenitev v Logarski dolini [Pleistocene glaciation in the Logar Valley]. Dela 40:25–38. https://doi.org/10.4312/dela.40.2.25-38
Teržan J (1967) Ob 65 letnici PD Ruše [The sixty-fifth anniversary of the Ruše Alpine Club]. Planinski Vestnik 67(2):50
Triglav National Park (2024). Javni zavod Triglavski narodni park, Bled. https://www.tnp.si/en/park/area/triglavski-narodni-park/. Accessed 1 Jul 2024
Uredba o nacionalnim parkovima [Decree on national parks] (1938) Službeni list 434/71. Službene novine 128/1939. Beograd
Uredba o Regijskem parku Pohorje [Decree on Pohorje Regional Park] (2024) Uradni list Republike Slovenije 9/2024. Ljubljana
Vintar Mally K, Bobovnik N, Lampič B, Kušar S (2020) Analiza socialno-ekonomskih vplivov projektnih aktivnosti na lokalno gospodarstvo in prebivalstvo: v okviru izvajanja projekta "Ohranjanje in upravljanje suhih travišč v Vzhodni Sloveniji" [Analysis of the socioeconomic impacts of project activities on the local economy and population: in the context of implementing the project "Conservation and management of dry grasslands in Eastern Slovenia"]. Univerza v Ljubljani, Ljubljana
Zakon o narodnih parkih [National Parks Act] (1959) Uradni list Ljudske Republike Slovenije 6/1959. Ljubljana
Zakon o ohranjanju narave [Nature Conservation Act] (1999) Uradni list Republike Slovenije 56/1999. Ljubljana
Zakon o Triglavskem narodnem parku [Triglav National Park Act] (1981) Uradni list Socialistične Republike Slovenije 17/1981. Ljubljana
Zakon o Triglavskem narodnem parku [Triglav National Park Act] (2010) Uradni list Republike Slovenije 52/2010. Ljubljana
Zavarovana območja v Sloveniji = Protected areas of Slovenia (2008). Ministrstvo za okolje in prostor, Ljubljana
Žiberna M (2016) Usklajevalec interesov [Interest coordinator]. Natl Geogr Slov 11(5):130–133
Žiberna I, Zajc P (2020) Geografske značilnosti Pohorja [Geographical characteristics of Pohorje]. Proteus 83(2–5):104–114
Zorn M, Smrekar A, Skoberne P, Šmuc A, Brancelj A, Dakskobler I, Poljanec A, Peršolja B, Erhartič B, Ferk M, Hrvatin M, Komac B, Ribeiro D (2015) Dolina Triglavskih jezer [The Triglav Lakes Valley]. Geografija Slovenije, vol 32. Založba ZRC, Ljubljana. https://doi.org/10.3986/9789610503620
Zupančič B (1998) Padavine [Precipitation]. In: Fridl J, Kladnik D, Orožen Adamič M, Perko D (eds) Geografski atlas Slovenije. DZS, Ljubljana, pp 98–99

27 Temperature Conditions and Vegetation in Dinaric Frost Hollows

Matej Ogrin, Gojko Nikolić, Domen Svetlin, Pavle Cikovac, and Carola Küfmann

Abstract

Topographic depressions in high mountain karst regions are among the coldest known areas not only in a regional context but also on a global scale, which is why they are also referred to as frost hollows. Although knowledge of thermal conditions in depressions is generally well established, scientific research on microclimates and periglacial ecosystems in glaciokarstic depressions of the High Dinaric Mountains is scarce and suffers from a lack of continuous data collection. Due to the prevailing natural chemical weathering processes in the limestone areas, karst landscapes with a large number of karst depressions, e.g., dolines, uvalas, and poljes, are characteristic in this region. They provide specific types of microclimates with adequate ecological conditions for cold-adapted species and are interesting geo- and biodiversity sites. Their occurrence indicates migrations during the cold phases of the Pleistocene glacial cycles, in the reverse direction along the north–south axis to the Balkans. The paper deals with temperature conditions and combined plant species as ecological indicators of biological feedbacks to the exceptional frost hollow microclimates in periglacial areas in the Dinaric Mountains of Montenegro. The study area covers the altitudes between 1,200 and 2,026 m in the mountain karst plateau of Sinjajevina Mountains, the Orjen massif, and the Durmitor Mountains. The related fieldwork analyses and continuous measurements were carried out in representative plots in the study area. Strikingly, the results show that the absolute lowest temperatures for the whole Montenegro are found in Durmitor Mountains.

Keywords

Glaciokarst · Microclimate · Cold-air pool · Warm stage refugia · Arcto-alpines · Bio- and geodiversity · Dinaric Mountains · Montenegro

M. Ogrin (✉)
Department of Geography, Faculty of Arts, University of Ljubljana, Ljubljana, Slovenia
e-mail: matej.ogrin@ff.uni-lj.si

G. Nikolić
Department of Geography, Faculty of Philosophy, University of Montenegro, Nikšić, Montenegro
e-mail: gojkorn@t-com.me

D. Svetlin
Envirodual d.o.o., Slovenske Konjice, Slovenia
e-mail: domen.svetlin@envirodual.com

P. Cikovac · C. Küfmann
Department of Geography, University of Munich, Munich, Germany
e-mail: pavle.cikovac@campus.lmu.de

C. Küfmann
e-mail: carola.kuefmann@geographie.uni-muenchen.de

M. Zorn et al. (eds.), *Discourses on Mountains of Montenegro and Slovenia*, Historical Geography and Geosciences, https://doi.org/10.1007/978-3-031-99017-5_27

27.1 Introduction

Relief has a major influence on climatic and weather diversity, so the different topography is often reflected in the different microclimates of an area (Ogrin and Kozamernik 2020). Frost hollows are a typical representative of the synergy between relief forms and meteorological phenomena, which have a great influence on temperature conditions at the local scale and, consequently, on other areas of living and non-living nature. In many places, they represent an important element of geodiversity and influence biodiversity. The term frost hollow refers to a concave relief shape in which temperatures are lower than outside the frost hollow at the same elevation (Trošt 2008). The mechanism of nocturnal cooling in frost hollows has already been described in the scientific literature (e.g., Sauberer and Dirmhirn 1954; Whiteman et al. 2004; Ogrin et al. 2006; Ogrin 2007; Ortar 2011; Dorninger 2016) and in a related article in this monograph, so we will not revisit it here. The key factor in the occurrence of frost hollows is the presence of concave relief and the formation of cold-air pools (CAP). This is often the case in karst areas where water dissolves the limestone through corrosion and drains vertically into the ground, forming depressions. These are dolines, collapse dolines, uvalas, and poljes. Concave relief forms are also found in non-karst areas with glacial or periglacial transformation. Concave relief forms often occur behind frontal moraines, and nivation hollows can also form in the area of periglacial impacts (Natek 2007). In volcanic topography, concavities are also represented by volcanic craters.

The phenomenon of frost hollows in the Alps has been described several times, mainly with climatological or meteorological focus, such as nocturnal cooling (e.g., Whiteman et al. 2004; Ogrin and Ogrin 2005; Zängl 2005; Steinacker et al. 2007; Trošt 2008; Ortar 2011; Svetlin 2020). Interestingly, research on frost hollows in Dinaric Mountains focuses more on vegetation and ecological conditions (e.g., Gams 1972; 2004; Martinčič 1975, 1977; Cikovac and Hölzle 2018; Cikovac and Ljubisavljević 2020). In this article we focus on climatological, vegetation, and ecological conditions of Dinaric frost hollows with emphasis on selected frost hollow sites in Montenegrin mountains of the Durmitor, Sinjajevina, and Orjen massifs. No less than 75% of the territory of Montenegro is covered by carbonate rocks, mostly limestone (Nikolić and Bušković 2012). These include Lovćen, Orjen, Rumija, Durmitor Mountains, Morača Mountains, Maganik, Sinjajevina, Komovi, Prokletije Mountains, Piva Mountains, and others, where many karst depressions of different sizes exist, and where CAPs are regularly formed by radiative processes under clear sky conditions. As a result, the minimum and average temperatures here are much lower than outside these areas (Ogrin et al. 2018). The temperature conditions of the Dinaric frost hollows have been much less studied than those of the Alpine frost hollows (e.g., Sauberer and Dirmhirn 1954; Whiteman et al. 2004; Ogrin and Ogrin 2005; Ogrin et al. 2006; Ogrin 2007; Steinacker et al. 2007; Dorninger 2016). Ortar et al. (2010) presented climatological conditions in Valoviti do on Durmitor and Reovce on Orjen. The first measurements confirmed that Valoviti do is a very favorable location for nocturnal cooling, so the measurements were continued. Ogrin et al. (2018) prepared an analysis of temperature conditions in the Valoviti do in comparison to meteorological mountain stations in Montenegro. In this article (Fig. 27.1), these findings are upgraded with climatological data from measurements in frost hollows in the Valoviti do (Figs. 27.2 and 27.3) and Planinica on the Sinjajevina (Figs. 27.4 and 27.5), and placed in the context of vegetation and ecological conditions based on observations in the frost hollow Borovi do (Figs. 27.7 and 27.8) and Opuvani do (Figs. 27.6 and 27.10) on the Orjen massif.

Fig. 27.1 The area of research (*Sources* EEA EU-DEM v1.1 2016; Eurostat 2020). (Author of the map Domen Svetlin)

Fig. 27.2 Bottom of frost hollow Valoviti do with a measurement instrument. (Photo by Matej Ogrin)

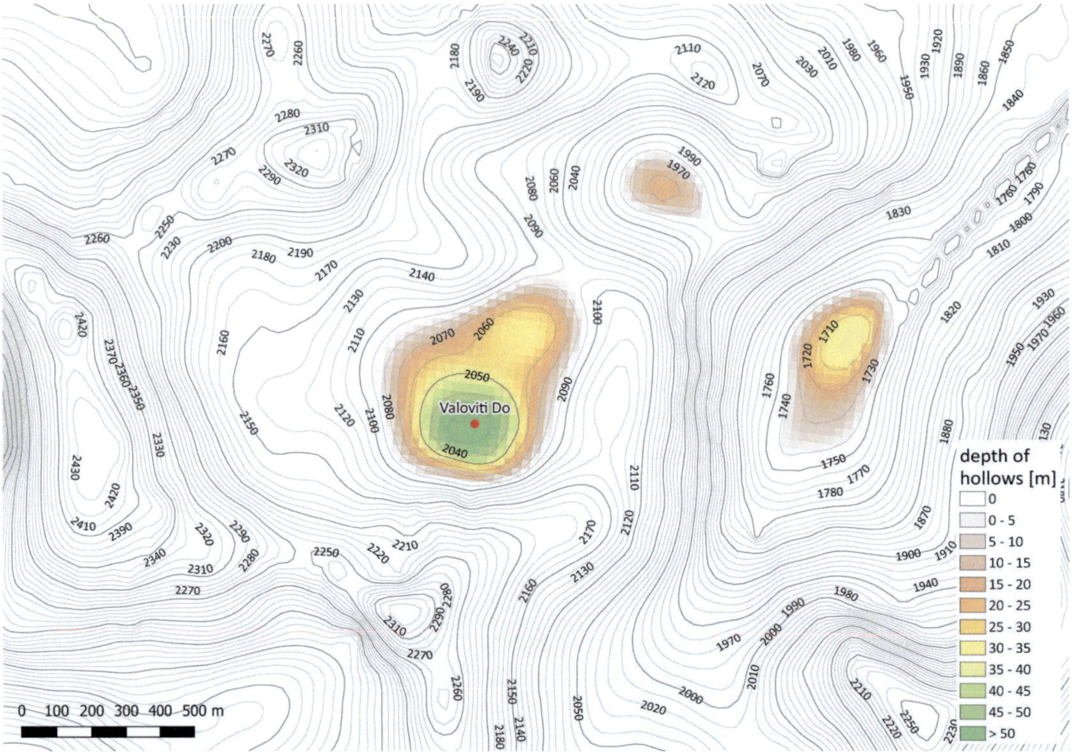

Fig. 27.3 Topography of Valoviti do (*Sources* EEA EU-DEM v1.1 2016; Eurostat 2020). (Author of the map Domen Svetlin)

Fig. 27.4 Summer view on frost hollow Planinica with measuring instrument on the pole. (Photo by Matej Ogrin)

Fig. 27.5 Topography of Planinica (*Sources* EEA EU-DEM v1.1 2016; Eurostat 2020). (Author of the map Domen Svetlin)

27.2 Temperatures in Frost Hollows of Valoviti Do and Planinca as an Indicator of Temperature Conditions of Frost Hollows in Dinaric Mountains

27.2.1 Study Area

27.2.1.1 Valoviti Do (Durmitor)

Durmitor is a mountain range in northern Montenegro and, along with Prokletije, the highest mountain massif in Montenegro. The highest peak is Bobotov Kuk (2,523 m), and the mountain range is part of Dinaric Mountains, which stretches from Slovenia to the Prokletije Mountains on the border between Montenegro and Albania. It is composed of carbonate rocks (Đurović 2009), mainly limestones formed during the Alpine orogeny (Morley 2007). Due to the geological composition and high altitude, there are a large number of karst, glacial, and periglacial relief forms in this area (Annys et al. 2014).

Valoviti do (Figs. 27.1–27.3) is a high karst depression in the central part of Durmitor. The bottom is at an altitude of 2,018 m, and the depression is surrounded in the west by the highest peaks of Durmitor, such as Bobotov Kuk (2,523 m), Bezimeni vrh (2,487 m), and Lučin vrh (2,396 m), in the south the depression closes Minin Bogaz (2,387 m), and in the east Pasovi and Sagorele ploče (2,120 m). The eastern edge of the hollow is also the lowest and represents the outflow of cold air at an altitude of almost 2,100 m. The bottom of the hollow is heavily fragmented with slope material, and the hollow is regularly filled with avalanches triggered by the slope of Minin Bogaz, forming snowfields in late summer. The bottom of the hollow is about 220 m long and 300 m wide. The depth of frost hollow is 59 m and sky view factor 0.9 (Table 27.1).

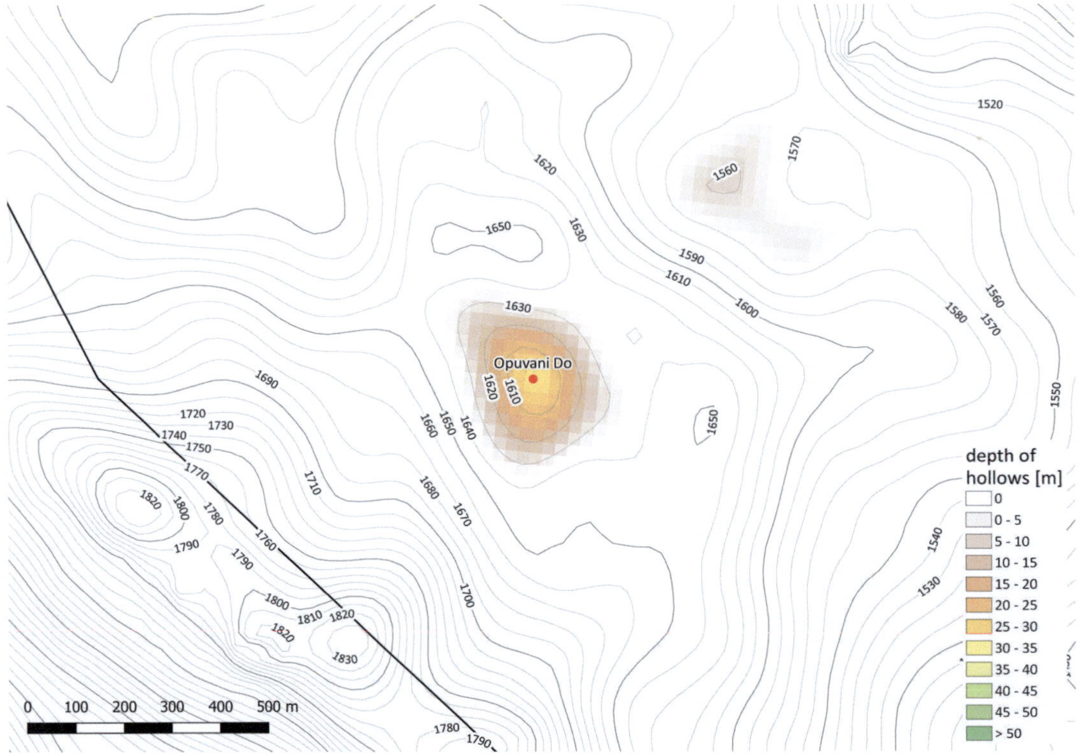

Fig. 27.6 Topography of Opuvani do (*Sources* EEA EU-DEM v1.1 2016; Eurostat 2020). (Author of the map Domen Svetlin)

27.2.1.2 Planinica (Sinjajevina Plateau)

Sinjajevina is the highest limestone plateau in Montenegro, located in the northern part of central Montenegro, with its northwestern parts reaching about 1,600 m above sea level, while the central and southeastern parts are higher, reaching about 1,900 m above sea level. Six peaks exceed 2,000 m above sea level, the highest being Babin zub (2,253 m above sea level). The composition of the limestone allows for numerous karst phenomena, and the surface is riddled with karst cavities. Many of them are pronounced frost hollows, the potential of which has not yet been explored. The frost hollow Planinica (Figs. 27.1, 27.4, and 27.5) is located at an altitude of 1,631 m in the central part of the plateau, about 5.5 km horizontally from the settlement Dobrilovina and 3 km west of the Zabojsko Lake in the area called Planinica. It is a shallow and very open frost hollow, with depth from the lowest point 9.5 m and sky view factor 0.99 (Table 27.1).

27.2.1.3 Opuvani Do and Borovi Do (Orjen Mountains)

Orjen is a mountain massif that represents the border area of Montenegro with Croatia and Bosnia and Herzegovina. It reaches an altitude of 1,893 m (Mt. Veliki Kabao). In the northwest, it extends to the Konavle polje, in the north to southern Herzegovina, Trebinje, and Grabsko poljes, in the southeast and east to the Grahovsko and Dragaljsko poljes, and in the south to the Bay of Kotor. As a typical Dinaric mountain massif, Orjen is built of limestone and represents a typical mountain karst surface, which has been glacially overformed in some places (Stepišnik et al. 2009). The frost hollows Borovi do (1,411 m; Figs. 27.1 and 27.7) and Opuvani do (1,604 m; Figs. 27.1 and 27.6) are located in the northern part of the

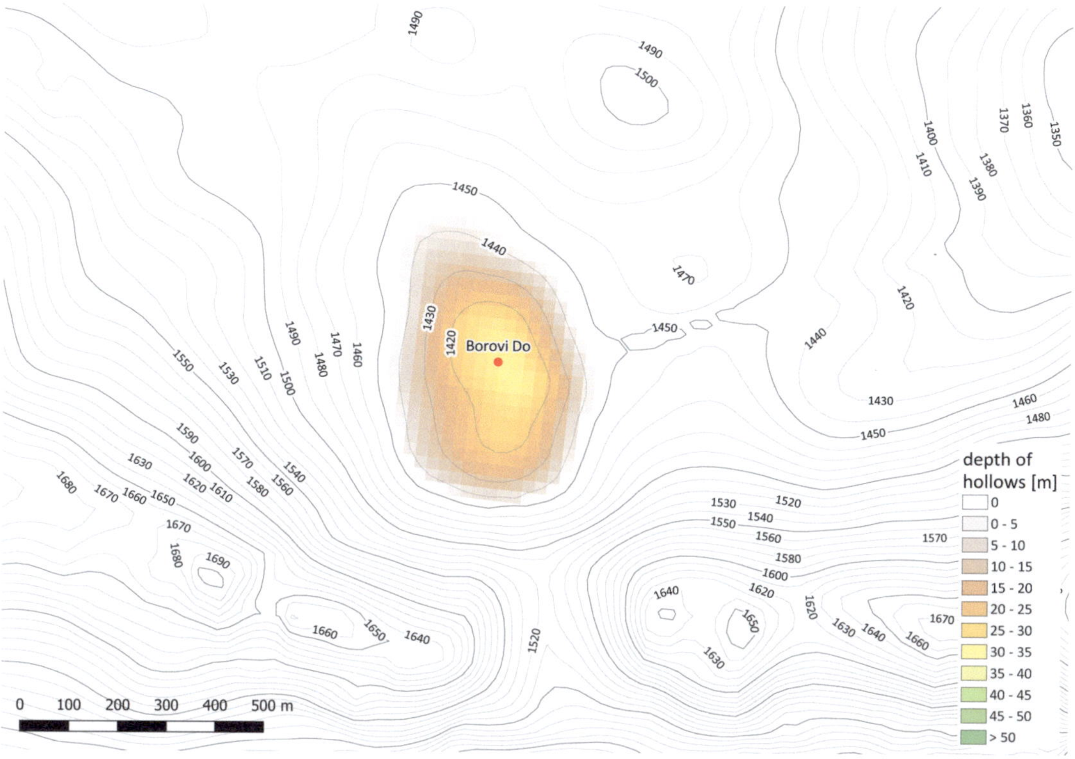

Fig. 27.7 Topography of Borovi do (*Sources* EEA EU-DEM v1.1 2016; Eurostat 2020). (Author of the map Domen Svetlin)

Fig. 27.8 The pan shaped Borovi do (1,470 m) frost hollow on Mt. Orjen measures 450 m across. (Photo by Pavle Cikovac)

Table 27.1 Metadata of researched frost hollows (for locations, see Fig. 27.1)

Frost hollow	Broader area	Latitude	Longitude	Depth (m)	Altitude of the deepest point (m)	Sky view factor (mean)	Data logger type	Measurement period
Valoviti do	Orjen	43.131880	19.048830	59.0	2,026.0	0.900	Dallas Semiconductor i- button, Madgetech Cryo-Temp	2007–2008, 2010–2013, 2014–2015
Opuvani do	Orjen	42.602917	18.548390	31.0	1,603.7	0.962		
Planinica	Sinjajevina	43.010560	19.338330	9.5	1,630.8	0.990	Dallas Semiconductor i- button	
Borovi do	Durmitor	42.582070	18.570549	34.9	1,411.0	0.935		

massif and are two of numerous frost hollows in this area. The depth of Opuvani do is 31 m, while Borovi do is 4 m deeper (35 m). Mean sky view factor in Opuvani do is 0.962 and in Borovi do 0.935 (Table 27.1).

27.2.2 Methodology of Measurements

For an automated reading of daily temperatures, digital temperature data loggers were installed in frost hollows. The devices recorded temperatures at a predetermined time and at a selected time step. The measurements were conducted in the cold season, covering the period from the end of October until end of March; in some cases they were extended to April. Temperature readings from the Valoviti do were obtained in 2007/2008, 2010/2011, 2011/2012, 2012/2013, and 2014/2015, and in Planinica for 2010/2011, 2011/2012, 2012/2013, and 2014/2015. The temperature readings were set to intervals of 30 min during 2007/2008, to 15 min during 2010/2011, 2011/2012, 2014/2015, and to 10 min during 2012/2013. In Valoviti do in 2012/2013 season, the Madgetech CryoTemp Data Logger measuring instrument was used. The Dallas Semiconductor i-button recorders were used in other years in Valoviti do and Planinica. The Dallas Semiconductor recorder obtains temperatures down to – 41 °C; the error interval is –0.9 to + 1.6 °C at the lowest temperatures and –0.7 to + 1.2° C at –30 °C (Vertačnik and Sinjur 2013), while the measurement range of the Data Logger CryoTemp is down to –86 °C and the accuracy is ± 1° C. The accuracy of the measurements was 0.5 ºC.

A temperature recorder was installed on a pole, with probes protected by a specially designed radiation shield for the protection from the weather events (see also Ogrin 2007). The radiation shield was designed to meet World Meteorological Organization standards as much as possible—it is made of materials with low thermal capacity and conductivity and high light reflectance. The radiation shield with the temperature recording device was placed in the lowest part of the frost hollow at a height of 4–5 m above the ground, so that it would not be effected from high snow cover.

Data from November to March were included in the analysis. For Valoviti do, the 2012/2013 data set is incomplete because an avalanche overturned and buried the recorder in February. At Valoviti do, measurements were taken on 28 days in November 2007 (3–30 November 2007) and on only 25 days in March 2011 (1–25 March 2011).

27.2.3 Temperature Conditions in Frost Hollows of Valoviti Do and Planinica

The analysis included minimum temperatures in both frost hollows and the number of days with characteristic minimum temperatures. Data from the meteorological station Žabljak, the highest meteorological station in Montenegro, located 7 km from Valoviti do, were also used for comparison (Table 27.2).

A comparison of the minimum temperatures of the two Dinaric frost hollows and the highest and coldest stations within the network of National Meteorological and Seismological Service of Montenegro in Žabljak (Durmitor) shows more extreme temperature conditions in the frost hollows. Of particular interest are the temperatures from the frost hollow Planinica, which lies at an altitude of only 150 m above the altitude of the meteorological station at Žabljak. During four winter measurement periods, the average minimum temperatures on Planinica were up to 3.9 °C lower than the temperature in Žabljak. Valoviti do, which lies at an altitude 418 m above Planinica, was on average 4.1 °C colder than Žabljak or only 0.2 °C colder than Planinica for the same period. Thus, the average minimum temperatures of the considered period show very similar conditions for both frost hollows. The average minimum temperature in Valoviti do was –8.8 °C and in Planinica –8.6 °C.

Tables 27.3, 27.4, and 27.5 show the number of characteristic days with minimum temperatures in both frost hollows. Interestingly, the number of days with minimum temperature below 0 °C is higher in Valoviti do (8.4 days), and at minimum temperatures below –10 °C the ratio turns in favor of Planinica frost hollow: 3.7 days at minimum temperature < –10 °C, 5.2 days at minimum temperature < –20 °C, and 1.9 days at minimum temperature < –30° C. The cooling potential in frost hollows is influenced by several factors, and elevation is only one of them. In general, higher elevation frost hollows have a higher cooling potential due to lower average humidity and generally lower air temperatures (Dorninger 2016).

However, the shape of the frost hollow and the surroundings is also important (Ortar 2011). On the other hand, the cooling potential is not necessarily related to average temperatures or average minimum temperatures or to the frequency of occurrence of a CAP, which is a basic condition

Table 27.2 Average minimum air temperatures (°C) in the Valoviti do and Planinica frost hollows and at the Žabljak (Durmitor) meteorological station in the period from November to March and average minimum temperatures for each season

Measurement period	Study area	November	December	January	February	March	Season (°C)
2007/2008	Žabljak (1,450 m)	–5.4	–7.1	–6.2	–6.2	–3	–5.6
	Valoviti do (2,018 m)	–8	–15.2	–10.9	–11.4	–7	–10.5
2010/2011	Žabljak (1,450 m)	1.6	–5.6	–6.9	–6.6	–4.5	–4.4
	Planinica (1,631 m)	0.8	–8.4	–12.9	–14.6	–7.4	–8.5
	Valoviti do (2,018 m)	–0.2	–10.3	–12.8	–14.9	–14.9	–10.6
2011/2012	Žabljak (1,450 m)	–2.6	–5	–9.5	–11.8	–3.2	–6.4
	Planinica (1,631 m)	–9.2	–9.1	–15.6	–14.4	–6.7	–11.0
	Valoviti do (2,018 m)	–4.8	–11.1	–13.9	–13.4	–8.6	–10.4
2012/2013	Žabljak (1,450 m)	1.4	–6.9	–6.2	–4.7	–3.1	–3.9
	Planinica (1,631 m)	–0.9	–11.3	–10.3	–9.4	–7.5	–7.9
	Valoviti do (2,018 m)	–1.1	–10	–10.4	/	/	–7.2
2014/2015	Žabljak (1,450 m)	1.9	–4.4	–6.1	–6.9	–4.1	–3.9
	Planinica (1,631 m)	–1	–7.4	–9.2	–11.8	/	–7.4
	Valoviti do (2,018 m)	–0.3	–7	–8.6	–13.1	/	–7.3

Table 27.3 Average number of days by months and total average number of days with characteristic temperatures in the Planinica frost hollow during the winter seasons 2010/2011, 2011/2012, 2012/2013, and 2014/2015 (*only three measurement seasons)

Planinica (1,631 m)	November	December	January	February*	March	Σ (November–March)
< 0 °C	15.7	24.0	29.8	27.7	19.3	116.6
< –10 °C	3.7	12.3	15.7	18.3	9.3	57.3
< –20 °C	0.0	5.7	6.8	8.0	3.7	23.0
< –30 °C	0.0	1.00	2.20	1.3	0.7	5.2

Table 27.4 Average number of days by months and total average number of days with characteristic temperatures in Valoviti do in the winter seasons 2010/2011, 2011/2012, 2012/2013, and 2014/2015 (*only three measurement seasons)

Valoviti do (2,018 m)	November	December	January	February*	Σ (November–February)
< 0 °C	18.3	28.5	30.5	28.3	105.6
< –10 °C	1.0	12.5	15.8	17.0	46.3
< –20 °C	0.0	5.0	4.0	6.3	15.3
< –30 °C	0.0	0.5	0.8	1.3	2.6

Table 27.5 Average number of days with characteristic temperatures in the Valoviti do and Planinica during the winter seasons (November–February) 2010/2011, 2011/2012, 2012/2013, and 2014/2015

	Valoviti do (2,018)	Planinica (1,631 m)
< 0 °C	105.6	97.2
< –10 °C	46.3	50.0
< –20 °C	15.3	20.5
< –30 °C	2.6	4.5

for low temperatures in frost hollows (Ogrin et al. 2018). In our cases, Valoviti do appears to be more frequently ventilated, which indicates a lower frequency of lower minimum temperatures (below –10 °C). However, in the four winter seasons, the cooling potential in Valoviti do proved to be higher than in Planinica, as the absolute minimum in Valoviti do was –40.0 °C (4 January 2011) and in Planinica –38, 5 °C (13 December 2012). The absolute minimum of the considered period in Valoviti do is also the lowest documented temperature measured in Montenegro.

27.3 Frost Hollow Vegetation in Glaciokarstic Basins

The temperature conditions are important for other ecological conditions. Temperature affects both biotic (e.g., vegetation, decomposers) and abiotic nature (e.g., chemical, mechanical weathering) in frost hollows. Since limestone is the predominant rock in Montenegro and on Dinaric Mountains, there are many frost hollows, so the knowledge of ecological conditions in these environments has a significant impact on the bio- and geodiversity of the whole Dinaric Mountains.

Newly detected, highly fragmented occurrences of glacial relicts in the Southeastern Dinaric Mountains have challenged previous assumptions on the existence of topographically based warm-stage refugia under Mediterranean climates (Cikovac and Ljubisavljević 2020). They are associated with specific karst topographic landforms (e.g., dolines, collapse dolines, uvalas, poljes) with enclosed atmospheres of frost hollows that have the ability to create stable cold-functional environments. Frost hollows have comparably lower minimum temperatures, provide higher ground moisture availability, and sustain longer snow cover through the summer ablation period, which is prolonged through wind and topography-controlled snow distribution (Horvat et al. 1974; Cikovac and Ljubisavljević 2020).

Thus, karst depressions enclose local areas with continental temperature regimes which modify landscape terrain morphology by seasonal diurnal freeze–thaw cycles. Under such microclimatic conditions they can offer resilient habitats for essentially arcto-alpine communities, even at sites in the Mediterranean climatic zones with relatively warm winters (i.e., sub-Adriatic Dinaric Mountains).

In limestone areas in Montenegro karst is well developed, which is a consequence from the amount of precipitous waters, which is greater than in most of Europe (Ducić et al. 2012). It lays the foundation to a great deal of karst depression development and not much surface runoff, even though the slopes are steep. As the Mesozoic limestones are at 1,000–2,500 m, a great number of karst depressions lies above tree lines, where they provide snow-rich and cold microhabitats in the Dinaric-alpine region (Lakušić 1970; Matvejev and Puncer 1989). In Mediterranean climate type areas in Montenegro, where mountain ranges are not high enough to reach the alpine life zone, alpine habitats occur at spatial scales of karst depressions from a few hundred meters across (Fig. 27.8) (Cikovac and Hölzle 2018). As these depressions could hold possible warm stage reserves for adaptations to the changing of modern ecosystems, scientific studies in global change and biodiversity-related topics have recently shifted their interest to frost hollows that could play a critical role in climate resilience for a specific set of temperature sensitive communities (i.e., climatic niches) (Cikovac and Ljubisavljević 2020). Such local spots with potential to buffer from ambient climate in a warming world are regularly established in topographic concavities of the Dinaric karst in Montenegro (Cikovac and Hölzle 2018).

27.3.1 Vegetation-Frost Hollow Interactions

Studies on diurnal mountain wind circulations exist since the nineteenth century (Zardi and Whiteman 2013). Studies in frost hollow vegetation in the Northwestern Dinaric Mountains root in the first half of the nineteenth century (Horvat 1953). They served as classic sites to formulate concepts of frost hollow vegetation phenomena (Horvat 1961). However, frost hollows in the physiographic area of Montenegro have been studied for a mere period of a couple of years (Ogrin et al. 2018); and although physical dynamics of frost hollow atmospheres (e.g., formation of CAP) are well documented, understanding of ecological responses is still lacking (Cikovac and Hölzle 2018).

Beck (1906) tied "vegetation inversion" (Ger. *Höhenstufen-Umkehr*) as a conceptual interpretation of doline vegetation to temperature inversion. However, the basic cause of azonally distributed taxa in dolines goes back to landscape-vegetation functional feedbacks. Microclimatological research suggests that a systematic measurement of the common air temperature of meteorological observations does not refer to the real microclimatic condition in itself to draw a real picture of the local distribution of compact habit alpine plants (Körner 2021). This is supported by systematic analysis of phytocoenoses in the Viljinska ponikva doline in the Northwestern Dinaric Mountains (Antonić et al. 1997), where no link between the local arrangement of phytocoenoses and air temperature was revealed. Thus "vegetation inversion" is a misnomer to isotherms of temperature in the distribution of alpine life forms above climatic treelines.

27.3.2 Ecological Gradients in Dinaric Frost Hollows

Snow load distribution patterns and snow depth variation in Dinaric karst depressions are most important structured by land surface properties and accumulation of snow patch through

Table 27.6 N, C, $CaCO_3$ and pH-values of soils from tall herb (1), dwarf shrub heath (2), tussock grasslands (3), and snowbed communities (4) in the Opuvani do doline (Mt. Orjen)

Location	N (%)	C_{tot} (%)	C_{org} (%)	$CaCO_3$ (%)	Soil Ph	Fe (%)
Tall herb (1)	0.95	10.22	10.16	0.54	4.71	–
Dryadetum octopetalae (2)	0.38	10.45	5.81	72.38	7.09	–
Oxytropidion dinaricae (3)	1.00	13.32	10.60	22.65	7.25	1.4
Salicion retusae (4)	1.38	19.58	19.35	1.85	6.76	1.4

windblown redistribution (Horvat et al. 1974). Snow cover insulation avoids critical frost damages in freezing sensitive organisms in Dinaric snowbed and dwarf scrub mountain tundra communities (e.g., *Salicetum retusae*, *Dryadetum octopetalae*), which thrive on active periglacial substrates. Wind exposed slopes and rigdes are dominated by Dinaric calcicolous tussock grasslands in the *Oxytropidion dinaricae* alliance (i.e., *Helianthemo alpestre-Caricetum kitaibelianae*). Tall-herb meadow vegetation in the *Adenostyllion alliariae* alliance and montane tall grass communities (i.e., *Deschampsietum subalpinum*) grow in concavities at bottoms, indicating oligotrophic, water-rich soils. Warmer and sheltered doline habitats are dominated by oro-Mediterranian tussock grasslands in the *Seslerion robustae* alliance. The enhanced snow accumulation in basins consequently regulates runoff fractions from snow and water with partitioning of lateral water discharge through snowmelt runoff channels and redistribution of nutrients and carbon to low-lying areas (Table 27.6).

Thus, spatial heterogeneity of snow cover has causal impacts in spatial variations of Dinaric plant-community mosaics, which are basically dependent on convergent doline environments (e.g., *Allietum schoenoprasii*, *Plantaginetum atratae*, *Rumicetum alpinae*), and from accumulation of water and soil, which helps to retain moisture during the dry summer period.

27.3.3 Phytogeographic Aspects of Dinaric Frost Hollows

Dinaric frost hollow habitats are dominated by shade-intolerant compact alpines that lack competition to tall grass and trees. Frequent subzero temperatures during the active growing phenological phase lean competitive advantages toward cold-affiliated compact organisms, while less cold-tolerant species will be constrained or would have limited concentration (Modrić-Surina and Surina 2010).

Dolines and other karst depressions in the Southeastern Dinaric Mountains create with their specific buffered microclimate strong topographical control on solifluction-vegetation interactions with reflections on community levels and species composition (Mucina 2015). Their vegetation in Montenegro is dominated by glacial relicts of arctic-alpine (e.g., *Dryas octopetala*, *Saxifraga oppositifolia*, *Oxytropis campestris* agg.), boreo-montane (e.g., *Deschampsia caespitosa*, *Dactylorhiza viridis*, and *Allium schoenoprasum*), and Dinaric endemic (e.g., *Viola calcarata* ssp. *zoysii*, *Festuca violacea* agg., *Armeria canescens*) affiliation (Cikovac and Hölzle 2018). Biogeographic connections to neighboring mountain ranges are expressed by South-European frigophilic and calcicolous snowbed species (e.g., *Plantago atrata*, *Heliosperma pusillum* s.l., *Salix retusa*, *Potentilla crantzi*, *Androsace villosa*, *Leontodon hispidus* spp. *hispidus*, *Carex kitaibelliana*, *Scabiosa silenifolia*, and *Crepis aurea* ssp. *glabrescens*) (Stevanović et al. 2009). Floristically similar are snowbeds from karst concavities in Slovenia and Croatia in the *Drepanoclado uncinati-Heliospermetum pusillae* community (Modrić-Surina and Surina 2010). However, frost hollows in Montenegro have a comparably higher species (e.g., *Edraianthus serpyllifolius*, *Aquilegia dinarica*, *Saxifraga prenja*, *Androsace federici-augusti*) and vegetational diversity (i.e., *Amphoricarpion neumayeri*, *Seslerion robustae*) (Antonić and Lovrić 1996).

27.3.4 Microclimatic Aspects of Different Frost Hollows

Stably stratified air layers in terrain constrictions are created by near surface cooling and fillings with heavy cold-air mass (Dorninger 2016). Deeper and more persistent stagnant cold-air masses are built in topographically sheltered concavities (Fig. 27.9) with consequent elimination to synoptic and mesoscale flows. This consideration in the climatic sensitivity to both wind direction and local topography is observed in frost hollows in the Orjen range (e.g., Borovi do; Fig. 27.8), where alpine and arctic-alpine vegetation exists in frost hollows unaffected by strong valley and upslope winds. Its indication might be transferred to the Durmitor massif, where depressions are arranged in ventilated U-shaped valley (i.e., Velika kalica valley). Unsurprisingly, it is from the Valoviti do doline, which lies on a sheltered high-plateau, and is roughly 60–100 m deep and 750 m across, where the lowest temperature minima in Montenegro was registered ever (Ogrin et al. 2018).

27.3.5 Case Studies on Mt. Orjen

First and foremost, this analysis is a detailed empirical study of the frost hollow vegetation in the Orjen range, but there are several good reasons for believing that the selected field

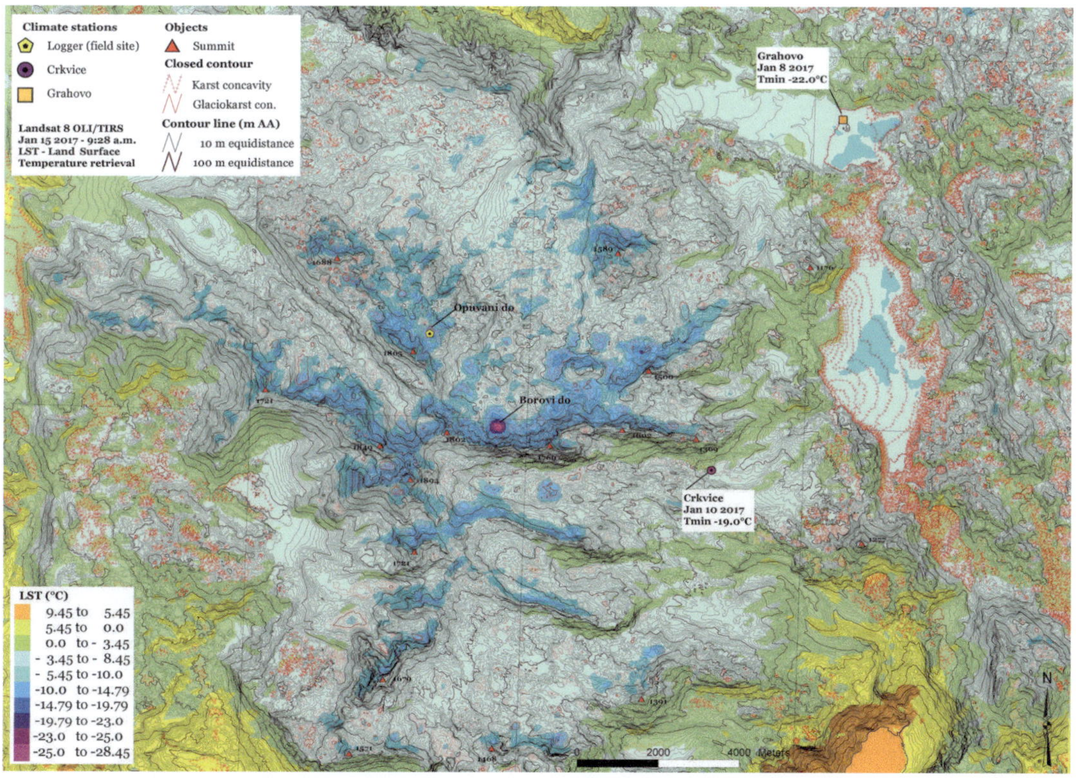

Fig. 27.9 Retrieved Land Surface Temperature (LST) from a Landsat 8 satellite image on January 15 2017. T-values are given in °C. LST was calculated through surface-temperature retrieval algorithms in ENVI 5.1, for which multispectral (OLI) and thermal (TIRS) data were transformed to radiance and reflectance values, emissivity, NDVI, and proportion of vegetation (Avdan and Jovanovska 2016). During the extreme cold in January 2017, the Borovi do doline in the Orjen range held the (presumable) cold minimum for Montenegro on January 15 2017. In situ measurements of January minima air temperature are given from official meteorological stations (Godišnjak ... 2019). (Author of the map Pavle Cikovac)

sites are an ideal starting point for charting the characteristics of frost hollow vegetation for the whole of Montenegro. Firstly, the altitudinal zonation is across biogeographic realms of the Mediterranean and temperate zone. Secondly, the species composition is markedly of the oro-Mediterranean type (Stevanović et al. 2009). This makes it manageable to distinguish between local oro-Arctic functional vegetation types in frost hollows, embedded in the oro-Mediterranean life zone, but, at the same time, also includes the basic outfit of the typical vegetational organization and species pool of calcicolous Southeastern Dinaric alpine communities. This aspect is important, as the sheer number, scope, and complexity in the range of Southeastern Dinaric depressions often make in-depth empirical studies of the entire area unfeasible.

27.3.5.1 Frost Hollow Microrefugia

The Montengrin High-Karst physiographic area has gone through complex landscape evolution. Pre-Pleistocene dolines were vigorously glacierized during the last glacial cycle and are today modified by active periglacial processes (Stepišnik et al. 2009; Veress 2017). The Opuvani do doline on Mt. Orjen is a lateral elongated doline downstream the north oriented tributary hanging valley from the northern headwall of the Velika Jastrebica (1,865 m) (Fig. 27.10). Slopes are characterized by periglacial talus and periglacially molded bedrock with incised linear gullies and runnels in the upper sediments of deep cryoturbated soil layers. They are characterized by mountain aven (*Dryas octopetala*) communities (*Scabioso silenifoliae-Dryadetum octopealae)*. Snowbeds of the *Salicetea herbaceae* alliance thrive in areas with the longest lying snow patches. They are represented with the stunted willow (*Salix retusa*) community (*Salicetum retusae*). The mountain rib grass (*Plantago atrata*) snowbed wetland community (*Plantaginetum atratae*) emerges from late melting snow on level areas with water accumulation on fine-grained superficial deposits. Similarly, snowmelt runoff catchments are dominated by chive (*Allium schoenoprasum*) communities (*Allietum schoenoprasi*) (Cikovac and Hölzle 2018). Frost-sensitive tall herb communities (*Rumicetum alpinae*) occupy damp water-sucked and wind-protected low-lying basin areas on nutrient-rich soils.

Altogether 13 arctic-alpine species grow in the Opuvani do (Fig. 27.10), and comparably 22 arctic-alpine species are found in the Valoviti do on Durmitor (Fig. 27.2), from which 9 were not registered in the Opuvani do (Vukojičić 2008). Frost hollows are refugias for the highly fragmented distribution of the alpine salamander

Fig. 27.10 An elongated doline Opuvani do 1,584 m) with periglacially molded bedrock and meltwater runnels in cryoturbated soils. (Photo by Pavle Cikovac)

(*Salamandra atra*) (Cikovac and Ljubisavljevic 2020). They are also main habitats for Lepidopteran larvae (e.g., *Boloria pales*, *Boloria graeca*, *Pyrgus andromedae*) in high alpine butterfly species dependent on arctic-alpine plants (*Viola*, *Dryas*) and topographic concavities (e.g., *Agriades dardanus*, *Erebia gorge*) (Verovnik 2002).

27.3.6 Climate Change and Frost Hollow Refugia

Dinaric frost hollows provide an important physical factor in creating cold-functional habitats that maintain fundamental ecosystem services by safeguarding high species diversity at southern outposts of arctic and alpine plants. They can exist in regions where cold-adapted species groups currently have no suitable climate (e.g., Mt. Orjen). Thus, they could provide warm stage refugia for relicts with fragmented populations (Šunje et al. 2021) so they contribute to increase of biodiversity on local level. Frost hollows are relevant for snow accumulation and preservation. They interact with snowmelt runoff regimes, and increase N deposition from partitioning of snow loads which buffer against regional species losses (Körner 2021). Thereby they affect the seasonal start of ablation, plant phenology, and site selective regimes in physiological plant trait evolution and adaptation to soil-surface textures. Higher ground moistures than in surrounding landscapes add to comparably greater environmental heterogeneity, geodiversity, and species diversity.

27.4 Conclusion

Measurements in frost hollows of Southeastern Dinaric Mountains confirmed conclusions from alpine case studies (e.g., Sauberer and Dirmhirn 1954; Whiteman et al. 2004; Ogrin and Ogrin 2005; Ogrin et al. 2006; Ogrin 2007; Steinacker et al. 2007; Dorninger 2016) that diurnal cold-air accumulation in topographic concavities can lead to extreme and long frost periods. Thereby it mitigates thawing and the onset of warmer seasons.

Several years of measurements in the Dinaric frost hollows of Montenegro have shown that the cooling potential is similar to that in alpine frost hollows. Higher altitude frost hollows have a higher potential for absolute minima, but the frequency and duration of CAPs depends more on depression morphology than on altitude. This was also evident in our cases, as CAPs are more persistent in the Planinica frost hollow than in Valoviti do.

Periodic freeze–thaw processes enhance periglacial processes, frost creep, solifluction, and frost weathering (Cikovac and Hölzle 2018), with fundamental relevance for providing valuable ecosystem functions to arcto-alpines with complex trait responses to substrate textures from intense soil movements (e.g., cryoturbation and solifluction). Arcto-alpines at low latitude distributions at the southern margins of their geographic ranges in the maritime Mediterranean Southeastern Dinaric Mountains frequently persist only in these local favorable habitats outside the Arctic cold biome. The evidence of microrefugia in the Mediterranean climate type Dinaric Mountains approves to the potential importance of frost hollows in ameliorating warming temperatures, which locally might diminish and mitigate the effects of regional climate change. Current predictions caution that long-term warming leaves possibilities for cold-air refugia to resist, reflect, or disappear. They could at least hold interim solutions to resist deep ecosystem perturbations and vegetation assemblage conversions (Cikovac and Ljubisavljević 2020). Knowledge of temperature and vegetation conditions in frost hollows opens a new chapter in the study of the ecological characteristics of frost hollows and their role in assessing the evaluation of bio- and geodiversity of mountain areas.

Acknowledgements The authors would like to thank Golub Culafić (National Meteorological and Seismological Service of Montenegro) and Dragan Burić (Department of Geography, Faculty of Philosophy, University of Montenegro, Nikšić) for providing data. We would also like to thank students from the Department of Geography at the Faculty of Arts, University of Ljubljana, as well as Gregor

Vertačnik, Iztok Sinjur, Dejan Medojević, Darko Ogrin, Dušan Plut, Boštjan Rogelj, Tajan Trobec, Blaž Repe, and Dejan Cigale for their research assistance.

References

Annys K, Frankl A, Spalević V, Čurović M, Borota D, Nyssen J (2014) Geomorphology of the Durmitor Mountains and surrounding plateau Jezerska Površ (Montenegro). J Maps 10(4):600–611. https://doi.org/10.1080/17445647.2014.909338

Antonić O, Lovrić AŽ (1996) Numerical analysis of vegetation complexes and community diversity of major coastal Dinaric mountains. J Veg Sci 7(1):73–80. https://doi.org/10.2307/3236418

Antonić O, Kušan V, Hrašovec B (1997) Microclimatic and topoclimatic differences between the phytocoenoses in the Viljska ponikva sinkhole, Mt. Risnjak, Croatia. Hrvatski Meteorološki Časopis 32:37–49

Avdan U, Jovanovska G (2016) Algorithm for automated mapping of land surface temperature using LANDSAT 8 satellite data. J Sens 2:1–8. https://doi.org/10.1155/2016/1480307

Beck G (1906) Die Umkehrung der Pflanzenregionen in den Dolinen des Karstes. Sitzungsberichte der Akademie der Wissenschaften: mathematisch-naturwissenschaftliche Klasse 115(1):3–20

Cikovac P, Hölzle I (2018) Glacial relicts in the Mediterranean Dinarides: a phenomenon of cold-air pool microclimates. Botanica Serbica 42(1):74

Cikovac P, Ljubisavljević K (2020) Another isolated relic population of the Alpine salamander (*Salamandra atra* Laureti, 1768) (Amphibia: Caudata: Salamandridae) in the Balkans. Russian Journal of Herpetology 27(2):109–112. https://doi.org/10.30906/1026-2296-2020-27-2-109-112

Đjurović P (2009) Reconstruction of the Pleistocene glaciers of Mount Durmitor in Montenegro. Acta Geogr Slov 49(2):263–289. https://doi.org/10.3986/AGS49202

Dorninger M (2016) Topografische und meteorologische Faktoren für extrem tiefe Temperaturminima in Kaltluftseen. Promet 98:43–58

Ducić V, Luković J, Burić D, Stanojević M (2012) Precipitation extremes in the wettest Mediterranean region (Krivošije) and associated atmospheric circulation types. Nat Hazard 12(3):687–697. https://doi.org/10.5194/nhess-12-687-2012

Gams I (1972) Prispevek k mikroklimatologiji vrtač in kraških polj [A contribution to the microclimatology of the karst dolinas and poljes]. Geografski Zbornik 13:5–72

Gams I (2004) Kras v Sloveniji v prostoru in času [Karst in Slovenia in space and time]. Založba ZRC, Ljubljana

Godišnjak meteoroloških i hidroloških podataka 2017 [Yearbook of meteorological and hydrological data 2017] (2019). Zavod za hidrometeorologiju i seizmologiju Crne Gore, Podgorica

Horvat I (1953) Vegetacija ponikava: prilog biljnoj geografiji krša [The vegetation of dolines: a supplement to the vegetation geography of karst]. Geografski Glasnik 14–15:1–25

Horvat I (1961) Die Pflanzenwelt der Karst-Ponikven, eine besondere Vegetationserscheinung. Phyton 9:268–283

Horvat I, Glavač V, Ellenberg H (1974) Vegetation Südosteuropas. Gustav Fischer, Stuttgart

Körner C (2021) Alpine plant life: functional plant ecology of high mountain ecosystems. Springer, Berlin. https://doi.org/10.1007/978-3-642-18970-8

Lakušić R (1970) Die Vegetation der südöstlichen Dinariden. Vegetatio 21:321–373. https://doi.org/10.1007/BF02269665

Martinčič A (1975) Vpliv človeka na nastanek nekaterih mrazišč v Trnovskem gozdu [Human influence on the formation of some frost holows on the Trnovski gozd Plateau]. Razprave: posebna številka. Simpozij: meteorologija - gospodarstvo, vol 1. Društvo meteorologov Slovenije, Ljubljana, pp 35–43

Martinčič A (1977) Prispevek k poznavanju ekologije mrazišč v Sloveniji: botanično ekološka skica [Contribution to the knowledge of the ecology of frost holows in Slovenia: a botanical-ecological sketch]. Razprave IV Razreda SAZU 20(5):227–317

Matvejev DS, Puncer JI (1989) Karta bioma: predeli Jugoslavije i njihova zaštita [Map of biomes: landscapes of Yugoslavia and their protection]. Prirodnjački Muzej u Beogradu, Posebna Izdanja 36:1–76

Modrić-Surina Ž, Surina B (2010) Snowbed vegetation in Croatia: phytosociology, ecology and conservation status. Plant Biosyst 144(4):747–768. https://doi.org/10.1080/11263504.2010.502716

Morley MW (2007) Mediterranean Quaternary rockshelter sediment records: a multi-proxy approach to environmental reconstruction. PhD Thesis, University of Manchester, Manchester

Mucina V (2015) *Muscario-Scillion nivalis*: the mystery of an elusive alliance. In: Chytrý M, Daniëls F, Di Pietro R, Koroleva N, Mucina V (eds) Nomenclature adjustments and new syntaxa of the Arctic Alpine and oro-Mediterranean vegetation. Hacquetia 14(2): 284–286. https://doi.org/10.1515/hacq-2015-0004

Natek K (2007) Periglacial landforms in the Pohorje mountains. Dela 27:247–263. https://doi.org/10.4312/dela.27

Nikolić G, Bušković V (2012) Prostorni plan posebne namjene za Durmitorsko područje: poglavlje 2.3.1: prirodne karakteristike [Special purpose spatial plan for the Durmitor area: chapter 2.3.1: natural characteristics]. IBI Group i CAU Centar za arhitekturu i urbanizam-Podgorica, Podgorica

Ogrin M (2007) The minimum temperatures in the winter 2006/2007 in Slovenian frost holows and cold basins. Dela 28:221–237. https://doi.org/10.4312/dela.28

Ogrin M, Kozamernik E (2020) Vertical precipitation gradients: a case study of Alpine valleys of northwestern Slovenia. Theoret Appl Climatol 140:401–409. https://doi.org/10.1007/s00704-019-03051-z

Ogrin D, Ogrin M (2005) Predhodno poročilo o raziskovanju minimalnih temperatur v mraziščih pozimi 2004/2005 [Preliminary report about measuring extreme nocturnal temperatures in concave relief shapes in winter 2004/2005]. Dela 23:221–233. https://doi.org/10.4312/dela.23

Ogrin M, Sinjur I, Ogrin D (2006) Minimalne temperature v slovenskih mraziščih pozimi 2005/2006 [The minimum temperatures of cold air pools in winter 2005/2006]. Geografski Obzornik 53(2):4–12

Ogrin M, Nikolić G, Ogrin D, Trobec T (2018) An investigation of winter minimum temperatures in the mountains of Montenegro—a case study from the karst depression of Valoviti Do and selected mountain stations of northern Montenegro. Geographica Pannonica 22(4):241–252. https://doi.org/10.5937/gp22-18017

Ortar J (2011) Tipizacija in regionalizacija Slovenskih mrazišč [Typification and regionalisation of frost hollows in Slovenia]. BSc Thesis, University of Ljubljana, Ljubljana

Ortar J, Ogrin M, Vertačnik G, Sinjur I (2010) Primerjava temperaturnih razmer v mraziščih Reovce (Orjen), Valoviti Do (Durmitor), Luknja in Mrzla Komna (obe Julijske Alpe) v meteorološki zimi 2007/2008 [Comparison of temperature conditions in frost hollows Reovice (Orjen), Valoviti Do (Durmitor), Luknja (Julian Alps) and Mrzla Komna (Julian Alps) in meteorological winter 2007/2008]. In: Geoekologija XXI vijeka: teorijski i aplikativni zadaci: zbornik referata. Filozofski fakultet, Nikšić, pp 553–561

Sauberer F, Dirmhirn I (1954) Über die Entstehung der extremen Temperaturminima in der Doline Gstettner-Alm. Archiv Für Meteorologie, Geophysik und Bioklimatologie B5:307–326. https://doi.org/10.1007/BF02242757

Steinacker R, Whiteman CD, Dorninger M, Pospichal B, Eisenbach S, Holzer AM, Weihs P, Mursch-Radlgruber E, Baumann K (2007) A sinkhole field experiment in the Eastern Alps. Bull Am Meteor Soc 88(5):701–716. https://doi.org/10.1175/BAMS-88-5-701

Stepišnik U, Ferk M, Kodelja B, Medenjak G, Mihevc A, Natek K, Žebre M (2009) Glaciokarst of western Orjen, Montenegro. Cave Karst Sci 36(1):21–28

Stevanović V, Vukojičić S, Šinžar-Sekulić J, Lazarević M, Tomović G, Tan K (2009) Distribution and diversity of arctic-alpine species in the Balkans. Plant Syst Evol 283:219–235. https://doi.org/10.1007/s00606-009-0230-4

Šunje E, Bermejo AZ, Van Damme R, Backelajau T, Poskić N, Lukić-Bilela L, Kalamujić Stroil B (2021) Genetic diversity and differentiation of alpine salamanders from the Dinarides—an evolutionary perspective with insights for species conservation. Salamandra 57:75–88

Svetlin D (2020) Analiza temperaturnih razmer v mraziščih Komne [An analysis of temperature conditions in frost hollows on the Komna plateau]. MSc Thesis, University of Ljubljana, Ljubljana

Trošt A (2008) Mrazišča na Komni [Frost hollows in Komna Plateau]. BSc Thesis, University of Ljubljana, Ljubljana

Veress M (2017) Solution doline development on glaciokarst in alpine and Dinaric areas. Earth Sci Rev 173:31–48. https://doi.org/10.1016/j.earscirev.2017.08.006

Verovnik R (2002) The presence of *Plebeius pyrenaicus dardanus* (Freyer, 1845) (Lepidoptera: Lycaeinidae) on Mt Čvrsnica in Bosnia and Herzegovina, with notes on the butterfly fauna of this mountain. Entomologist's Gazette 55:29–34

Vertačnik G, Sinjur I (2013) Merilna napaka temperature zraka v aluminijasto-plastičnem zaklonu Slovenskega meteorološkega foruma [Air temperature measurement error in aluminium-plastic radiation shield of the Slovenian meteorological forum]. Vetrnica 5:62–75

Vukojičić S (2008) Glacijalni relikti u orofitskoj flori Srbije, Crne Gore i Makedonije [Glacial relicts in the orophytic flora of Serbia, Montenegro and Macedonia]. PhD Thesis, Univesity of Belgrade, Belgrade

Whiteman CD, Haiden T, Pospichal B, Eisenbach S, Steinacker R (2004) Minimum temperatures, diurnal temperature ranges, and temperature inversions in limestone sinkholes of different sizes and shapes. J Appl Meteorol 43(8):1224–1236. https://doi.org/10.1175/1520-0450(2004)043%3c1224:MTDTRA%3e2.0.CO;2

Zängl G (2005) Formation of extreme cold-air pools in elevated sinkholes: an idealized numerical process study. Mon Weather Rev 133(4):925–941. https://doi.org/10.1175/MWR2895.1

Zardi D, Whiteman CD (2013) Diurnal mountain wind systems. In: Chow FK, De Wekker SFJ, Snyder BJ (eds) (2013) Mountain weather research and forecasting: recent progress and current challenges. Springer, Dordrecht, pp 35–119. https://doi.org/10.1007/978-94-007-4098-3_2

Frost Hollows as Interesting Areas of Geo- and Biodiversity in the Alps

28

Domen Svetlin, Matej Ogrin, and Janez Kermavnar

Abstract

Frost hollows are areas where cold air pools frequently form due to relief, resulting in specific microclimatic conditions. The different microclimate also manifests itself in specific ecological conditions where only adapted plant communities develop. These unique areas are of high conservation importance as they can provide suitable habitats for cold-adapted plant species that are not found in the surrounding landscape. Such microrefugia are known to maintain relict populations of some endemic species. It can be concluded that frost hollows are special areas in terms of both geodiversity and biodiversity. The chapter discusses the climatological and vegetation characteristics of frost hollows from the Slovenian alpine areas.

Keywords

Mountain climate · Topoclimate · Temperature inversion · Vegetation inversion · Cold air pools · Mountain species Adaptation

28.1 Introduction

Alpine frost hollows are climate-specific areas and represent an interesting area of bio- and geodiversity. Varying temperature conditions influence many physical processes, such as pedogenesis, decomposition of organic matter, soil moisture conditions, corrosion and mechanical weathering of rocks, thickness and duration of snow cover, and the like. In frost hollows with a persistent and pronounced temperature inversion, vegetation inversion also occurs, which is a good example of the influence of geodiversity on biodiversity. Morphologically, frost hollows are concave relief forms, they can also be called relief depressions which enable the basic condition of the frost hollow, i.e., the formation temperature inversion and cold air pools in the evening and at night.

The definition of a frost hollow was initially established as a depressional karst form, e.g., dolines or collapse dolines, in which a temperature inversion and a vegetation inversion occur. Martinčič (1975, 1977) defines a frost hollow as a special habitat that differs from its immediate

D. Svetlin
Envirodual d.o.o., Slovenske Konjice, Slovenia
e-mail: domen.svetlin@envirodual.com

M. Ogrin (✉)
Department of Geography, Faculty of Arts, University of Ljubljana , Ljubljana, Slovenia
e-mail: matej.ogrin@ff.uni-lj.si

J. Kermavnar
Department of Forest Ecology, Slovenian Forestry Institute, Ljubljana, Slovenia
e-mail: janez.kermavnar@gozdis.si

M. Zorn et al. (eds.), *Discourses on Mountains of Montenegro and Slovenia*, Historical Geography and Geosciences, https://doi.org/10.1007/978-3-031-99017-5_28

surroundings in botanical, zoological, and climatological terms. These definitions point to the influence of the frost hollow on biodiversity, more specifically on vegetation. Later, when research in frost hollows focused on climatological and meteorological processes, other definitions emerged, such as that the frost hollow is an area where on clear and windless nights the temperature falls much lower than in the surrounding area at a similar altitude (Trošt 2008).

Undoubtedly, frost hollows are also special areas of geodiversity. At this point, it should be emphasized that the concept of geodiversity is treated holistically, i.e., as the diversity of abiotic nature and not only as the diversity of geological or geomorphological features of space. Although geodiversity is also commonly defined as diversity of abiotic nature (Prosser 2002; Gray 2013), the term geodiversity is now mostly used in the context of geological and geomorphological, and sometimes pedological, interpretations (e.g., Sharples 1993, 2002; Kiernan 1994, 1996, 1997; Dixon 1995; Eberhard 1997; Stanley 2000; Stepišnik and Repe 2015; Stepišnik and Trenchovska 2016; Stepišnik et al. 2017). We are not aware of research that would also address other abiotic phenomena in the context of geodiversity, e.g., in the hydrosphere or atmosphere.

In this paper we focus on the phenomenon of frost hollows as important areas of biodiversity and geodiversity of alpine areas in Slovenia, where the abiotic part of the natural environment is understood more broadly than only in geological and geomorphological terms and includes atmospheric phenomena and processes.

28.2 Frost Hollow as a Climatological and Geomorphological Phenomenon

The basic meteorological phenomenon that defines a frost hollow is temperature inversion. Under certain atmospheric and surface conditions, the air temperature in the near-surface layer of the frost hollow is significantly lower than in the higher layers. Temperature inversion results from the negative energy balance of the surface when it emits more energy in the evening and at night than it receives and therefore cools down. (Fig. 28.1). The surface cools the air directly above the ground. If this process takes place on the slopes, the cooled air directly above the slopes begins to flow toward the bottom of the frost hollow, and in its place less cold air flows from areas farther from the slopes. Thus, during the night, cold air flows to the bottom of the frost hollow. When the cold air reaches the bottom of the frost hollow, it continues to cool at the bottom. At the same time, cooled air flows in from the slopes, which is warmer due to the shorter cooling time and accumulates above the "older" and colder air at the bottom of the frost hollow. Thus, a cold air pool is formed during the night, reaching to the lowest point of the top of the frost hollow (Fig. 28.2). The cold air pool persists until it is dissipated by solar radiation (thermal dissipation) or the intrusion of wind turbulence (dynamic dissipation). In summer, a cold air pool can dissipate daily; in winter, it can persist for days or even weeks.

At frost hollows in karst areas, a cold air pool can also form when air escapes from the karst bedrock (Mihevc 2021), but this cooling mechanism is possible only when the ground is not covered with snow and when the karst bedrock is colder than the air in the frost hollows. At elevation of Komna (app. 1,350–1,700 m) frost hollows, this is the case during the day from June to November. Thus, this mechanism cannot contribute to extreme winter lows.

28.3 Conditions for Low Temperatures in Frost Hollows

28.3.1 Meteorological Conditions

Temperature of Air Mass

A lower initial air temperature usually means that a lower minimum temperature will be reached. The intensity of longwave radiation from the atmosphere also depends on air temperature (Ortar 2011). Low initial air temperatures

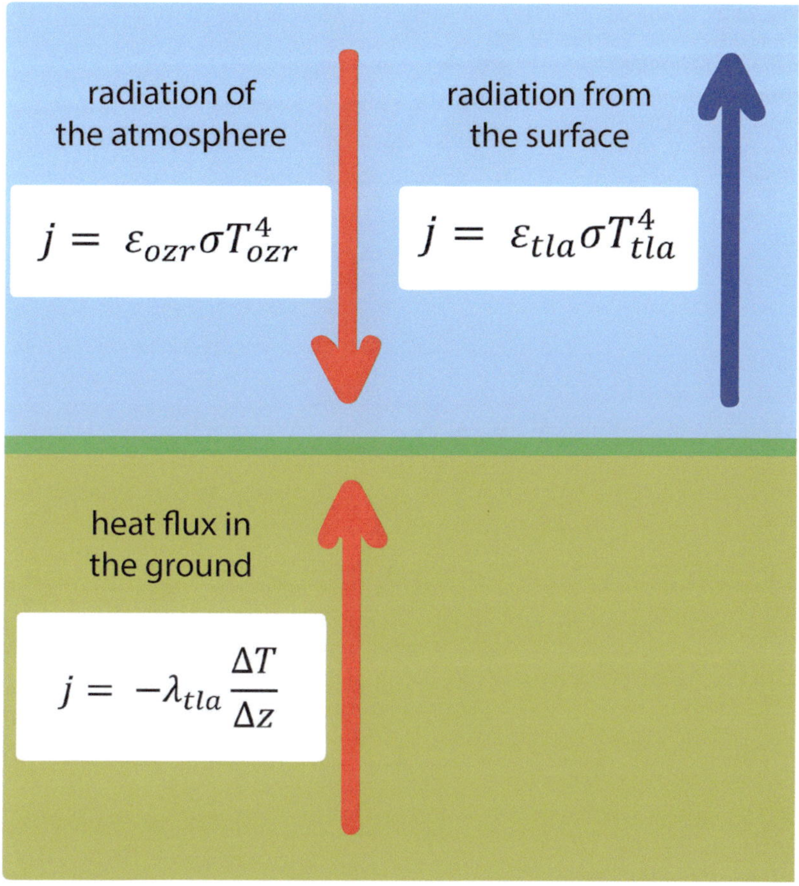

Fig. 28.1 The prerequisite for the occurrence of a radiative temperature inversion is a negative surface heat balance. The losses due to longwave radiation from the ground must be greater than the heat input from the ground and the heat input from the atmosphere. (Figure by Domen Svetlin)

in the free atmosphere in Slovenia are associated with advection of polar continental air masses from the north, northeast, or east; at the same time a blockade of the general western air circulation must be established. The pronounced area of high pressure over Scandinavia and northwestern Russia in winter often allows the advection of cold air from the northeast or east via Eastern Europe to the Central Europe, the Alps, and Balkan Peninsula. Traditionally, episodes of such advection are also associated with the Siberian anticyclone, whose influence on the weather in the Alps has been very rare in recent decades.

Wind

As the wind occurs at ground level, the layer of cooled air mixes with the warmer air higher up due to turbulence, so the formation of ground temperature inversion is difficult or completely impossible with stronger winds. If the general winds in the area of the frost hollow are relatively weak, a cold air pool can still form and continue to cool. If a ground layer of cold air has formed, it is more difficult for the overlying winds to penetrate the depressions with cooled air due to static stability. In general, shallower depressions are more prone to turbulent mixing than deeper depressions, which are more closed and where the inversion layer may be thicker. Particularly favorable conditions for radiative cooling are provided by a pronounced area of high pressure over Central Europe with weak winds, which expands and strengthens after the passage of the cold front (Whiteman et al. 2004; Ogrin et al. 2006; Trošt 2008).

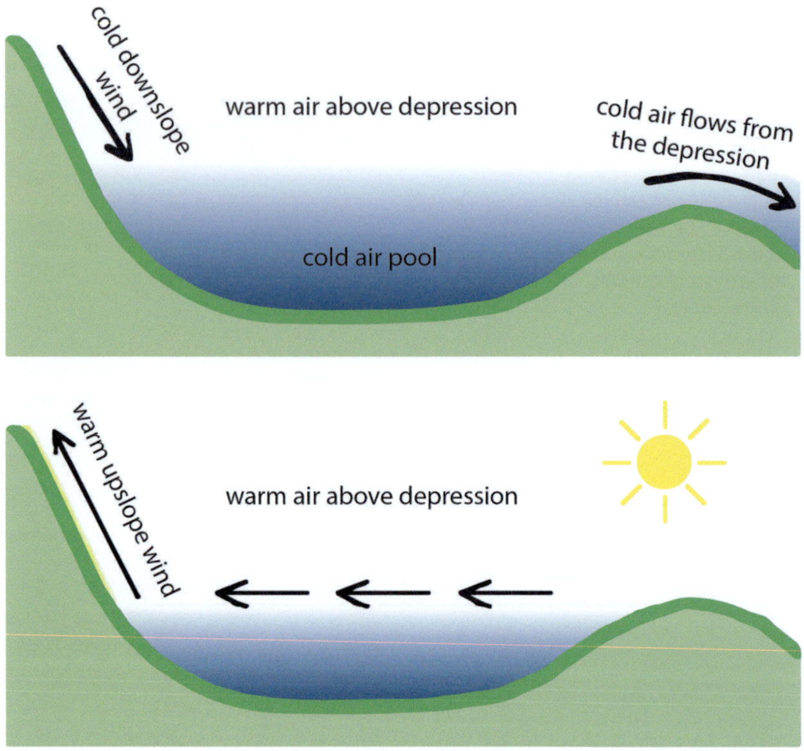

Fig. 28.2 Nocturnal formation (above) and late morning dissipation (below) of the cold air pool in the frost hollow. (Figure by Ana Seifert, authors archive)

Air Humidity

Of all the greenhouse gases, water vapor has the most important effect in emitting radiant energy from the atmosphere. The drier the air, the lower its emissivity. The saturation vapor pressure of water vapor, which limits the amount of water vapor in the atmosphere, decreases with decreasing temperature and increasing altitude. Roughly speaking, the amount of water vapor in the atmosphere decreases with altitude, so the emissivity of the atmosphere also decreases (Staley and Jurica 1972; Henderson 1994). Until cloud droplets or ice crystals form, increased relative humidity in the higher layers of the troposphere has no significant effect on the overall atmospheric emissivity (Ortar 2011).

28.3.2 Altitude

As altitude increases, the atmosphere becomes rarer and contains less and less water vapor and other greenhouse gases. The effect of atmospheric radiation is therefore less the higher the frost hollows. Higher altitude generally means lower temperatures at the onset of cooling, since temperature generally decreases with altitude. Thus, under favorable other conditions, the decrease in temperature and humidity with elevation allows higher elevation frost hollows to cool more than those at lower elevations (Ogrin et al. 2006). Wind speed usually increases with the altitude which is less favorable from the point of view of pronounced temperature changes, especially for shallower frost hollows. Thick snow cover reduces ground roughness, which reduces friction with the ground and increases wind speed at the ground (Vrhovec et al. 2006).

28.3.3 Topography

Due to complex topographic influences, such as colder downslope wind, reduced sky view factor, altered wind circulation and turbulence, heat storage in sunny slopes, and reduced insulation time, in mountain relief the dynamics of formation, maintenance, and dissipation of temperature inversion are different from those in lowlands. In the evening, longwave radiation from the surface causes a layer of cold air to form just above the surface, and density differences lead to the movement of cooled air down the slopes to lower areas. In its place comes warmer air from the higher areas, which cools above the ground and flows down the slope. This creates a downslope wind, also known as a katabatic wind. Cold air can flow down the slope, but it can't flow out of the frost hollows in a calm atmosphere, so it stays in them and builds up. This is the main reason for the formation of cold air pools which are filled with cooled air by slope winds. During radiant weather at night and in the morning, temperatures in concave relief are always lower than in the surrounding area, and inversions are more pronounced than in valleys, basins, and plains (Geiger 1966).

The concave shape of dolines, karst poljes, and other karst depressions protects the air at the bottom inside the depression from mixing with the surrounding air once a stable inversion layer has formed there. Thus, the layer of cooled air formed early in the night can more easily resist the turbulence at the top of the cold air pool that forms at the boundary between higher air currents due to shear, even if it is weak. As the temperature difference between the air at the bottom and the air at the upper edge of the depression increases, the direct downward flow of air is limited due to the static stability of the inversion. Since air temperatures are lowest at the bottom of the depression, the air there is almost completely isolated from the higher, warmer layers. Weak crosswinds can penetrate the higher parts of the basin, but they cannot directly mix the air at the bottom (Whiteman et al. 2004).

Another consequence of the concave relief of the frost hollows is the reduced proportion of visible sky. Because of the surrounding slopes, less sky is visible from the bottom of the frost hollow than from a flat or convex surface. This means that more longwave radiation from the slopes is also directed downward to the bottom of the frost hollow, reducing the total emitted longwave radiation. The downward longwave radiation and upward flux of perceived heat from deeper ground layers are the only processes that counteract radiative surface cooling. The surface temperature approaches the equilibrium temperature asymptotically during the night. As it approaches the equilibrium value, the loss of net longwave surface radiation (radiation balance) is nearly offset by the upward heat flux from the ground, so that further surface cooling is very small (Whiteman et al. 2004). The more open and shallow the frost hollow, the greater the proportion of visible sky and the more undisturbed the heat can radiate toward the sky. Shallower frost hollows, on the other hand, are more exposed to turbulence and wind effects, which reduces the number of cases of undisturbed cooling (Ogrin et al. 2006).

28.3.4 Surface and Vegetation

Lower thermal conductivity of soil means less emission of longwave radiation to the atmosphere. Higher thermal conductivity of rocky surface or higher thermal capacity of moist and grassy soils contributes to slower cooling of air and higher minimum temperatures reached than if the ground were covered with a blanket of snow. Events with extreme cooling in frost hollows were always associated with the presence of fresh snow cover (Zängl 2005).

In the Grünloch (Austria) frost hollow, the measured temperature differences between the bottom of the hollow and the edge of the hollow were between 25 and 30 K when fresh snow cover was present, and between 10 and 15 K without snow cover (Sauberer and Dirmhirn 1954, 1956). Fresh snow contains about 50 to 95% of air, which rests between the ice crystals and prevents heat transfer from the ground. Thus, when cooled by radiation, snow acts as an excellent insulator and prevents the flow of heat from the ground to the atmosphere, which would otherwise counteract the effect of radiative heat loss from the surface. The ground remains warmer than it would be without snow cover, and the air above the snow can cool more (Ogrin et al. 2006). In hollows covered with shrubs or forests, the temperatures do not drop as low as in those without higher vegetation cover. Trees, especially compared to snow cover, have a much lower albedo, so they absorb more radiation during the day and then emit it during cooling, while at the same time providing a barrier to longwave radiation from the ground by partially obscuring the sky. The higher the tree density on the frost hollow surface, the higher the longwave radiation from tree trunks and canopies and the coverage of the sky, and the lower the radiative losses from the surface. Deciduous tree species cover the sky less in winter than coniferous trees, so radiative cooling is even more inhibited in a coniferous forest than in a deciduous forest. The emissivity of tree canopies is high, ranging from 0.97 to 0.99 (DeWalle and Rango 2008).

28.4 Temperature Conditions in Frost Hollow Luknja on Komna Plateau as a Distinct Representative of an Alpine Frost Hollow

As an example of a distinct frost hollow in Slovenian Alps, we present measurements from the Luknja frost hollow at Komna Plateau, where we also measured temperatures outside the frost hollow (Dom na Komni measuring site) in its immediate vicinity (Fig. 28.3), allowing a direct comparison of conditions in the frost hollow and in the environment outside the hollow.

The Luknja frost hollow (Figs. 28.3, 28.4, and 28.6) is located in the central part of Spodnja Komna Plateau (Julian Alps) at an altitude of

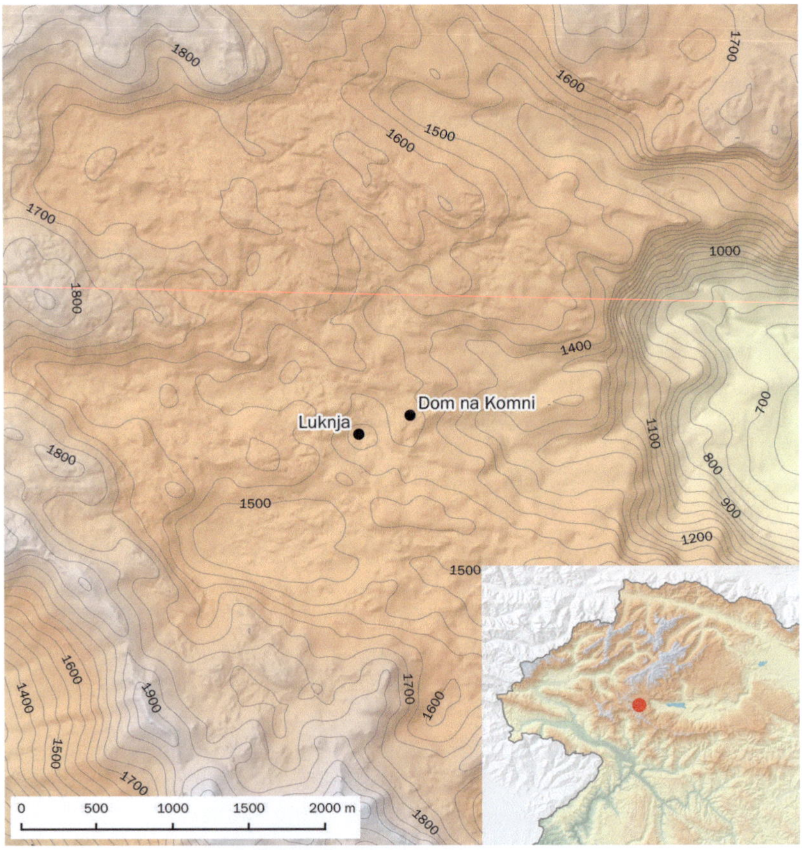

Fig. 28.3 Komna Plateau with locations of measuring sites (black dot). (Author of the map Domen Svetlin)

Fig. 28.4 The Luknja frost hollow in winter from south. (Photo by Matej Ogrin)

Fig. 28.5 Measuring site Dom na Komni outside the frost hollow. (Photo by Domen Svetlin)

1,427 to 1,483 m, about 400 m as the crow flies west to southwest of the mountain hut on Komna (*Dom na Komni*). The 56.5 m deep hollow is one of the deepest frost hollows of Komna plateau; on the bottom of the hollow there are many smaller dolines, which are only a few meters deep and wide. Most of the hollow is overgrown with dwarf pine, and only at the bottom there are some grassy clearings. The measurement site in the frost hollow was placed on a grassy area about 26 m west to northwest of the deepest point of the frost hollow at an elevation of 1,429.5 m. Measurements at the bottom of the hollow began on December 12, 2005, and initially took place in a classic radiation shield. At the beginning of July 2006, the shelter was replaced by a Stevenson screen.

The Dom na Komni monitoring site (Figs. 28.3 and 28.5) is located on a small ridge about 70 m northwest of mountain hut on Komna Plateau at an elevation of 1,521 m. At the beginning of the measurements on September 22, 2006, the measuring site, where the classical shield was used, was for some time at another microsite nearby (at a small spruce tree); on November 20, 2009, a Stevenson screen was set at the current location. Temperature measurements at this location are still ongoing. The monitoring site is primarily a reference station for temperature measurements outside the frost hollows, as its location is completely different in terms of the occurrence of ground temperature inversion, despite its proximity. The location of the measuring station on an open and slightly convex surface is closer in terms of temperature to the condition in the ground layer of the free atmosphere, as there is no accumulation of the cooled air or this effect is not so pronounced; at the same time, the wind exposure and thus the air mixing are greater.

The temperature conditions in the frost hollow differ significantly from those outside the frost hollow in some parameters (Table 28.1). Thus, of course, the greatest differences are in the absolute minimum temperatures (19.7 °C) and in the average minimum temperature (5.3 °C). The average daily temperature is also significantly lower, the difference being 2.5 °C. The average number of cool days (Tmin < 0 °C) per year is 75 days higher in the frost hollow, or negative temperatures occur on 57% of the days. There is a smaller difference in the number of icy days (Tmax < 0 °C), which is only 9 days per year, and the difference in the number of cold days (Tmin < –10 °C) increases sharply, because it is up to 54 days. In the frost hollow there are even 73 such days, i.e., every fifth day. Outside the frost hollow the temperatures never fall below –30 °C; in the frost hollow this happens on average on 3 days per year.

The differences in temperature amplitude are also interesting. The average temperature amplitude is 4.7 °C higher in the frost hollow than outside, and the absolute maximum temperature amplitude is 14.8 °C higher in the frost hollow. Temperature amplitudes above 20 °C occur 12% of the time in the frost hollow, while outside the frost hollow such days do not occur; a day with a temperature amplitude above 30 °C can also

Table 28.1 Temperature conditions in Luknja frost hollow and outside it on na measuring site Dom na Komni in period 2006–2015 (data source: Slovenian Meteorological Forum)

Parameter	Dom na Komni	Luknja
Average annual temperature (°C)	5.0	2.5
Average daily Tmax (°C)	9.3	8.7
Average daily Tmin (°C)	1.6	–3.7
Absolute Tmax (°C)	28.0	28.0
Absolute Tmin (°C)	–20.0	–39.7
Average annual number of days with Tmin < 0 °C (cool days)	135	210
Average annual number of days with Tmax < 0 °C (icy days)	40	49
Average annual number of days with Tmin < –10 °C (cold days)	19	73
Average annual number of days with Tmin < –20 °C	0	24
Average annual number of days with Tmin < –30 °C	0	3
Average annual number of days with Tmax < –10 °C	1	3
Average annual number of days with Tmax < –20 °C	0	0
Average annual number if days with Tmax ≥ 25 °C (warm days)	5	4
Average annual temperature amplitude (°C)	7.7	12.4
Absolute highest temperature amplitude (°C)	20.0	34.8
Average annual number of days with temperature amplitude < 10 °C	254	140
Average annual number of days with temperature amplitude > 20 °C	0	45
Average annual number of days with temperature amplitude > 30 °C	0	1

occur in the frost hollow, which is not the case outside. On the other hand, there is almost no difference in maximum temperature between the frost hollow and the surrounding area, which is understandable since there are no conditions for lower temperatures in the middle of the day in the frost hollow, when in most cases temperature inversion dissipates. Because the bottom of the frost hollow is lower and less ventilated (greater impact of the surface), maximum temperatures may even be higher than outside the frost hollow. The average daily maximum temperature is only 0.6 °C lower in the frost hollow, while the absolute maximum is the same at both sites. It can be concluded that the temperature conditions in the frost hollow are more extreme than in its surroundings. The difference is due to the lower minimum temperatures, while the maximum temperatures in the frost hollow are very similar to those outside.

28.5 Ecological Conditions, Flora, and Vegetation in Alpine Frost Hollows as a Result of Lower Temperatures

Specific microclimates, described in previous chapters, induce unique abiotic site conditions in which only adapted plant communities develop. Cold air accumulation has ecologically important consequences for the development of biota in such adverse habitats. Lower temperatures at the bottom of frost hollows due to cold air accumulation result in overall more stressful conditions for plant establishment, growth, and reproduction. A whole range of biological processes in the ecosystem are related to the lower temperatures, with shorter growing season being the most noticeable. Plant species therefore show rapid development during a short time window of more favorable ecological conditions. However, the main determinants of vegetation development are not only mean temperatures, but also microclimatic extremes. These act as a significant selective filter for plant species (Körner 2003). Frost events caused by cold spells, which can also occur in the summer months, have a major influence on plant survival. The harshest environmental conditions are found at the bottom of frost hollows. At the same time, concave terrain depressions are characterized by pronounced gradients in ecological factors (temperature, light, soil moisture) due to differences in local topography. The cooling effect depends mainly upon the dimension (depth, diameter) and geomorphology (shape, orientation) of depressions (Whiteman et al. 2004; Kobal et al. 2015; Čarni et al. 2022). Large-sized frost hollows generally show greater potential for persistent cold air pool compared to smaller depressions. Slope aspect and inclination are also important factors. North-facing slopes receive less solar radiation compared to south-facing slopes. Such differences impose prominent spatial partitioning of plant communities within frost hollows (Bátori et al. 2009, 2017; Čarni et al. 2022). Soil mineral nutrients, water availability, and microclimate change over short distances which translates into fine-scale mosaic of varying life conditions. Lower temperatures affect soil process. The decomposition rate of plant litter is lower at colder sites, which has important implications for organic matter formation, nutrient cycling, and nutrient availability to plant species (Aerts 2006). The formation of raw humus may contribute to the occurrence of acidophilic floral elements despite carbonate parent material. Freezing temperatures make water unavailable to plants.

Temperature inversion in frost hollows is accompanied by vegetation inversion (Fig. 28.6), where plant communities with higher thermal requirements on the edges and slopes of the hollows are gradually replaced downslope by more cold-tolerant vegetation toward the bottom (Bátori et al. 2014, 2019). Cold air pool and temperature inversion drive broad-scale patterns in vegetation and community assembly in frost hollows. The frigophilous character of vegetation weakens with increasing elevation, i.e., with distance from the bottom of the frost hollow. In the Alps, dwarf pines (*Pinus mugo*) indicate the presence of extreme conditions too unfavorable for the growth of tree species. Tree species such as European beech (*Fagus sylvatica*) are burned by late frosts, leaving only the Norway spruce

Fig. 28.6 The Luknja frost hollow is a typical alpine frost hollow. The inversion of the vegetation is remarkable: spruces (*Picea abies*), larches (*Larix decidua*), and beeches (*Fagus sylvatica*) grow at the edge; only spruces in the middle and dwarf pine (*Pinus mugo*) at the bottom. In many alpine frost hollows, the original vegetation has been removed by grazing and the vegetation inversion is formed by selective succession. (Photo by Matej Ogrin)

(*Picea abies*). Thus, bottoms of frost hollows are occupied by treeless vegetation composed predominantly by dwarf shrubs, alpine flowering plants, non-woody frigophilous scree vegetation, graminoids (grasses, sedges), and hygrophilous bryophytes. The boundaries between adjacent communities are often sharp and differ not only in terms of distinct species composition but also in contrasting plant forms and functions. Bottoms of frost hollows are extreme environments for tree growth (Dy and Payette 2007). However, treeless clearings at the bottoms of some frost hollows in the Alps or other mountainous areas are not merely a result of adverse conditions but can also be interpreted as a consequence of profound anthropogenic activities in the past (Fig. 28.7; Martinčič 1975). Woody plants generally have weaker cold tolerance than herbaceous vegetation, and their altitudinal limits are strongly controlled by cold sensitivity (D'Odorico et al. 2013). Bottoms of frost hollows are characterized by long winters in which deposited snow persists into the summer. Under these harsh conditions, the length of the growing season is short (about three to four months), but it can also be shorter depending on elevation and snowfall patterns. Such habitats provide suitable conditions for rare plants with specific developmental (phenological) adaptations. For instance, bottoms of frost hollows may be favorable for life cycle completion of alpine snow-bed clonal herb *Soldanella pusilla (Primulaceae)* due to long-lasting snowpack. In this case, the snow cover provides protection from deep soil frosts, thus protecting plant's growth and reproductive structures from freezing (Wheeler et al. 2014). Similarly, *Androsace alpina* (*Primulaceae*), a taxon endemic to the Alps, prefers scree areas, which can also be found at the bottom of frost hollows.

In terms of vegetation ecology and phytosociology, frost hollow vegetation is a classic example of azonal vegetation. This means that extreme cold significantly reduces the influence of macroclimate, so that low temperatures and associated temperature amplitudes become the limiting factor. Vegetation-microclimate feedbacks (D'Odorico et al. 2013) can be observed at the bottom of frost hollows where treeless vegetation (mostly formed by tussock grasses and cushion plants) cannot reduce net longwave radiation at clear nights, resulting in greater radiative cooling compared to higher parts of depressions where tree canopy cover induces a buffering effect and prevents additional cooling of the air. Sparse

Fig. 28.7 Medvedova konta frost hollow on the Pokljuka Plateau in Julian Alps. Although it is a pronounced frost hollow, the vegetation inversion is not formed. The clearing at the bottom of the frost hollow is of anthropogenic origin. (Photo by Uroš Stepišnik, authors archive)

forests are more susceptible to loss energy than closed-canopy stands (Martinčič 1975).

From an ecological, floristic, and vegetational perspective, cold air pool sites are interesting because they provide suitable environments for high-altitude plant species which can thus occur also at lower elevations. Botanists, phytosociologists, plant ecologists, and foresters have frequently studied this phenomenon particularly in large karst depressions of Dinaric mountains (e.g., Zupančič 1980) and less so in the Alps. Terrain depressions as characteristic topographic situation and diagnostic landforms of karst landscapes (Kobal et al. 2015) were examined in different setup contexts, respectively. Bátori et al. (2012) reported the importance of forested dolines in preserving relict, endemic, and high mountain plant species in *Fagus sylvatica* forests. Similar conclusions were made by Vilović et al. (2019).

In adverse environments with extreme site conditions and lack of disturbance-driven dynamics (Kammer and Möhl 2002), temporal community-level turnover tends to be low, but such stability is counterbalanced by high spatial asynchrony due to diverse microtopography and fine-scale changes in environmental conditions. In general, frost hollows represent one of the most stressful habitats due to cold conditions, where only cold-adapted plant species can survive. Communities are formed by plants with specific functional traits that reflect their adaptations to the harsh abiotic environment and ultimately capture the allocation trade-offs between coping with environmental stress and competitive ability (Körner 2016). Plants growing at high altitudes and frost hollows are considered typical stress-tolerators with conservative strategies (e.g., small stature, low specific leaf area, and high leaf dry matter content). These life history traits imply their intolerance to biotic competition. Owing to stressful conditions, these alpine plant communities tend to be relatively species-poor, but at the same time include a large number of unique species with narrow ecological niches (Good et al. 2019). Community species richness within the alpine life zone is generally controlled by stress intensity (Kammer and Möhl 2002).

Plant species in alpine habitats are exposed to several limiting constraints: low mean temperatures, thermal extremes, short growing season, scarcity of soil, low nutrient levels, strong daily thermal amplitudes, and related ecosystem processes (e.g., slow decomposition of organic matter). Plants of extreme environments have evolved numerous adaptations and mechanisms in response to these selective pressures that enable their survival (Sakai and Larcher 1987; Lütz 2012). Adaptations to challenging physical environments are manifold and can be detected on morphological, anatomical, phenological, and reproductive levels (Billings 1974; Körner et al. 1989). Life cycles of alpine plants are threatened by the high uncertainty whether flowering and fruiting, germination, and establishment can be successfully completed.

There is a relatively high degree of similarity between the vegetation of high alpine mountains and the most stressful parts of frost hollows, i.e., bottoms. However, important differences in physical characteristics between the alpine environment and the ecological conditions found in frost hollows suggest that these two habitats cannot simply be considered equivalent. Because of the concave terrain, bottoms of the frost hollows are characterized by less windy and more humid conditions in contrast to alpine slopes and especially ridges (strong winds, lower humidity, higher solar radiation). Nevertheless, some common patterns are expressed, such as the high proportion of long-lived species that rely on vegetative reproduction (via rhizomes, stolons, or bulbils) for population dynamics. Clonality and lateral spread seem to occur more frequently in cold environments compared to other, less stressful occasions (Klimešová and Doležal 2011). Relatively few life forms are present in alpine vegetation: evergreen or deciduous prostrate shrubs, short-stemmed herbaceous perennials, cushion plants, rosette plants, lichens, and mosses. Succulent plants with fleshy storage organs indicate water-limiting conditions. Annual plants are very rare because their life cycle depends on flowering and fruiting (Larcher et al. 2010). The vegetational dimension of frost hollows, bearing some botanical and floristic peculiarities, should be described from multiple (i.e., taxonomic, ecological and functional) aspects (Good et al. 2019). Lim et al. (2017) showed a shift in ecological strategies in frost hollows. Shrubs that are shorter and have smaller leaves may be better at avoiding the risk of frost damage. Their study shows how fine-scale turnover in shrub species composition around frost hollows is related to plant functional traits.

Mountain habitats play an important role from the point of view of biodiversity and its protection, as they have a unique flora and harbor relict vascular plant and bryophyte species. Alpine vegetation is highly susceptible to climate warming and drying (Theurillat and Guisan 2001). Populations of cold-adapted species confined to alpine habitats and frost hollows are highly vulnerable to climate change, while their habitats tend to shrink (Plasse and Payette 2015). Frost hollows are of high conservation importance as they can provide suitable habitats for cold-adapted plant species that are not found in the surrounding landscape (Bátori et al. 2012, 2014, 2017). These unique areas have the potential to act as refugia in times of global change. Such microrefugia are known to maintain relict populations, and many endemic taxa of European conservation concern can be found in these vulnerable systems. The relatively high degree of specialization and naturalness (i.e., preserved species composition due to lower anthropogenic impacts in the past) of such areas should further facilitate our conservation efforts. The climate-induced decrease of snow cover often leads to frost damage to vegetation, which creates gaps suitable for the establishment of new colonizing species. One of the species groups that benefits from warming is the more thermophilic tall herbs, which are more common in the subalpine zone but have recently been actively spreading (Grabherr et al. 2010). However, in mountainous regions, topological differences at the microscale can strongly influence microclimate and counteract the average effects of climate warming (Lampei et al. 2019). Decoupling local microclimate from regional macroclimatic conditions plays an important role in maintaining certain plant communities in concave landforms (Lenoir et al. 2017).

28.6 Conclusions

Alpine frost hollows are especially common in alpine karst areas, although they occur in other areas as well. They are among the coldest areas in the Alps and in Europe, and the temperature at the bottom may fall below –50 °C (Trošt 2008). Specific relief conditions influence the occurrence of a specific microclimate and specific soil conditions, which often leads to the occurrence of vegetation inversion. From the point of view of climatic conditions, frost hollows are extreme environments, which is reflected mainly in significantly lower minimum temperatures, while the maximum temperatures do not differ significantly from the ambient temperature. Only cold-adapted plant communities are able to thrive in harsh conditions, with specific plant life-history strategies and functional traits reflecting their adaptations to stressful environments. Alpine frost hollows are also home to many relict and endemic species with narrow ecological niches. These additionally increase the overall biodiversity value of such unique alpine habitats.

Undoubtedly, frost hollows are climatological, vegetational, and geomorphological phenomena that contribute to both biotic and abiotic diversity. Geomorphological features (such as concavity) enable the maintenance of a cold air pool that is long-lasting in some frost hollows and short-lived in others. The longer a cold air pool is present, the more likely it is to form specific soil conditions and vegetation inversion. Unlike geomorphological forms, which are permanent, and vegetation inversion, which is also a long-term phenomenon, the climatological phenomenon lasts much shorter. In most frost hollows, the cold air pool dissipates daily in the warm half of the year, and in winter usually only in the advective type of weather. From a climatological point of view, frost hollows can therefore be considered as phenomena rather than forms, representing an element of geodiversity, but in a different context from geological, geomorphological, and vegetation forms. Something similar can be observed with some hydrological forms, which are also not permanent. Such examples are torrents, intermittent lakes, and high tides, which, despite the limited time of their occurrence, make an important contribution to geo- and biodiversity.

Acknowledgements The authors thank Iztok Sinjur, Gregor Vertačnik, and other members of Slovenian Meteorological Forum for their help with the measurements on Komna Plateau.

References

Aerts R (2006) The freezer defrosting: global warming and litter decomposition rates in cold biomes. J Ecol 94(4):713–724. https://doi.org/10.1111/j.1365-2745.2006.01142.x

Bátori Z, Csiky J, Erdős L, Morschhauser T, Török P, Körmöczi L (2009) Vegetation of the dolines in Mecsek mountains (South Hungary) in relation to the local plant communities. Acta Carsologica 38(2–3):237–252. https://doi.org/10.3986/ac.v38i2-3.125

Bátori Z, Körmöczi L, Erdős L, Zalatnai M, Csiky J (2012) Importance of karst sinkholes in preserving relict, mountain, and wet-woodland plant species under sub-Mediterranean climate: a case study from southern Hungary. J Cave Karst Stud 74(1):127–134. https://doi.org/10.4311/2011LSC0216

Bátori Z, Csiky J, Farkas T, Vojtkó AE, Erdős L, Kovács D, Wirth T, Körmöczi L, Vojtkó A (2014) The conservation value of karst dolines for vascular plants in woodland habitats of Hungary: refugia and climate change. Int J Speleol 43(1):15–26

Bátori Z, Vojtkó A, Farkas T, Szabó A, Havadtői K, Vojtkó AE, Tölgyesi C, Cseh V, Erdős L, Maák IE, Keppel G (2017) Large- and small-scale environmental factors drive distributions of cool-adapted plants in karstic microrefugia. Ann Bot 119(2):301–309. https://doi.org/10.1093/aob/mcw233

Bátori Z, Vojtkó A, Maák IE, Lőrinczi G, Farkas T, Kántor N, Tanács E, Kiss PJ, Juhász O, Módra G, Tölgyesi C, Erdős L, Aguilon DJ, Keppel G (2019) Karst dolines provide diverse microhabitats for different functional groups in multiple phyla. Scientific Reports 9. https://doi.org/10.1038/s41598-019-43603-x

Billings WD (1974) Adaptations and origins of alpine plants. Arct Alp Res 6(2):129–142. https://doi.org/10.2307/1550081

Čarni A, Čonč Š, Breg Valjavec M (2022) Landform-vegetation units in karstic depressions (dolines) evaluated by indicator plant species and Ellenberg indicator values. Ecological Indicators 135. https://doi.org/10.1016/j.ecolind.2022.108572

D'Odorico P, He Y, Collins S, De Wekker SFJ, Engel V, Fuentes JD (2013) Vegetation-microclimate feedbacks in woodland-grassland ecotones. Glob Ecol Biogeogr 22(4):364–379. https://doi.org/10.1111/geb.12000

DeWalle DR, Rango A (2008) Principles of snow hydrology. Cambridge University Press, Cambridge

Dixon G (1995) Aspects of geoconservation in Tasmania: a preliminary review of significant earth features. Parks and Wildlife Service, Hobart

Dy G, Payette S (2007) Frost hollows of the boreal forest as extreme environments for black spruce tree growth. Can J for Res 37(2):492–504. https://doi.org/10.1139/X06-235

Eberhard R (ed) (1997) Pattern and process: towards a regional approach to national estate assessment of geodiversity. Australian Heritage Commission, Canberra

Geiger R (1966) The climate near the ground. Harvard University Press, Cambridge
Good M, Morgan JW, Venn S, Green P (2019) Timing of snowmelt affect species composition via plant strategy filtering. Basic Appl Ecol 35:54–62. https://doi.org/10.1016/j.baae.2019.01.004
Grabherr G, Gottfried M, Pauli H (2010) Climate change impacts in alpine environments. Geogr Compass 4(8):1133–1153. https://doi.org/10.1111/j.1749-8198.2010.00356.x
Gray M (2013) Geodiversity: valuing and conserving abiotic nature. Wiley, Chichester
Henderson JR (1994) Remote measurement of ground temperature and emissivity. In: Proceedings: international symposium on optics, imaging, and instrumentation. San Diego. https://doi.org/10.1117/12.188681
Kammer PM, Möhl A (2002) Factors controlling species richness in alpine plant communities: an assessment of the importance of stress and disturbance. Arct Antarct Alp Res 34(4):398–407. https://doi.org/10.1080/15230430.2002.12003510
Kiernan K (1994) The geoconservation significance of lake pedder and its contribution to geodiversity. Unpublished Report to the Lake Pedder Study Group. University of Tasmania, Hobart
Kiernan K (1996) The conservation of glacial landforms. Forest Practices Unit, Hobart
Kiernan K (1997) The conservation of landforms of coastal origin. Forest Practices Board, Hobart
Klimešová J, Doležal J (2011) Are clonal plants more frequent in cold environments than elsewhere? Plant Ecolog Divers 4(4):373–378. https://doi.org/10.1080/17550874.2011.586734
Kobal M, Bertoncelj I, Pirotti F, Dakskobler I, Kutnar L (2015) Using lidar data to analyse sinkhole characteristics relevant for understory vegetation under forest cover–case study of a high karst area in the Dinaric mountains. PLoS ONE 10(3). https://doi.org/10.1371/journal.pone.0122070
Körner C (2003) Alpine plant life—functional plant ecology of high mountain ecosystems. Springer, Berlin, Heidelberg. https://doi.org/10.1007/978-3-642-18970-8
Körner C, Neumayer M, Pelaez Menendez-Riedl S, Smeets-Scheel A (1989) Functional morphology of mountain plants. Flora 182(5–6):353–383. https://doi.org/10.1016/S0367-2530(17)30426-7
Körner C (2016) Plant adaptation to cold climates. F1000Res 5:2769–2774. https://doi.org/10.12688/f1000research.9107.1
Lampei C, Wunder J, Wilhalm T, Schmid KJ (2019) Microclimate predicts frost hardiness of alpine *Arabidopsis thaliana* populations better than elevation. Ecol Evol 9(23):13017–13029. https://doi.org/10.1002/ece3.5659
Larcher W, Kainmüller C, Wagner J (2010) Survival types of high mountain plants under extreme temperatures. Flora 205(1):3–18. https://doi.org/10.1016/j.flora.2008.12.005
Lenoir J, Hattab T, Pierre G (2017) Climatic microrefugia under anthropogenic climate change: implications for species redistribution. Ecography 40(2):253–266. https://doi.org/10.1111/ecog.02788
Lim FKS, Pollock LJ, Vesk PA (2017) The role of plant functional traits in shrub distribution around alpine frost hollows. J Veg Sci 28(3):585–594. https://doi.org/10.1111/jvs.12517
Lütz C (2012) Plants in alpine regions—cell physiology of adaptation and survival strategies. Springer, Wien. https://doi.org/10.1007/978-3-7091-0136-0
Martinčič A (1975) Vpliv človeka na nastanek nekaterih mrazišč v Trnovskem gozdu [Human impact on the formation of some frost hollows in the Trnovski gozd plateau]. Razprave: posebna številka. Simpozij: meteorologija - gospodarstvo, vol 1. Društvo meteorologov Slovenije, Ljubljana, pp 35–43
Martinčič A (1977) Prispevek k poznavanju ekologije mrazišč v Sloveniji: botanično ekološka skica [Contribution to the knowledge of the ecology of frost hollows in Slovenia: a botanical-ecological sketch]. Razprave IV Razreda SAZU 20(5):227–317
Mihevc A (2021) Ice in caves and its effect on thermal inversion and permafrost in the case of the Velika ledena jama v Paradani, Smrekova draga and neighbouring dolines. Acta Carsologica 50(2–3):241–251. https://doi.org/10.3986/ac.v50i2-3.10495
Ogrin M, Sinjur I, Ogrin D (2006) Minimalne temperature v slovenskih mraziščih pozimi 2005/2006 [Minimum tempertures in Slovenian frost hollows in winter 2005/2006]. Geografski Obzornik 53(2):4–12
Ortar J (2011) Tipizacija in regionalizacija Slovenskih mrazišč [Typification and regionalisation of frost hollows in Slovenia]. BSc Thesis, University of Ljubljana, Ljubljana
Plasse C, Payette S (2015) Frost hollows of the boreal forest: a spatiotemporal perspective. J Ecol 103(3):669–678. https://doi.org/10.1111/1365-2745.12399
Prosser C (2002) Terms of endearment. Earth Herit 17:12–13
Sakai A, Larcher W (1987) Frost survival of plants—responses adaptation to freezing stress. Springer, Berlin, Heidelberg. https://doi.org/10.1007/978-3-642-71745-1
Sauberer F, Dirmhirn I (1954) Über die Entstehung der extremen Temperaturminima in der Doline Gstettner-Alm. Archiv Für Meteorologie, Geophysik und Bioklimatologie B5:307–326. https://doi.org/10.1007/BF02242757
Sauberer F, Dirmhirn I (1956) Weitere Untersuchungen über die kaltluftansammungen in der Doline Gstettner-Alm bei Lunz im Niederösterreich. Wetter Leben 8:187–196
Sharples C (1993) A methodology for the identification of significant landforms and geological sites for geoconservation purposes. Forestry Commission Tasmania, Hobart
Sharples C (2002) Concepts and principles of geoconservation. Tasmanian Parks and Wildlife Service, Hobart

Staley DO, Jurica GM (1972) Effective atmospheric emissivity under clear skies. J Appl Meteorol Climatol 11(2):349–356. https://doi.org/10.1175/1520-0450(1972)011%3c0349:EAEUCS%3e2.0.CO;2

Stanley M (2000) Geodiversity. Earth heritage 14:15–18

Stepišnik U, Repe B (2015) Identifikacija vročih točk geodiverzitete na primeru krajinskega parka Rakov Škocjan [Identification of geodiversity hotspots on example of the Rakov Škocjan landscape park]. Dela 44:45–62. https://doi.org/10.4312/dela.44.45-62

Stepišnik U, Trenchovska A (2016) A proposal of quantitative geodiversity evaluation model in the example of Upper Pivka Karst, Slovenia. Dela 46:53–65. https://doi.org/10.4312/dela.46.2.41-65

Stepišnik U, Ilc Klun M, Repe B (2017) Vrednotenje izobraževalnega potenciala geodiverzitete na primeru Cerkniškega polja [Assessment of educational potential of geodiversity on example of Cerknica polje, Slovenia]. Dela 47:5–21. https://doi.org/10.4312/dela.47.1.5-39

Theurillat JP, Guisan A (2001) Potential impact of climate change on vegetation in the European Alps: a review. Clim Chang 50:77–109. https://doi.org/10.1023/A:1010632015572

Trošt A (2008) Mrazišča na Komni [Frost hollows in Komna Plateau]. BSc Thesis, University of Ljubljana, Ljubljana

Vilović T, Buzjak S, Buzjak M (2019) Floristic and microclimatic features of the Sovljak doline (Mt. Velika Kapela, Croatia). Šumarski List 143(1–2):35–43. https://doi.org/10.31298/sl.143.1-2.4

Vrhovec T, Kastelec D, Petkovšek Z (2006) Vreme in podnebje v gorah [Weather and climate in the mountains]. Tehniška založba Slovenije, Ljubljana

Wheeler JA, Hoch G, Cortés AJ, Sedlacek J, Wipf S, Rixen C (2014) Increased spring freezing vulnerability for alpine shrubs under early snowmelt. Oecologia 175:219–229. https://doi.org/10.1007/s00442-013-2872-8

Whiteman CD, Haiden T, Pospichal B, Eisenbach S, Steinacker R (2004) Minimum temperatures, diurnal temperature ranges, and temperature inversions in limestone sinkholes of different sizes and shapes. J Appl Meteorol 43(8):1224–1236. https://doi.org/10.1175/1520-0450(2004)043%3c1224:MTDTRA%3e2.0.CO;2

Zängl G (2005) Formation of extreme cold-air pools in elevated sinkholes: an idealized numerical porocess study. Mon Weather Rev 133(4):925–941. https://doi.org/10.1175/MWR2895.1

Zupančič M (1980) Smrekovi gozdovi v mraziščih Dinarskega gorstva Slovenije [Spruce forests in the frost hollows of the Dinaric Alps of Slovenia]. Dela IV Razreda SAZU, vol 24. SAZU, Ljubljana

Part VII
Conclusion

Between Slovenia and Montenegro 29

Drago Perko and Matija Zorn

Abstract

The Dinaric Alps are a large European geographical unit connecting Slovenia and Montenegro, whose capitals, Ljubljana and Podgorica, which are 553 km apart as the crow flies, are located right next to their northwestern and southeastern edges, respectively. The two countries are also connected by the Adriatic with its geographical and cultural Mediterranean influences. In addition, they are also linked historically, having belonged, either in full or in part, to the same larger states, such as the Republic of Venice, France's Illyrian Provinces, the Habsburg Monarchy, and Yugoslavia. Slovenia has been an independent country since 1991 and Montenegro since 2006. Slovenia was the first country to open its embassy in Montenegro.

Keywords

Geography · History · Landscapes · Karst surface · Ljubljana · Podgorica · Dinaric Alps

D. Perko · M. Zorn (✉)
ZRC SAZU Anton Melik Geographical Institute, Ljubljana, Slovenia
e-mail: matija.zorn@zrc-sazu.si

D. Perko
e-mail: drago.perko@zrc-sazu.si

29.1 From Slovenia to Montenegro

Over the course of history, parts of what are now Slovenia and Montenegro belonged to the same state at various times, and for a short time Ljubljana was even the capital of part of today's Montenegro. Therefore, the two countries seem to be closer than they actually are in geographical terms. Ljubljana, the capital of Slovenia, and Podgorica, the capital of Montenegro, are 553 km apart as the crow flies, with as many as eleven other European capitals being closer to Ljubljana (i.e., Zagreb: 117 km; Vienna: 278 km; the City of San Marino: 286 km; Bratislava: 305 km; Budapest: 381 km; Sarajevo: 393 km; Vaduz: 400 km; Prague: 447 km; Rome: 489 km; Vatican City: 490 km; and Bern: 549 km), and six being closer to Podgorica (Tirana: 131 km; Sarajevo: 173 km; Skopje: 185 km; Belgrade: 279 km; Sofia: 334 km; and Zagreb: 458 km). The road distance between Ljubljana and Podgorica is nearly one and a half times longer than the air distance.

Slovenia and Montenegro are among the smaller countries in the world (ranked 154th and 161st in terms of size), but Slovenia (20,273 km^2) is nearly twice as large as Montenegro (13,812 km^2), its population is nearly 3.5 times larger (around two million compared to just under 620,000), and its population density is over twice that of Montenegro (105 compared to 45 people per km^2). The two countries also

M. Zorn et al. (eds.), *Discourses on Mountains of Montenegro and Slovenia*,
Historical Geography and Geosciences, https://doi.org/10.1007/978-3-031-99017-5_29

differ in terms of GDP per capita, which in Slovenia is nearly three times higher (approximately €27,000 compared to less than €10,000). There are also major differences in the structure of tourism. In terms of overnight stays, seaside resorts predominate in Montenegro (94.3% of all overnight stays), while overnight stays in Slovenia are more spread across different types of tourist resorts (mountain resorts predominate with 31.1% of all overnight stays) (Table 29.1).

The highest peaks in both countries rise over 2,500 m in elevation. They are a little higher in Slovenia, even though the average elevation of Montenegro (approximately 1,000 m) is nearly 500 m higher than that of Slovenia (approximately 550 m). Just under half of Slovenia is over 500 m in elevation, compared to over four-fifths of Montenegro. A major difference can also be observed in the length of the two countries' coastlines, with Montenegro's coast (316 km) being nearly seven times longer than that of Slovenia's (47 km). The divide between the Black Sea and the Adriatic runs through both countries; approximately four-fifths of Slovenia's watercourses drain to the Black Sea, compared to over half of Montenegro's. Both countries are also known for their karst landscape with an abundance of surface and subterranean karst features; this reflects a strong presence of carbonate rock, which covers over three-fifths of Montenegro and just under half of Slovenia. Both countries have similar forest coverage (approximately 60% of the surface; Table 29.1).

In terms of landscape, four European macroregions meet in Slovenia (i.e., the Alps, Dinaric Alps, Pannonian Basin, and Mediterranean) compared to two in Montenegro (i.e., the Dinaric Alps and Mediterranean). This is reflected in the high landscape diversity of both countries, which is confirmed by the Landscape Diversity of Europe map (Fig. 29.1) and the European Countries survey (Ciglič and Perko 2013; Perko et al. 2017). In terms of landscape diversity across European countries, Slovenia is the most diverse country in Europe on average, a first-class European landscape hotspot, and Montenegro is in third place.

Geography is what not only distinguishes the two countries from one another, but also links them together. This is greatly contributed to by the Dinaric Alps and the Adriatic, two large European geographical units, which run parallel in a northwest–southeast direction from Slovenia to Montenegro, connecting them.

29.2 Geographical Links

From the perspective of cultural geography, Slovenians and Montenegrins are closest in terms of language because they both speak a South Slavic language. In other terms, Slovenians largely belong to Central Europe with its predominant Roman Catholicism and Latin and German cultural influences, whereas Montenegrins belong to southeastern Europe with its predominant Eastern Orthodoxy and Byzantine and Turkish influences. The coastal areas of both countries are also under the strong influence of the Mediterranean natural and cultural features.

Location on the Mediterranean or Adriatic coast is the first connecting geographical point between the two countries, followed by the Dinaric Alps, which start in Slovenia and end in the wider area of Montenegro (or vice versa).

29.2.1 The Dinaric Alps

The Dinaric Alps (Sln. *Dinarsko gorovje*, Cro. and Bos. *Dinarsko gorje*, Mtg. and Srb. *Dinarske planine* or *Dinarski Alpi*, It. *Alpi Dinariche*, Alb. *Alpet Dinarike*) were named after the nearly 100 km long Dinara mountain range on the border between Croatia and Bosnia and Herzegovina (Fig. 29.2). The highest peak in the range is Mount Troglav (1,913 m) in Herzegovina, followed by Mount Sinjal (1,831 m; Fig. 29.3), which is the highest mountain in Croatia and is often marked on maps as Dinara, that is, with the name of the entire mountain range (Fig. 29.3).

The Dinaric Alps include various others mountain ranges, such as the Velebit (1,758 m) and Biokovo (1,762 m) ranges in Croatia (Figs. 29.4 and 29.5), and the Accursed Mountains (Mtg.

Table 29.1 Some comparative data on Slovenia and Montenegro

	Slovenia	Montenegro
Area, km^2 (world rank by area)	20,273 (154)	13,812 (161)
Population (2023)	2,120,937	616,695
Population density (inhabitants/ km^2)	105	45
Capital city (population, 2023)	Ljubljana (296,228)	Podgorica (173,024)
Highest peak (mountain range; m)	Triglav (Julian Alps; 2,864)	Zla Kolata (Prokletije Mountains; 2,534)
Average height above sea level (m)	556.4	c. 1,050
Area (%) below 500 m	52.3	15.8
Area (%) above 500 m	47.7	84.2
Area (%) above 1,000 m	11.5	57.0
Area (%) above 1,500 m	2.5	18.5
Area (%) above 2,000 m	0.4	1.1
Number of neighboring countries (list)	4 (Italy, Austria, Hungary, Croatia)	5 (Croatia, Bosnia and Herzegovina, Serbia, Kosovo, Albania)
Length of land border (km)	1,322	622
Length of coastline (km)	47	316
GDP per capita, € (2022)	27,040	9,598
UNESCO world heritage sites: number, list	5 – Škocjan Caves (*natural site*) – Ancient and Primeval Beech Forests of the Carpathians and Other Regions of Europe (*natural site, transboundary*) – Prehistoric Pile Dwellings around the Alps (*cultural site, transboundary*) – Heritage of Mercury (*cultural site, transboundary*) – The Works of Jože Plečnik in Ljubljana–Human Centered Urban Design (*cultural site*)	4 – Natural and Culturo-Historical Region of Kotor (*cultural site*) – Durmitor National Park (*natural site*) - Stećci Medieval Tombstone Graveyards (*cultural site, transboundary*) – Venetian Works of Defense between the sixteenth and seventeenth Centuries (*cultural site, transboundary*)
National parks: number, list, area of country (%)	1 – Triglav National Park 4.1	5 – Biogradska Gora National Park – Durmitor National Park – Lovćen National Park – Prokletije National Park – Lake Skadar National Park 7.7
European macroregions	– Alps – Dinaric Alps – Pannonian Basin – Mediterranean	– Dinaric Alps – Mediterranean
River Basin (%)	Black Sea (Danube) Basin (81.0) Adriatic Sea Basin (19.0)	Black Sea (Danube) Basin (54.6) Adriatic Sea Basin (45.4)

(continued)

Table 29.1 (continued)

	Slovenia	Montenegro
Longest river (km)	Sava River (221)	Tara River (156)
Largest natural permanent lake (km^2)	Lake Bohinj (3.28)	Lake Skadar (370–540; c. 60% belongs to Montenegro)
Carbonate rock (limestones and dolomites, % area of the country)	c. 43	> 60
Forest (% area of the country, 2022)	59.1	59.9
Tourism: overnight stays (%, 2023)	– Mountain resorts: 31.1 – Seaside resorts: 19.1 – Capital: 14.3 – Health (spa) resorts: 19.6 – Other: 16.0	– Mountain resorts: 2.0 – Seaside resorts: 94.3 – Capital: 2.4 – Other: 1.3

Sources Radojičić 2008; Uhan et al. 2010; Pešić et al. 2019; Sekulić and Radulović 2019; Hrvatin et al. 2020; Monstat 2023, 2024a, 2024b; Britannica 2024; Montenegro.travel 2024; National ... 2024; Radulović 2024; SiStat 2024a, 2024b, 2024c, 2024e; Slovenia.info 2024; Triglav ... 2024; UNESCO 2024; Zavod ... 2024

Prokletije), which also feature the highest peak in the entire Dinaric Alps, 2,694 m Mount Jezercë (Alb. *Maja e Jezercës*) in northern Albania (Fig. 29.6).

The Dinaric Alps cover an area of nearly 120,000 km^2. They can be divided into three parallel belts (Fig. 29.2) stretching from the northwest to the southeast between Slovenia (Fig. 29.7) and Albania (Fig. 29.8). The Mediterranean belt (also the Adriatic, maritime, outer, or southwestern belt) extends along the Adriatic coast, roughly between Trieste, Italy (Fig. 29.9) and Ulcinj, Montenegro (Fig. 29.10). It is over 800 km long and covers an area of approximately 24,000 km^2, which corresponds to 21% or just under one-fifth of the entire mountain range. The Pannonian belt (also the continental, inner, or northeastern belt) runs along the southern edge of the Pannonian Basin, roughly between the Krka River in Slovenia (Fig. 29.11) and the Ibar River in Serbia (Fig. 29.12). It is 900 km long and covers an area of approximately 47,000 km^2, corresponding to 41% or over two-fifths of the entire range. The central belt lies between them, stretching roughly between the Soča River in Slovenia (Fig. 29.13) and the Drin River in Albania (Fig. 29.14). It is over 900 km long and covers approximately 43,000 km^2, which equals 38% or just under two-fifths of the entire range.

The Dinaric Alps are among the largest mountain ranges in Europe. They are very rugged because they are largely composed of limestone and dolomite, which are corroded by water and folded and broken up by tectonics. They have also been partly reshaped by glaciers. Three-fifths (59.0%) of the mountain range is karst landscapes with terrain that is difficult to traverse and numerous caves (Zwitter et al. 2024). The rare surface rivers, such as the Kolpa, Neretva, Tara, and Drin, cut deep canyons between the karst plateaus, separating them and providing space for roads and other transport infrastructure. The Dinaric Alps are known for their natural subterranean and surface tourist sites, such as Škocjan Caves (Fig. 29.15), the Plitvice Lakes (Fig. 29.16), the Cetina Valley (Fig. 29.17), and Lake Skadar (Fig. 29.18).

29.2.2 The Adriatic Sea

The Adriatic Sea (Sln. *Jadransko morje*, Cro. and Mtg. *Jadransko more*, It. *Mare Adriatico*, and Alb. *Deti Adriatik*) is an elongated gulf in the Mediterranean Sea between the Apennine and Balkan peninsulas. Its border with the Ionian Sea as defined by the International Hydrographic Organization runs through the Strait of Otranto, from the Cape of Santa Maria di Leuca in Italy in the

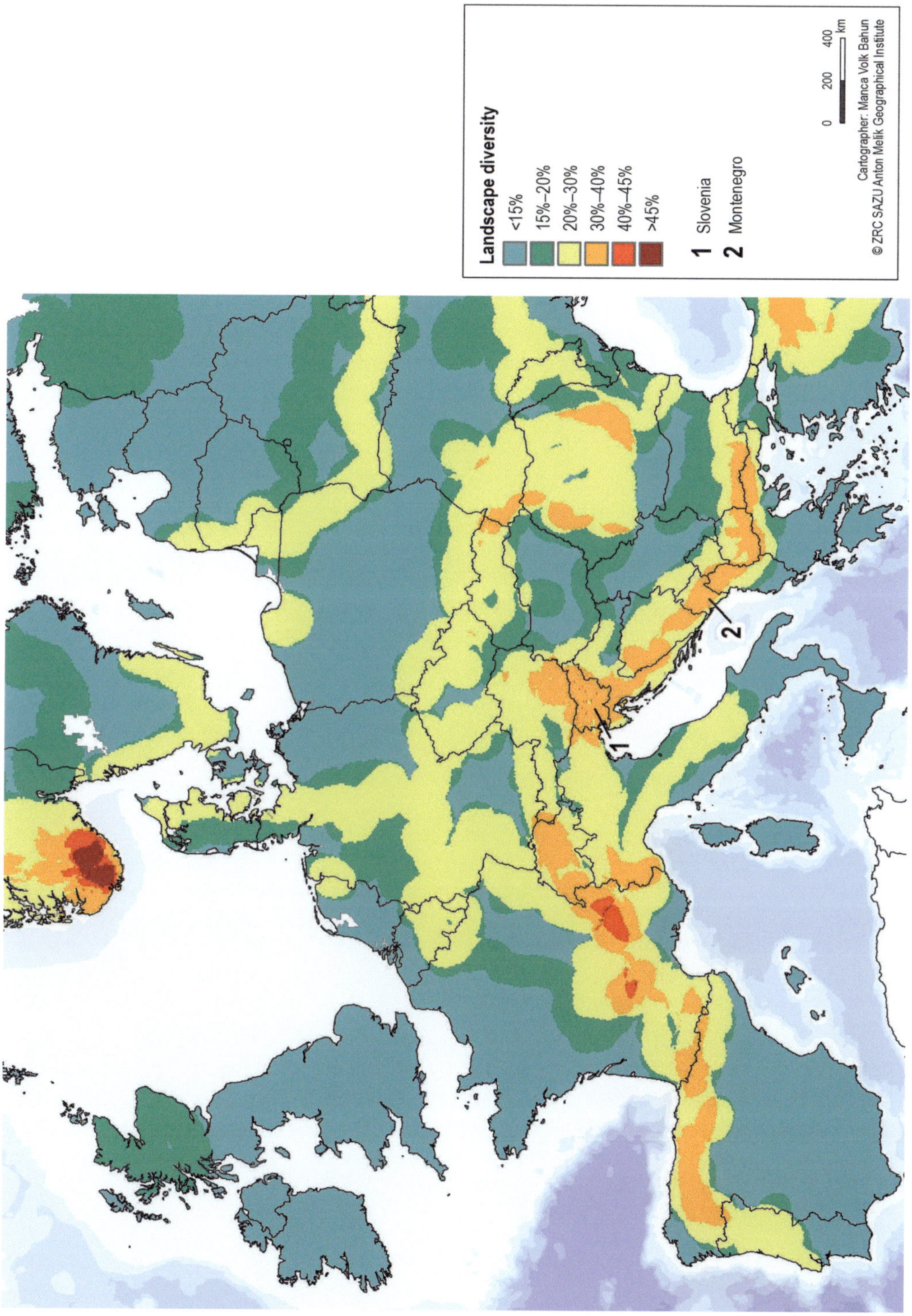

Fig. 29.1 Europe's landscape diversity map shows the percentage of landscape types occurring in each cell within a 50 km radius around it, out of all possible landscape types of the landscape classifications considered (adapted from Ciglič and Perko 2013)

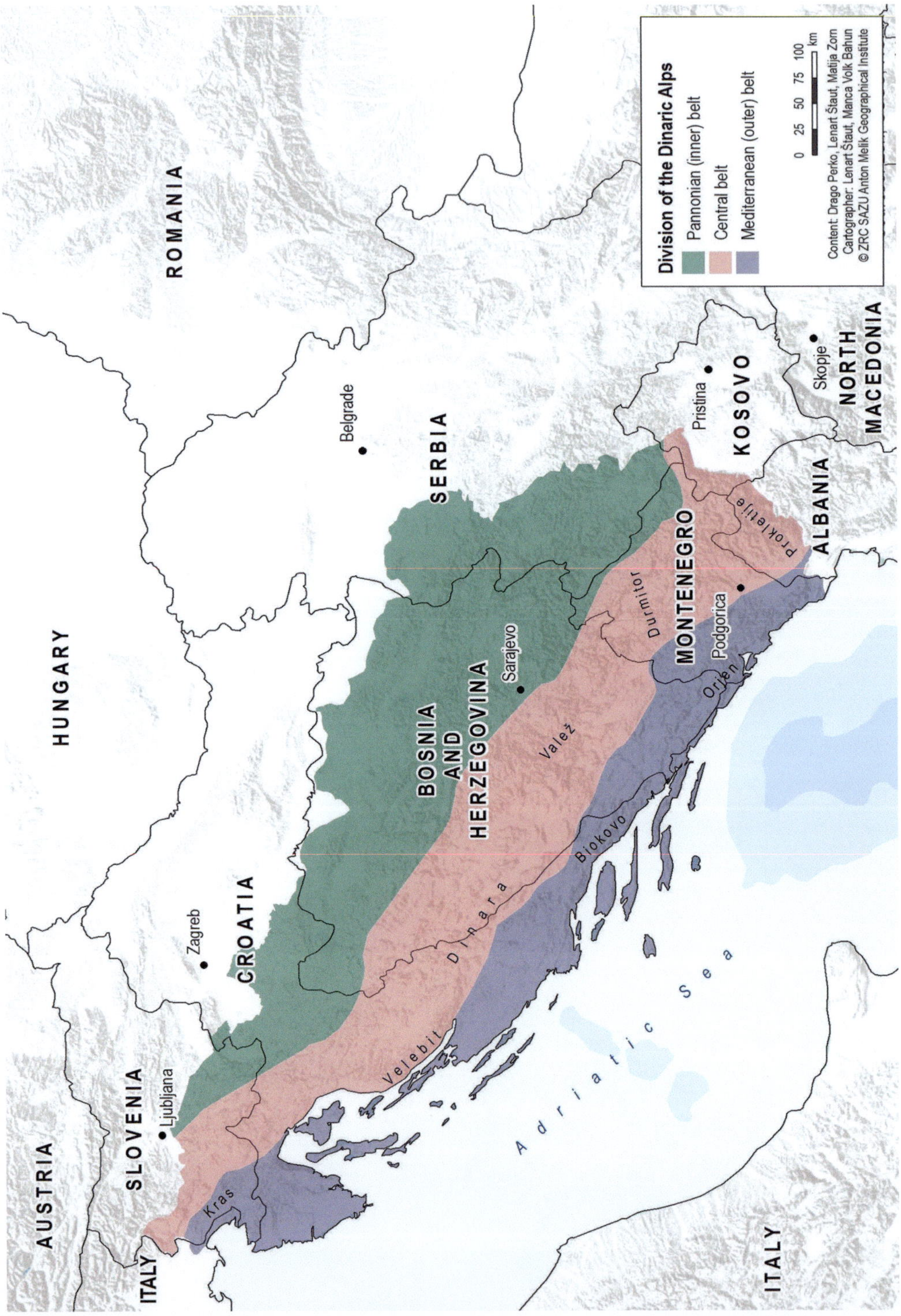

Fig. 29.2 The Dinaric Alps can be divided into three belts stretching from the northwest to the southeast: the Mediterranean (also the Adriatic, maritime, outer, or southwestern) belt, the central belt, and the Pannonian (also the continental, inner, or northeastern) belt

Fig. 29.3 The Dinaric Alps were named after Mount Dinara (a.k.a. Mount Sinjal, 1,831 m), the highest Croatian mountain, where the Cetina River has its headwaters. (Photo by Dario Bajurin, Adobe Stock)

Fig. 29.4 Velebit (highest peak: *Vaganski vrh* "Vagan Peak," 1,757 m) is a karst mountain range 145 km long and up to 30 km wide, extending along Croatia's Adriatic coast. (Photo by Nino Pavišić, Adobe Stock)

west along the northern coast of the Greek island of Corfu to the mouth of the Vivari Channel in Albania in the east (International ... 1953). Within this scope, it is just over 800 km long and 150 to 200 km wide, covering an area of 138,600 km^2.

Slovenia has two types of Mediterranean landscapes that differ from one another primarily by their rock composition: the Mediterranean low hills and the Mediterranean plateaus (Perko and Ciglič 2020; Perko et al. 2021). The former are characterized by a predominance of flysch and closely alternating small valleys and ridges, and the former are dominated by carbonate rock, especially limestone, and a highly fragmented karst surface.

Fig. 29.5 Biokovo (highest peak: *Sveti Jure* "Saint George Peak," 1,762 m) is a karst mountain range along Croatia's coast between the Cetina and Neretva rivers. It has been a nature park since 1981. (Photo by Mahey, Adobe Stock)

Fig. 29.6 Mount Jezercë (Alb. *Maja e Jezercës*, derived from Slavic *jezerce* "small lake") rises to 2,694 m and is the highest mountain in the Dinaric Alps. (Photo by Lenart Štaut, GIAM ZRC SAZU archive)

The situation is similar in Montenegro (Ministarstvo ... 2015), only that there the flysch belt runs even closer to the coast than in Slovenia. In both countries, the influence of the Mediterranean climate extends into the interior: in Slovenia, it primarily spreads through the Soča Valley into the heart of the Julian Alps, and in Montenegro it spreads from Lake Skadar through the Zeta and Morača valleys toward the karst plateaus and mountains.

The rugged Adriatic coast is also characterized by bays (Figs. 29.19 and 29.20), with Kvarner Bay as the largest one, peninsulas, such as Istria (the largest), numerous islands, such as the picturesque Kornati islands (Fig. 29.21), traditional Mediterranean towns, such as Koper, Pula, Split, Dubrovnik, and Kotor, and modern ports, such as Koper (Fig. 29.22), Rijeka, and Bar (Fig. 29.23).

Fig. 29.7 The border between the Alps and the Dinaric Alps also runs through the tectonic valley of the Idrijca River, a tributary of the Soča River, where the little town of Idrija is located with its well-known former mercury mine. Mercury heritage in Idrija has been a UNESCO world heritage site since 2012. (Photo by Matej Garbajs, GIAM ZRC SAZU archive)

Fig. 29.8 The Drin River in Albania is roughly the southeastern border of the Dinaric Alps. After it flows down from the mountains at Shkodër, it meanders toward the Adriatic Sea. (Photo by dudlajzov, Adobe Stock)

29.3 Historical Links

The first state that included both the territory of what is now Slovenia and Montenegro was the Roman Empire (Fig. 29.24). After the collapse of the Roman Empire, its territories were no longer part of the same state until the fifteenth century, when parts of Slovenia (i.e., Slovenian Istria and part of the Soča Valley until the sixteenth century) and Montenegro (the Montenegrin Littoral) came under the rule of the Republic of Venice (Veliki ... 1999: 89; Bajt and Vidic 2011). When Napoleon conquered the Republic of Venice in 1797, its territory was divided

Fig. 29.9 The port of Trieste in Italy lies on the northwestern edge of the Adriatic belt of the Dinaric Alps. (Photo by Hans und Christa Ede, Adobe Stock)

Fig. 29.10 The town of Ulcinj in Montenegro lies on the southeastern edge of the Adriatic belt of the Dinaric Alps. (Photo by Dmitrii, Adobe Stock)

between France and the Habsburg Monarchy. The latter had already acquired most of what is now Slovenia before that, and at that point it also obtained Istria; from what is now Montenegro, it won the Montenegrin Littoral (Andrijašević and Rastoder 2006). From then to the 1991 dissolution of Yugoslavia, parts of what are now Slovenia and Montenegro belonged to the same state, except for a short period between 1813 and 1814, when Montenegrin forces also controlled the Montenegrin Littoral (Andrijašević and Rastoder 2006), and during the Second World War.

Fig. 29.11 Flowing below the plateau-like Gorjanci Mountains in Slovenia, the Krka River follows the northwestern edge of the Pannonian belt of the Dinaric Alps. (Photo by Jure Tičar, GIAM ZRC SAZU archive)

Fig. 29.12 In southwestern Serbia, the Ibar River, which has its source in Montenegro, forms the southeastern edge of the Pannonian belt of the Dinaric Alps. (Photo by Milan, Adobe Stock)

29.3.1 The Illyrian Provinces

The Illyrian Provinces were established by Napoleon on October 14, 1809, when he signed a peace treaty with the Habsburg Monarchy. This political and administrative unit included parts of several Habsburg hereditary lands, the eastern territories of the former Republic of Venice (i.e., Venetian Istria, Dalmatia, and the Montenegrin Littoral), and the dissolved Republic of Ragusa. The Illyrian Provinces were part of the French Empire, but they were never fully integrated into it, primarily due to their geographical remoteness (Bajt and Vidic 2011). Legally, they were completely independent of Napoleon's Kingdom of Italy (Gašperič et al. 2012). Their capital was Ljubljana, which had never before been a capital of any large political and administrative unit (the Illyrian Provinces covered an area of around 55,000 km^2; Bajt and Vidic 2011).

Fig. 29.13 Slovenia's picturesque town of Most na Soči lies on the northwesternmost edge of the central belt of the Dinaric Alps. (Photo by marcin jucha, Adobe Stock)

Fig. 29.14 The Drin Valley (the photo shows the Fierzë Reservoir near the Albanian town of Kukës) meanders along the southeastern edge of the central belt of the Dinaric Alps. (Photo by marketanovakova, Adobe Stock)

The new state hence brought together the territories of what are now western, central, and southern Slovenia, which were part of the provinces of Carniola and Istria, and of what is now the Montenegrin coast, which belonged to the Province of Ragusa. Within the Province of Carniola (*Province de la Carniole*), today's Slovenian territory, with a population of 370,340, was divided into the Ljubljana (*Laybach*), Novo Mesto (*Neustadt*), and Postojna (*Adelsberg*) districts, and within the Province of Istria (*Province de l'Istrie*) the Slovenian territories covered the Koper (*Capo d'Istria*) district with the Koper and Piran (*Pirano*) cantons, with a total population of 20,956. The Province of Ragusa (*Province de Raguse*) was divided into three districts. Today's Montenegrin Littoral was part of the Kotor (*Cattaro*) district, with a population of 31,507. This district was further divided into three cantons: Kotor (*Cattaro*), Herceg Novi (*Castelnuovo*), and Budva (*Budua*; Gašperič et al. 2012) (Fig. 29.25).

The Illyrian Provinces were dissolved in the fall of 1813, when they were occupied by

Fig. 29.15 Škocjan Caves (Sln. *Škocjanske jame*) lie below Slovenia's Karst region. The Reka River flows through the canyon, which is up to 146 m deep, resurfacing as the Timavo River in Italy after 34 km and emptying into the Adriatic. The caves have been a UNESCO world heritage site since 1986. (Photo by Andreas, Adobe Stock)

Fig. 29.16 Croatia's Plitvice Lakes (Cro. *Plitvička jezera*) area has been a national park since 1949 and a UNESCO world heritage site since 1979. The park is composed of sixteen cascading small lakes on dolomite and limestone, covering an area of around 200 hectares. They are just over 50 km from the Adriatic coast. (Photo by Dimitar Chilov Wirestock Creators, Adobe Stock)

the Habsburg Monarchy, which subsequently acquired their entire territory through peace treaties (Šumrada 1990).

29.3.2 The Habsburg Monarchy

Based on the agreements reached by the Sixth Coalition and the later Treaty of Paris, signed in May 1814, the entire territory of the Illyrian Provinces came under the Habsburg Monarchy (Andrijašević and Rastoder 2006). Their northern part, which included the ethnic Slovenian territories, became part of the Kingdom of Illyria (1816–1849; Žontar 1990). In turn, the Montenegrin Littoral became part of the Kingdom of Dalmatia (1815–1918). The Habsburg Monarchy also won the remaining parts of what is now Slovenia—that is, the Slovenian part of the Duchy of Styria, and Prekmurje as part of the Kingdom of Hungary (Fig. 29.26; Bajt and Vidic 2011). In 1919, what is now Slovenia had a population of 1,321,098, which corresponded to sixty-five people per km^2 (SiStat 2024d). The Kotor political

Fig. 29.17 The Cetina River is the longest river in Dalmatia (101 km). In places it cut deep canyons into the karst surface. Several hydroelectric plants operate on it, and the ruins of several ancient and medieval fortresses can be found above the river. It flows into the Adriatic at Omiš. (Photo by Branko, Adobe Stock)

Fig. 29.18 Lake Skadar (Mtg. *Skadarsko jezero*, Alb. *Liqeni i Shkodrës*) lies on the southeastern edge of the Dinaric Alps, on the border between Montenegro and Albania. It is the largest lake in the Balkans. Its area varies between 370 and 540 km^2, depending on the water level, and it is 5 to 10 m deep. The Bojana River flows from it, emptying into the Adriatic Sea after 30 km. (Photo by Florian Scholl, Adobe Stock)

district (Ger. *politischer Bezirk Cattaro*), which covered an area of 674 km^2, had a population of 40,582, corresponding to sixty people per km^2 (Spezialortsrepertorium ... 1919).

29.3.3 Yugoslavia

After the First World War (and until 1941, when Slovenian ethnic territory was divided among the Axis powers) most of what is now Slovenia and the entire territory of what is now Montenegro belonged to the new Kingdom of the Serbs, Croats, and Slovenes, which was renamed the Kingdom of Yugoslavia in 1929 (Fig. 29.27). It covered approximately three-quarters of today's Slovenia (15,849 km^2; Statistički ... 1934); the remaining part belonged to the Kingdom of Italy. After 1929, Slovenian territory was administratively part of the Drava Banovina (i.e., province; *Dravska banovina*), whose capital was in Ljubljana. The newly established kingdom included the entire territory of what is now Montenegro, which was an administrative part of the Zeta Banovina (*Zetska banovina*), whose seat was at Cetinje. This province was nearly twice the

Fig. 29.19 In the north the Bay of Piran is dominated by the town of Piran and in the south, at the mouth of the Dragonja River, there are the Sečovlje Saltpans, a Ramsar site since 1993 and partly still in operation today. (Photo by Roland, Adobe Stock)

Fig. 29.20 The Bay of Kotor (Mtg. *Boka kotorska*) is a ria of the Boka River, which flowed from now heavily karstified Mount Orjen (1895 m) during the Pleistocene. The Kotor area has been a UNESCO world heritage site since 1979. (Photo by Mike Mareen, Adobe Stock)

size (30,997 km^2; Statistički ... 1934) of today's Montenegro (Table 29.1) because it also encompassed parts of what are now Croatia, Bosnia and Herzegovina, Serbia, and Kosovo. The highest peak of the Drava Banovina, Mount Triglav (2,864 m), was also the highest mountain of the entire kingdom, and the highest peak of the Zeta Banovina was Mount Gjeravica (Alb. *Gjeravicë*, Mtg. *Đeravica*; 2,656 m; Statistički ... 1934), which is the highest mountain of today's Kosovo. Compared to today's forest coverage in both countries (Table 29.1), forest only covered 9.42% of the Drava Banovina and 15.47% of the Zeta Banovina in 1931 (Statistički ... 1934). In 1931, the population of the Drava Banovina was 1,152,596 (8.2% of the total population of the Kingdom of Yugoslavia), compared to 1,397,650 in the entire territory of what is now Slovenia (SiStat 2024d). The Zeta Banovina had a population of 937,280 (6.7% of the total Yugoslav population) in 1931 (Statistički ... 1934), but the population of what is now Montenegro was only 360,044 (2.6% of the total population in the Kingdom of Yugoslavia; Premović and Doderović 2020). Compared to today (Table 29.1), population density was lower: seventy-three people per

Fig. 29.21 The Kornati archipelago is composed of around 150 islands and cliffs, which are largely made of limestone and thus karstified. It stretches 35 km along the central part of Croatia's Adriatic coast between Zadar and Šibenik. It was declared a national park in 1980. (Photo by Lukas Gojda, Adobe Stock)

Fig. 29.22 Koper, known as *Capris* in Roman times, lies on the northwestern coast of Istria. Its medieval historic center stands on a former island. The town is the largest port in Slovenia and the largest container port in the northern Adriatic region. (Photo by Andrew Mayovskyy, Adobe Stock)

km^2 in the Drava Banovina and sixty-nine people per km^2 in what is now Slovenia, and thirty people per km^2 in the Zeta Banovina and only twenty-six people per km^2 in what is now Montenegro.

After the Second World War, when the Kingdom of Yugoslavia was officially dissolved, Slovenia and Montenegro became two of the six people's republics of the Federative People's Republic of Yugoslavia (1945–1963) or, later on, socialist republics as part of the Socialist Federative Republic of Yugoslavia (1963–1991). The people's republics had their own insignia, governments, people's assemblies, and legislation. It was envisaged that the borders between the republics would be defined by laws approved by the federal people's assembly, but that never happened. This continues to cause border disputes between the Yugoslav successor states (Bajt and Vidic 2011). With the final establishment of Yugoslavia's national borders (Slovenia's borders were established in 1954, when the border with Italy was set), the size of both republics was also defined (Fig. 29.28). Both countries retained this size in full even after the collapse of Yugoslavia,

Fig. 29.23 Bar is the largest port in Montenegro, separated from nearby Lake Skadar only by Mount Rumija (1,593 m). The port was heavily damaged in a 1979 earthquake, but it resumed operations in 1983, after reconstruction was completed. (Photo by bigguns, Adobe Stock)

Fig. 29.24 Certain Roman settlements on a late-medieval copy of a map showing the road network of the Roman Empire, known as *Tabula Peutingeriana.* Above and to the right of the Istria (*Isteria*) Peninsula, in what is now Slovenia, towns such as *Longatico* (now Logatec), *Nauporto* (Vrhnika), *Emona* (Ljubljana), *Celeia* (Celje), and *Petauione* (Ptuj) can be identified (upper rectangle), and above what is now Puglia and the Adratic, across a wider area of what is now Montenegro (Mijušković 1969), places such as *Resinum* (Risan), *Vicinium* (Ulcinj), and *Batua* (Budva) can be recognized, as well as *Scobre* (Shkodër) in Albania (lower rectangle). (GIAM ZRC SAZU archive)

Fig. 29.25 Detail from a map of the Illyrian Provinces (Malte-Brun and Lapie 1812). (David Rumsey Map Center, Stanford Libraries)

Fig. 29.26 Detail from a map of Austria-Hungary showing the western part of the state, including the provinces with a Slovenian population and the Montenegrin Littoral (Orožen 1902). (GIAM ZRC SAZU archive)

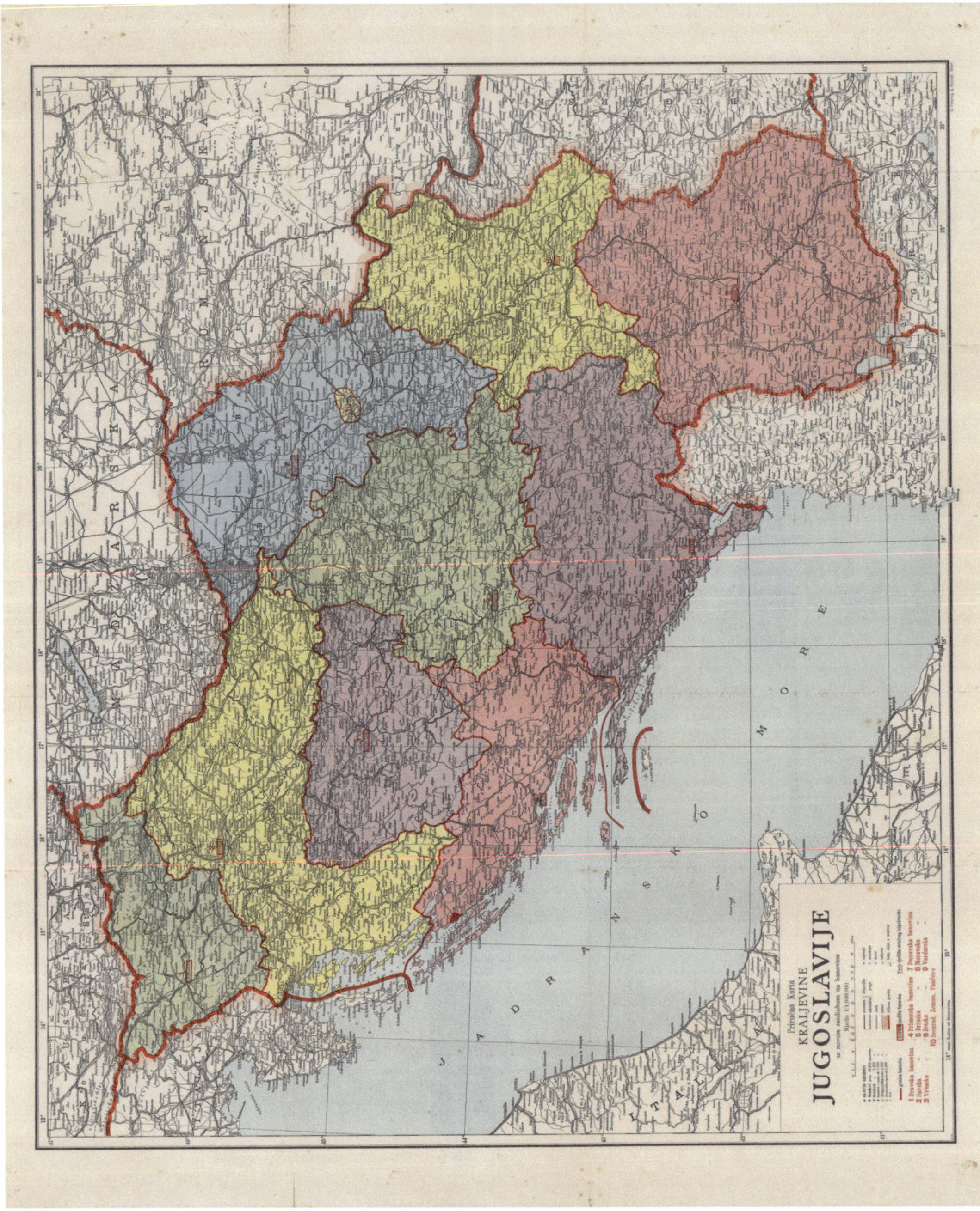

Fig. 29.27 Map from the 1930s showing the administrative division of the Kingdom of Yugoslavia into provinces. All of Montenegro was part of the Zeta Banovina, and the northwestern, central, and eastern parts of what is now Slovenia belonged to the Drava Banovina. (GIAM ZRC SAZU archive)

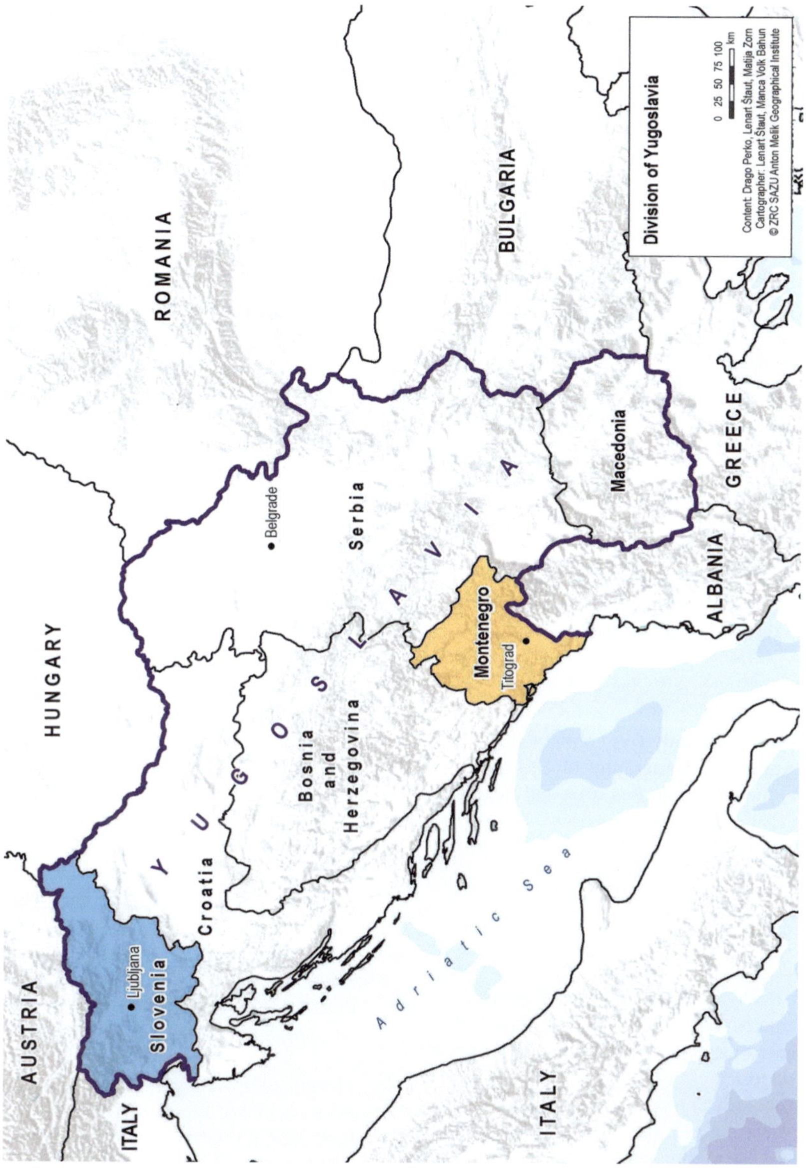

Fig. 29.28 Slovenia and Montenegro were two of the six republics constituting Yugoslavia between 1945 and 1991. The borders of both, now independent states, are nearly the same as when they were still part of Yugoslavia

or their declaration of independence (Slovenia in 1991 and Montenegro in 2006).

In 1991, when Yugoslavia broke up, Slovenia had a population of 1,965,986, corresponding to ninety-seven people per km^2, and Montenegro had a population of 615,035, corresponding to forty-five people per km^2 (Doderović and Ivanović 2018). The coats of arms of both Yugoslav republics featured their "holy" mountains—that is, Slovenia's Mount Triglav (Mikša 2018) and Montenegro's Mount Lovćen—as well as the sea or watercourses (Fig. 29.29). This shows the great national importance of mountains for both Slovenia and Montenegro. Slovenia retained these symbols in its national insignia (the coat of arms and flag) even after its independence.

Fig. 29.29 The coats of arms of the socialist republics of Slovenia (left) and Montenegro (right) highlight Mount Triglav (Slovenia) and Mount Lovćen (Montenegro) as supreme national symbols as well as the importance of the sea or watercourses

Acknowledgements This study was conducted as part of the research program The Geography of Slovenia (P6-0101), financed by the Slovenian Research and Innovation Agency.

References

Andrijašević ŽM, Rastoder Š (2006) Istorija Crne Gore: od najstarijih vremena do 2003 [History of Montenegro: from the earliest times to 2003]. Centar za iseljenike Crne Gore, Podgorica

Bajt D, Vidic M (eds) (2011) Slovenski zgodovinski atlas [Slovenian historical atlas]. Nova revija, Ljubljana

Britannica (2024) List of the world's largest countries and dependencies by area. https://www.britannica.com/topic/list-of-the-total-areas-of-the-worlds-countries-dependencies-and-territories-2130540. Accessed 3 May 2024

Ciglič R, Perko D (2013) Europe's landscape hotspots. Acta Geogr Slov 53(1):117–139. https://doi.org/10.3986/AGS53106

Doderović M, Ivanović Z (2018) Stanovništvo Crne Gore [Population of Montenegro]. Matica 19(74):121–140

Gašperič P, Orožen Adamič M, Šumrada J (2012) Zemljevid Ilirskih provinc iz leta 1812 = Carte des Provinces illyriennes de 1812 [Map of the Illyrian provinces from 1812]. Založba ZRC, Ljubljana. https://doi.org/10.3986/9789612543907

Hrvatin M, Pavšek M, Topole M (2020). Waters of Slovenia. In: Perko D, Ciglič R, Zorn M (eds) The geography of Slovenia: small but diverse. Springer, Cham, pp 59–70. https://doi.org/10.1007/978-3-030-14066-3_4

International Hydrographic Organization (1953) Limits of oceans and seas. IHO, Monte Carlo

Malte-Brun C, Lapie P (1812) Provinces Illyriennes, Servie et Bosnie. In: Atlas Complet Du Precis De La Geographie Universelle De M. Malte-Brun. Chez Francois Buisson, Paris

Mijušković S (1969) O rimskoj cesti kroz Boku Kotorsku [On the Roman road through Boka Kotorska]. Boka: zbornik radova iz nauke, kulture i umjetnosti 1:33–51

Mikša P (2018) Kako je Triglav pristal v grbu Republike Slovenije [How Triglav ended up on the coat of arms of the Republic of Slovenia]. In: Zorn M, Mikša P, Lačen Benedičič I, Ogrin M, Kunstelj AM (eds) Triglav 240. Založba ZRC, Ljubljana, pp 15–28. https://doi.org/10.3986/9789610500841

Ministarstvo održivog razvoja i turizma (2015) Mapiranje i tipologija predjela Crne Gore [Mapping and typology of the subdivision of Montenegro]. Ministarstvo održivog razvoja i turizma Crne Gore, Podgorica

Monstat (2023) Montenegro in figures 2023. Statistical Office of Montenegro, Podgorica

Monstat (2024a) Šumarstvo—podaci [Forestry—data]. Uprava za statistiku, Podgorica. https://www.monstat.org/eng/index.php. Accessed 3 May 2024

Monstat (2024b) Arrivals and overnights of tourists, total. Uprava za statistiku, Podgorica. https://monstat.org/eng/novosti.php?id=3884. Accessed 3 May 2024

Montenegro.travel (2024) Facts about Montenegro. https://www.montenegro.travel/en/plan-your-stay-in-montenegro/useful-tips-and-facts/facts-about-montenegro. Accessed 3 May 2024
National parks of Montenegro (2024). https://nparkovi.me/en/. Accessed 3 May 2024
Orožen F (1902) Zemljepisni atlas za ljudske šole s slovenskim učnim jezikom [Geographical atlas for elementary schools with Slovene as the language of instruction]. Hölzel, Dunaj
Perko D, Hrvatin M, Ciglič R (2017) Determination of landscape hotspots in Slovenia. Acta Geogr Slov 57(1):7–29. https://doi.org/10.3986/AGS.4618
Perko D, Ciglič R, Hrvatin M (2021) Landscape macrotypologies and microtypologies of Slovenia. Acta Geogr Slov 61(3):7–89. https://doi.org/10.3986/AGS.10384
Perko D, Ciglič R (2020) Slovenia's landscapes. In: Perko D, Ciglič R, Zorn M (eds) The geography of Slovenia: small but diverse. Springer, Cham, pp 211–225. https://doi.org/10.1007/978-3-030-14066-3_14
Pešić V, Paunović M, Kostianoy AG (2019) The rivers of Montenegro: introductory remarks. In: Pešić V, Paunović M, Kostianoy A (eds) The rivers of Montenegro. The handbook of environmental chemistry, vol 93. Springer, Cham, pp 1–12. https://doi.org/10.1007/698_2019_416
Premović M, Doderović M (2020) Demographic and historical features of Montenegro. Near East Historical Review 10(1):1–11. https://doi.org/10.29228/nehrreview.41554
Radojičić B (2008) Geografija Črne Gore: knjiga I, prirodna osnova [Geography of Montenegro: vol 1, natural basis]. Dukljanska akademija nauka i umjetnosti, Podgorica
Radulović MM (2024) Hydrogeological characteristics of Montenegro. In: Barovic G (eds) Speleology of Montenegro. Springer, Cham, pp 11–19. https://doi.org/10.1007/978-3-031-49375-1_2
Sekulić G, Radulović MM (2019) The hydrology and hydrogeology of Montenegro. In: Pešić V, Paunović M, Kostianoy A (eds) The rivers of Montenegro. The handbook of environmental chemistry, vol 93. Springer, Cham, pp 13–42. https://doi.org/10.1007/698_2019_413
SiStat (2024a) Višinski pasovi in nakloni terena [Altitude zones and inclination of the terrain]. Statistični urad Republike Slovenije, Ljubljana. https://www.stat.si/. Accessed 3 May 2024
SiStat (2024b) Dolžina državne meje [Length of the state border]. Statistični urad Republike Slovenije, Ljubljana. https://www.stat.si/. Accessed 3 May 2024
SiStat (2024c) Gross domestic product, Slovenia, annually. Statistični urad Republike Slovenije, Ljubljana. https://www.stat.si/. Accessed 3 May 2024
SiStat (2024d) Prebivalstvo in gospodinjstva, Slovenija, popisi 1857–1991 [Population and households, Slovenia, censuses 1857–1991]. Statistični urad Republike Slovenije, Ljubljana. https://www.stat.si/doc/pub/rr776-2002/1/t01-01-00.htm. Accessed 3 May 2024
Slovenia.info (2024) Facts about Slovenia. https://www.slovenia.info/en/plan-your-trip/facts-about-slovenia. Accessed 3 May 2024
Spezialortsrepertorium der österreichischen Länder: Bearbeitet auf Grund der Ergebnisse der Volkszählung vom 31. Dezember 1910 (1919) XII. Dalmatien. Statistischen Zentralkommission, Wien
Statistički godišnjak 1931 = Annuaire statistique 1931 [Statistical Yearbook 1931] (1934). Opšta državna statistika, Beograd
Šumrada J (1990) Ilirske province [Illyrian provinces]. In: Javornik M (ed) Enciklopedija Slovenije, vol. 4. Mladinjska knjiga, Ljubljana, pp 110–111
Triglav National Park (2024). https://www.tnp.si/en/park/area/triglavski-narodni-park/. Accessed 3 May 2024
Uhan J, Dobnikar Tehovnik M, Pavlič U (eds) (2010) Vode v Sloveniji [Waters in Slovenia]. Agencija Republike Slovenije za okolje, Ljubljana
UNESCO (2024) World heritage list. https://whc.unesco.org/en/list/&order=country. Accessed 3 May 2024
Veliki atlas svetovne zgodovine [The great atlas of world history] (1999). DZS, Ljubljana
Zavod za gozdove Slovenije (2023) Poročilo Zavoda za gozdove Slovenije o gozdovih za leto 2022 [Slovenian forest service's forest report 2022]. ZGS, Ljubljana
Žontar J (1990) Ilirsko kraljestvo [Kingdom of Illyria]. In: Javornik M (ed) Enciklopedija Slovenije, vol. 4. Mladinjska knjiga, Ljubljana, p 111
Zwitter Ž, Fuerst-Bjeliš B, Mrgić J, Petrić H, Zorn M (2024) Some insights into the environmental history of the Dinaric Karst. In: Fuerst-Bjeliš B, Mrgić J, Petrić H, Zorn M, Zwitter Ž (eds) Environmental histories of the Dinaric Karst. Springer, Cham, pp 1–25. https://doi.org/10.1007/978-3-031-56089-7_1

FSC
www.fsc.org
MIX
Papier aus verantwortungsvollen Quellen
Paper from responsible sources
FSC® C105338

If you have any concerns about our products,
you can contact us on
ProductSafety@springernature.com

In case Publisher is established outside the EU,
the EU authorized representative is:
Springer Nature Customer Service Center GmbH
Europaplatz 3, 69115 Heidelberg, Germany

Printed by Libri Plureos GmbH
in Hamburg, Germany